BACK TO THE GALAXY

J. H. Oort

1900–1992

Photograph by Loek Zuyderduin, courtesy of Sterrewacht Leiden

AIP CONFERENCE PROCEEDINGS 278

BACK TO THE GALAXY

COLLEGE PARK, MD 1992

EDITORS:

STEPHEN S. HOLT
FRANCES VERTER

NASA/GODDARD
SPACE FLIGHT CENTER

AIP
American Institute of Physics   **New York**

L.C. Catalog Card No. 93-71543
ISBN 1-56396-227-6
DOE CONF-9210330

Printed in the United States of America.

Preface

The first October Astrophysics Conference in Maryland was held two years ago. It was entitled "After the First Three Minutes," and was devoted to the study of the universe at $z>10^3$. Last year's conference, "Testing the AGN Paradigm," localized the topic to $z<10$. This year we have brought the conference "Back to the Galaxy."

These conferences are organized by astrophysicists at the Goddard Space Flight Center and the University of Maryland. The topic for each conference is selected by a permanent committee of senior scientific staff with the help of an International Advisory Committee, the current membership of which is:

Roger Blandford, Pasadena   *Sir Martin Rees*, Cambridge
Arnon Dar, Haifa   *Vera Rubin*, Washington
Alan Dressler, Pasadena   *Joseph Silk*, Berkeley
Riccardo Giacconi, Baltimore   *Rashid Sunyaev*, Moscow
Stephen Holt, Greenbelt   *Alex Szalay*, Budapest
Frank Kerr, College Park   *Yasuo Tanaka*, Tokyo
Richard McCray, Boulder   *Scott Tremaine*, Toronto
James Peebles, Princeton   *Joachim Trümper*, Munich

We have begun each of the conferences on the Monday corresponding to the federal observance of Columbus Day. This year we beat the 7-to-1 odds and began the meeting on October 12, the "true" Columbus Day. It was a very special Columbus Day, indeed, in the year of the 500th anniversary of Columbus' discovery voyage to the New World.

As in previous years, the conference began with two invited summaries reviewing the general subject (this year's were delivered by *Frank Kerr* and *Jerry Ostriker*). The conference then proceeded through the next two days with a series of non-parallelled sessions, each devoted to a specific topic and led by a distinguished session chair. Each of these sessions featured two or three invited talks and an extensive discussion period, and may have also included one or two short contributions "promoted" from the poster papers by the session chair. All the chairs should be commended for keeping the activities lively and the speakers to their allotted times.

Eli Dwek   *Ivan King*   *Vera Rubin*
Neil Gehrels   *Dick McCray*   *Charles Townes*
Namir Kassim   *Mort Roberts*   *Fran Verter*

The concluding rapporteur talks after lunch on Wednesday were delivered by *Scott Tremaine* and *Carl Heiles*.

These Proceedings are meant to be a cogent record of the entire meeting. All of the invited speakers provided us with manuscripts, and shorter written contributions were accepted for the poster papers. This year there are two new features, both of which are required if these Proceedings are to be truly representative of the conference. Because of the great interest among the conference participants generated by *Martin Harwit*'s banquet speech, were are including his thoughtful observations on the anticipated reduction in the gradient (if not the amount) of federal support for astronomy during the next few years. The other new feature is the inclusion of color plate pages in order to do justice to some of the graphics that were presented in the posters as well as in the invited talks.

All the members of the Scientific Organizing Committee deserve credit for the final product:

Chuck Bennett	*Frank Kerr,*
Leo Blitz	*Dave Thompson*
Jordan Goodman	*Fran Verter*
Mike Hauser	*Stuart Vogel*
Namir Kassim	*Rosie Wyse*

but special recognition should be given to *Chuck Bennett* and *Leo Blitz* for their leadership in the design of the scientific program, and to *John Trasco, Fran Verter* and *Paula Webber* for assuring the smooth operation of the conference.

These Proceedings represent the best available summary of the current state of our knowledge of the large-scale properties of the Milky Way. Virtually all the new relevant data are presented here in one form or another, including those from the current space-borne observatories like the *Cosmic Background Explorer (COBE)*, The *Röntgen Satellit (Rosat)*, and the *Compton Gamma Ray Observatory (CGRO)*. Interestingly, there were two "old refrains" which kept emerging during the conference. One was that the need for astrometry remains essential for significant new progress to be made in the understanding of the dynamics of the galaxy. The other was that much of the current conventional wisdom was anticipated before the outset of observations from space, and that no one was more important to the understanding of the structure of our galaxy than was *Jan Oort*. Just a few days after our conference was concluded, we received word that *Jan Oort* had passed away at the age of 92. We respectfully dedicate these Proceedings to his memory.

Steve Holt
December 1992

Note added: Fran Verter wishes to thank Bill Miranda, for his
encouragement, and our new daughter Shai Miranda Verter,
who expedited the editing of these proceedings by sleeping
through the night.

TABLE OF CONTENTS

LOCAL ENVIRONMENT

INFRARED EMISSION

INTERSTELLAR MEDIUM

BANQUET SPEECH

Back to Earth

BACK TO EARTH

Martin Harwit
Laboratory for Astrophysics
National Air and Space Museum
Washington, DC 20560

Perhaps I should begin with an apology. After dinner talks are sometimes expected to be entertaining. But a number of concerns have worried me for such a long time, that I thought I might be excused for presenting them. With this conference's emphasis on dealing with matters closer at hand than the remote universe and the earliest times, I thought it might not be inappropriate to bring all of us back to earth for a few minutes and talk about the future shape of astronomy and astrophysics as seen in a broader context.

I am fortunate to have come into our field when young astrophysicists were at a premium, jobs were plenty, promotions rapid, and grants easy to come by. I remember Nancy Roman, who ran NASA's Astrophysics Program calling me up one day to ask whether I'd like to start up an infrared astronomical rocket program at Cornell, much like the program I had helped to initiate at the Naval Research Laboratory the year before, and wanting to know how much it would cost. I, of course, said I'd be delighted and the start-up would require about a quarter of a million dollars.

Soon thereafter, the grant money arrived, and Tommy Gold, my department chair worried aloud that I might be too inexperienced to spend it, after only one year as an Assistant Professor during which I'd had the enormous salary of $7,000.

Those were the days of discovery: The microwave background, infrared galaxies, pulsars, superluminal sources, x-ray stars and the gamma ray background. It was the sixties and anything seemed possible.

That's no longer true today, less than thirty years later. Morale in our field, and perhaps in all of basic research seems to be at an all time low, possibly because of outmoded expectations that may now be unjustified.

The American Institute of Physics has been publishing figures on job opportunities for young college graduates and PhDs in recent months, and the job market for this latest addition to our ranks is bleak. All of us know this even without reading the statistics.

And the dearth of jobs is not only in basic research. One of my sons is an applied physicist with a recent PhD from Stanford. While he has been able to secure a pretty good job in industry, he tells me that many of his classmates have found nothing for which their training qualifies them.

What should we be doing about this? The answers aren't easy, nor are they likely to be popular. But I am certain of one thing. They won't be anything like the op-ed piece the Washington Post carried on June 2 of this year. The article's headline reads " Too Many Scientists? Don't Believe It".

The writer asserts,

To hear Congress, the media, policy gurus and even a few scientists tell it, we will soon face a glut of scientists and engineers in America.

He goes on to say,

Unfortunately the stridency and nature of their arguments .... threatens to undermine our national commitment to improving science education at all levels.... The most important factor affecting our long-term production of scientists is the tragic inadequacy of our primary and secondary mathematics education programs. The results of this educational disaster are felt at every point along the path to postgraduate and professional degrees.

We have two statements here. First, that we need more scientists; second that the inadequacy of our school system threatens to prevent us from producing adequate numbers. I believe that neither statement is correct.

Given the state of the job market, how are we to take this op-ed piece, particularly given that it is signed by James J. Duderstadt, President of the University of Michigan and, more germane to the issue, also the Chairman of the National Science Board, the body responsible for overseeing the policies governing the National Science Foundation?

Is the National Science Board so out of touch with the market? Or, possibly worse, does President Duderstadt reflect a community that is out of touch with national realities? Either prospect would be dismaying.

Let's look at our own community:

Just before the Bahcall Committee Report was to come out last year, John Bahcall called me to ask whether I'd be willing to write a review of the study in the **Air & Space Smithsonian** magazine, which has a circulation of about 400,000. He thought the publicity would be good for astronomy.

I said I didn't know what the Report advocated, since I hadn't had any part in it, and that I didn't want to write a destructive review if it turned out that I didn't like it. But John didn't seem worried and said he'd have an advance copy sent to me. As it was, I did write a reasonably positive review, because I felt the choice of recommended new facilities had been thoughtfully selected. I sent John a copy ahead of time as a courtesy, got a return message that he saw nothing that was wrong or misrepresented, and then went to press with it.

Still, the study had me worried:

Let me read the relevant portions from the review I wrote. I believe they are accurate:

For ground-based research, the report, i.e.the Bahcall Committee Report, assigns the highest priority to supporting the field's infrastructure and to the growth of research grants.....Grants to young astronomers have suffered a decline stemming from the report's single most startling finding: between 1980 and 1990 the number of active astronomers in the United States grew by 42 percent..... Mindful of current tight federal budgets, the report nevertheless recommends an expansion of research grants, one aim being the support of graduate students and young postdoctoral researchers -- a reasonable request, if the goal is further expansion of the field. But does this request imply the need for an equal growth in the number of astronomers during the 1990s? How large a growth rate should be encouraged? How many astronomers and how much research can the nation responsibly support?

As far as I am concerned, none of us have been willing to take a serious look at this question, with the possible exception of Robert White, President of the National Academy of Engineering, who gave a thought-provoking interview in MIT's **Technology Review** a couple of years ago. He is quoted there as saying,

We appear to have too many scientists and engineers chasing too few research dollars...Are there too many scientists and engineers or is there too little money?

He goes on to make the first serious attempt I have seen to look at the question in an objective way. He says,

It is not sufficient to claim entitlement to as much money as is necessary to support every good scientist.... Science and engineering research, like any other activity in this country, has a social purpose, and it must justify expenditures in ways that can be understood and lead to the social and economic betterment of the country.

White's statement has a ring of truth to it, and we should not dismiss it lightly.

Whether or not we believe that the country now has enough scientists, can be disputed, but let us at least agree that an unbridled growth cannot go on forever. Sooner or later, there will be too many scientists. If not now, then at least pretty soon. A growth rate of 42% per decade simply cannot be sustained forever. It is unlikely that the economy will ever be able to afford a research budget that runs over a few percent of federal expenditures. We already are at that threshold now, and it just does not seem reasonable to expect the rate of research support to dramatically increase.

Nor can we expect to gain a great deal from international collaborations, that is, by having other countries help pay for the research. Europe and Japan already are actively engaged in large scale research efforts. We might expect some economies through international streamlining of efforts, but that will not go all that far.

So, what should we do to keep astronomical research specifically, and science more generally as exciting, in this country, as it has been in the past?

Let me suggest three points, though these are by no means exhaustive, and many of you will be able to think of several additional tacks one might take. I'll mention these in order of increasing difficulty of implementation, though none of them are particularly easy.

One approach is to recognize the separability of front-line university research from graduate student and post-doctoral education. In the national laboratories and centers, that doesn't come as news. But in the universities, the structure of research groups still centers on the production of PhDs.

In government laboratories and in industry, scientists to a much larger extent work with professional research assistants rather than students. The same could be done in our universities.

That would permit university research groups to continue their high-caliber research, but the number of graduate students could be reduced to the point where our production of new astronomers roughly matched the rise in the gross national product. That way we would not need to worry about a mushrooming over-production of young astronomers who had no hope of becoming accommodated in the job market.

That's a much fairer way of dealing with graduate students, instead of raising unwarranted hopes of a career after investing years in working toward a PhD.

But we also can do more. A second way is through education in our public schools and colleges. By that I do not mean what James Duderstadt has referred to when he writes,

Nationwide, 10 times as many students major in accounting as in mathematics;

and he adds, with evident dismay,

last year, at my own University of Michigan, we graduated more than 1,300 students planning to enter law school but only 22 in mathematics.

That may be too many lawyers, but if the market can't absorb even 22 mathematicians, should we be worrying that the University of Michigan hasn't trained more of them?

I do not think so.

What I do believe we can do, is to provide far better science education in our schools and colleges, aimed at non-scientists. Not in order to make them into junior chemists or biologists or physicists or even astronomers, but to give them an understanding for what science can and cannot do.

Most of our citizens have no idea whatsoever about the way science works. They see it as a sort of crank that you turn to get out truths. That leaves no room for comprehending how and why scientific disputes arise, and why two scientists can differ on what may or may not be taken to be factual: For example, is there global warming or is there not? Does it constitute a threat or doesn't it? Can we delay doing something about the growing ozone hole and will further delay cost more in the long run than prompt action?

These are questions to which we need answers. People in Congress need to understand the nature of our arguments, and yet they lack the necessary background to appreciate how scientists reason, how to assess risks, and how to judge probabilities.

A simple case concerns asbestos: If people working in asbestos factories face an increased risk of death from cancer, after exposure for several decades, does that mean that all asbestos tiles must be removed from schools across the entire nation? It may, if the tiles flake a lot. But if they don't, it may be unimportant. Rather than mandating the cost of asbestos replacement, we might then decide to put money into paying higher teacher salaries to attract better teachers, particularly in science. The Congress needs to be able to judge such trade-offs, rather than condemning anything related to asbestos out of hand, without further evaluation.

The science education offered by our schools should provide the nation with a better understanding of what science is and what it does. For the moment we do not need to worry all that much about going to special lengths to also have these schools train future scientists. Right now, we already appear to be training a sufficiently large number to fill the nation's needs.

Third, we will need to look, far more seriously than we ever have before, at how we have actually been carrying out our own work. We hardly understand that at all, despite the existence by now of some first rate studies, such as Robert Smith's book **The Space Telescope**, published just before the telescope's launch. Smith's study describes the three-cornered disputes between scientists, engineers and managers, which continued unresolved for decades. And these make understandable how this costly telescope could have been left untested for serious potential flaws. The book documents the managerial techniques used on this large scale project, gets one thinking about how such methods might be improved, and could ultimately lead to better ways of handling such undertakings that require the efforts of enormous teams for periods lasting a generation.

We need to better understand the ways in which we work, because there are many myths around, and we need to establish clarity in their stead. The most glaring myth, perhaps, is a belief that somehow astronomers can fend for themselves -- that we depend on no one else, especially not the military. If the planning documents produced each decade -- the Greenstein Committee Report, the Field Committee Report, and the Bahcall Committee Report -- are any indication, astronomy and military research have hardly anything in common.

Yet, when I wrote the book **Cosmic Discovery** more than a decade ago, I was struck by the fact that every major post-World-War-II astronomical discovery had been made possible only by virtue of instrumentation we had inherited from the military. We may feel uncomfortable with that fact, but to me it seems indisputable. Whether it was the discovery of radio galaxies with techniques developed for radar during the War, or the realization, through early rocket observations, that X-ray galaxies exist, or the dramatic detection of gamma-ray bursts by the military Vela satellites -- we all depended on equipment or information handed down by the military. And that continues today, as we plan the Space Infrared Telescope Facility, SIRTF, or take advantage in ground-base optical astronomy of active and adaptive optics pioneered by defense research.

Where there is no economic or defense imperative, and industrial or military research is reduced, astronomers and astrophysicists will need to be prepared to carry out costly research in certain critical areas. That will require intelligent planning. In fields where the military has not been interested to date, progress has been noticeably slower than elsewhere. A prime example is the relativity experiment, Gravity-Probe-B. This heroic effort spearheaded first by William Fairbank, and now by Francis Everitt, has been in progress for three decades and still awaits launching. Most of us feel it is an important experiment and that it needs to be carried out. But the slow pace of progress is symptomatic of what it may be like to do research without the support of military advance research easing the way.

As a first step towards coping with major impending changes in the way we work, we will need to understand the extent to which we have been dependent on the military and others, so that we might be in a position to plan astronomical progress intelligently. And for that, we'll need to pay far more attention than we have, to books by scholars like Robert Smith, Karl Hufbauer, Walter McDougal, and others who have studied, and to a reasonable extent captured, the nature of modern astronomical research.

In conclusion then, three thoughts that I've tried to emphasize. First, let's try to understand the actual needs for astronomy in our society, in terms of our utility to society and in the longer run to civilization. That's what Robert White would recommend. Second, let's make sure we have an informed citizenry, which understands what science can do as well as its limitations. We could only be helped if every Congressman was able to come up with that kind of assessment. And third, let's try to understand our own working methods better, and our dependence on research and technologies springing up in the military and industry.

All three factors are crucial. Maybe we won't be able to bring back the spirit of the heady 1960s. But if we begin to shed better light on the nature of the process, it will assuredly help us to strengthen astronomical research.

I believe that is something all of us would wish to support.

An Artist's Rendition
of the Milky Way Galaxy

AN ARTIST'S RENDITION OF THE MILKY WAY GALAXY

Jon Lomberg
P.O. Box 207 Honaunau, Hawaii 96726

ABSTRACT

The most up-to-date information on the large scale structure of the Milky Way has been incorporated in a painting that presents in image of the Galaxy in an oblique view from a bit more than twice the solar distance from the center.

1. THE PAINTING

Plate 2 shows a painting by astronomical artist Jon Lomberg, commissioned by the National Air and Space Museum for its exhibition "Where Next, Columbus?", which opened in December, 1992. The original painting is a 6' × 8' canvas, on display at the museum.

The painting resulted from a collaborative research effort by the artist and Jeff Goldstein of the Laboratory for Astrophysics at the National Air and Space Museum. They were guided by the advice of Leo Blitz of the University of Maryland. Woodruff Sullivan of the University of Washington also consulted on the project.

The painting attempts to represent an accurate synthesis of various data sets combining current best estimates of large scale Galactic structure with known locations of almost 300 individual objects. Known clusters and nebulae in the Galactic disk are identified by purple or green dots as described below. Imaginary objects have been placed in plausible locations to suggest the vast numbers of undiscovered objects that must surely exist. Since the intent of this painting is to convey the range and positions of various kinds of Galactic features, modifications have been made to the brightness of objects to permit their inclusion. However, the relative brightness of objects in the painting is realistic.

The perspective chosen for this view is that of an observer located 18.8 kiloparsecs from the Galactic center and 10 degrees above the Galactic plane. The locations and sizes of the various components have been accurately computed for this view. The Sun and the Galactic center are aligned on the vertical axis of the painting. The position of the Sun is just above and to the right of the California Nebula.

The detail in the painting is extremely fine to accommodate the mural dimensions, and some details will not be visible in the figure unless it is viewed with a magnifying glass.

2. SPIRAL STRUCTURE

The Sun is located between the Sagittarius and Perseus arms, in a spur running diagonally between them. The Norma arm and the 3 kpc arm lie further in. The spiral structure has been filled out with plausible spurs and structure on the far side of the center. An asymmetrical "bar-like" structure is visible as a subtle elongation of the halo slightly to the left of the vertical axis.

3. GLOBULAR CLUSTERS

With one exception, all globular clusters are mapped in true locations. All clusters have been enlarged by about 50% in order to make them more visible, but the relative sizes are approximately accurate. The large cluster at the extreme left is an imaginary cluster, included so the structure of globular clusters would be clearer to the Museum visitor.

4. NEBULAE AND OB ASSOCIATIONS

Two sources have been used to identify the location and size of nebulae and OB associations: the Sharpless (1959) catalogue of optical HII regions and the survey of very large radio-bright HII regions compiled by Lockman (1979). Objects from the Sharpless catalog are marked with a purple dot beside them, and objects from the Lockman survey are marked with a green dot. The artist has placed a visible HII region in the locations found in the Lockman survey. (These purple and green dots are readily visible in the original artwork but may be hard to see in the figure.)

Many familiar, nearby nebulae are depicted in the painting. The Air and Space Museum's Laboratory for Astrophysics mapped their locations and calculated their sizes, and the artist used photographs as the source of the renditions here. The brightness of the nebulae has been slightly exaggerated for the sake of clarity. One of the chief aims of this painting is to show where familiar objects are located in the Galaxy. Therefore, the nebulae have been painted as they appear from Earth.

Barnard's Loop is shown as a large, faint ring around the Orion Nebula. A number of other similar loops have been imagined in other locations around the Galaxy (*e.g.*Blitz 1992).

5. DUST

In a few places (*e.g.* the Coal Sack nebula) the artist has mapped actual dust clouds, but most of the dust in this painting is imaginary, plausibly placed along the inner edge of arms and spurs. Dust in the Sagittarius arm has been arranged to suggest the look of the actual dust lanes seen from Earth in the direction of the Galactic center.

6. FAMILIAR OBJECTS DEPICTED IN MILKY WAY PAINTING

1) Orion Spur (location of Sun)
2) Sagittarius Spiral Arm
3) Perseus Spiral Arm
4) Omega Centauri Globular Cluster
5) Orion Nebula
6) Lagoon Nebula
7) Veil Nebula (Supernova remnant)
8) Coal Sack Nebula (Dark Nebula)
9) North America Nebula

10) Eta Carina Nebula
11) Barnard's Loop (Supernova remnant)
12) Rosette Nebula
13) Cone Nebula
14) Crab Nebula
15) Norma Spiral Arm
16) Orion OB2 Association
17) Jewel Box Cluster
18) Eagle Nebula
19) 3 kiloparsec Spiral Arm
20) Omega Nebula
21) California Nebula
22) Bubble Nebula
23) Gum Nebula (Supernova remnant)
24) Trifid Nebula

ACKNOWLEDGEMENTS

Painting © 1992 by Jon Lomberg and National Air and Space Museum.

REFERENCES

Blitz, L. 1992, Nature, 357, 361
Lockman, F.J. 1979, ApJ, 232, 701
Sharpless, S. 1959, ApJS, 4, 257

COLOR

PLATES

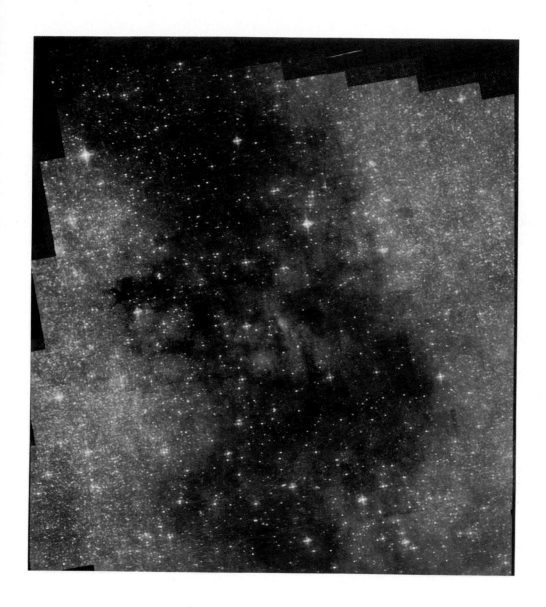

Plate 1. A false-color composite of two Galactic center images obtained with the near-infrared camera described by Ueno *et al.* in this volume. The color coding is blue for H band and red for K′ band filters. The coordinates and orientation of this image are shown in Fig. 2 of Ueno *et al..* This composite reveals a ring-like structure in the obscuration around the Galactic center.

Plate 2. This is a painting by astronomical artist Jon Lomberg, commissioned by the National Air and Space Museum, which uses the most up-to-date information on the large scale structure of the Milky Way to construct an image of the Galaxy in visible light. Pictured here is the view seen by an observer located slightly above the disk, at twice the solar distance from the Galactic center. Objects which appear in the painting are described in the article by Lomberg in this volume.

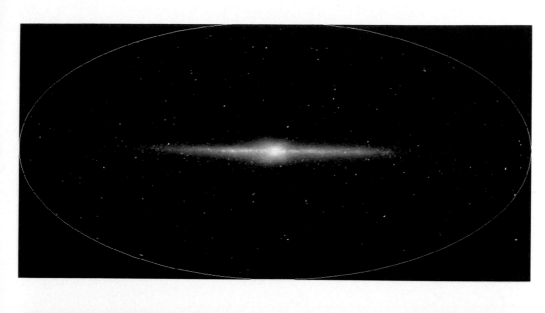

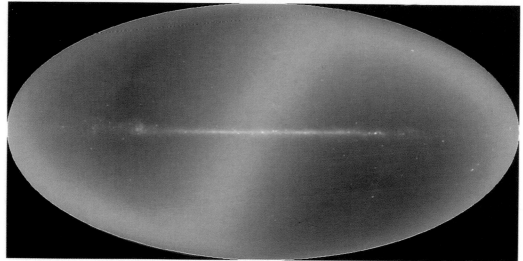

Plate 3. (*Top:*) A full sky, false-color image in galactic coordinates of the photometric intensity observed by the DIRBE instrument on the *COBE* satellite at 1.25, 2.2, and 3.5 μm (blue, green, and red respectively). (*Bottom:*) A full sky, false-color image in galactic coordinates of the photometric intensity observed by the DIRBE instrument on the *COBE* satellite at 4.9, 12, and 25 μm (blue, green, and red respectively). In both images, the solar elongation angle is 90° for all pixels. They are presented in this volume by Hauser.

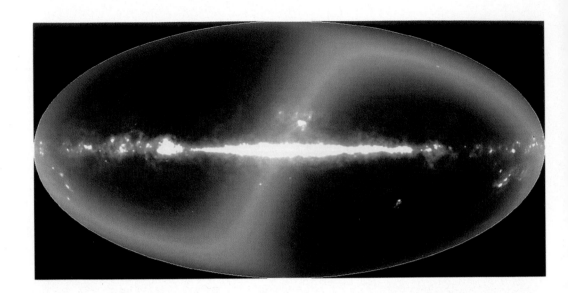

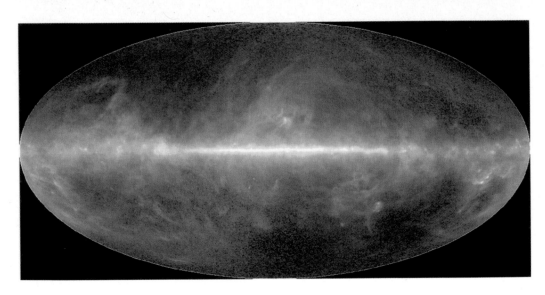

Plate 4. (*Top:*) A full sky, false-color image in galactic coordinates of the photometric intensity observed by the DIRBE instrument on the *COBE* satellite at 25, 60, and 100 μm (blue, green, and red respectively). (*Bottom:*) A full sky false-color image in galactic coordinates of the photometric intensity observed by the DIRBE instrument on the *COBE* satellite at 100, 140, and 240 μm (blue, green, and red respectively). In both images, the solar elongation angle is 90° for all pixels. They are presented in this volume by Hauser.

COBE FIRAS [C II] (158 μm) Line Intensity

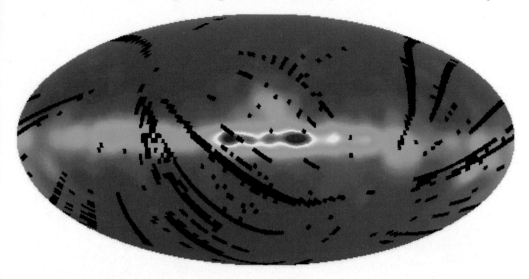

COBE FIRAS [N II] (205 μm) Line Intensity

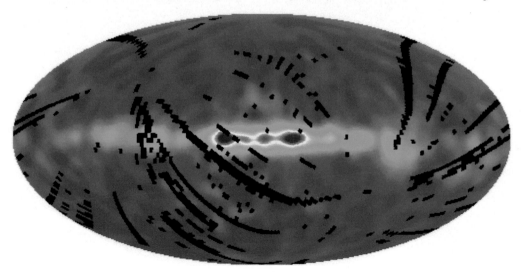

Plate 5. (*Top:*) A full sky map in galactic coordinates of the spectral line intensity of the 158 μm C II transition as observed by the FIRAS instrument on the *COBE* satellite. (Note that Plate 7 is a map of the C II transition observed in the Galactic plane by BICE). (*Bottom:*) A full sky map in galactic coordinates of the spectral line intensity of the 205 μm N II transition as observed by the FIRAS instrument on the *COBE* satellite. These maps are presented in this volume by Bennett & Hinshaw.

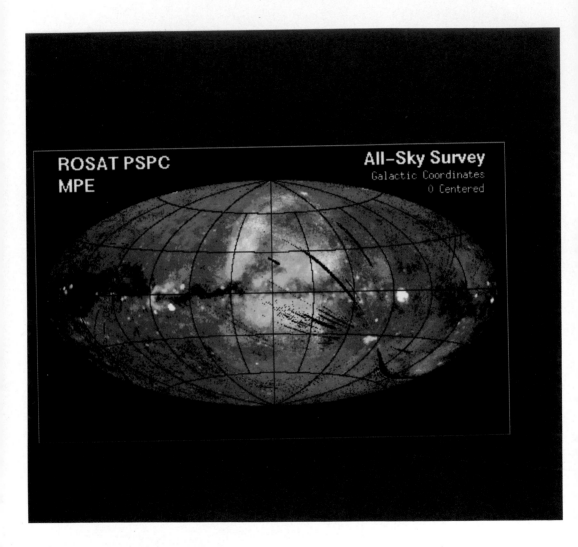

Plate 6. Three-color composite map of the *ROSAT* X–ray Sky Survey of the Max Planck Institute for Extraterrestrial Physics (Snowden *et al.* 1993b). Red shows the 1/4-keV intensity, green is 3/4-keV, and blue is 1.5 keV. This is a preliminary version with 2° angular resolution; the intrinsic resolution of the sky survey is about 3 arc minutes. This work is discussed in the article by McCammon in this volume.

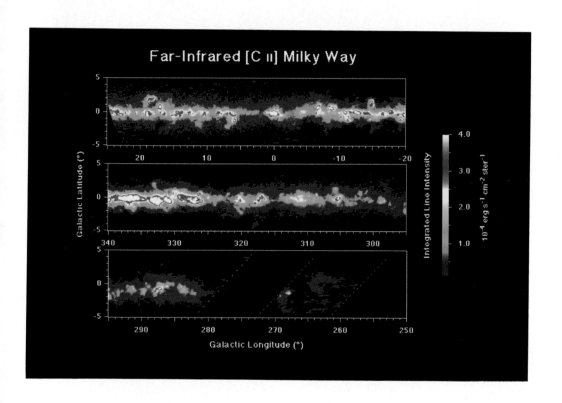

Plate 7. The far-infrared (158 μm) [C II] line emission map of the Galactic plane observed by the Balloon-borne Infrared Carbon Explorer (BICE). This data is discussed in this volume by Nakagawa *et al.* and by Doi *et al.*. The effective resolution is 15 arcmin. (Note that Plate 5 is a full sky map of the [C II] transition observed by the FIRAS instrument on the *COBE* satellite.)

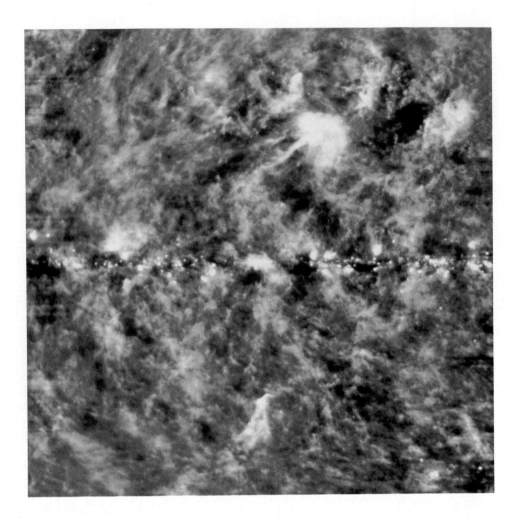

N
↑

Plate 8. Fine-scale structure in the 100 μm emission for the 60° × 60° field covering the Galactic center, bulge, molecular ring, and nearby regions. The smoothly-varying midplane-dominated emission has been eliminated by dividing out a smoothed version of the original IRAS 100 μm mosaic. The nearby Ophiucus and Lupus dark clouds are clearly evident to the north of the Galactic plane, and the Corona Australis dark cloud is prominent to the south. The remaining fine-scale features probably represent a superposed mix of nearby and more distant filaments and shells. These structures are discussed by Waller & Boulanger and Boulanger & Waller in this volume. For comparison, a different filtering of the 100 μm emission of the first Galactic quadrant is presented in Plate 12.

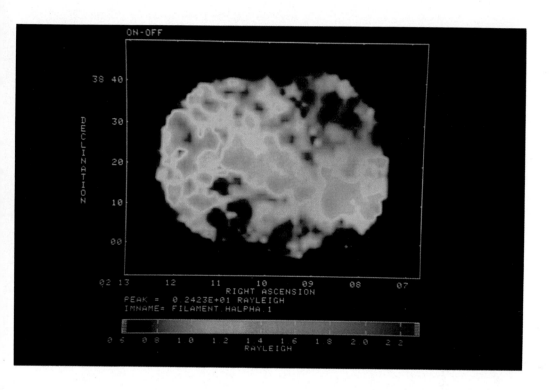

late 9. A high angular resolution (3′), narrow radial velocity interval (-56 m s^{-1}; LSR) map of the $H\alpha$ background within an approximately 1° diameter eld centered at 2^h10^m +38°20′ (1950) ($l = 140.°0$, $b = $-21.°6) (see Fig. 1 of eynolds in this volume). The image shows small scale structure associated with portion of a faint (emission measure ≈ 2 cm^{-6} pc), narrow ($\approx 1°$) filament ionized gas that stretches more than 15° across the sky (Ogden & Reynolds)85). The filament fills the oval shaped field of this map. This map was)tained by subtracting an off-line image with the pass band centered at -112 n s^{-1} from the on-line image centered at -62 km s^{-1} (Kung, McCullough, eynolds, & Tufte, in preparation).

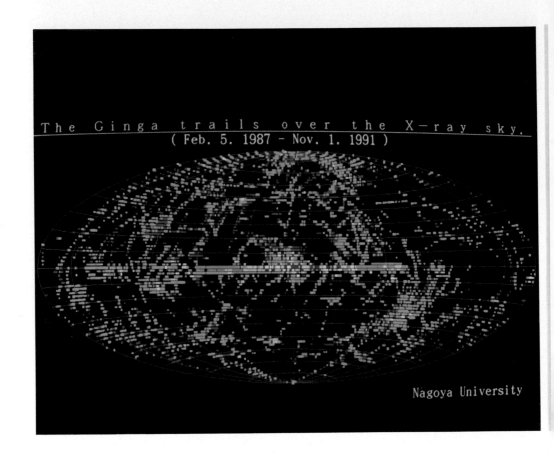

Plate 10. The Ginga satellite trails over the sky in the X-ray energy range of 2 – 10 keV. In this plate, the hue indicates the intensity of the X-ray detection in units of counts/sec/LAC beam. The color coding is blue (10 – 15), light blue (15 – 20), green (20 – 50), yellow (50 – 100), red (100 – 500), and white (> 500). Plate courtesy of H. Awaki, discussed in this volume by Koyama & Yamauchi.

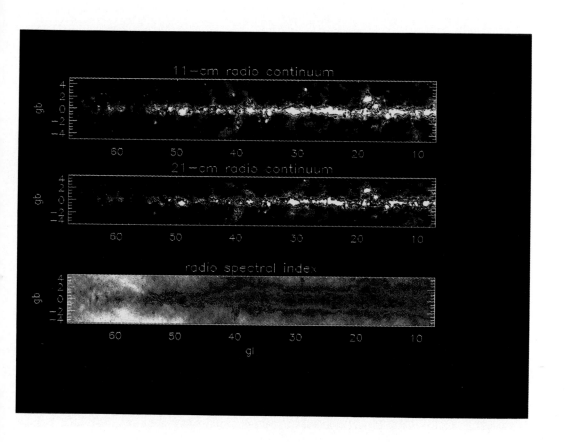

Plate 11. Median-filtered radio continuum maps at 11 cm *(top)* and 21 cm *(middle)*, and the radio spectral index *(bottom)*, for the first Galactic quadrant. Worms are evident in the radio continuum maps a structures protruding from the Galactic midplane. The spectral index map shows variations from predominantly thermal in the midplane (dark red) to nonthermal at high latitudes and high longitudes. Note that the radio spectral index pertains to the total emission (worm+source+background); the worms are much fainter than the radio background, so they only cause small changes in the index. This image is discussed in the article by Reach, Heiles, & Koo in this volume.

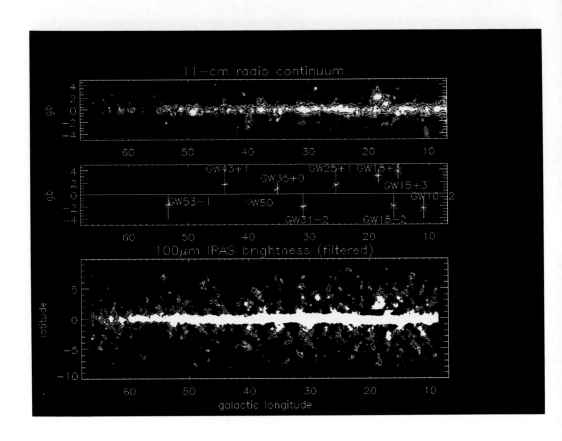

Plate 12. Median-filtered infrared surface brightness maps at 100μm *(bottom)* together with a worm-finder chart *(middle)* and a reproduction of the radio 11 cm map *(top)* for the first Galactic quadrant. All the worms in Table 1 of the article by Reach, Heiles, & Koo in this volume are included on the finder chart some are too small to be clearly seen in this global view. The position of the bright radio supernova remnant W50 is indicated on the finder chart because it looks like a worm on the radio maps.

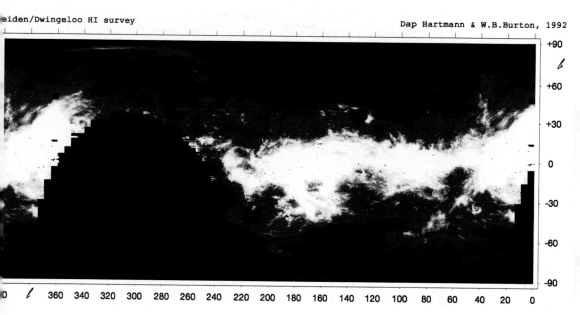

Plate 13. Distribution on the plane of the sky of HI column densities integrated over the range of velocities $-300 \leq v_{\mathrm{LSR}} \leq +300$ km s^{-1}. This large velocity range incorporates essentially all of the HI emission associated with the Galaxy, excepting weak emission from some high–velocity–cloud objects at extreme velocities. This large range obviously suppresses substantial kinematic information. Much structural information is also smoothed away, because of the inherent kinematic variations which lend a quite feathery appearance to moment maps calculated over narrow, *i.e.* $\Delta v \leq 5$ km s^{-1}, intervals. The false–color coding represents N(HI) in the range $0 - 2500 \times 10^{18}$ atoms per cm^{-2}. (Data from the Leiden/Dwingeloo 21–cm survey of galactic HI; Dap Hartmann and W.B. Burton, Leiden Observatory.)

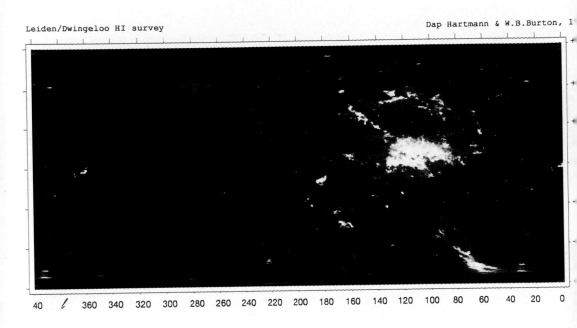

Plate 14. Distribution on the plane of the sky of HI column densities integrate
over $-375 \leq v_{\mathrm{LSR}} \leq -125$ km s^{-1}. This range of velocities incorporates mo:
emission from the high–velocity–cloud objects at negative velocities, althoug
some weak emission is found at more extreme negative velocities. Some isolate
knots in this map are contributed by external galaxies. The large patch o
emission lying several degrees above the galactic equator and covering much o
the second longitude quadrant is contributed by the outer reaches of the Milk
Way. The narrow pattern emanating from this patch is the HVC Complex A; th
pattern arching over the patch is the HVC Complex C. The emission extendir
toward the south galactic pole in the first quadrant comes from the Magellan
Stream. HVC complexes show much kinematic structure which is lost under th
large range of integration. Some of the HVC patterns may be traced across th
arbitrary boundary of -125 km s^{-1}. The color coding represents N(HI) in tl
range $0 - 75 \times 10^{18}$ atoms per cm^{-2}. (Data from the Leiden/Dwingeloo 21–c:
survey of galactic HI; Dap Hartmann and W.B. Burton, Leiden Observatory.

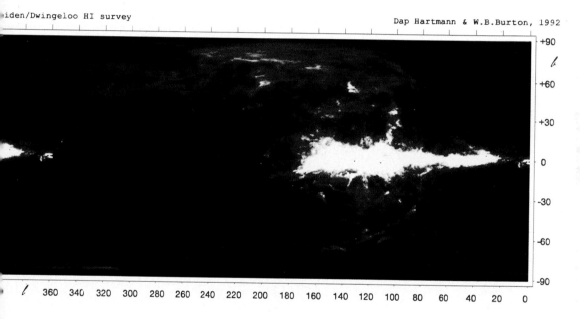

late 15. Distribution on the plane of the sky of HI column densities integrated ver the range of velocities $-50 \leq v_{\mathrm{LSR}} \leq -45$ km s^{-1}. The detailed topography galactic emission at such velocities remains largely unstudied; this range of locities lies outside that entering the Heiles–Habing survey. These velocities clude some of the emission from the anomalous intermediate–velocity–cloud jects, as well as some of the emission from the conventionally–defined galac- c gas layer. The patchwork pattern results from the low–level ($T_b \leq 0.4$K) ntamination from radiation entering the sidelobes of the telescope for which e data represented here have yet to be corrected. The false–color coding rep- sents N(HI) in the range $0 - 100 \times 10^{18}$ atoms per cm^{-2}. (Data from the iden/Dwingeloo 21–cm survey of galactic HI; Dap Hartmann and W.B. Bur- n, Leiden Observatory.)

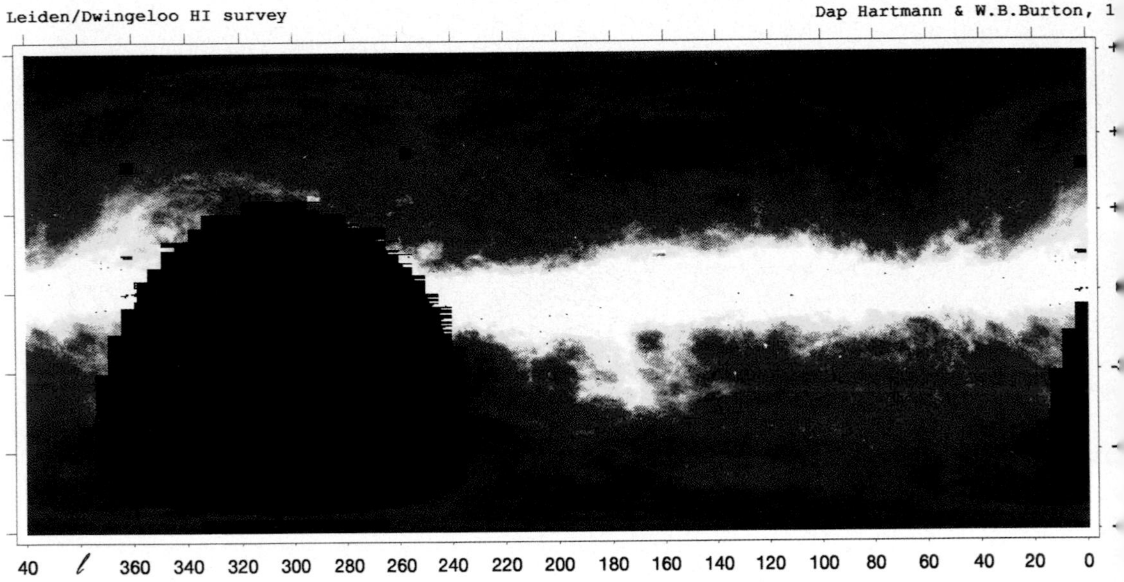

Plate 16. Distribution on the plane of the sky of HI column densities integrate over the range of velocities $0 \leq v_{\mathrm{LSR}} \leq +5$ km s^{-1}. This range of velocities "permitted" at longitudes in the first and third longitude quadrants, and mild "forbidden" elsewhere. Much, but certainly not all, of the emission represente in this map is contributed by HI gas lying within several hundred pc from th Sun. Some of the structural features shown here have characteristic veloci scales actually less than the 5 km s^{-1} extent of the integration; moment ma representing a smaller velocity range typically show a more feathery structur The false–color coding represents N(HI) in the range $0 - 500 \times 10^{18}$ atoms p cm^{-2}. (Data from the Leiden/Dwingeloo 21–cm survey of galactic HI; Da Hartmann and W.B. Burton, Leiden Observatory.)

INTRODUCTORY

OVERVIEWS

THE MILKY WAY: A TYPICAL SPIRAL GALAXY?

FRANK J. KERR
University of Maryland / USRA

I was asked to begin the proceedings by setting our Galaxy in context relative to the extragalactic universe. For the rest of the conference, other galaxies will be mentioned rarely, by people wanting to make special points.

What do we know? Firstly, our Galaxy is a field galaxy and not in a tight cluster. It has small satellites, and a number of small neighbors, but there are no large galaxies nearby. Secondly, it is a large galaxy - one of the largest in fact, but its exact ranking depends on the choice of the Hubble constant, which determines distances to other galaxies, and thus their sizes. It has been known for a long time that it is a very flattened system, but until early in this century we were thought to be approximately in the center. This view was changed when the effects of interstellar extinction were recognized.

The best evidence for the Milky Way being a _typical galaxy_ comes from a wide view towards the center. Whole-sky optical views started to show similarities to an edge-on galaxy, but the really strong evidence has come from infra-red views, especially that resulting form the DIRBE instrument on the COBE satellite. This well-known view shows the Milky Way system as a typical very thin and flat edge-on galaxy, with a widening in the central few kiloparsecs, corresponding to the budge. The galaxies most mentioned as similar to our own are NGC891 and NGC4565, which are both edge-on and with a similar sized bulge. The similarity is very striking indeed.

These comparisons show that we are in a typical galaxy, but are we in a typical _spiral_ galaxy? The first evidence for spiral structure in our Galaxy was obtained by Morgan, Sharpless and Osterbrock (1952), and caused great excitement when it was produced at the Rome IAU General Assembly in 1952. They plotted the positions of all the stellar associations and clusters known at the time, and found a clear indication of three pieces of trailing spiral arms, although of course limited to a small area near the Sun.

Since then, there has been a long succession of studies, trying to understand the spiral pattern. To cover the whole Galaxy, radio methods are of course best, in order to penetrate the interstellar dust, and spectral lines give the possibility of measuring radial velocities and thus of estimating distances by making use of our knowledge of the velocity pattern. Hence the greatest effort has gone into studies of the 21-cm HI line and the 3-mm CO line. One of the earliest attempts was the well-known diagram by Oort, Kerr and Westerhout (1958) dating from the late fifties. This was based on 21-cm observations from Leiden and Sydney, to cover the two hemispheres, and showed a well-developed spiral pattern with lengthy arms extending over the whole Galactic disk. The trailing arms in this pattern displayed a pitch angle of about 7^0, and the diagram was affected by an apparent asymmetry in the motions as between the two sides in the region inside the solar circle. The same study revealed the warp in the outer parts of the gas layer, as well as leading to the definition of the new system of galactic coordinates.

We now know that we were too optimistic then in the interpretation of the data, because of the impossibility of allowing properly for the presence of noncircular velocity components when we aim to go from the measured radial velocities to the distances of the various arm segments. Many studies of the spiral pattern have been made since then, using various constituents as spiral arm traces, but it is interesting to note that the rate of publication of new diagrams and new papers on the subject has been slowing down, with some lengthy reviews ending without the authors feeling confident about showing any spiral diagram at all (e.g. Burton 1992). The main difficulty arises from the fact that distances to objects far away in the Galaxy are mostly derived kinematically, i.e. distances to objects are estimated from their measured radial velocities, assuming that one knows the galactic rotation pattern. However, the motions are not entirely circular, and quite small noncircular components can produce substantial errors in the derived distances. This effect, coupled with the existence of a distance ambiguity inside a circle through the Sun (there can always be two distances corresponding to a particular radial velocity), together with the difficulty of deciding how to join together multiple segments of arms into long, continuous structures, makes the drawing of the spiral pattern a very difficult exercise. It is a good illustration of the difficulty that spiral diagrams for the Galaxy in the literature cover the range from 2 to 13 arms, with pitch angles varying from 5^0 to 25^0.

For the region inside the solar circle, the most certain evidence for the existence of a large-scale spiral pattern is found in the longitude-velocity plane for HI or CO observations. Lengthy interstellar structures can be identified in such an l-v diagram. The problem lies in the transformation to a real map, because of the difficulties mentioned above. Spiral arms are certainly present, but they are not easy to delineate, and no really acceptable diagram of the overall spiral pattern exists at the present time.

We can do better outside the solar circle, where there is no distance ambiguity, although we must still contend with the existence of noncircular velocity components. Combining HI observations from Parkes and Hat Creek shows the presence of long, continuous structures (Henderson et al. 1982; Kulkarni et al. 1982). The outer patterns on the two sides of the Sun-center line show an interesting asymmetry, giving more evidence of the way that lack of knowledge of the velocity distribution can affect the appearance of the spiral pattern. The apparent asymmetry could be due to a real structural asymmetry, it could be entirely due to the reflection of an asymmetry in the velocity pattern, or it could be a mixture of the two effects.

CO observations have provided another approach to the spiral problem, with similarities and differences. CO has one advantage over HI that it is not so ubiquitous in the Galactic disk, and therefore there is more chance of separating the arms or segments of arms. However, the overall conclusions are very similar. We can see lengthy features and small segments, but there is the same difficulty in drawing an overall pattern.

Is there any possibility of improving knowledge of the spiral pattern? We are unlikely to better the understanding of the velocities, at least for the smaller-scale features, but astrometry is being revolutionized and might open up alternative approaches to the distance problem. We will be able to get better distances to optical and radio sources in the Galaxy from Hipparcos, HST, and radio and optical interferometers.

Francoise Combes (1991) has asked the interesting question of how we would get on if we were in the disk of a very well-defined spiral like M51 and tried to work out the spiral pattern from HI or CO observations. She worked out a longitude-velocity diagram from CO observations of M51, if viewed from an interior position analogous to our own, and then attempted a spiral map. This exercise did not lead to a clear-cut spiral. We can turn the argument around and say that the difficulty of seeing a well-defined pattern in our own

Galaxy does not definitely rule out the possibility that we do have as clear a spiral as M51 does.

In other galaxies, the spiral pattern ranges from a very regular style, as in M101 or NGC4622 to a very irregular or "flocculent" type, as in NCG253. Our Galaxy probably lies somewhere between these extremes in style, but the degree of regularity is difficult to determine.

If we look for other characteristics of typical spirals which are reproduced in our Galaxy, we should especially mention a flat rotation curve, with a velocity a bit over 200 km/sec, and a warp in the outer parts (Kerr 1957; Burke 1957). Various people, such as de Vaucouleurs, have discussed the possibility that we live in a barred galaxy. Recently, Blitz and Spergel (1991) have produced more definite evidence from HI and CO studies that our Galaxy has a pronounced bar, of intermediate strength as far as bars go.

Accepting that the Galaxy is a spiral, we want to consider what is its Hubble type. Although we cannot delineate the spiral pattern very well, we can get clues from various parameters. The rotation velocity puts the type between Sb and Sc, the probable pitch angle of 7^0-10^0 fits a type near Sb, as does the extent of the central bulge. Table 1 lists a number of criteria which have been used by various authors to give evidence of the Hubble type for our Galaxy.

de Vaucouleurs and his coworkers have developed estimates of various photometric parameters in a way which could be compared with similar quantities for other galaxies. Table 2 (by de Vaucouleurs and Pence 1978) presents comparisons of this type between our Galaxy and the average of four galaxies in the Sbc region of the classification scheme. The two sets of numbers are very similar.

Overall, we say that our Galaxy is a typical spiral galaxy but we cannot say very much about the detailed spiral pattern. The Hubble type seems to be in the region of Sb or Sbc. The degree of regularity of the pattern is not well known.

REFERENCES

Blitz, L. and Spergel, D. N. 1991, **379**,631
Burke, B. F. 1957, **62**, 90
Burton, W. B. 1992, "The Astron. and Astrophys. Encyclopedia", ed.
    S. P. Maran, van Nostrand Reinhold, 204
Combes, F. 1991, Ann. Rev. Astron. Astrophysics **29**, 195.
Hendenson, A. P., Jackson, P.D., and Kerr, F. J. 1982, Astrophys. J.,
    263:116.
Kerr, F. J. 1957, Astron. J., **62**, 93.
Kulkarni, S. R., Blitz, L., and Heiles, C. (1982), Astrophys. J.
    Letters, **259**, L63
Morgan, W. W., Sharpless, S., and Osterbrock, D. E. (1952), Astron. J.,
    57, 3.
Oort, J. H., Kerr, F. J., and Westerhout, G. (1958), Mon. Not. R. Astron. Soc.,
    118, 379.
de Vaucouleurs, G., and Pence, W. D. (1978), Astron. J., **83**, 1163.
Wevers, B. M. H. R. (1984), Ph. D. Thesis, Groningen.

Table 1. Some indicators of Hubble type

Criterion	Indicated Type
Extent of bulge	Sb
Bulge to disk ratio (de Vaucouleurs)	Sbc
Rotation velocity	Sbc
Azimuthal profiles - disk colors (Wevers)	Sb
Ratio of surface brightness in B band to	
HI surface density	Sb
CO central depression	Sb

Table 2. Photometric Parameters of the Galaxy and the
Average of Four Spirals **

Parameter		Average	Galaxy
Isophotal diameter,	D_0 (kpc)	22.7	23.0
Effective diameter,	D_e (kpc)	10.3	10.2
Inner ring diameter,	$D(r)$(kpc)	5.75	6.:
Absolute B magnitude,M_T°		-20.0	-20.1
Color index, $(B - V) \, _T^{\circ}$		0.51	0.53
Spheroidal fraction, K_I (B)		(0.33)+	0.34
Luminosity index,Λ		0.67	0.71
Mean effective surface brightness*,$m_e^{!}$		22.08	22.06

**NGC 1073, SB(rs)c II; NGC 4303, SAB(rs)bc I: NGC 5921, SB(rs)bc I-II;
NGC 6744, SAB(r)bc II. +NGC 6744 only. *B mag sec $^{-2}$.

CURRENT ISSUES IN THE STUDY OF THE LARGE-SCALE PROPERTIES OF THE MILKY WAY

Jeremiah P. Ostriker
Dept. Astrophysical Sciences, Princeton Univ., Princeton, NJ 08544
JPO@astro.princeton.edu

Frances Verter
MC 685, NASA Goddard Space Flight Center, Greenbelt, MD 20771
verter@stars.gsfc.nasa.gov

1. INTRODUCTION

The intention of this invited review is to pose questions, not present answers. For our own, best-studied, Galactic system, there is a great deal that we do not know; and of what we do know, there is a great deal which we do not understand. Working from the inside of the Galaxy outwards, these are some of the questions that we continue to ask. We will not understand galaxy formation and the universe, until we understand our own Galaxy.

2. GALACTIC CENTER (0.1 - 10 pc)

1) Are there one or more black holes at the Galactic center?

A black hole of mass $\sim 10^6 M_\odot$ could be responsible for the high orbital velocities in the center of the Galaxy (Lacy 1989) and might drive the magnetic field in the dramatic nonthermal filaments (Yusef-Zadeh 1989). These and other outstanding puzzles in the study of the Galactic center are reviewed by Genzel & Townes (1987), and by Townes (1989).

But the existence of a black hole should create an accretion disk, with characteristic emission from infalling material (Lynden-Bell & Rees 1971; Blandford & Rees 1992). If there is a black hole in the Galactic center, then why doesn't the nucleus of the Milky Way look like the centers of AGN galaxies (Lo 1989; Phinney 1989; Ozernoy 1989; but see Mirabel *et al.* 1992)?

2) What is the distribution of normal stars in the Galactic center?

Even in the near-infrared, the center of the Galaxy is veiled by murky extinction (see the review by Rieke in this volume). Is the peak of the central luminosity profile a sharp cusp around a central point mass, or is it more gradual like the nucleus of M31? Estimates of the core radius of the Galactic light profile have ranged from ≤ 0.08 pc (Allen, Hyland, & Jones 1983) to 0.6 pc (Rieke & Lebofsky 1987). By comparison, the core radius of M31 is 1.4 pc (Light, Danielson, & Schwarzschild 1974), and the tightest globular cluster has a core radius of 0.2 pc (Peterson & King 1975).

In the nuclear region of the Galaxy, is there ongoing star formation? The possibility that a burst of star formation occured in the Galactic center a few million years ago is based on the existence of M supergiants (Lebofsky, Rieke, & Tokunaga 1982) and a WN9 star (Allen, Hyland, & Hillier 1990).

3. GALACTIC BULGE / SPHEROID ($10^2 - 10^3$ pc)

1) Does the Galaxy have a triaxial bulge or an inner bar or both?

Blitz & Spergel (1991a) argue that a slowly rotating triaxial spheroid, extending beyond the solar circle, can explain large-scale asymmetries in the distribution and motion of HI in the outer disk, as well as offsets between the velocity centroids of various Galaxy components. An alternative interpretation of these motions is a lopsided distortion of the outer Galaxy (Kuijken 1991a). These two models are reviewed by Spergel in this volume. Metzger & Schechter claim in this volume that the opposite motions of the stars and gas in the outer Galaxy favor the lopsided distortion model. The triaxial spheroid model can be tested by surveys of OH/IR stars in the bulge, using the method of Habing (1988) to select stellar candidates.

Blitz & Spergel (1991b) argue further that an inner bar, located well inside the molecular ring and oriented nearly perpendicular to their triaxial spheroid (see their Fig. 1), is implied by 2.4 μm photometry of the Galactic center (Matsumoto *et al.* 1982). Weinberg (1992) also interprets his IRAS 12 μm survey of AGB variables as providing a direct map of an inner bar (see his Figs. 2 & 3 in this volume). Binney *et al.* (1991) argue independently for an inner bar to explain the orbital dynamics of molecular and atomic gas near the Galactic center. The predicted observational signatures of an inner bar are asymmetries in the stellar photometry of the Galactic center, described by Blitz in this volume, and noncircular motions of gas in the Galactic center, described by Binney in this volume. The best test of the predicted near-infrared photometry will be made by data from the DIRBE instrument onboard the COBE satellite; preliminary results are reported by Weiland *et al.* in this volume.

Photometric observations of external galaxies suggest that spherical or axisymmetric bulges may be the exception rather than the rule (Zaritsky & Lo 1986). However, "peanut-shaped" or even "boxy" isophotes can be produced by more than one mechanism, including bar instabilities (Raha *et al.* 1991) and/or satellite accretion (Statler 1988).

2) What is the age of the bulge/spheroid?

The oldest identifiable population in the Galaxy are globular star clusters. Recent advances in our understanding of horizontal branch stars, coupled with CCD photometry of globular clusters, have made it possible to study the ages of globular clusters to markedly greater precision and distance than before (Bolte 1992; Lee 1992a). Lee (1992b) concludes that the oldest stars in the Galactic bulge, RR Lyrae, are $\sim$ 16 Gyr old. This must be a lower limit on the age of the Universe. Ongoing measurement programs that are capable of pinpointing globular cluster ages to $\leq$ 1 Gyr will further refine our understanding of the timescale of galaxy formation (Bolte 1992).

It now appears that the long-sought "second parameter" (after overall metallicity) which controls the distribution of horizontal branch colors in globular clusters is the cluster age (Bolte 1992; Lee 1992a). This identification allows bulge stars to be 1.3 $\pm$ 0.3 Gyr older than the halo population and yet more metal rich (Lee 1992b), removing a long-standing puzzle about the relative ages of the bulge and halo (see the review by Larson 1990).

When the optics of the Hubble Space Telescope are fully corrected, it should be possible to study the ages of stellar populations out to the edges of the Galactic halo and in virtually all Local Group galaxies (Bolte 1992; Lee 1992a).

3) Are globular clusters truly rare near the Galactic center?

One expects that they should be: Within ~ 2 kpc of the Galactic center, large globular clusters will tend to spiral into the nucleus by dynamical friction, and small clusters will tend to be disrupted by gravitational shocks from passage through the Galactic disk and bulge (Aguilar, Hut, & Ostriker 1988). Thus, the use of globular clusters to trace the early history of the Galaxy must allow for their subsequent evolution.

Do the observations confirm these predictions? The present distribution of globular clusters appears less centrally peaked than the underlying spheroidal population (Webbink 1985), and their velocity ellipsoid is nearly isotropic (Harris & Racine 1979), unlike the highly radial velocity dispersions of extreme population II stars. However, predictions of the past globular cluster population are strongly dependent on the assumed Galactic model (Aguilar, Hut, & Ostriker 1988).

4. GALACTIC SPHEROID / INNER DISK (1 - 5 kpc)

1) Is the Milky Way a barred spiral?

If seen from outside, would the Milky Way look like the prototypical barred spiral NGC1300 (Sandage 1961)?

Classic barred spiral models are generated by an oval distortion in the distribution of bulge stars that predate the disk (Sanders 1977). By comparison, the inner bar of Binney et al. (1991) and Blitz & Spergel (1991b) has the high pattern speed and peanut-shaped isophotes predicted for a bar that forms by a buckling instability of the disk (Raha et al. 1991).

The presence of a bar in the Galaxy would provide a driving mechanism for density waves that could generate the spiral arms (Roberts, Roberts, & Shu 1975). But SB spirals are observed to have stronger arm patterns than SA spirals (Kormendy & Norman 1979), so if the Galaxy is a barred spiral, then why is it so difficult to identify its spiral pattern (see the review by Kerr in this volume)?

2) Is the 3 kpc "expanding arm" actually streaming motions of gas in a bar?

Radio observations of gas in the inner Galaxy can be interpreted either as an expanding arm at a radius of 3 kpc (Oort 1977) or as gas streaming along a rapidly rotating bar with corotation radius near 3 kpc (Liszt & Burton 1980; Binney et al. 1991). The expanding arm hypothesis has the difficulty that any triggering explosion(s) would be more likely to expand out of the Galactic plane rather than into the disk (e.g., Mac Low & McCray 1988). The bar hypothesis has the advantage that bars occur naturally by dynamical instability in disk galaxies (Ostriker & Peebles 1973).

The existence of a rotating bar asymmetry in the potential of the inner Galaxy would alter the orbits of globular clusters, but would not strongly increase their destruction rate (Long, Ostriker, & Aguilar 1992).

3) Is the spheroid tipped with respect to the disk?

The gas disk itself is not perfectly flat, but has a tilt in the inner 3 kpc that mirrors the outer warp, perhaps due to cosmic infall (Ostriker & Binney 1989). In addition, models of the inner bar require that it be inclined a few degrees out of the Galactic plane (Binney et al. 1991; Blitz & Spergel 1991b). A similar tilt is seen between the bulge and disk of M31 (Ciardullo et al. 1988). However, the preliminary DIRBE results reported by Weiland *et al.* in this volume show that the apparent tilt of the Galactic bulge which is seen at a single near-infrared wavelength is revealed to be an extinction effect when photometric maps at several wavelengths are compared.

4) Can a self-consistent stellar model be contructed for the bar, bulge, and disk of the Galaxy?

The phenomena of triaxial structure, figure rotation, and matter streaming are coupled to each other by the requirement that the observed motions occur on predicted orbit streamlines. For example, Vietri (1986) has shown that if he assumes the gas in the Galactic center is all on stable, non-intersecting orbits, then the bulge must be triaxial and counter-rotating with respect to the disk.

If triaxial components of the Galaxy are confirmed, including some combination of a spheroid and bar, then the next issue is to constrain their rotation rates. Existing kinematic models of gas motions in the inner Galaxy, reviewed by Binney in this volume, have differed greatly in their estimation of pattern speeds and the radii at which resonances occur.

5. BODY OF THE GALAXY: THE FACTS (5 - 20 kpc)

We separate the discussion of the main body of the Galaxy into two pieces: those facts that we don't know (this section), and those facts that we do know but don't understand (next section).

1) Gas: Is the net gas flow into or out of the Galaxy?

In the fields of cosmolgy and chemical evolution, the prevailing paradigm is to regard gas as falling into galaxies. But in the study of the interstellar medium and its evolution, the paradigm is to view gas as being blown out of galaxies. Both can occur, as in the "galactic fountain" model of the interstellar medium (Bregman 1980; Houck & Bregman 1990). Thus, it may be that the net flow direction is time variable.

For example, a gas infall of 1 $M_\odot$ yr^{-1} should lead to an X-ray luminosity of 10^{41} erg s^{-1}; this is not seen today for most galaxies (Fabbiano 1989). Therefore, the net inflow of gas that is associated with galaxy formation cannot continue indefinitely.

2) Disk: Is there dark matter in the disk?

Ever since Oort (1932,1965) first solved the Possion equation to obtain the mass density in the solar neighborhood, it has been debated whether stars and gas can account for all of the gravitational mass. Bahcall (1984a,b) and coworkers have found that the mass to light ratio of known objects in the solar neighborhood, $2.6^{+1.9}_{-1.2}$ $M_\odot/L_\odot$ (Bahcall, Flynn, & Gould 1992) requires dark matter, whereas Kuijken and coworkers find a mass to light ratio, 1.0 ± 0.3 $M_\odot/L_\odot$ that is consistent with no dark matter in the disk near the Sun (Kuijken & Gilmore 1989; Kuijken 1991b). The census of dark material in the solar neighborhood, in the form of binary companions, faint degenerate dwarfs, etc., is so uncertain that the required "dark matter" may be nothing more exotic than ordinary objects which have been incompletely assessed.

3) <u>Thick Disk:</u> Does one exist?

What is meant by "thick"? In the past decade, a laborious program of counting stars at high latitude and determining their absolute magnitudes by photometric parallax has improved our knowledge of the local distribution of stellar luminosity and mass as a function of distance from the Galactic plane. Out to $z \sim 1$ kpc, the density law is dominated by the old disk, with exponential scale height ~ 300 pc; beyond this lies a "thick disk" of older, more metal-poor stars that can be described by an exponential of scale height $\sim 1 - 1.5$ kpc (Gilmore & Reid 1983).

Is the thick disk a separate stellar component or the high-latitude tail of the distribution of disk stars? All of the parameters describing the thick disk remain poorly determined, because their mean offset from the properties of thin disk stars is much less than the dispersion in these quantities. Perhaps with more observations, more model components will be needed, until ultimately we view the distributions of stellar age and location as a gradual merging of disk and halo.

Are there young stars far from the Galactic plane, other than runaways? Searches for star formation among high-latitude molecular clouds (Magnani, Caillault, & Armus 1990) and IRAS point sources (Paley et al. 1991) have been almost totally negative. But the weakness in these searches is that almost all the atomic and molecular cirrus that we can see at high latitude is very local (mean distance 100 pc; Magnani, Blitz, & Mundy 1985).

4) <u>Halo:</u> What is the Galactic halo made of; how big is it?

The concept that galaxies reside in wells of dark matter has gone from heresy to paradigm (Tremaine 1992). The suggested dark contents of the Galactic halo include brown dwarfs, stellar remnants, black holes, and various mysterious elementary particles.

The modeled extent of the Galactic halo ranges from 50 kpc (Little & Tremaine 1987) to a joint halo encompassing both the Galaxy and M31 (Peebles et al. 1989). This is related to the question of the mass of the Galaxy, which is reviewed by Fich & Tremaine (1991).

The typical isolated giant late-type spiral has a dark halo that extends $\gtrsim$ 200 kpc and contains $\gtrsim 10^{12}$ $M_\odot$ (Zaritsky et al. 1993).

5) <u>Spiral Structure:</u> What are the demographics of the arm contents?

Does the Galaxy have azimuthal fluctuations in the density of old disk stars, like the red arms in M51 (Zwicky 1955, Schweizer 1976)? So far, the detected fluctuations in the luminosity distribution of the Galactic plane can be explained by extinction (Kent, Dame, & Fazio 1991).

Can studies of spiral arms be used to constrain molecular cloud lifetimes? Gas dynamics are essential to maintain spiral structure, which would only exist as transient reactions in a purely stellar system (Sanders 1977). If giant molecular clouds only occur in spiral arms, then these clouds must be continually forming and breaking up as they orbit the disk. In principle, it should be possible to constrain the timescale of these processes. In practice, it is very difficult to model the hydrodynamics of the interstellar medium effectively, because it contains structure on all scales (Scalo 1990). Efforts to overcome this obstacle are reviewed by Binney in this volume.

6. BODY OF THE GALAXY: PHYSICS AND EVOLUTION (5 - 20 kpc)

1) Gas: What regulates star formation?

Another way of looking at this question is to ask what regulates the internal and external velocity dispersions of molecular clouds.

The external velocity dispersion of molecular clouds plays a role in regulating star formation because it governs the cloud collision rate. This dispersion is roughly constant over three orders of magnitude in cloud mass, indicating that the massive clouds do not experience equipartition of energy with the smaller clouds (Stark & Brand 1989). The velocity dispersion among small clouds is consistent with the amount of kinetic energy input by supernova shocks (McKee & Ostriker 1977), although improved fluid dynamic models are needed to demonstrate how this energy can be converted into bulk cloud motions. The velocity dispersion among giant molecular clouds can be explained by gravitational scattering (Gammie, Ostriker, & Jog 1991).

The internal velocity dispersion of molecular clouds plays a role in regulating star formation because it supports the cloud against gravitational collapse. The empirical mass-luminosity relation of molecular clouds is consistent with Virial equilibrium, but may only be an artifact of their empirical size-linewidth relation (Maloney 1990). The internal kinetic energy evidenced by the linewidths of molecular clouds has been attributed to turbulent motions (Kleiner & Dickman 1987; Falgarone, Puget, & Perault 1992), but here theory falls behind observation, because there are very few models of turbulent cascades which can be compared to the available data (Shore 1992).

On large scales in external galaxies, Kennicutt (1989) notes that the star formation rate is proportional to a weak power of the total gas surface density, as originally proposed by Schmidt (1959), provided the density exceeds a critical threshold. This threshold appears to be associated with the onset of gravitational instabilities leading to the growth of large-scale density perturbations (*i.e.* clouds) in the disk (Toomre 1964; Cowie 1981).

2) Disk & Halo: How are they related?

Two basic variants for the birth scenario of the Galaxy are the smooth collapse of a single protogalactic cloud (Eggen, Lynden-Bell, & Sandage 1962), versus the chaotic condensation and merger of subunits within a protogalactic cloud (Searle & Zinn 1978). The mean variation and dispersion of globular cluster ages as a function of Galactocentric distance are used to decide between these two scenarios. The latest studies find that clusters in the inner halo (Galactocentric R < 8 kpc) are relatively old and have an age spread of $\approx$ 1.5 Gyr, while clusters in the outer halo (R $\geq$ 8 kpc) are younger and have an age spread of $\approx$ 4.5 Gyr (see Fig. 6, Lee 1992a). These results imply that the formation of the globular cluster system occured from the inside out, over a prolonged time, consistent with the merger of many protogalactic gas clouds. The Galactic disk began to form 3 – 4 Gyr after the formation of the oldest halo population (Lee 1992a).

Is the disk stable? The susceptibility of a disk galaxy to various dynamical instabilities depends on the amount and distribution of dark matter in its halo (*e.g.*, Ostriker & Peebles 1973).

3) Thick Disk: How could it have formed?

Gilmore, Wyse, & Kuijken (1989) review scenarios for the origin(s) of the thick disk. We boil down the alternatives to three basic possibilities:

One is that the thick disk pre-dates the body of the galaxy, and is from the settling phase of galaxy formation. If this is true, further observations should confirm that stellar ages in the thick disk are intermediate between extreme population II stars and old disk stars.

A second is that the thick disk post-dates the body of the galaxy, being a high-latitude tail of the thin disk. Some examples of dynamical mechanisms that could increase the stellar velocity dispersion out of the plane enough to create a thick disk are the decay of the disk buckling instability (Raha *et al.* 1991), scattering of disk stars by halo black holes (Lacey & Ostriker 1985), or satellite accretion into the thin disk (Quinn & Goodman 1986; Toth & Ostriker 1992).

The third possible origin is that debris from satellite infall preferentially populates orbits in the thick disk (Statler 1988). The distinction between the options listed in the second and third categories requires observations which can determine whether the thick disk is kinematically distinct.

7. GLOBAL EVOLUTIONARY QUESTIONS

1) A consistent temporal account is needed for the formation of the major Galaxy components: bulge, halo, and disk.

Our basic concept of the collapse of a proto-galactic cloud and its evolution into first spheroidal and later disk components has not changed very much since the seminal paper of Eggen, Lynden-Bell, & Sandage (1962). Yet, none of the proposed models and scenarios really *works* in detail at explaining the observed correlations between the properties and ages of the various Galaxy components (see the review by Gilmore, Wyse, & Kuijken 1989).

Our efforts to deduce the formation history of the Galaxy from the properties of its present day components will always be hampered by those physical processes which act on the parameters used as historical tracers so as to rearrange them or reset their clocks. For instance, the fossil record encased in stellar orbits has been partially scrambled by the process of violent relaxation in the early history of the Galaxy (Lynden-Bell 1967), followed by secular scattering events (Wielen 1977). Stellar orbits may also have been altered by past mergers and cannibalism. The timescale of Galaxy formation can be significantly altered by the role of dissipational processes in the proto-Galactic nebula (*e.g.*, Fall & Efstathiou 1980; Gunn 1982). Finally, the use of metallicity abundances to clock past star formation can be thrown off by inhomogeneities in the rates of enrichment and/or collapse (Searle & Zinn 1978).

2) Evolution of globular cluster systems.

In the inner Galaxy, globular clusters are destroyed primarily by bulge shocking, while at the Galactic radius of the Sun, evaporation is the major destruction mechanism, followed by disk shocking and dynamical friction. Aguilar, Hut, & Ostriker (1988) estimate it is possible that a large fraction of the original population of globular clusters has perished.

How many globular clusters did the Galaxy have originally? Is the Galactic spheroid composed of stars from clusters that were disrupted within the cental 2 – 3 kpc of the Galaxy?

3) How are stars accelerated from the Galactic plane?

Stars are formed in a thin layer of molecular gas, but the velocity dispersion of stars in the solar neighborhood is observed to increase with age. This is believed to be caused by stochastic accelerations from scattering events over the life of the star (Spitzer & Schwarzschild 1951,1953; Wielen 1977). The rate of dynamical heating and orbit diffusion can be derived for specific populations of scattering bodies (Lacey 1984; Jenkins & Binney 1990), but so far none of the models quite fit the observations. Possible scatterers include stellar clusters, giant molecular clouds, spiral arms, black holes, and satellite interactions.

Photometric observations of edge-on spirals are consistent with the hypothesis that old disk stars undergo secular acceleration (van der Kruit & Searle 1982).

4) How much merging and infall of small satellite galaxies occured in the early history of the Milky Way?

Several lines of evidence suggest that galaxy mergers were much more prevalent in the past than today: In the recent past, IRAS has revealed large numbers of IR-luminous galaxy mergers (Sanders *et al.* 1988). In the distant past, galaxy counts (Cowie, Songaila, & Hu 1991) and QSO absorption line systems (Lanzetta 1992; Yanny 1992) suggest that there were large numbers of star-forming dwarf galaxies that are gone today. Models of galaxy mergers indicate that they can provide an explanation for the creation of (abnormal) ellipticals (Toomre 1977; White 1978; but see Ostriker 1980), cD galaxies (Hausman & Ostriker 1978), and starbursts (Barnes & Hernquist 1991). Moreover, mergers are even more common in galaxy groups than clusters (Barnes 1985; Mamon 1987). In this volume, Byrd *et al.* simulate the orbital evolution of the Local Group over a period of 30 billion years. Nonetheless, Toth & Ostriker (1992) argue that in our Galaxy and other spirals with similar disk scaleheight and velocity dispersion, the contribution of merging to the mass in the inner Galaxy has been less than 4% over the past 5 billion years.

ACKNOWLEDGEMENTS

FV is supported by grant LTSA-89-100 from the NASA Long-Term Space Astrophysics Research Program.

REFERENCES

Aguilar, L. A., Hut, P., & Ostriker, J. P. 1988, ApJ, 335, 720
Allen, D. A., Hyland, A. R., & Hillier, D. J. 1990, MNRAS, 244, 706
Allen, D. A., Hyland, A. R., & Jones, T. J. 1983, MNRAS, 204, 1145
Bahcall, J. N. 1984a, ApJ, 276, 169
Bahcall, J. N. 1984b, ApJ, 287, 926
Bahcall, J. N., Flynn, C., & Gould, A. 1992, ApJ, 389, 234
Barnes, J. 1985, MNRAS, 215, 517
Barnes, J. E., & Hernquist, L. E. 1991, ApJ, 370, L65
Binney, J., Gerhard, O. E., Stark, A. A., Bally, J., & Uchida, K. I. 1991, MN-
RAS, 252, 210
Blandford, R. D., & Rees, M. J. 1992, in Testing the AGN Paradigm, ed. S. S.
Holt, S. G. Neff, & C. M. Urry (New York: American Institute of Physics),
3

Blitz, L., & Spergel, D. N. 1991a, ApJ, 370, 205
Blitz, L., & Spergel, D. N. 1991b, ApJ, 379, 631
Bolte, M. 1992, PASP, 104, 794
Bregman, J. N. 1980, ApJ, 236, 577
Ciardullo, R., Rubin, V. C., Jacoby, G. N., Ford, H. C., & Ford, K. C. 1988,
      AJ, 95, 438
Cowie, L. L. 1981, ApJ, 245, 66
Cowie, L. L., Songaila, A., & Hu, E. M. 1991, Nature, 354, 460
Eggen, O. J., Lynden-Bell, D., & Sandage, A. 1962, ApJ, 136, 748
Fabbiano, G. 1989, ARA&A, 27, 87
Falgarone, E., Puget, J. L., & Perault, M. 1992, A&A, 257, 715
Fall, S. M., & Efstathiou, G. 1980, MNRAS, 193, 189
Fich, M., & Tremaine, S. 1991, ARA&A, 29, 409
Gammie, C. F., Ostriker, J. P., & Jog, C. J. 1991, ApJ, 378, 565
Genzel, R., & Townes, C. H. 1987, ARA&A, 25, 377
Gilmore, G., & Reid, N. 1983, MNRAS, 202, 1025
Gilmore, G., Wyse, R. F. G., & Kuijken, K. 1989, ARA&A, 27, 555
Gunn, J. E. 1982, in Astrophysical Cosmology, ed. H. A. Bruck, G. V. Coyne,
      & M. S. Longair (Vatican City: Pontif. Acad. Sci.) 233
Habing, H. J. 1988, A&A, 200, 40
Harris, W., & Racine, R. 1979, ARA&A, 17, 241
Hausman, M. A., & Ostriker, J. P. 1978, ApJ, 224, 320
Houck, J. C., & Bregman, J. N. 1990, ApJ, 352, 506
Jenkins, A., & Binney, J. 1990, MNRAS, 245, 305
Kennicutt, R. C. Jr. 1989, ApJ, 344, 685
Kent, S. M., Dame, T. M., & Fazio, G. 1991, ApJ, 378, 131
Kleiner, S. C., & Dickman, R. L. 1987, ApJ, 312, 837
Kormendy, J. & Norman, C. A. 1979, ApJ, 233, 539
Kuijken, K. 1991a, in Warped Disks and Inclined Rings Around Galaxies, ed. S.
      Casertano, P. Sackett, & F. Briggs (Cambridge: Cambridge Univ. Press)
      159
Kuijken, K. 1991b, ApJ, 372, 125
Kuijken, K., & Gilmore, G. 1989, MNRAS, 239, 605
Lacey, C. G. 1984, MNRAS, 208, 687
Lacey, C. G., & Ostriker, J. P. 1985, ApJ 299, 633
Lacy, J. H. 1989, in The Center of The Galaxy, proc. IAU Symp. 136, ed. M.
      Morris (Dordrecht: Kluwer) 493
Lanzetta, K. M. 1992, PASP, 104, 835
Larson, R. B. 1990, PASP, 102, 709
Lebofsky, M. J., Rieke, G. H., & Tokunaga, A. T. 1982, ApJ, 263, 736
Lee, Y.-W. 1992a, PASP, 104, 798
Lee, Y.-W. 1992b, AJ, 104, 1780
Light, E. S., Danielson, R. E., & Schwarzschild, M. 1974, ApJ, 194, 257
Liszt, H. S., & Burton, W. B. 1980, ApJ, 236, 779
Little, B., & Tremaine, S. 1987, ApJ, 320, 493
Lo, K. Y. 1989, in The Center of The Galaxy, proc. IAU Symp. 136, ed. M.
      Morris (Dordrecht: Kluwer) 527
Long, K., Ostriker, J. P., & Aguilar, L. 1992, ApJ, 388, 362
Lynden-Bell, D. 1967, MNRAS, 136, 101
Lynden-Bell, D., & Rees, M. J. 1971, MNRAS, 152, 461
Mac Low, M.-M., & McCray, R. 1988, ApJ, 324, 776
Magnani, L., Blitz, L., & Mundy, L. 1985, ApJ, 295, 402

Magnani, L., Caillault, J.-P., & Armus, L. 1990, ApJ, 357, 602

Maloney, P. 1990, ApJ, 348, L9

Mamon, G. A. 1987, ApJ, 321, 622

Matsumoto, T. et al. 1982, in The Galactic Center, ed. G. Riegler & R. Blandford (New York: American Institute of Physics), 48

McKee, C. F., & Ostriker, J. P. 1977, ApJ, 218, 148

Mirabel, I. F., Rodriguez, L. F., Cordier, B., Paul, J., & Lebrun, F. 1992, Nature, 358, 215

Oort, J. H. 1932, Bull. Astron. Inst. Netherlands, 6, 349

Oort, J. H. 1965, in Galactic Structure, ed. A. Blaauw & M. Schmidt (Chicago: Univ. Chicago Press) 455

Oort, J. H. 1977, ARA&A, 15, 245

Ostriker, E. C., & Binney, J. J. 1989, MNRAS, 237, 785

Ostriker, J. P. 1980, Comm. Ap., 6, 177

Ostriker, J. P., & Peebles, P. J. E. 1973, ApJ, 186, 467

Ozernoy, L. M. 1989, in The Center of The Galaxy, proc. IAU Symp. 136, ed. M. Morris (Dordrecht: Kluwer) 555

Paley, E. S., Low, F. J., McGraw, J. T., Cutri, R. M., & Rix, H.-W. 1991, ApJ, 376, 335

Peebles, P. J. E., Melott, A. L., Holmes, M. R., & Jiang, L. R. 1989, ApJ, 345, 108

Peterson, C. J., & King, I. R. 1975, AJ, 80, 427

Phinney, E. S. 1989, in The Center of The Galaxy, proc. IAU Symp. 136, ed. M. Morris (Dordrecht: Kluwer) 543

Quinn, P. J., & Goodman, J. 1986, ApJ, 309, 472

Raha, N., Sellwood, J. A., James, R. A., & Kahn, F. D. 1991, Nature, 352, 411

Rieke, G. H., & Lebofsky, M. J. 1987, in The Galactic Center, ed. D. Backer (New York: American Institute of Physics), 91

Roberts, W. W. Jr., Roberts, M. S., & Shu, F. H. 1975, ApJ, 196, 381

Sandage, A. 1961, The Hubble Atlas of Galaxies (Washington, D.C.: Carnegie Institution Publication 618)

Sanders, D. B., Soifer, B. T, Elias, J. H., Madore, B. F., Matthews, K., Neugebauer, G., & Scoville, N. Z. 1988, ApJ, 325, 74

Sanders, R. H. 1977, ApJ, 217, 916

Scalo, J. M. 1990, in Physical Processes in Fragmentation and Star Formation, ed. R. Capuzzo-Dulcetta, C. Chiosi, & A. DeFazio (Dordrecht: Kluwer) 151

Schmidt, M. 1959, ApJ, 129, 243

Schweizer, F. 1976, ApJS, 31, 313

Searle, L., & Zinn, R. 1978, ApJ, 225, 357

Shore, S. N. 1992, An Introduction to Astrophysical Hydrodynamics (San Diego: Academic Press)

Spitzer, L. Jr., & Schwarzschild, M. 1951, ApJ, 114, 385

Spitzer, L. Jr., & Schwarzschild, M. 1953, ApJ, 118, 106

Stark, A. A., & Brand, J. 1989, ApJ, 339, 763

Statler, T. S. 1988, ApJ, 331, 71

Toomre, A. 1964, ApJ, 139, 1217

Toomre, A. 1977, in Evolution of Galaxies and Stellar Populations, ed. B. M. Tinsley & R. B. Larson (New Haven: Yale Univ. Press) 401

Toth, G., & Ostriker, J. P. 1992, ApJ, 389, 5

Townes, C. H. 1989, in The Center of The Galaxy, proc. IAU Symp. 136, ed. M. Morris (Dordrecht: Kluwer) 1

Tremaine, S. 1992, Physics Today, 45, 28

van der Kruit, P. C., & Searle, L. 1982, A&A, 110, 61

Vietri, M. 1986, ApJ, 306, 48

Webbink, R. F. 1985, in IAU Symp. 113, Dynamics of Star Clusters, eds. J. Goodman & P. Hut (Dordrecht: Reidel), 541

Weiland, J. L., Hauser, M. G., Kelsall, T., Dwek, E., Moseley, S. H., Silverberg, R. F., Mitra, M., Odegard, N. P., Spiesman, W. J., Sodroski, T. J., Stemwedel, S. W., Freudenreich, H. T., & Lisse, C. M. 1993, this volume

Weinberg, M. D. 1992, ApJ, 384, 81

White, S. D. M. 1978, MNRAS, 184, 185

Wielen, R. 1977, A&A, 60, 263

Yanny, B. 1992, PASP, 104, 840

Yusef-Zadeh, F. 1989, in The Center of The Galaxy, proc. IAU Symp. 136, ed. M. Morris (Dordrecht: Kluwer) 243

Zaritsky, D., & Lo, K.-Y. 1986, ApJ, 303, 66

Zaritsky, D., Smith, R., Frenk, C. S., White, S. D. M. 1993, to appear in March ApJ

Zwicky, F. 1955, PASP, 67, 232

GALACTIC

CENTER

THE CENTRAL ENGINE AND ACTIVITY AT THE GALACTIC CENTER

Mark Morris

Department of Astronomy, UCLA, Los Angeles, CA 90024, USA

ABSTRACT

Many kinds of activity are in evidence in the Galactic center, and there are different kinds on different scales from several A.U. to a few hundred parsecs, although some of them might be related to a common cause. This review provides a brief summary of the various known forms of non-standard activity there, and of the structures which accompany them. It then examines the question of the identity of the central engine; is it a single massive black hole, a cluster of stellar-mass black holes, or something else? The total mass of the underluminous matter in the inner few tenths of a parsec is consistent with either of these hypotheses. Finally, the nature of the luminous objects in the central parsec is considered. The pros and cons of the hypotheses that they are massive young stars, collisionally merged stars, or stellar remnants with new atmospheres are discussed.

I. INTRODUCTION

The title of this paper begs the question. Is there really the implied analogy to active galactic nuclei? If activity is defined as anything that cannot easily be attributed to common processes such as star formation, or gravitational reaction to the central mass concentration, then there is indeed a great deal of curious activity, although presently with two or three orders of magnitude less energy than is manifested in a Seyfert nucleus. This review first describes the forms of activity that have been documented at various scales in the galactic center, then considers the nature of the central engine. Finally, it offers speculations about how some of this activity may have come about. The implications for other spiral galaxies having a central concentration of interstellar gas are evident, inasmuch as the processes discussed here are necessarily generalizable. However, the implications for active galaxies are discussed elsewhere (Morris 1993a, 1993b).

II. FORMS OF ACTIVITY

The following is a compendium of the non-standard forms of galactic center activity, and the structures that are associated with them. They are arranged in descending order of physical scale. The busy galactic center arena is radically different when viewed at different scales, and the structures and activities on one scale are often quite unrelated to those at another scale. The notion of a "central engine" is encountered only on subparsec scales.

A. Large-Scale (> 10 pc)

Jets. The largest scale feature that has been noted is a putative low-energy jet of enormous length: over 4 kpc. Sofue *et al.* (1989), who call it the "Galactic Center Spur", extracted it from the 408-MHz database of Haslam *et al.* 1982 and confirmed it at higher frequencies. It consists of a narrow tongue of radio emission, seen only at positive latitudes, which is perpendicular the galactic plane and which appears to intersect the plane at or near the galactic center. This feature has not yet been followed closer to the galactic center than several hundred parsecs because of its relatively low intensity, and there is no evidence yet that it is not a fortuitously placed local spur like the North Polar Spur, so the significance of the Galactic Center Spur remains uncertain.

A much smaller, low-frequency "jet" was previously seen in a 160-MHz map of the galactic center by Yusef-Zadeh *et al.* 1986. The ridge of emission begins on, or close to, the nucleus, and extends $\gtrsim$ 30 pc to only *negative* latitudes, so it might be the counterjet to the "Galactic Center Spur". Like the larger structure, however, its identification as a jet from the nucleus remains to be convincingly demonstrated.

The Galactic Center Lobe (GCL). Radio continuum images of the inner few degrees of the Galaxy reveal a pair of extended ridges of centimeter-wave radio emission which extend out of the galactic plane toward positive latitudes (Altenhoff *et al.* 1978; Sofue and Handa 1984). The GCL straddles the galactic nucleus, but is not symmetric about it; the eastern leg lies at $l \simeq 0.2°$ while the western leg is at $l \simeq$ -0.5°. Sofue (1985) has argued that the two ridges are connected at about $b = 1°$ (150 pc), giving the overall appearance of a limb-brightened, closed cylindrical "lobe" above the galactic center. This picture inspired a number of models in which an energy release at the nucleus, or a magnetic field twisted by galactic rotation, drives a vertical gas outflow (see Shibata 1989 for a summary). However, Tsuboi *et al.* (1986) and Uchida *et al.* (1990, 1993) have argued that the two legs of the GCL have completely dissimilar characteristics, so the existence of the "lobe" may be illusory. Rather, it appears that there are two unrelated radio ridges: the eastern ridge is a continuation to high latitudes of the nonthermal, polarized emission in the unique magnetic structure known as the "Radio Arc", while the western ridge is a thermal emitter with an infrared and molecular counterpart and it is probably a large-scale shock.

The Expanding Molecular Ring (EMR). A longitude-velocity (L-V) plot for molecular gas in the galactic plane, taken from the ATT Bell Labs survey of ^{12}CO (Bally *et al.* 1993), is shown in figure 1. For this figure, the emission is integrated over all latitudes $|b| > 15'$ in order to reduce confusion with intervening material lying close to the galactic plane. In addition to the pervasive emission from local gas emitting near zero velocity, and the extended gas in the "3-kpc expanding arm" which emits near -60 km s^{-1}, one can clearly see the broad velocity range covered by the complex distribution of molecular gas lying within a few hundred parsecs of the galactic nucleus. Gas orbiting with pure circular rotation would be found in this diagram between the zero-velocity horizontal and the approximately diagonal locus representing the rotation curve. Indeed, this portion of the diagram is filled with emission, and the extensions to the upper left and lower right of the emission complex probably represent the concentration of emitting gas toward the rotation curve. In addition, however, a large fraction of the gas emits above the rotation curve in this diagram, and some is even located in the quadrants that are "forbidden" relative to pure circular rotation (the upper right and lower left). Clearly, there is a considerable non-circular component to gas motions in the galactic center.

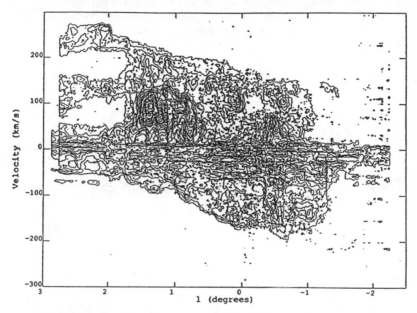

Figure 1: Longitude-velocity contour plot for ^{12}CO J $= 1 - 0$ emission from the galactic center. The emission is averaged over all Galactic latitudes satisfying $0.4° > |b| > 0.15°$. The data are from the AT&T Bell Labs survey (Uchida 1992; Bally *et al.* 1993).

Much of the non-circular motion appears as an elliptical locus in the L-V plane. This was first noticed by Scoville (1972), who interpreted the ellipse in terms of an expanding ring of molecular gas (see also Kaifu *et al.* 1972). More recent analyses based on this hypothesis (Bania 1977; Güsten & Downes 1980) indicate that the EMR, with a current total mass of $\sim 2 \times 10^6$ $M_\odot$, would have a radius of 150 pc and would be expanding radially with a velocity of 160 km s^{-1}. The EMR hypothesis requires a tremendous energy input ($\gtrsim 10^{55}$ ergs) about 10^6 years ago. This may turn out to be quite a strong constraint, since other clues for such an energy release are lacking (unless the GCL is indeed an explosively produced structure). Also, the plane containing the ring is tilted by about 15° with respect to the Galactic plane, so either the explosion was asymmetric, or the original gaseous disk in which the explosion took place was itself tilted.

An alternative explanation for the elliptical locus of emission in the L-V plane is that it corresponds to the response of gas to a strong bar potential in the inner Galaxy. Binney *et al.* (1991) have argued for this hypothesis, finding that they can more-or-less account for the shape of the EMR feature in the L-V plane if they place all of the gas in the innermost, self-intersecting x_2 orbit. Actually, the predicted locus in the L-V plane is a parallelogram rather than an ellipse, and this appears to agree rather well with the L-V plot for emission at low galactic latitudes ($|b| < 15'$). The idea that our Galaxy has a central bar has received support from other quarters (Blitz & Spergel 1992; Blitz, these proceedings), so the bar interpretation must be important at some level. Nonetheless, it is curious that high- latitude CO emission agrees with the classical EMR hypothesis, while low-latitude CO emission supports the bar-induced dynamics hypothesis. Problems facing the bar hypothesis are summarized by Uchida (1992) and Uchida *et al.* (1993).

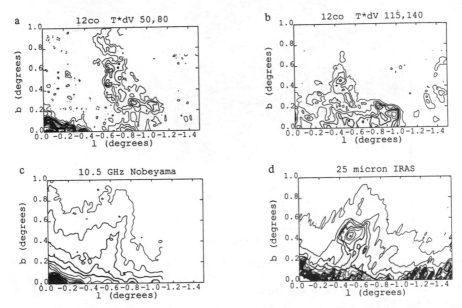

Figure 2: CO emission near AFGL5376 integrated over two velocity ranges (50 to 80 km s^{-1} and 115 to 140 km s^{-1}), compared to the radio continuum and 25-μm IRAS images (Sofue and Handa 1984; Uchida *et al.* 1993).

Large-Scale Shocks. In molecular line surveys that have been made of the Galactic center region, many peculiar streams, or kinematically continuous complexes of gas can be identified (*e.g.*, Bally *et al.* 1988, 1993), and their motion is often markedly non-circular. These streams inevitably undergo shocks as they encounter each other, as they encounter the bar shock, or as they encounter ambient gas moving normally with the sense of galactic rotation. In principle, those shocks can be quite violent and have a large physical scale. They therefore represent an important way in which energy, in the form of kinematic motions of clouds, is released as luminous energy. A well-documented case is the ridge associated with the strong 25-μm source AFGL5376. Here, a molecular system with a highly forbidden velocity is colliding with another system, producing perhaps two apparent shocks (Uchida *et al.* 1990, 1993) along a 90-pc front (figure

2). This vertical shock can be identified with the western leg of the GCL, so it is one of the more prominent features of the galactic center region. Other shocks are undoubtedly present, and will ultimately be identified as velocity discontinuities in the molecular streams, or as sources of shock-induced infrared lines (*e.g.*, 2-μm H_2 lines, or far-infrared, atomic fine-structure lines).

The origin of the peculiar motions of galactic center gas streams is not yet understood. Some appear to be associated with the EMR; the shock associated with AFGL5376, for example, has been ascribed to the expansion of the EMR (Uchida *et al.* 1990). Others might correspond to ejection from the nucleus (Oort 1977), or to infall of gas clouds from the bulge and halo.

Localized Star Formation Episodes. Starbursts potentially provide one of the most important forms of activity for a galactic nucleus, but in our Galactic center, star formation appears to be at a low ebb (Morris 1989, 1993a). While there are prominent star formation complexes (notably Sgr B2, Sgr C, and G0.5-0.0), they do not constitute starbursts. Indeed, given the $3 - 10 \times 10^7$ $M_\odot$ of gas within ± 300 pc of the galactic center (Güsten 1989; Cox and Laureijs 1989), there is no more star formation evidenced in the Galactic center than there is in a comparable mass of gas in the Galactic disk, and possibly less (Morris 1989).

However, it is entirely possible that spectacular starbursts occured in the past history of the Galaxy. The expansion of the EMR, for example (if that is the appropriate interpretation), could have been caused by a recent nuclear starburst involving $10^4 - 10^5$ supernovae. At least one major star formation event did take place recently within the inner few hundred parsecs: the 75-pc diameter superbubble G359.1-0.5, formerly regarded as a single supernova remnant, is surrounded by an apparently expanding ring or a shell of molecular gas. Models of the energetics and kinematics of this structure suggest that $\gtrsim 10^2$ supernovae occured there within the last few million years (Uchida *et al.* 1992a,b). Such localized depositions of energy may substantially alter the local velocity field of galactic center gas, but they are not of the scale usually attributed to an "active" galaxy.

Magnetic Activity. Some of the radio-emitting structures seen in the galactic center are magnetic in character, and they raise the question of whether the magnetic field is one of the dominant determinants of the manifestations of "activity" near the galactic center. The most prominent magnetic feature is the radio Arc, which consists of a bundle of non-thermally emitting, parallel linear filaments oriented perpendicular to the galactic plane (Yusef-Zadeh *et al.* 1984). Tsuboi *et al.* (1986) showed with polarization measurements that the filaments trace out the direction of the magnetic field lines. In addition, there are a number of isolated filaments, or "threads" within a degree of the Galactic plane, all oriented approximately perpendicular to the plane (Morris and Yusef-Zadeh 1985; Yusef- Zadeh 1989; Gray *et al.* 1991). In spite of the relative tumult of the galactic center environment, all of the extremely narrow filaments have only a gentle, uniform curvature. Only one thread – the one most distant from the galactic center – shows any hint of a distortion or irregularity (Gray *et al.* 1991). The uniformity of the threads can be used to set a limit on the rigidity of the field, and it is found that milligauss fields are needed to prevent clouds and velocity inhomogeneities from buffetting and distorting the filaments (Morris 1990).

The confinement of such filaments presents an interesting problem; without an external pressure, the filaments, which would presumably have a force-free magnetic field configuration, would expand freely on a short time scale and disappear. Indeed, observations of diffuse X-rays and the 6.7-keV iron line indicate that the kinetic gas pressure of the intercloud medium of the Galactic center appears to be at least 2 or 3 orders of magnitude higher than the gas pressure in the local disk (Kawai *et al.* 1988; Yamauchi *et al.* 1990; Spergel and Blitz 1992). Even so, this high pressure falls short of that necessary to confine the filaments by almost two orders of magnitude. If the filaments are to be confined, then it is preferable to invoke a milligauss field having a dipolar geometry which pervades the inner 100 – 200 pc of the Galaxy. The filaments then reduce to being those flux tubes which happen to be illuminated by the deposition of relativistic particles. This hypothesis leaves us with the inference that the magnetic field is the dominant source of pressure in the galactic center, and that the field can be dynamically important for interstellar gas clouds. It also leaves open the question of the origin of the relativistic particles, inasmuch as no obvious sources are found along the length of the filaments. A variety of acceleration mechanisms can be imagined, however, including field line reconnection, diffusive shock acceleration, and acceleration by an electric field induced by the movement of a conducting cloud through the magnetic field (Morris and Yusef- Zadeh 1989; Rosso and Pelletier 1993).

With such a strong field, several forms of magnetic activity are possible. The field will clearly affect gas flows, it will ionize the surfaces of clouds moving through it at sufficient speeds and induce current flows (Morris and Yusef-Zadeh 1989), and it may help cause an inward migration of gas toward the galactic center by exchanging angular momentum with the gas, and transferring it wherever the field lines happen to be anchored (halo gas? the outer disk?). Of course, the inward migration of gas will lead to other forms of activity as that gas arrives at the central engine (discussed below).

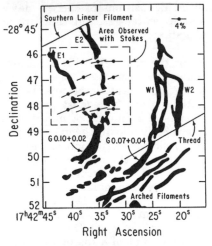

Figure 3: Far-IR polarization E-vectors superimposed on a sketch of the radio emission comprising the "arched filaments" of the Arc (Morris *et al.* 1992). According to the standard interpretation of thermal emission from magnetically aligned dust grains, the field is perpendicular to the polarization vectors.

A hint of the importance of magnetic fields inside of dense clouds is provided by observations of the polarization of far-infrared emission from aligned dust grains in the galactic center clouds. The percent polarization is generally higher in the galactic center clouds which have been observed than it is anywhere else. Thus, the field in the 1 – 10 pc circumnuclear disk (described below) appears to be dominated by an azimuthal component, and the total field may be important for extracting angular momentum from the disk and driving a wind (Hildebrand *et al.* 1990, 1993; Wardle and Königl 1990). Also, the field in an anomalous-velocity cloud underlying part of the Radio Arc is found to be surprisingly uniform over a large region, possibly as a result of dynamical shear of the cloud as it orbits close to the galactic center (figure 3, from Morris *et al.* 1992). Such polarization measurements have only just begun, and one can expect that a great deal more can be learned about the geometry of magnetic fields in clouds as the University of Chicago far-IR Polarimeter Array (STOKES) is used in coming years. The one drawback of this method is that the polarization cannot be used to derive a direct estimate of the strength of the magnetic field.

B. Intermediate Scale (1 – 10 pc)

The radio complex coincident with the galactic nucleus, Sgr A, dominates the activity at intermediate scales. It consists of a nonthermal shell source, Sgr A East, surrounded by a halo which appears to be a mix of thermal and nonthermal emission. The halo of Sgr A East is also evident in dust emission observed at $\lambda 1.3$ mm (Mezger *et al.* 1989). Within the projected periphery of the shell source lies Sgr A West, the thermal source coinciding with the nucleus. Studies carried out at low radio frequencies reveal that optically thick Sgr A West absorbs the bright, background nonthermal emission from Sgr A East, so it therefore lies in front of the bulk of the shell source (Yusef-Zadeh & Morris 1987; Pedlar *et al.* 1989).

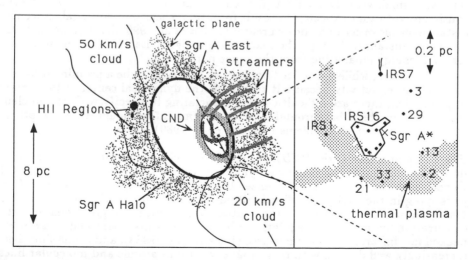

Figure 4: Schematic diagram illustrating the relative placement of the various structures described in the text.

Sagittarius A East. It is perhaps only a coincidence that a shell source encompasses the dynamical center of the Galaxy (figure 4), given that it lies somewhat behind the center. The possibility that Sgr A East is a supernova remnant has been considered for some time (Jones 1974; Ekers *et al.* 1975), but if so, it is unusual in some respects (Ekers *et al.* 1983; Goss *et al.* 1983; Mezger *et al.* 1989). An alternative hypothesis which has been considered is that Sgr A East be a bubble of relativistic gas emanating from the nucleus. Inasmuch as Sgr A East is not centered on the galactic nucleus, either along the line of sight or in the plane of the sky, this hypothesis requires that there be a relative motion between the medium into which the bubble is expanding and the source of relativistic gas at the nucleus (Yusef-Zadeh & Morris 1987; Mezger *et al.* 1989). It may be problematical for this hypothesis that the circumnuclear disk, which surrounds the nucleus and would presumably impede the equatorial expansion of relativistic gas from the nucleus (figure 4), has no perceptible effect on the shape or the brightness of the shell.

Two other pieces of evidence allow one to constrain the location and the nature of Sgr A East. First, the shell is elongated along the Galactic plane with an aspect ratio of about 1.5. An initially spherical bubble expanding into a static, stratified gas layer would naively be expected to be elongated along the density gradient, *i.e.*, perpendicular to the Galactic plane. The observed opposite elongation can probably be attributed to shear of the bubble in a differentially rotating medium, but if so, it means that the expansion velocity of the bubble is not much larger than the shear velocity. The magnitude of the shear depends on the proximity of Sgr A East and West, which is not well known, but it is probably no larger than $\sim$100 km s^{-1}. With a shell radius of about 3 pc, this gives a lower limit on the shell's expansion time of 3×10^4 years. The second piece of evidence is the rather striking interaction of the shell with the prominent "50 km s^{-1}" molecular cloud (Goss *et al.* 1985; Genzel *et al.* 1990; Ho *et al.* 1991; Serabyn *et al.* 1992). The 50 km s^{-1} cloud, and by association Sgr A East, is unlikely to be very close to the dynamical center, because its relatively small velocity gradient (5–6 km s^{-1} arcmin^{-1}; Serabyn *et al.* 1992) is consistent with that expected to result from tidal shear only for a galactocentric distance on the order of 50 pc or greater (Zylka *et al.* 1990; Serabyn *et al.* 1992). At such a distance, the 6-pc shell constituting Sgr A East cannot be accounted for with energy emanating from the inner parsec.

Therefore, while Sgr A East appears at first sight to be a prominent form of galactic center activity because it surrounds the dynamical center of the Galaxy in projection, there are difficulties with associating it directly with the nucleus. Indeed, it may be a background supernova remnant, although that hypothesis is also not without its problems (Mezger *et al.* 1989).

The Circumnuclear Disk (CND). A lumpy disk with a central cavity of about 1 pc radius surrounds the Galaxy's dynamical center (reviewed by Güsten 1987; Genzel and Townes 1987; and Genzel 1989). Fortunately for the study of objects lying in the cavity, the disk is tilted by about 20° relative to the galactic plane, allowing us to peer into the central parsec without a great deal of extra obscuration by the disk. The relatively hot CND was first identified as a double-peaked far- infrared continuum source (Becklin *et al.* 1983), and it has since been increasingly well defined with maps made in various atomic and molecular lines.

The total disk mass, estimated to be several times 10^4 M$_\odot$, is presumably supplied by infall of matter from mass-losing stars in the halo, by the inward migration of gas from the galactic plane as a result of various angular momentum loss processes, and possibly by gas streams from nearby, tidally disrupted clouds (Okumura *et al.* 1991). While some have claimed that there is a substantial inward radial component to gas motions in the CND, a recent limit of 20 km s^{-1} on such a motion (Jackson *et al.* 1993) limits the present mass infall rate through the disk to about 0.1 M$_\odot$ yr^{-1}. Some of the gas may leave the disk surface as a magnetically driven wind (Wardle and Königl 1990), but there is still room for a substantial accretion into the central cavity and onto any objects residing at the center. Indeed, the "northern arm", a stream of ionized gas within Sgr A West, and the associated reservoir of atomic gas, indicate a mass inflow rate into the cavity of 3 × 10^{-2} M$_\odot$ yr^{-1} (Jackson *et al.* 1993). Whether or not this stream originates in the CND is still open to question. In any case, the gas inflow rate onto the center is important for the present discussion inasmuch as it supplies the fuel for the central engine, *if* a central engine actually exists. [Another gas stream identified by Ho *et al.* (1991) and Sandqvist *et al.* (1989) may bring gas directly into the center, bypassing the CND in the process. This claim is based primarily on the placement and geometry of the stream; the physical connection to the nucleus remains to be demonstrated.]

The fuel supply coming through the CND does not appear to be constant. The CND is, according to Genzel (1989), an unstable configuration that may expand inward to fill the central cavity on a time scale not much longer than a few times 10^5 yr. One might imagine, then, that the mass accretion rate onto the nucleus will be much larger at some point in the future than it is at present.

The Radio Streamers. Yusef-Zadeh and Morris (1987) pointed out that part of the halo of Sgr A consists of a collection of radio continuum streamers oriented perpendicular to the Galactic plane. They appear to originate within Sgr A West and they are perceptible only at positive latitudes. The interpretation of these features is based on the interaction of a strong wind from objects in the core with the CND (indeed, this wind can help account for the sharp inner edge to the CND). In fact, the distance along the galactic plane from which the streamers appear to arise is comparable to the inside diameter, 2 pc, of the CND. The wind from the nucleus is deflected away from the galactic plane by gas in the CND, and in the process, it mixes with the more diffuse component of the disk gas and entrains it in the flow. Yusef-Zadeh and Morris (1987), and Yusef-Zadeh and Wardle (1993) argue that the filamentary character of the streamers is caused by a strong, poloidal field that directs the flow parallel to the field and out of the plane. Therefore, the streamers may be the largest-scale manifestation of a nuclear wind, although no kinematic information is yet available.

C. Small Scale ($\lesssim$ 1 pc)

The inner parsec of the Galaxy – wherein lies the core of the central star cluster, a collection of unusual luminous objects, and a unique nonthermal radio source – is essentially guaranteed of having something that can be called a "central engine". The question is what that engine is capable of.

IRS16, the Infrared Cluster, the Broad-Line Region, and the Galactic Center Wind. Within a few arcseconds, the dynamical center of the Galaxy is coincident with a collection of luminous, relatively blue objects known collectively as IRS16.

Eckart *et al.* (1992) report a total of about 24 point-like near-IR components in this cluster. Their luminosities are individually on the order of 10^5 $L_\odot$, but are rather uncertain because of large reddening corrections. These objects are candidates for being the source of ultraviolet photons required to explain the ionization of gas in Sgr A West (Rieke *et al.* 1989), although they might not be able to supply enough flux (Tollestrup *et al.* 1989). The ionization state of the gas limits the effective temperature of the source of ionization to ~35,000 K (Lacy *et al.* 1982), corresponding to a main sequence O9 star. However, the components of IRS16 are too luminous to be main sequence O9 stars. The standard interpretation is that they be a cluster of evolved blue supergiants which resulted from a relatively recent star formation event in the inner parsec.

The IRS16 cluster is coincident with a "broad-line region" seen in near-IR lines of helium and hydrogen (Hall *et al.* 1982). Lines having widths of ~700 km s^{-1} were found to be spread over a region several arcseconds in extent (Geballe *et al.* 1987, 1991). It appears from recent studies, however, that the broad lines come from a collection of at least a dozen discrete sources. Many of the bright, near-IR components of IRS16, as well as a number of other luminous objects in the vicinity, appear to be helium emission-line stars with Wolf-Rayet characteristics (Allen *et al.* 1990; Krabbe *et al.* 1991). Interpreted as massive stars, these objects support the notion that a recent burst of star formation led to a coeval cluster of stars of curiously comparable mass.

A corollary of this is that the central stellar cluster produces a global wind resulting from the collective winds of the emission-line stars, and possibly also from other objects, such as Sgr A*. At the moment, this wind constitutes one of the more prominent forms of "activity" in the galactic center. It has a few notable effects on its surroundings, in addition to the radio "streamers" mentioned above: (1) as it strikes the inner edge of the circumnuclear disk, it shocks the gas and gives rise to near-IR H$_2$ emission (Gatley *et al.* 1984, 1986. An alternative, radiative model for the H$_2$ emission may not be ruled out, however). (2) The ram pressure of the wind may play a role in keeping the cavity within the CND relatively devoid of gas, and it can help account for the relatively sharp inner edge to the CND (although a dynamically explanation for the sharp inner edge was offered by Duschl 1989). (3) The red supergiant IRS7, which apparently lies close to the source of the wind, has developed a cometary "tail" as either its stellar wind, or the outer layers of its loosely bound atmosphere, are blown radially outwards from the galactic center (Yusef-Zadeh & Morris 1991; Serabyn *et al.* 1991, Yusef-Zadeh & Melia 1992). The galactic center wind might also facilitate the disruption of red giant atmospheres in the central cluster (*cf.*,.Sellgren *et al.* 1990) once tidal interactions with passing stars sufficiently distort or levitate parts of the atmosphere.

The "Missing" Red Giants. One of the most interesting discoveries in the central parsec is that the relative absorption strength of the 2.3-μm CO bandhead, which results from the collective light of the large number of cool giants concentrated there, drops abruptly at galactocentric radii less than about 0.6 pc (Sellgren *et al.* 1990). This could imply a new population of objects luminous at 2.3-μm, but without any intrinsic CO absorption, or it may mean that the atmospheres of red giants in this dense region are strongly perturbed by tidal collisions with field stars. Indeed, the predicted collision rates between giants and field stars in the inner parsec are large enough to strip the atmospheres of a large fraction of the red giants there (Lacy *et al.* 1982; Phinney 1989).

Sgr A*. The compact, variable radio source Sgr A*, also located at the Galaxy's dynamical center, has long had the status of being the most popular candidate for the carrier of the few million solar masses inferred dynamically for the central few tenths of a parsec (Genzel and Townes 1987; Serabyn *et al.* 1988; McGinn *et al.* 1989). The radio spectrum, and limits at other wavelengths, have been modelled by Melia (1992) and Melia *et al.* 1992) in terms of a $\sim 10^6$ $M_\odot$ black hole, but Ozernoy (these proceedings) argues that a mass as small as 500 $M_\odot$ is sufficient (see also Allen and Sanders 1986). Obviously, this debate is crucial to the question of whether this object is potentially a "central engine".

The properties of Sgr A* are reviewed by Lo (1989). Since that review, a new spectral component has been recognized, and it may turn out to be quite important for understanding the nature of Sgr A* when it is more thoroughly observed. Zylka *et al.* (1992) reported a pronounced excess at submillimeter wavelengths above the flat or slowly rising spectrum from centimeter to millimeter wavelengths. This excess has been confirmed with interferometric observations at $\lambda 1.4$ mm. It remains to be seen whether it is thermal emission from a rather extended, massive accretion disk, or nonthermal emission from the inner magnetosphere around a black hole. Either hypothesis would have profound implications for the nature, and mass, of the object. Another recent finding is that Sgr A* is coincident with a near-IR source with relatively blue colors (Eckart *et al.* 1992). Given the high density of 2-μm sources in this region, the coincidence may just be a chance superposition, but if the near-IR source is really associated with Sgr A*, then a "missing luminosity" problem with accretion disk models for Sgr A* will be diminished.

III. The Nature of the Central Engine

The several million solar masses in the central parsec are not easily attributable to stars unless the stellar core has a rather unlikely mass distribution (*e.g.*, Serabyn and Lacy 1985). It is probably comprised of underluminous matter, and several forms can be imagined: very cold dark matter particles, stellar remnants, or a small number (one?) of supermassive black holes. Besides a gravitational field, there is no other evidence for dark matter there, and it is difficult to imagine how dark matter particles could have been dissipative enough to sink into the central parsec unless they have stellar masses and could thus dissipate their energy by dynamical friction.

Stellar remnants, on the other hand, *will* sink to the core if they are massive enough; in particular, if they are more massive than typical field stars ($\sim 1\ M_\odot$). It is straightforward to estimate how much mass in black hole and neutron star remnants might have piled up within the central stellar core during the lifetime of the Galaxy. The stellar density distribution, as judged by the distribution of 2-μm light, depends on galactocentric radius, r, as $\rho_*(r) \propto r^{-1.8}$ all the way from the central core out to a few hundred parsecs (Becklin and Neugebauer 1968; Matsumoto *et al.* 1982; Allen *et al.* 1983). Morris (1993a) has used this dependence, normalized to give the correct mass distribution (*cf.*, Genzel & Townes 1987), to determine how much gravitational settling, or mass segregation, has taken place. Various assumptions are necessary, including the form of the initial mass function, the lower and upper cutoff masses, and the mass above which a star leaves a black hole remnant (typically assumed to be about 50 $M_\odot$). For a wide range of assumptions centered on standard values, but with a larger-than-usual lower mass cutoff to the IMF, the calculation gives a total mass of remnants in the core which is comparable to the mass of the dynamically-inferred non-luminous matter. Black holes, assumed to have a typical mass of 10 $M_\odot$, dominate the remnant mass. Neutron stars, even with their higher production rate, are not as important because they are far less gravitationally differentiated than black holes.

Therefore, one possibility for the central engine is that it is a dense cluster of $\sim 10^5$ stellar-mass black holes. This possibility was previously considered for active galactic nuclei by Weedman (1983) and by Pacholczyk & Stoeger (1986, with several follow-up papers in subsequent years). Such an environment offers obvious possibilities for "activity" based on accretion energy if gas is channeled into the core. The size of the cluster of remnants is dictated by energy exchange with field stars in the central stellar core. If energy equipartition applies, the black hole cluster core would be about 1/3 the size of the stellar core, or ~ 0.1 pc, to within a factor of 2. Interestingly, this is about the size of IRS16.

The cluster of remnants at the core is inevitable as long as suitably massive stars form. The persistence of such a cluster against collapse to a single, super-massive object, however, is an open question. There is a competition between various instabilities leading to collapse, on the one hand, and, on the other hand, the reheating of the cluster by gravitational encounters involving binary stars, in which the binaries are hardened in the collision (discussed by Morris 1993a). The outcome of this competition has not yet been determined for conditions appropriate to the galactic center, but the formation of a supermassive black hole may be inevitable (Quinlan & Shapiro 1987, 1989, 1990). Even if part of this cluster of remnants coalesced into a single massive object, there should still be a large number of stellar mass black holes in the entourage of that object, as they are continuously migrating inwards.

A single supermassive black hole is a very familiar and oft-considered central engine. If accreting at the Eddington rate, a 2 or 3×10^6 $M_\odot$ black hole would generate energy at a rate approaching that of a Seyfert nucleus. In the Galactic center, however, such rates would occur only under 3 circumstances:

1) when the black hole tidally disrupts and swallows a star, which it may do every 10^4 years (Sanders & van Oosterom 1984; Phinney 1989), it will have an impulse of accretion energy lasting ~ 100 years.

2) If one of the dense clouds orbiting the Galactic center on a highly elliptical orbit were to have sufficiently small angular momentum (because of torques exerted by the magnetic field or by a bar), it could pass directly over the central accretor(s) and furnish an accretion disk with a 10^5-year supply of matter, for an accretion luminosity $\gtrsim 10^{44}$ ergs s^{-1} (Sanders 1981; Bottema & Sanders 1986).

3) The circumnuclear disk is apparently a transient phenomenon (Genzel & Townes 1987; Phinney 1989). It accumulates matter at an unknown rate, but the inner edge of the disk is impeded from migrating inwards, perhaps because of the nuclear wind, as suggested above. If this description is apt, this is an inherently unstable situation. At some point, the disk will have too much mass to be held at bay, and it will collapse catastrophically onto the nucleus, giving rise to a situation very much like that encountered in the previous possibility. Note that the latter two circumstances would equally well power a cluster of stellar mass black holes, although the emergent radiation spectrum would be quite different (e.g., Stoeger et al. 1992).

IV. THE NATURE OF THE LUMINOUS OBJECTS IN THE CORE

If the luminous objects in the inner few tenths of a parsec are massive young stars, they had to have formed at their presently observed radii. There would have been insufficient time for them to have sunk to the core as a result of dynamical friction. However, it appears to be very difficult to form stars at such radii. Morris (1989; 1993a) has argued that strong tidal forces, large magnetic fields, and large internal motions in clouds near the Galactic center all act to inhibit star formation by raising the Jeans mass to an untenably high value. Even cloud collapse provoked by external compression – the apparently dominant mode of star formation in at least the inner 10 pc or so – is unlikely to overcome the magnetic pressures resisting collapse, except under particularly fortuitous magnetic geometries.

A few alternatives to young stars have been considered for the nature of the luminous, central objects: the formation of massive stars by stellar coalescence, and accretion-powered stellar remnants with optically thick atmospheres. At the high densities of the central stellar core, stars have a better-than-even chance of merging with another star in a Hubble time (Phinney 1989). Indeed, within a radius of 0.2 pc, there is a collision about every 2.5×10^5 years. Once stellar masses start rising as a result of mergers, their cross section rises, and subsequent mergers occur more quickly (Quinlan & Shapiro 1990). However, it remains to be seen with a detailed calculation whether this process can produce a few dozen objects, each possessing a few tens of solar masses, as would be required to account for IRS16.

If the central mass is comprised of a cluster of stellar-mass black holes, then another possibility arises. The occasional collision between a black hole remnant and a red giant could lead to the transfer of a substantial amount of matter to the remnant, depending in detail on the parameters of the collision. This endows the remnant with an optically thick atmosphere, or an accretion disk so thick that it is, for all practical purposes, quasi-spherical. The long-term stability of such an object is not known; for as long as it lasts, however, it could conceivably mimic a hot star with a high enough luminosity to drive a wind. If such objects can stably maintain their atmospheres for 10^6 - 10^7 years, then their production rate would be sufficiently high to account for the population of luminous objects in IRS16 (Morris 1993a).

V. SUMMARY

"Activity" is a relative term. The activities listed in this review are those that one does not find anywhere in the Galaxy except in the Galactic center. Yet the level of activity of our Galactic nucleus pales in comparison with that seen in AGN's. Nonetheless, the elements comprising the galactic center activity are all in place, and the differences with AGN's may only be based on the magnitude of one of the following quantities: the mass of the central black hole, or cluster of black holes, the total mass of interstellar gas, the relative amplitude of the central bar distortion, and the strength of the magnetic field. Alternatively, the difference between the Galactic center and other galactic nuclei could be one of temporal phase; the activity level could rise as gas accretes onto the nucleus, and then fall again as the resulting activity temporarily halts the accretion process. In that view, we are between energetic events in an unstable cycle. The remnants of the past events would then be readable in terms of the dynamics of the gas. At the moment, however, we have insufficient information to unambiguously decide whether the Galactic nucleus has fits of extreme activity, or whether it remains eternally in its present, relatively placid state.

REFERENCES

Allen, D.A., Hyland, A.R. & Hillier, D.J. 1990, MNRAS, 244, 706.

Allen, D.A., Hyland, A.R., Jones, T.J. 1983, MNRAS, 204, 1145.

Allen, D.A. & Sanders, R.H. 1986, Nature, 319, 191.

Altenhoff, W.J., Downes, D., Pauls, T.A., & Schraml, J. 1978, A&AS, 35, 23.

Bally, J., Stark, A.A., Wilson, R.W. & Henkel, C. 1988, ApJ, 324, 223.

Bally, J. *et al.* 1993, in preparation.

Bania, T.M. 1977, ApJ, 216, 381.

Becklin, E.E. & Neugebauer, G. 1968, ApJ, 151, 145.

Becklin, E.E., Gatley, I. & Werner, M.W. 1982, ApJ, 258, 134.

Binney, J., Gerhard, O.E., Stark, A.A., Bally, J. & Uchida, K.I. 1991, MNRAS, 252, 210.

Bottema, R. & Sanders, R.H. 1986, A&A, 158, 297.

Cox, P. & Laureijs, R. 1989, in "The Center of the Galaxy" ed: M. Morris (Dordrecht: Kluwer) p.121.

Duschl, W.J. 1989, MNRAS, 240, 219.

Eckart, A., Genzel, R., Krabbe, A., Hofmann, R., van der Werf, P.P. and Drapatz, S. 1992, Nature, 355, 526.

Ekers, R., Goss, W.M., Schwarz, U.J., Downes, D. & Rogstad, D.H. 1975, A&A, 43, 159.

Ekers, R., van Gorkom, J., Schwarz, U. & Goss, W.M. 1983, A&A, 122, 143.

Gatley, I, Jones, T.J., Hyland, A.R., Beattie, D.H. & Lee, T.J. 1984, MNRAS, 210, 565.

Gatley, I. *et al.* 1986, MNRAS, 222, 299.

Geballe, T.R., Krisciunas, K., Bailey, J.A. & Wade, R. 1991, ApJL, 370, L73.

Geballe, T.R., Wade, R., Krisciunas, K., Gatley, I. & Bird, M.C. 1987, ApJ, 320, 562.

Genzel, R. 1989, in "The Center of the Galaxy" ed: M. Morris (Dordrecht: Kluwer) p.393.

Genzel, R., Stacey, G.J., Harris, A.I., Townes, C.H., Geis, N., Graf, U.U., Poglitsch, A. & Stutzki, J. 1990, ApJ, 356, 160.

Genzel, R. & Townes, C.H. 1987, Ann. Rev. Astron. Ap., 25, 377.

Goss, W.M., Schwarz, U.J., van Gorkom, J.H. & Ekers, R.D. 1985, MNRAS, 215, 69p.

Gray, A.D., Cram, L.E., Ekers, R.D. & Goss, W.M. 1991, Nature, 353, 237.

Güsten, R. 1987, in AIP Conf. Proc. No. 155: "The Galactic Center" ed: D.C. Backer (AIP: New York) p. 19.

Güsten, R. 1989, in "The Center of the Galaxy" ed: M. Morris (Dordrecht:

Kluwer) p.89.

Güsten, R. & Downes, D. 1980, A&A, 154, 25.

Hall, D.N.B., Kleinmann, S.G. & Scoville, N.Z. 1982, ApJL, 262, L63.

Haslam, C.G.T., Salter, C.J., Stoffel, H. & Wilson, W.E. 1982, A&AS, 47, 1.

Hildebrand, R.H., Gonatas, D.P., Platt, S.R., Wu, X.D., Davidson, J.A., Werner, M.W., Novak, G. & Morris, M. 1990, ApJ, 362, 114.

Hildebrand, R.H., Davidson, J.A., Dotson, J., Figer, D.F., Novak, G., Platt, S.R. & Tao, L. 1993, preprint.

Ho, P.T.P., Ho, L.C. Szczepanski, J.C., Jackson, J.M., Armstrong, J.T. & Barrett, A.H. 1991, Nature, 350, 309.

Jackson,, J.M., Geis, N., Genzel, R., Harris, A.I., Madden, S.C., Poglitsch, A., Stacey, G.J. & Townes, C.H. 1993, ApJ, 402, 173.

Jones, T.W. 1974, A&A, 30, 37.

Kaifu, N., Kato, T. & Iguchi, T. 1972, Nature Phys. Sci., 238, 105.

Kawai, N., Fenimore, E.E., Middleditch, J., Cruddace, R.G., Fritz,G.G., & Snyder, W.A. 1988, ApJ, 330, 130.

Krabbe, A., Genzel, R., Drapatz, S. & Rotaciuc, V. 1991, ApJL, 382, L19.

Lacy, J.H., Townes, C.H. & Hollenbach, D.J. 1982, ApJ, 262, 120.

Lo, K.Y. 1989, in "The Center of the Galaxy" ed: M. Morris, p. 527.

Matsumoto, T., Hayakawa, S., Koizumi, H. and Murakami, H. 1982, in The Galactic Center, AIP Conf. Proc. No. 83, eds: G.R. Riegler and R.D. Blandford, AIP, New York.

McGinn, M.T., Sellgren, K., Becklin, E.E. & Hall, D.N.B. 1989, ApJ, 338, 824.

Melia, F. 1992, ApJL, 387, L25.

Melia, F., Jokipii, J.R. & Narayanan, A. 1992, ApJL, 395, L87.

Mezger, P.G., Zylka, R., Salter, C.J., Wink, J.E., Chini, R., Kreysa, E. & Tuffs, R. 1989, A&A, 209, 337.

Morris, M. 1989, in "The Center of the Galaxy" ed: M. Morris (Dordrecht: Kluwer) p. 171.

Morris, M. 1990, in "Galactic and Extragalactic Magnetic Fields , IAU Symp. 140, eds: Beck, Kronberg, and Wielebinski (Kluwer: Dordrecht) p.361.

Morris, M. 1993a, ApJ, in press.

Morris, M. 1993b, in "First Light in the Universe: Stars or QSO's?", eds: B. Rocca-Volmerange, M. Dennefeld, B. Guiderdoni, & J. Tran Thanh Van, Editions Frontiŕes, in press.

Morris, M., Davidson, J.A., Werner, M., Dotson, J., Figer, D.F., Hildebrand, R., Novak, G. & Platt, S. 1992, ApJL, 399, L63.

Morris, M. & Yusef-Zadeh, F. 1985, AJ, 90, 2511.

Morris, M. & Yusef-Zadeh, F. 1989, ApJ, 343, 703.

Okumura, S.K., Ishiguro, M., Fomalont, E.B., Hasegawa, T., Kasuga, T., Morita, K.-I., Kawabe, R. & Kobayashi, H. 1991, ApJ, 378, 127.

Oort, J.H. 1977, Ann. Rev. Astron. Ap., 15, 295.

Pacholczyk, A.G. & Stoeger, W.R. 1986, ApJ, 303, 76.

Pedlar, A., Anantharamaiah, K.R., Ekers, R.D., Goss, W.M., van Gorkom, J.H., Schwarz, U.J. & Zhao, J.-H. 1989, ApJ, 342, 769.

Phinney, E.S. 1989, in "The Center of the Galaxy," ed: M. Morris, Dordrecht:

Kluwer, p. 543.

Quinlan, G.D. & Shapiro, S.L. 1987, ApJ, 321, 199.

Quinlan, G.D. & Shapiro, S.L. 1989, ApJ, 343, 725.

Quinlan, G.D. & Shapiro, S.L. 1990, ApJ, 356, 483.

Rosso, F. & Pelletier, G. 1993, A&A, submitted

Sanders, R.H. & van Oosterom, W. 1984, A&A, 131, 267.

Sandqvist, Aa., Karlson, R. & Whiteoak, J.B. 1989, in "The Center of the Galaxy," ed: M. Morris, Dordrecht: Kluwer, p. 421.

Scoville, N.Z. 1972, ApJL, 175, L127.

Sellgren, K., McGinn, M.T., Becklin, E.E. & Hall, D.N.B. 1990, ApJ, 359, 112.

Serabyn, E., Carlstrom, J.E. & Scoville, N.Z. 199x, ApJL, 401, L87.

Serabyn, E. & Lacy, J.H. 1985, ApJ, 293, 445.

Serabyn, E., Lacy, J.H. & Achtermann, J.M. 1991, ApJ, 378, 557.

Serabyn, E., Lacy, J.H. & Achtermann, J.M. 1992, ApJ, 395, 166.

Serabyn, E., Lacy, J.H., Townes, C.H. & Bharat, R. 1988, ApJ, 326, 171.

Shibata, K. 1989, in "The Center of the Galaxy," ed: M. Morris, Dordrecht: Kluwer, p. 313.

Sofue, Y. 1985, PASJ, 37, 697.

Sofue, Y. & Handa, T. 1984, Nature, 310, 568.

Spergel, D.N. & Blitz, L. 1992, Nature, 357, 665.

Stoeger, W.R., Pacholczyk, A.G. & Stepinski, T.F. 1992, ApJ, 391, 550.

Tollestrup, E.V., Capps, R.W. & Becklin, E.E. 1989, AJ, 98, 204.

Tsuboi, M., Inoue, M., Handa, T., Tabara, H., Kato, T., Sofue, Y. & Kaifu, N. 1986, AJ, 92, 818.

Uchida, K.I. 1992, PhD Thesis, Dept. of Astronomy, Univ. of California at Los Angeles.

Uchida, K.I., Morris, M., Bally, J., Pound, M. & Yusef-Zadeh, F. 1992a, ApJ, 398, 128.

Uchida, K.I., Morris, M. & Yusef-Zadeh, F. 1992b, AJ, 104, 1533.

Uchida, K.I., Morris, M. & Serabyn, E. 1990, ApJ, 351, 443.

Uchida, K.I., Morris, M., Serabyn, E. & Bally, J. 1993, ApJ, submitted.

Wardle, M. & Königl, A. 1990, ApJ, 362, 120.

Weedman, D.W. 1983, ApJ, 266, 479.

Yamauchi, S., Kawada, M., Koyama, K., Kunieda, H., Tawara, Y. & Hatsukada, I. 1990, ApJ, 365, 532.

Yusef-Zadeh, F. 1989, in "The Center of the Galaxy," ed: M. Morris, Dordrecht: Kluwer, p.243.

Yusef-Zadeh, F. & Melia, F. 1992, ApJL, 385, L41.

Yusef-Zadeh, F. & Morris, M. 1987, ApJ, 320, 545.

Yusef-Zadeh, F. & Morris, M. 1991, ApJL, 371, L59.

Yusef-Zadeh, F., Morris, M. & Chance, D. 1984, Nature, 310, 557.

Yusef-Zadeh, F., Morris, M., Slee, O.B. & Nelson, G.J. 1986, ApJL, 300, L47.

Yusef-Zadeh, F. & Wardle, M. 1993, submitted to ApJ.

Zylka, R., Mezger, P.G. & Wink, J.E. 1990, A&A, 234, 133.

Zylka, R., Mezger, P.G. & Lesch, L. 1992, A&A, 261, 119.

STELLAR POPULATIONS AT THE GALACTIC CENTER

Marcia J. Rieke
Steward Observatory, University of Arizona, Tucson, AZ 85721.

Email ID
mrieke@as.arizona.edu

ABSTRACT

Near-infrared imaging and spectroscopy have revealed a rich collection of stars whose character are now being understood as the result of several epochs of star formation. The collection of objects known as IRS16 may be intermediate temperature supergiants which need not contribute significantly to the ionizing radiation present in the central few parsecs.

1. INTRODUCTION

What kinds of stars do we find in the nucleus of the Milky Way? Are the stars old and similar to the bulge stars studied by Frogel and Whitford (1987) and others or are the stars younger and more closely related to populations found in the disk? Are there any unusual stars? How much of the ionizing flux and luminosity are provided by the stellar population as opposed to a black hole? The desire to answer these questions has driven the study of the Galactic Center at near-infrared wavelengths where the stellar population can be observed.

When we consider how to analyze the stellar population at the Galactic Center, we must remember how heavy the interstellar extinction is along this line of sight through the galaxy. The value of A_V is approximately 30 magnitudes. This translates to 8.19 magnitudes at J(1.25μm), 5.25 magnitudes at H(1.6μm), and 3.36 magnitudes at K(2.2μm) using the extinction law of Rieke and Lebofsky (1985). Note that the dust is optically thick even at K, and worse yet, if there is an uncertainty of only 5% in the extinction law at J, the dereddened magnitude is uncertain by 0.4 magnitudes! The heavy level of extinction drives one to observing in the infrared, and virtually all traditional tools used in studying stellar populations such as spectral type and metallicity indicators cannot be utilized. Even the rough metallicity indicator J-K cannot be used for Galactic Center stars.

What types of stars would one expect to find at the Galactic Center? One population that must exist at some level is a population similar to that seen in Baade's Window (BW), in others words, an extension of the bulge population into the center. BW lies at a projected distance of about 500 pc from the center, and this population is undoubtedly similar to some Galactic Center stars. The BW and other bulge populations have been extensively studied in the infrared (Frogel and Whitford 1987, Frogel *et al.* 1990, Terndrup *et al.* 1991). The metallicity of the BW K giants has been measured by Rich (1990). The picture which emerges from these studies is one of an old, metal-rich population. The most luminous stars, if moved to the Galactic Center, would have apparent magnitudes of $K \approx 8.6$.

Another population which could exist at the Galactic Center would be one which is associated with the molecular material near the center. A molecular ring with an inner radius of 1.7pc surrounds the Galactic Center. Genzel (1989) has reviewed the properties of this ring which indicate that at the current time, this material is too hot and has too high a velocity dispersion to be a likely location for star formation. No sites of current star formation have been observed in this ring, but whether any of the stars observed near the Center could have formed from ring material is an open question. Kinematic studies may be able to demonstrate whether there is a connection between the ring and stars, but the number of stars with kinematic data available now is too small to look for this connection.

2.CHARACTERISTICS OF STARS FROM INFRARED SPECTROSCOPY

The existence of late-type stars at the Galactic Center has been known since 1976 when Neugebauer *et al.* observed the brightest 2μm using a low-resolution infrared spectrometer. These authors surmised that IRS7, the brightest 2μm source at the Galactic Center, is an M supergiant. Subsequent work at higher spectral resolution (Treffers *et al.* 1976, Wollman *et al.* 1980, Lebofsky *et al.* 1982) confirmed the nature of IRS7 as an M supergiant, and revealed that several other sources are late-type stars, possibly supergiants. The supergiant nature of IRS7 indicates that it must be a massive star; comparison with tracks from Maeder and Meynet (1988, 1989) indicate that IRS7 has a progenitor of at least 20 $M_\odot$ and an age of about 10 million years. Note that the now ample evidence of IRS7's interactions with the Galactic Center environment (Rieke and Rieke 1989, Geballe *et al.* 1989, Serabyn *et al.* 1991, Yusef-Zadeh and Melia 1992) demonstrates the proximity of IRS7 to the center.

These studies also showed that some sources including the IRS16 complex are not late-type stars because of the lack of CO absorption at 2.3μm. Both Brγ and HeI(2.06μm) emission were also seen, but this emission is distributed widely in the central 2pc, and is not necessarily associated with any of the discrete 2μm sources. The suggestion by Krabbe *et al.* 1991 that the Galactic Center contains a cluster of HeI emission stars must be viewed cautiously in view of the widespread distribution of the HeI emission which is demonstrated by their detection of HeI(2.06μm) from sources such as IRS17 which show CO absorption (Rieke and Rieke 1988) characteristic of cool stars. The strongest limit that can be placed on the spectral types of the IRS16 components based on their IR spectra indicates that they must be earlier than type G. A similar statement holds for the 2μm spectra of the sources most prominent at 10μm.

Another type of star has been discovered at the Galactic Center by Allen *et al.* 1990. This star exhibits Brγ and HeI mentioned above and an additional HeI line at 2.113 μm. A high resolution spectrum of the HeI(2.06 μm) line shows a FWHM of 750 km/sec. Allen *et al.* conclude from the line ratios, line widths, and the star's absolute magnitude that it is very similar to the WN9 and Ofpe stars seen in the Magellanic Clouds; there is no other example known of this stellar type in the Milky Way. Allen *et al.* suggest an age of approximately 5 million years for this star.

3.CHARACTERISTICS OF STARS FROM INFRARED IMAGING

Imaging at H and K has revealed a dense concentration of stars with typical observed K magnitudes of 7-12 extending over a region at least 15pc x 15pc in size (Rieke 1987). This central concentration is very apparent in the large scale images of Glass *et al.* 1987. Within this region is a central cluster of stars located within the molecular ring; to date all of the blue sources (*i.e.*ones without CO bands and with dereddened near-infrared spectra which are Rayleigh-Jeans) lie in this cluster. From the two-color imaging, this entire region suffers about 30 magnitudes of extinction which when coupled with the appearance of the Glass *et al.* image makes a strong case for these stars to be physically close to the Galactic Center.

From mid-infrared fine structure lines (Serabyn and Lacy 1985), a limit of $35,000°K$ can be placed on ionizing sources at the Galactic Center. This translates into a spectral type of no earlier than O9 for main sequence stars. An O9V star at the Galactic Center would have $K \approx 13.3$. Such stars have not yet been detected. Note that emission is detected between the discrete stars seen at the Center; this emission is the output of many stars, perhaps as many as 1000 / sq. arc sec (Bailey 1980). This diffuse stellar emission has a brightness of about $K \approx 13$/sq. arc sec which will make detection of the main sequence difficult until this region is studied with resolution enhancement techniques.

Equipped with H and K images of the Galactic Center, one can proceed to extract a luminosity function. The basic prescription is to extract magnitudes using a profile fitting program such as DAOPHOT or DOPHOT. The H and K magnitudes can then be compared to estimate the extinction assuming that late-type stars predominate. After dereddening, the K magnitudes can be converted to bolometric magnitudes using a bolometric correction and a distance modulus to the Galactic Center. Haller *et al.* (1993a) argue that the average extinction corresponds to $A_V = 31.8$ and that the average dereddened H-K color is 0.3, the average color in Baade's Window. The bolometric correction to K appropriate to giants and supergiants with this H-K color is 3 magnitudes. Haller *et al.* also show that the derived bolometric magnitudes will change by at most 10% if the intrinsic H-K varies from 0.1 to 0.5 with A_V varying from 35 to 29 magnitudes. A distance modulus of 14.2 (R_o=7kpc) has been used for ease in comparing the Galactic Center luminosity function with that derived for BW and other bulge fields by Frogel *et al.* 1990.

Figure 1 shows the resulting luminosity function normalized to have equal numbers in the range M_{bol}=-4.2 to -2.2 for the Galactic Center data and the bulge data. The input K data came from an image of the Galactic Center taken on May 10, 1992, using a NICMOS3 256x256 array with 0.6"/pixel. The seeing disk had a FWHM of 1.6". These data are of substantially higher quality than the data presented in Haller *et al.* because of the better sampling of the PSF, and because the data were acquired over a much shorter period of time. The Galactic Center has an excess of stars brighter than $M_{bol} \leq -4.2$.

Another aspect of the Galactic Center stellar population can be studied by repeat imaging. Because we know that most of the individual stars that we can detect at the center are red giants or supergiants, it is very likely that some of them vary. Haller *et al.* (1993b) have discovered 60 variables in the same 15pc x 15pc field used in deriving the Galactic Center luminosity function. These stars have near-infrared colors on average redder than those observed for the non-variables which supports the suggestion that they are long period variables. The bolometric magnitudes of the variables average -4.9, considerably brighter than the average of - 4.2 for long period variables in the bulge and of -4.1 in globular clusters (Frogel *et al.* 1990). The brightness of the Center variables suggests that they may be an intermediate age population ($10^8 - 10^9$ yrs) which may be related to the OH/IR stars near the Center whose radio spectra indicate a similar age (Winnberg *et al.* 1985).

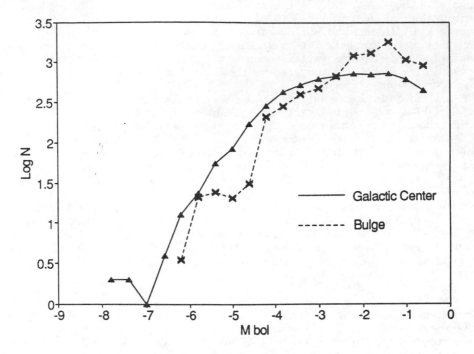

Fig. 1. Comparison of Galactic Center and bulge luminosity functions. The bulge data are taken from Frogel *et al.* (1990) while the Galactic Center data are described in the text.

4. STAR FORMATION AT THE GALACTIC CENTER

We can identify at least three distinct stellar populations based on the discussion above: 1) Stars with ages less than 50-100 million years. IRS7, the Allen *et al.* WN9 star, and probably the bright 10 μm sources and the components of IRS16 fall into this category. 2) Stars with ages of $10^8 - 10^9$ years. The Galactic Center variables and OH/IR stars fall into this category. 3) Stars with ages of 5-10 Gyrs. The extension of the bulge population falls into this category. In the case of IRS7 and the WN9 star, the masses of their precursors must be at least $20 M_\odot$. To form such massive stars from main sequence or giant stars from the bulge population would require an implausibly large number of stellar mergers, so one must understand how star formation can occur in the Galactic Center environment. Less certain is the status of the IRS16 components. These stars must have temperatures and luminosities which place them in the B-G supergiant category since lunar occultations have demonstrated that they are not unresolved stellar clusters (Simons *et al.* 1990, Simon *et al.* 1990).

Tamblyn and Rieke (1993) have investigated whether one episode of star formation could account for all of the young objects seen at the Center including the IRS16 sources. They have modelled the population within 1 pc of the Center, and since they were making a plausibility argument, they have used constraints making it as difficult as possible for star formation to match the observed properties. They assumed that a burst of star formation could use at most 4×10^5 $M_\odot$ with the remainder of the central mass apportioned between a black hole or other unseen matter and the observed stars and gas. They have also required that the burst provide 10^{50} ionizing photons/sec (Mezger and Wink, 1986), no more than 25% of which can come from stars with $T_{eff} > 37,000°K$. The mix of stellar types that the burst was required to produce consisted of at least 4 stars matching the properties of the IRS16 components and 7 to 15 red supergiants. They were able to construct bursts which matched all of these properties, but only if the age of the burst was restricted to the range $6\text{-}9 \times 10^6$ years. Younger bursts were excluded because the ionizing radiation would be too hot and no red supergiants would be seen. Older bursts were excluded because too little ionizing flux was produced and too many red supergiants as compared to IRS16-type stars were produced. In this model the IRS16 stars have $T_{eff} \approx 14,000°K$ and are late B to early A-type supergiants. They produce very little of the ionizing flux which is instead supplied by main sequence stars and Wolf-Rayet stars.

ACKNOWLEDGEMENTS

This work was supported by the National Science Foundation grant AST-911442. I would also like to acknowledge helpful discussions with G. Rieke, P. Tamblyn, and J. Haller.

REFERENCES

Allen, D. A., Hyland, A. R., & Hillier, D. J. 1990, MNRAS, 244, 706

Bailey, M. E. 1980, MNRAS, 190, 217

Frogel, J. A., Terndrup, D. M., Blanco, V. M., & Whitford, A. E. 1990, ApJ, 353, 494

Frogel, J. A., & Whitford, A. E. 1987, ApJ, 320, 199

Geballe, T., Baas, F., & Wade, R. 1989, A&A, 208, 255

Genzel, R. 1989, in The Center of the Galaxy(IAU Symp. No. 136), ed. M. Morris (Dordrecht : Kluwer), 393

Glass, I. S., Catchpole, R. M., & Whitelock, P. A. 1987, MNRAS, 227, 373

Haller,J., Rieke, M. J., & Rieke, G. H. 1993a, in preparation

Haller, J., Rieke, M. J., Rieke, G. H., & Speer, J. 1993b, in preparation

Krabbe, A., Genzel, R., Drapatz, S., & Rotaciuc, V. 1991, ApJ, 382, L19

Lebofsky, M. J., Rieke, G. H., & Tokunaga, A. T. 1982, ApJ, 263, 736

Maeder, A., & Meynet, G. 1988, A&A Supp, 76, 411

Maeder, A., & Meynet, G. 1989, A&A, 210, 155

Mezger, P. G., & Wink, J. E. 1986, A&A, 157, 252

Neugebauer, G., Becklin, E. E., Matthews, K., & Wynn-Williams, C. G. 1976, ApJ, 205, L139

Rich, R. M. 1990, ApJ, 362, 604

Rieke, G. H., & Lebofsky, M. J. 1985, ApJ, 288, 618

Rieke, G. H., & Rieke, M. J. 1988, ApJ, 330, L33

Rieke, G. H., & Rieke, M. J. 1989, ApJ, 344, L5

Rieke, M. J. 1987, in Nearly Normal Galaxies, ed. S. M. Faber (New York : Springer-Verlag), 90

Serabyn, G., & Lacy, J. 1985, ApJ, 293, 445

Serabyn, G., Lacy, J. H., & Achtermann, J. M. 1991, ApJ, 378, 557
Simon, M., Chen, W.-P., Forrest, W. J., Garnett, J. D., Longmore, A. J., Gauer,
    T., & Dixon, R. I. 1990, ApJ, 360, 95
Simons, D. A., Hodapp, K.-W., & Becklin, E. E. 1990, ApJ, 360, 106
Tamblyn, P., & Rieke, G. H. 1993, ApJ, submitted
Terndrup, D. M., Frogel, J. A., & Whitford, A. E. 1991, ApJ, 378, 742
Treffers, R. R., Fink, U., Larson, H. P., & Gautier, T. N. 1976, ApJ, 209, L115
Winnberg, A., Baud, B., Matthews, H. E., Habing, H. J., & Olnon, F. M. 1985,
    ApJ, 291, L45
Wollman, E. R., Smith, H. A., & Larson, H. P. 1982, ApJ, 258, 560
Yusef-Zadeh, F., & Melia, F. 1992, ApJ, L65

$R_o(t)$: A BRIEF HISTORY OF OUR DISTANCE FROM THE GALACTIC CENTER

Virginia Trimble
Astronomy Department, University of Maryland, College Park MD 20742
and
Physics Department, University of California, Irvine CA 92717

ABSTRACT

When Shapley moved us out of the center of the galaxy, he did so with excessive vigor, to a distance of 20 kpc or more. Since then, published values of R_o have oscillated between large and small, with only margi nal evidence for convergence.

For more than a century after William Herschel (1785), astronomers lived essentially at the center of a galaxy not much more than 6000 LY across (see Illustration 18 in Jaki 1972; the poster version of this paper consisted largely of historical images of the Milky Way, which considerations of space and copyright preclude reproducing here). Her schel arrived at his result by counting stars as a function of apparent magnitude in various directions ("star gauging"), and according to Kopal (1971, p. 21) increased the diameter to 20,000 LY in 1806. The issue of whether the spiral nebulae might constitute other island uni verses was discussed sporadically throughout the 19th century, but was not the focus of anyone's research. Newcomb (1882; Jaki illustration 19), for instance, put the "region of the nebulae" immediately above and below a Herschel-like disk.

Cornelius Easton's (1900, picture in Berendzen p. 56) galaxy was also small and sun-centered, but he was the first to give the Milky Way spiral arms. An honest examination of the sky forced him to displace the center of the spiral pattern away from us by more than half the galactic radius in the direction of Cygnus, and his drawing gives the impression of a man struggling with the truth and losing. Parsecs gradually replaced LY as the unit of choice between 1900 and 1920. Schwarzschild's (1910) galaxy was 10 kpc across, 2 kpc thick, and sun centered, while Eddington (1912, picture in Whitney p. 196) put us 60 LY above the center of the galactic plane. Hoskin (1976) summarizes a number of other diameter determinations between 15 and 30 kLY derived essentially from star counts during the first two decades of the 20th century.

Shapley (1918, 1919, and earlier references therein) took an entirely different approach, assumed that the globular clusters traced out the skeleton of galactic structure, and found distances to them from B stars, Cepheids, and other individual stars. There is a certain youthful exuberance to his distances - 67 kpc for NGC 7006 and 13.9 kpc even for M 3. Inevitably, the centroid of his cluster distribution fell far from the sun, at R_0 = 13-25 kpc. The 1919 paper settled on 20 kpc, and a total diameter at least three times that. Sha pley's universe had no room for anything outside this enormous galaxy, and he remained a single-system man at least to the end of the 20's. Pannekoek (1919) concurred in placing the sun far off center, but in a smaller galaxy (R_o = 40-60,000 LY; d = 80-120,000 LY).

The Curtis-Shapley debate (Shapley & Curtis 1921) was supposed to have been about the distance scale of the universe. In practice, the arguments for and against the existence of other, comparable galactic systems were thoroughly admixed. The critical point is that Curtis's Milky Way was only 10 kpc across, with the sun at R_0 = 3 kpc, and so could co-exist with other island universes, in a way that Shapley's ten-times bigger galaxy could not.

Meanwhile, Kapteyn and van Rhijn (1920; Kapteyn 1922) were counting stars more precisely than they had ever been counted before - but totally neglecting interstellar absorption. The first result was R_0 = 0 and d = 24,000 pc; the second, R_o = 3 kpc, d = 17,000 pc (picture, Berendzen p. 24). But Shapley's result seems to have dominated most people's thinking very quickly: Sir Harold Spencer Jones (1923, General Astronomy), Sir James Jeans (1927, Astronomy and Cosmogony), and Russell, Dugan, and Stewart (1927, Vol. 2) all place the galactic center 20 kpc away. Jeans describes the Milky Way and other spirals as having the relationship of a cake to a bunch of bisquits. Kapteyn's work is described as referring to a local stellar subsystem.

Trumpler (1930, picture Berendzen p. 93) made a valiant attempt to declare both parties correct: his drawing shows a coordinate system centered on the sun at the middle of a slightly-tilted 10 kpc "Kapteyn Universe," but globular clusters scattered over an 80 kpc spheroid, centered about 18 kpc away from us!

Oort's discovery of galactic rotation quickly led to a new calibration of distance scales. His first version (Oort 1927, including picture) reported 6300 ± 2000 pc, soon revised upward to 10 kpc (Oort 1932, including picture). This value was widely used over the next 20 years (e.g. Bok 1937, with several interesting pictures).

Baade (1953), however, looked again at the globular clusters and their RR Lyrae variables and settled on R_0 = 8.16 kpc. This value (as 8.2 kpc) was generally accepted as the standard for reducing galactic rotation curve data over the next decade (Westerhout 1956; Kerr 1962, including pictures). N.G. Roman (pr. comm. 1992), who attended the symposium where Baade shrank the galaxy, describes herself as having gone to college at 10 kpc and to graduate school at 8.2 kpc.

The present author did precisely the opposite, for in 1963 Oort (1963, cf. Schmidt 1965) moved us back out to 10 kpc. And there the official IAU set of galactic rotation constants kept us until the 1985 General Assembly in Bangalore, where the Commission on Galactic Structure (cf. Kerr & Lynden-Bell 1986) voted to reduce R_o to 8.5 kpc. That number is the average of a long table, containing numbers as big as 11 kpc and as small as 6 kpc. Recent trends have been perhaps toward the small end of the range (see e.g. RAS 1990). In other words, our picture of the galaxy and the rest of the universe is somewhat closer to that advocated by Curtis than to Shapley's version, though Shapley is generally regarded as having won the debate.

ACKNOWLEDGEMENTS: Many of the images and numbers for this paper were collected in Peridier Library of the astronomy department, Univ. of Texas, Austin, while the author was Beatrice M. Tinsley visiting professor there in spring 1992.

REFERENCES

Baade, W. 1953, in Symposium on Astrophysics (U. Michigan) p. 25
Berendzen, R. et al. 1976, Man Discovers the Galaxies (NY Science
    History Publications)
Bok, B.J. 1937, The Distribution of Stars in Space (U. Chicago Press)
Easton, C. 1900, ApJ 12, 136
Eddington, A.S. 1914, Stellar Movements (London, Macmillan) p. 31
Herschel, W. 1785, Phil. Trans. Roy. Soc. 75, pt. 1
Hoskin, M.A. 1976, J. Hist. Astron. 17, 169
Kapteyn, J.A. 1922, ApJ 55, 65
Kapteyn, J.A. & P.J. van Rhijn 1920, ApJ 52, 23
Kerr, F.J. 1962, MNRAS 123, 327
Kerr, F.J. & D. Lynden-Bell 1986, MNRAS 222, 1023
Kopal, Z. 1971, Widening Horizons (NY, Taplinger)
Jaki, S. 1972, The Milky Way (NY, Science History Publications)
Newcomb, S. 1882, Popular Astronomy 4th ed. (NY Harper & Bros.) p 493
Oort, J.A. 1927, BAN 4, 79
Oort, J.A. 1932, BAN 6, 279
Oort, J.A. 1964, IAU Inf. Bull. 11
Oort, J. & A. van Woerkom 1941, BAN 9, 185
Pannekoek, A. 1919, MNRAS 79, 500
RAS 1990, Observatory 111, 67 (discussion session)
Schmidt, M. 1965, in Galactic Structure (U. Chicago Press), p. 513
Schwarzschild, K. 1910, AN 185, 81
Seeliger, H. 1920, Muenchen Ber. p. 87
Shapley, H. 1918, ApJ 48, 176
Shapley, H. & M.B. Shapley 1919, ApJ 50, 116
Shapley, H. & H.D. Curtis, 1921, Bull. Nat. Res. Council 2, 171
Trumpler, R. 1930, LOB 14, 154 (No. 420)
Westerhout, G. 1956, BAN 13, 201
Whitney, C.A. 1971, The Discovery of our Galaxy (NY, Knopf)

High Resolution Study of the -190 km s^{-1} Molecular Gas in the Galactic Center; IRAM Observations of CO J=1→0 and J=2→1 in the Central 80"

J. M. Marr[1], C. Lemme[2], T. A. Pauls[3],
M. C. H. Wright[4], T. L. Wilson[2], A. L. Rudolph[5]

[1] Haverford College
[2] Max-Planck-Institute für Radioastronomie
[3] Naval Research Laboratory
[4] University of California, Berkeley
[5] NASA Ames Research Center

ABSTRACT

We have mapped the CO emission in J=2→1 and J=1→0 in a 140" by 100" region around Sgr A West at radial velocities from –250 to –170 km s^{-1} with a resolution of 11". Our maps reveal that the –190 km s^{-1} molecular feature reported by Liszt (1991) appears to end at the brightest clump in the circumnuclear molecular ring (Güsten et al.1987) and makes a sharp bend about a point 50" due West of the southern end of the ring. Additionally, the blueshift increases with increasing distance from the ring and Galactic center.

1. Introduction

In low resolution CO observations of the Galactic center region, Liszt (1991) mapped a feature at –190 km s^{-1} and found it to extend about 9' northwest from the Galactic center. Liszt noted that this molecular gas feature runs perpendicular to the Galactic plane with one end at the Galactic center and therefore has the appearance of a molecular gas "jet." However, three problems plague this interpretation: the large blueshift suggests that the gas does not actually move perpendicular to the Galactic plane; the high velocity requires a significant acceleration mechanism that does not destroy molecules; and this feature appears to miss the Galactic center in Liszt's map by about an arcminute.

Interferometric observations of Sgr A West in HCO$^+$ (Marr et al. 1992) and H$_2$CO (Pauls et al. 1992) have suggested that molecular gas at velocities from –170 to –210 km s^{-1} exists within the central 20" of the Galactic center. Because the HCO$^+$ and H$_2$CO were mapped in absorption, these latter studies traced only the gas in positions where continuum emission exists and could not reveal whether this gas is related to the extended feature mapped by Liszt.

We have, therefore, observed and mapped at high resolution the CO emission at these velocities in and around Sgr A West in order to determine the actual distribution of this highly blueshifted molecular gas in the central region.

2. Observations

Observations of the CO J=2→1 and J=1→0 lines were carried out on 18-20 January 1992 with the IRAM 30-m telescope near Granada, Spain. The angular resolution of this telescope is 11" at 1.3 mm and 22" at 2.6 mm. Three receivers were used simultaneously—two 1.3-mm SIS receivers and a single 3-mm SIS receiver. The noise temperatures were between 500 and 680K for the 1.3-mm receivers and between 900 and 1100K for the 2.6-mm receiver. For each 1.3-mm receiver, a 512 channel by 1 MHz spectrometer was used; for the 2.6-mm receiver, an acoustic optical spectrometer was employed.

Spectra at velocities from –170 to –250 km s^{-1}, relative to the Local Standard of Rest, were obtained employing position switching with a reference at (ΔRA, ΔDec)=(+150″, –150″), relative to the position of Sgr A*. The reference showed no CO emission in the velocity range from about –180 to –250 km s^{-1}. We first obtained the spectra in a 60″ by 60″ square grid around Sgr A* using a procedure that involved a chopper wheel calibration first, followed by a reference spectrum observation, and then five on-source spectra observations at positions spaced by 6″. Further from Sgr A*, the spacings were increased to 12″. A few single spectra were taken later at other positions using one on-source spectrum observation per reference observation. The integration time per on-source spectrum was 30 seconds and that for each reference observation was 60 seconds. On the average, each position was measured twice. The full extent of our maps covered a 140″ by 100″ region.

The data were reduced using the CLASS program package. Usually a third order baseline was removed from each spectrum; in a few cases, a fifth order baseline was required. The spectra for the two 1.3-mm receivers were reduced separately, since there were small differences in the calibration temperatures.

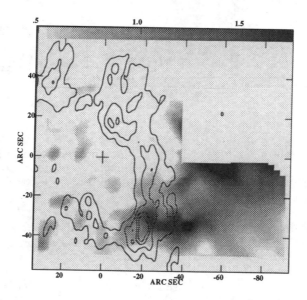

Figure 1. The CO J=2→1 emission from the Galactic center at velocity –190 km s^{-1} is shown in gray-scale overlaid with a contour map of the HCN emission from the circumnuclear ring (Güsten *et al.* 1987). The square in the upper right corner is blank because it contains no data. The coordinates are relative to Sgr A*, the position of which is indicated by the '+'. The displayed contours correspond to 24, 60, and 80% of the peak HCN intensity. The maximum flux-density for the gray-scale is 1.75 K km s^{-1}.

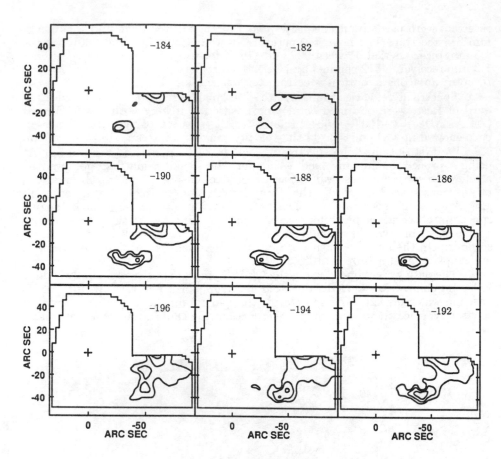

Figure 2. Contour maps of the CO J=2→1 emission at velocities from −182 to −196 km s^{-1} are shown in steps of 2 km s^{-1} per frame. As in Figure 1, Sgr A* is located at the '+' at position (0,0). The displayed contours correspond to brightness temperatures of 1.25, 1.50, 1.75 and 2.00 K km s^{-1}.

Figures 1 and 2 display maps of the CO J=2→1 emission at velocities from −182 to −196 km s^{-1}. The CO J=1→0 maps, which are not shown, look similar but have poorer resolution and so are less revealing.

3. Discussion

The −190 km s^{-1} feature mapped by Liszt (1991) enters the map in Figure 1 from the northwest in the cut-out square at the right and connects to the emission feature shown in Figure 1. The other end of this highly blueshifted molecular gas feature appears to coincide with the brightest clump and highest blueshifted ($v \approx 110$ km s^{-1}) part of the circumnuclear molecular ring. This positional coincidence suggests that this feature and the ring, especially the bright clump, are causally connected. Such a relation, if true, can alleviate the difficulty in explaining the large blueshift of the extended molecular feature, since an acceleration of only 80 km s^{-1} is needed. Furthermore, the mechanism responsible for kicking molecular gas out of the ring could also be the cause of the enhanced excitation of the molecules in the bright clump.

An alternative scenario is that the highly blueshifted molecular gas falls in toward the Galactic center from behind on a parabolic or hyperbolic orbit and collides with the circumnuclear molecular ring. The excitation of the molecules in the ring, then, are enhanced at the collision site yielding the bright clump.

The observed morphology and velocity structure of the –190 km s^{-1} gas, though, is not satisfied by either of these models. The observed morphology suggests that the gas' motion has a sharp bend about a point well away from the known significant masses in the region, assuming that the extended shape of the emission is due motion of the gas. The curvature of this path does not fit acceleration by a mass at the dynamical center (located at the '+').

The velocity structure, as shown in Figure 2, involves a blueshift that increases with increasing distance from the center, suggesting an acceleration outwards. The increase in blueshift could be due to a curved path of motion that curves toward the line of sight with increasing distance from the center. In either case an accelerating mechanism other than the gravitational field of the center is required.

ACKNOWLEDGEMENTS

We wish to thank Harvey Liszt for helpful discussions.

This research was partially supported by funds from the Margaret Cullinan Wray Charitable Lead Annuity Trust.

REFERENCES

Güsten, R., Genzel, R., Wright, M. C. H., Jaffe, D. T., Stutzki, J., & Harris, A. I. 1987, ApJ, 318, 124
Liszt, H. S. 1991, BAAS, Observatory Reports
Marr, J. M., Pauls T., Rudolph. A., Wright, M. C. H., & Backer, D. C. 1992, ApJ, 400, in press
Pauls, T., Johnston, K. J., Wilson, T. L., Marr, J. M., & Rudolph, A. 1992, ApJ, in press

CO (J=2–1) OBSERVATIONS OF THE MOLECULAR CLOUD COMPLEX IN THE GALACTIC CENTER

TOMOHARU OKA[1], TETSUO HASEGAWA[2]

MASAHIKO HAYASHI[1], TOSHIHIRO HANDA[2]

SEI'ICHI SAKAMOTO[1]

1) Department of Astronomy, Faculty of Sience, University of Tokyo, Bunkyo, Tokyo 113, Japan
2) Institute of Astronomy, Faculty of Sience, University of Tokyo, Mitaka, Tokyo 181, Japan

ABSTRACT

We report a large scale mapping observation of the Galactic center region in the CO (J=2–1) line using the Tokyo-NRO 60cm survey telescope. Distribution of the CO (J=2–1) emission in the l-V plane suggests that molecular clouds forms a huge complex (*Nuclear Molecular cloud Complex, NMC*). Tracers of star formation activities in the last 10^6–10^8 years show that star formation has occured in a ring $\sim$ 100 pc in radius. Relative to this *Star Forming Ring*, the molecular gas is distributed mainly on the positive longitude side. This may indicate that much of the gas in NMC is in transient orbit to fall into the star forming ring or to the nucleus in the near future.

Molecular gas in our Galaxy is concentrated in the inner a few hundred parsecs (*e.g.*, Liszt & Burton 1978), exhibits complicated kinematics (*e.g.*, Bally *et al.* 1987). It is known to be warm (30–60K) and dense (Morris *et al.* 1983; Bally *et al.* 1987, 1988). In order to illustrate the distribution and kinematics of the warmer and denser molecular gas in the galactic center region, we have started a program to make a large-scale map in the CO (J=2–1) line emission. In this paper, we report its initial results, focusing on the gas in the galactic plane.

Observations are carried out using the Tokyo-NRO 60cm survey telescope at Nobeyama with a $9' \pm 1'$ beam (FWHM). About 100 CO (J=2–1) spectra were taken along the galactic plane between $l = -3°$ and $3°$, and along strips perpendicular to the plane between $b = -0°.5$ and $0°.5$.

The distribution of the CO (J=2–1) line emission (Fig.1) shows a strong asymmetry; strong emission occurs mainly in the positive longitude with positive velocities. The data are compared directly with the CO (J=1–0) data from the Columbia survey (Bitran *et al.* 1987). The average J=2–1/J=1–0 line intensity ratio is 1.1, which is significantly higher than the ratio for clouds in the galactic disk (0.6 – 0.8; Hasegawa et al. 1992, Sakamoto *et al.* 1992). This means that the gas in the galactic center is denser and $X_{CO}/(dV/dR)$ is smaller.

Fig.1 recalls us a possibility to conceive the molecular clouds in this region as a huge single complex, which is $\sim$2°.5(375 pc) in l and $\sim$0°.3(45 pc) in b. We call it *the Nuclear Molecular cloud Complex (NMC)*. If we adopt the conversion factor in the galactic disk ($N_{H_2}/I_{CO} = 2.0 \times 10^{20}$ (cm)$^{-2}$(Kkms^{-1})$^{-1}$), the total mass of NMC is about 10^8 $M_\odot$.

Probes of star formation activities in the last 10^6–10^8 years, i.e., HII regions (filled circles in Fig.1, data from Pauls and Mezger 1975) and OH/IR stars (Lindqvist *et al.* 1991), show much more symmetric patterns in the *l-V* plane. This suggests that star formation has occured in a ring (or a rigid-rotating disk). The structure, *Star Forming Ring*, is $\sim$ 100 pc in radius and its rotation is approximately circular.

The strong asymmetry in the distribution of molecular gas relative to the Star Forming Ring may indicate that much of the gas in NMC is in transient orbits. Some gas may settle in the ring to fuel a starburst, or may fall even closer to the nucleus (like the 20kms^{-1}and 40kms^{-1} clouds, for example).

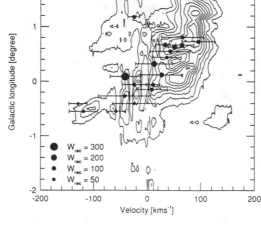

Figure 1—The CO (J=2–1) *l-V* diagram at b=0°. Filled circles show the positions, velocities, and integrated intensities (W_{rec} in KkHz) of the H109 α recombination line observed by Pauls and Mezger (1975). The bars represents the widths (FWHM) of the H109 α line.

REFERENCES

Bally, J., Stark, A. A., Wilson, R. W. 1987, *Ap. J. Suppl.*, **65**, 13
Bally, J., Stark, A. A., Wilson, R. W. 1988, *Ap. J.*, **324**, 223
Bitran, M. E. 1897, PhD Thesis, U. of Florida
Genzel, R. 1990, In *From Ground-Based to Space-Bourne Submm Astronomy.*
        , eds. N. Longdon, B. Kaldeich
Goldreich, P., Kwan, J. 1974, *Ap. J.*, **189**, 441
Hasegawa, T. *et al.* 1992, in preparation
Lindqvist, M., Winnberg, A., Habing, H. J., Matthews, H. E. 1991, *Astr. Ap. Suppl.*, **92**, 43
Liszt, H. S., Burton, W. B. 1978, *Ap. J.*, **226**, 790
Morris, M., Polish, N., Zuckerman, B., Kaifu, N. 1983, *Astr. J.*, **88**, 1228
Pauls, T., Mezger, P. G. 1975, *Astr. Ap.*, **44**, 259
Sakamoto, S., Hasegawa, T., Hayashi, M., Handa, T., Oka, T., 1992, *Ap. J.* submitted

DUST EMISSION IN THE GALACTIC CENTER

D. C. Lis
Downs Laboratory of Physics 320–47
California Institute of Technology, Pasadena, CA 91125

J. E. Carlstrom
Astronomy Department 105–24
California Institute of Technology, Pasadena, CA 91125

ABSTRACT

We conducted a survey of the 800 μm continuum emission from $\sim$1.5° × 0.2° area around the Galactic center. The 800 μm continuum emission, which traces temperature weighted column density of dust, shows a strong asymmetry between the positive and negative galactic longitudes, similar to that seen in the molecular data; the bulk of the emission occurs at the positive galactic longitudes. One of the most interesting features in the 800 μm map is a clumpy ridge of emission apparently connecting Sgr B2 and the Radio Continuum Arc. A molecular counterpart of this ridge can be seen in the CS $J = 2 \rightarrow 1$ data. The ridge contains a number of dust cores without identified far–infrared sources or compact H II regions. These dust cores may be unusually cold.

1. INTRODUCTION

The total infrared luminosity and the ionizing flux inferred from radio continuum observations are consistent with a normal rate of star formation per unit mass of molecular material in the Galactic center (Mezger & Pauls 1979; Güsten 1989). However, H_2O and OH masers commonly found in the sites of high–mass star formation are relatively rare in this region (Güsten & Downes 1983). Boissé et al. (1981) and Odenwald & Fazio (1984) suggested a decrease in the formation rate of stars with masses greater than $\sim$20 $M_\odot$ in the Galactic center. Morris (1989) also argued that the star formation could be currently suppressed in the Galactic center compared to the disk and that a substantial fraction of the far–infrared and radio continuum emission may be due to other processes.

To ascertain the contribution of high–mass star formation to the luminosity and ionization of the Galactic center region, we conducted a survey of the 800 μm continuum emission from $\sim$1.5° × 0.2° area of the nuclear disk. In a previous paper (Lis & Carlstrom 1993) we compared the morphology of the dust emission in this region to those of the far–infrared and radio continuum emission. In the known high–mass star forming regions, such as Sgr B2, the large–scale distribution of the submillimeter continuum emission, which traces the temperature weighted column density of dust, was found to correlate well with those of the radio continuum and far–infrared emission. However, in the area between Sgr A and Sgr B2 which includes the Radio Continuum Arc, we found a poor correlation between the dust emission and the other two tracers. This suggests that a substantial fraction of the far–infrared and radio continuum luminosity in this region may be produced by processes unrelated to high–mass star formation. In the present paper we compare the distribution of the dust emission with that of the molecular gas.

2. OBSERVATIONS

The data were taken in 1992 July using the bolometer of the Caltech Submillimeter Observatory on Mauna Kea, Hawaii (Lis, Carlstrom, & Keene 1991). Differential altazimuth images were reconstructed using the algorithm described by Emerson, Klein, & Haslam (1979) to recover the emission on angular scales larger than the chopper throw used during the observations ($\sim$3'). The area covered is $\sim$1.5° × 0.2°, elongated along the Galactic plane. The angular resolution of the final map interpolated on the regular grid in the equatorial coordinate system is $\sim$30″ (FWHM). The $\sim$218,000 flux measurements are equivalent to $\sim$18 hours of integration. The rms sensitivity of the final image is $\sim$1.5 Jy ($\sim$0.5 % of the peak flux density in a 30″ beam measured toward Sgr B2). Uranus was used for flux calibration and pointing of the telescope was checked by frequent observations of Sgr B2(N), the strongest compact source in the region.

3. COMPARISON OF THE DUST AND THE MOLECULAR EMISSION

A contour map of the 800 μm continuum emission from the nuclear disk is shown in Figure 1. The boundary of the region covered in the CSO survey is outlined by the solid line. The dashed line marks the position of the Galactic plane. There is a strong asymmetry in the intensity of the dust emission between the positive and negative galactic longitudes. The 800 μm emission peaks toward the Sgr B2(N) continuum source; about half of the 800 μm flux density in the region mapped ($\sim$11.6 kJy) comes from the Sgr B2 complex. The integrated flux density toward Sgr C is only $\sim$20% of that observed toward Sgr B2. A similar asymmetry has been observed in molecular data (e.g., Bally et al. 1987). In Sgr B2, approximately 30 % of the total molecular mass ($\sim$7 × 10^6 $M_\odot$; Lis & Goldsmith 1990) is traced by the dust emission.

One of the the most interesting features in the 800 μm map is the narrow, clumpy ridge of emission apparently connecting Sgr B2 and the radio continuum source G0.18–0.04 associated with the Radio Continuum Arc. No radio continuum emission or compact far–infrared sources are observed toward the dust cores in this ridge. This may be the first detection of *compact* dust cores in the Galactic center *without embedded far–infrared sources or H II regions*, indicating that these dust cores may be unusually cold. The total H_2 mass in the ridge based on our 800 μm data is $\sim$5 × 10^5 $M_\odot$. Limited CS $J = 5 \rightarrow 4$ data taken toward the ridge with the CSO indicate gas velocities between 0 and 40 km s^{-1}. Low angular resolution (100″) channel maps of the CS $J = 2 \rightarrow 1$ emission in this velocity range (Bally et al. 1987) show a morphology similar to that of the dust emission. The velocity of the gas associated with the dust ridge is similar to that of the molecular cloud associated with the radio continuum source G0.18–0.04 (Serabyn & Güsten 1991). This suggests that the ridge may

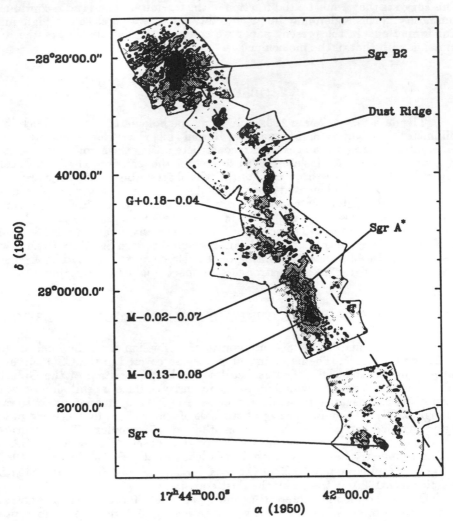

Figure 1. A contour map of the 800 μm continuum emission toward the nuclear disk. Contour levels are 5, 10, 15, and 20 Jy in a 30″ FWHM beam. The boundary of the region surveyed is outlined by the thin solid line. The dashed lines mark the position of the Galactic plane. The 800 μm continuum emission shows strong asymmetry between the positive and negative galactic longitudes, similar to that observed in molecular emission.

be kinematically connected to the Arc. High angular resolution spectroscopic studies of this region are required to determine the kinematics of the dense gas and the physical conditions in the compact dust cores.

Because the mean volume densities of the GMCs in the Galactic center are comparable to the critical density of the CS $J = 2 \rightarrow 1$ transition, the existing CO and CS surveys of the Galactic center (e.g., Bally et al. 1987; Heiligman 1987) show primarily the extended cloud envelopes rather than the high density cores. Lis et al. (1993) mapped the envelope and the core of the Sgr C molecular cloud in various transitions of CS and CO isotopes. The distribution of the dust continuum emission is resembled most closely by the CS $J = 5 \rightarrow 4$ emission, while the CS $J = 2 \rightarrow 1$ emission is dominated by the low density gas in the extended molecular cloud. A complete survey of the Galactic center region in the $J = 5 \rightarrow 4$ transition of CS, when compared with the dust continuum data, would be extremely useful for studying the distribution and kinematics of the dense gas and deriving the fraction of mass in the high–density cores.

ACKNOWLEDGEMENTS

This research has been supported by NSF grant AST 90–15755 to the Caltech Submillimeter Observatory. We thank K. Young for implementing the continuum on–the–fly mapping mode at the CSO.

REFERENCES

Bally, J., Stark, A. A., Wilson, R. W., & Henkel, C. 1987, ApJS, 65, 13.
Boissé, P., Gispert, A., Coron, N., Wijnberger, J. J., Serra, G., Ryter, C., & Puget, J. L. 1981, A&A, 94, 265.
Emerson, D.,T., Klein, I., & Haslam, C. G. T. 1979, A&A, 76, 92.
Güsten, R. 1989, in The Center of the Galaxy, ed. M. Morris (Dordrecht: Kluwer), p. 89.
Güsten, R., & Downes D. 1983, A&A, 117, 343.
Heiligman, G. M. 1987, ApJ, 314, 747.
Lis, D. C., & Carlstrom, J. E. 1993, in Sky Surveys: Protostars to Protogalaxies, ed. B. Soifer, in press.
Lis, D. C., Carlstrom, J. E., Goldsmith, P. F., & Bergin, E. A. 1993, in preparation.
Lis, D. C., Carlstrom, J. E., & Keene, J. 1991, ApJ, 380, 429.
Mezger, P. G., & Pauls, T. 1979, in The Large–Scale Characteristics of the Galaxy, ed. W. Burton, 357.
Morris, M. 1989, in The Center of the Galaxy, ed. M. Morris (Dordrecht: Kluwer), p. 171.
Odenwald, S. F., & Fazio, G. G. 1984, ApJ, 283, 601.
Serabyn, E., & Güsten R. 1991, A&A, 242, 376.

ON THE THERMAL INSTABILITY AROUND THE BOUNDARY OF H II REGION IN THE GALACTIC CENTER

Masa–aki Kondo
Senshu University
Higashi–mita, Tama–ku, Kawasaki–shi, Kanagawa, 214 Japan

ABSTRACT

The excitation of oscillation in the rotating disk of the Galactic center is considered. HII regions are usually thermally stable, because of the heating due to photoelectrons produced by ionizations of the UV radiation from central OB stars. The different point from the usual HII regions is that the Galactic center is the rotional HII region. The coupling between the rotational and non-adiabatic effects brings the instabilities of nonradial oscillations in the HII region of the Galactic center. Hence, it is possible to say that the spiral and bar-like structure is the trapped nonradial oscilations of gravity mode, excited by the heating of photoelectrons. Furthermore, the oscillation in the outer region surrounding the HII region is generated by the nonadiabatic resonance, where the epicyclic frequency is near to the Lamb frequency. This oscillation is expected to be the warp motion of the molecular ring.

I. INTRODUCTION

The structure of Galactic center seemes to be still puzzeled, although the detailed velocity and pattern have been observed by the radio continum and the infrared line of $NeII$ (cf. Brown and Liszt, 1984; Gentzel and Townes, 1987). The spiral pattern and bar has ben interpreted as free-falling of giant gaseous cloud, which would has lost its angular momentum by collisions against other turbulent molecular clouds on the boundary of HII region. In such case, however, more clouds would have been fallen into the center, because of the turbulent state around the HII region. If the spiral is a pattern, the loss of angular momentum is out of problem. Then, it is worth to consider the instability of corrugation wave, which has been proposed by Kato (1983) in the case of the accretion disks around black holes (cf. Lacy et al, 1991).

On the other hand, the various molecules of CO, HCN, H_2 have been observed outside HII region, forming the ring structure. Especially, the HCN ring shows the warp motion. Then, one will presume that the spiral pattern inside the HII region and the warp motion outside are connected through some excitation mechanism, which is examined by the analysis of stability in a rotating HII region. As detailed structure of the unperturbed state is not obtained yet, the qualitative characters will be discussed in the following stability analysis.

II. UNPERTURBED STATE

The unperturbed state is the steady rotational gaseous disk, where the accretion is not considered. The cylindrical coordinate of r, ϕ, z is used. The notations are usual ones.

Momentum equations: $\dfrac{1}{\rho_0}\dfrac{\partial p_0}{\partial r} = -g_r + r\Omega^2 \equiv g_{eff}$, $\dfrac{1}{\rho_0}\dfrac{\partial p_0}{\partial z} = -g_z$, (1)

Energy   equation : $div\boldsymbol{F}_0 = \Gamma_{ei} + \Gamma_d + \Gamma_x$,   (2)

Radiative equations: $div\boldsymbol{F}_\nu = -(\sigma_\nu + \sigma_\nu^d)J_\nu + \sigma_\nu B_\nu + \sigma_\nu^d B_\nu^d + \sum_j W_j L_j$,   (3)

$$\tfrac{1}{3}\nabla J_\nu = -(\sigma_\nu + \sigma_\nu^d)\boldsymbol{F}_\nu ,$$ (4)

where g_r , g_z and g_{eff} are the radial, vertical and effective gravitational forces, Ω the rotational angular frequency. In the energy equation, $\boldsymbol{F}_0$ is the integrated radiative flux, Γ_{ei}, Γ_d and Γ_x the heatings due to photoelectrons, viscous dissipations and X-ray photons (cf. Morris, 1988). In the radiative equations, $\boldsymbol{F}_\nu$ and J_ν are the radiative flux and intensity of frequency ν, σ_ν and σ_ν^d the gaseous and dust-grains absorption coefficients, B_ν and B_ν^d the emissions due to the gaseous and dust-grains temperatures, W_j and L_j the dilution factor and luminosity of the star j (cf. Spitzer, 1978). The observations of HeI and $NeII$ have shown that the UV photon sources are sufficient due to OB stars of $38000K$, which are clustering in $IRS16$.

Here, the above equations are not solved exactly, but they are used for to estimate the order of magnitude of the parameters appeared in the dispersion relation for perturbations.

III. PERTURBATION EQUATIONS

The effects of self-gravitation and viscosity are not considered, but the nona-diabatic effect is taken into account. Then, the perturbations equations are as follows, where perturbed quantities are denoted by the prime symbol:

$$\left(\frac{\partial}{\partial t} + \Omega\frac{\partial}{\partial\phi}\right)\rho' + \frac{1}{r}\frac{\partial}{\partial r}r\rho_0 v_r' + \frac{1}{r}\frac{\partial}{\partial\phi}\rho_0 v_\phi' + \frac{\partial}{\partial z}\rho_0 v_z' = 0 ,$$ (5)

$$\left(\frac{\partial}{\partial t} + \Omega\frac{\partial}{\partial\phi}\right)v_r' - 2\Omega v_\phi' = -\frac{1}{\rho_0}\frac{\partial p'}{\partial r} + \frac{\rho'}{\rho_0^2}\frac{\partial p_0}{\partial r} ,$$ (6)

$$\left(\frac{\partial}{\partial t} + \Omega\frac{\partial}{\partial\phi}\right)v_\phi' + \frac{\kappa^2}{2\Omega}v_r' = -\frac{1}{\rho_0 r}\frac{\partial p'}{\partial\phi} ,$$ (7)

$$\left(\frac{\partial}{\partial t} + \Omega\frac{\partial}{\partial\phi}\right)v_z' = -\frac{1}{\rho_0}\frac{\partial p'}{\partial z} + \frac{\rho'}{\rho_0^2}\frac{\partial p_0}{\partial z} ,$$ (8)

and

$$\left(\frac{\partial}{\partial t} + \Omega\frac{\partial}{\partial\phi}\right)(p' - c_s^2\rho') + c_s^2\left(\frac{N_r^2}{g_{eff}}v_r' + \frac{N_z^2}{g_z}v_z'\right) = \mathcal{D}_p p' + \mathcal{D}_\rho \rho' ,$$ (9)

where

$$\kappa^2 \equiv 4\Omega^2\left(1 + \frac{1}{2}\frac{\partial ln\Omega}{\partial ln r}\right) ,$$ (10)

$$N_r^2 \equiv g_{eff}\left(\frac{1}{\gamma}\frac{\partial ln p_0}{\partial r} - \frac{\partial ln \rho_0}{\partial r}\right) , \qquad N_z^2 \equiv g_z\left(\frac{1}{\gamma}\frac{\partial ln p_0}{\partial z} - \frac{\partial ln \rho_0}{\partial z}\right) ,$$ (11)

$$\mathcal{D}_p \equiv \frac{1}{\delta p}\delta[-div\boldsymbol{F} + \Gamma_{ei} + \Gamma_d + \Gamma_x] , \text{ and } \mathcal{D}_\rho \equiv \frac{1}{c_s^2}\frac{1}{\delta\rho}\delta[-div\boldsymbol{F} + \Gamma_{ei} + \Gamma_d + \Gamma_x] .$$ (12)

Here δ means the variation. κ is the epicyclic frequency, N_r and N_z the radial and vertical Brunt–Väisäla frequencies.

Now, we assume a perturbation q' takes the wave form of $exp\ i(\omega t + m\phi + \ell z)$.

After some manupilation, we have the canonical equations concerning the radial derivative. Under the further assumption of the radial wave form of $exp(ikr)$, we can get the simplified dispersion relation:

$$[\bar{\omega}^2 - \kappa^2 + N_r^2 \frac{\bar{\omega}}{\bar{\omega} - i\mathcal{D}_\rho}][(\bar{\omega}^2 + K_{zz})\frac{\bar{\omega} + i\mathcal{D}_p}{\bar{\omega} - i\mathcal{D}_\rho} - L^2)] = k^2 c_s^2 \bar{\omega}^2 . \tag{13}$$

where $\bar{\omega} = \omega - m\Omega$, $K_{zz} = \partial g_z/\partial z$, and L^2 is the Lamb frequency given by

$$L^2 \equiv c_s^2(\frac{m^2}{r^2} + \ell^2) , \tag{14}$$

Precisely speaking, k must be determined by the quantized condition of giving k indicates the wavenumber contained in the trapped region (cf. Unno etal, 1989).

In the adiabatic case of $\mathcal{D}_p = \mathcal{D}_\rho = 0$, ω is denoted by ω_0, which is determined by the following dispersion relation:

$$[\bar{\omega}_0^2 - \kappa^2 + N_r^2][\bar{\omega}_0^2 - L^2 + K_{zz}] = k^2 c_s^2 \bar{\omega}_0^2 . \tag{15}$$

The oscillations with real ω_0 exist in the region satisfying the following condition:

$$(\omega_0 - m\Omega)^2 > \kappa^2 - N_r^2 \quad \text{or} \quad (\omega_0 - m\Omega)^2 < L^2 - K_{zz} . \tag{16}$$

Here, the oscillations satisfying the former condition is the sound mode and the latter be the gravity mode. In the following, we will consider the case that $\Omega = V_0/r$, because the rough treatment of dipersion relation will be mitigated. The normalization is done such that r and frequencies are in the units of the radius R of HII region and V_0/R respectively.

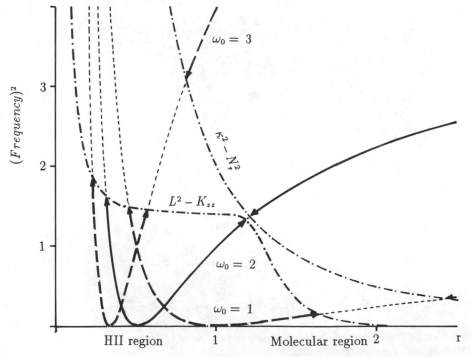

Fig. The propagation diagram. the rotaional anguler frequency is assumed to be proportional to $1/r$.

In the figure, the schematic distributions of $\kappa^2 - N_r^2$ and $L^2 - K_{zz}$ are shown by the dot-dashed curves. The dashed curves indicate $\bar{\omega}_0^2 = (\omega_0 - \Omega)^2$ in the case of $m = 1$. The bold portions indicate the waves trapped in the inner region, satisfying the latter of condition(16), and those in the outer region, satisfying the former condition. The oscillations indicated by the bold dashed curves may be trapped, but they are less excited than the ones indicated by the bold full curves, if the nonadiabatic effects discussed later are taken into accout.

Then, to see the non-adiabatic effects, $\mathcal{D}_p$ and $\mathcal{D}_\rho$ are assumed to be small, so that the adiabatic frequency ω_0 changes to $\omega_0 + \omega_1$. From equation (16), ω_1 is derived as follows:

$$\omega_1 = \frac{-ik^2 c_s^2 \bar{\omega}_0^2}{2[\bar{\omega}_0^4 - (\kappa^2 - N_r^2)(L^2 - K_{zz})]}[(\mathcal{D}_p + \mathcal{D}_\rho)\frac{\bar{\omega}_0^2 + K_{zz}}{\bar{\omega}_0^2 - L^2 + K_{zz}} + \mathcal{D}_\rho \frac{N_r^2}{\bar{\omega}_0^2 - \kappa^2 + N_r^2}] , (17)$$

This equation should be the integrated over the trapped region obtained from the adiabatic consideration (cf. Lynden-Bell and Ostriker, 1967), so that the terms appearing in equation (17) are interpreted as the means in the trapped region.

Since the heating due to the photoelectrons due to UV radiation is dominat in the HII region, $\mathcal{D}_p + \mathcal{D}_\rho$ and $\mathcal{D}_\rho$ are positive (cf. Osterbrock, 1988). In the outer molecular ring, $\mathcal{D}_p + \mathcal{D}_\rho$ and $\mathcal{D}_\rho$ are also positive, although cooling due to atoms and molecules is effective.

In the static state of $\kappa = N_r = K_{zz} = L = 0$, the condition of the thermal instability of condensation mode is $\mathcal{D}_\rho < 0$, and that of sound mode is $\mathcal{D}_p + \mathcal{D}_\rho < 0$. These conditions correspond to those obtained by Field (1969), where the independent variables are T and ρ. Hence, this state is thermally stable.

However,in the rotational case with nonzero κ, N_r, K_{zz} and L, the positive $\mathcal{D}_p$ and $\mathcal{D}_\rho$ let $i\omega_1$ positive, so that oscillations are unstable where N_r^2 is also positive in the HII region. The most unstable mode is determined by the term that the magnitude of the denominator in equation (17) is as possible as small. Especially, if $\kappa^2 - N_r^2$ and $L^2 - K_{zz}$ are degenerate, ω_1 becomes very large like in a resonance case. Such situation occurs under the condition that $\ell \sim r\kappa/c_s$, which is constant in the isothermal state with $\Omega \sim 1/r$. This result means that vertical and radial motions are coupled into the warp motion. If c_s is the sound velocity of gase, ℓ is very large, so that the warp is not so intensive. Rather, vertical propagation will occur like a bar. However, if c_s is given by the turbulent motion, ℓ is near to k, so that the large scale warp motion will be generated.

REFERENCES

Brown, R. L. and Liszt, H. S., 1984, *Ann. Rev. Astron. Astrophys.* **22**, 223.

Field, G., 1965, *Astrophys. J.,* **142**, 531.

Gentzel, R and Townes, C. H., 1987, *Ann. Rev. Astron. Astrophys.* **25**, 377.

Kato, S., 1983, *Publ. Astron. Soc. Japan,* **35**, 249.

Lacy, J. H., Achermann, J. M. and Serabyn, E., 1991, *Astrophys. J.,* **380**, L71.

Lynden-Bell, D. and Ostriker, J. P., 1967, *Mon. Not. R. astr. Soc.,* **136**, 293.

Morris, M., 1987,*The Center of the Galaxy,* ed. M.Morris,(Kluwer Acad. Press), 171.

Osterbrock, D., 1989, *Astrophysic of Gaseous Nebulae and Active Galactic Nuclei,* (Univ. Science Books).

Spitzer, L., 1978, *Physical Processes in the Interstellar Medium,* (John Wiley Press).

Unno, W., Osaki, Y, Ando, H, Saio, H. and Shibahasi, H., 1989, *Nonradial Oscillations of Stars,* (Univ.of Tokyo Pess).

ENERGETICS AND MORPHOLOGY OF THE GALACTIC CENTER INFRARED SOURCES MODELED FROM 4.8 - 12.4 μm ARRAY IMAGES

Dan Gezari and Eli Dwek

NASA/Goddard Space Flight Center, Infrared Astrophysics Branch Code 685

We have modeled the dust emission from the central parsec of the Galaxy at seven wavelengths between 4.8 and 12.4 μm using new diffraction-limited (1 arcsec) array images (see Gezari 1992, Gezari and Dwek 1993). A full description of the array camera system is presented by Gezari et al. (1992). The fundamental mechanism responsible for the high infrared luminosity of the Galactic Center region is still a mystery. One school of thought contends that a recent episode of massive O star formation has occured (e.g. Rieke and Lebofsky 1982, Allen et al. 1990) which would account for the high observed luminosity and ionizing radiation. Another argues that the luminosity and mass concentration are evidence for a "central engine" associated with Sgr A*, and that this object may be a massive black hole (e.g. Rees 1982) surrounded by an accretion disk.

A complete set of array camera images been obtained for the central 15 arcsec array field of view (0.6 parsec at 8.5 kpc). The 12.4 μm image is shown in Figure 1a (see Gezari 1992 for additional data). The array images have been used to derive the physical properties in Sgr A West with a multi-composition dust model, in which we assumed that all the sources are radially symmetric shells of either silicate or amorphous carbon dust. Detailed radiative transfer models (Wolfire and Churchill 1987) show that the stellar luminosity is absorbed in the inner layers of the shell and re-radiated at near-infrared wavelengths. The re-radiated emission is very well described by a single-temperature dust shell (assuming r^{-2} radial density structure). The observed energy distribution I_ν from a source is then simply given by

$$I_\nu = [1 - e^{-\tau}] \, \tau_S(\nu) \, B_\nu(T)$$

where $B_\nu(T)$ is the Planck function at the dust temperature T, $\tau_S(\nu)$ is the optical depth of the emitting dust, and τ is the opacity of dust along the line-of-sight which gives rise to the observed extinction. The cool grains responsible for the extinction are taken to consist of a characteristic mixture of interstellar silicate and graphite dust. In our calculations we used the optical constants of Draine & Lee (1984) to calculate the emissivities and opacities of the silicate and graphite dust, and those of Edoh (1983) to calculate the emissivities of amorphous carbon. We solved the simple radiative transfer equation by fitting the spectrum of each 1/4 arcsec resolution element (pixel) within the region, obtained from the "stack" of aligned array images.

The compact IRS sources are found to be local temperature maxima (IRS1 = 300K, IRS3 = 500K for the carbon grain source case) as seen in Figure 1b. For the silicate grain source case the spatial distribution is basically the same but the temperatures are a factor of 3 higher. IRS 1, 3, 10, 13 and 21 are all seen to be compact local temperature peaks. IRS 13 is relatively quite warm, however, IRS 2 is not evident as a temperature feature in the east-west bar. The fact that the mid-infrared IRS objects in the observed field coincide with unambiguous, compact, symmetrical temperature peaks is evidence that the IRS objects are internally heated by individual luminosity sources. The intriguing warm rim seen along the inside of the infrared ridge (directly west of IRS1) is correlated with two Helium I line stars observed by Krabbe et al. (1992), as discussed below.

© 1993 American Institute of Physics

An interesting result of the model is the opacity distribution of the emitting dust, which shows the appearance of a shell-like structure around IRS 1 in the carbon grain case (as seen in the "image" of the opacity of emitting grains, Figure 2a). This structure is also evident but weaker in the silicate grain model results. The shell-like structure is suggestive of depletion of dust surrounding an imbedded luminosity source. The modeled mid-infrared luminosity distribution for the carbon grain case (Figure 2b) shows peak values of 6 x 10^4 L_0 arcsec^{-2} at IRS1 and about 1 x 10^5 at IRS3. The derived luminosities in the silicate source case are generally almost an order of magnitude higher. The total luminosity integrated over the 15 arcsec (~0.6 parsec) region modeled is 1 x 10^6 L_0 for the carbon source case and 7 x 10^6 L_0 for silicates. These luminosities are conservative since they do not include the contribution from cold grains radiating at far-infrared wavelengths. The results of this modeling program are presented in detail by Gezari and Dwek (1993).

Krabbe et al. (1992) have recently detected a group of compact sources in the 2.06 μm emission line of Helium I, in a 40 arcsec^2 diameter field centered on Sgr A*. They are identified as HeI emission line stars (late-type blue supergiants or Wolf-Rayet stars) with individual luminosities in the range 0.1 - 2 x 10^6 L_0 resulting in a total luminosity of 1.2 x 10^7 L_0 for the ~40 arcsec cluster (Krabbe et al. 1992). Significant correlations have been found in a detailed comparison (Gezari 1992) between the positions of these HeI line stars and either structure in the 12.4 μm array image or features in the derived dust color temperature "image". The HeI stars coincide with the compact 4.8 or 12.4 μm sources IRS 1, 2, 3, 9, 13 and 21 in the 12.4 μm image, as well as with dust color temperature peaks. HeI stars are also correlated with regions which appear to be voids or cavities in the emitting dust. These source correlations place the HeI star cluster physically in the Galactic Center infrared source complex, not simply coincident in the line-of-sight.

We conclude that the compact IRS sources are hot spots in the dense Galactic Center dust clouds and are heated internally by luminous HeI emission line stars imbedded in the emitting dust material. Considering the uncertainties of the observations and the model results, radiation from the HeI emission line stars could essentially account for the observed luminosity of Sgr A West. The weight of observational evidence for the origin of the ~10^7 L_0 luminosity of the Galactic Center thus seems to be falling on the side of heating by luminous young stars embedded within and between the dense dust clouds, without invoking an exotic "central engine".

Allen, D. A., Hyland, A. R., and Hillier, D. J. 1990, M.N.R.A.S., 244, 706.
Draine, B. T. and Lee, H. M. 1984, Ap. J., 285, 89.
Edoh O. 1983, Ph. D. thesis, University of Arizona.
Gezari, D. Y. 1992, "The Center, Disk, and Bulge of the Galaxy", ed. L. Blitz, Kluwer Academic Publishers (Dordrecht), 23.
Gezari, D. Y. and Dwek, E. 1992, (in preparation for Ap. J.).
Gezari, D. Y., W. Folz, L. Woods and Varosi, F. 1992, P.A.S.P., 104, 191.
Krabbe, A., Genzel, R., Drapatz, S. and Rotaciuc, V. 1992, submitted to Ap. J. (Letters).
Rees, M. J. 1982, "The Galactic Center", ed. G. Riegler and R. Blanford, American Institute of Physics, Conf. Series, 83, 166.
Rieke, G. H. and Lebofsky, M. J. 1982, "The Galactic Center", ed. G. Riegler and R. Blanford, American Institute of Physics, Conf. Series, 83, 194.
Wolfire, M. G. and Churchwell, E. 1987, Ap. J., 315, 315.

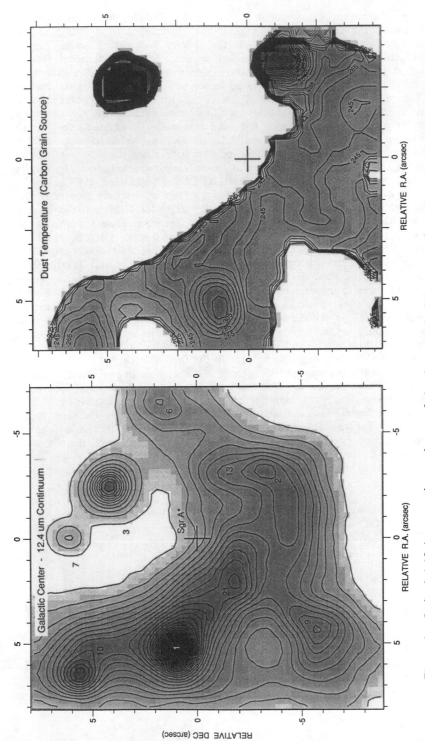

Figure 1 - Left: (a) 12.4 μm continuum image of the the central ~0.7 parsecs of the Galactic Center complex made with the 58 × 62 array camera (Gezari *et al.* 1992) at the 3-m NASA/IRTF Telescope. The compact IRS sources are numbered (peak brightness at IRS1 = 16 Jy arcsec⁻², lowest coutour [3σ] = 0.3 Jy arcsec⁻²). *Right: (b)* Modeled temperature distribution of emitting carbon grains (IRS1 = 300K, IRS3 = 500K). Source temperatures for silicate grains are typically 2 – 3 times higher.

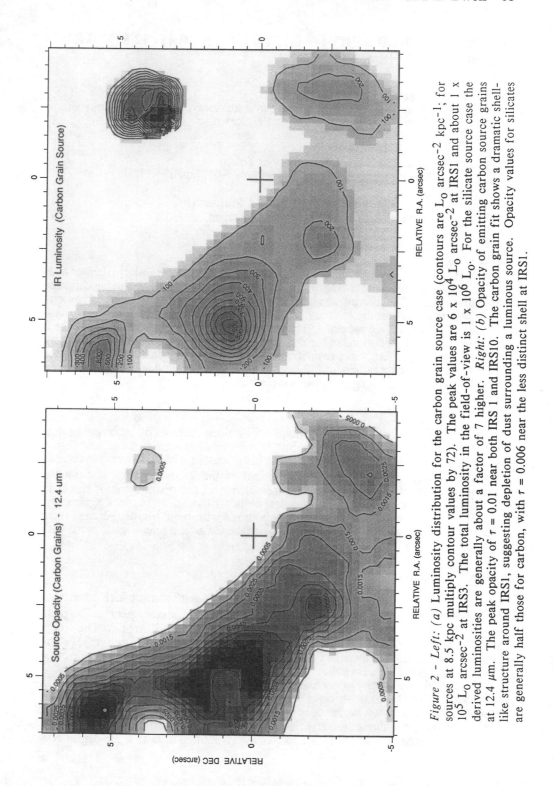

Figure 2 – Left: (a) Luminosity distribution for the carbon grain source case (contours are $L_\odot$ arcsec^{-2} kpc^{-1}; for sources at 8.5 kpc multiply contour values by 72). The peak values are 6 x 10^4 $L_\odot$ arcsec^{-2} at IRS1 and about 1 x 10^5 $L_\odot$ arcsec^{-2} at IRS3. The total luminosity in the field-of-view is 1 x 10^6 $L_\odot$. For the silicate source case the derived luminosities are generally about a factor of 7 higher. Right: (b) Opacity of emitting carbon source grains at 12.4 μm. The peak opacity of $\tau = 0.01$ near both IRS 1 and IRS10. The carbon grain fit shows a dramatic shell-like structure around IRS1, suggesting depletion of dust surrounding a luminous source. Opacity values for silicates are generally half those for carbon, with $\tau = 0.006$ near the less distinct shell at IRS1.

NEAR INFRARED SURVEY OF THE CENTRAL 12 SQUARE DEGREE OF THE GALAXY

Munetaka Ueno
Department of Earth Science and Astronomy, University of Tokyo
Komaba, Meguro-ku, Tokyo 153, Japan

Takashi Ichikawa
Kiso Observatory, University of Tokyo
Mitake-mura, Kiso-gun, Nagano, 397-01, Japan

Shuji Sato, Yasumasa Kasaba and Masanao Ito
National Astronomical Observatory Japan
Osawa, Mitaka, Tokyo 181, Japan

ABSTRACT.   We obtained a near infrared survey of the Galactic center at H and K' bands and detected a ring-like structure of the obscuration surrounding the Galactic nucleus. The survey was performed by a special purpose infrared camera, which was mounted on the University Hawaii's 61 cm telescope atop Mauna Kea, Hawaii in June 1991. We present an infrared color map of the Galactic center, which clearly shows the distributions of the dust clouds. Even in near infrared wavelengths, the Galactic plane suffers heavy extinction by a number of dark patches. The darklane of the Galactic disk spreads within 2 degree along the Galactic equator, and whose extinction is typically 14 mag at visible wavelength.

1. INTRODUCTION

The center of our Galaxy is completely hidden from us by intervening dust clouds at photographic wavelength but is visible at infrared, radio and X-ray wavelengths. The dust extinction at 2 μm is only one-tenth of that at visible band in magnitude scale.[1] The dominant component of the 2 μm radiation is from the photospheres of stars. We can make observations of the distribution of the stars around the Galactic nucleus and of the dust clouds silhouetting in black against the concentration of the bulge stars. The observations with large beams (~1 deg) from space delineated brightness distributions around the Galactic nucleus while those with small beams detected discrete sources in a limited area[2,3].

2. OBSERVATIONS

We made a wide field infrared camera for the present survey, which consists of 25 cm Newtonian telescope and NAOJ's 512 x 512 PtSi infrared camera[4]. The plate scale is 6.1 x 9.4 arcsec per pixel, giving a field of 40.2 x 52.3 arcmin. The filters used were H and K' (central wavelength 1.65 and 2.15 μm respectively). The telescope was attached to the side flange of the University of Hawaii's 61 cm telescope on top of the Mauna Kea, Hawaii.

The Galactic center survey was performed in 35 quadrants at H and K' band in June 1991. Two frames, which are shifted slightly relative to each other, with exposure times of 240 sec, were added for each quadrant. Data reduction is done in a usual manner; dark subtraction, flat fielding and mosaicing, using IRAF system.** Total field covered in this survey is approximately 3.2 deg in E-W x 3.6 deg in N-S direction. The limiting magnitudes obtained by the sky background noise were 12.0 mag at K' band and 13.0 mag at H band respectively, however the sensitivities of the concentrated regions are limited by the confusion of the stars.

3. RESULTS

Plate 1. shows the infrared false-color image combined from the H band (blue) and K' band (red) data. The Galactic plane is strongly obscured at H band, which is displayed as a red belt in the present map, while the Galactic center is quite prominent at K' band (red). The most remarkable thing is discovery of a ring-like structure of obscuration threading through the dark clouds surrounding the Galactic nucleus. The size of the dust ring is approximately 50 arcmin (150 pc at 10 Kpc) and which is tilted by 35° with respect to the Galactic equator. The edges of the ring coincide to the dark clouds, DC2 and DC3 identified in the former work[5]. The DC2 is resolved into three dark patches in the present survey, the easternmost of which coincides to an H II region, RCW142[6]. Other parts of the dust-ring are identified with 2.6 cm radio continuum sources[7], OH or H_2O maser sources and IRAS sources[8]. The close association of the dark clouds with radio continuum indicates these knots to be regions of OB star formation, while the presence of masers proves to be indicators of proto-stellar objects. Thus the dust-ring might be a site of current star formation. The well-known radio lobes[9] must be distinguished by the near-infrared map. The southwest lobe delineates the obscuration in the K band map and is supposed to be a thermal origin, while the north lobe has no coincidence with any infrared features and supposed to be a non-thermal origin.

Within the bounds of the present survey we can derive a typical extinction since the (H-K) color is insensitive to the stellar population but sensitive to the extinction because a typical stellar has a constant (H-K) color. The extinction of the Galactic disk is determined to be 14 mag in average with 2 degree width along the Galactic equator and has a range of 10-50 mag. (Figure 2)

It would be of interest to note that there are no conspicuous clusters of stars expect for the bright infrared cluster; AFGL2004[10,11]. The detection limit is enough deep to detect a typical globular cluster lying at the distance of the Galactic center with a typical extinction (Av~20mag). The number of globular clusters is estimated to be 20 under the effect of dynamical friction within 3 deg of the center of the Galaxy[12]. Our survey may give a restriction on the formation of globular cluster around the Galactic nucleus. However, more deep infrared survey is necessary to draw a conclusion of this problem.

We develop a new infrared survey system, which uses a 1040 x 1040 PtSi infrared camera and 32 cm Newtonian telescope. We plan to continue this survey to the outer region of the Milky way in 1993.

ACKNOWLEDGMENT.   We thank the day crews of University of Hawaii's telescope for their great supports to attach the survey system on the telescope and Dr. A. Pickles for the supports of the computer networks. **IRAF is distributed by the National Optical Astronomy Observatories, which is operated by the Association of Universities for Research in Astronomy, Inc. (AURA) under cooperative agreement with the National Science Foundation.

1.   Rieke, G. & Lebofsky, M. *Astrophys. J.* **288**, 618

2.   Storey, J. W. V. & Allen, D. A. *Mon. Not. R. Astr. Soc.* **204**, 1153 (1983)

3.   Glass, I. S., Catchpole, R. M. & Whitelock, P.A. *Mon. Not. R. Astr. Soc.* **227**, 373 (1987)

4.   Ueno, M. *et al*. *Proceeding of SPIE conference on Infrared Technology XVIII* Andresen,
     B.F. & Shepherd, F.D., 1992

5.   Hiromoto, N. *et.al*. *Astr. Astrophys*. **139**, 309 (1984)

6.   Gardner, F. F., & Whiteoak, J. B. *Mon. Not. R. Astr. Soc*. **171**, 29P (1975)

7.   Handa, T. *et.al*. *Publ. Astron. Soc. Japan* **39**, 709 (1987)

8.   Beichman, C. A. *et.al*. *IRAS Explanatory Supplement* 1985, JPL D-1855 (Jet
     Propulsion Laboratory, Pasadena).

9.   Sofue, Y., *Publ. Astron. Soc. Japan* **37**, 697 (1985)

10. Okuda, H. *et.al*. *Astrophys. J*. **351**, 89 (1990)

11. Nagata, T. *et.al*. *Astrophys. J*. **351**, 83 (1990)

12. Oort *Astrophys. J*. **218**, L97 (1991)

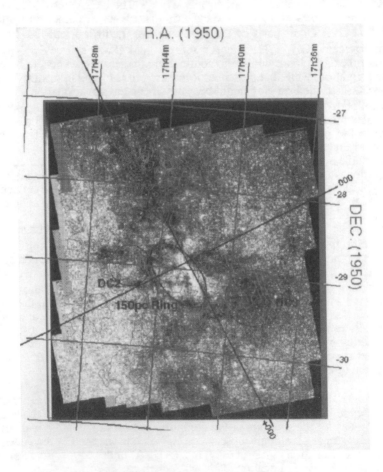

Figure 2. The extinction map overlaid on the K' band map. The contour interval is 5 mag and the lowest contour is 10 mag. The dark lane is wound to the south of the Galactic center. The 150 pc ring and the DC2 and DC3 are superimposed in the map.

THE MILKY WAY: AN ACTIVE GALAXY?

William C. Oelfke
2319 Huron Tr., Maitland, FL 32751

ABSTRACT

The behavior and interactions of one or more core black holes is described for the mass density environment of the galaxy core and comparisons made between this environment and that of a large spherical stellar system typical of an active galaxy. Data is presented for a range of galaxy types in local galaxy clusters that indicates jetting with kinetic energy greater than 10^{55} ergs from a collapsed object or system of objects in the galaxy core may have occurred during the very early phase of development of the Milky Way but is unlikely in later epochs.

INTRODUCTION

A comparison is drawn between the expected activity of the Milky Way core and that of a typical giant elliptical, M87, based on the assumption that the central engine is comprised of a single or multiple black hole system. The gravitational interaction, orbital decay and accretion properties of binary black holes in the known core environments of spiral and elliptical galaxies are sufficient to model many of the properties demonstrated by these two uniquely different types of active galaxies. Of all spiral galaxies approximately 1%, the Seyfert class, exhibit heightened core activity and that activity does not appear to be bipolar in nature. Collision-induced core compression is believed to be responsible for triggering this heightened core activity which appears to be transient rather than periodic. In contrast approximately 1% of all large elliptical galaxies exhibit bipolar jetting resulting in the production of bright radio lobes. Many of these radio galaxies have left remnants of previous jetting that indicate a possible periodic nature to this giant elliptical core activity. These two strikengly different core activities may result: in the case of spirals, from accretion around single Schwarzschild black holes in the cores where binaries decay and merge rapidly; and in the case of ellipticals, from accretion around multiple Kerr black holes in the cores where binaries become hardened and remain in stable orbits for times approacing the Hubble time.

ORBITAL DECAY OF CORE BLACK HOLES

The Model presented by Begelman, et al. (1980) for the decay of binary black holes due to dynamical friction, gas damping and gravitational radiation is used for the comparison of binary black hole decay modes in spiral and elliptical galaxy cores. In this model the decay time due to dynamical friction for $r \geq 1$ pc is

$$T_{DF} = (6 \times 10^6 / \log N)(v_0 / 300 \text{ km/s})(r_c / 100\text{pc})^2 (10^8 \, M_0 / m) \text{ yr,} \qquad (1)$$

due to dynamical friction with loss-cone depletion for $1\text{pc} \geq r \geq 0.01\text{pc}$ is

$$T_h = T_{DF} \, \text{Max}[(r_h / r) \ln N, (M / M_c)^2 \, N], \qquad (2)$$

and due to gravitational radiation for $r \leq 0.01$ pc is

$$T_{GR} = 10^4 \, (r / 0.01 \text{ pc})^{5/2} (M / 10^9 \, M_0)^{-3/2} \text{ yr.} \qquad (3)$$

If a significant amount of gas and dust is present in the galactic core as in the case of spiral galaxies then orbital decay due to gas damping continues after dynamical friction declines and T_h = 10^8 years. Since for galactic core environments T_{DF} and T_{GR} are of order 10^6 years, it is unlikely that black hole binaries exist for more than 10^8 years in spiral galaxies, and that a single, merged Schwarzschild core black hole is found at the core of most spirals whereas Kerr binaries may exist for times approaching the Hubble time in the cores of large elliptical galaxies.

For the Milky Way Galaxy, core activity may be dominated by steady accretion by a low angular momentum, 3 x 10^6 M_o, core black hole at a rate of 9 x 10^{-4} M_o per year, producing the observed, steady level of activity of 2 x 10^{36} Watts. Impulsive (Seyfert) core activity and possibly jetting from individual Kerr black holes in decaying binaries could have occured during initial protogalaxy collapse and may occur again as a result of core compression or merger induced by galaxy collision, however the current level of activity may be representative of the sustained output of the Milky Way core. The lack of significant Lense-Thirring "frame dragging" around this low angular momentum core black hole would greatly inhibit the central engine's ability to form highly focused, stable bipolar jets and therefore one would expect the galaxy to generate core activity that is mostly isotropic.

In contrast the active core of the giant elliptical M87 may represent accretion by multiple 10^8 M_o Kerr black holes where bipolar jetting and subsequent increase in core potential sustains a radial core oscillation of 10^8 year period (Smith 1992). The formation of a compression cusp and a slowly precessing, highly focused, bipolar jet represents a 5% impulsive core duty cycle which is repeated each cycle for 10^{10} years at which time the core binaries will have decayed to radii of less than 0.01 pc. During this final stage of rapid binary decay due to gravitational radiation damping, high precession rates of individual Kerr black holes prevent self focusing of bipolar jets with the result that bright radio lobes are formed. If M87 is assumed to be typical of all large elliptical galaxies then one would expect to see cusps at the cores of less than 5%, and radio lobes associated with 1% of any representative sample. In addition one would expect to see evidence of the radio-quiet jetting that occured during the 10^{10} years that the core black hole binaries were hardened.

The author (Oelfke, 1992) reports that within galactic clusters in the sextant from 10 to 14 hours right ascension, a total of 332 bright galaxies form 15 unique periodic series of extended galaxy pairs each of which leads to a large elliptical within its respective cluster. The positions of pair centers are consistent with locations of periodic jetting from these moving progenitors of sufficient energy to trigger galaxy formation. In each of these 15 series, the principal axes of the moving ellipticals are aligned with their respective transverse velocities indicating that each progenitor is moving through an intergalactic medium and displays an elliptical isophote elongated by ram pressure. Bipolar jetting from the core engines of large ellipticals appears to dictate the structure of galaxy clusters (Trevese, 1992) and may represent the principal mechanism for establishing the large scale structrue of the visible universe (Struble, 1985).

REFERENCES

Begelman, M. C., Blandford, R. D., & Rees, M. J. 1980, Nature, 287, 307
Oelfke, W. C. 1992, J. Ap. & Ast. (Submitted Aug. 11, 1992)
Smith, B. F. & Miller, R. H. "Galactic Oscillations and Internal Dynamics" (This publication)
Struble, M. F. & Peebles, P. J. E. 1985, AJ, 90, 582
Trevese, D., Cirimele, G., & Flin, P. 1992, AJ, 104, 935

EVALUATING THE MASS OF A PUTATIVE BLACK HOLE AT THE GALACTIC CENTER

L. M. Ozernoy[†]

Lab. for High Energy Astrophys., NASA/GSFC, Greenbelt, MD 20771

Email: ozernoy@heavax.gsfc.nasa.gov

ABSTRACT

A wide-spread belief in a $\sim 10^6$ M$_\odot$ mass for a putative black hole at the Galactic center is based on the approaches which employ the gravitational field of a point mass at such large distances that the contribution of any distributed matter is not negligible, and therefore the figure above cannot be considered as a reliable one. A method used in the present work employs the radiation spectrum of the black hole, *i.e.* it is based on a *local* approach. Approximate analytical expressions are obtained to estimate the mass of the black hole from its spectral luminosity assumed to result from the Bondi accretion of the wind from a nearby group of hot massive stars, IRS 16. To fit the radio spectrum of Sgr A*, $M_{bh} \simeq 5 \cdot 10^2$ M$_\odot$ is required; a 10^6 M$_\odot$ black hole would result in an electron temperature only marginally relativistic, which is incompatible with the observed (synchrotron) spectrum. A comparatively low value for the black hole mass is consistent with the upper limits to Sgr A* mass found earlier.

1. INTRODUCTION

There is a long-term controversy on whether there is or not a supermassive black hole (BH) at the Galactic center (for recent reviews, see Genzel & Townes 1987, Phinney 1989, Ozernoy 1989). A wide-spread belief in a $\sim 10^6$ M$_\odot$ mass for a putative BH at the Galactic center is based on the approaches which employ the gravitational field of a point mass at such large distances that the contribution of any distributed matter is not negligible. For instance, all the previous approaches to fitting of the distribution of ionized gas in the central parsec, both kinematically and spatially, suffer from non-accounting for non-gravitational forces far from the assumed point mass. As Yusef-Zadeh & Wardle (1992) and Chevalier (1992) have pointed out, the wind from IRS 16 is able to distort appreciably the velocity pattern in the ionized gas streamers around Sgr A*. Besides, in a recent attempt to treat the mini-spiral as a density wave in the gravitational potential of a point mass, Lacy *et al.* (1991) have misleadingly neglected all but one ionized gas streamers so there is no surprise that a point mass gave them the best fit. Stars as a tracer of the gravitational field are not susceptible to magnetic and other non-gravitational forces. Meanwhile a $\sim 10^6$ M$_\odot$ BH is consistent but not demanded by kinematics of stars in the central parsec or so (McGinn *et al.* 1989).

Because of an ambiguity of the dynamical fits, I have tried for a long time to employ a completely different approach bearing in mind to constrain the BH mass from a variety of involved physical processes. Four different methods – (i) tidal disruption of stars by a BH; (ii) displacement between Sgr A* and IRS 16; (iii) electron-positron pair production by a BH via electromagnetic cascade; and

---

[†] NAS/NRC Senior Research Associate

(iv) wind diagnostics of Sgr A* – all resulted in upper limits to the BH mass, which depending on the method employed range between $\sim 30 - 100$ M$_\odot$ and $2 \cdot 10^4$ M$_\odot$ (see Ozernoy 1992a and refs. therein). Although these methods have some flaws, all of them unambiguously indicate for a putative BH at the Galactic center to have a mass $M \ll 10^6$ M$_\odot$.

Recently Sanders (1992) has pointed out that the tidal forces from a point mass of a 10^6 M$_\odot$ would make impossible the formation of young stars in the vicinity of Sgr A*, contrary to what is seemingly observed. Unfortunately, this argument to the case against a supermassive BH cannot be reliably quantified since it is hardly possible to separate the tidal fields of a point mass and the core stars around it. An additional problem, as Morris (1992) has suggested, is that unusual conditions for star formation at the Galactic center make it very difficult to assess the density of protostars.

In the present work, we make a next step and employ a *local* approach trying to find the mass of Sgr A* from its radiation spectrum assuming that the latter is due to accretion onto a BH. The basis for this is given by an earlier suggestion (Ozernoy 1989) that the Bondi case approximates accretion onto Sgr A* that is fed by the wind from a nearby group of hot massive stars, IRS 16. As the boundary conditions for this accretion can be set up from the observational data on the wind, the BH mass can be derived from a consistent accretion theory. Such a theory is outlined in Sec. 2 (a brief account of this approach was earlier presented in Ozernoy 1992b). Sec. 3 contains discussion and comparison with some recent work on the subject.

2. WIND ACCRETION ONTO SGR A*

Numerous data (*e.g.* Geballe *et al.* 1991, Krabbe *et al.* 1991) indicate the following values for the parameters of the IRS 16 wind: $n_\infty \simeq 10^4$ cm^{-3}, $v_\infty \simeq$ 700 km/s, $\dot{M}_{ej} \simeq 2 \cdot 10^{-3}$ M$_\odot$/yr, where n_∞ is gas number density in the wind far from Sgr A* , v_∞ is velocity of the wind, and $\dot{M}_{ej}$ is the resulting power of the wind.

The BH intersepts a fraction of $\dot{M}_{ej}$ that depends on the BH mass M and also on n_∞ and v_∞ but does not depend on a distance from the source of the wind provided it is large enough. The Bondi accretion rate can be written as $\dot{M}_a \simeq 10^{-4}$ M$_\odot$/yr $M_6^2 n_4 v_{700}^{-3} \equiv 10^{-4} A M_6^2$ M$_\odot$/yr, where $M \equiv M_6 \cdot 10^6$ M$_\odot$, $n \equiv n_4 \cdot 10^4$ cm^{-3}, and $v \equiv v_{700} \cdot 700$ km/s. It is assumed hereinafter that the accretion parameter $A \simeq 1$.

In Bondi accretion onto a BH, once n_∞ and v_∞ are known, the accretion rate, $\dot{M}_a \propto M^2 n_\infty v_\infty^{-3}$, is determined solely by the BH mass M. We employ the standard theory of spherical accretion (*e.g.* Ipser & Price 1982) modified by taking into account the decoupling of electrons and ions at small distances from the BH where the energy exchange time between them, $t_{ei} \propto T_e^{3/2}/n$, exceeds the inflow time, $t \sim r/v$. Since the density in the flow varies as $n = 2.5 \cdot 10^{12}$ cm$^{-3} \beta_v \dot{m} M_6^{-1} (r/r_g)^{-3/2}$, where $r_g = 2GM/c^2$ is the gravitational radius and $\beta_v \equiv v/v_{ff}$, v_{ff} being the free-fall velocity, the decoupling happens at the distances $r \lesssim r_{dec}$, where $r_{dec} \simeq 2.6 \cdot 10^3 \dot{m}^{-2/3} r_g = 2.1 \cdot 10^4 M_6^{-2/3} A^{-2/3} r_g$ and the Coulomb logarithm $\Lambda = 20$ is adopted.

Further analysis employs some features from the theory of turbulent spherical accretion (Maraschi *et al.* 1982, Colpi *et al.* 1984).

Heating of the accreting plasma is due to dissipation of turbulence and magnetic fields. The magnetic field in the accretion flow can be evaluated by assuming equipartition of the magnetic energy density either with electron energy density (Model 1) or with kinetic energy density (Model 2).

• Model 1. Assuming that $B^2/8\pi = \beta'_B nkT_e$ with $\beta'_B = \text{const} \sim 1$ and bearing in mind that the electron temperature of the accreting plasma varies with r as $T_e(r) = T_p(r) \simeq 0.1\, m_p c^2 (r/r_g)$ at $r \geq r_{dec}$ and $T_e(r) = \text{const} \simeq T_p(r_{dec})$ at $r \leq r_{dec}$, one gets for the magnetic field:

$$B \simeq 2.0 \cdot 10^4 \text{ G } \left(\frac{\beta'_B}{\beta_v}\right) A^{1/2} \begin{cases} (r/r_g)^{-5/4} & \text{at } x > x_{dec} \\ (r/r_g)^{3/4} x_{dec}^{-1/2} & \text{at } x < x_{dec}. \end{cases}$$

• Model 2. Assuming that $B^2/8\pi = \beta''_B \rho v^2/2$ and substituting ρ and v as given above, one gets $B \simeq 4.6 \cdot 10^4 \text{ G } (\beta_v \beta''_B)^{1/2} A^{1/2}(r/r_g)^{-5/4}$.

Having compared B in Models 1 and 2, one can see that the value of the magnetic field in the accretion flow is rather insensitive to various equipartition models, at least at sufficiently large radii from the BH.

Thus, the heating of the accretion flow due to dissipation of turbulence and magnetic fields is given by $\Gamma \simeq 4 \cdot 10^6 \, \beta_v \beta_d A M_6^{-1} (r_g/r)^4$ erg/cm^{3}s, where $\beta_d \sim 1$ is a numerical coefficient absorbing uncertainties in the dissipation process. One can see that the heating rate turns out to be basically a function of M only.

Cooling of the accretion flow is due to mainly cyclo-synchrotron emission and can be presented as (Ipser & Price 1982):

$$\Lambda = 2\pi f \int_{-1}^{+1} d\cos\theta \int_0^{\infty} j_\nu d\nu \simeq 1 \cdot 10^6 \, \beta_B f A^2 \left(\frac{T_e}{10^9 \text{ K}}\right)^2 \left(\frac{r_g}{r}\right)^4 \frac{\text{erg}}{\text{cm}^3\text{s}},$$

where a factor f accounts for an optically thin part of synchrotron radiation. The value of Λ is a function of T_e and M. In the steady-state situation, the thermal balance in the accretion flow ($\Gamma = \Lambda$), is established and at the distances of interest, $r \gtrsim r_{dec}$, the value of T_e turns out to be a function of M only, with a weak dependence upon the uncertainties associated with the dissipative heating:

$$T_e \simeq 5.2 \cdot 10^9 \text{ K } M_6^{-1/2} (\beta_v \beta_d/\beta_B)^{1/2} f^{-1/2} A^{-1/2}.$$

An optically thin synhrotron spectral luminosity is given by

$$L_{\nu_\infty} = 8\pi^2 \int_{-1}^{\cos\theta_{max}} d\cos\theta \int_{r_g}^{\infty} dr r^2 e^{-\tau_{\nu'}} j_{\nu'} \left(1 - \beta^2\right)/(1 - \beta\cos\theta)^2$$

(Ipser & Price 1982). It is easy to show that the accretion radiation falls into a comparatively narrow band from cm to submm wave lengths. Evaluation of the last integral yields $L_{\nu_\infty} \simeq 2.8 \cdot 10^{32} \, M_6^3 \beta_B^{1/2} \beta_v A^{3/2} \left(\nu/10^{10} \text{ Hz}\right)^{1/3}$ erg/s Hz.

The observed spectral radio luminosity of Sgr A* (Lo 1989) is an optically thin, synchrotron emission of relativistic electrons with Lorents-factor $\gamma \sim 10^2 -$

10^3 and can be presented by $L_{obs} \simeq 10^{22.5} \left(\nu/10^{10} \text{ Hz}\right)^{0.3}$ erg/s Hz. Evidently, L_{ν_∞} gives a fair representation of the observed Sgr A* radio emission if

$$M \simeq 0.5 \cdot 10^3 \text{ M}_\odot \beta_B^{-1/6} \beta_v^{-1/3} A^{-1/2},$$

with a rather weak dependencies on the uncertainties involved.

3. DISCUSSION

The results of the above calculations are in contradiction with those of Melia (1992) and Melia *et al.* (1992) who have used a similar aproach but reached at a different value of $M \approx 10^6$ M$_\odot$. Meanwhile it is straighforward to see that such a high value of M would make the temperature of the accretion flow given above only marginally relativistic for electrons, instead of providing them the Lorents-factor $\gamma \sim 10^2 - 10^3$ required by the radio data. The value of M as large as 10^6 M$_\odot$ fails to explain the synchrotron reabsorption of Sgr A* emission at $\nu \leq 10^9$ Hz, whereas $M \simeq 500$ M$_\odot$ is consistent with it.

4. CONCLUSION

An approximate analytical expression for the spectral luminosity of Sgr A* is obtained to estimate the mass of the underlying putative BH. To fit the radio spectrum of Sgr A*, $M \simeq 5 \cdot 10^2$ M$_\odot$ is required; a 10^6 M$_\odot$ BH would result in an electron temperature only marginally relativistic, which is incompatible with the observed (synchrotron) spectrum. A comparatively low value for the BH mass is consistent with the upper limits to Sgr A* mass found earlier (Ozernoy 1992 and refs. therein).

REFERENCES

Chevalier, R. 1992, ApJ,397, L 39
Colpi, M. *et al.* 1984, ApJ, 280, 319
Geballe, T.R. *et al.* 1991, ApJ, 320, 562
Genzel, R. & Townes, C.H. 1987, Ann. Rev. Astron. Astrophys., 25, 377
Ipser, J.R. & Price, R.H. 1982, ApJ, 255, 654
Krabbe, A. *et al.* 1991, ApJ, 382, L19
Lacy, J.H., Achterman, J.M., & Serabyn, E. 1991, ApJ, 380, L71
Lo, K.Y. 1989, in The Center of the Galaxy (Proc. IAU Symp. No. 136), ed.
     M. Morris, Kluwer Acad. Publ., p. 527
Maraschi, L., Roasio, R., & Treves, A. 1982, ApJ, 253, 312
Melia, F. 1992, ApJ, 387, L25
Melia, F., Jokipii, J.R., & Narayanan, A. 1992, ApJ, 395, L87
McGinn, M.T. *et al.* 1989, Ap J., 338, 824
Morris, M. 1992, Preprint
Ozernoy, L.M. 1989, in The Center of the Galaxy (Proc. IAU Symp. No. 136),
     ed. M. Morris, Kluwer Acad. Publ., p. 555
Ozernoy, L.M. 1992a, in Testing the AGN Paradigm. Eds. S.S. Holt et al., AIP
     Conf. Proc., 254, 40 and 44
Ozernoy, L.M. 1992b, BAAS, 24, 746
Phinney, E.S. 1989, in The Center of the Galaxy (Proc. IAU Symp. No. 136),
     ed. M. Morris, Kluwer Acad. Publ., p. 543
Sanders, R. 1992, Nature, 359, 131
Yusef-Zadeh, F. & Wardle, M. 1992, ApJ (in press)

ULTRA HOT GAS IN THE GALAXY:
POSSIBLE ORIGINS AND IMPLICATIONS

L. Ozernoy [†], L. Titarchuk [†], R. Ramaty
Lab. for High Energy Astrophysics, NASA/GSFC, Greenbelt, MD 20771
Email: ozernoy@lheavx.gsfc.nasa.gov   titarchuk@lheavx.gsfc.nasa.gov
ramaty@lheavx.gsfc.nasa.gov

ABSTRACT

A remarkable enhancement of the 6.7 KeV line emission and associated thermal continuum emission from the Galactic ridge toward the Galactic center discovered recently by *Ginga* satellite opens new opportunity of exploring high-energy phenomena in the Galaxy. This emission is thought to originate in a hot optically thin gas, and the central enhancement seems to be associated with an ultra hot ($T \approx 10^8$ K), rarefied [$n \sim (3-6) \cdot 10^{-2}$ cm^{-3}] interior of the 200-pc expanding molecular ring in the Galactic center. Assuming that this interior is a superbubble resulting from an adiabatic explosion, we analyse the dynamics of the explosive phenomenon. Some implications of the presence of a cold (molecular) and a warm (radio emitting) gas between us and the ultra hot superbubble, such as fluorescence of iron lines in the cold gas and scattering of iron lines in warm (hot) gas are discussed.

1. INTRODUCTION

Both continuum (Skinner *et al.* 1987) and emission line observations (Koyama *et al.* 1989) have revealed the presence of a high-temperature ($T \sim 10$ KeV) plasma at the Galactic center. Gas of such high temperature had previously been found only in clusters of galaxies and in some active galactic nuclei. In the Galaxy, earlier X-ray imaging with *Einstein* and *ROSAT* has uncovered a very hot coronal component of the interstellar gas: it has a temperature in the range $T \sim 10^{6.3}-10^7$ K, significantly hotter than the coronal component ($T \sim 10^5-10^6$ K) that was found in UV resonance lines and in very soft ($\lesssim 0.25$ KeV) X-ray background (*e.g.* Wang 1992). Therefore the novel, 10^8 K component of ISM could be called *ultra hot*.

Recent studies of the ultra hot gas by the *Ginga* satellite have found that an intense 6.7 KeV Fe line and associated thermal continuum emission from the Galactic ridge have a significant enhancement toward an extended ($\sim 1.8°$) central region of the Galaxy (Koyama *et al.* 1990). An extensive study of this region resulted in the following parameters of the emission (Yamauchi *et al.* 1990, Yamauchi 1990): characteristic radius of the region filled with hot gas, $R \approx 150$ pc; temperature of hot gas, $T \approx 10^8$ K; total luminosity, $L \approx 4 \cdot 10^{37}$ erg/s (2 − 10 KeV); line luminosity $L_{\text{Fe}} \approx 2.1 \cdot 10^{36}$ erg/s; and equivalent width of the line, $E.W. = 600 \pm 70$ eV.

From the emission measure, the density of hot plasma is evaluated to be $n \sim (0.03 − 0.06)$ cm^{-3} (Yamauchi et al. 1990) (without incorporating a filling factor). This implies that the thermal energy density of the gas is $w_t \sim (5 − 10) \cdot 10^{-10}$ erg/cm^3, the total thermal energy $W_t \sim (4 − 8) \cdot 10^{53}$ erg so that the cooling time, (3 - 6)$\cdot 10^8$ yr, is much less than the age of the Galaxy.

[†] NAS/NRC Senior Research Associate

Moreover, this gas is highly non-stationary: its characteristic expansion time, $\sim 2 \cdot 10^5$ yr, indicates a very recent origin.

In this paper, we intend to discuss the possible origins of the ultra hot gas and some novel approaches to further studying its parameters. A more detailed presentation of these results will be given elsewhere (Ozernoy et al. 1993).

2. POSSIBLE ORIGINS OF ULTRA HOT GAS

The emission is thought to originate in a hot optically thin gas, and the central enhancement seems to be associated with an ultra hot ($T \approx 10^8$ K), rarefied [$n \sim (3 - 6) \cdot 10^{-2}$ cm^{-3}] interior of the 200-pc expanding molecular ring in the Galactic center (Sofue 1990). We assume that this interior is a fossil hot superbubble resulting from an adiabatic explosion (some other options will be discussed later on). Below, we analyze the dynamics of the explosive phenomenon. We employ the observed size, temperature, and density of the superbubble to derive the characteristic parameters of the explosion, such as the power, the time elapsed since the onset of the explosion, and the initial gas density.

The evolution of a hot bubble is described by the well-known Sedov-Teylor self-similar solution. The radius of the bubble is given by (Castor et al. 1975, Weaver et al. 1977) as $R = 270$ pc $\left(L_{38} t_7^3 / n_{i0}\right)^{1/5}$, where $L_{38} = L/10^{38}$ erg/s, $t_7 = t/10^7$ yr, and $n_{i0} = n/1$ cm^{-3}. Here L is the power of the explosion, which is normalized so as to release, in terms of supernova explosions, 10^{51} erg each $3.2 \cdot 10^5$ yr. The temperature and density distributions in the interior of the bubble, if one neglects the structure of the latter, are as follows: $T = 3.5 \cdot 10^6$ K $L_{38}^{8/35} t_7^{-6/35} n_{i0}^{2/35}$, and $n = 4.0 \cdot 10^{-3}$ cm$^{-3} L_{38}^{6/35} t_7^{-22/35} n_{i0}^{19/35}$. Then the above three equations can be regarded as giving the three observable values (R, T and n) as functions of the three unknown parameters – initial gas density n_i, time elapsed since the onset of the expansion, t, and the power (kinetic luminosity), L. If one takes (Yamauchi et al. 1990) $R = 135$ pc, $n = (3 - 6) \cdot 10^{-2}$ cm^{-3}, and $T = 10^8$ K, the solution of the above set of equations is as follows: $L = 6.5 \cdot 10^{42}$ erg/s, $t = (4.2 - 8.4) \cdot 10^3$ yr, and $n_i = (1.5 - 12) \cdot 10^{-4}$ cm^{-3}.

The characteristic energy release during the explosion phase is $Lt = (0.9 - 1.7) \cdot 10^{54}$ erg. In terms of multiple supernova explosions, the required supernova rate would be as high as 0.2 yr^{-1}, if the energy release per supernova is 10^{51} erg. Such a supernova rate does not seem to be enormous, given the fact that many starburst galaxies demonstrate the star formation rate to be comparable or even exceeding the above value. As for the Galactic nucleus, this would imply that it was in a starburst phase some time ago. A relatively young stellar population at the Galactic center (Lebofsky et al. 1982) and especially IRS 16, a group of hot, massive stars in the central parsec of the Galaxy (more exactly, in a fraction of it) is probably the remnant of this starburst. Generally, the Galactic nucleus seems to be a tracer of a past starburst, rather than a remnant of Seyfert galaxy activity (Ozernoy 1992).

A problem with the multiple supernova model is that the evidence for a supernova rate at the Galactic center as high as ~ 1 SN/5 yr is lacking so far. An alternative scenario might incorporate an explosion a while ago of a single very massive star (like η Carinae). We shall not discuss this alternative in more detail here, addressing the reader to Sofue 1990, Ozernoy 1992.

The free-free emisivity of the ultra hot gas at the Galactic center is about 10^{-26} $(n/3 \cdot 10^{-2}$ cm$^{-3})$ erg s^{-1} cm^{-3}. It is of interest that Skibo & Ramaty (1993) assuming an injection spectrum of relativistic electrons necessary to explain the diffuse Galactic low energy gamma ray continuum have found a local power input of about $2 \cdot 10^{-26}$ erg s^{-1} cm^{-3} due to electrons of $E > 50$ KeV. Therefore those relativistic electrons, in principle, could heat the gas bearing in mind that its cooling time is rather short. The total power input into the Galaxy has been found to be $4 \cdot 10^{41}$ erg/s, which exceeds by a factor of 10 the total power supplied to the nuclear cosmic rays unless the electrons occupy the central 200 ps or so, instead of being distributed more or less uniformly throughout the Galaxy. The possibility of heating the ultra hot Galactic gas by means of the soft (~ 100 KeV) tail of cosmic ray electrons is worth of further study.

3. IMPLEMENTING IRON LINES

Further progress in inderstanding of the origin of the ultra hot gas could be reached by implementing such tools as iron lines, which are invaluable sources of information on the physical state of the gas. For instance, the equivalent width of the iron lines depends both on the temperature and iron abundance, which in principle can be separately determined while combined with additional data. The equivalent width of the iron lines is given by (Ozernoy *et al.* 1993):

$$E.W. = 830 \text{ eV} \frac{N_H/N_e}{0.85} \frac{f'}{0.44} \frac{\eta_Z}{0.45} Y_{Fe},$$ (1)

where f' is an effective oscillator strength for Fe XXVI (Mewe et al. 1985), Y_{Fe} is a local iron abundance, and η_Z is the Fe XXVI species concentration as a function of $T_8 \equiv T/10^8$ K, which we derived as a numerical interpolation from the Mewe (1990) data to be $\eta_Z \approx 1.2\sqrt{T_8} \exp\left(-\sqrt{T_8}\right)$. It can be seen that the observed equivalent width of K$_\alpha$-line, $E.W. = 600 \pm 70$ eV, is consistent with Eq. (1) at reasonable values of T and Y_{Fe}.

Three ionization stages of iron coexist at $T \sim 10^8$ K: He-like Fe XXV, H-like Fe XXVI, and the bare nucleus Fe XXVII. Three types of lines are associated with the presence of these species: K$_\alpha$, L$_\alpha$ (a resonance line), and L$_c$, respectively. If, along with ultra hot gas, there is cold gas too, some of the above lines might be seen by fluorescence; besides, the presence of hot gas would make them seen by means of scattering. Below, we calculate the corresponding intensities of these lines, both for fluorescence and scattering.

(i) *Fluorescence of iron lines by cold gas.*
It can be shown that the equivalent width of the K$_\alpha$-line is given by:

$$E.W. = \frac{F_{K_\alpha}}{I_c} \approx 5 \cdot 10^{-3} \left(\frac{f}{0.1}\right)^{1/3} \frac{\tau_0}{\mu} Y_{Fe} \epsilon_K \exp\left(\frac{\epsilon_K}{10 \text{ KeV}}\right),$$ (2)

which is about 70 eV (τ_0/μ) when $f = 0.1$ and $Y_{Fe} = 1$. Here F_{K_α} is the line radiation flux outgoing at the angle $\theta =$ arc cos μ with the normal to the source assumed to be a flat layer of the optical thickness τ_0, and $I_c \propto \exp(-\epsilon_K/10$ KeV) is the intensity of the continuum.

Radio observations of the central region of the Galaxy indicate the density of warm ($T \sim 10^4$ K) gas to be $n \sim 10^4$ cm^{-3} (Mezger *et al.* 1990, Zylka *et al.* 1990). Although both the electron density and its filling factor still need to be accurately determined, there are serious reasons to believe that the factor (τ_0/μ) in Eq. (2) might be of the order of unity or so in the region surrounding the ultra hot gas. Therefore, searches for fluorescent K_α-line in the central region of the Galaxy would be of great interest. We note in passing that in an extreme (probably, unlikely) case when $\tau_0 \gtrsim 1$ the K_α-line would not be seen and, instead, a broad absorption feature at the position of the K series might be observed.

(ii) *Scattering of iron lines by warm (hot) gas.*

Along with direct components of iron lines, the scattered components might also be seen unless the optical depth in scattering is too small or too large. The ratio of intensities of scattered and direct components of iron lines is expected to be of the order of $\tau_0 \ln \tau_0^{-1/2}$, *i.e.* might be significant unless τ_0 is too small.

4. CONCLUSIONS

In this paper, we have discussed several options for the origin of an ultra hot ($T \approx 10^8$ K) gas in the central 200 pc of the Galaxy. It appears that this gas might be a fossil, hot superbubble resulting from a recent explosion at the Galactic center, either of multiple supernova phenomenon or a single very massive object (like η Carinae). Heating of the gas by non-thermal electrons is an other option that needs to be thoroughly analyzed. More data on the distribution of the ultra hot gas are needed in order to discriminate between these possibilities. New approaches to studying the iron lines from this gas considered above would hopefully be helpful in resolving those issues.

Acknowledgement. L.O. and L.T. acknowledge a NRC/NAS Research Associateship.

REFERENCES

Castor, J., McCray, R., & Weaver, R. 1975, ApJ 200, L107
Koyama, K. *et al.* 1989, Nature 339, 603
Koyama, K. *et al.* 1990, Nature 343, 148
Lebofsky, M.J., Rieke, G.H., & Tokunaga, A.T. 1982, ApJ 263, 736
Mewe, R. 1985, A&A Suppl. Ser. 62, 197
Mewe, R. 1990, in Physical Processes in Hot Cosmic Plasmas. Eds. W. Brinkman
    et al. Kluwer Acad. Publ., p. 39
Mezger, P.G., Wink, J.E., & Zylka, R. 1990, A&A 228, 95
Ozernoy, L.M. 1992, in Testing the AGN Paradigm. Eds. S.S. Holt, S.G. Neff,
    & C.M. Urry. AIP Conf. Proc. 254, 44
Ozernoy, L. M., Titarchuk, L.G., & Ramaty, R. ApJ 1993 (to be submitted)
Skibo, J.G. & Ramaty, R. 1993, A&A Suppl. Ser. (in press)
Skinner, G.K. *et al.* 1987, Nature 330, 544
Sofue, Y. 1990, Astro. Lett. and Communications 28, 1
Titarchuk, L.G. 1987, Astrophysics 26, 97
Wang, Q. 1992, in The ESO/EIPC Workshop on Star Forming Galaxies and
    Their Interstellar Medium. Eds. J. Fraco et al. (in press)
Weaver, R. *et al.* 1977, ApJ 218, 377
Yamauchi, S. 1990. Thesis (ISAS Research Note No. 474)
Yamauchi, S. *et al.* 1990, ApJ 365, 532
Zylka, R., Mezger, P.G., & Wink, J.E. 1990, A&A 234, 133

ANNIHILATION AND BACKSCATTERED LINES

IN BLACK HOLE CANDIDATES

Ph. Durouchoux and P. Wallyn
Service d'Astrophysique, Centre d'Etudes Nucléaires de Saclay
91191 Gif sur Yvette Cedex FRANCE.

ABSTRACT

On May 22nd 1989, HEXAGONE observed a broad region (19°FWHM) in the direction of the center of the Galaxy, and detected a narrow 511 keV line (Durouchoux et al. 1992) and also a broad emission around 170 keV (Smith et al. 1992) interpreted as backscattered annihilation photons coming from the 1E1740.7-2942 hard X-ray source. On the other hand, the french SIGMA experiment aboard the GRANAT satellite observed the Nova Muscae source on Jan 20-21 1991 and detected in addition to a line around the annihilation energy, two features which appeared successively around 102 keV and 170 keV (Goldwurm et al 1992) . Considering a new transient source EXS1737.9-2952 (Grindlay et al. 1992) which showed a bump around 102 keV a week before the HEXAGONE observation, we are questioning here if the 170 keV bump detected by HEXAGONE was also emitted by EXS source. We present a model where 102 keV and 170 keV double and single backscattered lines are explained in terms of supercritical states occuring in the accretion disk surrounding a black hole; we verify the model with a Monte Carlo simulation which accounts for the observations and mainly their timing sequence of appearance. We conclude by indicating that not only an annihilation line but also backscattered features might be signatures for black hole candidates.

I- INTRODUCTION

A high resolution gamma-ray balloon borne spectrometer HEXAGONE was launched from Alice Springs (Australia) in May 22nd 1989. The target was the Galactic Center region and we detected a narrow 511 keV line (Durouchoux et al. 1992) and a broad emission around 170 keV (Smith et al. 1992). It is fairly well established that the 511 keV annihilation line has two origins (Lingenfelter & Ramaty 1989): a steady diffuse emission, which follows more or less the type I supernova distribution along the Galactic plane (e. g. Skibo et al. 1992), and a variable point source, assumed to be coming from positrons emitted by 1E1740.7-2942 (1E hereinafter) and annihilating in a nearby cold molecular cloud (Bally & Leventhal 1991, and Mirabel et al. 1991). Here we question the validity of these assumptions: analysing carefully the balloon data successfully gathered during spring 1989 (GRIP, EXITE, POKER and HEXAGONE), we investigate the possibility that the bumps (around 170 keV, 102 keV and 72 keV) which appear in some of the spectra (and not in others) can be explained if we assume that 1E is not the positron source which produce the backscattered photons. Among the four balloon experiments mentioned above, the first three have positioning cababilities, whereas the fourth has high energy resolution. GRIP detected on April 12 th the 1E source in its normal state, EXITE (May 9 th) found a "new" source: EXS 1737.9-2952 (40' West of 1E, EXS hereinafter) which exhibits a feature in the range 83-111 keV, and also a bump in the soft X-ray band (20-30 keV).
Furthermore, a bump around 170 keV in the spectrum corresponding to a large region of the center of the Galaxy has been detected on May 23 rd (Matteson et al. 1991) and interpreted as backscattered 511 keV photons (Lingenfelter & Hua 1991, LH hereinafter)

We must underline the similarities in the behaviour of EXS and the recently observed low mass X-ray binary: Nova Muscae. As shown in fig 1, the observations monitored by the GRANAT observatory in 3-1300 keV band on January 20/21 exhibit features around 102 and 170 keV and a line around 480 keV. We are trying to elaborate a model which takes into account the timing variation of the single 511 keV backscattering bump (170 keV) and the double backscattered (102 keV).

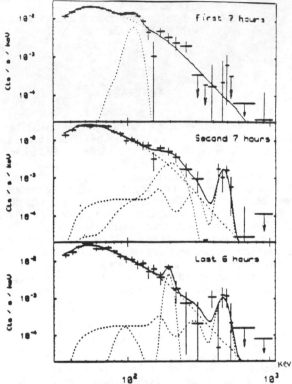

Fig 1: Nova Muscae spectra measured at different times (Goldwurm et al. 1992).

II- A MODEL FOR THE BACKSCATTERED LINES

By examining the timing sequence in Nova Muscae, one must point out that the 102 keV feature appears prior the 170 keV bump, as well as in EXITE and HEXAGONE GC observations. Moreover, in the case of Nova Muscae a narrow possibly redshifted annihilation line (centroid at 471±30 keV, line width of 30 ± 30 keV, during the last sub-session) is observed and requires a cold region (T < 10^8 K, Goldwurm et al. 1992) to produce the annihilation of the positrons. Such a region must be separated from the hot pair plasma region which produces the pairs. These results also emphasize the separation between the positron creation and annihilation regions. Shakura & Sunyaev (1973), have investigated the evolution of an accretion disk around a black hole related to a variation of the accretion rate of matter. During supercritical inflows of matter, a strong anisotropic mass outflow, in a small cone near the axis of disk rotation due to radiation pressure over gravitational attraction, can be observed. The matter flows out in a spiral with a speed decreasing when matter moves away from the compact source (see Figure 9 in Shakura & Sunyaev 1973). They expect the matter to flow away from a region close to 7 Schwarzschild radii from the central source.

We thus propose that Nova Muscae and the 1989 observations can be tentatively explained in a roughly similar framework (the important timescale difference between Nova Muscae - 3 sequences of roughly 7h - and EXS - typically two weeks between EXITE and HEXAGONE observations- lets here the possibility of more than one 511 keV burst in the former source): a high activity state in the inner part of the accretion disk (following a variation in the accretion rate) creates the conditions for a hot plasma (T>10^9 K) and the subsequent emission of pairs.

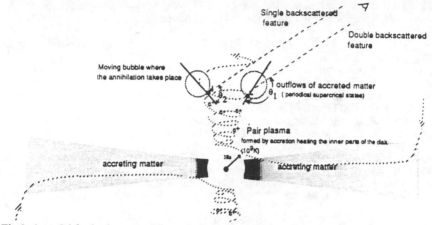

Fig 2: A model for backscattered lines: during a high activity of the central source (accretion rate variation) an outflow of the accreted matter takes place and a hot pair plasma is formed. The interactions of the positrons emitted in a narrow beam with the spiraling outflow of matter are able to explain the observed features; (for more details, see Wallyn & Durouchoux 1992)

This stream of electron-positron particles will create a hot spot in the outflowing matter when it penetrates this region. If the geometry allows the matter to lie between the hot spot and the observer, the annihilation line is obscured and the double backscattered line prominent .

III·MONTE CARLO SIMULATION

In order to verify our model, we performed a Monte Carlo simulation based on a simple geometry,

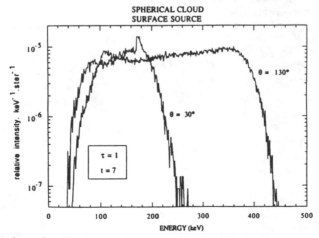

Fig 3: Comparison of two backscattered spectra obtained at different times (different θ)

consisting of a spherical cloud with an opacity defined in terms of the compton scattering optical depth τ at 511 keV and a monoenergetic and anisotropic 511 keV point source set to the surface of the cloud. Since the cloud, where the backscattered lines are formed, is spiralling due to angular momentum of the accreted matter, time evolution is related to the observing angle (see fig 2)

We present in fig 3 the backscattered spectrum created in the cloud, obtained at different times (which means with different observing angles). It is therefore possible to observe first ($\theta = \theta_1$) a bump around 102 keV and then ($\theta = \theta_2$) a feature at 170 keV, when decreasing the value of θ, sequence which is present both in EXS and Nova Muscae sources.

IV-CONCLUSION

In order to explain the different features seen in the GC in spring 1989, as well as the evolution of these features versus time, we developed a model involving an evolutionary geometry and propose that a unique source EXS, is the progenitor of both the 102 and 170 keV features detected by EXITE and HEXAGONE (and could be responsible of the features around 170 keV previously detected in the Galactic center region by high energy experiments during the last decade). This model can also be applied to Nova Muscae source, giving a view of the timing sequence.

A unique signature which could allow discrimination of Galactic black hole candidates from other gamma-ray emitters is not yet available (see Liang 1992 for a review). We tentatively propose the presence of such backscattered annihilation features associated with a variable power law spectrum as a new black hole signature.

REFERENCES

Bally, J., & Leventhal, M., 1991, Nature, 353, 234

Durouchoux, Ph., et al., 1992, Astr. Ap. Suppl. Ser., in press

Goldwurm, A., et al. 1992, Astr. Ap. Suppl. Ser., in press.

Liang, E. P., 1992, in Compton Observatory Science Workshop,
        eds. C. Shrader, N. Gehrels and B. Dennis, NASA/GSFC, p 173

Lingenfelter, R. E., & Hua, X., 1991, Ap.J., 381, 426

Lingenfelter, R. E., & Ramaty, R., 1989, Ap. J., 343, 686

Matteson, J. L., et al., 1991, in proceedings on Gamma Ray Line in Astrophysics eds. Ph. Durouchoux and N. Prantzos. (New York: AIP 232), p 45.

Mirabel, I. F., Morris, M., Wink, J., Paul, J., & Cordier, B., 1991,
        Astr. Ap. (Letters), 251, L43.

Shakura, N. I., & Sunyaev, R. A., Astr. Ap., 1973, 24, 337.

Skibo, J. G., Ramaty, R., & Leventhal, M., 1992, submitted to Ap. J

Smith, D., et al. 1992, Astr. Ap. Suppl. Ser., in press.

Wallyn, P., & Durouchoux, Ph. 1992, submitted to Ap. J

THE STRONG MAGNETIC FIELD (SMF) CENTRAL ENGINE

HOWARD D. GREYBER
GREYBER ASSOCIATES, 10123 FALLS ROAD, POTOMAC, MD 20854, U.S.A.

ABSTRACT

The Strong Magnetic Field Central Engine model (SMF) has been developed since 1961 to explain the variety of morphology and energetics of objects of galactic dimension. It is suggested that in the very center of our own Milky Way Galaxy, a newly discovered object, GZ-A, is the remnant of our own weak central engine, but a prototype of those in far more energetic active galactic nuclei, such as in Seyferts and quasars.

INTRODUCTION

For decades a minority point of view has argued against the galactic dynamo model, and for a primordial magnetic field existing before the galaxies were formed. Since in the pregalactic, precollapse giant cloud the Debye length is small compared to other lengths of interest, the precollapse cloud is a *plasma cloud* and electromagnetic effects dominate, even if the ionization is as low as 10%. Theorists have ruled that the **ONLY** large, basic components of a galactic size object are conventional stars, white dwarfs, neutron stars and black holes. *But does* the Universe (over 95% plasma) *obey* !?

The Strong Magnetic Field model (SMF) was created in 1961 originally to explain spiral arms, then extended for radio galaxies, quasars and their jets, and AGN in general.[1-13] SMF argues that when a pregalactic plasma cloud, with a large-scale poloidal magnetic field inside, collapses under gravitation, an extremely intense, relativistic, gravitationally bound current loop (GBCL) is formed around the central object, presumably a black hole. A very slender plasma toroid surrounds the extremely slender current loop. The two are bound together by the well-known Maxwell "frozen-in" field condition, and, in turn, the slim toroidal mass is gravitationally bound to the black hole. SMF agrees with the insightful comment by Martin Rees that "the phenomena of quasars and radio galaxies cannot be understood until they are placed in the general context of galactic evolution".[14]

Furthermore, SMF has shown that large scale turbulence in an isolated plasma cloud will prepare, *naturally over time,* by the processes of dissipation, a toroidal volume current with its dipole-like magnetic field, suitable for amplification.[3] Spiral galaxies (found isolated or in loose clusters) follow this path. However the ordinary elliptical galaxies (found in rich clusters) do *not,* having collided and cancelled out poloidal effects. The obvious exceptions are the most *massive* galaxies in a tight cluster, the "giant ellipticals" (often strong radio sources), since their poloidal development by dissipation processes is too strong to be cancelled out by interaction.[3]

As described back in 1966, one examines the dissipation term in the following equation easily derivable from electromagnetic theory, and sees that for long times, and

$$\frac{\partial H}{\partial t} = \text{curl} (v \times H) + \eta \nabla^2 H$$

for finite electrical resistivity, η , the magnetic energy dissipation goes as $\eta k^2 H$, where k is the wave number. Similarly in the theory of turbulence, the dissipation term goes as $\gamma\, k^2 u(k)$, where γ is the viscosity and u(k) is the Fourier velocity component whose square is proportional to the energy in wave number k. Clearly, the small wavelengths dissipate *fastest*, before the largest wavelength. Thus, after a time far less than 10^9 years, the non-poloidal components die out in this bounded cloud volume, leaving a dipole-like field suitable for amplification under gravitational collapse.[3]

Gravitational collapse means that the myriads of individual filamentary current loops making up the toroidal volume current are forced to coalesce into a GBCL anchored around the black hole.[8] The bursting force of the strong dipole magnetic field is in stable equilibrium with the gravitational attraction between the slender toroidal plasma and the black hole (notice the different ratios for the two figures with different morphology).[8]

One can see such a current loop forming in the figures in a very fine old paper by Mestel and Strittmatter, although they did not notice this.[15] A crude underestimate of the field amplification can be done assuming flux conservation.[8] However, a large calculation is needed, using particle-in-cell (PIC) simulation of gravitational collapse with a toroidal volume current, with the latest 3D electromagnetic particle codes.

DISCUSSION AND CONCLUSIONS

A reasonable deduction from SMF is that, even in our old tired Milky Way spiral galaxy, there may remain a remnant of the GBCL created when the Galaxy was formed (although extremely weak compared to those in AGN and quasars). The evidence is (a) the magnetic field close to Sag A* is observed to be perpendicular to the galactic plane, and (b) the radio arcs observed by Morris and Yusef-Zadeh - both suggesting a weak dipole magnetic field at the galactic center, and therefore a weak GBCL around Sag A*.

Recently M. R. Rosa, et al, have made a far-red (980 nm) high precision study of the Galactic Center region.[16] They discovered a new source, GZ-A, which is the closest non-radio source to the fiducial center of the Galaxy known to date, *and also has no infrared counterpart*. My conjecture is that GZ-A may be *the relatively weak local GBCL in our Galaxy shining at us in the plane* by virtue of its Cherenkov radiation. The GBCL signature to look for is the polarization of the Cherenkov radiation, and also the shape of the intensity versus frequency curve for Cherenkov radiation.[17]

Hopefully these critical observations will be made in the not too distant future. Of course GBCLs should exist in M31, M82, other nearby spirals, etc., but the problem of resolution at those distances makes observation with today's technology doubtful.

It is a pleasure to acknowledge the late Donald H. Menzel and John A. Wheeler for advice and comments, and Gart Westerhout for permission to use the U.S.N.O. library.

REFERENCES

1. Greyber, H. D. June 1, 1962, Air Force Office of Scientific Research, Report No. 2958
2. Greyber, H. D 1964, Quasistellar Sources and Gravitational Collapse, ed. I. Robinson et al, U. of Chicago Press, Chapter 31
3. Greyber, H. D. 1967, Instabilite Gravitationelle et Formation des Etoiles, des Galaxies et de Leurs Structures Caracteristiques, Memoirs Royal Society of Sciences of Liege, XV, 189
4. Greyber, H. D. 1967, ibid. XV, 197
5. Greyber, H. D. 1967, Publications Astronomical Society of the Pacific, 79, 341
6. Greyber, H. D. 1984, 11th Texas Symposium, Annals New York Academy of Sciences, 422, 353
7. Greyber, H. D. 1988, Supermassive Black Holes, ed. M. Kafatos, Cambridge University Press, p. 360
8. Greyber, H. D. 1989, Comments on Astrophysics, 13, 201
9. Greyber, H. D. 1989, The Center of the Galaxy, ed. Mark Morris, Kluwer Academic Press, p. 335
10. Greyber , H. D. 1990, 14th Texas Symposium, Annals New York Academy of Sciences, 571, 239
11. Greyber, H. D. 1992, Testing the AGN Paradigm, eds. S. Holt et al, A.I.P. Conference. Proc. 254, 467

12. Greyber, H. D. 1993, in STScI Preprint, "Astrophysical Jets"
13. Greyber, H. D. 1993, in Proceedings of the Compton Gamma Ray Observatory Conference, Oct. 1992
14. Rees, M. J. 1986, Highlights of Modern Astrophysics, eds. S. Shapiro & S. Teukolsky, Wiley, p. 163
15. Mestel, L. & Strittmatter, P. 1967, Monthly Notices Royal Astronomical Society, 137, 95
16. Rosa, M. R. et al 1992, Astronomy & Astrophysics, 257, 515-522
17. Landau, L. D. & Lifshitz, E. M. 1960, Electrodynamics of Continuous Media, Pergamon Press, p. 357

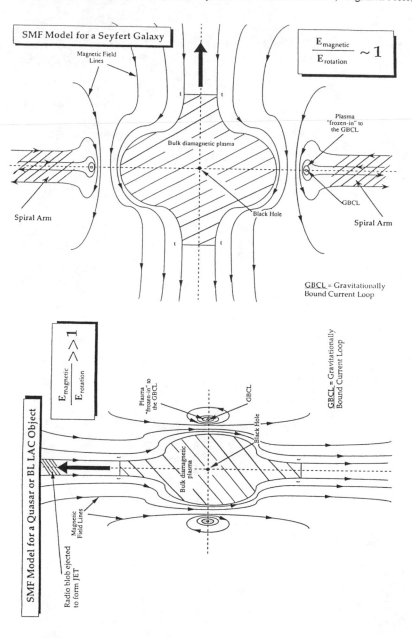

BULGE/BAR

DYNAMICS OF THE GALACTIC BULGE-BAR

James Binney
Theoretical Physics, Keble Road, Oxford OX1 3NP, U.K.

Ortwin Gerhard
Landessternwarte, Königstuhl, D-6900 Heidelberg 1, Germany.

ABSTRACT

The non-circular velocities of gas in the inner few degrees of our Galaxy would arise naturally if the Milky Way had a small bar at its centre. Recent studies suggest that this bar is more rapidly rotating than had been previously suggested; corotation probably lies near $R = 2.4\,\mathrm{kpc}$. The circular speed associated with the Galaxy's monopole component must rise as $v_c \sim R^{1/8}$ in the inner kiloparsec or so. It seems likely that the bar is the peanut-shaped object seen by COBE, and formed by spontaneous thickening of a bar-unstable disk. The lop-sidedness of the distribution of molecular gas at $|l| \lesssim 2°$ suggests that the gas flow at the Centre is unsteady. This unsteadiness may be an important clue to the correct technique for simulating the large-scale dynamics of the interstellar medium, which displays a wealth of unresolvable structure.

1. INTRODUCTION

From at least the work of Burton & Liszt (1978) and Sinha (1979) it has been known that both HI and CO surveys of gas in the central few degrees of the Galaxy showed that a lot of gas is moving at 'forbidden velocities', i.e., velocities that would not occur along the given line of sight if all gas moved on circular orbits. Two possible explanations of this phenomenon are: (i) gas at the Centre has been driven outwards by one or more explosions; (ii) gas is moving on roughly elliptical streamlines because the Galaxy's potential is significantly non-axisymmetric.

To many people the second of these explanations is more attractive than the first for at least two reasons

1. The force of an explosion at the Centre would be expected to drive gas predominantly upwards and downwards out of the plane, rather than outwards within the plane. Hence the overall energy of the explosion should considerably exceed that associated with the observed planar motions. Since the latter involve $\sim 10^8\,M_\odot$ moving at velocities in excess of $50\,\mathrm{km\,s^{-1}}$, the total energy must exceed that released into the ISM by 10,000 supernovae; Sanders & Prendergast (1974) conclude from hydrodynamical simulations that as much as 10^{58} erg would be required because of radiative cooling.

2. Since the work of Miller & Prendergast (1968) we have known that initially cold, self-gravitating disks invariably form a tumbling bar. Moreover, roughly a half of all disk galaxies are classified as barred, and a number of those classified as un-barred have turned out on close examination to have a small barred structure at the centre (e.g., Sellwood & Wilkinson 1992). So *a priori* it would not be surprising if the Milky Way were to have a central bar.

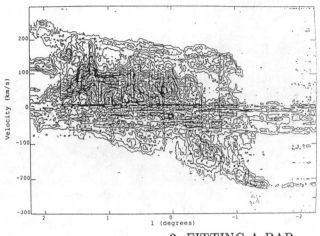

Figure 1. (l,v) digram of ^{12}CO $J = 1 \rightarrow 0$ averaged over $|b| < 0.1°$. The contours (regularly spaced at intervals of 1 K) show a striking parallelogram. (From Binney *et al.*, 1991).

2. FITTING A BAR

The key parameter of a bar is its pattern speed ω_p. Indeed, studies of n-body bars and gas flows in bars (e.g. Sellwood 1981, Athanassoula 1991) reveal that bars extend to roughly the corotation radius, R_{CR}, divided by 1.2. Hence, once one knows the bar's pattern speed, one can estimate its length quite accurately.

Binney, Gerhard, Stark, Bally & Uchida (1991) have argued that the key to deducing the pattern speed of the Galaxy's bar is a striking parallelogram in the (l,v) plane that has been seen in surveys of ^{12}CO (Bania 1977) and ^{13}CO (Heiligman 1987, Bally *et al.* 1987). Binney *et al.* argue that this structure, which is shown in Figure 1, is formed in a shock associated with the cusped orbit depicted in the upper right panel of Figure 2: this figure shows the principal closed orbits in the plane of a rotating bar. The outermost orbits are elongated parallel to the long axis of the bar, and are mostly occupied by HI clouds. These clouds gradually drift inwards, onto more elongated orbits. Eventually they reach the cusped orbit, encounter powerful shocks, in which most of the gas condenses into molecular form, and then plunge onto the outermost of the small central orbits – these orbits are elongated perpendicular to the bar.

The lower panels of Figure 2b show the traces in the (l,v) plane of the orbits plotted in the upper panels, when the bar is viewed from a line of sight inclined by 16° to the bar's long axis. The trace of the cusped orbit forms a parallelogram like that of Figure 1. Since the chosen line of sight runs parallel to two of the cusped orbit's straight sections, on which a cloud's velocity changes considerably as its galactocentric distance changes, two of the edges of the parallelogram in the lower right panel of Figure 2 run parallel to the v axis, as in the observed parallelogram of Figure 1. Requiring the theoretical and observed (l,v) plots to match in this respect determines the 16° angle of our line of sight in Figure 2, while from the roughly ±2° angular extent of the observed parallelogram in Figure 1 we deduce the linear scale of Figure 2: we require that the cusped orbit extend to $R \simeq 500$ pc, and this, for the given potential, in turn implies $\omega_p = 63\,\mathrm{km\,s^{-1}}/\mathrm{kpc}$ and $R_{CR} = 2.4$ kpc.

In the interpretation of Binney *et al.*, the parallelogram determines the Sun's viewing angle and the size of the cusped orbit rather directly. It determines R_{CR} much more indirectly, however, because the position of corotation depends on the shape of the Galaxy's circular-speed curve in the region $R \lesssim 2.5$ kpc. In

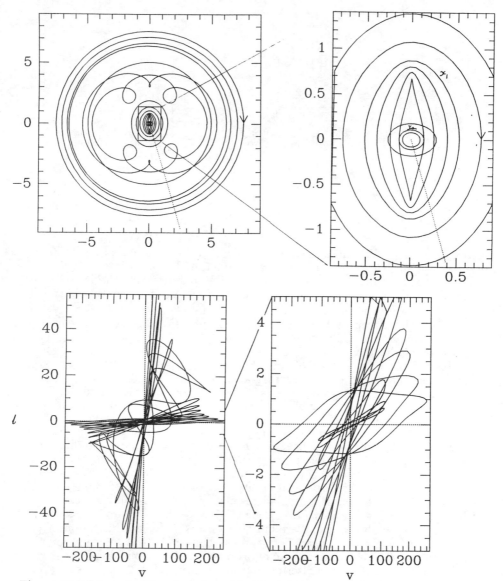

Figure 2. Orbits in the Binney *et al.* model of the galactic potential. The long axis of the bar runs vertically. The top panels show the structure in real space; they are labelled in kpc and show the line of sight from the Sun as a dotted line. The lower panels show the corresponding (l, v) plots.

the presence of a bar it is important to be clear what one means by 'the circular-speed curve', for this will not coincide with the speed of any actual orbits. We define the 'circular speed' $v_c(R)$ to be the circular speed at R provided by the monopole component of the potential; we have $v_c^2(R) = GM(R)/R$ precisely. The potential used to obtain Figure 2 has been matched to the HI data of

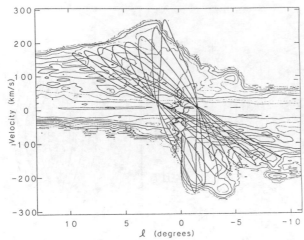

Figure 3. Contours of 21-cm intensity in the (l, v) plot from Liszt & Burton (1980). The traces of some closed orbits in the Binney *et al.* potential are superimposed.

Liszt & Burton (1980) through Figure 3. Here the (l, v) traces of orbits in the potential are superimposed on the Liszt & Burton HI contours and $v_c(R)$ has been adjusted so that the envelopes of the HI contours and the (l, v) traces match. Although the terminal velocity we are matching falls sharply outwards from $l = 2°$, the circular speed we require to match this *rises* outwards, as $v_c \propto R^{1/8}$. If this interpretation is correct, the sharp central peak in conventional rotation curves of the Galaxy (e.g. Clemens 1985) is an artifact of the type discussed by Gerhard & Vietri (1986).

Most of the CO, and all of the CS, emission from $|l| < 2°$ is associated not with the parallelogram, but with giant molecular clouds such as Sgr B2. In the Binney *et al.* model these are on the innermost orbits plotted in Figure 3a—these perpendicularly aligned orbits are called x_2 orbits by Contopoulos and Papayannopoulos (1980).

Previous attempts to interpret non-circular motions in the Galactic disk in terms of the dynamics of a bar have focused on the '3 kpc expanding arm' and from it inferred a smaller pattern speed than that of Binney *et al.* For example, Sanders (1979) and Mulder & Liem (1986) concluded that $R_{CR} \simeq R_0$, the solar radius. Can the 3 kpc arm be fitted into the scheme of Binney *et al.*? If $R_{CR} = 2.4$ kpc, the outer Lindblad resonance (OLR) would lie near 4 kpc, and from the work of Schwarz (1981) one might expect a pile-up of gas similar to the great molecular ring, at around this radius. The accumulation of gas near the OLR arises from the tendency of the bar's potential to pump angular momentum into clouds on near resonant orbits at the OLR, pushing these clouds outwards into material that is filtering in from greater radii. Thus the effect seen by Schwarz has a sound physical basis and may be expected to be robust. On the other hand, Schwarz's bar extended all the way to the OLR, rather than terminating at $R_{CR}/1.2$ as we now believe it should, with the result that the non-axisymmetric component of his potential at the OLR was unrealistically large.

It remains to be seen whether an evacuated region around corotation and something like a giant molecular ring will form naturally in the Binney *et al.* model. But if they do, the 3 kpc arm may find a natural interpretation as a streamer of gas that is moving through the evacuated region on a highly non-circular orbit after being ejected from the molecular ring, perhaps by a supernova-driven bubble.

An alternative explanation of the great ring of molecular clouds is offered by Dame (this conference): Dame suggests that the 'ring' may really be just the high-density parts of a pair of bisymmetric arms which emanate from the ends of the bar.

Long, Ostriker & Aguilar (1992) have investigated the effects on the orbits of globular clusters of a bar that is just a little bit larger and more slowly rotating than that proposed by Binney et $al.$: Long et $al.$ adopt $\omega_p = 40\,\mathrm{km\,s^{-1}}/\mathrm{kpc}$. They find that such a rapidly rotating bar only slightly increases the rate at which bulge shocking destroys globular clusters at $r \simeq 2.5\,\mathrm{kpc}$.

3. PROBLEMS WITH THE MODEL

So far we have concentrated on the successes of the Binney et $al.$ model. However, it does have significant problems. In particular:

1. Three quarters of the molecular gas inside $|l| = 2°$ lies at $l > 0$. Moreover, the parallelogram in Figure 1 is significantly displaced towards positive longitudes. Why is this?

2. If the parallelogram were the (l, v) trace of the cusped orbit, its extent in v would match the top of the HI envelope plotted in Figure 3, since the cusped orbit is simply the last in the sequence of elongated orbits whose (l, v) traces have been fitted to the HI contours. Actually the parallelogram extends only to $v \simeq 220\,\mathrm{km\,s^{-1}}$ while the HI envelope extends to $v \simeq 270\,\mathrm{km\,s^{-1}}$.

In some measure the displacement of the parallelogram towards positive longitudes may be explained as a simple perspective effect. Indeed, in Figure 2b the (l, v) traces can be seen to be displaced to positive l by about $0.3°$ for just this reason. Unfortunately, the observed parallelogram seems to be more lop-sided than this. Moreover, perspective cannot explain why three quarters of the molecular gas is at positive longitudes. Overall one cannot escape the conclusion that something important is missing from the model.

The Binney et $al.$ model is based on the naïve expectation that clouds move on closed orbits. But of course clouds are susceptible to hydrodynamic forces in addition to gravitational ones. It is precisely these hydrodynamical forces which drive clouds towards closed orbits. But in the neighbourhood of a resonance, where the available closed orbits are liable to intersect, strong shocks will form and hydrodynamic forces become non-negligible. Hence in the neighbourhood of resonances, clouds may deviate significantly from the closed orbits assumed by Binney et $al.$. Can these hydrodynamic effects, coupled with the self-gravity of the cloud medium, explain the lop-sidedness of the molecular gas, and the mis-match between the velocity scales of the parallelogram and the HI contours?

4. HYDRODYNAMICS OF INTERSTELLAR GAS

It is not at all obvious how the large-scale dynamics of interstellar gas should be modelled, since observations show that the ISM has structure on all observable scales. We can have no realistic hope of modelling all this structure, from sub-parsec scales through to kiloparsecs-long features. We must seek a phenomenological representation of the effective dynamics on the scales that are of interest to us—in the present case, scales ranging from tens of parsecs up through a kiloparsec or so. Of course a similar requirement for a phenomenological representation of the macroscopic effects of the interactions of myriads of molecules underlies standard hydrodynamics. This analogy frequently induces people to assume without much reflection that the dynamics of the ISM can be modelled with the standard equations of hydrodynamics. But this assumption is not well founded.

Indeed, if one argues that the ISM *is* only a fluid composed of molecules, so it must be describable by standard hydrodynamics, then one is obliged to follow its dynamics at such high resolution that the density of molecules does not change significantly from one resolution element to another. Since this is impossible, it is tempting to apply the chain of reasoning that leads from dynamics on a molecular scale to the Navier–Stokes equations on the macroscopic scale, to elementary units that would ordinarily be regarded as of macroscopic scale, such as Bok globules or other cloudlets.

In this way one might hope to derive a set of equations that describe the evolution of the density, velocity, etc., of an ensemble of cloudlets. It would be a grave mistake to assume that these macroscopic equations for the large-scale dynamics of a cloud ensemble are simply the Navier–Stokes equations. For the microphysics of cloudlets differs from that of molecules in important respects: (i) their encounters are significantly inelastic; (ii) mass will often be exchanged between cloudlets during an encounter, and (iii) magnetic fields and the dispersive effects of star formation are likely to play some role. Moreover, the cloudlets of cold gas must be embedded in a diffuse, hotter medium, and it is likely that there are also non-negligible exchanges of mass and momentum between clouds and the embedding fluid.

In fact, it is far from clear that *any* system of partial differential equations governs the macroscopically averaged variables of the cloud ensemble; there is no logical necessity for this. Even if such equations do exist, it seems unlikely that they will closely resemble the Navier–Stokes equations, given that the microphysics of clouds differs so significantly from that of molecules. In these circumstances it is no advantage for a numerical scheme to be securely founded on the Navier–Stokes equations, and becomes simply irrelevant whether the scheme works well for test problems such as shock tubes and the Sedov solution. Instead we must seek from first principles a credible way of simulating the dynamics of a cloud ensemble.

Nevertheless, most studies of the dynamics of gas in barred potentials have employed a scheme based on the Navier–Stokes equations. These include grid schemes (van Albada, 1985; Athanassoula, 1984, 1991; Mulder & Liem, 1986), beam schemes (Sanders & Huntley, 1976; Schempp, 1982), and the currently popular smooth-particle hydrodynamics (SPH) algorithm (Wada & Habe, 1992, Friedli & Benz 1992).

The results from all hydrodynamic simulations show that away from resonances, gas streamlines do shadow closed orbits. In the vicinity of a resonance, the flow lines are less nearly closed and, unless the bar is weak or of low central concentration, shocks form. In the case of the ILR these shocks appear downstream of the bar's long axis, or, since the gas is rotating faster than the bar, on the bar's leading edge. Figure 4 shows that for a bar of moderate strength and central concentration, the shocks form roughly along two of the four straight segments of the cusped orbit. In the Binney *et al.* model, the line of sight from the Sun runs nearly parallel to these shocks, which then form the vertical edges of the parallelogram in the CO (l, v) plane. The peak velocity achieved on these straight segments is smaller than the peak velocity along the cusped orbit, because only a fraction of the gas in the shock has fallen all the way from the cusp; much material joins the shock between the cusp and where it runs tangent to a

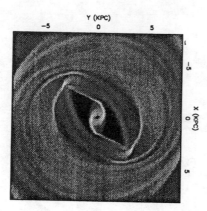

Figure 4. Contours of gas density in a hydro-dynamical simulation of Athanassoula (1991).

circle about the Centre. It remains to be seen whether this difference between the peak velocity along the shock and the peak velocity of the cusped orbit agrees with the difference between the velocity extent of the parallelogram and the peak of the HI terminal-velocity curve.

Once gas falls inside the cusped orbit, shocks rapidly bleed it of energy and angular momentum, and it soon falls into the region in which it can move quiescently on one of the x_2 orbits that fit inside the cusped orbit. Hence the surface density is low in the entire region between the cusped orbit and the largest populated x_2 orbit. This explains why most of the galactic-centre molecular material is concentrated in the giant clouds, which in the Binney *et al.* model are on x_2 orbits.

Grid schemes are generally designed to converge rapidly on a quasi steady-state, and sometimes explicitly assume two-fold symmetry. Hence the lop-sidedness of the data, which must be associated with unsteadiness in the flow, lies beyond their scope. Moreover, particle-based algorithms such as the beam scheme and SPH have high effective transport coefficients, so in them noise tends rapidly to damp away.

Jenkins (1992) has recently simulated the Binney *et al.* model with a 'sticky particle' scheme of the type used in simulations of stellar accretion disks (Larson, 1978; Whitehurst 1988). In this scheme the force between particles is given by

$$\mathbf{F}(r,v) = K\frac{\hat{\mathbf{r}}}{r} \times \begin{cases} 1 + v^2/v_0^2, & v < 0, \ r < r_{\max} \\ 1, & v \geq 0, \ r < r_{\max} \end{cases}, \tag{1}$$

where $v = \mathbf{v}\cdot\hat{\mathbf{r}}$ is the velocity of the forcing particle away from the particle under consideration, and K, v_0 and $r_{\max}$ are constants. Thus closely packed particles repel each other, mimicking a pressure force. Encounters at speed $v \ll v_0$ are elastic, while faster encounters are inelastic.

Figure 5 shows the evolution of a simulation started with all particles moving on precisely closed orbits. Particles that start out on orbits at all close to the cusped orbit, quickly fall in and settle on an x_2 orbit, leaving the vicinity of the cusped orbit strongly depopulated. In the (l,v) plane, this depopulation manifests itself in a terminal-velocity envelope which does not rise strongly near $l = 2°$, like the observed one.

This depopulation of orbits near the cusped one appears to be caused by the mutual repulsion of the particles in the high-density regions apparent along the major axis in the top left panel of Figure 5a. In real life, cloudlets do not

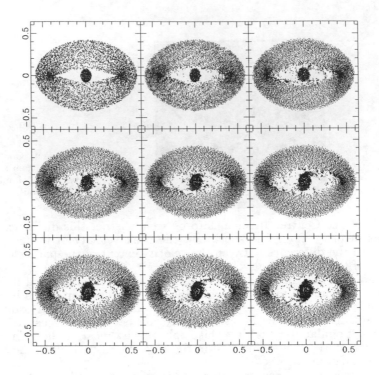

Figure 5. Evolution of a sticky-particle simulation of gas flow in the Binney *et al.* model from Jenkins (1992). The interaction scheme is given by equation (1).

repel each other; on the contrary, they must initially attract one another, even if star formation may subsequently cause massive clouds formed in mergers, to disrupt or suffer heavy mass loss. Therefore we have experimented with an alternative sticky particle scheme, in which the force on a particle at $(\mathbf{x}, \mathbf{v})$ from its neighbours is given by

$$\mathbf{F}(\mathbf{x}, \mathbf{v}) = K \sum_{\text{neighbours } i} (\mathbf{v}_i - \mathbf{v}) \exp\left(-|\mathbf{x}_i - \mathbf{x}|^2 / h^2\right). \tag{2}$$

This scheme is motivated by the idea that cloudlets exchange mass during close encounters. The amount of mass exchanged is assumed to vary Gaussianly with their separation. Figure 6 shows the evolution from the same initial conditions as Figure 6 under this new force law. Two features are noteworthy: (i) orbits near the cusped orbit are no longer rapidly depopulated, and (ii) the flow is now much more unsteady and lop-sided.

Although these experiments are still preliminary and much remains to be done before we can hope for a satisfactory understanding of gas at the Centre, they do illustrate the importance of trying different schemes for simulating interstellar gas, and they suggest that the flow near the centre may be as unsteady and lop-sided as the observations suggest.

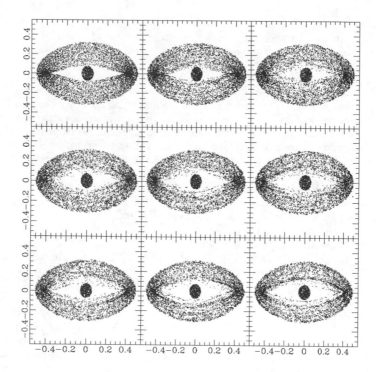

Figure 6. Evolution of a sticky-particle simulation of gas flow in the Binney *et al.* model using the interaction scheme of equation (2).

VERTICAL STRUCTURE

Fully three-dimensional studies of bar formation have become an industry only in the last few years, although a key fact has been in the literature since 1981, when Combes & Sanders observed a peanut-shaped bulge forming from a bar-unstable *n*-body disk. Combes *et al.* (1990) have now examined this phenomenon in detail. They find significant numbers of bulge stars are on orbits trapped around closed orbits that make two vertical and radial oscillations for every complete rotation about the centre in the bar's rotating frame. One says that these orbits are trapped around the $\omega_R : \omega_\phi : \omega_z = 2{:}1{:}2$ resonance. Raha *et al.* (1991) argue that peanut-bulges form suddenly through a collective 'firehose' instability in the original planar bar, while Combes *et al.* (1990) and Friedli & Pfenniger (1990) suggest that the bar thickens as a result of stars individually suffering resonant acceleration in the vertical direction. Certainly the suddenness of the thickening observed in the Raha *et al.* models, and the loss by the system of equatorial symmetry as it begins to thicken, suggest that an instability of the firehose type is involved. On the other hand, Combes *et al.* show a model in which the peanut-bulge begins to grow only 4 Gyr after the formation of the planar bar, leaving sufficient time for asymmetric resonant scattering to become important.

What do we know of the vertical structure near the centre of our Galaxy? Sellwood (this meeting) has shown how well one of these n-body peanuts can mimic the $2\,\mu$ surface photometry reported by the COBE–DIRBE team. By contrast, if one supposes that the Disk continues with constant scale height $z_0 \simeq 165\,\mathrm{pc}$ in to $R < 1\,\mathrm{kpc}$, one finds that the vertical frequency in the Binney *et al.* potential would be twice the radial frequency at $R \simeq 1\,\mathrm{kpc}$, and the central orbital structure would be dominated by orbits trapped around the 2:1:4 resonance, rather than the 2:1:2 resonance, which appears to dominate the n-body models. In short, the half-thickness of the Galactic bar must be substantially greater than 160 pc if it is to resemble one of the n-body peanuts, as the COBE photometry suggests.

Burton & Liszt (1978) and Sinha (1979) already reported that at $R \lesssim 2\,\mathrm{kpc}$, the HI appears not to lie exactly in the Plane. Liszt & Burton (1978, 1980) developed a model in which gas moves on elliptical orbits in a plane tilted by 24° with respect to the Plane; the line of nodes in which the former intersects the latter makes an angle of 57° with the line of sight to the Centre, and the long axis of the elliptical orbits makes an angle of 51° with the line of sight. The nearer end of the bar lies at positive longitudes as in the Binney *et al.* model.

Although Heiligman (1987) concluded that the ^{13}CO distribution is also tilted (though to a smaller degree), the suggestion that the gas lies in a tilted plane has remained controversial, in part because at $|l| \leq 2°$ the ^{12}CO distribution appears to peak near the line $b = 0$. Recently Blitz has pointed out that even in the Bally *et al.* data, material with forbidden velocities is most conspicuous above the plane at negative longitudes, and below the plane at positive longitudes, and that this is just what one would expect if the gas were on tilted, elliptical orbits since the forbidden velocities occur near the ends of the bar.

Can these out-of-plane motions be explained in terms of the instability strips that surround certain vertical resonances (Binney, 1981; Heisler, Merritt & Schwarzschild, 1982, Pfenniger 1985), and in particular the 2:1:2 resonance which appears to play an important rôle in n-body peanuts? The answer to this question is not yet in. Friedli & Udry (1992) have investigated closed orbits in three-dimensional barred potentials, including those of n-body simulations, and find that the closed orbits to which gas settles need not be confined to a plane. In particular it appears possible for gas to settle to the figure-of-eight shaped 2:1:2 resonant closed orbits, but the symmetry of these orbits is not suited to the generation of a tilt. It looks as if we still have an insufficiently clear enough picture of what resonances are most important, and how the strengths of their instability strips are affected by such factors as the roundedness of the bar's ends.

REFERENCES

Athanassoula, E. 1984. *Phys. Rep.*, **114**, 319

Athanassoula, E. 1991. In *Dynamics of Disk Galaxies*, ed. B. Sundelius, p. 71 (Göteborg: Chalmers University)

Bally, J., Stark, A. A., Wilson, R. W. & Henkel, C., 1987. *ApJSuppl*, **65**, 13

Bania, T. M. 1977. *ApJ*, **216**, 381

Binney, J. J., Gerhard, O. E., Stark, A. A., Bally, J., Uchida, K. I. 1991. *MNRAS*, **252**, 210

Binney, J. J. 1981. *MNRAS*, **196**, 455

Burton, W. B. & Liszt, H. S. 1978. *ApJ*, **225**, 815
Clemens, D. P. 1985. *ApJ*, **295**, 422
Combes, F., Debbasch, F., Friedli, D. & Pfenniger, D. 1990. *A&A*, **233**, 82
Combes, F. & Sanders, R. H. 1981. *A&A*, **96**, 164
Contopoulos, G. & Papayannopoulos, Th. D. 1980. *A&A*, **92**, 33
Friedli, D. & Benz, W. 1992. *A&A*, **00**, 00
Friedli, D. & Pfenniger, D. 1990. In *Bulges of galaxies*, eds. B. J. Jarvis & D. M. Tendrup (ESO, Munich), p. 265
Friedli, D. & Udry, S. 1992 *IAU Symposium 153*, ed. H. Dejoghe (Dordrecht: Reidel)
Gerhard, O. E., Vietri, M. 1986. *MNRAS*, **223**, 377
Heiligman, G. 1987. *ApJ*, **314**, 747
Heisler, J., Merritt, D. & Schwarzschild, M. 1982. *ApJ*, **258**, 490
Jenkins, A. R. 1992. D.Phil. thesis, University of Oxford
Larson, R. B. 1978. *J. Comp. Phys.*, **27**, 397
Liszt, H. S. & Burton, W. B. 1978. *ApJ*, **226**, 790
Liszt, H. S. & Burton, W. B. 1980. *ApJ*, **236**, 779
Long, K., Ostriker, J. P. & Aguilar, L. 1992. *ApJ*, **388**, 362
Miller, R. H. & Prendergast, K. H. 1968. *ApJ*, **151**, 699
Mulder, W. A. & Liem, B. T. 1986. *A&A*, **157**, 148
Pfenniger, D. 1985. *A&A*, **150**, 112
Raha, N., Sellwood, J. A., James, R. A. & Kahn, F. D. 1991. *Nature*, **352**, 411
Sanders, R. H. 1979. In IAU Symposium 84 *Large Scale Characteristics of the Galaxy*, ed. W. B. Burton
Sanders, R. H. & Huntley, 1976. *ApJ*, **209**, 53
Sanders, R. H. & Prendergast, K. H. 1974. *ApJ*, **188**, 489
Schempp, W. V. 1982. *ApJ*, **258**, 96
Schwarz, M. P. 1981. *ApJ*, **247**, 77
Schwarz, M. P. 1985. *MNRAS*, **209**, 53
Sellwood, J. A. 1981. *A&A*, **99**, 362
Sellwood, J. A. & Wilkinson, A. 1992. *Rep. Prog. Phys.*, **00**, 00
Sinha, R. P. 1979. Ph.D. thesis, University of Maryland
van Albada, G. D. 1985. *A&A*, **142**, 491
Wada, K. & Habe, A. 1992. *MNRAS*, **258**, 82
Whitehurst, R. 1988. *MNRAS*, **233**, 529

Photometric Evidence for the Bar

Leo Blitz
Astronomy Department, University of Maryland, College Park, MD, 20742

Email ID
blitz@astro.umd.edu

ABSTRACT

Good evidence for the existence of a bar at the center of the Milky Way was available nearly twenty years ago, but it is only in the last several years that its existence has become compelling. Previous evidence for the existence of a bar is reviewed, and the recent photometric evidence is discussed. The system of metal-rich globular clusters is shown to exhibit an elongation and orientation consistent with the bar determined by other means. It is argued that the reason that the bar was not previously embraced is that it conflicted with the prevailing idea that the non-circular motions of the gas near the center is due to explosions, which had an influential proponent in Jan Oort.

1. SOME HISTORICAL NOTES

To first order, the motion of the atomic hydrogen gas in the disk of the Milky Way implies that the gravitational potential is axisymmetric. However, it had been known since the first 21-cm line surveys that the region near the Galactic Center is peculiar. In an axisymmetric galaxy, gas motions are circular; toward $l = 0°$, the velocities would be perpendicular to the line of sight, and the line widths should reflect the velocity dispersion of the gas, about $4 - 7 \text{ km s}^{-1}$. However, linewidths in excess of 300 km s^{-1} are observed in many locations, and non-circular velocities are seen over a longitude extent of at least 20°. Furthermore, certain velocities are forbidden in a circular velocity field, namely negative velocities at positive longitudes, and positive velocities at negative longitudes. Position–velocity maps of atomic hydrogen clearly show, however, that 'forbidden' velocities in excess of 100 km s^{-1} are clearly observed within several degrees of the Galactic Center. De Vaucouleurs (1964) noticed that the non-circular velocities observed in the Milky Way are similar to those of the gas near the centers of barred spiral galaxies; he suggested on that basis, that the Milky Way is actually a barred spiral.

Previous explanations of the non-circular HI motions were based on the supposition that they are due to expansion and rotation, rather than steady state orbits (Rougoor and Oort 1960; Rougoor 1964) in spite of the large kinetic energy required to accelerate the gas to the observed velocites. Scoville (1972) invoked a similar origin for the motion of a smaller "expanding" ring structure he identified in CO. As early as 1975, however, Peters showed that it was possible to produce the large-scale morphology of the longitude-velocity plots of HI without invoking any expansion. He did the first detailed kinematic modeling of the gas showing that the emission is consisent with gas moving on elliptical streamlines, presumably in response to forcing by a bar. He assumed that the atomic hydrogen within 15° of the Galactic Center is constrained to move on ellipses with a common orientation to the Sun-Center line. The modeling produced simulated position-velocity diagrams with all of the qualitative features of the 21-cm data. The orientation of the ellipses was constrained by the kinematics so that the part of the orbit closest to the Sun is in the first galactic quadrant; Peters found that the best match to the data were obtained with a position angle of 45°±15°. The differences between Peters' results, and the bar inferred from more recent observations is cosmetic.

Several other investigations quickly followed that gave strong support to the idea that the Milky way is a barred spiral. The first, by Cohen and Few (1976) found the CO ring identified by Scoville is mimicked in OH and they argued that the peaks of the OH distribution within about a degree and a half of the Galactic Center are situated in longitude-velocity space as if they were on an elliptical orbit with the same orientation demanded by Peters, but with a larger axial ratio. They showed, however, that an expanding and circularly rotating ring of gas would produce a nearly identical kinematic signature. Thus it was not possible to distinguish between expansion and elliptical orbits from the OH data alone. In the same year, Sanders and Huntley (1976) did some hydrodynamical simulations of gas distributions in a rapidly rotating bar-like potential and showed that one could obtain non-circular gas velocities of the correct magnitude, although they did not attempt a detailed simulation of HI in the $l - v$ plane. Finally, Burton and Liszt (1978, 1980) and Liszt and Burton (1980), looked at atomic hydrogen and CO and included the latitude distribution of the gas. They confirmed the previous results of Peters, and Cohen and Few, but added the new wrinkle that the atomic hydrogen and very likely the CO as well, is orbiting at some angle inclined to the galactic plane. Their first two papers included circular rotation and expansion as separate kinematic motions, but at the suggestion of Ostriker, the third paper showed that the description of the data are equally good with elliptical streamlines.

It would seem that the kinematic modeling and the results of the hydrodynamic simulations would have been strong evidence that the Milky Way is a barred spiral, but in spite of the analytical successes, not only was there no consensus that the Milky Way has a bar, but there had been little mention of this possibility in the literature subsequent to the Burton and Liszt (1980) model. A notable exception is the work of Gerhard and Vietri (1986), in which it was argued that the detailed shape of the HI and CO terminal velocity curves could be reconciled with the bulge density distribution inferred from infrared observations, and the local density of spheroid stars only if the bulge is non-axisymmetric and the resulting potential is triaxial. Even this work did not seem to have the expected impact on the community, especially among the observers. Why was it, given the weight of the evidence, that there no strong consensus developed that the Milky Way is a barred spiral?

It appears that a combination of exciting new results in astronomy and weight of opinion of one of the world's most influential astronomers made the expansion hypothesis seductively attractive. Consider what was going on in the years just prior to Peters' seminal analysis. Radio interferometry , and especially the Westerbork synthesis array were providing increasingly clear images of the radio jets from active galaxies. X-ray satellites were gathering data suggesting that particular sources might actually be black holes. In 1971 Lynden-Bell and Rees published a paper suggesting that black holes might be the engines that power the high luminosities of quasars and are the sources of explosive outflows in their host galaxies. Significantly they suggested that a black hole at the center of the Milky Way could power the luminosity of the center and would be manifested as an unresolved point source of low frequency radio radiation. In 1971 Ekers and Lynden-Bell identified a compact radio source at the center of the Galaxy, and in 1974, Balick and Brown established that its size is less than 0.″1; the source has become known as Sgr A*. Ever since that discovery, there has been a major effort to discern whether Sgr A* actually marks the presence of a black hole.

The idea that a black hole powers the energetic activity and luminosity at the Galactic Center fit in quite comfortably with the ideas about explosive activity generating the non-circular motions propounded by Oort. In 1977, he published an influential review on the Center of the Galaxy where he reviewed the relevant literature up to that time. Although Oort discussed both gravitational and explosive origins for the non-circular motions of the HI at the center, one is left with the unmistakable impression that he strongly favored the explosion hypothesis. In his sole reference to Peters (1975), Oort says that he, "has *tried* [emphasis mine] to represent the HI velocity longitude diagram for the central region by gas moving in concentric elliptical orbits with axial ratios of about 0.5. He suggested that such orbits might be sustained by a bar-like perturbation, but no quantitative dynamical foundation was given." In fact, the agreement between Peters' kinematic model and the data are quite impressive given the simplicity of the modeling. Given that no mechanism was identified to provide the necessary 10^{55-56} ergs necessary to power the the radial motions required by the outflow hypothesis, and that the idea of a black hole at the Galactic Center received more favorable attention in Oort's review, it is hard to conclude that Oort felt that both ideas were equally meritortious. The powerful influence that Oort had on the astronomical community as a whole, as well as the influence that this particular review had on ideas about the Galactic Center, ultimately stymied the bar hypothesis. It was not until a flurry of observational papers in the last few years, all giving consistent results, as well as an important dynamical analysis by Binney *et al.* (1991) which provided a good physical basis for the kinematics, that the tide seems to have turned.

2. PHOTOMETRIC EVIDENCE

There have been at least five studies published within the last two years suggesting that the distribution of stars in the bulge is not axisymmetric. They may be divided into two catagories: investigations of total luminosity of bulge stars, and investigations of the distribution of individual stars identified with the bulge. All show that the distribution of stars in the bulge is non-axisymmetric, and the orientation of the long axis of the distribution is the same in each case.

Diffuse Emission

The first of these studies, by Blitz and Spergel (1991), is an analysis of the balloon observations of the Galactic Center by Matsumoto *et al.* (1983). The observations are at a wavelength of 2.2 μm of the inner 13° around the Galactic Center and have a resolution of about 0.5°. At 2.2 μm, the emission is dominated by red giants, and the extinction is a tenth of what it is in the optical. Thus, at a sufficiently large angular distance from the galactic midplane, the near infrared emisison is dominated by bulge stars, and the extinction due to dust is sufficiently small that it can be shown to be negligible for determining the degree of axisymmetry of the bulge population. Blitz and Spergel estimated the mean foreground extinction from CO and HI maps of the galactic plane, and assumed that the dust-to-gas ratio has a standard value along the line of sight. They found that if the analysis of the bulge emission is limited to latitudes greater than about 3° from the midplane, foreground emission is insufficient to affect their general conclusions.

By analyzing the low extinction emission above and below the plane, Blitz and Spergel were able to make a quantitative assessment of the shape of the bulge. If the bulge is axisymmetric, lines of sight that make the same angle on either side of the center will appear equally bright. On the other hand, if the bulge is elongated and its long axis is at some angle oblique to the line of sight (see Figure 1), perspective effects cause equivalent lines of sight on either side of the center to have systematically different surface brightnesses. For most lines of sight, the near side of the bar will appear to be brighter than the far side because lines of sight on the near side traverse paths with a higher volume density of stars. The near side will also appear to be thicker than the far side. However, close to the center, lines of sight on the far side are actually brighter because the stellar densities sampled are almost the same, but the lines of sight on the far side traverse a longer path of stars. This counterintuitive result requires that the differential surface brightness of the two halves of an obliquely oriented bar show a sign reversal near the center, and that the photometrically determined near side have a larger angular thickness. An axisymmetric bulge, as would be found in a normal (non-barred) galaxy, would show no such effects.

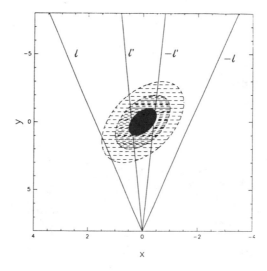

Fig. 1. The orientation of an elliptical distribution of stars projected onto the galactic plane as viewed from the Sun. The different shadings represent different stellar surface densities. Note that close to the center, a line of sight through the bar on the far side actually traverses a longer path length than on the near side.

Figure 2 shows the results of folding the data of Matsumoto *et al.* along the line $l = 0°$ and subtracting the emission at negative longitudes from that at positive longitudes. The figure shows that over most longitudes, the emission is systematicaly brighter at positive longitudes than at negative longitudes, that the relative differential brightness increases with increasing longitude, and that there is a location near $l = 0°$ where the relative surface brightness changes sign. Furthermore, the percentage difference is consistent with the parameters Blitz and Spergel used to define the bar. The emission at positive longitudes (the photometrically determined near side of the bar) also has a larger angular scale height than the far side. All of these effects are consistent with a non-axisymmetric bar-like distribution of stars similar to that shown in Figure 1, and inconsistent with an axisymmetric distribution.

The DIRBE experiment on the COBE satellite has collected a data set similar to that of Matsumoto *et al.* The full COBE data set is, however, much more sensitive, better calibrated, and are also done at several different wavelength bands so that an explicit correction for extinction is possible. The analysis of a small subset of the COBE data is presented elsewhere in this volume (Weiland *et al.*), and the results are found to be consistent with the Blitz and Spergel results. However, where Blitz and Spergel find a small tilt to the the bar, the COBE results suggest that the tilt is due to the effects of uncorrected extinction.

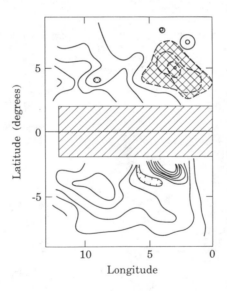

Fig. 2. Absolute difference between negative and positive longitudes of the Matsumoto *et al.* (1982) data. The lowest positive contour and the contour intervals are in steps of $0.3 \times 10^{-10} W cm^{-2} \mu m^{-1} sr^{-1}$. Negative contours are indicated by cross-hatching.

Stars

There have been three published analyses of the distribution of bulge stars. All of them use different subsets of the IRAS point source catalogue, and all find results consistent with the geometry of the bar determined from diffuse light observations.

The largest sample of stars was used in an elegant analysis by Weinberg (1992). He identified 3170 AGB stars in a flux limited sample that is large enough to determine the surface density distribution of the stars in the plane of the galaxy. The stars were selected on the basis of their variability, but they also had to satisfy a color criterion. Weinberg made bolometric and extinction corrections in order to determine apparent magnitudes for the stars, and assumed that the stars then have a well-defined mean bolometric luminosity with zero dispersion. The effects of this assumption were tested with numerical simulations. The main thrust of Weinberg's work borrows from a theorem in quantum mechanics: if a wave function is known for one postion, it is then determined for all space; one need only determine the translation operator to determine the wave function for any other position. Weinberg determined the components of the 'wave function' for the distribution of stars in a Sun-centered coordinate system, and then transposed that distribution to a coordinate system with an origin at the Galactic Center.

The positions of the stars that met his selection criteria were plotted in projection on the galactic plane, and the surface density distribution in a Sun-centered coordinate system was determined. The distribution was then decomposed into a Fourier-Bessel series where the coefficients of the first ten terms were calculated. Weinberg then derived the translation operator and transformed the series into a Galactic Center based coordinate system. Keeping only the even terms of the series, the density distribution of stars is shown in Figure 3. There is a distinct non-axisymmetric component to the distribution that Weinberg identifies with the galactic bar. The long axis of of the distribution has a position angle of 36° pointing to the first galactic quadrant in a manner similar to that implied by the diffuse infrared emission. Weinberg's bar is larger than that inferred from other studies (about 4 kpc semi-major axis compared to 2 kpc), and it is unclear to what this difference might be attributed.

A second study is that of Whitelock and Catchpole (1992) in which suspected Mira variables identified from the IRAS point source catalogue were selected in two strips between $-8° < b < -7°$ and $7° < b < 8°$. The stars were subsequently monitored for several years in the near infrared to confirm or reject the initial identification, and and ultimately a listing with 113 firm identifications was produced. It was found that the the longitude distribution of the Miras is asymmetric with respect to the Galactic Center, with more Miras on the positive longitude side, as would be expected from a bar with the long axis pointing to the first quadrant. In addition, the distance modulus for stars on the positive longitude side of the distribution is distinctly smaller than that on the negative longitude side.

Whitelock and Catchpole then made a model for the stellar distribution in which the density distribution was assumed to be ellipsoidal in three dimensions, and then compared the results projected on the sky to the data. They found a good fit with a 4:1 axis ratio, and a stellar surface density scale height of 212 pc. The angle the bar makes with the Sun-center line is 45°. Their model and data are shown in Figure 4.

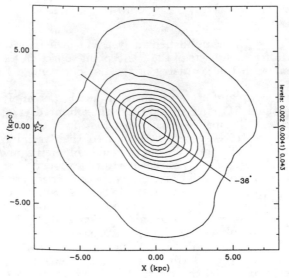

Fig. 3 Contour diagram of the reconstructed density distribution of AGB variables on the galactic center of Weinberg (1992). Only even terms are included.

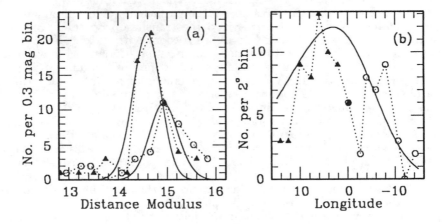

Fig. 4. a) The distribution of IRAS Miras in distance modulus from the work of Whitelock and Catchpole (1992). The open circles and closed triangles represent those with negative and positive longitudes respectively. b) The longitude distribution of the same sources. The solid lines are the best fitting models (prolate ellipsoid).

The third photometric investigation is that of Nakada *et al.* (1991), the first of the stellar analyses to be published. They selected stars from the IRAS point source catalogue with red 25μm/12μm flux ratios, and within a specified 12μm flux range toward the Galactic Center. They limited their sample to $|b| > 3°$ to avoid galactic plane absorption; their sample contained 50 sources. As was true for the Whitelock and Catchpole sample, they find more sources at positive longitudes, and the sources are systematically brighter. Nakada *et al.* concluded that the distribution of sources is consistent with the bar suggested by Blitz and Spergel.

Globular Clusters

The globular cluster system has long been known to have a metal-rich subsystem that is centrally concentrated about the center of the Galaxy. Following a suggestion of Jeff Goldstein of the Air and Space Museum, Figure 5 shows the distribution of globular clusters located within a radius of 4 kpc of the center of the Milky Way. This sample does not take reddening into account, but nonetheless shows some elongation along an axis similar to that found for the diffuse infrared emisison and the distribution of evolved stars. The dashed line in the figure is a fit to the data with a position angle of 25°. The uncertainty in the fit is 8°. Although distance errors would tend to make the distribution appear elongated, distance errors alone would not produce the skewed distribution shown. Furthermore, since most of the clusters are far from the galactic plane, extinction effects should not be a major contributor to the distribution of clusters; which are in any event presumably accounted for in the determination of the cluster distances. Nevertheless, a proper analysis requires a detailed and explicit analysis of the effects of extinction.

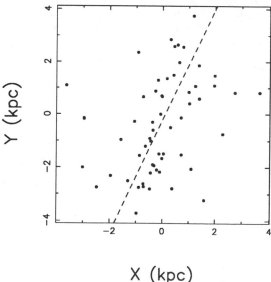

Fig. 5. The distribution of globular clusters within a 4 kpc box centered on the Galactic Center projected on the galactic plane. The Sun is at 0,8. The dashed line is a least squares fit to the data of a line forced to go through the Galactic Center. This figure was prepared by Peter Teuben of the University of Maryland.

ACKNOWLEDGEMENTS

Partial support for this work comes from a grant from the National Science Foundation.

REFERENCES

Balick, B., & Brown, R.L. 1974, ApJ, 194, 265
Blitz, L. & Spergel, D.N. 1991, ApJ, 379, 631
Cohen, R.J. & Few, R.W. 1976, MNRAS, 176, 495
de Vaucouleurs, G. 1964, in *IAU Symposium 20: The Galaxy and the Magellanic Clouds*, eds Kerr, F. J. & Rodgers, A. W., Sydney: Australian Academy of Science, p. 195
Ekers, R.D., & Lynden-Bell, D. 1971, ApLett, 9, 189
Gerhard, O.E., & Vietri, M. 1986, MNRAS 223, 377
Lynden-Bell, D., & Rees, M.J. 1971, MNRAS, 152, 461
Matsumoto, T., *et al.* 1982, in *The Galactic Center*, G. Riegler & R. Blandford, eds., American Institute of Physics:New York, p.48
Nakada, Y., Deguchi, S., Hashimoto, O., Izumiura, H., Onaka, T., Sekiguchi, K., & Yamamura, I. 1991, *Nature*, **353**, 140.
Oort, J.H. 1977, ARAA, 15, 295
Peters, W.L. 1975, ApJ 195, 617
Rougoor, G.W., & Oort, J.H. 1960, Proc. Nat Acad. Sci, 46, 1
Rougoor, G.W. 1964, BAN, 17, 31
Scoville, N.Z. 1972, ApJL, 175, 125
Weinberg, M.D. 1992, ApJ, 384, 81
Whitelock P. & Catchpole, R. 1992, in *The Center, Bulge and Disk of the Milky Way*, L. Blitz, ed. Kluwer:Dordrecht p.103

X-RAY HALO FROM THE GALACTIC BULGE

-THE 6.7 KEV LINE MAPPING-

K. KOYAMA AND S. YAMAUCHI

Department of Physics, Faculty of Science, Kyoto University,
Sakyo-ku, Kyoto, 606-01

ABSTRACT

The 6.7 keV line is the K-shell resonance transition from a Helium like iron (Fe XXV) and is the most intense line in an optically thin cosmic plasma with a temperature ranging from 10^7 to 10^8 K. Therefore optically thin hot plasma in our Galaxy can be selected by the 6.7 keV line. This paper report on the 6.7 keV line map and discuss on the nature of the hot gas components in our Galaxy.

OBSERVATIONS AND RESULTS

Observations were made with the Large Area Counter (LAC) onboard Japanese Astronomical satellite *Ginga*. Since the filed of view of the LAC has a y-axis direction, we can survey the Galactic plane with the $1° \times 2°$ beam size. Although, the LAC is not a all-sky survey instrument, we have covered about 1/3 of the sky during the Ginga life time. The 6.7 keV line intensity map of the sky is given in figure 1(color plate page). From the map we found excess diffuse emission along the galactic inner disk. In particular, the excess shows prominent peak at the Galactic center and is largely extended around it.

a) Overall Distribution of the 6.7 keV Line

Typical X-ray spectra of the diffuse excess emission on the Galactic plane are shown in Plate 10. We found a strong iron line at about 6.7 keV. To determine the iron line intensity, these X-ray spectra were fitted with a model of a thermal bremsstrahlung, with a narrow emission line. This model generally gives acceptable fits except for the regions where the confusion of bright binary X-ray sources is significant. The center energy and equivalent width of the iron line are plotted in figure 2 as a function of the best-fit continuum temperature. Solid lines indicate the theoretical values of thin thermal plasma emission with cosmic abundance in ionization equilibrium.

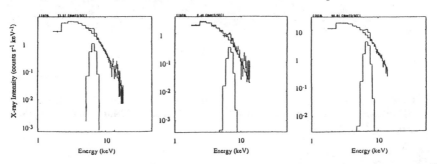

Fig. 2. X-ray spectra from the diffuse Galactic emission at the 4-kpc arm (left), the Galactic bulge (middle) and the Galactic center (right).

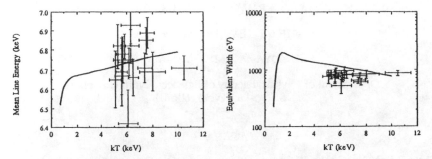

Fig. 3. Line energies(left) and equivalent widths (right) as a function of the best-fit continuum
temperature. Solid lines are the theoretical values of thin thermal plasma emission
with cosmic abundance.

From the best-fit flux of the iron line, we made a contour map of the iron line in
figure 3, where blank regions are due to the limited number of scan paths. For
comparison, the 6.7 keV line distribution along the Galactic plane (Galactic latitude ~ 0°)
is given in figure 4   together with the scan profile in the 1.1 - 18.5 keV energy band.
This figure demonstrates clear difference between the X-ray sky of the 6.7 keV line and
that of the continuum energy band.

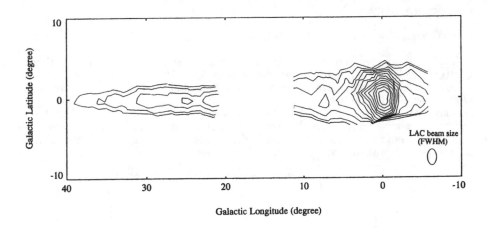

Fig. 4 Two-dimensional distribution of 6.7 keV iron line emission in - 6° ≤ l ≤ 11° and 20° ≤ l ≤
40° regions. The contour levels are 1.0, 1.2, 1.6, 2.0, 2.5, 3.2, 4.0, 5.0, 6.3, 7.9, 10.0,
12.6, and 15.8 photons s^{-1} (LAC beam)$^{-1}$.

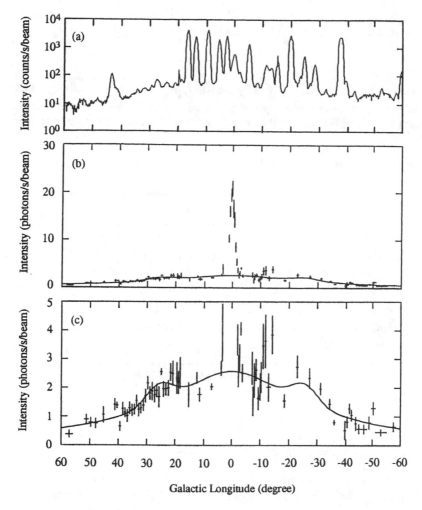

Fig. 5 Intensity distribution of the 1.1 - 18.5 keV band (a) and the 6.7 keV iron line (b)
along the Galactic plane. Enlarged plot of (b) is given in (c). The solid line in Figure
5-(b), (c) is the intensity profile expected from the galactic disk plus "4-kpc arm"
model (see text and Table 1). Vertical Error bars are at 1 σ level.

b) Galactic Plane Emission

To evaluate the 6.7 keV line distribution quantitatively, we modeled the two-dimensional distribution of the 6.7 keV line flux. We assumed that the volume emissivity of 6.7 keV line (η) is given as a function of the galactocentric radius (R) and the distance from the Galactic plane (z). Since the 6.7 keV line distribution shows tail structure to a large longitude, we applied an exponential disk model. Then the data shows systematic excess emission above this model curve at around $l = \pm 30°$. Therefore we added a "Galactic arm" component at this position assuming a Gaussian shape emissivity :

$$\eta(R, z) = \left\{ \eta_d \exp\left(-\frac{R}{R_d}\right) + \eta_a \exp\left(-\frac{(R - R_a)^2}{2w^2}\right) \right\} \times \exp\left(-\frac{|z|}{z_s}\right)$$

The best-fit model profile is given by the solid line in figure 4. Although this model is still not acceptable, improvement of χ^2 value indicates that the arm component is significant with a confidence level of more than 99%. The best-fit model parameters are listed in Table 1.

Table 1 Best-fit parameters for the Galactic plane emission

Parameter	Value
η_d (photons s^{-1} cm^{-3})	$(1.2 \pm 0.1) \times 10^{-21}$
R_d (kpc)	3.6 ± 0.2
η_a (photons s^{-1} cm^{-3})	$(2.6 \pm 0.5) \times 10^{-22}$
R_a (kpc)	4.0 ± 0.2
w (kpc)	0.51 ± 0.12
z_s (kpc)	0.10 ± 0.02
Reduced χ^2 / d.o.f.	1.958 / 126

Note.- Quoted errors are are single-parameter 90 % confidence limits.

The position of the "Galactic arm" is derived to be at 4 kpc from the Galactic center, and hence may be referred to as the "4-kpc arm". This region coincides with the "5-kpc arm" of CO line named when the nominal distance of the Galactic center was 10 kpc instead of the present value of 8.5 kpc.

With this best-fit model, we estimated the 6.7 keV line luminosity's from the Galactic plane to be 6.6 $\times 10^{36}$ ergs s^{-1}. Scale height of the disk emission is about 100 pc.

Since the iron line luminosity is about 10 % of the 2-10 keV luminosity, we can estimate that 2 - 10 keV X-ray band luminosity is about 10^{38} ergs s^{-1}, This value is consistent with the previous observation with *HEAO-A, EXOSAT* and *Tenma* (Worrall et al. 1982; Warwick et al. 1985; Koyama et al. 1986).

c) Galactic Center Emission

We found prominent Galactic center emission which is extending larger than that of the collimator response.

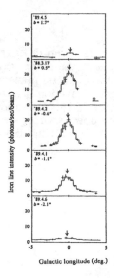
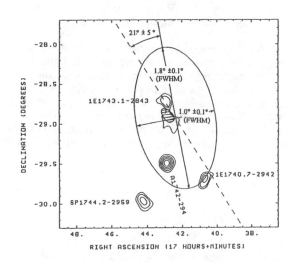

Fig. 6 (right) The Galactic longitude distribution of the iron line intensity. The solid lines show the best-fit curve with the model given in the text. The arrows indicate the peak position (Yamauchi et al. 1990).

Fig. 7 (left) The structure of the Galactic center plasma(solid line) superposed on the SPARTAN 1 point sources(Kawai et al. 1988).

Figure 5 shows scan profiles with different elevation angle from the Galactic plane. Since the intensity peak position shows systematic shift as a function of the elevation angle, the Galactic center emission is not spherically symmetric. We fit the scan profile with the model of two-dimensional Gaussian:

$$S(l, b) = S_{center} \left(\frac{1}{2\pi\sigma_l\,\sigma_b}\right) \exp\left[- \left\{\frac{(x - l_0)^2 + (y - b_0)^2}{2\sigma_l^2\,\sigma_b^2}\right\}\right] \quad \text{(photons s}^{-1}\text{ deg}^{-2})$$

$$x = l \times \cos\theta + b \times \sin\theta$$
$$y = -l \times \sin\theta + b \times \cos\theta$$

The best-fit structure and parameters are shown in figure 6. The total luminosity of iron line is estimated to be 1.1×10^{36} ergs s^{-1}

d) Excess Emissions near the Galactic Bulge

The iron line intensity distribution near the galactic center is given in figure 7 by closed circle as a function of the galactic latitude, while that of the plane region is shown by open circles. The open circles followed well to the Galactic plane model with the scale height of 100 pc. However the Galactic center emission shows clear excess at a large latitude. Here we call this new component as the Galactic bulge emission. Since the bulge region is relatively free from the point source, we found diffuse emissions even with the continuum energy band and found that the excess emission is extending to more than 10° from the Galactic center.

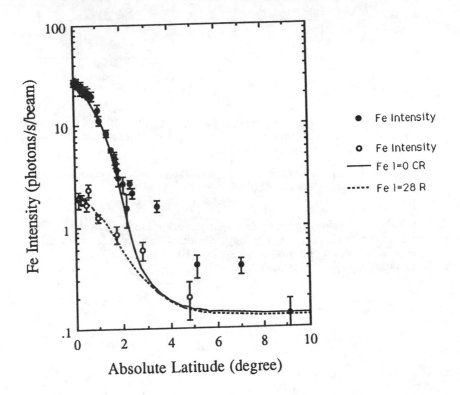

Fig. 8   Iron line flux as a function of the galactic latitude near the galactic center of $l = 0°$ and Galactic arm at $l = 28°$. Solid lines indicate the model distribution of the Galactic center and plane model described in the text.

Although the data coverage of the bulge region is far from complete, we still fit the model with a surface brightness of two-dimensional Gaussian:

$$S(l, b) = S_0 \times \exp\left[-\left(\frac{l}{x}\right)^2 - \left(\frac{b}{y}\right)^2 \right]$$

The overall distribution is large elliptical shape with the best-fit parameters given in Table 2. The total iron line luminosity of the bulge component is estimated to be 2.8×10^{36} ergs s^{-1}.

Table 2 Best-fit parameters for the Galactic bulge emission

Parameter	Value
S_0 (photons s^{-1} cm^{-2} sr^{-1})..	0.55 ± 0.10
x (degree)..	12 ± 2
y (degree)..	5 ± 1
Reduced χ^2 / d.o.f.	$2.436 / 24$

Note.- Quoted errors are single-parameter 90 % confidence limits.

DISCUSSION

The galactic hot plasma is similar to the near infrared map taken with *COBE* satellite. The near infrared map may traces the star distribution. However the scale height of the hot plasma of about 100 pc may be smaller than that of late type stars. If we compare the young object such as SNRs and CO cloud, we find similar distribution for the Galactic plane emission. Thus Koyama, Ikeuchi &Tomisaka (1986) argued that, in the Galactic plane, there would be more than 10^3 unidentified SNRs, which have rather low surface brightness in X-rays as well as in radio bands, and hence they were not detected previously. One example of such SNRs may be RX04591+5197, a new SNR detected with the soft X-ray telescope onboard *ROSAT* (Pfeffermann et al. 1991). This SNR was also detected in the hard X-ray band with the *Ginga* satellite. If we assume that many radio-weak SNRs similar to RX04591+5197 are lying in the Galactic plane, they would be hardly detected as each SNR by radio band but would contribute to the diffuse X-rays and the 6.7 keV lines. Since the soft X-rays from the SNRs in the Galactic plane would be largely absorbed, X-ray spectra of the integrated emission of the SNRs would become harder than unabsorbed spectra.

The Galactic center emission with iron line is more prominent than that of the CO line (Dame et al. 1987) . Also no bulge comments of CO line is reported. Thus new puzzles arise for the origin of the hot plasma at the Galactic center.

Since the *Ginga* field of view is large, whether the extended iron line emission is diffuse or clumpy or integrated point source emission is unclear. Therefore we estimate several physical parameters as a function of filling factor (f).   If X-ray emission is completely diffuse, the filling factor (f) is unity. In this case we obtained drastic results. For example, the Galactic center emission has total energy of about 10^{54} erg, which is 1000 times of typical supernova explosion. If we divide the diameter of the center plasma by the sound velocity, we can estimate that the dynamical time scale of the Galactic center plasma is only 10^5 years. In other word, the Galactic center plasma may be produced by an extraordinary explosion within 10^5 year (Koyama et al. 1989, Yamauchi et al. 1990).   This rather fantastic idea may be supported by the Galactic center 1.8 Mev γ–line emission from ^{26}Al (von Ballmoos et al. 1987).

Since the gas density is very low. the cooling time scale is larger than 10^8 years. Therefore if there were same kind of energetic explosion before 10^5 year , the temperature of hot gas is still larger than the escape velocity of the Galactic gravity. Thus the hot gas would escape from the Galactic center to the Galactic bulge. This would be one possibility for the origin of the Galactic bulge emission.

REFERENCES

Dame, T. M. et al. 1987, ApJ., 322, 706.
Kawai N, et al. 1988, ApJ., 330, 130.
Koyama, K., Makishima, K., Tanaka, Y., & Tsunemi, H. 1986, PASJ, 38, 121.
Koyama, K., Ikeuchi, S., & Tomisaka, K. 1986,   PASJ, 38, 503.
Koyama, K., Awaki, H., Kunieda, H., Takano, S., Tawara, Y., Yamauchi, S.,
    Hatsukade, I., & Nagase, F. 1989, Nature , 339, 603.
Pfeffermann, E., Aschenbach, B., & Predehl, P. 1991, A. & ApJ., 246, L28.
von Ballmoos et al. ApJ. 318, 654.
Warwick, R. S., Turner, M. J. L., Watson, M. G., & Willingale, R.,
    1985 Nature, 317, 218.
Worrall, D. M., Marshall, F. E., Boldt, E. A., & Swank, J. H. 1982, ApJ., 255, 111.
Yamauchi, S., Kawada, M., Koyama, K., Kunieda, H., Tawara, Y., & Hatsukade, I.
    1990, ApJ., 365, 532.

ON THE ROTATION AND PROPER MOTIONS
OF THE BULGE STARS

HongSheng Zhao
Columbia University, Astronomy Department, NY, NY 10027

1. INTRODUCTION

The bulge of the Milky Way is observed to have significant flattening and stellar rotation, especially for the metal rich population. Kent *et al.*(1991) find a 400 pc scale length and a flattening of $\epsilon \equiv 1 - c/a = .4 \pm .2$ for the 2μ light. Similar flattening and scale height are found by other infraed studies (Harmon & Gilmore 1988). Miras in the inner kpc are observed have a linear regression of radial velocity vs longitude, with the slope ω in the range of $7 - 14km/s/deg$ (Catchpole 1990; Menzies 1990) Kinnman *et al.*(1990) find that planetary nebulae rotate in the bulge with a speed $\approx 100km/s$. Minniti *et al.*(1992) find that K giants rotate with a speed of $50km/s$ at a field $(l, b) = (8, 7)$.

If our bulge is, like extragalatic bulges, rotation flattened, then it should satisfy the well known relation between the V_{rot}/σ and the flattening for systems with isotropic intrinsic dispersions (Binney 1978; Komendey & Illingworth 1982). As a rough estimate of the the averaged value of V_{rot}/σ of the whole bulge, one can take the maximum dispersion $\sigma = 120km/s$ near the center of the bulge and the maximum observed rotation speed of the bulge $V_{rot} = 100km/s$. One finds $V_{rot}/\sigma = .8$, which is also that needed to fully explain the flattening of the bulge with $\epsilon = .4$.

In this paper, we discuss an independent constraint on the rotation from the proper motion of the bulge stars, and the metallicity dependence of the bulge rotation.

2. MODEL

Recent proper motion data in Baade's Window show an apparent anisotropy between the two proper motions. Spaenhauer *et al.*(1992) found the dispersion in the plane $\sigma_l = 115 \pm 4(R_0/7.5kpc)km/s$ and $\sigma_b = 100 \pm 4(R_0/7.5kpc)km/s$, with the former 2σ greater than the latter; R_0 is the galactocentric distance.

The apparent anisotropy is due to rotation of the bulge with isotropic intrinsic dispersion. At Baade's Window's position, which is on the minor axis, rotation is in the plane and perpendicular to the line of sight. The proper motion perpendicular to the plane is purely due to random motion, while the observed proper motion in the plane is a line of sight integration of both the random motion and the rotation of the stars, the later broadens the dispersion. So for a minor axis field,

$$\langle \sigma_l^2 \rangle = \langle \sigma_b^2 \rangle + \langle \omega^2 r^2 \rangle, \qquad (1)$$

where quantities in the brackets are density weighted line of sight integrations. Using a Kent density model (Kent *et al.*1990) and substituting in the proper motion dispersions found by Spaenhauer *et al.*(1992), we find the the solid body rotation speed to be $\omega = 8(R_0/7.5kpc)km/s/deg$, which agrees with the values found by direct observation of various bulge tracers, $7 - 14km/s/deg$. Increasing R_0 improves the agreement.

Large surveys of the stars in the bulge which yields both proper motion and abundance, is necessary to see the metallicity dependence of the bulge rotation. Here we show a simple conversion from the observed proper motion anisotropy to the V_{rot}/σ in Fig.1. Under the assumption that the intrinsic dispersion is isotropic, we have

$$V_{rot}/\sigma = \gamma(\sigma_l^2/\sigma_b^2 - 1)^{1/2}, \tag{2}$$

where $\gamma = 1$ if both σ_l/σ_b and V_{rot}/σ are averages in a minor axis field (solid line in Fig.1), or $\gamma = \sqrt{2}$ if both quantities are averages over the whole bulge (dotted line in Fig.1). Bigger proper motion anisotropy implies bigger rotation. The metal rich population both rotates faster and has lower dispersion than the metal poor population, hence it should show bigger proper motion anisotropy. Such a trend is already shown by the error bars in Fig.1, where the metal rich subsample has a bigger proper motion ratio than the whole sample of Spaenhauer *et al.*(1992).

3. DISCUSSION

The recent *COBE* data show that the bulge has a boxy shape and the boxyness is in agreement with the projection of a triaxial bar. Such a bar is also inferred from the non-circular motion of the gas; the corotation radius is at $2.4kpc$, which corresponds to a pattern rotation speed around $10km/s/deg$ (Binney *et al.*1991). If the internal streaming is small, our interpretation of proper motion anisotropy is not affected. If the internal streaming is not small, then proper motion perpendicular to the plane can be broadened by minor axis rotation, and the radial velocity can be broadened by non-circular streaming. Compared with metal poor population, the metal rich component should stream faster and concentrate more towards the plane. A dynamical model based on the Kent density distribution, modified into a triaxial picture, is developed to account for kinematics of different metallicity groups in the bulge (Zhao *et al.*1993).

ACKNOWLEDGEMENTS

The author wishs to thank Mike Rich and James Applegate for discussions which lead to this work.

REFERENCES

Binney, J. 1978, MNRAS, 183, 501
Binney, J., Gerhard, O. E., Stark, A. A., Bally, J.,& Uchida, K. I. 1991, MN-
    RAS, 252, 210
Catchpole, R. M. 1990, in ESO-CTIO Workshop on Bulges of Galaxies, Eds.
    Javis, B. J. & Terndrup, D. M.
Menzies, J. W. 1990, in ESO-CTIO Workshop on Bulges of Galaxies, Eds.
    Javis, B. J. & Terndrup, D. M.
Kent, S. E., Dame, T. M., & Fazio, G. 1991, ApJ, 178, 131
Kinnman, T.D., Feast, M. W., & Lasker, B. M. 1988, A. J., 95, 804
Kormendy,J.,& Illingworth, G. 1982, ApJ, 256, 460
Minniti, D., White, S. D. M. , Olszenski, E., & Hill,J. 1992, ApJ., 393, L47
Spaenhauer, A. , Jones, B. F. ,& Whitford, A. E. 1992, AJ, 103, 297
Wyse, R. F. G.,& Gilmore, G. 1989 Comments on Astrophysics, 13, 125
Zhao, H. S., Rich, R. M., & Applegate, J. H. 1993, in preparation

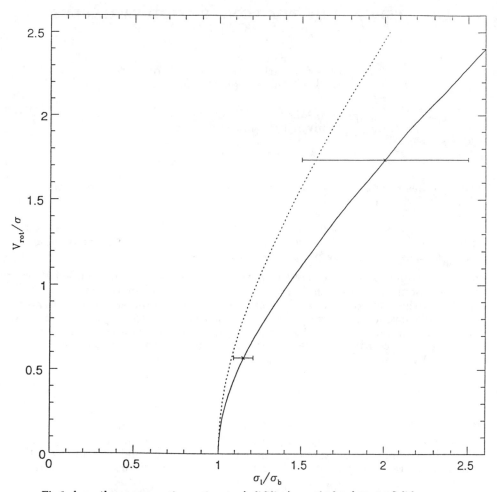

Fig.1 shows the proper motion ratio σ_l/σ_b (solid line), vs. the local V_{rot}/σ. Solid line is for the case that both σ_l/σ_b and V_{rot}/σ are averages in a minor axis field, and the dotted line is for the case that both quantities are averages over the whole bulge. The error bars are for the whole sample and the metal rich subsample of Spaenhauer et al.(1992). Compared with the metal poor population, the metal rich population both rotates faster and has lower dispersion, hence it should show bigger proper motion anisotropy.

THE RESPONSE OF THE LONG-PERIOD VARIABLES
TO A GALACTIC BAR

Kevin Long
University of Massachusetts; SUNY Brockport
Email: kevin@falcon.phast.umass.edu

Martin D. Weinberg
University of Massachusetts

ABSTRACT

Weinberg has recently detected a large-scale quadrupole bar in the spatial distribution of the long-period variables. To understand the underlying potential producing this feature, we have simulated the dynamical response of the old Galactic disk to several models of a Galactic bar. In each of our trial potentials, we integrated the orbits of an ensemble of test particles having initial kinematics similar to those of the long-period variables.

From each simulation, we selected a spatially-limited sample and applied Weinberg's harmonic analysis technique to reconstruct the multipole moments of the density; in this way, our simulations can be directly compared to the observed feature. We have found several barred potentials that can roughly reproduce the detected quadrupole feature. To further constrain the parameters of a Galactic bar, we plan to continue testing the successful bar potentials against observed stellar and gas kinematics and structural morphology.

1. GALACTIC POTENTIAL

We used a three-component model of the axisymmetric potential, consisting of a Miyamoto-Nagai (1975) disk, a Hernquist (1990) spheroid, and logarithmic halo. The component parameters were chosen to approximately reproduce the Bahcall-Soniera (1980) model. To represent the bar, we used a quadrupole potential

$$\Phi(r, \theta) = \frac{fGQ_0 r^2}{[r_b{}^2 + r^2]^{5/2}} P_2(\cos \theta),$$

where θ is the angle from the long axis of the bar. The bar strength was specified as a fraction f of the characteristic quadrupole moment $Q_0 = \frac{v_c(r_b)^2 r_b{}^3}{G}$.

2. SIMULATIONS

The test particles representing the LPVs were initially distributed axisymmetrically. A preliminary kinematic model was generated, in which the local velocity ellipsoid was used everywhere. A consistent kinematic model was then generated by allowing the preliminary model to phase mix in the axisymmetric potential.

We integrated orbits in several models with different final quadrupole strengths

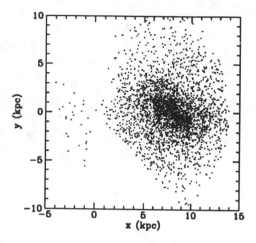

Figure 1. Observable sample of test particles from our simulation having bar length $r_b = 3$ and bar strength $f = 0.4$, projected onto the galactic plane. The origin is the position of the Sun. Stars beyond 14 kpc are excluded due to the flux limit of Weinberg's sample; the wedges devoid of stars are longitude ranges in which variables could not be identified reliably.

f and bar lengths r_b. The bars were grown to their maximum strengths over a timescale of one bar rotation period. The corotation radius was $r_c = 1.5r_b$ for all runs.

3. ANALYSIS

From each ensemble of test particles, we derived a spatially-limited sample viewed from the vantage point of the Sun. The bar position angle was assumed to be 35°. The sample from the $r_b = 3.0$, $f = 0.4$ run is shown in Figure 1. The density distribution was reconstructed with Weinberg's (1992) harmonic analysis technique. In Figure 2 we compare the reconstructed $m = 2$ density components of our simulated bar to Weinberg's bar.

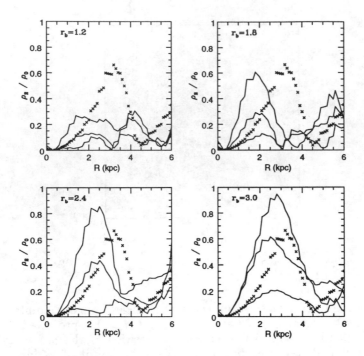

Figure 2. Quadrupole density components, normalized to the axisymmetric component, *vs.* galactocentric radius for our simulations and for Weinberg's sample. Each graph shows the results of simulations with a single bar length, indicated on the figure. The different solid curves are for different bar strengths $f = 0.4, 0.8, 1.0$; the crosses show Weinberg's model. The increase beyond 5 kpc is spurious; it is due to the small number of stars at large radii and the distance limit of the sample. No error bars are shown because of the difficulty in estimating errors in Weinberg's method.

4. CONCLUSIONS

Our models with $r_b = 2.4$ and 3.0, $f = 0.4$ and 0.8 roughly reproduce the magnitude and length of the observed bar.

Interestingly, we were unable to reproduce the observed bar with a small-scale (r_b=1.2 kpc), rapidly rotating bar potential. Our small-scale potential does cause a large-scale "anti-bar" outside corotation, but its amplitude is too small to match the observations.

We plan to try additional bar parameters, and to continue testing our successful models against other observational data, such as gas kinematics.

ACKNOWLEDGEMENTS

This work was supported by NASA grant NAG 5-1999 and by startup funds from the State University of New York.

REFERENCES

Bahcall, J. and Soniera, R. 1980, Apj.Suppl, 44, 73.
Hernquist, L 1990, ApJ., 356, 359.
Miyamoto, M., and Nagai, R. 1975, PASJ, 27, 533.
Weinberg, M.D. 1992. ApJ, 384, 81.

GALACTIC ELLIPTICAL STREAMLINES AND THEIR LARGE-SCALE ORGANIZATION

K. Rohlfs, H. Kampmann
Astronomical Institute of the Ruhr University,
Postfach 102148, W-4630 Bochum, Germany

Email ID
astrorub@ruba.rz.ruhr-uni-bochum.dbp.de

ABSTRACT

Interstellar gas moving in elliptical streamlines is used to model the observed terminal velocities of neutral hydrogen gas in the I. and IV. galactic quadrant. The excentricity of the orbits decreases with increasing r, and they are well aligned with a position angle of 40^0 for $r \leq 3$ kpc indicating the existence of a bar in that region. Plotting the orbits in the galactic plane spiral-arm-like features are marked by orbital crowding effects. This spiral arm system is 4-armed for $3 \leq r \leq 6$ kpc, and two-armed outside of this and it agrees reasonably well in position with the spiral structure outlined by Georgelin and Georgelin (1976). The strong peak of the galactic rotation curve found amongst others by Rohlfs and Kreitschmann (1987) near $r = 500$ pc is seen to be caused by the alignment of the elliptical orbits and not to be a feature of the dynamics.

1. THE MODEL

The large-scale kinematics of the galactic interstellar gas is reproduced to within 20 km/s by galactic rotation in circular orbits. Observation of the terminal velocity in 21-cm line profiles show, however, systematic differences for the $|v_{term}|$ at the co-longitudes, l and $360^0 - l$, $0^0 \leq l \leq 90^0$ which are not compatible with such circular orbits. These velocity differences are coherent over a wide range of longitudes and therefore must be caused by some large-scale features of the gas motion.

Orbits of the gas particles therefore will be non-circular, and assuming them to be non-intersecting in order to avoid excessive shocks caused by cloud collisions, streaming in closed, concentric elliptical orbits is the most simple generalization that comes into mind. Indeed do the computations by Contopoulos and Papayannopoulos (1980) or Binney and Tremaine (1987) for flat, non-axisymmetric systems show the existence of two families of elongated, non-intersecting closed orbits (x_1 and x_2 orbit family). Another welcome feature of these orbits is that the angular momentum varies only by a few percent along any of them.

Observational data for the terminal velocity of galactic disk neutral hydrogen gas has been obtained by Rohlfs and Kreitschmann (1987) by fitting of synthetic line shapes to the high velocity part of the observed line profiles, and these data have been augmented by additional 21-cm line measurements obtained with the 64-m telescope at Parkes, Australia for the longitude range $350^0 \leq l \leq 10^0$ in 1987. This data set then was reproduced by fitting elliptical orbits with the parameters a, b – major and minor axis, α – position angle of major axis, and h – the angular momentum appropiate for the orbit. We varied a, b and α until the orbital radial velocities for the longitudes l and $360^0 - l$ approximated the observed $v_{term}(l)$ and $v_{term}(360^0 - l)$ within close limits. In this the angular momentum h was chosen to be equal to that of a circular orbit with the radius $\frac{1}{2}(a + b)$.

2. RESULTS

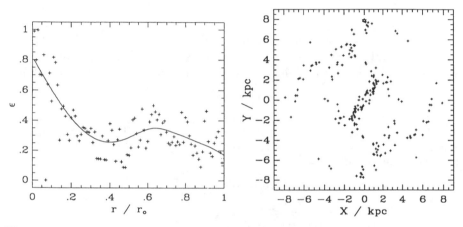

Fig. 1: Excentricity ϵ of the orbits as a function of the distance from the galactic center

Fig. 2: Apogalactic points in the galactic plane. The orbits for $r < 3\,\mathrm{kpc}$ are aligned at position angle $\alpha \simeq 40^0$ indicating a bar

The parameters of these orbits show a systematic variation with the galactic radius; Fig. 1 gives their excentricity $\epsilon = \sqrt{1 - b^2/a^2}$, and Fig. 2 plots the positions of the apogalactic points in the galactic plane. The inner orbits with $r/r_0 < 0.35$ are well aligned with a position angle of about $\alpha \simeq 40^0$ and they seem to indicate the existence of a bar reaching roughly to $3\,\mathrm{kpc}$.

If the large-scale trend of the orbital parameters is extracted from Fig. 1 and Fig. 2 by fitting smoothing splines to ϵ and α the observed values for v_{term} are well reproduced (Fig. 3); the resulting orbital angular momentum is shown in Fig. 4. This diagram gives the orbital velocity at the apogalactic point of each orbit. It is interesting that the strong peak and the sharp decline of the galactic rotation velocity for $400 \leq r \leq 1000\,\mathrm{pc}$ as found by Rohlfs and Kreitschmann (1987) for circular orbits is not present here. The large radial velocity extremes visible in the gas are not caused by large values of the galactic radial force-field but are caused by the orientation of strongly elliptical streamlines for the gas.

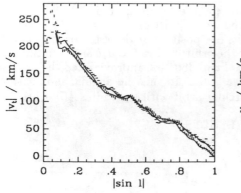

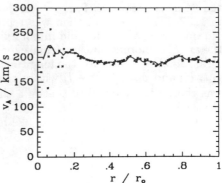

Fig. 3: The observed terminal velocity (ı : I. quadrant, – : IV. quadrant) and its reproduction by the elliptical orbit model

Fig. 4: The velocity at the apogalactic points in the elliptical orbits, the full drawn curve is derived from the interpolating spline of ϵ and α

Another interesting feature of the large-scale organization of the elliptical orbits becomes visible if the orbits themselves are plotted (Fig. 5). Orbital crowding mark two strong spiral arms in the range $6 \leq r \leq 8\,\text{kpc}$ while in the inner 6 kpc four arms are indicated.

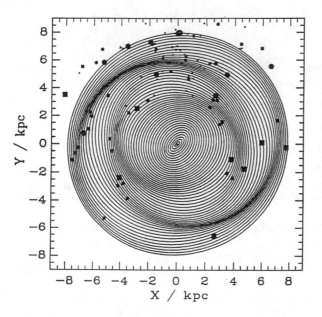

Fig. 5: Elliptical streaming orbits in the galactic plane. The points are positions of large galactic HII regions as determined by Georgelin and Georgelin (1976). The lsr is located at X = 0 kpc, Y = 7.9 kpc.

Finally, although only the high-velocity part of the observed line profiles has been used to determine the velocity field, this field can be used to compute the shape of the full line profile for the gas in the inner Galaxy (provided the gas density distribution is given). Again, the velocity field derived from elliptical orbits result in a much improved shape of the synthetic line profile.

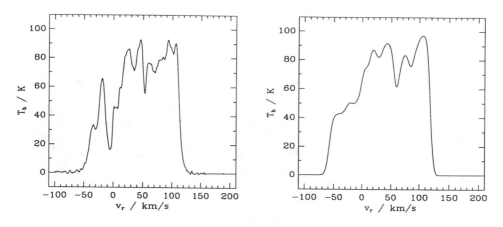

Fig. 6: Observed and computed 21-cm line profile for $l = 30^0$

3. CONCLUSIONS

Just as the kinematics of galactic rotation in circular orbits served as the basis for axisymmetric galactic mass models, we believe that the gas streaming in elliptical orbits can faciliate the construction of non-axisymmetric galactic models, in which a central bar is a major feature. At least some of the general properties for the orbits should be represented by the elliptical orbits used here. But obviously a dynamical model together with detailed orbit computations will be needed if a satisfactory explanation of the observations can be generally accepted.

REFERENCES

Binney, J., & Tremaine, S. 1987, Galactic Dynamics, (Princeton : Princeton University Press)
Contopoulos, G., & Papayannapoulos, T. 1980, A&A, 92, 33
Georgelin, Y.M., & Georgelin, Y.P. 1976, A&A, 49, 57
Rohlfs, K., & Kreitschmann, J. 1987, A&A, 178, 95

SPIRAL DENSITY WAVES RESONANTLY EXCITED
BY A RAPIDLY ROTATING BAR

Chi Yuan

Physics Department, City College of New York, New York, NY 10031

Email ID
yuan@sci.ccny.cuny.edu

ABSTRACT

Recent observations at millimter wavelengths have revealed bar-spiral structures in the central regions of several nearby spiral galaxies. Moreover, new
infarred data and even COBE data seem to support the idea that there is a bar
in the central regions of the Milky Way. All of these new findings have rekindled
the interest of bar-driven spirals again. In this short report, we first describe a
bar-driving mechanism which, we believe, is responsible for generating the spiral
waves and then present new results from an improved non-linear theory. They
comfirm our previous conclusion that a minor oval distortion or an uneven distribution of mass, rotating at a typical speed in the center can excite a nonlinear
wave in close resemblance to the observed 3-kpc arm.

1. INTRODUCTION

Recent analyses of infrared observations suggest that there exists a bar
in the central regions of the Milky Way (Blitz and Spergel 1991; Weinberg
1991). COBE data reported in this conference also seem to support this picture
(Wieland 1992). Furthermore, high resolution observations at millimeter wavelengths have also revealed the existence of bar-spiral structure in the center of
several nearby galaxies (e.g.Planesas et al.1991; Turner & Hurt 1992). These
findings have rekindled the interest in the bar-driven spirals, in particular, in
connection to the 3-kpc arm.

In this report, I shall give a brief account of the past development on
this subject. The emphasis will be placed on the resonance excitation mechanism, which, we believe, is reponsible for generating this bar-spiral phenomenon.
Based on this idea, a non-linear theory has been developed, in which the effects
of self- gravitation and viscosity of the gas are fully included. Some new results
of the theory, in application to the 3-kpc arm, will be presented. We confirm
the previous conclusion (Yuan 1984; Yuan & Cheng 1988) that a minor oval
distortion in the center, turning at a typical central angular speed, can excite a
spiral wave which matches both the observed shape and kinematics of the 3-kpc
arm.

2. RESONANCE EXCITATION

The idea of exciting of spiral density waves through a resonance mechanism
is not new. The idea has been studied and used successfully to describe the wavy
structure of Saturn's rings and the formation of the gaps in the rings (Goldreich

126   © 1993 American Institute of Physics

and Tremaine 1978). In that case, the waves are excited by a satellite of Saturn. The satellite provides the periodic forcing, to which ring particles at the inner Lindblad resonances will respond resonantly. As a result, long trailing density waves are generated and propagate outward in the rings, carrying excess angular momentum (negative) with them. By removing angular momentum from the disk to the satellite, a gap will eventually form at a resonance and the spiral waves will appear near the edge of the ring disk exterior to the gap. This indeed is exactly what has been observed. (See Lissauer 1990)

The same mechanism also works for the bar-driving case in a disk galaxy. Now the gravitational field associated with the oval distortion will play the role of the satellite, exerting periodic force on the gaseous disk of the Galaxy. In response to this periodic forcing, the gas at the outer Lindblad resonance is excited and organized into long trailing waves, which will propagate toward the center. Unlike the Saturn's rings problem, the gas in the galactic disk, which has a relatively high value of Q, the Toomre's stability parameter, will not allow the long trailing waves to travel any significant distance inward before sending them out as reflected short trailing waves at the Q-barrier. The outgoing short trailing waves are the manefestation of the 3-kpc arm or the spirals next to the bar observed by radio astronomers. It should be pointed out that the amplitude of the waves will be attenuated quickly by viscosity in the gas cloud medium as they move out. Usually, the waves cannot go much beyond one wavelength. This is about 1.5 kpc for the Milky Way. Therefore, the 3-kpc arm will not interfere with the grand design spiral structure outside.

3. THE 3-KPC ARM

The 3-kpc arm is an arm-like feature in the central region of the Milky Way, with a spectacular 53 km/sec expansion velocity, discovered in the 21-cm line obervations more than thirty years ago (Rougour and Oort 1960). Several theories were proposed. Most of them are based on an expulsion hypothesis (van der Kruit 1971; Sanders and Pendergast 1974). These theories all have problems with the total energy needed in the explosions (expected to be recurrent) and the replenishment of mass loss in gas in the central regions. A theory which is free from these shortcomings was proposed by Yuan (1984), in which the existence of a minor oval distortion in the galactic central regions is assumed. It is shown that an oval distortion, rotating at a typical angular speed near the galactic center and providing a bar field about 5% of the mean field at a distance of 3 kpc from the center, can excite a pair of spiral density waves, in close resemblance to the observed 3-kpc arm, both in physical shape and in kinematics.

The fact that the gaseous disk can respond to a bar potential in forming spiral waves was first found by Sanders and Huntley (1976) in their numerical simulation of gas flows in the galactic disk, using a hydrodynamical code. But, the physical mechanism behind these results was not understood until the work of Yuan (1984). Yuan's work too has some limitations. His approach cannot be easily extended to include the effects of self-gravitation and viscosity. This difficulty was resolved by adopting another non- linear approach formulated by Shu, Yuan and Lissauer (1985) in treating the non-linear waves in Saturn's rings. In this approach, asymptotic approximation in the limit of tightly wound spiral pattern was used to reduce the problem from two dimensions to one dimension. The Poison equation is solved, in this approximation, such that the spiral potential is expressed as a singular integral of the surface density. Thus, the governing equation becomes a second order integro-differential equation. This complicated equation is further simplified with heurestic, yet physically sound argument, into pure differential form and solved with straight-forward numerical integration (Yuan and Cheng 1989;1991).

The results of the improved theory comfirm and strengthen our earlier investigations. The physical parameters adopted in the calculation are essentially the same as in Yuan (1984), except the perturbation field strength is only a half of the previous value. Thus, the oval distortion provides a field just about 5% of the mean field near the outer Lindblad resonance, which is taken at the galacto-centric distance of 3 kpc. This oval distortion rotates at pattern speed equal to 118 km/sec-kpc, corresponding to a co-rotation radius of 1.6 kpc, with all scales adopted from the 1965 Schmidt model. The randam veocity of the gas is taken to be 10 km/sec, and the kinematic viscosity is 0.4 kpc-km/sec. The surface density of the gas in the unperturbed state is 5.4 $M_\odot/pc^2$. The excited spiral density waves are shown in the left frame of Figure 1, in which the surface density in contour and the radial expansion velocity in arrow are plotted. The maximum radial velocity in this case is equal to 53 km/sec, and we choose the orientation such that it points towards the sun. Such an orientation requires that the major axis of the oval distortion lies in the first and third quadrants (Yuan and Cheng 1991). Later, this orientation was found in general agreement with the results of Blitz and Spergel (1991).

The 3-kpc arm, however, doesn't have a counterpart on the other side of the Milky Way, as our theoretical model might suggest. There is the so-called 135 km/sec arm in the fourth quadrant. Its velocity is too large and its gas mass is too small to be regarded as the other spiral arm of the two-arm system. Furthermore, its distance may be a lot closer to the center than the 3-kpc arm (Rougoor and Oort 1960). Naturally, we may ask: Can a one-arm spiral pattern be excited with the same mechanism? The answer is yes. An uneven distribution of mass in the center may result in a one-arm spiral wave at 3 kpc from the center. For this case, we need a pattern speed of 169 km/sec-kpc, corresponding to corotation at 1.2 kpc. The same perturbation field is also required to ensure an expansion velocity equal to 53 km/sec. The result is shown on the right side of Figure 1. The mass concentration of the uneven distribution, in this case, is located in the fourth quadrant.

4. CONCLUDING REMARKS

Recent high resolution observations of nearby galaxies offer a opportunity to test the resonance excitation mechanism. A galaxy with a minor oval distortion (or an uneven distribution of mass) in the central regions, may be expected to display two-arm (or one-arm) spiral pattern in gas. The gas in the spiral arms is highly compressed. In the case of isothermal gas, the compression can be as high as 60 to 1, whereas in the case of polytropic gas, with the index equal to 5/3, the compression is about 10 to 1, in our calculations of the 3-kpc arm problem. In such a region of high compression, we expect to find concentration of molecular clouds. This is indeed the case observed in the center of several external galaxies (e.g. Turner & Hurt 1992). Since the gas associated with the spiral arms is streaming out with high radial velocity, it would be extremely interesting to look for such radial streaming motions along the minor axis in the high velocity resolution observations in the future.

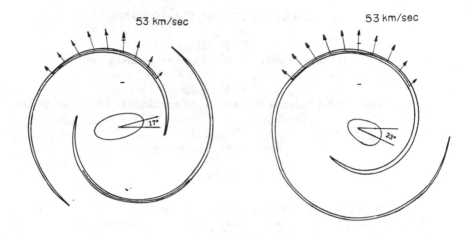

Fig. 1. Contour maps for surface density.

ACKNOWLEDGEMENTS

This work is supported in part by NASA-Ames Unversity Consortium grant NCA2-695 and PSC-CUNY Award 663372. Because of death in the family, the paper was not duely presented. I appreciate the organizers of the conference to allow me to present it in this written version.

REFERENCES

Blitz, L. & Spergel, D. 1991, ApJ, 370, 205
Goldreich, P. & Tremaine, S. 1978, Icarus, 34,227
Lissauer, J. J. 1990, in Dynamics of Astrophysical Discs, ed. J.A. Sellwood
    (Cambridge:Cambridge University Press)
Planesas, P., Scoville, N. Z. & Myers, S.T. 1991, ApJ, 369,364
Rougoor, G. W., & Oort, J. H. 1960, Proc. Nat. Aca. Sci., 46,1
Sanders, R. H. & Huntley, J. M. 1976, ApJ, 209, 54
Sanders, R. H. & Pendergast, K. H. 1974, ApJ, 188, 489
Shu, F. H., Yuan, C. & Lissauer, J. 1985, ApJ, 291, 456
Turner, J. L. & Hurt, R. L. 1992, ApJ, 384, 72
van der Kruit, P. C. 1971, AA, 13, 405
Weinberg, M. 1992, ApJ, 384, 81
Wieland, J. 1992 in this volumer
Yuan, C. 1984, ApJ, 281, 600
Yuan, C. & Cheng, Y. 1988, in The Outer Galaxy, Ed. L. Blitz & J. Lockman
    (Berlin: Sringer), p.144
Yuan, C. $ Cheng, Y. 1989, ApJ, 340, 201
Yuan, C. & Cheng, Y. 1991, ApJ, 376, 104

GALACTIC OSCILLATIONS

R. H. Miller
Astronomy Center, 5640 Ellis Ave., Chicago 60637

B. F. Smith
MS 245-3, NASA Ames Research Center, Moffett Field, CA 94035

Email ID
smith@pan.arc.nasa.gov

ABSTRACT

Several oscillations which have been identified in spherical galaxy models are likely to be present in our Galaxy. These are normal mode oscillations in a stable system. Each has its own distinct period and spatial form, and each rings without detectable damping through a Hubble time. The most important are: (1) a simple radial pulsation (fundamental mode), in which all parts of the galaxy move inward or outward with the same phase; (2) a second spherically symmetrical radial mode with one node, so material inside the node moves outward when material outside moves inward.

Numerical experiments suggest that normal mode oscillations like these may be present in nearly all galaxies at a considerably higher amplitude than has previously been thought. Amplitudes (variations in total KE, for example) can be as large as 10 percent of equilibrium values, and periods are around $50 - 300$ Myrs in a typical galaxy. This is long enough for gas trapped near the center to cool during an oscillation cycle, triggering star formation activity. Second mode oscillations could drive galactic bulges through bursts of star formation.

Possible observational consequences of these modes for our Galaxy are discussed.

1. DISTURBANCES

Fairly large systematic motions near the centers of galaxies reported earlier (Miller and Smith 1988, 1992) can produce effects often seen in the observations. The 50 km/sec velocity difference between galaxy and nucleus recently reported in $M87$ (Jarvis and Peletier 1991) is one such effect. Off center nuclei are another. Bajaja *et al.* (1984) reported that "the center of the HI distribution lies about 300 pc westward of the coincident optical and radio continuum nuclei" in the Sombrero galaxy, NGC 4594. They commented that this asymmetry is smaller than that which they usually find in spiral galaxies. Other examples are given in Miller and Smith (1992). These center motions are local disturbances as the nucleus orbits around the galaxy's mass centroid. They naturally raise the question whether global motions that affect the galaxy as a whole are also present and whether these might be found experimentally. Galactic oscillations were found as we pursued that question.

The fundamental mode shows up as a continuing oscillation in the total kinetic energy with a period around 300 Million years. Large amplitude oscillations (22% peak-to-peak fractional variations in the kinetic energy) showed no detectable damping over a Hubble time. A 5% decrease in amplitude could have been detected easily, but none was seen. This sets a lower limit to the damping time at around 100 billion years.

The second radial mode is spherically symmetrical with a node, and its period ranges from 100 − 150 million years (it is *not* commensurate with the fundamental mode). The node is a few kpc from the center, and the mode is strongest in the inner parts of the galaxy. The second radial mode drives variations in the central density as large as a factor of three in some of our experiments. This mode was identified as an instability by Hénon (1973), but here it is a normal mode in a stable galaxy.

Two other modes were identified as well, but they are not as important.

2. APPLICATION TO OUR GALAXY

Consider the disk component of the Galaxy as orbiting within a spherical dark halo. Disk material is carried along with normal mode oscillations in that halo.

2.1 The Fundamental Mode

The fundamental mode, in this picture, would be most apparent observationally in two different ways. The magnitude and sign of the effects depend both on the strength of the oscillation and on its phase at the present time.

The first of these effects is an inward or outward motion of the local standard of rest with respect to the Galactic center. Based on our experiments, we might anticipate a velocity as large as 10 − 15 km/sec. Observations bearing on this point give ambiguous results: see *e.g.* Blitz and Spergel (1991) or Kuijken and Tremaine (1991). Velocities inferred from HI absorption in the direction of the Galactic center (Radhakrishnan and Sarma 1980) are also ambiguous because HI densities are low over nearly half the path to the center.

The second effect is that the mode produces non-circular velocities in the Galaxy. Departures from circular motions would be greater in outer parts of the Galaxy. Burton (1988) noted some evidence for non-circular motions. There is also a weak positive velocity signal reaching to 40 km/sec essentially in the plane in the (v, b) diagram showing the anticenter region $b = 180°$ in Burton's (1988) Fig. 7.2. Both of these could be taken as evidence for the fundamental mode. The positive velocity implies that the present phase of the fundamental mode oscillation has outward-directed velocities (expanding). A signal with strong negative velocity, slightly above the plane in that same figure, is a high-velocity cloud.

The fundamental mode can be recognized in external galaxies because it distorts velocity maps to make them look warped. It would be interpreted as a kinematic warp. Unfortunately, that test cannot be applied to our Galaxy.

2.2 The Second Radial Mode

It is tempting to associate galactic bulges with the second radial mode. Gas from the interstellar medium, compressed during the compressive phase of that mode, can lead to a burst of star formation. This is nice because it leaves a footprint without requiring that the oscillation be observed at any particular phase. The association is again based on properties of external galaxies: this time it is the large variations of bulge size and of bulge luminosity within galaxies of the same Hubble type. It also accounts for rapid rotation of bulges (unlike the elliptical galaxies), the confinement of the bulge near the center (inside the node of this mode), and for different physical properties within the bulge, such as a hole or a depression in HI densities and the presence of hot gases.

The bulge (or spheroid) of our Galaxy also seems to be pretty much confined within the 3 − 4 kpc radius of an HI hole, with stars of different kinematical properties and abundances, and so on. A bulge might be interpreted as observational evidence for the second radial mode. This is clearly a topic that needs further study.

2.3 Center Motions

Center motions can also cause a radial velocity or a proper motion of the Galactic center. Errors assigned to the very small proper motion in the observations of Backer and Sramek (1987) still allow a velocity on the order of 25 km/sec in the plane of the sky. The object associated with the Galactic center may well be orbiting the Galaxy's mass centroid at a distance of some tens of parsecs and with velocites on the order of 25 − 40 km/sec, if evidence from external galaxies is applied to our Galaxy.

REFERENCES

Backer, D. C., & Sramek, R. A. 1987, in The Galactic Center, Proceedings of a Symposium Honoring C. H. Townes, AIP Conference Proceedings 155, ed. D. C. Backer (New York: American Institute of Physics) p. 163.

Bajaja, E., van der Berg, G., Faber, S. M., Gallagher, J. S., Knapp, G. R., & Shane, W. W. 1984, A&A, 141, 309.

Blitz, L., & Spergel, D. N. 1991, ApJ, 370, 205, Fiche 49-F11.

Burton, W. B. 1988, in Galactic and Extragalactic Radio Astronomy, eds. G. L. Verschuur & K. I. Kellerman (New York: Springer Verlag) p. 295.

Hénon, M. 1973, A&A, 24, 229.

Jarvis, B. J., & Peletier, R. F. 1991. A&A, 247, 313.

Kuijken, K., & Tremaine, S. 1991, in Dynamics of Disk Galaxies, ed. Sundelius, B. (Göteborg Sweden: Department of Astronomy, Göteborgs University) p. 71.

Miller, R. H. & Smith, B. F. 1988, in Applied Mathematics, Fluid Mechanics, Astrophysics, a Symposium to Honor C. C. Lin, eds. D. J. Benney, F. H. Shu, & Chi Yuan (Singapore: World Scientific) p. 366.

Miller, R. H. & Smith, B. F. 1992. ApJ, 393, 508, Fiche 122–B7 (includes a videotape).

Radhakrishnan, V. & Sarma, N. V. G. 1980, A&A, 85, 249.

A BAR MODEL FOR THE GALACTIC BULGE

J A SELLWOOD

Department of Physics and Astronomy, Rutgers University
P O Box 849, Piscataway, NJ 08855-0849

ABSTRACT

Bars in galaxies acquire a pronounced peanut shape when seen from the side. It is shown that the peanut-shaped bar in a 3-D N-body model has an appearance closely resembling the COBE image of the Milky Way bulge, when viewed from within the disc. Moreover, the projected velocity distribution is not far from isotropic, in agreement with modern kinematic data. Although the model is not a perfect match to the Milky Way, it exhibits a straightforward mechanism for the formation of the observed shape of the Milky Way "bulge" and provides a further argument in favour of a bar in the Galaxy.

1. FORMATION OF THE N-BODY BAR

A number of 3-D N-body simulations of barred disc galaxies have been reported in recent years (Combes & Sanders 1981, Combes *et al.* 1990 and Raha *et al.* 1991) in which the bar *always* acquired a box-peanut shape when viewed from the side. Here, I compare a model which exhibits this behaviour with the COBE surface photometry and recent kinematic data from the Milky Way bulge.

The initial bar unstable model contains 50K particles chosen so as to create a disc with a Kuz'min-Toomre density profile in the plane and to have a small but finite thickness. This component contains 70% of the total mass, the remaining 30% is in a rigid Plummer sphere which has half the scale length of that of the disc. (Raha *et al.* showed that the behaviour studied here is not greatly affected by this rigid bulge/halo approximation.) The disc particles were given initial velocities to create an equilibrium $Q = 1.2$ disc. The model was evolved using a 3-D Cartesian particle-mesh code having $127^2 \times 31$ cubic grid cells (double the resolution in each dimension of that used by Raha *et al.*).

Figure 1 shows that the model undergoes a normal bar instability, as expected, which saturates before time 80, though the disc continues to display weak spiral activity and the bar to intensify further up to time 120. The initial bar rotation period is 20 time units, but slows to about 26 units by the end. The bar gradually bends out of the plane, reaching a maximum around time 140. After this time, the bar abruptly regains symmetry about the disc plane, but is significantly fatter normal to the plane and noticeably shorter. The z-thickness of the bar is greatest at its ends, giving it a pronounced peanut shape when viewed from the side. The thickness of the disc at larger radii is not much affected by the bending of the bar, however. This new model confirms much of the behaviour seen previously using codes of lower spatial resolution. The peanut shape of the object is visible from viewing angles in the plane which are greater than about 30° to the bar major axis, but at even smaller angles, the object still appears boxy (*e.g.* time 160).

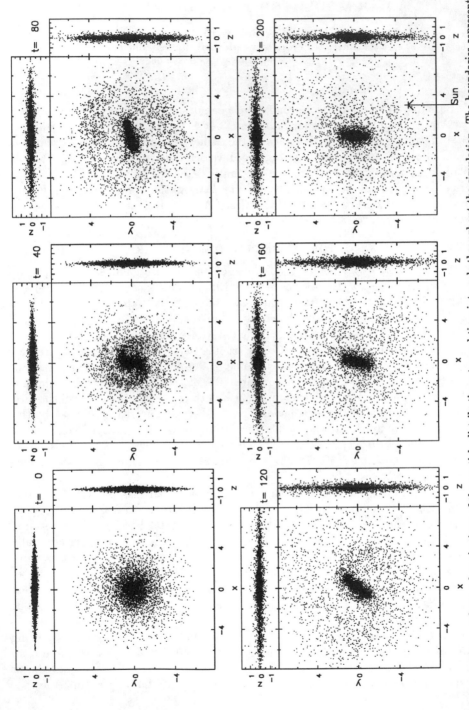

Figure 1 Three orthogonal projections of the particle distribution at equal time intervals throughout the simulation. The boundaries represent the grid edges and the length units are those of the KT disc. Notice the formation of the peanut shape sometime after the bar has formed.

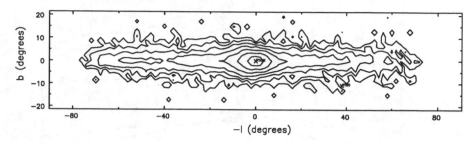

Figure 2 The projected surface brightness as "observed" towards the inner galaxy from the position marked in the last frame of Figure 1. The contours levels are spaced by factors of 3.

2. PROJECTED DENSITY FROM WITHIN THE DISC

Figure 2 shows contours of "surface brightness" towards the inner galaxy from the point marked in the last frame. The particles were projected into a raster of 65 × 15 bins and weighted by the inverse square of their distance from the observer, except those closer than half the distance to the centre were discarded. The viewing angle is 30° to the bar major axis; if this angle is increased, the bulge acquires too much of a peanut shape to be consistent with the extinction corrected COBE image (Weiland *et al.*, this meeting), while if it is reduced, the asymmetry between the two sides becomes more pronounced. The distance from the centre was chosen in order that the box-like feature in the third contour down from the peak subtends an angle of approximately 10°. (Equating this distance to 8 kpc and setting the velocity unit at 300 km s^{-1}, then the time unit of the simulation is 4.33 Myr and the total mass is 2.77×10^{10} M$_\odot$.) The resemblance to the Weiland *et al.* COBE photometry is quite striking, although the asymmetry in the *b* extent of the bar between the near and far sides is perhaps too strong.

3. KINEMATICS

Observations of proper motions of bulge stars in Baade's window (Spaenhauer *et al.* 1992) and also radial velocities (Terndrup 1992, Rich *et al.* in preparation) suggest that the Milky Way bulge has a nearly isotropic velocity distribution.

Data from the model are shown in Figure 3, which gives the dispersions of particle velocities along the line of sight and in the perpendicular component parallel to the galactic plane in the central few degrees. (The dispersion normal to the plane is not shown, but is substantially lower than for either component in the plane. Part of this anisotropy can be attributed to the grid-based numerical technique, in which forces are weaker than Newtonian at short range, having a disproportionately large effect on the vertical restoring force.) From this viewing angle, the perpendicular component is a little larger than the line-of-sight component; the difference would be less from a smaller viewing angle, however.

4. DISCUSSION

It is interesting to compare this model with other bar models proposed for the Milky Way. The position angle of the bar major axis to the viewing direction

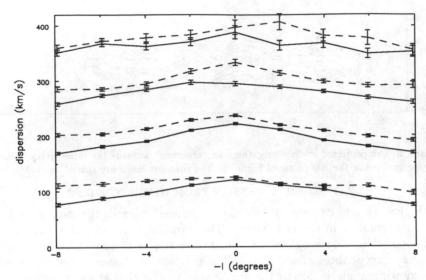

Figure 3 The line-of-sight components (full drawn) and perpendicular components (dashed) over a range of longitudes at fixed latitude. All particles within 2 kpc of the galactic centre distance contribute equally and data from above and below the plane combined to improve statistics. The bottom pair of curves are for $b = 0$, higher pairs are for b values increasing by $2°$ and offset vertically by 100 km/s each time for clarity.

is close to the $36 \pm 10°$ suggested by Weinberg (1992), but is considerably larger than the $16 \pm 2°$ favoured by Binney *et al.* (1991). On the other hand, the pattern speed of the bar in the simulation places co-rotation at about 2.6 kpc, in good agreement with the 2.4 kpc favoured by Binney *et al.*, but the semi-major axis is considerably less than Weinberg's estimate of 5 kpc. Blitz & Spergel (1991 and this meeting), from an analysis of the balloon borne near IR surface photometry, claim to see a short bar at a still larger angle, but again in the same quadrant. Whitelock (1992) finds a longitudinal asymmetry in the distribution of IRAS selected Mira variables, which she also attributes to a bar. All these strands of evidence indicate a bar-like feature pointing in a similar direction and the case for the COBE "bulge" feature being a bar now appears quite strong.

REFERENCES

Binney, J., Gerhard, O. E., Stark, A. A., Bally, J., & Uchida, K. I. 1991, MNRAS 252, 210
Blitz, L. & Spergel, D. N. 1991, ApJ 379, 631
Combes, F. & Sanders, R. H. 1981, A&A 96, 164
Combes, F., Debbasch, F., Friedli, D., & Pfenniger, D. 1990, A&A 233, 82
Raha, N., Sellwood, J. A., James, R. A. & Kahn, F. D. 1991, Nature 352, 411
Spaenhauer, A., Jones, B. F. & Whitford, A. E. 1992, AJ 103, 297
Terndrup, D. M., 1992, To appear in IAU Symposium 153, Galactic Bulges, eds H. Habing & H. Dejonghe (Dordrecht: Kluwer)
Weinberg, M. D. 1992, ApJ 384, 81
Whitelock, P. A. 1992, To appear in IAU Symposium 153, Galactic Bulges, eds H. Habing & H. Dejonghe (Dordrecht: Kluwer)

DIRBE OBSERVATIONS OF THE GALACTIC BULGE

J.L. Weiland[1], M.G. Hauser[2], T. Kelsall[3], E. Dwek[3], S.H. Moseley[3],
R.F. Silverberg[3], M. Mitra[1], N.P. Odegard[1], W.J. Spiesman[1],
T.J. Sodroski[4], S.W. Stemwedel[4], H.T. Freudenreich[5], C.M. Lisse[5]

[1]General Sciences Corp., Code 685.3, NASA/GSFC, Greenbelt, MD 20771
[2]Code 680, NASA/GSFC, Greenbelt, MD 20771
[3]Code 685.0, NASA/GSFC, Greenbelt, MD 20771
[4]Applied Research Corp., Code 685.3, NASA/GSFC, Greenbelt, MD 20771
[5]Hughes STX, Code 685.9, NASA/GSFC, Greenbelt, MD 20771

ABSTRACT

Near-infrared observations of the Galactic bulge from the Diffuse Infrared
Background Experiment (DIRBE) on-board NASA's Cosmic Background Explorer (COBE) satellite are discussed. Emphasis is placed on preliminary analysis of asymmetries in the brightness distribution of the bulge, and the implications for the presence of a stellar bar in the bulge region.

1. INTRODUCTION

Since interstellar extinction obscures a substantial fraction of the Galactic
bulge from observation at optical wavelengths, the morphology of the bulge is
more easily studied in the near-infrared. The bulge is evident in near-infrared
wavelengths as a roughly elliptically-shaped distribution of stars and interstellar
material centered within the flatter disk of the Galaxy.

Recently, Blitz & Spergel (1991) have used asymmetries in the bulge
brightness distribution at 2.2 μm to infer the presence of a stellar bar at the
center of the Galaxy, which is oriented at an angle to the observer in the plane,
and possibly also tilted out of the plane by a few degrees. The presence of a
bar has been postulated by others based on velocity data (e.g. de Vaucouleurs
1964, Lizst & Burton 1980) and IRAS late-type stellar population distributions
(e.g. Weinberg 1992, Whitelock & Catchpole 1992).

We discuss evidence for asymmetries in the bulge structure as seen in data
from the Diffuse Infrared Background Experiment (DIRBE), and the implications of our results for the stellar bar model of Blitz & Spergel (1991).

2. THE DIRBE INSTRUMENT AND OBSERVATIONS

The DIRBE instrument is a 10 spectral band absolute photometer with
wavelength coverage from 1 to 240 μm (Hauser et al. 1991). The DIRBE has
surveyed the entire sky with a 0.7° square field of view, systematically sampling
each pixel at solar elongation angles between 64° and 124°. The DIRBE data
have shown excellent relative photometric stability. A more complete description
of the DIRBE instrument and COBE mission is given by Boggess et al. (1992).

Using DIRBE observations spanning a 6 month interval, maps of the sky
interpolated to a solar elongation of 90° were created. An empirical fitting func-

tion (Hauser 1993) was used to remove foreground contamination from zodiacal dust from these maps. Estimated inaccuracies in this foreground removal for $\lambda \leq 3.5\mu m$ are $\leq 1 - 2\%$ of the median brightness of the bulge in the inner Galaxy region.

The Galactic bulge is clearly visible in the 1.25 to 4.9 μm DIRBE maps, within $|l| < 15°$ and $|b| < 10°$. In the 1.25 μm map, there is obvious evidence of a dust extinction lane running along the Galactic plane. The 2.2 μm map shows intensity levels and morphology similar to those observed by Matsumoto *et al.* (1982).

Longitudinal asymmetries in the near-infrared bulge brightness distribution at 1.25, 2.2, 3.5 and 4.9 μm can be discerned by 'folding' the intensity contours about $l = 0°$. At a fixed Galactic latitude, contours of the same intensity level for $l = 345°$ to $360°$ lie systematically inside those for $15°$ to $0°$, indicating asymmetry.

3. CORRECTION FOR EXTINCTION AT 2.2 μm

Extinction within the Galactic plane is still significant in the near-infrared, although considerably reduced compared to that at optical wavelengths. It is possible that some of the brightness asymmetries seen in the DIRBE data could be caused by patchy extinction, rather than by the underlying source distribution. Since the DIRBE data provide the first multi-wavelength near-infrared maps of the full Galactic plane, direct compensation for the effect of extinction can be made. As an example, we have performed the correction for the 2.2 μm DIRBE map.

The optical depth at 2.2 μm is estimated from the DIRBE data by two methods: 1) deriving the optical depth at 2.2 μm from the near-infrared DIRBE data as per Arendt *et al.* (1993), and 2) scaling the optical depth at 240 μm as determined by Sodroski *et al.* (1993) to 2.2 μm. The region within $|b| \leq 3°$ is excluded from analysis since the stellar light from the Galactic center region is too heavily attenuated to correct accurately. Both correction methods are in good agreement for $|b| > 3°$, yielding estimates for the 2.2 μm optical depth $\tau \lesssim 0.5$ in this region.

Even after an extinction correction is applied to the DIRBE data, some longitudinal asymmetries in the bulge brightness distribution remain. There are, however, structural differences in the appearance of the bulge between the corrected and uncorrected maps. In particular, extinction is chiefly responsible for the presence of the pronounced "peanut-shape" waist located near $l = 1°$, $b = 5°$ in the uncorrected DIRBE maps of the bulge at 1.25 μm and 2.2 μm. In the extinction-corrected maps, the pronounced "waist" disappears. This structure correlates strongly with features in the 2.2 μm optical depth map, and the near-infrared polarization (Arendt *et al.* 1993).

4. FRACTIONAL BRIGHTNESS DIFFERENCES AT 2.2 μm

A more quantitative way to characterize the asymmetries in the DIRBE maps is to compute difference maps between two halves folded either about $l = 0°$ or $b = 0°$ 'symmetry' planes. Following Blitz & Spergel (1991), we define the following fractional brightness differences:

about $l = 0°$: $\Delta I_l = [I(l,b) - I(-l,b)]/I(l,b)$
about $b = 0°$: $\Delta I_b = [I(l,b) - I(l,-b)]/I(l,b)$

DIRBE 2.2 μm fractional brightness differences about $l = 0°$ for data before and after extinction correction both show negative differences near $l = 0°$, become increasingly positive with increasing longitude out to 15°, and then decrease. The difference map created with data uncorrected for extinction possesses structure which is correlated with extinction: a highly negative feature appears near $l = 1°$, $b = 5°$, which we associated in the previous section with an extinction feature. Fractional differences about $l = 0°$ for extinction-corrected data are smoother and more latitudinally symmetric than the uncorrected data. Longitudinal asymmetries persist regardless of where the longitudinal symmetry plane is placed.

Difference maps created about $b = 0°$ show a general lack of structure beyond an overall latitudinal brightness displacement which is apparently the same for both the Galactic disk and bulge. Tests show that this gradient in brightness can be removed if the latitudinal symmetry plane is located not at $b = 0°$, but near $b = -0.16°$. This value is consistent with the assumption that the Sun is located roughly 10-20 pc above the Galactic plane (*e.g.* Djorgovsky & Sosin 1989, Freudenreich *et al.* 1993).

5. IMPLICATIONS FOR A STELLAR BAR

The emission from a triaxial stellar bar at the Galactic center may be modelled by parametrizing the stellar density distribution and choosing a geometrical orientation with respect to the observer. In the table below, the model predictions of Blitz & Spergel (1991) are compared with our findings for the observations at 2.2 μm:

Prediction	Verified by DIRBE?
If the bar is tilted so that the nearer half is in the positive longitude hemisphere, then the fractional brightness differences about $l = 0°$ should show an increasingly positive gradient as a function of longitude. The positive excess should be present at all latitudes.	*Yes*, although the gradient peaks and then drops off, presumably because light from the disk starts to dominate observed intensity.
Close to $l = 0°$, there should be a sign change in the fractional brightness differences about $l = 0°$.	*Yes.*
The near side of the bar should show a larger angular scale height. (Angular scale heights are computed assuming an exponential brightness distribution as a function of Galactic latitude).	*Yes.* Within $|l| < 7°$, scale heights at positive longitudes are larger than those at their corresponding co-longitudes.

Therefore, the DIRBE data are consistent with the signatures of a stellar bar with its near end in the first quadrant. It has been additionally suggested that the proposed stellar bar is tilted a few degrees out of the Galactic plane (Blitz & Spergel 1991, Liszt & Burton 1980). In this case, the fractional brightness difference maps are expected to show latitudinal asymmetries which are longitude dependent. There is no evidence for this effect in the extinction-corrected 2.2 μm DIRBE data.

6. SUMMARY

DIRBE data at 1.25, 2.2, 3.5 and 4.9 μm show longitudinal asymmetries in the Galactic bulge brightness distribution. The multi-wavelength nature of the DIRBE observations permits correction for extinction effects: even after correction, longitudinal asymmetries persist. However, extinction is responsible for some of the structure seen in the DIRBE near-infrared maps: in particular, the pronounced "peanut-shaped" waist in the northern part of the bulge.

The DIRBE fractional brightness differences and larger angular scale heights at positive longitudes are consistent with the signatures of a stellar bar with its near end in the first quadrant. This orientation is in agreement with the findings of previous studies (e.g. Liszt & Burton 1980, Blitz & Spergel 1991, Whitelock & Catchpole 1992). However, the extinction-corrected DIRBE data do not show the latitudinal asymmetries necessary to support the proposal that the stellar bar is also tilted a few degrees out of the Galactic plane.

ACKNOWLEDGEMENTS

The authors gratefully acknowledge the efforts of the DIRBE data processing and validation teams in producing the high-quality datasets used in this investigation. We also thank the COBE Science Working Group for helpful comments on this manuscript. COBE is supported by NASA's Astrophysics Division. Goddard Space Flight Center (GSFC), under the scientific guidance of the COBE Science Working Group, is responsible for the development and operation of COBE.

REFERENCES

Arendt, R.G., *et al.* 1993, this volume
Blitz, L. & Spergel, D. 1991, ApJ 379, 631
Boggess, N.W., *et al.* 1992, ApJ 397, 420
de Vaucouleurs, G. 1964, in "IAU Symposium 20: The Galaxy and the Magellanic Clouds", eds. F.J. Kerr & A.W. Rodgers (Australian Academy of Science: Sydney), p. 195
Djorgovsky, S. & Sosin, C. 1989, ApJ 341, L13
Freudenreich, H. *et al.* 1993, this volume
Hauser, M.G., *et al.* 1991, in "After the First Three Minutes", eds. S. Holt, C. Bennett & V. Trimble (AIP: New York), p. 161
Hauser, M.G. 1993, this volume
Liszt, H.S. & Burton, W.B. 1980, ApJ 236, 779
Matsumoto, T. *et al.* 1982, in "The Galactic Center", eds. G. Riegler & R. Blanford (AIP: New York), p. 48
Sodroski, T.J., *et al.* 1993, this volume
Weinberg, M.D. 1992 ApJ 384, 81
Whitelock, P. & Catchpole, R. 1992, in "The Center, Bulge and Disk of the Milky Way", ed. L. Blitz (Kluwer: Netherlands), p. 103

A STUDY OF THE ABUNDANCE DISTRIBUTIONS ALONG THE MINOR AXIS OF THE GALACTIC BULGE

NEIL D. TYSON

Princeton University Observatory, Princeton, New Jersey 08544 USA

ABSTRACT

I report the preliminary results of a project, in collaboration with R. Michael Rich (Columbia), where we derive heavy element abundances for hundreds of K-giants in seven windows of low extinction, along or near the minor axis of the Galactic bulge. By using the recently-calibrated Washington photometric filter system, the distribution function in [Fe/H] is determined for each field. Within 8° of the Galactic center (~ 1 kpc) our data are consistent with no gradient in the distribution of [Fe/H], which may hint to a dissipationless collapse, and/or sufficient mixing during the star-forming epoch when Fe was produced in the bulge. The mean abundance over this region is between two and five times solar. The form of these distributions is well-fitted by the simple (closed box) model of chemical evolution where the bulge is self-enriched by processing its original gas content to completion. Beyond 8° from the Galactic center, our data show that the mean of the abundance distributions drops precipitously. This is consistent with the notion that the inner bulge is chemically distinct from the halo. It may be possible to use kinematics to disentangle the two populations via a radial velocity survey.

1. OBSERVATIONS

We used the recently calibrated Washington photometric system (Geisler, et al. 1991) to obtain photometric abundances for hundreds of K-giants in seven low extinction windows to the Galactic bulge.

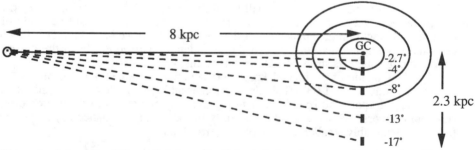

Figure 1. Schematic of lines of sight to the Galactic bulge, drawn roughly to scale. The concentric ellipses represent one, two, and three scale-heights in an exponential profile of the 2μm light (Kent et al 1991). For clarity, only five of the seven windows of low extinction that were selected for this study are shown. Omitted are the fields at $b = -6°$ and $b = -10°$. The Sun is taken to be 8 kpc from the Galactic center.

The data were obtained on the 0.9m telescope at CTIO and reduced using the DAOPhot. reduction package of Stetson (1987). Abundances accurate to ±0.25 dex in [Fe/H] were derived for each K-giant using the metallicity-sensitive indices of the Washington system. (For a recent review of the Washington system see Tyson 1991).

2. ABUNDANCE DISTRIBUTIONS

The shape of the stellar abundance distribution holds valuable information about the history of chemical enrichment in a system. The distributions for the three inner fields (i.e. $b = -2.7°, -4°, -6°$) bear a remarkable resemblance to the gaussian-convolved abundance distribution of the closed box, simple model of chemical evolution (Searle & Zinn 1978, also Rich 1990). Figure 2 displays the abundance distribution for the inner-most field of this study. Error bars represent the typical errors in precision for a single observation derived from the photometric uncertainties. There is the characteristic extended tail toward lower abundances and the relatively steep shoulder at high abundance. Consequently, our data suggest that the bulge abundance distributions, out to $b = -6°$, are consistent with a fully mixed, closed box, simple model of chemical evolution that has processed its gas to completion.

3. DISCUSSION

To supplement interpretation of the abundance distributions we can look at the run of the log mean abundance, $[< Fe / H >]$, as a function of Galactic latitude. This will constitute the first direct measurement of the latitude dependence of [Fe/H] in the Galactic bulge. Figure 3 presents these data. The errors are large, but two convincing features are immediately apparent:

[1] The inner latitudes stay at roughly constant mean abundance [Fe/H] $\approx 0.4 \pm 0.3$ (r < 1 kpc), which suggests that the bulge, within about 1 kpc, is a chemically well-mixed volume. There is a small trend for [<Fe/H>] to drop within $b = -6°$, but given the relatively large errors, it is not clear whether much should be made of this feature. The fact that the abundance distributions for the three inner latitudes show good agreement with the analytic form of the abundance distribution from the closed box simple model suggests strongly that the inner bulge is self-enriched.

[2] Outside of the inner 1 kpc, the mean abundance begin a precipitous 1.5 dex drop to latitude $b = -17.3°$. This drop is remarkable when we consider that it occurs across only 1 kpc. The 1σ error bars in Fig. 3 are computed from the photometric *and* zero-point errors. The statistical uncertainty in the mean abundance ($\sigma / N^{-1/2}$) for each field is considerably smaller than the plotted error bars.

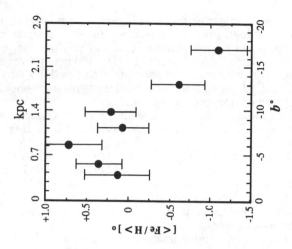

Figure 3. Galactic latitude dependence of the mean in the abundance distributions. Plotted is the error-corrected log of the mean abundance at all latitudes. The mean abundance does not show a trend downward until beyond ~ 1 kpc. The similarity among the inner latitudes in their mean abundance suggests strongly that the bulge, within about 1 kpc, followed common evolution. The similarity and shape of the abundance distributions for the inner fields suggests a common evolution described by the closed box simple model of chemical evolution.

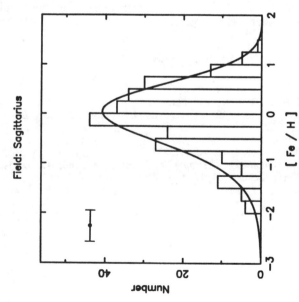

Figure 2. Abundance distribution for the Sagittarius field ($b = -2.7°$) overlaid with an error-convolved distribution expected for the closed box simple model of chemical evolution. There is remarkable agreement with the abundance distribution expected from the simple model of chemical evolution where a system has processed completely its original gas supply. The characteristic features are immediately apparent, such as the long low abundance tail, and the relatively steep high abundance shoulder. The error bar represents the errors in photometric precision.

4. FUTURE

One of the most useful aspects of the Washington system is its ability to flag interesting stars of unusually high (or low) abundance for follow-up spectroscopy. Some of these may be optical or normal binaries that will confuse the photometric interpretation (e.g. Twarog & Anthony-Twarog 1991). Others, however, will be genuinely interesting stars. The existence of super-metal rich ([Fe/H] > 0.5) stars is an exciting prospect that, if verified, poses theoretical challenges for our understanding of supernova yields, and stellar evolution models in general.

The idea of a kinematic dependence on abundance in the Galactic bulge is an exciting topic that demands much further study. Successful attack, of course, requires that one knows the radial velocity *and* the heavy element abundance for hundreds of stars. Ideally, one would like to target a high or low abundance subset of a distribution to explore the possible kinematic extremes. In collaboration with D. N. Spergel (Princeton), we have planned to obtain medium resolution spectra of the highest and lowest abundance stars from the abundance distributions of this study. Spergel has developed computer models of the Galactic bulge that allow one to derive observable quantities from segregated orbit families in phase space.

Answerable questions come to mind immediately: Can we further confirm the existence of a bar? If there is a bar, is the bar equally populated by stars of all abundances? Some low energy orbit families can evolve to become higher energy orbit families by perturbations from a triaxial gravitational potential. What would this look like in observable quantities? Can we measure minor axis rotation? By constraining the data to be along the bulge's minor axis we also constrain the explorable parameter space of the computer models by removing the rotational component. We expect this approach to be a powerful analytic tool that will enable interpretive leverage on abundance and velocity distributions that would not otherwise be possible.

REFERENCES

Geisler, Clariá, J. & Minniti, D. 1991, AJ 102, 1836
Kent, S. M., Dame, T. M., & Fazio, G. 1991, ApJ 299, 674
Rich, R. M. 1990, in *Bulges of Galaxies*, ed. B. J. Jarvis & D. M. Terndrup (Garching: ESO) p. 65
Searle, L. & Zinn R. 1978, ApJ 225, 357.
Stetson, P. 1987 PASP 99, 191.
Twarog, B. A. & Anthony-Twarog, B. J. 1991, AJ 101, 237
Tyson, N. 1991, in *Precision Photometry: Astrophysics of the Galaxy*, ed. A.G.Davis Philip, A. R.Upgren, & K. A. Janes, (Schenectady: L. Davis Press), p. 193

LOCAL
ENVIRONMENT

THE HOT IONIZED MEDIUM

Dan McCammon
Physics Department, University of Wisconsin, Madison, WI 53706

ABSTRACT

When early *ROSAT* results are included, no existing model can explain all the observed characteristics of the 1/4-keV diffuse background. It still seems likely, however, that most of the observed radiation at these energies is thermal emission from diffuse hot gas at temperatures near $10^{6.0}$ K, and that a large fraction of the Galactic disk within 100 pc of the Sun is filled with such gas. Further analysis of *ROSAT* observations should rapidly increase our understanding of this phase of the interstellar medium, and the prospect of high spectral resolution measurements promises more progress in the future.

1. INTRODUCTION

Existing data on the soft X–ray background come almost exclusively from proportional counters. A combination of the intrinsic energy resolution of these detectors supplemented by K-edge filters allows the energy range from 0.1 - 2 keV to be resolved into several overlapping bands, such as the ones shown in figure 1. This energy resolution makes it possible to distinguish different emission mechanisms and physical conditions in the source regions. Equally important for studies of the interstellar medium is the rapid variation of the interstellar absorption cross section with energy. Figure 2 shows that the soft X–ray interstellar mean free path at an average disk density spans the entire range of Galactic distance scales. Unfortunately, there is a large variation in cross section even within a band, but physically reasonable source spectra can be used as weighting functions to determine an effective cross section for each band. Table 1 gives typical values for the effective mean free paths. The difference between the three lowest energy bands turns out to be a very important factor in discriminating among models for the distribution of Galactic soft X–ray emission.

Figures 3 and 4 are maps of the sky in the C and M (M = M1+M2) bands, respectively. The sky looks quite different in these adjacent energy bands, implying distinct origins for the majority of the X–rays in each. The C and B band maps, on the other hand, look almost identical (McCammon *et al.* 1983), and the Be/B ratio is essentially constant for the 20% or so of the sky that has been measured in the Be band. The M band is discussed in §2, in part because some of it may indeed be due to hot interstellar gas, but primarily as a sort of cautionary parable, showing that the most reasonable answer is not necessarily the correct one. This will be useful to bear in mind while reading §3, which reviews existing models of the 1/4 keV (C, B, and Be band) background, and the apparent problems with each of them. Section 4 goes on to discuss prospects for further progress in our understanding of the role of hot ionized gas in the interstellar medium.

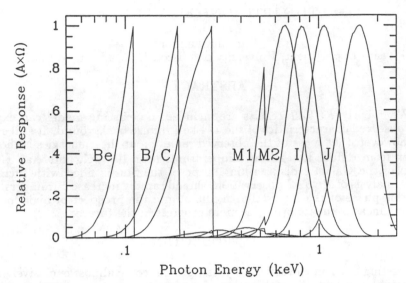

Fig. 1. Band response functions obtainable from proportional counters below 2 keV. The M1, M2, I, and J bands are defined primarily by using the pulse-height resolution of the counter. The Be, B and C bands are defined primarily by the transmission of a filter made of the corresponding element. (From McCammon & Sanders (1990).

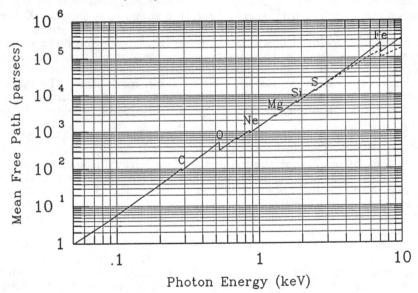

Fig. 2. Effective interstellar mean free path as a function of photon energy for an assumed interstellar gas density of 1 atom cm^{-3} with solar abundances. Data are from Morrison & McCammon (1983).

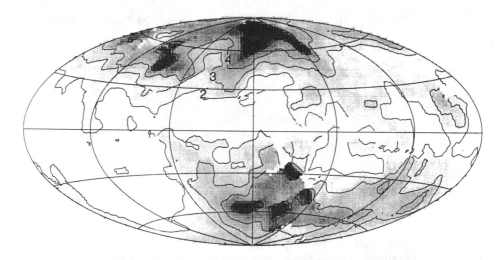

Fig. 3. C band (∼160 – 284 eV) map of the sky. The map is in Galactic coordinates with $l = 0°$ at the center and increasing to the left. Bright point sources have been removed. Data are from McCammon *et al.* (1983).

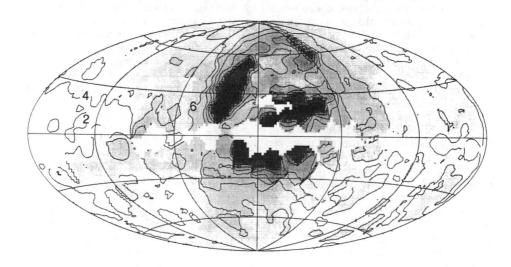

Fig. 4. M band (∼490 – ∼1190 eV) map of the sky. Otherwise same as figure 3.

Table 1: Band-averaged Effective Mean Free Paths

Band	Approx. Energy Limits (keV)	Effective Mean Free Path N_H (10^{20} cm^{-2})
Be	.07 – .111	0.10
B	.13 – .188	0.65
C	.16 – .284	1.3
M1	.44 – .93	12.
M2	.60 – 1.1	18.
I	.77 – 1.5	50.
J	1.1 – 2.2	130.

2. THE M BAND (0.5 - 1 KEV) BACKGROUND

The central part of the map is dominated by emission from Loop I, with a possible contribution from the M band Galactic bulge. There are a few other identifiable features, such as the Cygnus and Eridanus superbubbles, which are presumably similar to Loop I but more distant. The sky is otherwise nearly isotropic—a most surprising result in light of the ~1 kpc absorption mean free path in the Galactic plane for this energy band. An extragalactic source must vanish in the plane, where there are many optical depths of absorption, while a Galactic source should be more intense near the plane, where there is a much longer path length through the source distribution. Instead we see no variation at all in crossing the plane, particularly in the third quadrant. This feature was first noted on the Wisconsin rocket survey, and has persisted through the increasing angular resolution and statistical precision of later data: even the fine detail of the *ROSAT* maps shows no structure crossing the Galactic plane at most longitudes.

One's first inclination would be to ascribe such isotropic appearance to an inherently isotropic, and therefore local, source. It has proven very difficult to find a consistent local model, however (Edgar 1986), and recent *ROSAT* results have shown that the observed isotropy is apparently due to a remarkable but entirely accidental cancellation of inherently anisotropic components. The key observation is that at high latitudes at least 40% of the observed flux can be resolved into discrete sources that appear predominantly to be distant QSO's (Hasinger et al. 1993). The Galactic contribution to the remainder of the intensity seen at high latitudes is unknown, and may be very small: a reasonable extrapolation of the observed log N - log S relation for extragalactic sources could easily provide ~100% of the diffuse intensity, while stars are expected to contribute at most a few percent (Kashyap et al. 1991, Schmitt & Snowden 1990). On the other hand, there is some spectroscopic evidence for the existence of oxygen lines at high latitudes (Inoue et al. 1980; Schnopper et al. 1982), so it is still possible that some Galactic thermal source contributes a substantial fraction. This might be hot interstellar gas at temperatures of a few million degrees, but it does not appear to be part of the local interstellar medium region that produces the bulk of the 1/4-keV background: a *ROSAT* pointed observation of MBM 12, a molecular cloud at $l=159°$, $b=-34°$ with a well-determined distance of 65–75 pc (Hobbs, Blitz, & Magnani 1986), shows that essentially all of the M band X-rays are shadowed by the cloud, while most of the 1/4-keV diffuse flux originates in front of it (Snowden et al. 1993a).

In the Galactic plane, the large column densities of H I seem to preclude the possibility of any extragalactic contribution to the M band flux. The Galactic sources are not known, although M stars could provide a substantial fraction (*cf.* Kashyap *et al.* 1991). A superposition of supernova remnants and superbubbles could also contribute, but there may be problems with reproducing the observed smoothness. All that we really know is that at least 50% of the Galactic emission seen in the plane must fall off with increasing latitude at just the right rate to compensate for the increasing transmission of the extragalactic component.

3. THE C BAND (0.1 -0.28 KEV) BACKGROUND

The 1/4–keV emission at high latitudes is bright and very patchy, with large areas (tens of degrees across) that are strongly enhanced. The low- latitude flux is more uniform, and has an intensity about one-third of the average at high latitudes. The intensity variations are strongly (although imperfectly) anticorrelated with total H I column density, as shown in figure 5, implying that a substantial fraction of the emission lies beyond the H I. This is just what would be expected for an extragalactic or halo source. The behavior in the lower-energy B and Be bands can be used to check this model. Figure 6 shows the B/C band rate ratios observed over most of the sky, while figure 7 shows the relation between B and Be band intensities for a series of 15° x 15° fields distributed over a wide range of Galactic latitudes and B band intensities. Since the C:B:Be band ratios vary by less than 25% over the entire sky, while the

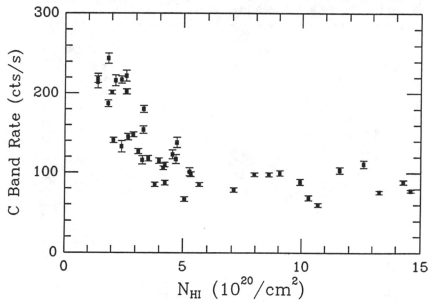

Fig. 5. C band rate from the Wisconsin survey plotted against total H I column density. Longitudes including Loop I (270° < l < 60°) have been excluded, as has a small region around the bright feature in Eridanus. Plotted values are averages over 22.5° in Galactic latitude by 30° in longitude. This X–ray *vs.* H I anticorrelation has the qualitative appearance of an absorbed distant source, but this probably is not the case. See text.

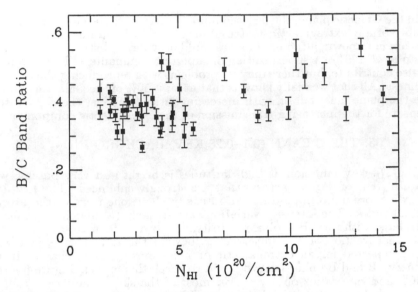

Fig. 6. Ratio of counting rates in the B and C bands plotted against total H I column density. The variations in this ratio are spatially coherent: all points with B/C > 0.4 are contiguous in a region roughly centered on $l = 160°$, $b = +15°$. Excluded regions are the same as in figure 5.

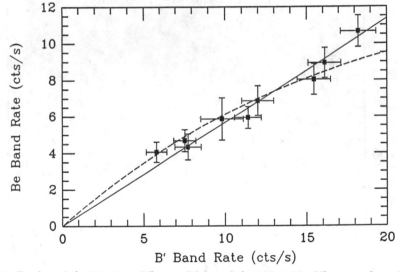

Fig. 7. Be-band ($\sim$70–111 eV) $vs.$ B'-band ($\sim$100–188 eV) rates for nine 15° x 15° fields widely distributed across the sky (Juda et $al.$ 1991). The straight line shows the relation expected if there is no intermixed absorbing gas. The dashed curve shows the relation expected for uniformly intermixed absorbing material with a total column density of $\sim$1.4 x 10^{19} cm^{-2} to the edge of the X–ray-emitting gas for the brightest point. This is a 2σ upper limit to what is allowed by the data.

total intensity varies by a factor of three and the effective absorption cross sections are in the ratio 1:2:10, it is not easy to explain the intensity variations as simple absorption of an extragalactic source. Extreme clumping of the H I could explain this, but does not appear to exist, at least in the total column density (Jahoda *et al.* 1985, 1986). So far, models in which the emission is intermixed with the H I are also inconsistent with these multi-color observations (Fried *et al.* 1980; Jakobsen & Kahn 1986; Kahn & Jakobsen 1988; Burrows 1989; Hirth *et al.* 1991; see also McCammon & Sanders 1990 and Burrows & Kraft 1993).

The mean free path for C band X–rays in interstellar material is only about 1×10^{20} H cm^{-2}, so the flux seen near the Galactic plane must originate rather close to the Sun, and this has been verified for one direction at least by the observation of MBM 12 described above. Because a local contribution is required, the simplest model would have it supply all of the observed flux. Unabsorbed emission from a 10^6 K plasma produces the correct C:B:Be band ratios and can produce the observed intensity if the hot gas fills a volume of typical extent 100 pc at a pressure $p/k \approx 10^4$ cm^{-3} K. The extent of this Local Bubble is such that variations in its radius could produce the observed negative correlation with N_H (Snowden *et al.* 1990). In this idealized model, all of the emission is closer than all of the H I, so there is no absorption and the intensity variations are naturally independent of photon energy.

This model is obviously oversimplified, but until recently it managed to fit all existing observations rather well. However, some of the first *ROSAT* observations showed that some small ($\sim 1°$) clouds in Draco shadowed half or more of the X–ray flux at high Galactic latitudes (Snowden *et al.* 1991; Burrows & Mendenhall 1991). In the one case where there is a good estimate of the cloud's distance, it is located at least 200 pc above the Galactic plane, placing half the observed X–ray emission beyond most of the H I—in a clear contradiction of the Snowden *et al.* model. There have long been theoretical expectations for such a hot Galactic corona (*e.g.* Spitzer 1956), but we are still left with the problem of how it can be that there are no appreciable changes in the energy spectrum with variations in the intervening column density. Additional shadowing observations of similar clouds nearby in Ursa Major show that the halo emission in this direction is at least a factor of three less than implied by the Draco measurements (Snowden *et al.* 1993c); if the Draco observations have demonstrated the existence of million-degree gas in the halo, it is necessarily very patchy on scales of 30° or more. A more detailed account of the impact of *ROSAT* observations on models of the 1/4-keV background can be found in Burrows & Mendenhall (1993).

4. THE FUTURE

We expect considerable progress in understanding the origins of the C and M band diffuse backgrounds as analysis of *ROSAT* data continues. Much more detailed all-sky maps are now being produced from the *ROSAT* X–ray sky survey of the Max Planck Institute for Extraterrestrial Physics. Plate 6 shows a color composite of the first versions of the 1/4, 3/4, and 1.5 keV maps from Snowden et al. (1993b). These maps use the *ROSAT* PSPC in a "light bucket" mode, and have 2° angular resolution. The resolution will be improved to a few arc minutes when position information from the PSPC is incorporated, and up to seven useful energy bands can be distinguished. The availability of data of this quality for the entire sky will enable many detailed studies of the correlation of X–ray emission with H I and CO observations to be made. Optical interstellar absorption measurements on the gas features can determine their distances and thus pin down the location of the X–ray emission. Analysis of the many *ROSAT* guest investigator observations of interstellar features will show even more detail of their interaction with the X–ray background. Images of normal galaxies could reveal the extent and true importance of million-degree gas in the disk—a difficult task for direct observation in our own Galaxy, since the X–rays produced by this gas can travel at most a few hundred parsecs.

Another prospect for progress (on a longer time scale) is observations with high spectral resolution. Thermal radiation at temperatures below 10^7 K is almost entirely in lines of the partially ionized heavy elements. Measuring the absorption of a narrow band of energies, or even of a single line, would eliminate the model dependencies introduced by band-averaged cross sections. Spectra that can resolve individual lines and blends can be used to learn a great deal about the detailed physical conditions in the hot gas: its temperature structure, its elemental abundances, and even some of its history. Figure 8 shows model spectra produced in three possible scenarios for the million-degree emission from the local interstellar medium. These are virtually indistinguishable using broad-band observations, but could be studied readily at ∼5 eV resolution.

It is symptomatic of our current lack of understanding of the interstellar medium that we cannot even say whether million-degree gas similar to that seen around the Sun occupies 10% or 90% of the volume of the Galactic disk—despite the enormous difference this would make in its effects on the structure and evolution of the Galaxy. With the rapid progress engendered by the unprecedented angular resolution and general quality of *ROSAT* observations of diffuse X–ray emission, and with the prospect of adding qualitatively new information from high resolution spectra, we can hope that this situation will soon improve.

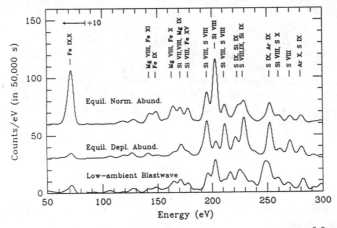

Fig. 8. Three models for thermal emission from a gas near $10^{6.0}$ K. The upper spectrum is for an equilibrium plasma with solar abundances in the gas phase. The middle spectrum also assumes equilibrium, but has the elemental depletions typically seen in dense clouds. The lower spectrum is from a time-dependent calculation of a supernova blast wave expanding into a low-density ambient, and shows strong non-equilibrium effects. The spectra have been convolved with an assumed 5 eV FWHM instrument resolution.

We thank Steve Snowden and the Max Planck Institute for Extraterrestrial Physics for the preparation of the three-color map from the *ROSAT* all-sky survey and for permission to include it here. This work was supported in part by NASA grants NAG 5-679 and NAG 5-1817.

REFERENCES

Burrows, D.N. 1989 , Ap. J., 340, 775

Burrows, D.N. & Kraft, R.P. 1993, Ap. J., submitted

Burrows, D. N. & Mendenhall, J.A. 1991, Nature, 351, 629

Burrows, D. N. & Mendenhall, J. A. 1993, 1992 COSPAR proceedings (preprint)

Edgar, R.J. 1986, Ap. J., 308, 389

Edwards, B.C. 1990, Ph.D. Thesis, University of Wisconsin, Madison

Fried, P.M., Nousek, J.A., Sanders, W.T., Kraushaar, W.L., 1980, Ap. J., 242, 987

Hasinger, G., Burg, R., Giacconi, R., Hartner, G., Schmidt, M., Trümper, J., Zamorini, G. 1993, preprint

Hirth, W., *et al.*, 1991, Astro. & Sp. Sci., 186, 211

Inoue, H., Koyama, K., Matsuoka, M., Ohashi, T., Tanaka, Y., Tsunemi, H. 1980, Ap. J., 238, 886

Hobbs, L.M., Blitz, L., & Magnani, L. 1986, Ap. J. (Letters), 306, L109

Jahoda, K., McCammon, D., Dickey, J.M. & Lockman, F.J. 1985, Ap. J., 290, 229

Jahoda, K., McCammon, D. & Lockman, F.J. 1986, Ap. J. (Letters), 311, L157

Jakobsen, P., & Kahn, S. M., 1986, Ap. J., 309, 682

Juda, M., Bloch, J.J., Edwards, B.C., McCammon, D., Sanders, W.T., Snowden, S.L., & Zhang, J. 1991, Ap. J., 367, 182

Kahn, S.M., & Jakobsen, P. 1988, Ap. J., 329, 406

Kashyap, V., Rosner, R.; Micela, G., Sciortino, S., Vaiana, G.S., & Harnden, F.R., Jr. 1992, Ap. J. 391, 667

McCammon, D., Burrows, D.N., Sanders, W.T. & Kraushaar, W.L. 1983, Ap. J., 269, 107

McCammon, D., & Sanders, W.T. 1990, Ann. Rev. Astr. Ap., 28, 657

Marshall, F.J. & Clark, G.W. 1984, Ap. J., 287, 633

Morrison, R., & McCammon, D. 1983, Ap. J., 270, 119

Schmitt, J.H.M.M. & Snowden, S.L. 1990, Ap. J., 361, 207

Schnopper, H.W., Devaille, J.P., Rocchia, R., Blondel, C., Cheron, C., *et al.* 1982, Ap. J., 253, 131

Snowden, S.L., Cox, D.P., McCammon, D., & Sanders, W.T. 1990, Ap. J., 354, 211

Snowden, S.L., McCammon, D., & Verter, F. 1993a, Ap. J. in press

Snowden, S.L., Hasinger, G., Freyberg, M.J., Schmitt, J.H.M.M., Trümper, J., Voges, W.H., McCammon, D., & Plucinsky, P.P. 1993b, in preparation

Snowden, S.L., Mebold, U., Hirth, W., Herbstmeier, U., & Schmitt, J.H.M.M. 1991, Science, 252, 1529

Snowden, S.L., *et al.* 1993c, in preparation

Spitzer, L. 1956, Ap. J., 124, 20

THE WARM IONIZED MEDIUM

R. J. Reynolds
Department of Physics, 1150 University Avenue
Madison, WI 53706

ABSTRACT

The warm ionized component of the interstellar medium is characterized by widespread regions of nearly fully ionized hydrogen with an electron density $n_e \simeq 0.1$ cm^{-3} and a temperature $T \simeq 8000$ K. The regions occupy at least 20% of the volume within a layer of half-thickness $H \simeq 1$ kpc and have a mass surface density approximately one-third that of the H I. The required ionizing power is 1×10^{-4} ergs s^{-1} per cm^2 of the Galactic disk near the solar circle. These results have been derived from a relatively small number of sight lines that sample a region of radius 3–4 kpc about the sun. The temperature and ionization state have been probed at low Galactic latitudes near the midplane, while the scale height, density, filling fraction, and power requirement are derived from data at high latitudes.

1. INTRODUCTION

Contrary to the traditional picture in which regions of fully ionized hydrogen are limited to "Strömgren spheres" near the Galactic midplane and within a few tens of parsecs of O stars, most of the interstellar H II is in fact located far from the midplane and far from O stars, in a widespread and massive ionized component. The existence of this component was first deduced by Hoyle & Ellis (1963) from the spatial and spectral characteristics of the Galactic synchrotron background at very low ($\nu \leq 5\ MHz$) radio frequencies. However, it was not until pulsars were discovered (Hewish et al. 1968) that widespread ionized hydrogen was generally recognized as a constituent of the interstellar medium. The pulsar dispersion measures revealed that the mean free electron density outside the traditional H II regions is about 0.03 cm^{-3}, approximately 100 times larger than that available from trace elements such as carbon ionized by ambient starlight. This was soon followed by the detection of faint diffuse optical line emission from the interstellar medium (e.g., Reynolds et al. 1971; Reynolds, Roesler, & Scherb 1973) which made possible more detailed investigations of the properties of this ionized gas, including its power consumption, temperature, ionization state, kinematics, and spatial distribution. Nearly all of the optical line studies have been carried out with the Wisconsin large-aperture Fabry-Perot spectrometer (Roesler et al. 1978).

The following is a review of some of the properties of this component derived from pulsar and optical emission line data, with an emphasis on the limitations and possible biases in the observations.

2. HYDROGEN IONIZATION RATE AND POWER REQUIREMENT

The intensity of the interstellar H$_\alpha$ background at high (i.e. $|b| > 15°$) Galactic latitudes provides a direct measurement of the hydrogen recombination (and thus ionization) rate within the diffuse ionized medium. Specifically, the recombination rate r within a cm^2 column in a direction with negligible interstellar extinction is given by $r = 4\pi I_\alpha/\epsilon$, where I_α is the H_α intensity in photons cm^{-2} s^{-1} sr^{-1} and ϵ is the average number of H_α photons produced per recombination. The value of ϵ changes slowly with electron temperature, decreasing from 0.71 at 100 K to 0.47 at 8000 K to 0.37 at 80,000 K (Martin 1988, and Pengally 1964; Case B). For a temperature near 8000 K (see Section 3 below) the average recombination rate $< r_G >$ per cm^2 of Galactic disk can therefore be determined from high latitude H_α observations by

$$< r_G > \simeq 53 < I_\alpha \sin|b| > \quad s^{-1}cm^{-2}. \tag{1}$$

A fully sampled sky survey of the interstellar H_α background does not exist yet. However, a selection of H_α scans toward 45 pulsars at $|b| > 15°$ have provided a good "random" sample of the sky at high latitudes and give $< I_\alpha \sin|b| > \simeq 9 \times 10^4$ photons cm^{-2} s^{-1} sr^{-1} (Reynolds 1992a). Thus $< r_G > \simeq 5 \times 10^6$ s^{-1} cm^{-2}, and at 13.6 eV per ionization the minimum power required to sustain this ionization is $< W > \simeq 1.0 \times 10^{-4}$ ergs s^{-1} per cm^2 of disk. This corresponds to approximately 16% of the Lyman continuum production rate from O stars (Abbott 1982) or 100% of the kinetic energy injected into the interstellar medium by supernovae (at 10^{51} ergs per 75 years within a Galactic radius of 12 kpc). The value of $< r_G >$ therefore tightly constrains the possible sources of the ionization. Of the known sources only the O stars comfortably exceed the minimum requirement, but it is not certain that the opacity of the interstellar H I allows Lyman continuum photons to penetrate sufficiently far from the stars to produce the widespread ionization (Reynolds 1992b; Miller & Cox 1993).

Since the scale height of the ionized gas is about 1 kpc (Section 5), these values correspond to a 3–4 kpc radius region of the disk around the sun. The examination of a substantially larger region of the Galaxy would require lines of sight at lower latitudes (i.e., $|b| < 15°$) and thus observations of hydrogen recombination lines that are not affected significantly by interstellar extinction, such as Brackett-α at 4 μm or much higher quantum number recombination lines at radio wavelengths. Unfortunately, current infrared and radio techniques are about three orders of magnitude less sensitive than the optical Fabry-Perot observations and thus are not capable of probing the diffuse ionized component.

Because interstellar extinction was neglected in the analysis of the high latitude H_α data, the derived values for $< r_G >$ and $< W >$ could be low by as much as 10–20%. Along a line of sight at $|b| = 30°$, for example, E(B-V) $\simeq 0.1$ (e.g., Burstein and Heiles 1982), corresponding to an optical depth due to both absorption and scattering of about 0.2 at H_α (Osterbrock 1979). However, forward scattering, which does not result in the loss of a photon from a diffuse source of emission, may account for one-half of this visual extinction (Mathis, Rumpl, and Nordsieck 1977). Thus the effective optical depth at $|b| = 30°$ may be only $\simeq$ 0.1, resulting in a loss of 10% of the H_α photons if the source is entirely behind the dust. On the other hand, interstellar grains at high latitude can *increase* the apparent H_α intensity by scattering into the field of view H_α photons originating from bright, traditional H II regions near the Galactic midplane (Jura 1979). This effect also must be relatively small because line intensity ratios in the interstellar background, specifically [S II] $\lambda6716$/H_α and [O III] $\lambda5007$/H_α, differ significantly from the ratios in the traditional H II regions (Reynolds 1985a, b, 1988). Furthermore, observations toward high latitude reflection nebulae that are presumably illuminated by the integrated light of the Galactic plane have shown no detectable scattered contribution of the H_α line. The available data suggest that scattered light from bright H II regions contributes $\lesssim 10\%$ to the H_α background (Reynolds, Scherb, & Roesler 1973; Reynolds 1990b).

3. TEMPERATURE AND NON-THERMAL VELOCITIES

The temperature T of the ionized gas has been estimated from the widths of the H_α, [N II] $\lambda 6584$, and [S II] $\lambda 6716$ emission lines. For example, if the H^+ and S^+ ions are well-mixed within the emitting regions then

$$T \simeq 22.5(W_H^2 - W_S^2 - 32) \ K, \tag{2}$$

where W_H and W_S are the full widths at half maximum of the interstellar H_α and [S II] line profiles, respectively, in units of km s^{-1}. A study of emission profiles in twenty "background" directions (Reynolds 1985a) provides an unweighted average value of $< T > = 7500 \pm 3200$ K, where 3200 K is the standard deviation of the individual measurements. The scatter appears to be due almost entirely to the measurement uncertainties. This average value is consistent with the lower limit of $T > 5500$ K derived from the forbidden line intensities (Reynolds 1989) and the temperature derived from [N II] and H_α line width comparison (Reynolds, Roesler, & Scherb 1977). The same line width analysis applied to twenty-two lines of sight toward "H II regions" associated with O stars and early B stars yields 6500 ± 500 K (Reynolds 1988).

Photoelectrons ejected from grains are a supplemental source of heat in the diffuse ionized medium, which could be important at densities $n_e \lesssim 0.1$ cm^{-3} (Reynolds & Cox 1992). For example, in regions where $n_e \approx 0.08$ cm^{-3} (see Section 6) grain heating is predicted to be approximately equal to hydrogen photoionization heating, resulting in a temperature that is approximately 2700 K higher than that in the denser, OB star H II regions. The available data do not provide evidence for such an elevated temperature, although it could easily be masked by the large measurement uncertainties. It must also be emphasized that [S II] and [N II] line widths can only be measured reliably where the line intensities are sufficiently bright, at low ($|b| \lesssim 25°$) Galactic latitudes. This biases the temperature measurements to gas near the Galactic midplane, where the densities are highest and, therefore, where the relative contribution of photoelectric grain heating is minimized. More accurate measurements of the electron temperature through the temperature sensitive line intensity ratio [N II] $\lambda 5755$/[N II] $\lambda 6584$ are possible in principle; however, the fact that the $\lambda 5755$ line is expected to have an intensity that is only 10^{-2} that of the H_α makes such measurements problematic at present.

The profiles of the emission lines have also been used to probe the non-thermal motions within the diffuse ionized component (Reynolds 1985a). The analysis shows that the most probable speed in the non-thermal radial velocity distribution along each of the twenty lines of sight ranges between about 12 km s^{-1} and 30 km s^{-1}. The large range appears to be real and is significantly higher than what is found toward isolated O and B star H II regions, where the values range between 5 km s^{-1} and 13 km s^{-1} (Reynolds 1988). The fraction of the non-thermal velocity that is due to *random* motion of the gas is not clear, since most of the directions with derived values in the range of 25–30 km s^{-1} have Galactic longitudes $117° \leq l \leq 164°$, where the contribution of Galactic differential rotation could be substantial (see Fig. 3 in Reynolds 1985a). At high latitudes, where the random vertical motions of the gas could be probed, [S II] and [N II] observations have not yet been made. However, the H_α line widths and radial velocities appear to be consistent with the non-thermal velocities derived at lower latitudes (Reynolds & Tufte, in preparation). Whether or not these H_α profiles have low-level, extended wings, such as those found in the 21 cm line profiles (Lockman & Gehman 1991), cannot be determined because of the relatively low signal-to-noise ratio of the high latitude H_α data.

4. IONIZATION STATE

The absence of detectable [O I] $\lambda6300$ and [N I] $\lambda5201$ implies that the diffuse ionized gas is confined to regions in which the hydrogen is nearly fully ionized. This conclusion is based upon high sensitivity searches in two diffuse background directions near the Galactic equator. Toward $l = 184°$ the intensity ratio [N I] $\lambda5201/$[N II] $\lambda6584 < 0.04$, which implies that within the [N II] emitting gas $n(N^+)/n(N°) > 4$ for $T > 6000$ K (Reynolds, Roesler, & Scherb 1977). Since the ionization potential of nitrogen is 0.9 eV above that of hydrogen, this suggests that the hydrogen also has a high fractional ionization. This is confirmed by the upper limit on the intensity ratio [O I] $\lambda6300/H_\alpha < 0.02$ toward a second background direction, $l = 114°$, b = 0° (Reynolds 1989). The ionization of oxygen is tied closely to that of hydrogen via a large charge exchange cross section for the reaction $H^+ + O° \rightleftharpoons H° + O^+$. The presence of H_α and [S II] coupled with the absence of [O I] implies that $T > 5400$ K and the hydrogen ionization ratio $n(H^+)/n(H°) > 2$ within the diffuse emitting gas; if $T = 8000$ K, then $n(H^+)/n(H°) > 15$.

The strong upper limit on the [O I] intensity also implies that less than 3% of the diffuse H_α emitted along this sight line can originate from partially ionized gas with $n(H^+)/n(H°) \simeq 0.1$, because within such a medium the intensity ratio [O I]$/H_\alpha > 0.7$ (for $T > 6000$ K). This clearly rules out penetrating radiation, including low energy cosmic ray protons, as the primary source of the diffuse ionization near the midplane.

High [S II] $\lambda6716/H_\alpha$ ($\simeq 0.25$–0.5) and low [O III] $\lambda5007/H_\alpha$ (< 0.05) intensity ratios appear to be characteristics of the diffuse ionized medium that clearly distinguish it from traditional O and B star H II regions (Reynolds 1985a, b, 1988). Because the ionization potential of S^+ is only 23.4 eV, while that of O^{++} is 35.1 eV, the ions probably are primarily in low stages of ionization (e.g., C^+, N^+, O^+, and S^+).

It must be emphasized that the forbidden-line observations have been limited to low Galactic latitudes, and that ionization and excitation conditions could vary with distance from the midplane. Just such a variation has in fact been observed within the extended ionized component of the edge-on spiral galaxy NGC 891 (Dettmar & Schulz 1992). Probing such effects in our Galaxy must await the construction of more sensitive instrumentation.

5. MASS AND SCALE HEIGHT

The mass surface density and extent of the ionized layer above the Galactic midplane have been estimated from pulsar dispersion measures. Distances have been determined for approximately 15% of the more that 500 known pulsars in the Galactic disk (see Frail & Weisberg 1990). These pulsars, with distances ranging from 130 pc (Gwinn et al. 1986) to more that 10 kpc, reveal ubiquitous ionization throughout the Galactic disk with a mean free electron density at the midplane $<n_e>_0 \simeq 0.03$ cm^{-3} (Lyne, Manchester, & Taylor 1985; Nordgren, Cordes, & Terzian 1992). This space-averaged value has changed little since pulsars were discovered more that two decades ago, although determining it to more than one significant figure is a matter of some controversy (see below).

Pulsars in globular clusters far from the Galactic midplane reveal the $|z|$-extent of the gas and provide a direct measurement of the total column density $N_\perp$ of H II perpendicular to the disk. Unfortunately, the number of samples is small. Twenty-nine pulsars have been detected within twelve globular clusters, but only five clusters are sufficiently far from the midplane (i.e., $|z| > 3$ kpc) to probe the entire H II column between the midplane and infinity. The resulting five values for the perpendicular column density $N_\perp \equiv N_{HII} \sin |b|$ range from $5.31 \pm 0.07 \times 10^{19}$ cm^{-2} to $9.52 \pm 0.01 \times 10^{19}$ cm^{-2} with an average value $N_\perp = 7.06 \times 10^{19}$ cm^{-2} (see Reynolds 1991a). This corresponds to a mass surface density for the ionized component of 1.6 M$_\odot$ pc^{-2} which includes a factor of 1.4 for helium. Nearly all ($\gtrsim 90\%$) of this can be attributed to warm ($\sim 10^4$ K) gas (Reynolds 1990a, 1992b). For these five lines of sight a comparison of the dispersion measures with the 21 cm H I column densities gives N(H II)/N(H I) column density ratios that range from 0.26 to 0.63 (Reynolds 1991a; Lockman, Langston, & Reynolds, in preparation).

The vertical extent of the diffuse ionized component can be calculated simply from the relationship $H = N_\perp / <n_e>_0$; for an exponential density distribution H is the scale height. Because the relatively small amount of hot ($\sim 10^6$ K) gas contributes to both $N_\perp$ and $<n_e>_0$, it should have little effect on the derived value of H for the warm gas. In this way, using essentially the same data, Nordgren, Cordes, and Terzian (1992; hereafter NCT) determine $H \simeq 670$ pc, while Reynolds (1991a) finds that $H \simeq 910$ pc. The difference is due almost entirely to the value adopted for $<n_e>_0$. The five high-$|z|$ globular clusters that provide the value for $N_\perp$ are at latitudes $27° \leq |b| \leq 80°$ and thus, for a gas layer half-thickness $H \sim 1$ kpc, sample a region within about 2 kpc of the sun. These high latitude lines of sight pass through only the diffuse component. On the other hand, the value for $<n_e>_0$ is usually derived from relatively low latitude sight lines that sample the disk out to 10 kpc from the sun. Therefore, the resulting value for $<n_e>_0$ could be affected by large scale variations in the mean density of the diffuse ionized gas over the Galactic disk. Frail & Weisberg (1990) have in fact found an increase in the density of the diffuse component toward the inner Galaxy. The derived value for $<n_e>_0$ also could be affected by the inclusion of lines of sight intersecting traditional H II regions that are too distant to be seen optically but nevertheless contribute significantly to the dispersion measure. In particular, there is a large concentration of H II regions in a thick annulus located approximately halfway between the Galactic center and the solar circle.

Lyne, Manchester, & Taylor (1985) addressed the question of H II region contamination and concluded that 38% (0.015 cm^{-3}) of the mean free electron density was due to H II regions within about 70 pc of the midplane with the remainder (0.025 cm^{-3}) due to the extended low density component. A value of 0.025 ± 0.005 cm^{-3} for the diffuse component was also found by Weisberg, Rankin, & Boriakoff (1980) using dispersion measure data for sight lines believed to be free of H II regions.

NCT addressed the problem of H II region contamination and a possible increase in density toward the inner Galaxy by rejecting pulsars with Galactic longitudes within 30° of the Galactic center and distances d > 5 kpc. From their sample (Table 1 in NCT) they derive $< n_e >_o = 0.033 \pm 0.003$ cm^{-3}. However, this value appears to depend on precisely which pulsars are rejected. For example, if the zone of rejection is widened slightly to $|l| < 41°$ and d > 3 kpc, which reduces the number of useable pulsars from 29 to 21, then the resulting value for $< n_e >_o$ is reduced to 0.026 cm^{-3}, which is not significantly different from that derived previously by Lyne et al. and Weisberg et al. The mean value of 0.050 cm^{-3} for the eight inner Galaxy pulsars rejected by the above criteria but retained by NCT strongly suggests that the NCT rejection criteria do not completely remove biases in the data resulting from higher densities in the inner Galaxy. If $< n_e >_o$ is defined by the nine pulsars in Table 1 of NCT that have d < 3 kpc, and therefore sample approximately the same portion of the disk as the high-$|z|$ globular cluster pulsars that define N$_\perp$, then $< n_e >_o \simeq 0.023$ cm^{-3} and H $\simeq 1000$ pc.

6. VOLUME FILLING FRACTION AND LOCAL DENSITY

It is only toward the four high-$|z|$ globular cluster pulsars accessible with the Wisconsin Fabry-Perot spectrometer that ratios of the pulsar dispersion measure (DM $\propto n_e$) and H$_\alpha$ emission measure (EM $\propto n_e^2$) provide accurate measurements of the clumping of the ionized gas within the interstellar medium (the fifth cluster has too low a declination). The data in these four directions indicate that the ionized gas is clumped into regions having an average electron density $n_e = 0.08$ cm^{-3} and a filling fraction f $\gtrsim 0.2$ of the volume within the 2 kpc thick layer of diffuse H II (Reynolds 1991b). If the density decreases exponentially with distance $|z|$ from the midplane, then $n_e \simeq 0.16$ cm^{-3} at z = 0.

Some analyses of the H$_\alpha$ and pulsar data suggest that the filling fraction increases from f ~ 0.1 at z = 0 to f > 0.4 at $|z| = 1$ kpc (Kulkarni & Heiles 1987; Reynolds 1991b). Given the uncertainties and the limited amount of data, such results must be considered tentative.

7. DISTRIBUTION ON THE SKY

High spectral resolution (12 km s^{-1}) Fabry-Perot scans have sampled the H$_\alpha$ background in a few hundred directions above declination $-20°$ with 0.°1 and 0.°8 diameter fields of view. The results indicate that the emission covers the sky. A low angular resolution, broad-band (-20 km s^{-1} $< V_{LSR} < +60$ km s^{-1}) map has been synthesized from these observations and is shown in the top half of Figure 1. The contours are in rayleighs, where 1 R $= 10^6/4\pi$ ph cm^{-2} s^{-1} sr^{-1}, corresponding to 2.41×10^{-7} ergs cm^{-2} s^{-1} sr^{-1} at H$_\alpha$. The extreme undersampling of the sky by the widely separated observation directions, which are spaced approximately 5–10° apart at $|b| \lesssim 20°$ and 10–20° apart at high b, restricts the information to large scale features. At $|b| \lesssim 5°$, interstellar extinction limits the effective sight line to a distance of 2–3 kpc and the diffuse H$_\alpha$ intensity to 3–10 R (Reynolds 1983). Local spiral structure is apparent; the intensity enhancement at $l < 30°$, $|b| < 10°$ is associated with the inner Sagittarius arm, and the emission at $60° < l < 240°$ is associated with the local Orion arm. Emission from the outer Perseus arm between $l = 90°$ and 150°, although present in the original H$_\alpha$ data, is excluded from this map by the choice of radial velocity interval. All three nearby arms are clearly seen in the radial velocity vs. longitude plots presented in Reynolds (1983). The relatively

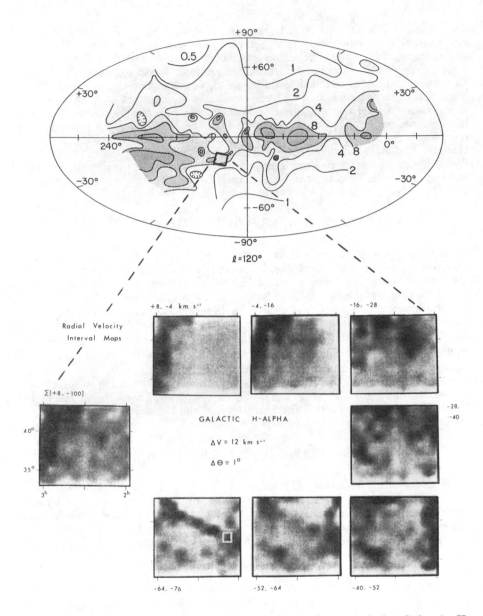

FIG. 1 *Top*: Coarse angular resolution ($\sim 10°$) map of the Galactic H_α background above declination $-20°$ within the radial velocity interval -20 km s^{-1} to $+60$ km s^{-1} (LSR). Contour values are in rayleighs, where 1 R $= 2.4 \times 10^{-7}$ ergs cm^{-2} s^{-1} sr^{-1} at H_α. The center of the Aitoff projection is at $l = 120°$, b $= 0°$. *Bottom*: Higher angular resolution ($\simeq 1°$), 12 km s^{-1} radial velocity interval maps of the H_α background within an approximately $11° \times 11°$ region centered at $l = 144°$, b $= -21°$ (see text). A $3'$ angular resolution image of the $\simeq 1$ square degree region outlined on the -64 to -76 km s^{-1} frame is shown in color plate 9.

small "bright spots" are associated with H II regions surrounding O stars and early B stars.

At higher latitudes the interstellar H_α intensities generally follow a csc $|b|$ law with $< I_\alpha \sin|b| > \simeq 1$ R. At a given latitude there are real intensity variations from direction to direction, which indicate the presence of smaller scale, unresolved angular structure. The standard deviation of $I_\alpha \sin|b|$ for the "random" sample of 45 background directions at $|b| > 15°$ (Reynolds 1992a) is 0.5 of the mean with the highest intensities approximately a factor of 3 above the lowest. Some of this scatter is the result of measurement uncertainties. At $|b| > 50°$ the mean intensity appears to be about one-half that predicted by the csc $|b|$ law at lower latitudes. The free electron column density should have a smoother distribution on the sky than the H_α. This expectation is consistent with the dispersion measures toward the high-$|z|$ globular cluster pulsars, the only lines of sight for which the total free electron column density is measured. The standard deviation of DM sin $|b|$ for these five widely spaced high latitude directions is only 0.23 of the mean.

To investigate the H_α intensity distribution on smaller angular scales, two approximately $11° \times 11°$ regions of the sky have been mapped on $1° \times 1°$ grids with a $0.°8$ beam. One of the maps is near the Galactic equator centered at $l = 213°$, b $= +3°$ (Reynolds 1987), and the other is at higher latitude, centered at $l = 144°$, b $= -21°$ (Reynolds 1980). The data for this second region are displayed in the lower portion of Figure 1 as a series of narrow-band H_α intensity maps, obtained by integrating the Galactic emission line profiles within seven 12 km s^{-1} radial velocity intervals between $+8$ km s^{-1} and -76 km s^{-1} (LSR). All of the maps are in the same orientation with right ascension increasing right to left from 2^h to 3^h and declination increasing upward from $+33°$ to $+43°$. On each map black represents the highest intensity, corresponding to an emission measure of about 23 cm^{-6} pc on the total intensity, "broad-band" map and to emission measures of 14, 6.8, 4.5, 2.9, 2.7, 2.5, and 1.8 cm^{-6} pc for each of the narrow velocity interval maps between $+8$ km s^{-1} and -76 km s^{-1}, respectively.

These $1°$ angular resolution pictures reveal that the H_α background has a complex spatial and kinematic structure with long filamentary regions and irregular patches superposed on a more diffuse background that does not show up well on this representation. Because the thermal width of hydrogen is 21 km s^{-1} (FWHM) at 10^4 K, there probably is not much additional structure that can be revealed by observations at higher spectral resolution. Much of the structure "averages out" through superposition on the broad-band map, which has only one "bright" feature along the eastern portion of the region. The mean value of $I_\alpha \sin|b|$ for the 121 directions is 3.1 x 10^{-7} ergs cm^{-2} s^{-1} sr^{-1} (corresponding to $<$EM sin $|b|$ $> = 2.9$ cm^{-6} pc), nearly identical to that for the random "all sky" sample discussed above. The standard deviation of $I_\alpha \sin|b|$ is 0.42 of the mean, which in this case is an accurate representation of the actual H_α intensity variations, since the measurement uncertainties are significantly less than the observed standard deviation.

The H_α emission appears to be diffuse at least on scales down to $3'$. Narrow-band (12 km s^{-1}) high angular resolution ($3'$) images of $0.°8$ diameter fields on the sky show that the emission does not resolve into small angular scale sources (Milster, Scherb, & Reynolds, in preparation). An exception is the long ($> 15°$), narrow ($\simeq 1°$), faint ($\simeq 1$ R) H_α filament on the -64 to -76 km s^{-1} map, originally investigated by Ogden & Reynolds (1985). When examined at $3'$ angular resolution (see color plate 9), a narrow ridge of enhancement about

6' wide with an EM $\simeq$ 3–4 cm^{-6} pc is found to run down the center of the filament (Kung, McCullough, Reynolds, & Tufte, in preparation). The intensity of this very narrow ridge is approximately one-half of the total (broad-band) interstellar H$_\alpha$ intensity in this part of the sky. The nature of this filament and its relationship to the more diffuse background are not yet known.

8. CONCLUSIONS

Faint optical line emission and pulsar dispersion measures are the principal sources of information about the warm ionized component of the interstellar medium. The derived properties are summarized in Table 1, which also lists the number of lines of sight for which measurements were made and the average latitude of those lines of sight.

Although many basic properties of the gas have been determined, the source of the ionization is not yet clear. Also, there is little understanding of the relationship between the ionized gas and other components of the interstellar medium, such as H I clouds and the hot "coronal" phase. This situation should be improved by a planned sky survey of the diffuse interstellar H$_\alpha$ emission at an angular resolution comparable to that of the H I surveys. Furthermore, the detection of extended, warm ionized gas in other galaxies (see the review by Dettmar 1992) has opened up the possibility of exploring the relationship of the warm ionized medium to galaxy type, star formation activity, extended non-thermal emission, and a variety of other global and local phenomena in galaxies.

This work was supported by the National Science Foundation through AST91-15703 and AST91-22701 and by the National Aeronautics and Space Administration through NAG5-674.

TABLE 1

MEASURED PARAMETERS OF THE IONIZED MEDIUM

Parameter	Value	Number of sight lines	$< |b| >$
Ionization rate, $< r_G >$	4.8 $\pm$0.4 $\times$ 10^6 s^{-1} cm^{-2}	45	37°
Temperature, T	7.5 $\pm$3.2 $\times$ 10^3 K	20	9°
Non-thermal speeds	12–30 km s^{-1}	20	9°
H II surface density	1.4 $\pm$0.2 $\times$ 10^{20} cm^{-2}	5	48°
Scale height, H	910[a](-320, +400) pc	5	48°
Volume fraction, f	> 0.2	4	49°
Local density, n_e	0.08[b] $\pm$ 0.02 cm^{-3}	4	49°
$n(H^+)/n(H^\circ)$	> 2	2	0°

[a]The value if $<n_e>_o$ = 0.025 (-0.05, +0.08) cm^{-3} (see text).
[b]The average density within ionized regions at $|z| \leq$ 910 pc.

REFERENCES

Abbott, D. C. 1982, ApJ, 263, 723

Burstein, D. & Heiles, C. 1982, AJ, 87, 1165

Dettmar, R. –J. 1992, Fund. of Cosmic Phy., 15, 143

Dettmar, R. –J. & Schulz, H. 1992, A&A, 254, L25

Frail, D. A., & Weisberg, J. M. 1990, AJ, 100, 743

Gwinn, C. R., Taylor, J. H., Weisberg, J. M., & Rawlings, L. A. 1986, AJ, 91, 338

Hewish, A., Bell, S. J., Pilkington, S. D., Scott, P. F., & Collins, R. A. 1968, Nature, 217, 709

Hoyle, F. and Ellis, G. R. A. 1963, Australian J. Phys, 16, 1

Jura, M. 1979, ApJ, 227, 798

Kulkarni, S. & Heiles, C. 1987, in Interstellar Processes, ed. D. J. Hollenbach & H. A. Thronson, Jr. (Dordrecht:Reidel), 87

Lockman, F. J. & Gehman, C. S. 1991, ApJ, 382, 182

Lyne, A. G., Manchester, R. N., & Taylor, J. H. 1985, MNRAS, 213, 613

Martin, P. G. 1988, ApJS, 66, 125

Mathis, J. S., Rumpl, W., & Nordsieck, K. H. 1977, ApJ, 217, 425

Miller, W. W., III & Cox, D. P. 1993, preprint

Nordgren, T. E., Cordes, J. M., & Terzian, Y. 1992, AJ, 104, 1465

Ogden, P. M. & Reynolds, R. J. 1985, ApJ, 290, 238

Osterbrock, D. E. 1989, in Astrophysics of Gaseous Nebulae and Active Galactic Nuclei (Mill Valley: University Science Books), p.204

Pengally, R. M. 1964 MNRAS, 127, 145

Reynolds, R. J. 1980, ApJ, 236, 153

Reynolds, R. J. 1983, ApJ, 268, 698

Reynolds, R. J. 1985a, ApJ, 294, 256

Reynolds, R. J. 1985b, ApJ, 298, L27

Reynolds, R. J. 1987, ApJ, 323, 118

Reynolds, R. J. 1988, ApJ, 333, 341

Reynolds, R. J. 1989, ApJ, 345, 811

Reynolds, R. J. 1990a, ApJ, 349, L17

Reynolds, R. J. 1990b in IAU Symposium No. 139, The Galactic and Extragalactic Background Radiation, ed. S. Bowyer and C. Leinert (Dordrecht:Kluwer), 157

Reynolds, R. J. 1991a, in IAU Symposium No. 144, The Interstellar Disk-Halo Connection in Galaxies, ed. H. Bloemem (Dordrecht:Kluwer), 67

Reynolds, R. J. 1991b, ApJ, 372, L17

Reynolds, R. J. 1992a, ApJ, 392, L35

Reynolds, R. J. 1992b, in Massive Stars: Their Lives in the Interstellar Medium, ed. J. Cassinelli & E. Churchwell, in press.

Reynolds, R. J. & Cox, D. P. 1992, ApJ, 298, L27

Reynolds, R. J., Roesler, F. L., Scherb, F. & Boldt, E. 1971, in The Gum Nebula and Related Problems, eds. S. P. Maran, J. C. Brandt, T. P. Stecher (NASA SP-332), p.169

Reynolds, R. J., Roesler, F. L., & Scherb, F. 1973, ApJ, 179, 651

Reynolds, R. J., Roesler, F. L., & Scherb, F. 1977, ApJ, 211, 115

Reynolds, R. J., Scherb, F., & Roesler, F. L. 1973, ApJ, 185, 869

Roesler, F. L., Reynolds, R. J., Scherb, F., & Ogden, P. M. 1978, in High Resolution Spectroscopy, ed. M. Hack (Trieste:Observatorio Astronomico), 600

Weisberg, J. M., Rankin, J. M., & Boriakoff, V. 1980, A&A, 88, 84

TURBULENCE IN THE CORONAL INTERSTELLAR MEDIUM

J.A. Phillips
Owens Valley Radio Observatory, Caltech 105-24, Pasadena, CA 91125

Andrew W. Clegg
Naval Research Laboratory, Code 7213, Washington, DC 20375

ABSTRACT

Low frequency scintillation observations of the nearby pulsar 0950+08 show that C_n^2, the amplitude of the electron density fluctuation spectrum, is $\simeq 10^{-4.5}$ m$^{-20/3}$ along the line of sight, a value which is 5 to 10 times smaller than C_n^2 measured towards other pulsars within a few kpc of Earth. Since the line of sight to PSR 0950+08 is contained mostly within the local X-ray bubble we can conclude that the coronal phase of the interstellar medium is weakly scattering.

1. SCATTERING TOWARDS PSR 0950+08

Density inhomogeneities in the interstellar plasma cause a variety of well-known propagation effects including angular broadening of radio images, temporal smearing of pulsar waveforms, and intensity scintillation of compact radio sources (see Rickett 1990 for a review). It is not known with certainty where the density fluctuations reside, but theoretical work on wave dissipation suggests the warm ionized medium, or the diffuse envelopes of HII regions (Spangler 1991; see also Spangler and Gwinn 1990). Efforts to identify the 'fluctiferous medium' (a term coined by Spangler) with a particular phase of the ISM have met with little success, largely because pulsars are distant objects and – on average – their lines of sight pass through an assortment of HII regions, cold clouds, supernova shells, etc. In such cases it is difficult to decide which object or region is responsible for the observed scintillation.

Studies of nearby pulsars may be the best way to understand density fluctuations in specific phases of the interstellar medium. PSR 0950+08 – one of the three nearest pulsars – is only 127±13 pc from the Earth, and the Earth-pulsar line of sight lies almost entirely within the local X-ray bubble (Fig. 1). The bubble is an elongated cavity in the local ISM filled with hot ($T \sim 10^5 - 10^6$ K), tenuous ($n_e \sim 0.01 - 0.005$ cm^{-3}) gas typical of the coronal phase of the interstellar medium. Scattering measurements of the pulsar would therefore provide an excellent probe of turbulence in the coronal phase of the interstellar medium.

Between 1989 and 1990 we obtained dynamic spectra of PSR 0950+08 at frequencies near 50 MHz (λ6m). The data acquisition and reduction are described in an earlier paper (Phillips & Clegg 1992). We chose to observe at meter wavelengths because we expected that the pulsar would be weakly scattered and that higher frequency measurements would be biased by radiometer bandwidths that are much smaller than the scintillation bandwidth (Fig. 2). Table 1 lists the observing parameters for our 50 MHz observations. Following standard practice we estimated the diffractive decorrelation time scale (τ_d) and frequency scale ($\Delta\nu_d$) from the two-dimensional autocorrelation functions of the dynamic spectra, where the dynamic spectrum is the observed pulsar flux density as a function of time and frequency. The values of τ_d and $\Delta\nu_d$ listed in Table 1 correspond to the half-width of the ACF along the time axis and the frequency axis, respectively.

The main goal of these observations is to estimate C_n^2, the amplitude of the electron density fluctuation power spectrum. For a power law model of interstellar density fluctuations the variance of electron density is related to C_n^2 by the equation $\delta n_e^2 \approx 0.7[C_n^2 \ell_o^{2/3}]$ cm^{-3}, where ℓ_o is the 'outer scale' in pc (see, e.g., Phillips & Clegg 1992). Cordes, Weisberg & Boriakoff (1985) have discussed in detail the estimation of C_n^2 from decorrelation bandwidth measurements, and based on their work we adopted the following estimator: $\hat{C}_n^2 = 2 \times 10^{-6} \nu^{11/3} D^{-11/6} \Delta\nu_d^{-5/6}$ m$^{-20/3}$, with ν in MHz, D in pc, and $\Delta\nu_d$ in kHz. The numerical coefficient and exponents in this equation assume a Kolmogorov spectrum of density fluctuations. The results from our data are listed in column (9) of Table 1. We estimate the errors in $\hat{C}_n^2$ to be $\sim 25\%$, based on the uncertainty in distance to PSR 0950+08 and the measurement errors in $\Delta\nu_d$.

Table 1: Dynamic Spectra and Scattering Parameters

Date	ν MHz	δt s	$\delta \nu$ kHz	T min	B MHz	$\Delta\nu_d$ kHz	τ_d s	$\log C_n^2$ m$^{-20/3}$	v_{iss} km s^{-1}
(1)	(2)	(3)	(4)	(5)	(6)	(7)	(8)	(9)	(10)
26-11-89	47	13	9.8	33	1.250	28 ± 2	279 ± 17	-4.6	50
01-01-90	51	13	4.9	14	0.625	49 ± 7	227 ± 32	-4.7	72
01-01-90	51	13	4.9	20	0.625	31 ± 4	372 ± 48	-4.5	35
27-01-91	51	12	4.9	20	1.250	19 ± 1	243 ± 15	-4.4	42

Symbol definitions: δt is the time resolution of the dynamic spectrum; $\delta\nu$ is the frequency resolution; T is the total observing time; B is the observing bandwidth; $\Delta\nu_d$ and τ_d are the diffractive scintillation bandwidth and time scales, respectively. The latter are defined as the half-width at half maximum of the dynamic spectrum's ACF, along the frequency axis ($\Delta\nu_d$) or the time axis (τ_d). The fractional errors in $\Delta\nu_d$ and τ_d were estimated from the formula $\epsilon = (\Delta\nu_d \tau_d / BT)^{1/2}$.

The derived values of C_n^2 for PSR 0950+08 are 5 to 10 times smaller than those obtained for other pulsars within a few kpc of Earth (Cordes et al. 1985). For example, three of the least strongly-scattered pulsars are 1937+21, 2016+28, and 2020+28. These are believed to lie in relatively quiescent regions between the local and Sagittarius spiral arms. Even so, they exhibit values of $\log \hat{C}_n^2$ close to -3.5, an order of magnitude larger than that of PSR 0950+08. We conclude then that coronal gas is very weakly scattering compared to other phases of the ISM.

Spangler and Gwinn (1990) have argued that the propagation speed of density irregularities may be associated with the Alfvén velocity in the scattering medium. Dewey et al. (1988) estimated the propagation speed of scattering irregularities for two relatively distant binary pulsars and found the velocity to be less than 25 km/s. For comparison, the Alfvén velocity is $\simeq 150$ km/s in the coronal phase, $\simeq$ 18 km/s in the warm ionized medium, and $\simeq$ 4 km/s in the diffuse envelopes of HII regions (Spangler and Gwinn 1990). The limits by Dewey et al. appear to rule out the coronal phase as a source of significant scattering, although they cannot distinguish between the WIM and envelopes of HII regions. If the scattering of PSR 0950+08 is indeed dominated by coronal gas in the local bubble, we would expect a relatively large propagation velocity commensurate with the expected Alfvén speed of $\sim$ 150 km/s. Column (10) of Table 1 shows the scintillation velocity at each epoch computed from the formula $v_{iss} = 1.06 \times 10^4 \sqrt{\Delta\nu_d D}/(\tau_d \nu)$ km/s where the symbols have the same meaning and units as in the equation for $\hat{C}_n^2$. The largest velocity in the table (72 km/s) is comparable to the expected Alfvén velocity of coronal gas, but there is an important caveat: Although the scintillation bandwidth is well determined in our data the scintillation time scale is not as well sampled. The derived velocities in Table 1 are inversely correlated with the observing time T, a fact that strongly suggests observational bias. We could improve the velocity estimates by taking data for a much longer period ($T \gtrsim$ 2 hr) to improve the sampling of intensity scintillations in time.

Of the data we have, the longest dynamic spectrum was 33 min, and the corresponding velocity was 50 km/s. The scintillation speed is a vector combination of the pulsar space velocity (v_{psr} = 24 km/s for 0950+08), the earth's orbital motion ($\mathbf{v}_\oplus$), and the drift velocity of the scattering material ($\mathbf{v}_{ism}$; see eq. [1] of Dewey et al. 1988). For a scattering screen midway along the line of sight the maximum contribution of the Earth-pulsar motion to v_{iss} is $\pm$27 km/s. The velocity of the scattering material could therefore have any value between $50 - 27 = 23$ km/s and $50 + 27 = 77$ km/s, depending on the relative orientation of $\mathbf{v}_{psr}$, $\mathbf{v}_\odot$, and $\mathbf{v}_{ism}$. The velocity range is not narrow enough to discriminate between the Alfvén velocity of the coronal gas and that of the warm ionized medium. If the lower velocity is correct it probably means that what little scattering there is comes from warm ionized gas outside the boundary of the X-ray bubble. On the other hand, the higher velocity would imply scattering by fast moving irregularities inside the bubble. Whichever is true, the conclusion that coronal gas is weakly scattering still holds.

REFERENCES

Cordes, J.M., Weisberg, J.M., and Boriakoff, V. 1985, ApJ, 288, 221

Cox, D.P. & Reynolds, R.J. 1987, ARA&A, 25, 303

Dewey, R.J., Cordes, J.M., Wolszczan, A, & Weisberg, J.M. 1988, in Radio Wave Scattering in the Interstellar Medium (eds. Cordes, J.M., Rickett, B.J. & Backer, D.C.) 217-221 (American Institute of Physics, New York)

Gwinn, C.R., Taylor, J.H., Weisberg, J.M. & Rawlings, L.A. 1986, AJ, 91, 338

Phillips, J.A. & Clegg, A.W. 1992, Nature, 360, 137

Rickett, B.J. 1990, ARA&A, 28, 561

Snowden, S.L., Cox, D.P., McCammon, D., & Sanders, W.T. 1990 ApJ, 354, 211

Spangler, S.R. 1988 in Radio Wave Scattering in the Interstellar Medium (eds. Cordes, J.M., Rickett, B.J. & Backer, D.C.) 66-69 (American Institute of Physics, New York)

Spangler, S.R. 1991, ApJ, 376, 540

Spangler, S.R. & Gwinn, C.R. 1990, ApJ, 353, L29

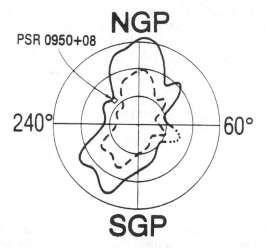

Fig. 1. PSR 0950+08 and the local X-ray bubble. The contours show the boundary of the local bubble inferred from the intensity of soft X-ray background emission (adapted from Fig. 6 of Snowden et al. 1990). The cross-section shown is through the north and south galactic poles along galactic longitude $l = 240°$ with the Sun located at the origin. The inner and outer contours represent two plausible models for the displacement of neutral material by the X-ray emitting gas. Three scale circles are shown with radii of 100, 200, and 300 pc. The location of PSR 0950+08 ($l = 228.9°$, $b = 43.7°$) is denoted by an open diamond. At least 70% of the line of sight to the pulsar passes through the bubble and, depending on the displacement model, 0950+08 could be completely within the nominal cavity boundary.

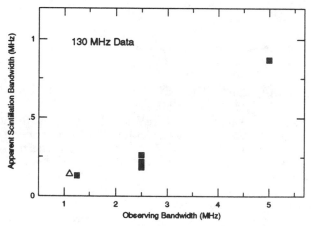

Fig. 2. The apparent scintillation bandwidth of PSR 0950+08 at observing frequencies near 130 MHz as a function of observing bandwidth. The trend suggests that the the scintillation bandwidth is poorly sampled at frequencies as low as 130 MHz because it is larger than or comparable to the observing bandwidth. The results reported in this paper are from 47 MHz data where the scintillation bandwidth is smaller than the observing bandwidth.

NEW INSIGHTS INTO THE STRUCTURE OF THE ISM PROVIDED BY ROSAT

Uwe Herbstmeier, Peter Moritz, Justina Engelmann
Jürgen Kerp, Gernot Westphalen

Radioastronomisches Institut der Universtiät Bonn,
Auf dem Hügel 71, D-W5300 Bonn 1, FRG
herbst@babsy.mpifr-bonn.mpg.de

1. INTRODUCTION

The concept of the Local Hot Bubble (LHB) was introduced (*e.g.* Snowden *et al.* 1990) to explain the observations of the diffuse soft X-ray background (SXRB) in the energy range from 0.078 to 0.284 keV (for a review, see McCammon & Sanders 1990). The structure of the LHB has been derived from the assumptions that the SXRB has little or no contributions from emitting regions outside the LHB, and that dense neutral clouds which absorb soft X-rays do not exist in significant quantities within the LHB (Cox & Reynolds 1987).

However observations with ROSAT (Trümper 1983) have shown that these basic assumptions for the LHB are no longer valid: Snowden *et al.* (1991), Burrows & Mendenhall (1991), and Lilienthal *et al.* (1992) have found shadows in the SXRB cast by clouds outside the LHB. These results prove that a large fraction of the SXRB is originating in the galactic halo. The detection of an elongated H I filament located in the LHB (Wennmacher *et al.* 1992 and Kerp *et al.* 1992) has shown that dense clouds are embedded in the local hot region. Therefore the ROSAT observations provide important datas for a better understanding of the 3-dimensional distribution of the hot plasma in relation to the other phases of the interstellar medium (ISM) (see also Mebold *et al.* 1993 and Hirth *et al.* 1992).

On the other hand the ROSAT results set up new perspectives for the determination of the structure of the Galaxy and the nature of it's ISM. Here several of these aspects are summarized.

2. THE MOLECULAR CONTENT OF INTERSTELLAR CLOUDS

Figure 1 shows the ROSAT 1/4 keV and the 100 μm IRAS survey images of the region of the Draco nebula, a molecular cloud at a large distance from the galactic plane ($z > 200$ pc, Lilienthal *et al.* 1991, Herbstmeier *et al.* 1992). We have compared the X-ray data with the far-infrared (FIR) emission of the cloud. The depth distribution of the shadow in Fig. 1 has been used to derive the total hydrogen column density distribution ($N(\mathrm{H})$), of the absorbing material (Moritz *et al.* in prep.). Here a smooth surface brightness has been assumed for the X-ray background of the Draco nebula as well as the absorption cross sections described by Snowden *et al.* (1991). The derived column density distribution has been compared to the 100 μm emission, I_{100}. Assuming a constant FIR-emissivity (a_{100}) for the whole nebula, the best fit value becomes

$$a_{100} = I_{100}/N(\mathrm{H}) = 1.1 \times 10^{-20} \text{ MJy ster}^{-1}\text{cm}^2.$$

This conversion factor is consistent with that derived for the same nebula from a comparison of I_{100} with the column density of atomic hydrogen $N(\text{H\,I}\,)$ and the velocity integral of the ^{12}CO J=1→0 line: Herbstmeier *et al.* (1992) have derived FIR-emissivities in the range 0.9×10^{-20} to 2.1×10^{-20} MJy ster^{-1}cm^2 varying across the Draco nebula.

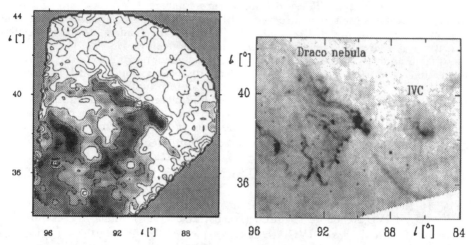

Fig. 1. left: The ROSAT 1/4 keV survey image of the region of the Draco nebula (provided by S. Snowden). The count rates range from $> 12 \times 10^{-4}$ (white) to $< 7 \times 10^{-4}$ counts(s arcmin^2)$^{-1}$ (black). Contours indicate the range from 8×10^{-4} to $> 18 \times 10^{-4}$ counts(s arcmin^2)$^{-1}$. Right: The same region at 100 μm wavelength (IRAS Sky Survey Atlas). The relative intensity peaks at > 5.5 MJy ster^{-1} (black).

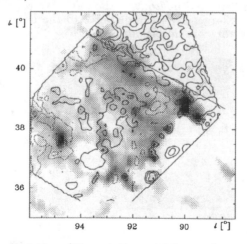

Fig. 2. The relative deviation $(C_{\text{obs}} - C_{\text{model}})/C_{\text{model}}$ (see text) overlayed on the IRAS 100 μm image. Negative values are represented by dashed contours. The contour distance is ± 0.1.

The conversion factor derived above does not fit, however, equally well for all parts of the nebula. Calculating the relative deviation $((C_{obs} - C_{model})/C_{model})$ of the observed count rate C_{obs} from our modelled count rate C_{model} we find the most striking deviations for the boundary towards low l and low b (galactic south-west (SW), see Fig. 2). The model count rate C_{model} has been determined from a smooth soft X-ray background, from the observed values of I_{100} and from the fitted value of a_{100}. These deviations are coincident with a chain of molecular clumps for which the observed depth of the shadow is smaller than that expected from our model. To explain the depth of the shadows of the molecular clumps a_{100} there is larger than the mean value given above.

The analysis of X-ray shadows is a new technique to derive the total column density of hydrogen nuclei for massive molecular clouds and to determine variations of the FIR-emissivity. This mass determination avoids the famous controversy about the conversion of the CO line integrals into the molecular column densities. Particularly interesting cases are those molecular clouds which can be observed as shadows against the harder X-ray background ($> 1 \, \mathrm{keV}$) close to the galactic plane with large total column densities ($> \sim 10^{22} \mathrm{cm}^{-2}$).

3. THE NATURE OF THE INTERMEDIATE VELOCITY CLOUDS

In the vicinty of the Draco nebula the head-tail shaped intermediate velocity cloud IVC 86+39-44 is found (Herbstmeier 1990). This IVC is possibly associated with the large IVC arc at $(l, b, v) \sim (110°, 40°, -50 \, \mathrm{km \, s}^{-1})$, located at a distance of $\sim 1 \pm 0.5$ kpc (Danly 1989). A comparison of the images in Fig. 1 and pointed observations with ROSAT clearly show that this cloud casts a shadow in the SXRB. From a preliminary analysis of the pointed ROSAT observations we derive the depth of the shadow to be $\sim 25 \%$. As the maximum hydrogen column density of IVC 86+39-44 is $\sim 2.5 \times 10^{20} \mathrm{cm}^{-2}$ the cloud is optically thick for the $1/4 \, \mathrm{keV}$ photons. A comparison of the shadow depth of the IVC with the values found for the Draco nebula shows that more soft X-ray photons are emitted in front of the IVC. We conclude that the distance to IVC 86+39-44 is larger than the distance to the Draco nebula, and that the IVC is emersed in the coronal plasma.

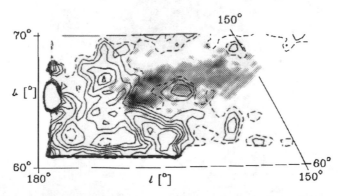

Fig. 3. The distribution of the H I column densities in the velocity range -126.3 to $-75.2 \, \mathrm{km s}^{-1}$ (gray scales) and the ROSAT $1/4 \, \mathrm{keV}$ survey image (contours, provided by S. Snowden) of the region of M I

4. THE GALACTIC CORONA, RELATED TO HIGH VELOCITY CLOUDS?

The positional association of some HVCs and diffuse soft X-ray spurs already discussed by Hirth *et al.* (1985) is confirmed by ROSAT survey data. An example for this is the area around the HVC called MI shown in Fig. 3. The distribution of the soft X-ray count rates in the area of MI is plotted as contours in the Aitoff projection overlayed on the image of the HI column densities of MI (Giovanelli *et al.* 1973, Westphalen in prep.). A comparison of the X-ray and the HI observations in the area of the HVC MI suggests that the X-ray emission looks like edge-brightening of the HVC.

If this suggestion is supported by further observations, it holds important clues to the understanding of the nature of soft X-ray emission in the galactic halo. An anticorrelation of X-ray and HI emission for MI can be interpreted in various ways. Attractive hypotheses are absorption of background emission by the neutral gas in the HVC, or the HVC as the densest inner part of a larger complex of gas whose outer parts are hot enough to emit soft X-ray photons, or alternatively, the soft X-ray enhancement as the bow wave due to the interaction of the infalling HVCs with the surrounding medium or a galactic wind.

ACKNOWLEDGEMENTS

We acknowledge the support for this work by Steve Snowden and others from the Max-Planck-Institut für Extraterrestrische Physik at Garching. This work was supported by the German space agency DARA under project no. 50 OR 9203.

REFERENCES

Burrows D.N., & Mendenhall J.A., 1991, Nat 351, 629
Cox D.P., & Reynolds R.J., 1987, ARA&A 25, 303
Danly L., 1989, ApJ 342, 78
Giovanelli R., Verschuur G.L., & Cram T.R., 1973, A&AS 12, 209
Herbstmeier U., 1990, in: Proc. of IAU Coll. 120 "Structure and dynamics of the interstellar medium" (eds. Tenorio-Tagle G., Moles M., Melnick J.), p. 458
Herbstmeier U., Heithausen A., & Mebold U., 1992, A&A in press
Hirth W., Mebold U., & Müller P., 1985, A&A 153, 249
Hirth W., Mebold U., Dahlem M., & Müller P., 1985, Ap&SS 186, 211
Kerp J., Herbstmeier U., & Mebold U., 1992, A&A submitted
Lilienthal D., Wennmacher A., Herbstmeier U., & Mebold U., 1991 A&A 250, 150
Lilienthal D., Hirth W., Mebold U., & de Boer K.S., 1992, A&A 255, 323
Mebold U., Kerp J., Herbstmeier U., Moritz P., Westphalen G., 1993, in: Proc. of the COSPAR/IAF Symp. "Recent results in X-ray and EUV astronomy" (eds. Trümper J., Bowyer S., *et al.*), submitted
McCammon D., & Sanders W.T., 1990, ARA&A 28, 657
Snowden S.L., Cox D.P., McCammon D., & Sanders W.T., 1990, ApJ 354, 211
Snowden S.L., Mebold U., Hirth W., Herbstmeier U., & Schmitt J.H.M.M., 1991, Sci 252, 1529
Trümper J., 1983, Adv. Sp. Res. 2(4), 241
Wennmacher A., Lilienthal D., & Herbstmeier U., 1992, A&A 261, L9

THE PROPERTIES OF THE LOCAL BUBBLE
FROM EUV OBSERVATIONS

R S Warwick, C R Barber, S T Hodgkin and J P Pye
Department of Physics and Astronomy, University of Leicester
Leicester, LE1 7RH, U.K.

1. INTRODUCTION

A catalogue of 384 relatively bright sources detected in the first-ever all sky survey in the extreme ultra-violet (EUV), carried out by the UK Wide Field Camera (WFC) on ROSAT, has recently been published (Pounds *et al.* 1993). Here we explore the statistical properties of the EUV source population, as represented by the WFC bright source catalogue, with a view to establishing details of the distribution of EUV-absorbing gas in the local interstellar medium (ISM).

The WFC catalogue lists a total of 335 sources detected in the short wavelength (60 – 140 Å) S1 filter observations and 262 sources detected in the longer wavelength (110 – 200 Å) S2 filter (239 sources were detected in both filters). These EUV sources are optically identified predominantly with white dwarf stars and active late-type stars of spectral types F,G,K and M (see Pounds *et al.* 1993); thus it is these two classes of source which are the most useful probes of the local ISM. From a compilation of distance estimates, we find that the late-type stars are seen in the EUV at distances ranging from a few pc up to ~ 100 pc, whereas the white dwarf detections extend out to about 300 pc.

2. MODELLING THE EUV SOURCE COUNTS

We have fitted a power-law relation ($dN/dC = KC^{-\gamma}$) to the binned differential counts for both the white dwarf stars and the late-type stars using a standard chi-squared minimization procedure. The white dwarf stars turn out to have an extremely flat source count in both survey filters (*i.e.* $\gamma \sim 1.4$). Similarly the counts for the late type stars are somewhat flatter than that expected for an unabsorbed uniform distribution of sources. The relatively shallow slope of the EUV source counts is almost certainly due to absorption of EUV photons by relatively cool, neutral (HI or HeI) or partially ionized (He II) gas in the local ISM.

For a nominally complete sample of sources, all at known distances, the observed form of the log N - log S curve can be readily compared to that expected for a specified absorption model. Initially we consider the simplest possible scenario in which the distribution of absorbing gas in the local ISM is taken to be uniform. In this case, in order to match the observed differential counts for the white dwarfs stars, an average gas density in the ISM of ~ 0.3 atoms cm^{-3} is required, whereas for the late-type stars the derived gas density is only ~ 0.05 atoms cm^{-3}. This disparity demonstrates that the assumption of a uniform distribution of absorbing gas is invalid. Since the white dwarf stars are detected at distances typically a factor ~ 5 greater than the late-type stars, we may infer that on average the white dwarfs suffer increased absorption outside of the region in which the bulk of late-type stars are found. This provides a strong impetus for modelling the source counts in terms of a local *bubble* within which the gas density is relatively low but which is bounded by a medium of much higher density.

We have modelled the global EUV source counts in terms of a spherically symmetric bubble/cavity centred on the solar system in which the interior gas density is n_i but outside of a radius, d_{bub}, the density rises to n_o. First we fixed n_i at a value of 0.05 atoms cm^{-3}, which is consistent with the results from the uniform absorption model for late-type stars. Excellent fits to the global differential counts of both the white dwarf and late-type stars are obtained with the exterior gas density set at $\gtrsim 5$ times the interior value. Specifically with $n_o = 0.5$ atoms cm^{-3} (*i.e.* 10 times the interior gas density) the best fitting bubble radius is 100 ± 25 pc for the S1 data and 80^{+10}_{-30} pc for the S2 sample. We have also investigated a model with the internal gas density set at a lower value of $n_i = 0.025$ atoms cm^{-3}, again with an exterior gas density $n_o = 0.5$ atoms cm^{-3}. In this case an excellent fit to the differential EUV source counts is obtained with a bubble radius ~ 80 pc. Hereafter we take as our "fiducial" model, a bubble with a radius of 80 pc, an interior gas density of 0.05 atoms cm^{-3}, and an exterior gas density of 0.50 atoms cm^{-3}.

3. THE SPATIAL DISTRIBUTION OF THE SOURCES

All-sky plots of the distribution of EUV sources in the WFC catalogue suggest marked deviations from isotropy (see Pounds *et al.* 1993). In order to investigate these effects we have applied a procedure, which compares the observed number of sources in a specified region of sky with the mean number predicted allowing for survey coverage. Figure 1 shows the resulting maps of the "fractional deviation" from a uniform distribution.

The deficit of sources, particularly white dwarf stars, in the general direction of the galactic centre is immediately apparent; Pounds *et al.* (1993) describe this feature in terms of a "wall" of absorption. We find that the paucity of sources in this region can be matched in the bubble model by reducing the bubble radius to ~ 10 pc.

A second feature evident in Fig. 1, is a marked excess of late-type stars in the southern part of the second/third galactic quadrant centred on $l \approx 210°$, $b \approx -30°$. This requires a reduction in the interior gas density, since the adjustment of the bubble radius beyond ~ 80 pc has relatively little impact on a stellar population observed typically within this distance. In fact, when we allow the interior gas density to fall to zero and extend the bubble radius to 100 pc, the predicted enhancement in source density matches that observed in the S2 sample, but falls a little short for the S1 sample. The effect of such a low-density interior for a bubble of radius 100 pc is to increase the predicted numbers of white dwarf stars by $\sim 80\%$, which is just compatible with the observed source density. Perhaps a more important point is that the lack of a *very strong excess* of white dwarf stars in the third galactic quadrant implies that the ultra-low density gas does not extend much beyond ~ 100 pc radius in this direction, at least not over a wide solid angle.

A final feature in Fig. 1 is the excess of white dwarf stars in a region of the northern sky in the second galactic quadrant (centred on $l \approx 150°$, $b \approx 45°$) where there is no corresponding enhancement in the number density of late-type stars. In terms of the bubble model this effect can be obtained by keeping the interior gas density fixed at the fiducial value but increasing the bubble radius; the observed peak enhancement of $\sim 80\%$ in the number density of white dwarf stars requires the bubble boundary to extend out to at least 120 pc.

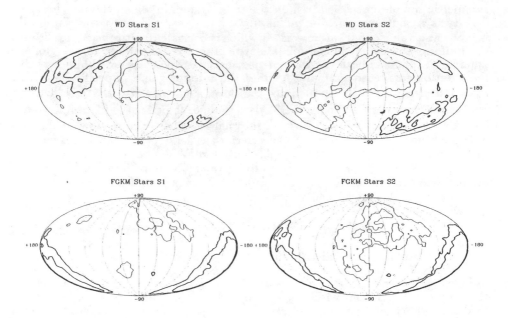

Figure 1. The "fractional deviation" from a uniform all-sky distribution of the white dwarf and late-type star stars in the WFC catalogue. The contour levels are set at 0.5, 0.3, -0.3, -0.5 (with the negative values shown by the fainter lines). The plots are in galactic coordinates centred on $l = 0°$, $b = 0°$.

4. DISCUSSION

Absorption line measurements in the optical/UV suggest that the Sun is located in a region roughly 100 pc in extent in which the neutral hydrogen density is anomalously low ($n_H \lesssim 0.1$ atoms cm^{-3}, e.g. Frisch and York 1983; Paresce 1984). Clearly this broad picture is now amply confirmed by the WFC all-sky survey.

The observed spatial distribution of EUV sources provides clear evidence for structure in the local ISM. The clearest deviation from isotropy in the EUV sky is the deficit of sources found in the general direction of the galactic centre, which we interpret as due to a relatively nearby (~ 10 pc) wall of absorption. The presence of relatively nearby interstellar clouds in this part of the sky is evident from the optical/UV absorption measurements of Frisch and York (1983) and Paresce (1984) and the polarization studies of Tinbergen (1982). More recently Frisch et al. (1990) have argued that in the galactic quadrants 1 and 4, the local fluff may well extend to 30 pc with a mean density ~ 0.1 atoms cm^{-3}. The picture which emerges is that beyond the very local fluff (~ 3 pc) there is a patchy distribution of clouds extending from ~ 5 pc to ~ 30 pc. These clouds are presumably intermixed with the hot plasma which produces the soft X-ray background. The wispy nature of the gas allows some relatively nearby late-type stars to be observed in the EUV, whilst the cumulative effect of this and more distant material is to present a column density sufficient to extinguish the vast majority of the white dwarf stars. This medium may in fact represent the nearside of the cool shell of the Loop I bubble (Cox and Reynolds 1987).

The second "interesting" region we identify in the EUV sky is the southern part of the second/third galactic quadrant centred on $l \approx 210°$, $b \approx -30°$. Here the observed excess of late-type stars is consistent with an ultra-low gas density in agreement with the findings of optical/UV absorption studies (e.g. Frisch and York 1983, 1991; Paresce 1984). Recently Welsh (1991) has reported a 'tunnel' in this direction of very low density ($n_H \ll 0.1$ cm^{-3}) cool gas, the tunnel being about 50 pc in diameter (*i.e.* subtending $\sim 10°$ diameter on the sky) and approximately 300 pc long. The absence in the WFC survey of a large excess of white dwarf stars in the third galactic quadrant indicates that the tunnel identified by Welsh (1991) is an exceptional feature and that the extent of the "EUV-window" in this part of the galaxy is more typically ~ 100 pc. Since, there is evidence for substantial amounts of HII gas in this quadrant (Gry, York and Vidal-Madjar 1985), a significant contribution to the EUV absorption beyond a 100 pc may originate in a HeI/HeII component.

The third region of interest identified in the EUV sky is in the northern part of the second galactic quadrant, centred on $l \approx 150°$, $b \approx 45°$. The excess of white dwarf stars but lack of an enhancement in the numbers of late type stars, suggests that the local bubble is particularly extended in this direction, but with a fairly typical interior gas density ($n_H \sim 0.05$ atoms cm^{-3}). This region coincides with one of the brightest features in the B-band map of soft X-ray background presented by McCammon *et al.* (1983). In terms of the displacement model of the soft X-ray background (Snowden *et al.* 1990), this again implies an extended path length through the local plasma-filled cavity.

REFERENCES

Cox,D.P. & Reynolds,R.J., 1987. Ann. Rev. Astron. Astrophys., 25, 303

Frisch,P.C. & York,D.G., 1983. Astrophys. J. Lett., 271, L59

Frisch,P.C.& York,D.G., 1991. In "Extreme Ultraviolet Astronomy", Eds. Malina & Bowyer, Pergamon Press, 322

Frisch,P.C., Welty,D.E., York,D.G. & Fowler,J.R., 1990. Astrophys. J., 357, 514

Gry,C., York,D.G. & Vidal-Madjar,A., 1985. Astrophys. J., 296, 593

McCammon,D., Burrows,D.N., Sanders,W.T. & Kraushaar,W.L., 1983. Astrophys. J., 269, 107

Paresce,F., 1984. Astron.J., 89, 1022

Pounds, K.A. et al. 1993. Mon. Not. R. astr. Soc., in press

Snowden,S.L., Cox,D., McCammon,D. & Sanders,W.T., 1990. Astrophys. J., 354, 211

Tinbergen,J. 1982. Astron. Astrophys., 105, 53

Welsh,B. 1991. Astrophys. J., 373, 556

OBSERVATIONS OF THE LOCAL INTERSTELLAR MEDIUM WITH THE HOPKINS ULTRAVIOLET TELESCOPE

Randy A. Kimble[1], Arthur F. Davidsen[2], Knox S. Long[3], and Paul D. Feldman[2]

[1]Laboratory for Astronomy and Solar Physics, Goddard Space Flight Center
[2]Center for Astrophysical Sciences, The Johns Hopkins University
[3]Space Telescope Science Institute

Email IDs
kimble@stars.gsfc.nasa.gov, afd@pha.jhu.edu, long@stsci.edu, pdf@pha.jhu.edu

ABSTRACT

During the Astro-1 space shuttle mission of December 1990, the Hopkins Ultraviolet Telescope was used to carry out an absorption study of the local interstellar medium (LISM). Through EUV observations of the hot DA white dwarf stars G191-B2B and HZ43, neutral hydrogen and neutral helium column densities have been determined along two lines of sight through the local interstellar cloud surrounding the Sun. The neutral hydrogen to helium ratios observed (in comparison to the cosmic abundance ratio of the two elements) provide an assessment of the relative ionization of the two species in the LISM. We find, in contrast to some previous indirect determinations, that hydrogen is *not* preferentially ionized compared with helium in the LISM and thus, that exotic sources of ionization are *not* required to explain the ionization state of the local cloud.

1. INTRODUCTION

Over the past twenty years or so, it has been established that our solar system currently lies within a small (few pc in radius), relatively cool, low density interstellar cloud, which is immersed in turn within a bubble of hot (million degree) gas extending for ~100 pc in most directions (*cf.* Cox & Reynolds 1987). As the various physical properties of the local cloud have been determined, a problem has arisen regarding the inferred ionization state of the local gas. Several lines of investigation suggest that hydrogen is substantially ionized within the local cloud, significantly more so than helium, implying a local neutral hydrogen to neutral helium ratio well below the cosmic abundance ratio of ~10.

However, theoretical modelling of the local cloud based on known sources of ionization (*e.g.* by Cheng & Bruhweiler 1990), predicts only a modest ionized fraction for hydrogen and actually a slightly higher ionized fraction for helium. This discrepancy is one of several motivations for a theory in which most of the mass of the universe consists of massive neutrinos which are unstable and decay slowly to photons that are capable of ionizing hydrogen but not helium (Sciama 1991). Alternatively, it has been suggested that the observed ionization is the result of a recent supernova within the local bubble. In either case, a proper understanding of our immediate interstellar neighborhood has far-reaching consequences.

We report here on the results of a LISM absorption study carried out using the EUV spectroscopic capability of the Hopkins Ultraviolet Telescope. (See Davidsen *et al.* 1992 for a description of HUT and its performance during the Astro-1 mission.) Using the continuum EUV emission of hot DA white dwarfs as a background, we have measured the absorption by neutral hydrogen and helium along two lines of sight through the local cloud and have thereby determined directly the relative ionization of the two species in the LISM.

2. HUT EUV OBSERVATIONS OF G191-B2B

The neutral hydrogen column density toward G191-B2B is derived by comparing the EUV spectrum obtained by HUT with the intrinsic flux expected from G191-B2B based on its spectral intensity in the far ultraviolet (where the interstellar attenuation is negligible) and its effective temperature (derived from Balmer line profile fitting). The neutral helium column is determined from the strength of the He I absorption edge at 504 Å. Comparing the relative neutral columns for hydrogen and helium with the standard cosmic abundance ratio then provides an assessment of the relative degree of ionization of the two species.

Carrying out such an analysis (see Kimble *et al.* 1993a for details), we find that the neutral hydrogen to neutral helium ratio in the local cloud toward G191-B2B is 11.6±1.0, close to, but slightly higher than, the canonical cosmic abundance ratio of 10 (see Figure 1). This result is in excellent agreement with the predictions of theoretical modelling of the local cloud and strongly contradicts the view that hydrogen is preferentially ionized in the LISM.

There is thus no need to invoke decaying neutrinos or recent supernova explosions to account for the ionization state of the local cloud. We address one of the arguments that originally motivated the notion of preferential ionization of local hydrogen below.

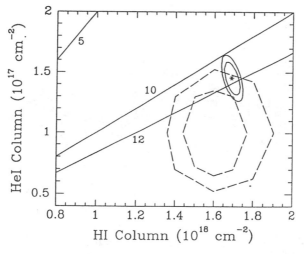

Fig. 1. Neutral hydrogen and helium columns toward G191-B2B. The solid contours are the HUT 90% and 99% confidence contours on the two columns, while the dashed contours are those of Green, Bowyer, & Jelinsky (1990). Also shown are lines of constant neutral hydrogen to neutral helium ratio.

3. HUT EUV OBSERVATIONS OF HZ43

HZ43 provided some of the motivation for the belief that hydrogen is preferentially ionized in the LISM, through a desire to reconcile EXOSAT and Voyager spectroscopy of this well-studied white dwarf with the belief that HZ43 contains a pure hydrogen atmosphere. If one attempts to account for the relatively low fluxes observed by EXOSAT in the 200 Å range by means of standard interstellar absorption, one finds that the neutral hydrogen column required would transmit far less flux in the 500-800 Å range than has been reported by Voyager (Holberg *et al.* 1980). As HZ43 shows no signs of helium or heavier element opacity in the photosphere which could suppress the intrinsic flux at EXOSAT wavelengths, Heise *et al.* (1988) suggested that perhaps the EXOSAT and Voyager spectra could be reconciled by anomalous interstellar absorption.

In this picture, a very low neutral hydrogen column permits consistency with the Voyager observations longward of 504 Å, while a much larger neutral helium column (which absorbs only shortward of 504 Å) serves to suppress the flux in the EXOSAT range. The relative columns correspond to a neutral hydrogen to neutral helium ratio of ~0.38 and require an ionization fraction of >96% for hydrogen in the local cloud.

The helium column required by this model would produce several optical depths of absorption at the helium edge, so EUV observations with HUT across the 504 Å edge provide a critical test of this hypothesis. Contrary to the prediction, the EUV spectrum obtained by HUT shows only a small drop in intensity across the 504 Å edge (Kimble *et al.* 1993b). Carrying out the same kind of model fitting analysis as described in the previous section, we derive the neutral hydrogen and neutral helium column densities shown in Figure 2.

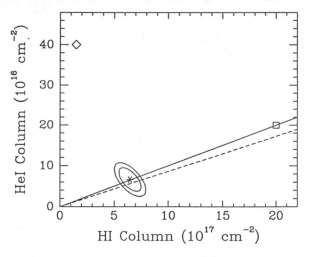

Fig. 2. Neutral hydrogen and neutral helium columns toward HZ43. The solid contours are the HUT 90% and 99% confidence limits on the two columns. The columns required by EXOSAT spectroscopy for normal (square) and anomalous (diamond) interstellar absorption are indicated. The solid line represents a H I/He I ratio of 10; the dashed line represents a ratio of 11.6, as found toward G191-B2B.

Though the neutral hydrogen to helium ratio is not tightly constrained by the low signal to noise HUT observation of HZ43, it is consistent with either the canonical cosmic abundance ratio of the elements or with the slightly higher hydrogen to helium ratio that was found toward G191-B2B. More significantly, it is clear that the low fluxes reported by EXOSAT can not be explained by either standard or anomalous interstellar absorption, as both models lie far from the confidence intervals permitted by the HUT observation, assuming a pure hydrogen atmosphere for HZ43. We conclude, therefore, that either there is a completely unexpected source of opacity present in the HZ43 photosphere that is suppressing the EUV flux at the shorter wavelengths, or there is a significant calibration error in the EXOSAT spectroscopy.

4. CONCLUSIONS

We have reported the results of an absorption study of the local interstellar medium, based on EUV observations of the hot DA white dwarfs G191-B2B and HZ43 made using the Hopkins Ultraviolet Telescope. The results indicate that, contrary to some previous indications, helium is slightly more ionized than hydrogen within the local cloud, in agreement with theoretical predictions based on known sources of ionization. No exotic sources of ionizing flux are required to sustain the local ionization fractions found by HUT. Furthermore, we find that the low fluxes reported by EXOSAT for HZ43 can not be explained by interstellar absorption. Either unexpected photospheric opacity or an EXOSAT calibration problem must be the cause.

ACKNOWLEDGEMENTS

We are grateful to the many people at the Johns Hopkins University and at NASA who contributed to the success of HUT and the Astro-1 mission. HUT is supported by NASA contract NAS 5-27000 to the Johns Hopkins University.

REFERENCES

Cheng, K.-P., & Bruhweiler, F. C. 1990, ApJ, 364, 573
Cox, D. P., & Reynolds, R. J. 1987, ARA&A, 25, 303
Davidsen, A. F., et al. 1992, 392, 264
Green, J., Jelinsky, P., & Bowyer, S. 1990, ApJ, 359, 499
Heise, J., Paerels, F. B. S., Bleeker, J. A. M., & Brinkman, A. C. 1988, ApJ, 334, 958
Holberg, J. B., Sandel, B. R., Forrester, W. T., Broadfoot, A. C., Shipman, H. L., & Barry, J. L. 1980, ApJ, 242, L119
Kimble, R. A., et al. 1993a, ApJ, 404, in press
Kimble, R. A., Davidsen, A. F., Long, K. S., & Feldman, P. D. 1993b, ApJ, submitted
Sciama, D. W. 1991, A&A, 245, 243

THE COOL NEUTRAL PHASE OF THE LOCAL ISM: COMPACT CLOUDS CARRY A MAJOR FRACTION OF THE MASS

Joaquín Trapero, John E. Beckman, Ricardo Génova
Instituto de Astrofísica de Canarias, 38200 La Laguna, Spain

Conal D. McKeith
Dept. of Pure and Applied Physics, The Queen's University, Belfast, U.K.

Ingemar Lundström
Lund Observatory, Sweden

1. INTRODUCTION: STRUCTURE IN THE LISM

The presence in the interstellar medium of co–existing phases at discrete, widely separate temperatures was predicted on general grounds by Spitzer (1968) and in the form of a detailed two–phase model ($T\sim10^2$ K and $T\sim10^4$ K) by Field, Goldsmith and Habing (1969). In the leading models of the 1970's which have not been superseded in their essentials, Cox and Smith (1974) and McKee and Ostriker (1977) predicted that the local ISM (LISM) should be dominated by hot, tenuous gas ($T>10^5$ K) heated by winds and supernovae from nearby OB associations, in which small, dense, cool clouds would float. There would also be stable interfaces at intermediate temperatures due to gas evaporating from these clouds into the hot environment. Since the advent of UV spectrographic satellites (Copernicus, IUE) in the 1970's and 80's, the presence of high temperature gas has been verified by the detection of ions such as C IV, Si IV and above all O VI in the LISM. However there has been much controversy especially about the relative importance of each phase: what fractions of the total volume and mass are occupied by the hot, warm and cool components.

In the present study we take a decisive step forward in elucidating the structure of the LISM, by identifying, locating, and quantifying a set of individual cold clouds within 200pc of the Sun. Using B stars, where possible fast rotators, as background illumination, we have accumulated high resolution, high S/N absorption spectra of the resonance lines $\lambda7699$Å of K I, and the D doublet of Na I. Combining our data with published LISM spectra towards neighboring objects, we can analyze the structure of the local phase, finding that it occupies less than 5% of the volume, but contributes the major fraction of the LISM mass. A characteristic cloud mass is of order $100\,M_\odot$. The clouds we have detected in the directions explored so far are being blown along in the interstellar wind flowing from the Sco–Cen OB complex.

2. THE OBSERVATIONS, AND THE DATA

For these studies we have used observations taken in both the northern and southern sky. In the northern hemisphere we observed from the Observatorio del Roque de los Muchachos, La Palma, Canary Islands, Spain, with three different configurations: the MES echelle on the INT 2.5m telescope, the QUBES echelle on the WHT 4.2m, and the IACUB echelle on the NOT 2.5m telescope. The

.resolution for all these studies is $\lambda/\Delta\lambda \sim 5 \times 10^4$, and with CCD detectors we obtained S/N ratios between 100 and 200. The southern spectra were obtained at the European Southern Observatory, La Silla, Chile (in collaboration with A. Ardeberg, H. Lindgren and E. Maurice), using the CES echelle at the CAT 1.5m telescope. The resolution used was $\lambda/\Delta\lambda \sim 1 \times 10^5$ and the S/N obtained was >250.

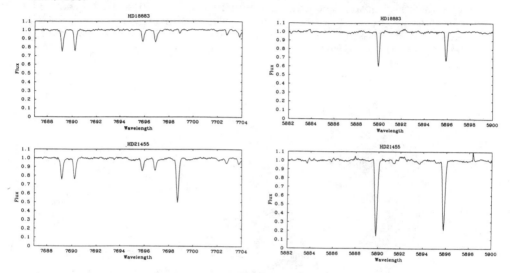

Fig. 1. Spectra of HD 18883 (upper) and HD 21455 (lower) in $\lambda 7699$Å K I (left) and Na I D doublet (right)

Data were reduced using the IRAF package, via standard steps: bias subtraction, flat–field division, precise wavelength calibration, and intensity normalization in the continuum. We could use telluric O_2 or H_2O absorption lines as fiducial wavelengths for calibration. In reducing the Na I spectra we had to take particular care to eliminate the telluric H_2O lines, which contaminate the spectrum and blend, especially, with the D_2 line. This telluric spectrum was removed using division by a spectrum synthesized according to the prescription of Lundström et al. (1991) on the basis of very precise measurements from previous observations.

In Fig 1 we illustrate LISM absorption along two lines of sight in both K I and Na I, towards stars in the 3rd Galactic quadrant. As expected from their abundances, the Na I lines are on the linear part of the curve of growth for smaller H I column densities, but also saturate at lower column densities than the K I feature. Thus one should use Na I to explore the range below $N(H I)=10^{20}$ cm^{-2}, and K I for the range above this value.

3. HOW TO DERIVE THE PARAMETERS OF INDIVIDUAL CLOUDS

Both in our data, and in those from the literature (see e.g. Stokes 1978) a typical LISM absorption spectrum in the K I or Na I resonance lines comprises a set of blended components, each with FWHM$\sim$2-3 km s^{-1}. Working in the

LISM, over path lengths <200pc, gives the advantage of few such components to de-compose.

Our technique is straightforward. We find a group of stars in a given region of the sky, within a radius of say 5-10°, whose distances we infer via spectroscopic parallax, with a reddening correction inferred from the LISM lines measured here. Identifying a velocity component in the absorption spectra, we note the stars in whose spectra the component does, or does not appear. The cloud in question lies in front of the stars with this given component, and we find each individualcloud to be well defined, with scales of a few parsecs in linear size perpendicular to the line of sight.

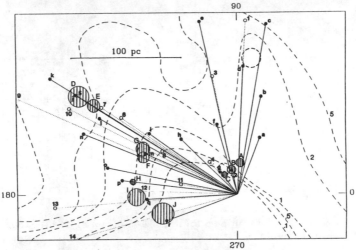

Fig. 2. Plan view in the Galactic plane of the cool clouds detected with K I. (see Trapero et al. 1992)

Our LISM lines give K I column densities, N(K I) from which we infer H I column densities N(H I) (see Trapero et al. 1992 for details and limitations), typically in the range 10^{20} to 10^{21} cm^{-2}. Assuming rough sphericity we can then infer particle number density, typically between 10 cm^{-3} and 50 cm^{-3}, and a cloud mass typically of order $100\,\mathrm{M_\odot}$. Assuming pressure equilibrium (the clouds are sub-Jeans in mass) and a canonical pressure of 3600 K cm^{-3}, in dimension p/k, we then deduce temperatures, which come out close to 100 K. The accuracy of a given determination depends, as well as on the assumptions, on the number of lines of sight in which the given spectral component is identified. The same technique using Na I samples the range 10^{19}-10^{20} cm^{-2} in N(H I). Figures 2 and 3 show the two regions close to the Galactic plane, which we have so far explored, indicating the locations, sizes, and heliocentric radial velocities of the cool clouds measured.

4. INFERENCES TO DATE

We can summarize our findings to date, as follows,

(a) Sharply varying LISM column density from one stellar line of sight to another nearby points strongly to the discrete nature of the cool clouds.

(b) They form a population in the "cold stability range" of the ISM, with T of order 100 K or less.

(c) Previously published local iso–column density maps centred on the Sun, do give a global picture of where the H I is concentrated, but give a misleadingly smooth local picture in the parameter space of diameter <10 pc where the clouds are found.

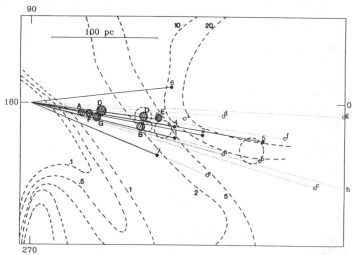

Fig. 3. Plan view in the Galactic plane of the cool clouds detected in the direction of Sco–Cen.

(d) We estimate the volume filling factor of these clouds in the region within 200pc of the Sun to be <5%, but their mass fraction is very high, probably more than 90% of the total LISM mass.

(e) The clouds detected so far tend to be flowing (not all of them, however) towards us from a direction l=340°, or away from us towards the opposing direction, with a heliocentric speed of $15\,\mathrm{km\,s^{-1}}$.

(f) Superposed on this flow, first detected by Crutcher (1982), is a random component of order $5\,\mathrm{km\,s^{-1}}$.

(g) At least one object within the present search has an H I column density of more than $2 \times 10^{21}\ \mathrm{cm^{-2}}$, and is hence optically thick in dust, and a probable candidate for a molecular core.

REFERENCES

Cox, D. P., & Smith, B. W. 1974, ApJ, 189, L105
Crutcher, R.M. 1982, ApJ 254
Field, G. B., Goldsmith, D. W., & Habing, H.J. 1969, ApJ, 155, L149
Lundström, I., Ardeberg, A., Maurice, E., & Lindgren, H. 1991, A&AS, 91, 199
McKee, C. F., & Ostriker, J.P. 1977, ApJ, 218, 148
Stokes, G.M. 1979, ApJS, 36, 115
Spitzer, L., Jr. 1968, in Diffuse Matter in Space, eds Interscience
Trapero, J., Beckman, J. E., Génova, R., & McKeith, C. D. 1992, ApJ, 394, 552

ANOMALOUS COSMIC RAYS
AND THE LOCAL INTERSTELLAR MEDIUM

James H. Adams, Jr. and Allan J. Tylka
E.O. Hulburt Center for Space Research
Code 7654, Naval Research Laboratory, Washington, DC 20375-5000

ABSTRACT

The *anomalous component of cosmic rays* comprises energetic particles which originate as neutral atoms in the Local Interstellar Medium (LISM). In 1991 it was discovered that these particles become trapped in the Earth's magnetic field, with lifetimes which concentrate the particles to ~500 times their interplanetary flux. This new "radiation belt" offers an opportunity for direct measurements of the isotopic composition of a number of elements from the LISM by a cosmic-ray instrument in low-Earth orbit. These measurements can be used to investigate how nucleosynthesis has evolved in the Galaxy and may offer clues to the circumstances surrounding the formation of the protosolar nebula.

1. ORIGIN OF THE ANOMALOUS COMPONENT

Cosmic rays have long been known to come from the Galaxy and occasionally from the Sun or the interplanetary medium. In the early 1970's, a third component of cosmic rays was discovered as a large "bump" in the spectra of He, N, and O at kinetic energies of ~10 MeV/nucleon. As spacecraft continued to move out through the heliosphere, the intensity of this "anomalous component" (hereafter AC) was observed to increase, indicating that these particles were not of solar origin. (See Webber (1989) for a review of the AC.)

Soon after the discovery of the AC, Fisk, Kozlovsky, & Ramaty (1974, hereafter *FKR*) proposed a theory for the origin of these particles. According to this theory, which is illustrated in Fig. 1, AC particles originate as neutral atoms in the Local Interstellar Medium (LISM). These atoms sweep into the heliosphere as the Solar System moves through interstellar space. At ~1-3 AU, these atoms become singly-ionized, either by photoionization or by charge-exchange with solar wind particles. They are then entrained in the solar wind, which carries them to an acceleration region (now identified as the solar wind termination shock (Pesses *et al.* 1981; Adams & Leising 1991; Jokipii 1992)), where their energies are raised from ~1 keV/nucleon to ~10 MeV/nucleon or higher. The particles then escape the acceleration region in all directions, with some heading back toward the Sun. Before and after acceleration, these particles pass through too little matter to be stripped of additional electrons, so they remain singly-ionized. Because of their large mass-to-charge ratio, these ions penetrate deeply into the heliosphere, even at these low energies where fully-ionized Galactic cosmic rays are turned back by the outflowing solar wind.

The *FKR* theory may seem somewhat elaborate, but some ingredients of the scenario, such as the inflow of interstellar neutrals (Weller & Meier 1974) and solar-wind pick-up (Mobius *et al.* 1985), have been independently established. Moreover, the theory makes two predictions about the AC, both of which have

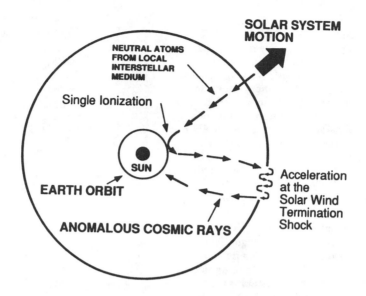

Fig. 1. Origin of anomalous cosmic rays.

now been experimentally confirmed.

The *FKR* theory first of all predicts that the AC should reflect the composition of the *neutral* LISM, resulting in an elemental composition vastly different from that of solar energetic particles and of Galactic cosmic rays. All presently available data on the AC are consistent with this prediction. In particular, the AC appears most strongly in elements with first ionization potential (FIP) ≥ 13.6 eV (since hydrogen atoms near the Sun absorb most UV photons with energy ≥ 13.6 eV). The elements H, He, N, O, Ne, and Ar all satisfy this requirement, and they have been observed in the AC. Since neutral atoms of all elements are presumably present to some degree in the LISM, the AC should also be present in elements with FIP < 13.6 eV, but at highly suppressed levels. The only such element observed so far is carbon (FIP $= 11.3$ eV), for which the AC appears as a very subtle enhancement in the observed spectrum.

Second, the *FKR* theory predicts that AC particles should be singly-ionized. Fisk called this the "definitive test" of theory, since this ionic charge state distinguishes AC ions from Galactic cosmic rays, solar energetic particles, and other energetic particles in the interplanetary medium, all of which are completely or nearly-completely stripped of electrons.

There have been numerous attempts to determine the charge state of the AC. (See Adams *et al.* 1991b for a review.) A definitive measurement of the average ionic charge state of AC oxygen was published in 1991 (Adams *et al.* 1991a). This work compared the AC oxygen flux measured *inside the magnetosphere* aboard Russian *Cosmos* satellites with simultaneous measurements *outside the magnetosphere* in interplanetary space from the US *IMP 8* satellite. Since an ion's ability to penetrate the Earth's magnetic field depends upon its momentum per unit charge, the average ionic charge state can be inferred from these measurements, as shown in Fig. 2. By combining data collected over a four-year period, the average ionic charge state of AC oxygen was measured to be $0.9^{+0.3}_{-0.2}$, confirming the *FKR* prediction.

Fig. 2. (from Adams *et al.* 1991a.) Ionic charge state of AC oxygen. Top panel shows flux measurements outside the magnetosphere from *IMP 8*. Curves show contributions of solar, anomalous, and Galactic particles to the observed flux. Bottom panel shows simultaneous measurements inside the magnetosphere aboard a *Cosmos* satellite. The curves show the expected fluxes (based on the *IMP 8* data) if the AC oxygen are either singly- (Q=+1) or fully-ionized (Q=+8). The data clearly favor Q=+1, as predicted.

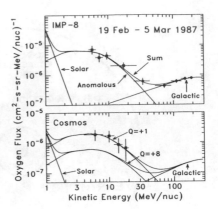

Fig. 3. (adapted from Grigorov *et al.* 1991.) History of geomagnetically-trapped oxygen flux on *Cosmos* satellites (left-hand side) and anomalous oxygen flux in interplanetary space on *IMP 8* (right-hand side). Note the different scales.

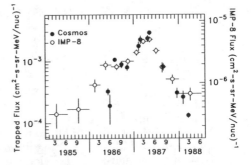

2. MAGNETOSPHERICALLY-TRAPPED ANOMALOUS COSMIC RAYS

Singly-ionized AC ions can penetrate deeply into the magnetosphere. Blake and Friesen (1977) predicted that AC ions can become geomagnetically trapped after being stripped of remaining orbital electrons in the residual atmosphere, with subsequent lifetimes (as limited by energy loss in the residual atmosphere) ranging from hours to months. These energetic particles therefore constitute a new "radiation belt", composed of nuclei from the LISM. The existence of trapped AC oxygen ions was confirmed by a US-Russian collaboration in 1991 (Grigorov *et al.* 1991). Fig. 3 shows the time history of the trapped oxygen flux (as measured on *Cosmos* satellites) and the interplanerary AC oxygen flux (measured on *IMP 8*). This strong temporal correlation, along with arguments on composition and energetics, demonstrated that the trapped particles came from the AC. Of particular note in Fig. 3 are the different flux axes on the right- and left- hand sides: the trapping process has concentrated the particle flux to ~500 times the interplanetary flux.

3. FUTURE STUDIES OF TRAPPED ANOMALOUS COSMIC RAYS

This large concentration factor makes trapped particles a potentially rewarding target for future measurements of the AC — and hence of the LISM. Trapping may also make it possible to extend observations of the AC to rare elements. (For example, although anomalous F, Kr, and Xe have not yet been observed, they all have FIP $\geq$ 13.6 eV and should therefore be present.) Because of selection biases introduced by ionization, acceleration, solar modulation, and trapping efficiency, trapped anomalous cosmic rays may be of limited usefulness in studying the *elemental* composition of the LISM. However, these effects should introduce only minimal distortions in relative *isotopic* abundances. Measurements of the isotopic composition of the anomalous component will therefore be particularly interesting (Mewaldt *et al.* 1984).

At present there are relatively few measurements of the isotopic composition of the LISM. Such measurements can contribute to our understanding of the history of nucleosynthesis in the Galaxy. In particular, by comparing isotopic ratios in different sources of matter, we can learn how nucleosynthesis has evolved as the Galaxy has aged and matter has been processed through successive generations of stars. The Solar System provides us with a sample of matter which is ~5 billion years old. Galactic cosmic rays, on the other hand, are generally believed to be a relatively young sample of matter, only ~10 million years old. The LISM provides us with a third sample of matter, older than Galactic cosmic rays and well separated in both time and space from the matter that collapsed to form the Solar System. Comparisons between the LISM and Galactic cosmic rays may also indicate to what extent Galactic cosmic rays are newly-synthesized matter or older interstellar matter swept up and accelerated by supernova shocks.

The isotopic composition of the LISM may also offer clues to the circumstances surrounding the formation of the Solar System. Comparisons of the isotopic compositions of the Solar System and the LISM may show that the Solar System is enriched in highly-processed material from a nearby supernova which triggered the formation of the protosolar nebula. This would be apparent, for example, in the ^{82}Kr/^{84}Kr ratio, since ^{84}Kr is due exclusively to the rapid neutron capture process which occurs only in supernovae while ^{82}Kr is a product of slow neutron capture which occurs in red giants.

Previous measurements of AC isotopic composition by instruments in interplanetary space have been limited by poor statistics (Mewaldt *et al.* 1984; Cummings *et al.* 1991). The *Cosmos* detectors, which discovered trapped AC oxygen ions, were unable to resolve isotopes. Cosmic-ray experiments aboard NASA's recently-launched *SAMPEX* satellite may study trapped AC ions. It is hoped that future missions will fully exploit this reservoir of interstellar matter available for study at Earth.

REFERENCES

Adams, J.H. Jr. *et al.* 1991a, ApJ Lett, 375, L45
Adams, J.H. Jr., Beahm, L.P. & Tylka, A.J. 1991b, ApJ, 377, 292
Adams, J.H. Jr. & Leising, M.D. 1991, Proc. 22nd ICRC (Dublin), 3, 304
Blake, J.B. & Friesen, L.M. 1977, Proc. 15th ICRC (Plovdiv), 2, 341
Cummings, A.C. *et al.* 1991, Proc. 22nd ICRC (Dublin), 3, 362
Fisk, L.A., Kozlovsky, B., & Ramaty, R. 1974, ApJ Lett, 190, L35
Grigorov, N.L. *et al.* 1991, Geophys Res Lett, 18, 1959
Jokipii, J.R. 1992, ApJ Lett, 393, L41
Mewaldt, R.A., Spalding, J.D., & Stone, E.C. 1984, ApJ, 283, 450
Mobius, E. *et al.* 1985, Nature, 318, 426
Pesses, M.E., Jokipii, J.R., & Eichler, D. 1981, ApJ Lett, 246, L85
Webber, W.R. 1989, AIP Conf. Proc., 183, 100
Weller, C.S. & Meier, R.R. 1974, ApJ, 193, 471

INFRARED

EMISSION

COLD DUST IN THE ISM?

Edward L. Wright

UCLA Dept. of Astronomy, Los Angeles CA 90024-1562

Email ID
wright@astro.ucla.edu

ABSTRACT

The spectrum of the Milky Way observed by the $COBE$ FIRAS instrument shows an excess of long wavelength emission when compared to a single temperature $\sigma \propto \nu^2$ dust model. This excess could be due to a massive population of large spherical grains or to a very small mass of fractal grains. Fractal dust grains with low equilibrium temperatures can produce the bulk of the IR emission by single photon thermal pulsing with peak temperatures of ≈ 20 K.

1. INTRODUCTION

The search for very cold dust in the ISM has long been motivated by the intriguing coincidence between the energy density of starlight in our galaxy with the energy density of the 2.7 K cosmic microwave background. Hoyle & Wickramasinghe (1969) suggested that a sufficiently high density of impurities in dust grains would allow their equilibrium temperature to be 3 K, so that dust grains could produce the microwave background. However, Purcell (1969) pointed out that the Kramers-Kronig relationship restricts the possible long wavelength absorption behavior of dust grains. The Kramers-Kronig restriction on long wavelength absorption can be relaxed for lower-dimensional grain shapes: the one-dimensional needles or whiskers (Rana 1979; Wright 1982; Hoyle & Wickramasinghe 1988) or fractal dust grains with dimensions between 1.7 and 2.5 (Wright 1987; Wright 1989).

One can also have very cold dust in the ISM by reducing the optical absorption efficiency instead of increasing the sub-mm emission efficiency. Rowan-Robinson (1992) has modelled the emission from galaxies using a mixture of grain types that includes spherical carbon grains with 30 μm radius. These grains have very low cross section per unit mass in the optical, and thus have equilibrium temperatures near 5 K. If any substantial amount of power were radiated by dust at 5 K, then these grains would have to dominate the mass budget for interstellar dust.

Microwave background anisotropy experiments have also provided measurements of the millimeter wave emission from the Milky Way, as shown in Lubin & Villela (1986). The spectrum of the galaxy obtained by Wright et al. (1991) using the FIRAS instrument on $COBE$ is another example.

The IRAS satellite provides a sensitive survey for almost all the sky at wavelengths $\lambda \leq 100$ μm. Paley et al. (1991) examined the IR and optical emission from a number of relatively isolated high latitude dust clouds and decided that three temperatures of dust were needed to explain their data: hot dust at 173 K, presumably transiently heated by single photon thermal pulsing (Purcell 1976); warm dust at 22.5 K; and a cold component with T < 15 K seen only in optical scattering with no $\lambda \leq 100\mu$m emission. Paley *et al.* found that these temperatures are very uniform, but that the ratio between the different components varied over the sky. Guhathakurta & Tyson (1989) found that the optical light from high latitude dust clouds is too red to be scattered light, and thus may be similar to the extended red emission (ERE) seen in reflection nebula by Witt & Schild (1988). The ERE is an extension of the 3 μm emission from thermally pulsing grains seen by Sellgren (1984).

2. OBSERVATIONS

Wright *et al.* (1991) measured the average spectrum of the Milky Way in the 1-90 cm^{-1} range and gave two fits to the continuum emission. One fit assumed two temperatures of dust with ν^2 emissivity laws:

$$g(\nu) \approx 0.00022(\nu/30 \text{ cm}^{-1})^2 [B_\nu(20.4) + 6.7B_\nu(4.77)] + S(\nu) \qquad (1)$$

where $S(\nu)$ models the synchrotron emission. This should not be taken as proving the existence of dust with $T = 5$ K, since an alternate fit with a single dust temperature is almost as good:

$$g(\nu) \approx 0.00016(\nu/30 \text{ cm}^{-1})^{1.65} B_\nu(23.3) + S(\nu) \qquad (2)$$

This form also matches observations of external galaxies by Carico *et al.* (1992).

The cold component in Eq(1) contributes only $\tau_c T_c^6 / \tau_w T_w^6 = 6.7(4.77/20.4)^6 = 0.1\%$ of the total power radiated by the warm dust. This very small fraction guarantees that the optical emission without IRAS 100 μm emission seen by Paley *et al.* is due to small transiently heated grains instead of large cold grains, since any component contributing only 0.1% of the high latitude optical cirrus would not be detectable by any observations made to date. This small fraction of the total power in the cold component also means that the large spherical grains proposed by Rowan-Robinson only require 12% of the cosmic carbon abundance.

3. THEORY

The ISM can be considered to contain a very dilute gas of dust grains, with density n, and polarizability $\alpha(\omega)$ (Purcell 1969). The dielectric constant of this medium is given by

$$\epsilon(\omega) = 1 + 4\pi n\alpha \qquad (3)$$

The propagating waves have an E-field that follows

$$E = E_\circ \exp(i(\omega/c)x\sqrt{\epsilon} - i\omega t) \qquad (4)$$

The absorption coefficient of the medium is

$$\kappa = 4\pi n\omega \text{Im}(\alpha(\omega))/c = n\sigma_{\text{ext}} \qquad (5)$$

so

$$\sigma_{\text{ext}} = 4\pi\omega\text{Im}(\alpha(\omega))/c \tag{6}$$

Now for a small spherical grain

$$\alpha = \frac{a^3(\epsilon_m - 1)}{(\epsilon_m + 2)} \tag{7}$$

(where ϵ_m is the dielectric constant of the grain material) which yields the standard Mie theory small sphere result.

The impulse response function for the electric dipole of the grain, when exposed to $E = \delta(t)$, is

$$d(t) = \frac{1}{2\pi}\int \alpha(\omega)e^{-i\omega t}\,d\omega \tag{8}$$

but $d(t) = 0$ for $t < 0$ by causality. This shows that α has no poles in the upper half plane $\text{Im}(\omega) > 0$. Thus the value of the integral

$$\oint \frac{\alpha(\omega)d\omega}{\omega - iy} = 2\pi i\alpha(iy) \tag{9}$$

for all $y > 0$ and is 0 for all $y < 0$. The principal value of the integral with $y = 0$ is

$$\int \frac{\alpha(\omega)d\omega}{\omega} = i\pi\alpha(0) \tag{10}$$

This gives the Kramers-Kronig relation between the DC polarizability and the integral of the absorption coefficient:

$$\alpha_{DC} = \frac{1}{\pi}\int \text{Im}(\alpha)d\omega/\omega = \frac{c}{2\pi^2}\int_0^\infty \sigma_{\text{abs}}\,d\omega/\omega^2 = \frac{1}{4\pi^3}\int_0^\infty \sigma_{\text{abs}}\,d\lambda \tag{11}$$

There is one further limitation on the polarizability α and hence the cross-section σ. Since the response $d(t)$ to a real electric field $E(t)$ must be real, one derives that $\alpha(-\omega) = \alpha(\omega)^*$. This in turn implies that σ is an even function of ω. Thus

$$\sigma(\omega) = A + B\omega^2 + C\omega^4 + ..., \tag{12}$$

and, unless $\alpha_{DC} = \infty$, we know from the Kramers-Kronig relation that $A = 0$. However, the Stone-Weierstrass approximation theorem guarantees that any continuous function can be approximated to any desired precision by a polynomial, and thus one can always approximate any $\sigma(\omega)$ as a series in ω^2. Still, one should clearly favor functional forms for σ that are manifestly functions of ω^2, such as the $\sigma_o/(1+(\omega_o/\omega)^2)$ appropriate for constant conductivity needles.

In order to have a large long wavelength absorption cross-section per unit mass of dust, one clearly needs to maximize α/V. Since $\alpha \propto L^3$, where L is the maximum dimension of the dust grain, α/V is maximized for grains with one long axis and two small axes: needles. In general for a fractal dimension D the volume is $V \propto (l_o)^3(L/l_o)^D$, where l_o is the size of the building block or monomer, so $\alpha/V \propto (L/l_o)^{D-3}$. Figure 1 shows α/V for the DLA and CL fractal grains considered by Wright (1989).

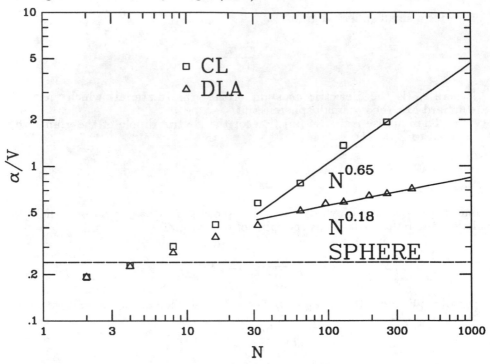

Fig. 1. Polarizability per volume *vs* number of subunits

4. THERMAL FLUCTUATIONS

Purcell (1976) found that for small interstellar grains the cooling time is about $400/T$ seconds. For small enough grains this is longer than the mean time between absorption of optical photons from the interstellar radiation field (ISRF). However, the cooling time depends on both the grain material and shape. Figure 2 shows the cooling times for graphite spheres, CL fractals and needles. Since graphite is such a good conductor, the graphite sphere behaves like a mirrored ball, reflecting all the IR radiation that hits it. This high reflectivity gives a low emissivity, and a long cooling time. On the other hand the fractal and needle have very short cooling times.

The ISRF can be reasonably well approximated by

$$I_\nu = 10^{-16} f_\nu \left(30766 B_\nu(2213) + 578 B_\nu(5280) + 1.14 B_\nu(24091)\right) \qquad (13)$$

where f_ν is 1 for $\lambda > 4000\text{Å}$ and 0.56 for shorter wavelengths, with an expected sharp drop to zero shortward of 911 Å. In this ISRF the dust grains in Figure 2 have equilibrium temperatures shown by the dots on the cooling time curves.

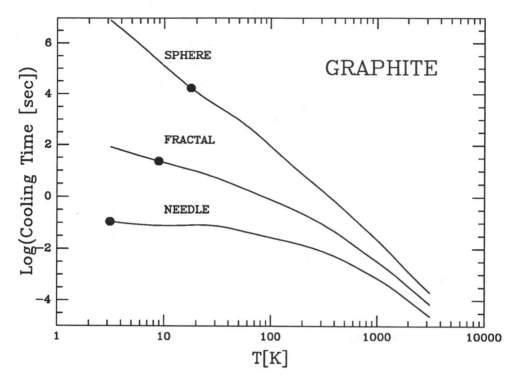

Fig. 2. Cooling time *vs* temperature for graphite grains

Because fractals and needles have such short cooling times, they will be far out of thermal equilibrium even for rather large masses. Figure 3 shows thermal time histories for a graphite sphere with a 100 Å radius, and a fractal and needle with the same mass as the sphere. Clearly the sphere is close to equilibrium, while the fractal shows large temperature swings, and the needle is entirely in the single photon limit. This means that the IR emission from fractals and needles is not determined by their equilibrium temperature, but rather by their enthalpy $U(T)$, their mass spectrum, and their absorption efficiency σ_ν.

The spectrum of the IR emission from thermally pulsing dust grains can be calculated using the method of Guhathakurta & Draine (1989). Figure 4 shows the *COBE* spectrum of the galaxy with the emission from a population of thermally pulsing DLA fractal grains made out of amorphous carbon, with equal total dust mass in logarithmic mass bins from 8,000 to 500,000 carbon atoms. The curve has been normalized so that it does not exceed the *COBE* spectrum at any wavelength. Extending the mass spectrum to smaller masses would clearly improve the fit at short wavelengths, and thus a population of thermally pulsing fractal grains can explain almost all of the IR spectrum of the galaxy. On the other hand, large fractal or needle-shaped grains can contribute at most a small fraction of the bolometric IR emission of the galaxy because their equilibrium temperatures are so low, and the *COBE* spectrum has such a small long wavelength excess.

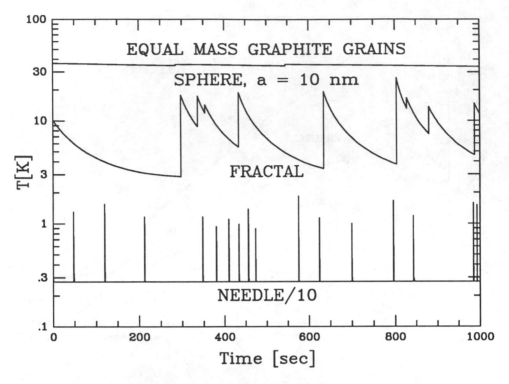

Fig. 3. Thermal time histories for graphite grains

5. CONCLUSIONS

Cold dust radiates only a small fraction (0.1 to 3%) of the total bolometric emission of IR dust. If the dust is cold because of a small optical σ/m then the mass fraction in the cold dust can still be substantial, as in the case of the 30 μm carbon spheres proposed by Rowan-Robinson (1992). In the case of fractal grains their low equilibrium temperature is caused by an increased sub-mm σ/m instead of a reduced optical σ/m. Thus if the fractal grains are in thermal equilibrium their mass fraction is limited to a few percent of the total dust mass. But thermally pulsing fractals with a suitable mass spectrum can produce close to 100% of the bolometric IR emission, and could thus make up close to 100% of the IS dust by mass. If so, the mass spectrum of the dust would have to vary from cloud to cloud to explain the results of Paley et al. (1991), and the high mass fraction would have to increase with galactocentric radius to explain the DIRBE results on the variation of the 240 μm to 100 μm color temperature that were presented at this conference by Sodroski et al. In this picture, most of the dust in the ISM is cold most of the time, but almost all of the IR emission is produced during brief warm intervals following the absorption of an optical photon.

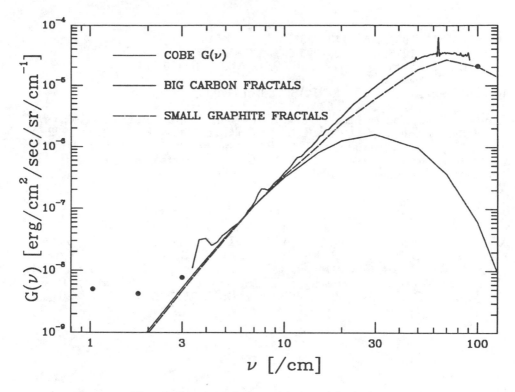

Fig. 4. Spectra of the galaxy and fractal dust

ACKNOWLEDGEMENTS

This work was supported in part by NASA Grant NAG 5-1309 to UCLA. *COBE* is supported by NASA's Astrophysics Division. Goddard Space Flight Center (GSFC), under the scientific guidance of the *COBE* Science Working Group, is responsible for the development and operation of *COBE*.

REFERENCES

Carico, D. P., Keene, J., Soifer, B. T. & Neugebauer, G. 1992, PASP, TBD, TBD

Draine, B. T. 1988, ApJ, 333, 848-872

Guhathakurta, P. & Draine, B. T. 1989, ApJ, 345, 230-244

Guhathakurta, P. & Tyson, J. A. 1989, ApJ, 346, 773-793

Hoyle, F. & Wickramasinghe, N. C. 1967, Nature, 214, 969-971

Hoyle, F. & Wickramasinghe, N. C. 1988, Ap. & Sp. Sci., 147, 245-256

Lubin, P. & Villela, T. 1986, in Cosmic Distances and Deviations from Universal Expansion, eds. B. F. Madore & R. B. Tully (Dordrecht: Reidel), pp. 169-175

Paley, E. S., Low, F. J., McGraw, J. T., Cutri, R. M. & Rix, H. 1991, ApJ, 376,

335
Purcell, E. M. 1969, ApJ, 158, 433
Purcell, E. M. 1976, ApJ, 206, 685-690
Purcell, E. M. & Pennypacker, C. R. 1973, ApJ, 186, 705-714
Rana, N. C. 1979, Ap.Space Sci., 66, 173
Rowan-Robinson, M. 1992, MNRAS, 258, 787-799
Sellgren, K. 1984, ApJ, 277, 623
Sodroski, T. J., Hauser, M. G., Dwek, E., Kelsall, T., Moseley, S. H., Silverberg, R. F., Boggess, N., Odegard, N. P., Weiland, J. L. & Franz, B. A. 1992, " Large Scale Physical Conditions in the Interstellar Medium from DIRBE Observations", presented at the "Back to the Galaxy" conference at U Md.
Witt, A. N. & Schild, R. E. 1988, ApJ, 325, 837
Wright, E. L. 1982, ApJ, 255, 401
Wright, E. L. 1987, ApJ, 320, 818
Wright, E. L. 1989, in Proceedings of IAU Colloquium 135 on Interstellar Dust, eds. L. Allamandola & X. A. M. Tielens, (Dordrecht: Kluwer), p. 337.
Wright, E. L., Mather, J. C., Bennett, C. L., Cheng, E. S., Shafer, R. A., Fixsen, D. J., Eplee, R. E. Jr., Isaacman, R. B., Read, S. M., Boggess, N. W., Gulkis, S. G., Hauser, M. G., Janssen, M., Kelsall, T., Lubin, P. M., Meyer, S. S., Moseley, S. H. Jr., Murdock, T. L., Silverberg, R. F., Smoot, G. F., Weiss, R., & Wilkinson, D. T., 1991, ApJ, 381, 200-209

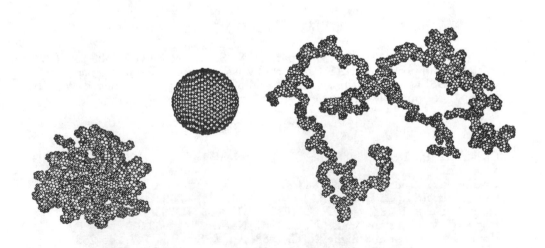

Fig. 5. Equal mass sphere (center), N = 1024 DLA fractal (left), & CL fractal (right).

COBE/DIRBE OBSERVATIONS
OF INFRARED EMISSION FROM STARS AND DUST

Michael G. Hauser
Code 680, NASA Goddard Space Flight Center, Greenbelt, MD 20771.

ABSTRACT

The Diffuse Infrared Background Experiment (DIRBE) on NASA's Cosmic Background Explorer (*COBE*) satellite has obtained low- angular resolution photometric maps of the full sky in ten broad spectral bands from 1.2 to 240 μm wavelength, and linear polarization maps at 1.2, 2.3, and 3.5 μm. These maps provide dramatic new views of the interplanetary dust cloud, integrated stellar light, and dust in the interstellar medium. An empirical function is used to separate the contribution from interplanetary dust, facilitating study of the stellar and interstellar emissions from the plane of the Milky Way. Initial results of such studies are presented in accompanying papers.

1. INTRODUCTION

The primary aim of the Diffuse Infrared Background Experiment is to conduct a definitive search for the isotropic cosmic infrared background radiation. This objective, and preliminary results of that search, have been discussed by Hauser et al. (1991). Additional objectives of the DIRBE are study of the diffuse interplanetary and interstellar media. To accomplish these objectives, the DIRBE was designed to obtain absolute brightness maps of the full sky in ten broad photometric bands from 1.2 to 240 μm wavelength. To facilitate study of the bright foreground contribution from interplanetary dust, linear polarization was also measured at 1.2, 2.2, and 3.5 μm, and all celestial directions were observed hundreds of times at all accessible elongation angles from the Sun in the range 64° to 124°.

A careful photometric reduction of the data from this unprecedentedly extensive infrared sky survey is currently in progress. In this paper, full sky images prepared from DIRBE data obtained during the initial five months of the *COBE* mission are presented. An empirical approach useful for separating the sky brightness contribution from interplanetary dust scattering and emission is described. Residual errors from this separation are small at low galactic latitude, facilitating studies of large scale galactic properties appropriate to the aims of this Conference. The specific investigations undertaken are described in accompanying papers at this Conference: stellar colors and extinction (Arendt *et al.* 1993); structure of the stellar bulge and implications for a stellar bar in the Milky Way (Weiland *et al.* 1993); large scale conditions in the interstellar medium (Sodroski *et al.* 1993); evidence for the galactic warp in the stellar and interstellar galactic components (Freudenreich *et al.* 1993); and a preliminary report on polarimetry with the DIRBE data (Berriman *et al.* 1993). The reader is referred to the above papers for the results of these DIRBE investigations. Implications of data from the *COBE* Far Infrared Absolute Spectrophotometer for the cooling of the interstellar gas are presented at this Conference by Bennett & Hinshaw (1993), and for the presence of very cold interstellar dust by Wright

(1993).

2. THE *COBE* MISSION AND DIRBE INSTRUMENT

The DIRBE was one of three instruments launched aboard the *COBE* satellite on Nov. 18, 1989. The *COBE* was placed in a circular, 900-km altitude, Sun-synchronous terminator orbit. The DIRBE was located inside a superfluid helium dewar which provided operating temperatures from 1.5 to 3 K for the various parts of the instrument from launch until expiration of the helium on Sept. 21, 1990. Since helium expiration, the interior of the dewar has stabilized near 50 K, cold enough for continued operation of the DIRBE from 1.2 to 4.9 μm; such operations are planned to continue until November 1993. In this orbit, the satellite is in full sunlight continuously except for parts of each orbit during a few month interval around the June solstice. During this 'eclipse' season, thermal conditions on the spacecraft are less stable than during the rest of the year. Data presented in this paper and the accompanying studies were obtained in the interval from Dec. 12, 1989, when instrument operation was optimized for standard orbital conditions, until May 1990, when the first 'eclipse season' began. A more detailed description of the *COBE* mission and its instruments has recently been given by Boggess *et al.* (1992).

The DIRBE instrument consists of a folded, off-axis Gregorian telescope with a 19-cm diameter primary mirror and a 10-spectral band absolute photometer and 3-band polarimeter. Sky brightness is measured at wavelengths of 1.2, 2.2, 3.5, 4.9, 12, 25, 60, 100, 140, and 240 μm, and linear polarization is measured at the three shortest wavelengths. The instrument rms sensitivity per field of view in 10 months is $\lambda I(\lambda) = (1.0, 0.9, 0.6, 0.5, 0.3, 0.4, 0.4, 0.1, 11.0, 4.0) \times 10^{-9}$ W m^{-2} sr^{-1}, respectively for the ten wavelength bands listed above. These levels are generally well below the total infrared sky brightness. The optical configuration (Magner 1987) has strong rejection of stray light from the Sun, Earth limb, Moon or other off-axis celestial radiation, or parts of the *COBE* payload (Evans, 1983; Evans and Breault 1983).

The DIRBE instrument measures absolute brightness by chopping between the sky signal and a zero flux internal reference using a tuning fork chopper. Instrumental offsets are measured by closing a cold shutter located at the prime focus. All spectral bands view the same instantaneous 0.7° x 0.7° field-of-view oriented at 30° from the spacecraft spin axis, which was maintained normally at 94° from the Sun and away from the Earth. This provides modulation of the solar elongation angle by 60° during each rotation, and mapping of 50% of the celestial sphere each day. As the orbit precesses 1° each day, the DIRBE slowly increases its sky coverage until the full sky is mapped each six months. The spiral scan yields hundreds of observations of each celestial direction every half year, with elongation angle varying over the full interval accessible at its ecliptic latitude in the range 64 to 124 degrees.

Four highly-reproducible internal radiative reference sources are used to stimulate all detectors when the shutter is closed to monitor the stability and linearity of the instrument response. Calibration of the photometric scale is obtained from observations of isolated bright celestial sources. The highly redundant sky sampling and frequent response checks will allow maintenance of precise photometric closure over the sky for the duration of the mission. The data obtained during the helium temperature phase of the mission are of excellent photometric quality, showing good sensitivity, stability, linearity, and stray light immunity. The measured sky brightnesses used for the studies presented at this Conference are conservatively estimated to have absolute photometric scale accuracy at each wavelength of 20%. Because a common celestial calibrator (Sirius) has been used from 1.2 to 12 μm, it is estimated that the accuracy of intensity ratios (colors) formed from these bands is of order 5%. Similarly, the 60 μm to 100 μm intensity ratio (common calibrator is Uranus) and 140 μm to 240 μm intensity ratio (common calibrator is Jupiter) are comparably accurate. The photometric consistency over the sky at each individual wavelength is better than 2%, except at 60 μm and 100 μm where non-linear detector effects have not yet been corrected.

3. THE APPEARANCE OF THE INFRARED SKY

Qualitatively, the initial DIRBE sky maps show the expected character of the infrared sky. This is shown in Figs. 1-4 (Color Plate numbers 3 and 4), each of which is a 3-band false color image of the full sky. Each image is a Mollweide projection in galactic coordinates, with the galactic center at the center of the image. In each case, the intensity at each wavelength is represented by the blue, green, or red intensity for the shortest to longest wavelength respectively. These images differ from earlier versions of the DIRBE data in two important respects: the data have been processed with the DIRBE pipeline software rather than 'quick-look' software, so that numerous segments of inferior data or processing artifacts have been eliminated and the full sky is included. Furthermore, the brightness shown at each sky pixel is the brightness observed when the solar elongation angle was 90°, so that there are no discontinuities in the images due to discreetly different elongations for adjacent sky pixels. This is the largest elongation angle for which the whole sky can be observed. The DIRBE, because of its dense sampling of each sky pixel at all possible elongation angles in the range 64° to 94°, has provided the first data set for which such images can be made.

Figure 1 shows a composite of the 1.2, 2.2, and 3.5 μm images. At 1.2 μm, stellar emission from the galactic plane and from isolated high latitude stars is prominent. Zodiacal scattered light from interplanetary dust is also present, but is so much fainter than starlight from the galactic disk that it is not apparent within the dynamic range displayed in Fig. 1. These two components continue to dominate out to 3.5 μm, though both become fainter as wavelength increases. Because extinction at these wavelengths is far less than in visible light, the disk and bulge stellar populations of the Milky Way are dramatically apparent in this image. Since extinction decreases from 1.2 to 3.5 μm wavelength, the image appears redder where there is more extinction, $e.g.$, in the plane of the disk and in discrete clouds in the northern part of the bulge.

Figure 2 shows a composite of the 4.9, 12, and 25 μm images. At 12 and 25 μm, emission from the interplanetary dust dominates the sky brightness everywhere but the central part of the galactic disk. This is evident in the bright S-shaped band, which is centered on the ecliptic plane. As with the scattered zodiacal light, the emitted contribution from the interplanetary dust decreases with increasing ecliptic latitude and with increasing solar elongation angle. The hard edges on the interplanetary dust emission on either side of the ecliptic plane are the result of emission from the zodiacal dust bands, first noted in the $IRAS$ data by Low $et\ al.$ (1984). Scattered light from these bands has also been detected at near-infrared wavelengths in the DIRBE data (Weiland $et\ al.$ 1991).

Figure 3 shows a composite of the 25, 60, and 100 μm images. At wavelengths of 60 μm and longer, emission from the interstellar medium dominates the galactic brightness, and the interplanetary dust emission becomes progressively less intense and hence appears blue in this image. The patchy infrared cirrus noted in IRAS data (Low *et al.* 1984) is evident at all wavelengths longer than 25 μm. Numerous nearby molecular clouds are evident in the image, such as the ρ Oph cloud North of the galactic center and the Orion complex below the plane at the right. Both Magellanic Clouds are also evident.

Figure 4 shows a composite of the 100, 140, and 240 μm images. The interplanetary dust and stellar contributions have nearly vanished, leaving a nearly uniform-temperature interstellar medium dominating the image. The series of images in Figs. 1-4 show clearly the changing character of the sky as source temperatures progress from thousands of degrees to tens of degrees. The DIRBE data will clearly be a valuable new resource for studies of the interplanetary medium and Galaxy as well as the search for the cosmic infrared background radiation.

4. REMOVAL OF THE INTERPLANETARY DUST SIGNAL

In order to carry out the quantitative studies of large- scale galactic properties described in the accompanying papers, an empirical procedure has been used to remove much of the foreground signal from the interplanetary dust (IPD). The procedure entails fitting an analytic function approximating the expected large- angular scale features of the IPD to the maps constructed from data at elongation angle 90°. Properties of the IPD cloud considered in choosing this function include the facts that the cloud is approximately an azimuthally symmetric distribution centered near the Sun and concentrated toward a symmetry surface, that this surface is inclined by a small angle with respect to the ecliptic plane, and that the dust density falls off rather rapidly with increasing distance from the Sun. This suggests that at high ecliptic latitude, β, the distribution should be close to a $\csc(\beta)$ distribution, which rounds off to a finite value as the symmetry surface is approached, and that the latitude of maximum brightness is not in general zero (though it is at that time of year when the Earth is in the line of nodes of the symmetry surface with the ecliptic plane). This suggests an expression for the intensity I at fixed elongation angle of the form:

$$I = I_o - k\{1 - \beta_c| \csc(\beta - \beta_o)|[1 - \exp(-x - \alpha x^2)]\}$$

where I_o = peak intensity, β_o = ecliptic latitude at maximum intensity, β_c = width parameter of the IPD brightness profile, k = intensity scale factor which defines the plane to pole intensity contrast, and $x = |(\beta - \beta_o)/\beta_c|$.

The parameter α can be chosen to minimize residuals. Though the value $\alpha = 1/2$ gives a profile with zero slope at $\beta = \beta_o$, one finds systematically smaller residuals for α values near 1/3. With the choice of $\alpha = 1/3$, the maximum residuals are typically a few percent of the peak IPD brightness and are dominated by the zodiacal band structure discovered in the *IRAS* data (Low *et al.* 1984).

This empirical function has been fitted separately to 1/4 -great circle map data segments at fixed ecliptic longitude from each ecliptic pole to the ecliptic latitude of maximum IPD intensity, β_o (recall that the solar elongation angle is fixed at 90° in these maps). The values of I_o and β_o in the ecliptic northern and southern segments of each 1/2 -great-circle at the same ecliptic longitude were required to be the same. This results in six free parameters for each half great-circle, and yields low residuals, particularly for the low galactic latitude regions studied here. Estimated residual contamination of the resulting 'galactic emission' maps by IPD signal depends on wavelength, and is noted in the accompanying papers. It might be noted that any constant contribution to the sky brightness (or instrument signal) over these data segments is included in the constant I_o, and does not appear in the residual 'galactic' emission maps.

5. CONCLUSIONS

The DIRBE instrument has provided extensive new data with many unique attributes showing the character of diffuse celestial emission from 1.2 to 240 μm wavelength. In particular, these data reveal large-scale properties of both the stellar and interstellar components of the Galaxy. A simple empirical fitting function permits removal of foreground solar system sky brightness contributions with small systematic residuals compared to galactic signals near the galactic plane. Some implications for the galactic bulge and disk are presented in five companion papers at this Conference.

ACKNOWLEDGEMENTS

The author gratefully acknowledges the contributions of his colleagues on the *COBE* Science Working Group to the accomplishments described here. He acknowledges particularly the contributions of N. Boggess, T. Kelsall, S. H. Moseley, R. F. Silverberg, and J. Weiland to this paper. The staff of the Cosmology Data Analysis Center, and particularly the *COBE* pipeline software development teams led by J. A. Skard and G. Toller, are thanked for their tireless efforts in producing the careful reduction of the data used in this investigation.

COBE is supported by NASA's Astrophysics Division. Goddard Space Flight Center (GSFC), under the scientific guidance of the *COBE* Science Working Group, is responsible for the development and operation of *COBE*.

REFERENCES

Arendt, R. G. *et al.* 1993, this volume
Bennett, C. L. & Hinshaw, G. 1993, this volume
Berriman, G. B. *et al.* 1993, this volume
Boggess, N. *et al.* 1992, ApJ, 397, 420
Evans, D. C. 1983, SPIE Proc., 384, 82
Evans, D. C.& Breault, R. P. 1983, SPIE Proc., 384, 90
Freudenreich, H. T. *et al.* 1993, this volume
Hauser, M. G. *et al.* 1991, in "After the First Three Minutes", eds. S. S. Holt,
    C. L. Bennett, V. Trimble, (AIP: New York), p. 161
Low, F. J. *et al.* 1984, ApJ, 278, L19
Magner, T. J. 1987, Opt. Eng., 26, 264
Sodroski, T. J. *et al.* 1993, this volume
Weiland, J. L. *et al.* 1991, BAAS, 23, 1313
Weiland, J. L. *et al.* 1993, this volume
Wright, Edward L. 1993, this volume

MAPPING THE ABSOLUTE BRIGHTNESS OF THE SKY AT LOW FREQUENCIES

G. De Amici, G. Smoot, M. Bensadoun, M. Limon, W. Vinje, C. Witebsky
Lawrence Berkeley Lab. and Space Sciences Lab., 50/351, Berkeley, CA 94720, USA

S. Torres, A. Umana, M. Becerra
Universidád de los Andes, Santa Fé de Bogotá, Colombia

Abstract

Uncertainties in the gain and zero-level of the existing radio and microwave frequency sky surveys often dominate the error budget in studies of anisotropies and spectral distortions of the Cosmic Background Radiation.

This paper discusses our existing prototype and the planned instruments and the observation techniques which are necessary for conducting an extended observational campaign to map the total sky brightness. Some experimental difficulties are outlined, and possible solutions are given.

1. Introduction

Measurements of the intensity and spatial distribution of the cosmic microwave background (CMB) radiation require the subtraction of the foreground galactic signals. The accuracy of the final result depends as much on the confidence with which this operation can be performed, as on the proper operation of the receiving instruments.

The galactic signal comes from the emission of relativistic electrons accelerated in the insterstellar magnetic field (synchrotron radiation), from thermal bremsstrahlung inside hydrogen clouds (HII or free-free radiation), and from thermal radiation of dust clouds (dust radiation). The intensity of each component changes with frequency in a unique way; this feature would disentangle one contribution from the other, given a sufficient number of measurements at different frequencies. Several excellent textbooks and articles deal with the theoretical aspects of the galactic emission components (Wrigth et al., 1991; Rybicki and Lightman, 1979; Longair, 1986). The dust component is dominant at the high end of the radio spectrum and in the IR region; its emissivity has a spectral index of about 1.6 (depending on the direction of observation) and is negligible for all measurements below about 50 GHz. At intermediate frequencies the thermal HII (free-free) emission is dominant, especially in the plane of the galaxy, where giant gas clouds provide the material and conditions for large concentrations of ionized hydrogen. The HII radiation has a spectral index of about -2.1, and is weakly dependent on the observing frequency and on the temperature of the electrons. At the low-frequency end of the spectrum and away from the galactic plane, the synchrotron radiation is dominant. The intensity is proportional to the depth of the electron cloud, the number density of the electrons, a power of the effective magnetic field, and a related power of the frequency. The exact value of the exponents depends on the energy spectral index of the electrons.

Recent and past measurements (Sironi et al., 1991 Bensadoun et al., 1993; Smoot et al. 1992; and references therein) of the spectrum and anisotropy of the CMB have skirted this problem by observing away from the galactic plane, toward areas where the HII signal, which is mostly associated with relatively dense and compact sources, is practically non-existent, except for a small and poorly measured diffuse signal, and the synchrotron radiation is at a minimum, and by using frequencies where the dust signal is negligibly small, and the extrapolation of the synchrotron radiation can be done with relative confidence.

Any accurate estimate of the synchrotron signal at mid- and high- frequencies (where the synchrotron signal is much smaller than the CMB, but still not negligible overall, and a significant component of the overall uncertainty) depends on the availability of a precise set of absolutely calibrated measurements at low frequencies (where the synchrotron signal is dominant).

2. Observational strategy

Several sky surveys have been conducted at frequencies ranging from 38 to 2300 MHz; with only one exception, these maps provide only partial sky coverage, and are affected by gain uncertainties and poorly known sidelobe contributions to the measured signal. An extended critique of the existing surveys has been made by Lawson et al (1987), and summarized and extended by De Amici et al (1993). Both studies have concluded that the published results are generally incompatible with each other, and large corrections need to be made to reconcile their quoted values. Even the best available survey, the whole-sky map by Haslam et al (1982), only marginally agrees with the absolutely calibrated measurement of the South Celestial Pole done at the same frequency by Price (1967). An upcoming map of the southern sky at 2.3 GHz (Jonas et al. priv. comm.) will provide excellent angular resolution, but will lack a reliable measurement of the zero level.

The need for a set of new, internally consistent maps is widely felt. The most desirable properties of these maps will be:

-constant angular resolution and beam pattern across frequencies, to allow for easy comparison of features as seen in different maps, and reduce the effect of beam pattern uncertainty

-total sky coverage, to force closure of the data

-absolute calibration of the zero level of the map to better than $(1 \text{ K}) * (v/408)^{-2.75}$, where v is the observing frequency in MHz, for $v \leq 1500$ MHz, and to better than 0.1 K for $v \geq 1500$ MHz

-accuracy of the gain level to better than 3%

-angular resolution of at least 3 degrees (for comparison with the pixelization of the COBE-DMR skymaps)

-sensitivity to the circularly polarized component (total intensity) of the galactic signal.

Taken together, the requirements on the gain and zero-level accuracy would imply a 0.04 K uncertainty in the extrapolated galactic emission at 1.4 GHz, a factor of three improvement over the 0.12 K uncertainty due to the presently available maps (Bensadoun et al 1993). These goals are not easily achieved. Constant angular resolution across frequencies requires carefully matched and frequency-scaled antennae. Total sky coverage with just one instrument can be obtained if the instrument is put into orbit: not an easy task for the antennae (20 m diameter) needed to obtain a 3° angular resolution at 408 MHz. Both absolute calibration and gain calibration require the use of quasi-free-space matched external loads at widely different temperatures; again a daunting manufacturing task for a large opening size antenna.

We believe that these results can be reached with a set of observations carefully distributed in longitude, so that complete sky coverage will not require any tipping of the antennae by more than 30° from local vertical. This limited zenith angle allows us to reduce the observing time needed at each location, while minimizing the contribution of the surrounding structures and terrain through the antenna sidelobes. Finally the antenna mount is of relatively simple construction if the zenith angle is limited. For a parabolic antenna, with extensions sidelobe screens, like the one used by the Berkeley team at the South Pole in 1991 (De Amici et al., 1993), the ground contribution for a 30° tilt has been measured to be less than 10 K. This value includes radiation diffracted over the shield, radiation reflected and scattered off the support legs for the feed antenna, radiation collected from the side and backlobe of the feed. With the selection of a feed antenna of better directivity, and improved radiometric properties, we expect to reduce the ground contribution to less than 3 K.

Because of the technique used to produce the existing maps (i.e. comparing measurements done with different instruments from different sites, and merging them together), there is a strong possibility that systematic errors might have been introduced; to minimize the uncertainties and corrections to the data, it is then desirable that the same instrument, or a very close duplicate, be used from each site. We have considered sites at several possible latitudes. While the apparent section of the sky seen from each site is approximatively constant, the real solid angle observed changes with latitude, giving the widest coverage and the most interesting targets to observers located near the Equator. We have started an international collaboration with research institutions in equatorial countries.

3. Calibration of the maps

The precise calibration of the maps is hampered by two, separate but related difficulties. First, the maps must be internally consistent; second, the zero level of the input signal must be determined with high accuracy.

The first problem arises from the fact that the radiometric properties of the receiver are not constant with time; even after the uttermost care has been taken to insulate the amplifiers and all the other components from the outside environment, radio receivers exhibit random gain changes at the 10^{-3} level over time scales of the order of one hour. The gain of the receivers will be periodically measured via a built-in constant reference temperature load (a solid state noise source); by monitoring the apparent antenna temperature of the reference, we will be able to track changes in the gain of the receiver. The experience gathered by the Berkeley group at the South Pole indicates that the receiver drifts can be modelled with good precision (less than 1.5% residual rms) if the noise source is fired every 40 s. The resulting loss of data (15% of the observing time) from the relatively high (10%) duty cycle is not an issue for this kind of experiment. A second generation receiver is expected to have even better stability; our goal is to model the gain changes to less than 1% rms.

The second problem comes from the fact that the zero level of the output scale of a radiometer can only be determined by comparing the antenna temperature of the sky against the antenna temperature of an external black-body-like load of well known thermodynamic temperature. In order to reduce the effects of systematic errors, an additional requirement is that the target temperature be very close to the temperature of the sky. A load with the required characteristics and of a size capable to fit a parabolic reflector antenna cannot be handled in the field, nor manufactured. A possible solution to the problem is to detach the feed antenna from the reflector, and use the feed alone to compare the signal from an appropriate target to the signal from the sky. Since the beam pattern of the feed antenna is much wider than the beam of the paraboloid, this technique requires convolving the measured sky signal to the feed beam pattern, with a consequent loss of accuracy. The insertion loss from the surface of the reflector antenna, and the effect of scatter off the structure supporting the feed must then be modelled and accounted for. The first effect is small at low frequency (for a clean reflector surface, we measured the antenna emission to be 0.08±0.03 K at 408 MHz), but the second one gives a non negligible contribution to the overall uncertainty. If there is scattering at the

5K level, we will have to measure and model it with a precision much better than 10%, in order to keep the experiment's error bars within the ±1K needed to substantially improve over the existing data sets. Another approach is to use a dedicated antenna with unobstructed aperture, whose beam pattern either closely matches the beam pattern of the parabolic reflector, or is accurately measured, and conduct a series of precise measurements of the temperature of the sky in a few selected areas. The sky signal is then compared to the signal from a waveguide load. By using an unobstructed aperture antenna, typically a horn either conical or rectangular, only the insertion loss of the walls and the sidelobes of the antenna, and the radiometric characteristics of the waveguide load, need to be modelled. At the frequency of 408 MHz, both the antenna walls and the waveguide load can be modelled with sufficient precision, that the accuracy of the sky measurement will limited only by our knowledge of the antenna sidelobes. We plan to use the AT&T horn-reflector antenna at Holmdel, NJ, (Crawford, 1961) as the calibration antenna. The main drawback of this approach is the necessity to accurately measure the insertion and reflection losses of the feed antenna. For the prototype unit deployed in 1991, the Berkeley group estimated that the feed antenna contributed about 20 K to the detected signal; this value will have to be improved to less than 5 K.

4. Conclusions

The desirability of a new and more accurate set of measurements of the galactic foreground emission is widely felt among the CMB community. A series of international collaborative efforts is being started to produce a series of maps of the microwave sky at frequencies ranging from 400 to 2500 MHz. Potential observing sites have been identified in California, Colombia and Antarctica; talks are underway with researchers in Brazil for a possible collaboration. Preliminary analysis and prototype tests indicate that a substantial improvement over the existing maps can be achieved within a few years from the beginning of the observing program.

This work has been supported in part by the US Department of Energy, by the US National Science Foundation, and by Colciencias of Colombia. We have drawn from the expertise and gratefully acknowledge the collaboration of J. Aymon, M. Bersanelli, H. Dougherty, D. Heine.

References

M. Bensadoun et al. to appear in: Ap. J., may 20, 1993
A.B. Crawford et al., Bell Sys. Tech. Jour., 1961, july issue, 1095
G. De Amici et al., in Proceedings "Observational Cosmology", Milano, 1993
C.G.T. Haslam et al., A.A.Suppl. 1982, 47, 1
K.D. Lawson, et al., M.N.R.A.S 1987, 225, 307
M.S. Longair, 1986, "High Energy Astrophysics", Cambridge Univ. Press, (Cambridge, UK)
R.M. Price, Aust. J. Phys. 1969, 22, 641
G.B. Rybicki and A.P. Lightman, 1979, "Radiative processes in astrophysics", Wiley & sons, (New York, USA)
G. Sironi et al., Ap.J. 1991, 378, 550
G. Smoot et al., Ap. J. Lett., 1992, 396, 1
E.L. Wrigth, et al., Ap. J., 1991, 381, 200

MODELLING GALACTIC MICROWAVE EMISSION
FROM 0.4 TO 50 GHZ

Chris Witebsky, George F. Smoot, Giovanni De Amici, and Jon Aymon
Lawrence Berkeley Laboratory and Space Sciences Laboratory,
University of California, Berkeley

Email: witebsky@smoot1.lbl.gov

ABSTRACT

We have developed a full-sky model of galactic microwave radiation at frequencies from 0.4 GHz to >50 GHz, with a resolution of ~2°. The primary components are free-free emission from ionized hydrogen and synchrotron emission from cosmic rays. (Dust emission is negligible in this frequency range.) A variety of data are used to estimate free-free emission from HII: the galactic plane and adjacent regions come from continuum surveys near 2.7 GHz; emission at higher galactic latitudes is approximated by a cosecant law. The synchrotron component is extrapolated from sky-survey data at 408 MHz. Both components vary with frequency. This model has been used to derive corrections for galactic emission in our measurements of the low-frequency CMBR spectrum.

1. INTRODUCTION

Modelling the microwave radiation from the galaxy serves several purposes. It helps to clarify our understanding of the distribution and characteristics of the processes that produce the radiation. In addition, galactic models are helpful in distinguishing the cosmic microwave background radiation (CMBR) or other extragalactic sources from galactic radiation. Several models designed for this application are discussed in Bennett *et al.* (1992) and papers cited therein.

Our model was originally developed for the analysis of CMBR anisotropy measurements, but it has also been used extensively to analyze low-frequency measurements of the CMBR spectrum (*e.g.* Bensadoun *et al.* 1992).

2. GALACTIC SPECTRUM

Galactic emission at microwave frequencies is made up of three components: synchrotron emission from cosmic-ray electrons in the galaxy's magnetic field, free-free emission (thermal bremsstrahlung) from ionized matter in HII regions and elsewhere, and thermal emission from interstellar dust. Typical spectra for the three components are plotted in Figure 1. Synchrotron emission decreases with frequency roughly as $\nu^{-2.7}$ to $\nu^{-3.1}$ over the range of interest (Banday & Wolfendale 1991). Free-free emission also decreases with frequency, but more gradually, approximately as $\nu^{-2.1}$. Dust emission increases approximately as $\nu^{1.65}$ (Wright *et al.* 1991), peaking in the IR.

It is evident from Figure 1 that dust emission is much weaker than synchrotron or free-free emission in the 0.4 to 50 GHz range. For this reason, dust emission will be omitted from further discussion.

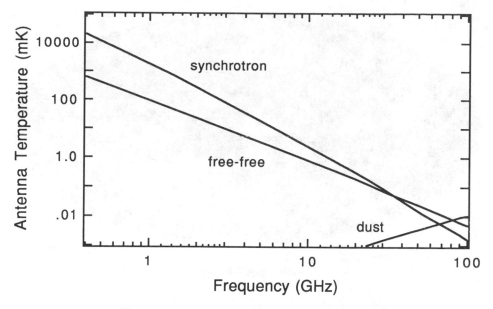

Fig. 1. Typical galactic emission spectra.

3. GALACTIC STRUCTURE

Synchrotron and free-free emission generate different distributions of radiation on the sky. Free-free emission is strongly concentrated near the galactic plane, with an additional weak diffuse component at high latitudes (Reynolds 1991). Synchrotron emission tends to be more dispersed and to have more structure on large angular scales.

4. MODELLING THE GALAXY

To model galactic emission, we combine frequency-scaled estimates of each component at each point on the sky. The combined result can then be convolved with an antenna pattern.

The estimated synchrotron component derives from the 408 MHz full-sky survey by Haslam *et al.* (1982), with small corrections to remove the estimated free-free component at that frequency. Frequency scaling is similar to Model C of Bennett *et al.* (1992).

Our model is unusual in that it attempts to reproduce the structural details of free-free emission at low galactic latitudes. To do so, it uses a list of HII source fluxes compiled from various galactic-plane surveys, mostly near 2.7 GHz (Fürst *et al.* 1990; Thomas and Day 1969a; Thomas and Day 1969b; Day, Thomas and Goss 1969; Beard 1966; Beard, Thomas and Day 1969; Goss and Day 1970; Beard and Kerr 1969; Day, Warne and Cooke 1970; Day, Caswell and Cooke 1972; Altenhoff *et al.* 1970; Wedker 1970; Felli and Churchwell 1972; Mills and Aller 1971). It also models a weak diffuse thermal component that varies with latitude as $\csc(b)$, derived by Smoot and Reach (1991) from measurements by Reynolds (1991). Figure 2 is a plot of the resulting distribution, frequency-

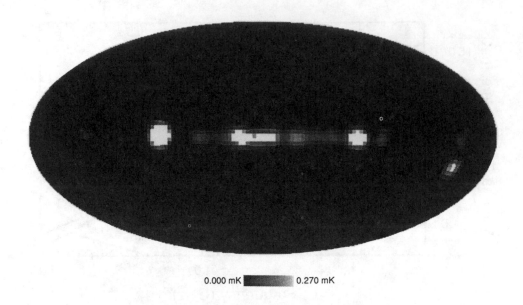

0.000 mK ▓▓▓▓▓▓▓ 0.270 mK

Fig. 2. HII model at 53 GHz, convolved with DMR gain pattern.

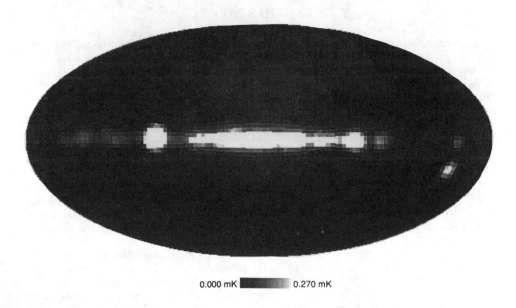

0.000 mK ▓▓▓▓▓▓▓ 0.270 mK

Fig. 3. Estimated galactic emission at 53 GHz, including synchrotron, free-free, and dust emission, convolved with DMR gain pattern.

scaled to 53 GHz and convolved with the gain pattern of the COBE differential microwave radiometer (DMR) antennas (Toral *et al.* 1989).

5. MODEL RESULTS

Figure 3 shows the estimated emission from the galaxy at 53 GHz, convolved with the DMR beam pattern. The distribution is similar to that of the DMR data with the dipole component removed (Smoot *et al.* 1992). We are currently analyzing the model maps to evaluate the residual discrepancies with the DMR data.

ACKNOWLEDGEMENTS

The COBE program is supported by Astrophysics Division of the NASA Office of Space Science and Applications. This work was supported in part by the Director, Office of Energy Research, Office of High Energy and Nuclear Physics, Division of High Energy Physics of the U.S. Department of Energy under Contract No. DE-AC03-76SF00098.

REFERENCES

Altenhoff, W.J., Downes, D., Goad, L., Maxwell, A. & Rinehart, R. 1970, A&AS, 1, 319
Banday, A.J. & Wolfendale, A.W. 1991, MNRAS, 248, 705
Beard, M. 1966, Aust. J. Phys., 19, 141
Beard, M. & Kerr, F.J. 1969, Aust. J. Phys., 22, 121
Beard, M., Thomas, B.MacA. & Day, G.A., 1969, Aust. J. Phys. Ap. Supp., 11, 27
Bennett, C.L. *et al.* 1992, ApJ, 396, L7
Bensadoun, M., Bersanelli, M., De Amici, G., Kogut, A., Limon, M., Levin, S.M., Smoot, G.F., & Witebsky, C. 1993, ApJ, 409
Day, G.A., Caswell, J.L. & Cooke, D.J. 1972, Aust. J. Phys., 25, 1
Day, G.A., Thomas, B.MacA. & Goss, W.M., 1969, Aust. J. Phys. Ap. Supp., 11, 11
Day, G.A., Warne, W.G. & Cooke, D.J. 1970, Aust. J. Phys. Ap. Supp., 13, 11
Felli, M. & Churchwell, E. 1972, A&AS, 5, 369
Fürst, E., Reich, W., Reich, P., & Reif, K. 1990, A&AS, 85, 805
Goss, W.M. & Day, G.A. 1970, Aust. J. Phys. Ap. Supp., 13, 3
Haslam, C.G.T., Salter, C.J., Stoffel, H. & Wilson, W.E. 1982, A&AS, 47, 1
Mills, B.Y. & Aller, C.H. 1971, Aust. J. Phys., 24, 609
Reynolds, R.J. 1991, ApJ, 372, L17
Smoot, G.F. *et al.* 1992, ApJ Lett, 396, 1
Smoot, G.F. & Reach, W. 1991 COBE Note Number 5037
Thomas, B.MacA. & Day, G.A. 1969a, Aust. J. Phys. Ap. Supp., 11, 3
Thomas, B.MacA. & Day, G.A. 1969b, Aust. J. Phys. Ap. Supp., 11, 19
Toral, M.A., Ratliff, R.B., Lecha, M.C., Maruschak, J.G., Bennett, C.L. & Smoot, G.F. 1989, IEEE T-AP, AP-37, 171
Wedker, H.J. 1970, A&A, 4, 378
Wright, E.L. *et al.* 1991, ApJ, 381, 200

POLARIMETRY WITH THE DIFFUSE INFRARED BACKGROUND EXPERIMENT ABOARD COBE

G. B. BERRIMAN
General Sciences Corporation, Code 685.3, NASA/Goddard Space Flight Center, Greenbelt, MD 20771

M. G. HAUSER, N. W. BOGGESS, T. KELSALL, S. H. MOSELEY, R. F. SILVERBERG
NASA/Goddard Space Flight Center, Greenbelt, MD 20771

T. L. MURDOCK
General Research Corporation, 5 Cherry Hill Drive, Danvers, MA 01923

ABSTRACT

This paper describes polarimetry of the Galactic plane and the zodiacal light, surveyed by the COBE Diffuse Infrared Background Experiment. Methods of analysis and the overall polarization characteristics of the diffuse sky are emphasized over interpretation. Results are illustrated by polarimetry at 2.2 μm obtained from 1990 August 5-12, when most of the Galactic plane was in view. Similar results are found at 1.2 μm and 3.5 μm.

I. INTRODUCTION

The Diffuse Infrared Background Experiment (DIRBE) is a cooled, off-axis Gregorian telescope with a 19 cm primary mirror, designed to operate as an infrared radiometer and polarimeter (Hauser et al. 1991; Boggess et al. 1992). It has performed the first linear polarization survey of the diffuse sky from 1.2 μm to 3.5 μm between November 1989 and September 1990.

Two large-scale sources of diffuse emission are expected to be polarized: the Galactic plane and the zodiacal light. The Galactic plane is the brightest structure in the sky in the near infrared. Its emission is largely due to starlight, polarized through absorption by magnetically aligned dust grains. Polarizations of less than ≈1% on average are to be expected (Martin and Whittet 1990). By contrast, the zodiacal light is much fainter than the Galactic plane, but its polarization, due to scattering of sunlight by grains close to the Earth, is expected to be as high as ≈10-15% (Leinert 1975). This paper outlines the construction and operation of the DIRBE polarimetry experiment, and describes the properties of the Stokes parameters obtained with it.

II. CONSTRUCTION OF THE DIRBE POLARIMETER

DIRBE measures polarizations at 1.2 μm, 2.2 μm and 3.5 μm in bandpasses that approximate the ground-based J, K and L' filters. The construction of the polarimeter is different from that of most ground-based polarimeters, where a single polarizer is used at all wavelengths and samples the signal by rotating about the optical axis of the telescope. DIRBE is instead equipped with a three channel polarimeter for each wavelength. Channel 'a' is a total intensity channel and channels 'b' and 'c' contain polarizers whose principal transmission axes are orthogonal. Each channel sees the same instantaneous field-of-view of 0.7° x 0.7°. The polarizers are fixed inside the DIRBE: those in the 'c' channels are always parallel to the instrument scan direction on the sky, while those in the 'b' channels always orthogonal to the scan direction. The orientations at which these polarizers measure the sky are therefore dictated by the motion of the instrument field-of-view across the sky.

III. MEASUREMENT OF POLARIZATIONS WITH DIRBE

Linear polarization is described by the Stokes parameters Q and U, measured as fractions of the total intensity. They are defined in ecliptic coordinates relative to the local line of longitude: Q is the component parallel or perpendicular to the local line of longitude, and U is the component at 45°E or 45°W of the line of longitude. In this convention, positive values of Q are parallel to the line of longitude, positive values of U are 45°E of the line of longitude.

The principal advantage of having the polarizers fixed inside DIRBE is that all instrumental polarization effects are eliminated: the only polarization signal measured by DIRBE is that attributable to the sky. As the spacecraft scans, DIRBE records the polarizer orientation and the intensities in each channel at a time resolution of 1/8 sec. The orientation, Θ, of the polarizer is defined as the angle between the local line of longitude and the polarizer axis in the 'c' channel projected on to the sky. It equals 0° when the polarizer is parallel to the local line of longitude, 90° when perpendicular to it. The "polarization signals" in channels 'c' and 'b', $y_{c,b}$, are related to the intensities via

$$y_c = (I_c/I_a)/k_c - 1; \quad y_b = 1 - (I_b/I_a)/k_b$$

where I_a, I_b and I_c are the intensities, and k_c and k_b are the "polarization coefficients" for the 'b' and the 'c' channels. The coefficients take account of the efficiencies of the polarizers and the throughput of the DIRBE optics. They are found from the mean intensity ratios I_c/I_a and I_b/I_a of unpolarized calibration stars, such as Sirius. The Stokes parameters are related to y and Θ via

$$y_{c,b} = Q_{c,b}\cos(2\Theta) + U_{c,b}\sin(2\Theta)$$

Observations of each sky pixel (approximately 0.32° on a side) must be accumulated over one week to obtain adequate signal-to-noise. What governs DIRBE's ability to measure Q and U is the distribution of Θ measured in one week, which is in turn governed by the DIRBE scan path. This important issue is amplified below.

IV. THE ORIENTATIONS OF THE POLARIZERS

Because the DIRBE field-of-view is inclined at 30° to the spin axis of the spacecraft, it traces out a helical path on the sky 60° wide, between solar elongation angles of 64° and 124°. Thus, the orientations change systematically across the scan path. At the edges of the scan path on the ecliptic equator, for example, the polarizers are perpendicular to the equator (i.e., parallel to the local line of longitude), and so have $\Theta \approx 0°$; these values increase progressively to $\Theta \approx 90°$ as the field-of-view rotates towards the middle of the scan path.

In one week, each sky pixel may be crossed while moving to the East or to the West of the local line of longitude, according to whether the field-of-view leads or follows the spin axis of the rotating spacecraft. Thus, successive orbits generally produce two distinct groups of orientations in each pixel. The range of Θ within each group is, however, only a few degrees because the COBE orbit precesses 1° per day.

An important consequence of this narrow range of Θ is that for most sky pixels, only one of the two Stokes parameters can be determined in one week. This applies to both the 'c' and 'b' channels: a consequence of having fixed orthogonal polarizers is that they both measure the same Stokes parameter; one channel does not, for example, measure Q while the other measures U. Q is only determined where $\Theta \approx 0°$ or $\pm 90°$, and so is generally only measured at the edges of the sky path and in the middle. Only U is determined near $\Theta = 45°$. Q and U are therefore measured alternately as the spacecraft scans across the sky.

V. RESULTS: POLARIZATION OF THE SKY AT 2.2 μm

a) The Data

This section summarizes the Stokes parameters obtained from 1990 August 5-12, when most of the Galactic plane was in view. They were found in each channel by performing a robust least-squares fit to all the data obtained in each sky pixel (excluding data contaminated by the Moon, the planets or the South Atlantic Anomaly). The two independent measures of Q or U given by the 'c' and 'b' channels are then averaged by weighting each value according to its uncertainty.

All results quoted here are preliminary because photometric closure corrections have not been applied: the values of the Stokes parameters will change by a few percent of those quoted here. Definitive weekly-averaged Stokes parameters Q and U for the whole mission will be part of the final data products released in June, 1994.

b) Discussion of Results

The 2.2 µm polarizations of the Galactic light and of the zodiacal light are very different. The zodiacal light is strongly polarized, with average Stokes parameters of Q=+10% and U=0%. The uncertainties per sky pixel are generally in the range σ_Q/Q and $\sigma_U/U = 1\%$ to 15%, the wide variation arising largely because stars contribute to the photometric noise in the large DIRBE beams.

Galactic polarizations are much smaller than the zodiacal polarizations. The two can be easily distinguished between galactic longitudes of $l=\pm 90°$, where the Galactic emission is very strong. Here, the average amplitudes of Q and U are approximately 0.7%, in accord with the results of ground-based polarimetry (Martin and Whittet 1990). Within $l=\pm 20°$ of the Galactic Center, the polarizations correlate with the 1.25 µm optical depth. Moreover, the Barnard 78 dark cloud complex 3° north of the Center is also strongly polarized (Arendt *et al.* 1993).

The uncertainties in the Galactic Stokes parameters are generally very small, with σ_Q/Q and $\sigma_U/U = 0.5\%$ per pixel. These values do not exhibit the wide fluctuations seen in the zodiacal light. This is because the many stars seen in each DIRBE beam make the signal appear spatially smooth on the scale of a DIRBE pixel. The direction of the polarization vector of the Galaxy is difficult to determine in this preliminary investigation: accurate determination will require the separation of the Galactic and zodiacal components. Analysis to perform this separation is planned.

ACKNOWLEDGEMENTS

The authors gratefully acknowledge the efforts of the DIRBE data processing and validation teams in producing the high-quality datasets used in this investigation. We also thank the COBE Science Working Group for helpful comments on this manuscript. COBE is supported by NASA's Astrophysics Division. Goddard Space Flight Center (GSFC), under the scientific guidance of the COBE Science Working Group, is responsible for the development and operation of COBE.

REFERENCES

Arendt, R. *et al.* 1993, this volume.
Boggess, N. W. *et al.* 1992, *Ap J*, **397**, 420.
Hauser, M. G. *et al.* 1991. In *After The First Three Minutes*, (AIP Conference Proceedings, Vol. 222). Holt, S. *et al.*, eds. p 161.
Leinert, Ch. 1975, *Space Science Reviews*, **18**, 281.
Martin, P. G. and Whittet, D. C. B. 1990, *Ap J*, **357**, 113.

LARGE-SCALE PHYSICAL CONDITIONS IN THE INTERSTELLAR MEDIUM FROM DIRBE OBSERVATIONS

T.J. Sodroski[1], M.G. Hauser[2], E. Dwek[3], T. Kelsall[3], S.H. Moseley[3],
R.F. Silverberg[3], N. Boggess[3], N. Odegard[4], J.L. Weiland[4], and B. Franz[1]

[1]Applied Research Corporation, Code 685.3,
NASA Goddard Space Flight Center, Greenbelt, MD 20771
[2]Code 680, NASA/Goddard Space Flight Center, Greenbelt, MD 20771
[3]Code 685, NASA/Goddard Space Flight Center, Greenbelt, MD 20771
[4]General Sciences Corporation, Code 685.3,
NASA/Goddard Space Flight Center, Greenbelt, MD 20771

ABSTRACT

Observations from the COBE Diffuse Infrared Background Experiment (DIRBE) of the 140 and 240 μm emission from the Galactic plane region ($|b| < 10°$) are analyzed. Assuming an isothermal dust distribution along each line of sight, maps of dust temperature, optical depth, and total far-infrared (FIR) brightness are derived. These quantities are combined with available ^{12}CO and HI data to produce maps of the gas-to-dust mass ratio and the FIR luminosity per hydrogen mass. The results differ from those of a similar analysis using IRAS 60 and 100 μm data (Sodroski et al. 1987), at least in part because emission from small, transiently heated grains does not contribute significantly at 140 or 240 μm.

A linear correlation analysis is used to decompose the 140 and 240 μm maps into components associated with the neutral atomic (HI), molecular (H_2), and extended low-density ionized (HII) gas phases of the interstellar medium. From the resulting emission components the large-scale physical conditions within each gas phase are derived. The implications of the results for the current picture of the Galactic large-scale FIR emission are discussed.

1. INTRODUCTION

The close agreement of the results of several independent IRAS studies of the Galactic large-scale 60 and 100 μm emission (Sodroski et al. 1986, 1987, 1989; Perault et al. 1989; Bloemen, Deul, & Thaddeus 1990; and references therein) has placed tight constraints on the large-scale physical conditions within the most massive gas phases (HI, H_2, and HII) of the interstellar medium and the distribution of FIR luminosity among these gas phase components. However, the limitations of these studies include the following: (1) IRAS may not be sensitive to a cold dust component which consitutes a large fraction of the total dust mass in the ISM. (2) There may be a significant contribution to the emission at 60 μm from small, transiently heated grains. Analysis of DIRBE observations at 140 and 240 μm can be combined with the results of the IRAS studies to arrive at a more complete interpretation of the Galactic large-scale FIR emission.

2. DATA SETS

The data used in this analysis comprise: (1) DIRBE (Hauser 1993) 140 and 240 μm observations at 0.7° resolution, from data averaged over 10 months of cryogenic operation. Zodiacal emission has not been subtracted. It is estimated to be less than 10% of the observed 140 μm intensity and less than 5% of the observed 240 μm intensity for all pixels along the Galactic plane (b = 0°); (2) Velocity-integrated HI 21-cm line observations at ~ 0.7° resolution. The data are from the surveys of Weaver & Williams (1973), Kerr et al. (1986), and Burton & Liszt (1983); (3) Velocity-integrated ^{12}CO (J = 1 → 0) line observations at 0.5° resolution. The data are from the surveys of the Goddard Institute for Space Studies (Dame et al. 1987); (4) 5 GHz radio-continuum observations of the Galactic plane region from the survey of Haynes, Caswell & Simons (1978). The data were smoothed to ~ 0.5° resolution.

3. DERIVED PHYSICAL CONDITIONS ASSUMING A SINGLE DUST TEMPERATURE ALONG EACH LINE OF SIGHT

Galactic plane region (|b| < 10°) maps of dust temperature, optical depth at 240 μm, and total integrated brightness over all FIR wavelengths (λ > 40 μm) were derived by assuming that the dust distribution along each line of sight is isothermal, and has a λ^{-2} emissivity law (see Sodroski et al. (1987) for details of this type of analysis). A map of the total H atom column density was derived from the HI and ^{12}CO surveys. These maps were then combined to derive maps of the gas-to-dust mass ratio and FIR luminosity per hydrogen mass. In order to derive the gas-to-dust mass ratio map, a dust mass absorption coefficient of 7.2 cm^2 g^{-1} at 240 μm was assumed (Hauser et al. 1984).

The 240 μm optical depth map has a maximum value of ~ 0.02. This shows that, averaged over the 0.7° DIRBE beam, the Galaxy is optically thin at wavelengths ≥ 140 μm. The total FIR brightness at b = 0° decreases by almost an order of magnitude from the inner (270° < ℓ < 90°) to the outer (90° < ℓ < 270°) Galaxy.

Longitude profiles of the dust temperature, gas-to-dust mass ratio, and FIR luminosity per hydrogen mass at b = 0° were derived. The profiles represent the mean properties within the latitude interval |b| < 0.75°. The dust temperature has a mean value over all longitudes of ~ 19 K, and decreases by ~ 20% from the inner to the outer Galaxy. We have also derived the latitude distribution of dust temperature from data averaged over 12 longitude bins. The latitude distribution of dust temperature shows a general trend of decreasing temperature with increasing latitude for the inner Galaxy, and increasing temperature with increasing latitude for the outer Galaxy. These results suggest that the dust temperature in the Galactic disk decreases significantly with increasing galactocentric distance. Assuming an axisymmetric disk, the radial distribution of dust temperature in the Galactic disk has been estimated by unfolding the 140 and 240 μm longitude profiles. A dust temperature exponential scale length of ~ 20 kpc is obtained.

The gas-to-dust mass ratio profile has a mean value of ~ 160, a value consistent within the uncertainties with the local canonical value of ~ 100. The ratio increases by ~ 35% from the inner to the outer Galaxy. However, this derived increase may not be real, since the assumption of a single dust temperature along each line of sight may lead to large errors in derived dust mass column density. The FIR luminosity per hydrogen mass has a mean value in the inner Galaxy of ~ 3 $L_{\odot}$ $M_{\odot}^{-1}$, and decreases by a factor of ~ 4 from the inner to the outer Galaxy.

Within a few degrees of longitude of the Galactic Center, the gas-to-dust mass ratio reverses its general trend and increases by a factor of $\sim 2 - 3$ above the inner Galaxy value. The FIR luminosity per hydrogen mass also reverses its general trend and decreases by a similar factor below the inner Galaxy value. One possible explanation of these results is that there is a significant deficiency of dust in the central region of the Galaxy (galactocentric distance < 500 pc) relative to the Galactic disk. Another possible explanation is that the ratio of H_2 column density to ^{12}CO intensity is significantly lower in the central region of the Galaxy than in the Galactic disk (Blitz et al. 1985), in which case the column density of molecular gas along lines of sight near the Galactic center has been overestimated.

The derived dust temperature and FIR luminosity per hydrogen mass profiles have important implications for our current picture of the Galactic large-scale FIR emission. They suggest that the equilibrium temperature of large dust grains in the interstellar medium, and therefore the ambient radiation field which heats these grains, decrease rapidly with increasing galactocentric distance. However, the mean dust temperature and FIR luminosity per hydrogen mass of star forming regions do not vary significantly from the inner to the outer Galaxy (Wouterloot, Brand, & Henkel 1988; Mead, Kutner, & Evans 1990). Therefore, our results suggest that the primary heating source of the dust in the Galactic disk is the general interstellar radiation field (Sodroski et al. 1989), which decreases by almost an order of magnitude from a galactocentric distance of 5 kpc to the solar circle (Cox, Krugel, & Mezger 1986). The derived longitudinal variations of dust temperature and FIR luminosity per hydrogen mass are consistent, within the uncertainties, with the corresponding variations predicted for dust heated by the general interstellar radiation field.

4. THE LARGE-SCALE PHYSICAL CONDITIONS WITHIN THE INDIVIDUAL GAS PHASES OF THE ISM: LINEAR DECOMPOSITION OF THE 140 AND 240 µm DATA

The large-scale physical conditions within the three most massive gas phases (H_2, HI, and HII) of the ISM have been inferred by applying the linear decomposition method of Sodroski et al. (1989) to the 140 and 240 µm data within the region $300° < \ell < 40°$, $|b| < 1.25°$. For the HI gas phase, the derived dust temperature and FIR luminosity per hydrogen mass profiles each decrease with increasing longitude in a manner consistent with that expected for dust heated by the general interstellar radiation field. A comparison of these results with the results of a similar analysis of the IRAS 60 and 100 µm data (Sodroski et al. 1989) shows that the relative contribution to the FIR intensity increases with increasing wavelength for the molecular component, whereas that of the ionized component decreases with increasing wavelength, consistent with the generally accepted view that dust associated with ionized gas has a higher equilibrium temperature than dust associated with molecular gas. The FIR luminosity per hydrogen mass profiles for the HI gas phase from the two decompositions are in close agreement. However, the dust temperature profile for the HI gas phase from the

IRAS decomposition shows no significant variation with longitude. We conclude that this derived constancy in dust temperature is probably a result of the contribution from small (< 10 Å), transiently heated grains to the 60 µm intensity, as suggested by Sodroski et al. (1987, 1989).

It is inferred from this analysis that the total FIR luminosity of the Galactic disk is ~ 1.5 x 10^{10} $L_{\odot}$, with 70%, 20%, and 10% from dust associated with neutral atomic hydrogen clouds, molecular clouds, and extended low-density HII regions, respectively. This result is consistent with the results of the IRAS analyses of the Galactic plane (Sodroski et al. 1986, 1987, 1989; Bloemen, Deul, & Thaddeus 1990; and references therein).

For future work it is intended to incorporate the kinematical information of the ^{12}CO and HI surveys into the analysis in order to derive the distributions of the large-scale physical conditions with galactocentric distance, and to do a more detailed study of the physical conditions within the central region of the Galaxy.

ACKNOWLEDGEMENTS

COBE is supported by NASA's Astrophysics Division. Goddard Space Flight Center (GSFC), under the scientific guidance of the COBE Science Working Group, is responsible for the development and operation of COBE.

REFERENCES

Blitz, L., Bloemen, J. B. G. M., Hermsen, W., & Bania, T. M. 1985, A&A, 143, 267
Bloemen, J. B. G. M., Deul. E. R., & Thaddeus, P. 1990, A&A, 233, 437
Burton, W. B., & Liszt, H. S. 1983, A&AS, 52, 63
Cox, P., Krugel, E., & Mezger, P. G. 1986, A&A, 155, 380
Dame, T. M., et al. 1987, ApJ, 322, 706
Hauser, M. G. 1993, this volume
Hauser, M. G., et al. 1984, ApJ, 285, 74
Haynes, R. F., Caswell, J. L., & Simons, L. V. S. 1978, Australia J. Phys., Ap. Suppl., 45, 1
Kerr, F. J., Bowers, P. F., Jackson, P. D., & Kerr, M. 1986, A&AS, 66, 373
Mead, K. N., Kutner, M. L., & Evans, N. J. 1990, ApJ, 354, 492
Pérault, M., Boulanger, F., Puget, J. L., & Falgarone, E. 1989, ApJ, submitted
Sodroski, T. J., Dwek, E., Hauser, M. G., & Kerr, F. J. 1986, in Star Formation in Galaxies, ed. C. J. Persson (Washington, DC: US Government Printing Office), p.99
----------. 1987, ApJ, 322, 101
----------. 1989, ApJ, 336, 762
Weaver, H., & Williams, D. R. W. 1973, A&AS, 8, 1.
Wouterloot, J. G. A., Brand, J., & Henkel, C. 1988, A&A, 191, 323

DIRBE OBSERVATIONS OF GALACTIC STELLAR POPULATIONS AND EXTINCTION

R. G. Arendt[1], M. G. Hauser[2], N. Boggess[3], E. Dwek[3], T. Kelsall[3],
S. H. Moseley[3], T. L. Murdock[4], R. F. Silverberg[3], G. B. Berriman[5],
J. L. Weiland[5], N. Odegard[5], and T. J. Sodroski[1]

[1]Applied Research Corp., Code 685.3, NASA Goddard Space Flight Center, Greenbelt, MD 20771
[2]Code 680, NASA Goddard Space Flight Center, Greenbelt, MD 20771
[3]Code 685, NASA Goddard Space Flight Center, Greenbelt, MD 20771
[4]General Research Corp., Technology Dept., 5 Cherry Hill Dr., Ste. 220, Danvers, MA 01923
[5]General Sciences Corp., Code 685.3, NASA Goddard Space Flight Center, Greenbelt, MD 20771

ABSTRACT

The short wavelength bands of the Diffuse Infrared Background Experiment (DIRBE) on NASA's Cosmic Background Explorer (COBE) mission allow us to examine the integrated near-infrared emission of various stellar populations within the Galaxy. The results of a preliminary analysis at 1.25, 2.2, 3.5, and 4.9 μm are presented. The characteristic colors of the bulge and disk populations are determined. The derivation of the line-of-sight extinction in these bands yields a mean interstellar reddening law. Comparison of the near-infrared extinction with that for the longer wavelength bands (140 and 240 μm) allows separation of the relatively nearby obscuration from that caused by more distant clouds.

1. INTRODUCTION

The near infrared region of the spectrum (1 – 5 μm) is an excellent window in which the stellar content of the Galaxy can be studied. At shorter wavelengths, extinction by dust prohibits observation of the full Galactic distribution of stars. At longer wavelengths, the diffuse Galactic light becomes dominated by the emission from dust in H II regions and in H I and molecular clouds.

The DIRBE instrument has mapped the full sky with a 0°.7 square beam at wavelengths from 1 to 240 μm (Hauser *et al.* 1991, Boggess *et al.* 1992), and has measured polarization from 1 to 3.5 μm (Berriman *et al.* 1993). DIRBE maps of the Galaxy at 1.25, 2.2, 3.5, and 4.9 μm clearly show a smooth Galactic disk of starlight surrounding an elliptical bulge. Extinction still significantly influences the appearance of the Galaxy at 1.25 μm (*e.g.* a dust lane is visible across the bulge of the Galaxy), but it quickly becomes less significant with increasing wavelengths in accord with the known properties of interstellar extinction (*e.g.* Rieke & Lebofsky 1985). It also polarizes starlight in the Galactic plane through selective absorption by magnetically aligned grains.

This paper examines the colors of the stellar populations and the characteristics of the interstellar extinction and polarization as observed in the short wavelength DIRBE bands. Questions regarding the spatial structure of the stellar bulge and the disk of the Galaxy are addressed in other papers in these proceedings (Weiland *et al.* 1993,

Freudenreich *et al.* 1993, Sodroski *et al.* 1993).

2. DATA

The data used here consist of observations spanning the six month interval from Dec 1989 to May 1990. All observations were then interpolated to 90° solar elongation, so that the signal contributed by interplanetary dust could be uniformly removed using an empirical fitting function (Hauser 1993). The estimated inaccuracies in our removal of this local emission are $\lesssim 1 - 2\%$ (at $1 - 4$ μm) and $\lesssim 5\%$ (at > 4 μm) of the median brightness of the inner Galaxy.

3. COLOR-COLOR DIAGRAMS

Color-color diagrams of I(1.25 μm) / I(2.2 μm) vs. I(2.2 μm) / I(3.5 μm) and of I(2.2 μm) / I(3.5 μm) vs. I(3.5 μm) / I(4.9 μm) were generated for various longitude ranges along the Galactic plane. The correlations between these colors were investigated by examining the colors in each DIRBE pixel (~0.32° square).

For the inner Galaxy, all three near-IR colors are linearly correlated with one another. The dispersion of points *along* these linear trends decreases towards the outer Galaxy and at higher latitudes. The range of colors is greatest for the I(1.25 μm) / I(2.2 μm) color and decreases with increasing wavelength. The correlations as a function of longitude in the 4th Galactic quadrant show good symmetry with those observed in the 1st quadrant. The Cygnus Region ($l = 70° - 90°$) and the Carina Region ($l = 270° - 290°$) show a greater range of colors than suggested by the general trend with longitude, and in the shorter wavelengths the trend has a significantly steeper slope.

4. EXTINCTION

The correlations between near-IR colors are strongly suggestive of extinction effects. We have quantified these trends by comparison with reddening lines calculated in three different ways: a) calculating a linear least squares fit to the data with the assumption that all sources are seen through the same amount of foreground extinction; b) using the Rieke-Lebofsky reddening law (Rieke & Lebofsky 1985) with the assumption of only foreground extinction; and c) using the Rieke-Lebofsky reddening law assuming that the sources of emission (stars) are embedded uniformly in an absorbing medium (dust).

The DIRBE-derived extinction law [method (a)] is consistent with the Rieke-Lebofsky reddening law [method(b)]. Other tabulated reddening laws (*e.g.* Mathis 1990, Koornneef 1983) do not fit the DIRBE data as well, particularly in the redder colors.

In the direction of the inner Galaxy and bulge ($70° < l < 290°$, $-5° < b < 5°$), the reddening is too large and too linear to be caused by a mixture of dust and stars along the line of sight [method (c)]. With uniformly mixed dust and stars, the observed colors will always be dominated by relatively nearby unreddened emission. Apparently the observed emission is actually dominated by that of the inner Galaxy and bulge as seen through up to 4 magnitudes of extinction (at 1.25 μm). The optical depth at 1.25 and 2.2 μm reaches 1.0 within ~2° of the Galactic plane at these longitudes. This means that the different wavelengths used here will sample different path lengths in the Galactic plane, complicating any detailed analysis.

It is noteworthy that none of the simple reddening models used here adequately describe the color-color diagrams of star-forming regions like Cygnus. Possible explanations for this are discussed below (§7).

5. INTRINSIC POPULATION COLORS OF THE INNER DISK AND BULGE

With the assumptions that the signal is predominantly from stars and that all the extinction lies in front of the stars which contribute to the short wavelength emission, we can examine lines of sight with low expected extinction and extrapolate reddening lines to $\tau = 0$ to find the intrinsic colors of the stellar populations along different lines of sight. In order to characterize the integrated emission of the stellar populations, the observed colors were compared with those of isolated bright stellar sources that were detected by DIRBE.

The unreddened Galactic light at $1.25 - 4.9\ \mu$m exhibits colors similar to those of late K and M giants and supergiants. The bulge stars appear to have $I(2.2\ \mu$m) $/ I(3.5\ \mu$m) color $\lesssim 5\%$ bluer than the disk stars, which is in accord with the finding that the bulge stars are ~ 400 K hotter than their disk counterparts (Terndrup, Frogel, & Whitford 1991). Difficulty in accurately determining the limit of $\tau \rightarrow 0$ prevents clear distinction between the $I(1.25\ \mu$m) $/ I(2.2\ \mu$m) colors of the bulge and disk populations.

6. MORE EXTINCTION AND POLARIZATION

From the ratio of $I(1.25\ \mu$m) $/ I(2.2\ \mu$m) we have constructed an optical depth ($\tau_{1.25}$) map of the Galaxy, neglecting the influence of any possible intrinsic source color differences between the inner and outer Galaxy or between the disk and the bulge. Features in this map are clearly anti-correlated with the intensity at $1.25\ \mu$m. The $\tau_{1.25}$ map also has been compared with the optical depth map at $240\ \mu$m constructed from the dust emission in the 140 and $240\ \mu$m DIRBE bands (Sodroski et al. 1993). Where $\tau_{1.25} \lesssim 1.0$, it is well correlated with the DIRBE $240\ \mu$m optical depth, implying that where the line of sight is optically thin at $1.25\ \mu$m, the same dust is observed in both absorption and emission. The DIRBE data indicate that $\tau_{1.25} / \tau_{240} \approx 0.0010$, which is roughly 2/3 of the value tabulated by Mathis (1990).

The $1.25\ \mu$m extinction map has also been compared with the DIRBE polarization maps at 1.25 and $2.2\ \mu$m (Berriman et al. 1993). Both wavelengths show similar gross polarization characteristics. In the Galactic plane, the amplitude of the polarization vector near the Galactic center ($l = 5° - 350°$) correlates with $\tau_{1.25}$. At $2.2\ \mu$m, the Galactic center is more strongly polarized than the rest of the plane. North of the Galactic plane ($l = 1°, b = 5°$), a distinct region of polarized light is seen at $1.25\ \mu$m and $2.2\ \mu$m, which correlates very well with high opacity structures in the $\tau_{1.25}$ map. This region is part of a complex of dust clouds near θ Oph which includes B 78 and other dark clouds. Extinction by dust in these clouds gives rise to the polarization and the apparent peanut shape of the bulge North of the galactic plane as observed at $1.25\ \mu$m (Hauser 1993).

7. CYGNUS AND OTHER STAR-FORMING REGIONS

The different character of the color-color diagrams in Cygnus and Carina is also seen in the ρ Oph region, Orion A, and the Rosette Nebula. Thus, the effect seems linked to star forming regions that are observed at locations that are not dominated by the bright light of the stellar Galactic disk and the bulge.

Simple modeling suggests that the trends can be characterized by various mixtures of stellar light and emission from very hot dust ($T_{dust} \approx 1000$ K). However, the high dust temperatures required would be hard to maintain if the grains were at equilibrium temperatures. Such hot dust would be found only very close to stars (« 1 pc), implying a very high density of very dusty stars. It seems more likely that the high color temperatures are due to small, transiently heated grains or PAHs such as observed in reflection nebulae (*e.g.* Sellgren, Werner, & Dinerstein 1983; Sellgren 1984). Models incorporating excess emission from very small grains and/or PAHs are under investigation.

On the basis of simple models it is found that the trends are not caused by stellar light mixed with line or continuum emission from ionized gas. Estimates of near-IR emission based on the observed Hα intensities of H II regions in Cygnus (Dickel, Wendker, & Bieritz 1969), suggest that the gaseous component of the ISM should not influence the DIRBE observations.

8. CONCLUSIONS

The reddening in the inner Galaxy can be effectively described by foreground extinction rather than extinction intermixed with the dominant stellar populations. This extinction strongly influences the appearance of the Galactic bulge at 1.25 μm and to lesser degrees at longer wavelengths. The extinction at 1.25 μm correlates with the 240 μm optical depth where $\tau_{1.25} \lesssim 1.0$, *e.g.* at latitudes $|b| \gtrsim 3°$. Infrared polarization signals are correlated with extinction, demonstrating that extinction gives rise to polarization of starlight from the Galactic bulge.

In the inner Galaxy ($|l| < 70°$), the integrated starlight seen by DIRBE at 1.25 – 4.9 μm matches the colors of late K and M giants. The bulge population seems to exhibit slightly different (bluer) 2.2 μm – 3.5 μm colors than the disk stars.

ACKNOWLEDGEMENTS

The authors gratefully acknowledge the efforts of the DIRBE data processing and validation teams in producing the high-quality datasets used in this investigation. We also thank the COBE Science Working Group for helpful comments on this manuscript.

COBE is supported by NASA's Astrophysics Division. Goddard Space Flight Center (GSFC), under the scientific guidance of the COBE Science Working Group, is responsible for the development and operation of COBE.

REFERENCES

Berriman, G. B. *et al.* 1993, this volume
Boggess, N. *et al.* 1992, ApJ, 397, 420
Dickel, H, R., Wendker, H., & Bieritz, J. H. 1969, A&A, 1, 270
Freudenreich, H. T. *et al.* 1993, this volume
Hauser, M. G. *et al.* 1991, in "After the First Three Minutes," eds. S. S. Holt, C. L. Bennett, V. Trimble, (AIP: New York), p. 161
Hauser, M. G. 1993, this volume
Koornneef, J. 1983, A&A, 128, 84
Mathis, J. S. 1990, ARAA, 28, 37
Rieke, G. H. & Lebofsky, M. J. 1985, ApJ, 288, 618
Sellgren, K., Werner, M. W., & Dinerstein, H. L. 1983, ApJ(Letters), 271, L13
Sellgren, K. 1984, ApJ, 277, 623
Sodroski, T. J. *et al.* 1993, this volume
Terndrup, D. M., Frogel, J. A., & Whitford, A. E. 1991, ApJ, 378, 742
Weiland, J. L. *et al.* 1993, this volume

FAR-INFRARED FINE STRUCTURE
AND ITS INTERSTELLAR COUNTERPARTS

Francois Boulanger
Institut d'Astrophysique Spatiale, B.P. 10, Route des Gatines,
91371 Verrières le Buisson Cedex, France.

Email ID
boulange@friap51.bitnet

William H. Waller
Code 681, NASA Goddard Space Flight Center, Greenbelt, MD 20771.

Email ID
waller@stars.gsfc.nasa.gov

ABSTRACT

Mosaics at 60 μm and 100 μm have been constructed of the 60° × 60° area covering the Galactic center, bulge, molecular ring, and nearby clouds of emitting dust. We examine the relationship between various tracers of the interstellar medium and the fine-scale structure that is evident in these mosaics. We pay special attention to the nearby Ophiucus cloud complex, a spectacular showcase for interactions between hot stars and dense interstellar matter. Here, the CO emission correlates well with much of the warm dust emission, whereas the Hα and FIR emission appear mostly uncorrelated. The warmest FIR colors are associated with the positions of young hot stars. Closer to the Galactic plane, the FIR fine structure is more likely tracing enhancements of warm dust in the more distant molecular ring and central bulge.

1. INTRODUCTION

In these proceedings, several new surveys of Galactic HI, FIR, x-ray, and radio-continuum emission are being presented. Comparisons between these surveys will no doubt yield exciting new results regarding the nature of the Galaxy's interstellar medium. Comparisons with pre-existing surveys will also be valuable. In particular, the data products recently created by IPAC from the IRAS mission database provide the best resolved coverage of the entire sky at FIR wavelengths. These newer, cleaner IRAS maps have a fully-sampled resolution of 4' × 4', and so are well-suited for comparisons with the relatively high-resolution CO, HI and optical maps which currently exist for small regions of the sky. They also add detail to the larger but lower resolution surveys of CO, HI, Hα, IR/FIR/sub-mm, radio-continuum, and x-ray emission that are currently available.

In this paper, we examine the central 60° × 60° of the Galaxy as mapped at 60 μm and 100 μm by IRAS. After elimination of the strong gradient in brightness towards the Galactic midplane, we compare the residual fine-scale structure with existing maps of CO and Hα emission. We pay special attention to the nearby Ophiucus cloud complex, a spectacular showcase for interactions between hot stars and dense interstellar matter.

2. MOSAICS AND PROCESSING OF THE FIR EMISSION

The 60 μm and 100 μm mosaics of the central 60° × 60° were constructed at IPAC from individual "plates" of the IRAS Sky Survey Atlas (ISSA). Most of these plates were within 50° of the Ecliptic plane and so were not part of the initial release on CD-ROM. Public release of the near-Ecliptic plates should occur by the time these proceedings are published, however. For all the plates, the Zodiacal light contribution had been modeled and removed. Residual contamination is negligible at 100 μm but more evident at 60 μm. After binning and mosaicing, the resulting pixel size is 3′ × 3′, slightly smaller than the nominal 4′ × 4′ resolution.

These mosaics are completely dominated by the strong gradient in brightness towards the Galactic midplane. To enhance structure, it necessary to remove the large-scale gradient. This was done by *median smoothing* the mosaics (with a window of size 15° × 0.05° in Galactic longitude and latitude) and subsequent *median filtering*. Figure 1 (Plate 8) shows the resulting *normalized residuals* after having divided the original 100 μm mosaic by its median-smoothed counterpart. In this particular representation, all of the *cosecant b* brightening towards the Galactic midplane is eliminated, thus revealing fine-scale structure at all latitudes. In addition to the median processed images, an image of the 60/100 μm color was made. This image highlights those regions which have been excited to warmer dust temperatures.

3. NEARBY CLOUDS

The most prominent structures outside the plane are the nearby clouds of Ophiuchus at positive latitudes and Corona Australis on the other side of the plane. To the west of the main Ophiuchus clouds are the Lupus clouds.

We have compared the FIR emission from the Ophiucus cloud system with the corresponding distributions of CO emission, Hα emission, and hot young stars (de Geus and Burton 1991). Figure 2 shows the examined field with the different components outlined. This figure should be compared with the Ophiucus portion of Figure 1 (Plate 8).

Much of the fine-scale FIR emission is associated with CO emission. However, there remain some FIR-emitting features with no CO counterparts. This is especially true near ζ Oph, ν Sco, τ Sco and π Sco. Most likely, these FIR features are associated with neutral gas that has been photo-dissociated by the nearby hot stars. The effect of the hot stars on the dust can be seen by overlaying Figure 2 on top of the image of 60/100 μm colors. This is shown in Figure 3. Nearly all of the color maxima coincide with a hot star associated with the complex.

The Hα emission seems to be uncorrelated with the 100 μm emission, the exceptions being near ν Sco and π Sco. Therefore, the contribution of ionized gas to the far-infrared emission is probably of less significance than that of the atomic and molecular phases.

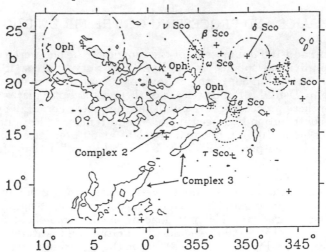

Fig. 2. CO, Hα, and hot stellar components in the Ophiucus and Scorpius region. Positions of the brightest stars (earlier than B2) in Upper Scorpius are plotted as plusses, together with the lowest contour of W(CO) (thick line). The dash-dotted lines outline the HII regions in Sharpless's catalogue (1959), and the dotted lines show the bright nebulae from Lynd's catalogue (1965). Taken from de Geuss and Burton (1991).

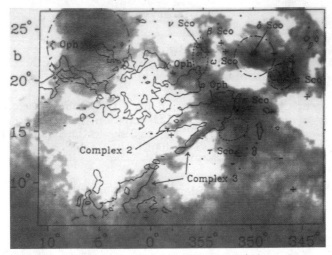

Fig. 3. The 60/100 μm intensity ratio ("color") for the Ophiucus and Scorpius region. Same field as in Fig. 3. Darker regions denote higher intensity ratios, e.g. "warmer colors."

4. STRUCTURE CLOSE TO THE PLANE

Structure close to the plane is too bright relative to the strong background to be of local origin (see Boulanger and Waller in these proceedings). It extends several degrees on each side of the plane. At the distance of the molecular ring it would correspond to a height above the plane of several hundred parsecs. There is no close correspondence between the 100 μm image of residuals and the 0.5° resolution CO map of Dame et al. (1987), even after median processing the CO emission to enhance its fine-scale structure. This is probably because the CO map is more sensitive to the cold nearby clouds (e.g. the Coalsack and those in the Aquila Rift) than the FIR map, whose near-plane emission is dominated by the more distant molecular ring and bulge. Support for this conclusion comes from the high FIR emissivity (relative to the HI emission and the visual extinction) that is measured in this direction. Both of these indices of FIR emissivity yield values ~10 times higher than is found in nearby clouds (cf. Deul 1988; Boulanger 1990; de Geus and Burton 1991).

5. CONCLUSIONS

1. The fine-scale FIR emission in the nearby Ophiucus cloud complex is well correlated with the CO emission and weakly correlated with the Hα emission.

2. The 60/100 μm colors in the Ophiucus/Scorpius region show strong maxima at the positions of the hot young stars. Photodissociation of molecular gas by these stars is probably responsible for the absence of CO counterparts to some of the FIR features.

3. The fine-scale FIR emission near the Galactic plane is too bright and warm to be local. It is probably tracing enhancements of warm dust in the more distant molecular ring and central bulge — where the exciting UV radiation fields are more intense (see Waller and Boulanger, these proceedings).

ACKNOWLEDGEMENTS

This research was funded in part by a NASA/ADP grant (NAG 5-1424) to StarStuff Inc. We also thank the folks at IPAC for their invaluable support.

REFERENCES

Boulanger, F. 1990, in The Galactic and Extragalactic Background Radiation, eds. S. Bowyer and C. Leinert, Reidel, p. 139

Dame, T. M., Ungerechts, H., Cohen, R. S., de Geus, E., Grenier, I., May, J., Murphy, D. C., Nyman, L. -Å., and Thaddeus, P. 1987, ApJ, 322, 706

de Geus, E. J. and Burton, W. B. 1991, A&A, 246, 559

Deul, E. 1988, Ph.D. Thesis, University of Leiden, Chap. 6

Lynds, B. T. 1965, ApJS, 12, 163

Sharpless, S. 1959, ApJS, 4, 257

A ROCKET-BORNE OBSERVATION OF DIFFUSE FAR-INFRARED EMISSION FROM INTERSTELLAR DUST

H. Matsuhara, M. Kawada, T. Matsumoto, S. Matsuura, and M. Tanaka
Department of Astrophysics, Nagoya University, Nagoya 464 Japan

J. Bock, V. V. Hristov, P. Mauskopf, P. L. Richards, and A. E. Lange
Department of Physics, University of California, Berkeley, CA 94720, U.S.A.

ABSTRACT

We report observations of diffuse far-infrared emission from interstellar dust at high Galactic latitudes using a rocket-borne, liquid-helium cooled telescope and photometer. The photometer has 5 spectral bands ranging from 90μm to 190μm with $0.5° \sim 0.65°$ FWHM fields-of-view. The detectors, stressed and unstressed Ge:Ga photoconductors, were cooled to 1K and coupled to charge integrating amplifiers to achieve high sensitivity. We observed regions near Ursa Major including infrared cirrus, high latitude molecular clouds, and the HI hole, a region of uniquely low neutral hydrogen column density. Preliminary analysis shows a strong correlation of the emission at 134, 154, and 186μm with 100μm IRAS data.

1. INTRODUCTION

The Infrared Astronomy Satellite (IRAS) discovered a component of far-infrared emission known as "infrared cirrus"(Low et al. 1984) extended over the high Galactic latitude sky. The infrared cirrus is spatially well correlated with the column density of neutral atomic hydrogen(HI) and is generally attributed to interstellar dust emission associated with HI gas(Boulanger and Pérault 1988). Observations at wavelengths longer than that of IRAS are important to determine the physical properties of the dust responsible for the infrared cirrus.

In order to study the nature of the infrared cirrus we have carried out sensitive far-infrared observations using a rocket-borne, liquid-helium cooled telescope equipped with a far-infrared photometer. The photometer also includes two narrow-band channels which are used for observation of diffuse [CII] ($^2P_{3/2} \rightarrow\ ^2P_{1/2}$: $\lambda = 157.7\mu$m) line emission(see Bock et al. of this conference).

2. INSTRUMENTATION AND OBSERVATIONS

The instrument consisted of a multi-channel photometer installed at the focus of a 10cm Ritchey-Chrétien telescope. The telescope had a specular forebaffle and an absorbing aftbaffle to achieve high off-axis rejection(Bock et al. 1992). The detectors were stressed Ge:Ga photoconductors mounted in compact stress rigs (Wang et al. 1987) except for the shortest wavelength channel, for which an unstressed Ge:Ga photoconductor was used. The detectors were cooled to 1K by a closed cycle ^{4}He refrigerator. At this temperature the detector dark current is negligible in comparison to the lowest photocurrent in flight.

The detectors were coupled to low-noise charge integrating amplifiers. All optical components used for the observations were cooled to 2K by superfluid helium, making instrumental emission negligible. A cold shutter was installed in front of the photometer entrance aperture to establish an absolute zero-level. The parameters of the far-infrared photometer are summarized in Table 1.

The instrument was installed on a single-stage, solid-fuel rocket, S-520-15 of the Institute of Space and Astronautical Science(ISAS). The instrument was launched at 1:00 JST on 2 Feb. 1992 from Kagoshima Space Center. The payload reached an apogee of 338km with a total flight time of about 10 minutes.

The payload was pointed in flight by an onboard three-axis control system with side-jets using dry N_2 gas. During most of the flight the telescope pointed at the Ursa Major HI hole($l \simeq 152°$, $b \simeq 52°$), a region of minimum HI column density(Jahoda, Lockman, and McCammon 1990). From 220sec to 310sec after launch the beam scanned through infrared cirrus and high latitude molecular clouds mapped in CO emission(de Vries, Heithausen, and Thaddeus 1987) and returned back to the HI hole along a path passing through the galaxy M82 which was used as an in-flight calibration source. The attitude control was accurate and stable to within 0.2°.

In Table 1 we summarize the flight performance of the instrument. The temperatures of the cryostat and the photometer housing remained close to 2K, and the detectors were stable at 1K. The in-flight sensitivities are estimated from the measured noise during the pointed observation of the HI hole.

3. PRELIMINARY RESULTS

In Fig.1 a-d we present the correlation of the observed far-infrared emission with IRAS 100μm emission(IRAS Sky Survey Atlas 1992) and with HI column density(Heiles 1975) from the scanning observations. There is a strong correlation with the IRAS data for all far-infrared channels. The correlations are very similar for regions with CO emission and without CO emission. The far- infrared color of the interstellar dust grains is the same for HI clouds as for clouds containing CO emission. However, as seen in Fig.1d, the far-infrared emission in regions with CO emission clearly exceeds that in regions without CO emission, as expected due to emission from dust grains associated with molecular hydrogen.

In Fig.2 we show the preliminary far-infrared spectrum of the IRAS correlated signal for a typical infrared cirrus cloud(IRAS$_{100\mu m}$ = 3MJysr^{-1}). At this stage of analysis all data points contain large error bars which are dominated by systematic errors. The spectrum of emission at the HI hole was also measured with high signal to noise, and will be discussed in a future paper.

We are grateful to Prof. H. Okuda and other launching staff of the ISAS for their kind launch support. We thank Prof. E. E. Haller and Dr. J. W. Beeman at University of California, Berkeley for providing stressed Ge:Ga detectors, and Dr. N. Hiromoto at the Communications Research Laboratory, Japan for providing unstressed Ge:Ga detectors. This work was supported in part by the ISAS, by the Daiko Science Foundation, by a grant-in-aid from the Ministry of Education, Science and Culture, in Japan, and by the National Aeronautics and Space Administration in the U.S.

REFERENCES

Bock, et al. 1992, in preparation.
Boulanger, F. & Pérault, M. 1988, ApJ, 330, 964
Heiles, C. 1975, ApJ Suppl. 20, 37
IRAS Sky Survey Atlas, 1992, (IPAC, California Institute of Technology)
Jahoda, K., Lockman, F. J., & McCammon, D. 1990, ApJ, 354, 184
Low, F. J., et al. 1984, ApJ, 278, L19
de Vries, H. W., Heithausen, A., & Thaddeus, P. 1987, ApJ, 319, 723
Wang, J. -Q., Richards, P. L., Beeman, J. W., & Haller, E. E. 1987, Appl. Opt.,
     26, 4767

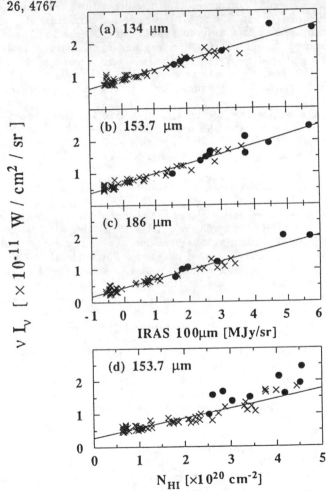

Fig. 1. The correlation of observed far-infrared intensity with IRAS 100μm(a-c) and with NHI(d) (Heiles 1975) are indicated by solid lines. Regions with appreciable CO emission(< 0.1K km/s, averaged over the beam) are shown as solid points, and are excluded from the fit.

Table.1 Parameters of the Far-Infrared Photometer

channel No.	1	2	3	4	5
$\lambda_c[\mu m]$	95	186	134	153.7	157.9
$\lambda/\Delta\lambda$ [1]	5	8	10	100	120
beam [°ϕ] [2]	0.5	0.5	0.5	0.65	0.65

detector	unstressed(ch,1) & stressed Ge:Ga
detector dark current	<10e/s@1K
read-out	charge integration amplifier
	1sec read-noise = 40-60e

=== In-flight performance ===

| Sensitivity [3] | 1.1×10⁻¹¹ 2.3×10⁻¹² 3.5×10⁻¹³ 1.6×10⁻¹² 3×10⁻¹⁴ |

$[W\ cm^{-2}sr^{-1}]$ (1sec,3σ)

detector temperature	0.9~1.0K
bath temperature	1.9~2.1K

1) $\Delta\lambda$ = equivalent square band widths.
2) Half-power beam widths.
3) For ch.1- 4, the limiting sensitivity in λI_λ is shown, while for ch.5 the sensitivity for the [CII] line intensity is shown.

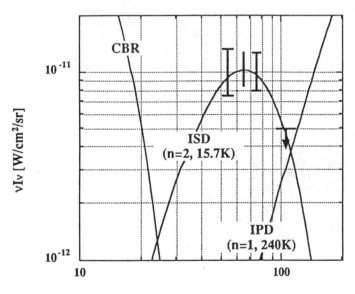

Fig. 2. Preliminary far-infrared spectrum(thick solid bars) of the infrared cirrus, normalized to a region in which the IRAS 100μm brightness is 3MJy/sr. Model spectra of the interstellar dust emission(ISD: $I_\nu \propto \nu^2 B_\nu(T = 15.7K)$, where $B_\nu(T)$ is the plank function), interplanetary dust emission(IPD, from Beichman and Helou 1991), and the 2.74K cosmic background radiation(CBR) are shown for comparison. The error bars are dominated by systematic uncertainty in the absolute gain, and may decrease with further analysis.

A STUDY OF THE INTERSTELLAR DUST DISTRIBUTION IN REGIONS OF LOW TOTAL COLUMN DENSITY

M. H. Jones

Astronomy Unit, School of Mathematical Sciences,
Queen Mary and Westfield College, Mile End Road, London E1 4NS

ABSTRACT

We present a study of the distribution of interstellar dust and HI, using 100μm and 21 cm measurements, for two regions of very low total column density. We find that down to the spatial resolution of the HI data (12'), there is very good agreement between dust and neutral gas features.

1. INTRODUCTION

It is well established that there is a strong correlation between the distribution of interstellar dust and HI on angular scales of several degrees over most of the sky with $|b| > 10°$ (Boulanger & Perault, 1988, and Rowan-Robinson *et al*, 1991). It is of interest to examine the relationship between dust and gas distributions on smaller scales, since this may provide insight into physical processes occurring in the interstellar medium.

Any such separation of dust and gas also has observational consequences. For example, in order to use the HI column density to estimate the optical depth to x-ray photons with energies above the carbon edge ($E \gtrsim 300eV$), it must be assumed that dust and atomic gas are well mixed on small scales, since a significant fraction of the metals are depleted onto grains. Similarly, the same condition applies when a measure of the dust column density (such 100μm flux - I_{100}) is used to estimate the photoelectric absorption below 300 eV.

In this paper, the dust and gas distributions are compared in two regions with very low total column density ($N_{HI} < 10^{20}$ H atoms cm^{-2}). These areas are of interest since they offer windows out of the Galaxy in the soft x-ray band ($E < 1$keV). An understanding of the relationship between the dust and gas distributions is required for the interpretation of soft x-ray observations made in these areas. A full description of this work is given by Jones *et al* (1992).

2. OBSERVATIONAL DATA

The areas studied are two of the very low column density regions described by Lockman, Jahoda & McCammon (1986). In equatorial coordinates (1950.0), the areas are: L3 ($\alpha = 13^h 27^m$ to $13^h 37^m$, $\delta = +37° 10'$ to $+39° 10'$); and, L7 ($\alpha = 10^h 01^m$ to $10^h 13^m$, $\delta = +50° 0'$ to $+56° 0'$). The mean N_{HI} values quoted by Lockman *et al* for these areas are 7.8×10^{19} and 7.6×10^{19} H atoms cm^{-2} respectively.

Since both areas are located away from obvious sources of heating, it is assumed that dust temperatures are reasonably constant, and consequently that I_{100} follows the dust density distribution. The distribution of HI was mapped at 21 cm using the Mark IA telescope (FWHM≈ 12′) at Jodrell Bank.

Maps of the 100μm emission were made from the IRAS Calibrated Reconstructed Detector Data (CRDD). The systematic 'striping' was reduced by using the model of Rowan-Robinson et al (1990) the predict the emission from the interplanetary dust cloud, and by forcing the zodiacally subtracted data to agree on scales of $\gtrsim 0.5°$ with the low-resolution maps of Rowan-Robinson et al (1991).

The 21 cm observations of the L7 field are described in detail by Willacy, Pedlar & Berry (1992). The L3 field was observed in a similar fashion. The observations were ordered such that consecutive pointings were made at increasing R.A. at constant declination. The spectra were corrected for stray radiation contamination using the method of Kalberla, Mebold & Reich (1980) adapted for the Mark IA by Stavely-Smith (1985).

3. RESULTS

In order to compare the 100μm and 21 cm maps, known far-IR point sources were removed from the 100μm data, and the maps were degraded in resolution to the beam width of the radio data. As an example, the resulting map for L3 is shown in figure 1. It is apparent that despite the fact that the map represents one of the lowest column density regions in the sky, the 100μm and 21 cm maps show a remarkable similarity.

For the L3 area (figure 1), both maps show a strip of low column density running East-West at the North of the map area, with minima in this strip lying near $13^h 35^m + 39° 0′$ and $13^h 33^m + 39° 0′$. South of this region there is a strip of higher column density material running East-West across the map at $\delta = + 38° 30′$. To the South of this region lies a hole with centre near $13^h 33^m + 37° 50′$. In the South East corner of the map there is a further minimum with centre near $13^h 36^m + 37° 15′$.

There are, however, several discrepancies. Of particular note are: the high 100μm intensity to the West of the strip of low column density around $13^h 28^m + 39° 0′$; the lack of an N_{HI} hole at $13^h 29^m + 38° 20′$; and the absence of an N_{HI} peak at $13^h 30^m + 37° 45′$. However, the N_{HI} map tends to have structures aligned along lines of constant δ which strongly suggests a systematic error on the days of observation. Similarly, the discrepant features in the 100μm map seem to be aligned along the scan direction of IRAS, again suggesting that systematic errors are responsible.

The L7 region (not shown) also shows many features common to both maps. Again, there are some discrepancies between the 100μm and N_{HI} maps, and the pattern of systematic error is similar to that seen in the L3 region.

In the light of these systematic errors it cannot be claimed that there is any significant separation of the dust and gas components. It appears that both maps agree to within about 2×10^{19} H atoms cm^{-2} or 0.2 MJy sr^{-1}. This corresponds to $\sim 30\%$ of the mean N_{HI} or I_{100} for these fields.

The correlation of N_{HI} and I_{100} (sampled in a similar way) is significant at at a confidence level $> 99\%$ for both fields. The ratio of 100μm intensity to HI column density (I_{100}/N_{HI}) is 1.13 and 1.02 MJy sr^{-1} $(10^{20}$ H atoms cm$^{-2})^{-1}$ for L3 and L7 respectively. This is higher than the mean value quoted by Boulanger & Perault (1986) for $|b| > 10°$, but lies within the 30% variation of I_{100}/N_{HI} which is commonly observed on scales of a few degrees at high $|b|$

(Rowan-Robinson *et al*, 1991)

As a further test of the similarity of the spatial distributions of 100μm intensity and N_{HI}, we compared the angular correlation functions (ACF), $A(\theta)$, for the 100μm and the 21 cm data. The random noise associated with independent measurements in both maps and point sources in the 100μm map suppresses the ACF away from $\theta = 0$ by a factor which is independent of the separation angle. So, if the dust and gas have the same underlying distribution, the ratio of the ACFs of the 100μm and the 21 cm maps will be a constant.

The ACFs of the 21 cm and the 100μm data for the L3 and L7 fields are shown in figure 2. For the L3 data, the ACF of the 21 cm data shows a small discontinuity between $\theta = 0$ and the curve beyond 6', due to the noise associated with independent measurements. At $\theta > 6'$ the 21 cm ACF is smooth and has an e-folding length of $\sim 45'$. Although the 100μm ACF appears to be dramatically different, this ACF has been distorted by the large variance at $\theta < 10'$ which arises from point-like fluctuation (background galaxies). The 21cm and 100μm ACFs of the L7 field (figure 2b) show greater similarity to one another than those of the L3 field. Again, the e-folding length of the ACF appears to be $\sim 45'$.

Figures 2c and 2d show the ratio of the ACFs (21cm/100μm) for the L3 and L7 fields respectively. The dashed lines show the weighted mean ratio, and it may be seen that the ratios of the ACFs are consistent with being constant on scales 12' to 60'. The constant value of this ratio indicates that the dust distribution follows the HI distribution over these scales in both fields.

4. CONCLUSIONS

In this paper we have compared dust and HI distributions in two regions of very low total column density ($N_{\text{HI}} < 10^{20}$ H atoms cm^{-2}). It is found that there is good spatial correlation of dust and gas features down to scales of $\sim 12'$. Assuming a typical distance to the neutral medium of ~ 150 pc (e.g. Cox & Reynolds, 1987) this then implies that any bulk separation of dust and gas components in these regions is less than ~ 0.6 pc. From the ACFs it is found that in common with previous studies of low column density regions, the structure of the neutral component of the ISM at scales $\lesssim 30'$ is dominated by smooth gradients rather than by discrete clumps.

REFERENCES

Boulanger, F., & Perault, M., 1988. ApJ 330, 964
Cox, D.P., & Reynolds, R.J., 1987. ARAA 25, 303
Jones, M.H., Rowan-Robinson, M., Branduardi-Raymont, G., Smith, P., Pedlar, A., & Willacy, K., MNRAS, submitted
Kalberla, P.M.W., Mebold, U., & Reich, W., 1980. A&A 82, 275
Lockman, F.J., Jahoda, K., & McCammon, D., 1986. ApJ 302, 432
Stavely-Smith, L.S., 1985. Ph.D. Thesis. University of Manchester.
Rowan-Robinson, M., Hughes, J., Vedi, K., & Walker, D.W., 1990. MNRAS 246, 273
Rowan-Robinson, M., Hughes, J., Jones, M., Leech, K., Vedi, K., & Walker, D., 1991. MNRAS 249, 729
Willacy, K., Pedlar, A., Berry, D., 1992. MNRAS, in press

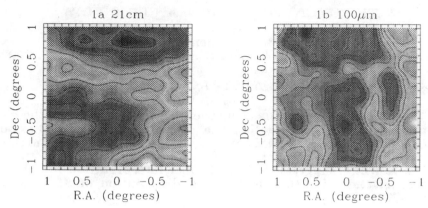

Fig. 1: Smoothed (a) 21 cm and (b) 100μm maps of the L3 region smoothed with a 2-D Gaussian filter with FWHM of 12'. Greyscales cover a range of: (a) 0.53 - 0.83 MJy sr^{-1} (contour interval of 0.05 MJy sr^{-1}); and, (b) (0.45 - 0.65)$\times10^{20}$ H atoms cm^{-2} (contour interval of 0.04×10^{20} H atoms cm^{-2}).

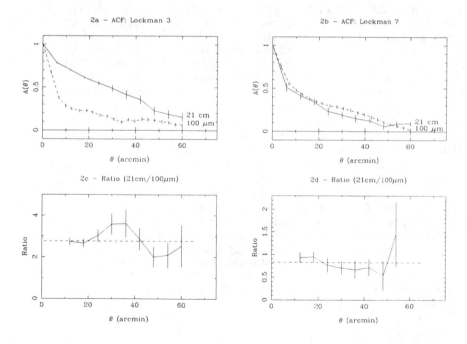

Fig. 2: The ACF of the 100μm (dashed line) and 21 cm (solid line) emission for; (a) the L3, and; (b) the L7 regions. (c) and (d): the ratio of the 21 cm and 100μm ACFs for L3 and L7 respectively. The dashed line indicates the mean value of the ratio.

THE CYGNUS-X REGION: AN IRAS VIEW

Sten F. Odenwald

BDM,Intl, Suite 340, 409 Third St. S.W., Washington,D.C. 20024

ABSTRACT

The Cygnus-X region has long been recognized as one of the more complex regions of the Galactic plane. It has been interpreted as both a single supergiant HII region, and as a chance piling up of unrelated SFRs along the line-of-sight. I will present a recent IRAS study of Cygnus-X which suggests that the former interpretation may be the most consistent with the number and distribution of SFRs within Cygnus-X.

1. INTRODUCTION

Since its discovery, Cygnus-X has been resolved into nearly 800 individual radio continuum peaks, star forming regions (SFRs), and numerous molecular clouds (Wendker, 1984; Cong, 1977). It can be readily identified as a gamma ray source (Hermsen, 1988) and figures prominently in both IRAS and COBE/FIRAS views of the Galactic plane (Habing, 1988; Bennett et $al.$, 1992). The 1-D velocity distribution of HI (Burton, 1985), CO (Leung & Thaddeus, 1992) and SFRs (Odenwald & Schwartz, 1993) reveals the dominant emission extending from 0 to -15 km sec^{-1}. Material between +20 and 0 km sec^{-1} is forbidden in this direction ($75° < l < 85°$), and is probably associated with the Cygnus Rift/Gould's Belt which form a dense obscuring screen beginning at $\approx$ 700 pc extending to 2 kpc (Bochkarev & Sitnik, 1985). Traces of an interarm gap appear between -20 and -35 km sec^{-1} followed by the Perseus Arm at - 35 to -50 km sec^{-1} ($\approx$ 8 kpc) and the Outer Arm at -60 to -80 km sec^{-1} ($\approx$ 12 kpc).

It is commonly assumed that Cygnus-X is simply the piling-up of unrelated HII regions and molecular clouds along the Local Arm. This interpretation is strongly favored by some (Bochkarev & Sitnik ,1985) since the distances to the most reliably known SFRs, despite their uncertainty, are seemingly incompatible with membership in a single complex. Only distances to OB associations (e.g. Cyg OB-2) and a small number of late-type supergiant stars are known with any precision from spectroscopic parallaxes. Less direct measures based on the presence of foreground absorbing features seen in H$_2$CO (Piepenbrink & Wendker, 1988), or assumed associations with optically visible OB stars (Cong, 1977) constitute the bulk of the Cygnus-X distance metrology. Consequently, no compelling case can be made for a 'one vs many' interpretation on the basis of distances alone. A detailed study by Piepenbrink and Wendker (1988) of the velocity distribution of the SFRs , however, does indicate that they are probably not part of an expanding 'superbubble' as was once thought (Abbott et $al.$, 1981).

2. THE VIEW FROM IRAS

Although IRAS does not provide further clues towards resolving the distance problem for Cygnus-X, it does provide several new elements to the story. A recent CO survey by Odenwald and Schwartz (1993) of 70 YSOs towards Cygnus-X for which $S(100\mu m) > 100$ Jy reveals that the majority are located in the Local Arm, with none detected in the Interarm Gap earlier than $\approx$ B3 ZAMS. The YSOs generally coincide with bright radio continuum peaks and are therefore at least B3 or earlier to be detected as radio sources. In view of the large number of radio sources, we may ask whether they are all SFRs, or whether a significant fraction correspond to inhomogeneities in the ionized ISM towards Cygnus-X?

The IRAS $100\mu m$ data (Figure 1) reveals three distinct components: 1) Discrete point sources; 2) Bright ridges, and; 3) an extended diffuse halo. The discrete sources correspond to the principal compact radio sources in this region identified by Downes and Rinehart (1966). The ridges coincide with regions of thermal radio emission and represent unresolved or partially-merged HII regions. They also coincide with the densest regions of CO emission (Odenwald, 1990). The $60{:}100\mu m$ temperature data in Figure 2 show that for λ^{-1} dust emission, the ridges are local temperature maxima with $38 < T_d < 42$ K within which still warmer 'hot spots' exist ($T_d > 43$ K) corresponding to individual SFRs such as DR 3, 6 and 15. This is consistent with earlier far-IR studies of the region (Campbell *et al.*, 1980). The extended far-IR halo has an equivalent dust temperature $T_d = 28 \pm 5$K, significantly warmer than for 'cirrus'-type dust heated by the ambient Galactic radiation field. The origin of its heating is unlikely to be the embedded SFRs that make-up the ridges since the local extinction within the ridges is too large for the UV radiation to escape and heat the large-scale halo component. Like the ridge emission, at least some of the extended component must also consist of merged HII regions from large numbers of SFRs.

To show this, consider a small sub-region near DR 10. The correspondence between the far-IR and radio emission shown in Figures 3 and 4 is generally good. Several prominent 'holes'(H1, H2, H3) can be identified in both maps, as can most of the bright radio peaks (A-J). DR 10, itself, corresponds to Peak C. The far-IR luminosities for these sources imply stars earlier than B3 ZAMS assuming d > 700 pc. The far-IR source FIR-1 ($L \approx 3000\ L_{\odot}$) has no radio counterpart and is possibly a protostar earlier than $\approx$ B2 ZAMS with no accompanying HII region. Radio source I, on the other hand, has no far-IR counterpart. The dust temperature map shows distinct maxima at each of the radio peaks confirming that they are internally-heated. This admittedly limited study of 5% of the Cygnus-X region, strengthens the view that the majority of the thermal radio sources identified by Wendker(1984) may, indeed, be SFRs and that Cygnus-X is an unusually dense region of star- forming activity.

The local surface density of HII regions is $\approx$ 7 kpc^{-2} (Crampton & Georgelin, 1975), which implies that to a maximum depth of 4 kpc the Cygnus-X region (1.4 kpc^2) should contain 10 HII regions if it were simply a chance superposition of SFRs along the Local Arm. This is 80-fold fewer than the number of radio peaks seen in Cygnus-X, implying that the 'chance superposition' model may not be completely viable. Cygnus-X, as a single giant star-forming complex, now seems a more likely possibility. At a distance of 2 kpc, Cygnus-X ($\approx 6^\circ$) has a linear diameter of 350pc, comparable to that of giant extragalactic HII regions such as 30 Dor (370 pc). A ring of optical filaments ($\approx 16^\circ$, d = 1000 pc) surrounds Cygnus-X, compared to a similar shells of HI/HII surrounding 30 Dor (Meaburn, 1979). Cygnus-X contains no fewer than 48 O-type stars, mostly in Cyg OB 1,2 and 9 (Bochkarev & Sitnik, 1985), compared to 114 for 30 Dor (Walborn, 1984). Galaxies such as NGC 1232 which are believed to resemble the Milky Way (Becker & Fenkart, 1970) also show large regions of star-forming activity within their arms, similar in size to Cygnus-X.

3. CONCLUSIONS

The IRAS data show that Cygnus-X is a collection of individual SFRs that correspond to possibly the majority of the radio/far-IR peaks seen in this direction. The surface density of these SFRs is too large to be consistent with a random superposition of SFRs along the Local Arm. Cygnus-X appears to have more in common with the large super-HII regions seen within the spiral arms of galaxies similar to the Milky Way.

REFERENCES

Abbott, D.C., Beiging, J.H. & Churchwell, E. 1981, ApJ 250, 645.
Becker, W. & Fenkart, R. 1970, in IAU Symposium No. 38, eds. W. Becker & G. Contopoulos (Reidel:Dordrecht), p. 205.
Bennett, C.L. et al. 1992, ApJ 396, L7.
Bochkarev, N.G. & Sitnik, T.G. 1985, Astr. Sp Sci 108, 237.
Burton, W.B. 1985, AASup. 62, 365.
Campbell, M.F. et al. 1980, ApJ 238, 122.
Cong, H. 1977, PhD Thesis Columbia University.
Crampton, D. & Georgelin, Y.M. 1975, AA 40, 317.
Downes, D. & Rinehart, R. 1966, ApJ 144, 937.
Habing, H.J. 1988, AA 200, 40
Hermsen, W. 1988, Sp. Sci. Rev. 49, 17
Leung, H.O. & Thaddeus, P. 1992, ApJ Suppl. 81, 267.
Meaburn, J. 1979, AA 75, 127.
Odenwald, S.F. 1990, BAAS v. 22, No. 4, p. 1298.
Odenwald, S.F. & Schwartz, P.R. 1992, ApJ 320
Piepenbrink, A. & Wendker, H. 1988, AA 191, 313.
Walborn, N.R. 1984, in IAU Symposium No. 108, eds. S. Van den Bergh & K. deBoer (Reidel:Dordrecht) p.243.
Wendker, H. 1984, AASup 58, 291.

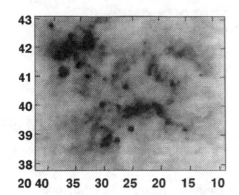

Figure 1 IRAS image of the Cygnus-X region at 100 μm between $20^h 40^m < \alpha < 20^h 10^m$ and $38^o < \delta < 43^o$. Gray scale range from 0.3 to 1.5 GJy/str. For discussion of principal features see Odenwald & Schwartz (1993).

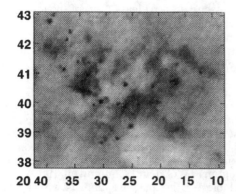

Figure 2 Dust temperature map of Cygnus-X region obtained from 60 and 100 μm IRAS images spanning a range $25 < T_d(K) < 45$. The dark regions have temperatures in excess of 38 K.

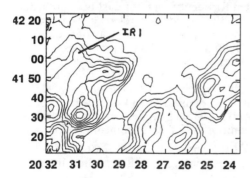

Figure 3 Dust temperature contour map of a region near DR 10. Contour intervals are 1 K over a range from 29 - 37 K.

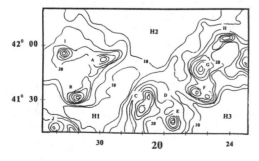

Figure 4 Radio emission at 4.8 GHz of DR 10 region from the survey by Wendker (1984). Contours at 10, 12, 18, 24, 36 and 42 in units of 0.05 K brightness temperature.

MASSIVE STARS EMBEDDED IN GMC'S IN THE SOUTHERN GALAXY

Leonardo Bronfman and Jorge May
Departamento de Astronomía, Universidad de Chile, Casilla 36-D, Santiago

ABSTRACT

Regions of massive star formation can be identified as IRAS point-like sources with characteristic FIR colors in the Galactic plane. The close association of these sources with molecular clouds provides a way of determining their kinematic distances and therefore luminosities. The mean radial distribution of the FIR luminosity generated by embedded massive stars in the IV Galactic quadrant is strongly peaked at the position of the molecular annulus. The FIR surface luminosity appears to be proportional to the 2nd power of the H_2 surface density in the Galactic disk. At least 13 % of the total FIR emission from massive star forming molecular clouds is produced in their dense cores, with typical sizes of 1-2 pc, in the close neighborhood of embedded OB stars. We present here a CS (2-1) map and near-infrared images of one of such regions, at 5.7 kpc of the Sun, associated with a strongly self-absorbed CO (1-0) profile.

1. GALACTIC DISTRIBUTION

Our study of the distribution of young massive stars in the Galaxy is based on the determination of the FIR luminosity of dust associated with dense molecular cores and heated by embedded stellar sources. Candidate regions of massive star formation are selected from the IRAS point source catalog based on their characteristic FIR colors, corresponding to those of compact H II regions (Wood and Churchwell 1989). The IRAS survey has the best sensitivity available for this purpose, and its wavelength bands at 60 μm and 100 μm are well suited for identifying hot dust near embedded stars. The close association of FIR sources with the dense cores of giant molecular clouds, which can be observed in millimeter-wave lines, provides a way of determining distances for the sources, and therefore their luminosities and other properties.

The sources discussed here were observed with SEST (Swedish-ESO Sub-millimeter Telescope) in the CS (2-1) transition (Bronfman *et al.* 1992), a good tracer of high density gas. The SEST beamwidth at the CS (2-1) frequency, 45", is comparable to the IRAS survey resolution, so dense gas and dust were probably sampled within the same spatial regions. About 350 such regions have been identified in the IV Galactic quadrant. The FIR luminosity for each point-like source was computed as $L = D^2 F$, where $F = 4\pi \Sigma \nu F\nu$ [$L_\odot$ pc^{-2}] is the "equivalent flux" defined by Boulanger and Perault (1988), representing the infrared flux integrated between 1 μm and 1000 μm; $F\nu$ is the flux at frequency ν; D is the distance of the parent molecular cloud, and the sum extends over all IRAS bands.

The kinematic distances of our sample of IRAS point-like sources, determined from the CS observations, allow a direct derivation of the radial distribution of the FIR luminosity of embedded massive stars. If we compare the face-on FIR surface luminosity associated with massive stars embedded in molecular clouds, averaged over 1 kpc wide rings in the Galactic disk, with the H_2 surface density derived from the Columbia CO Survey of the IV Galactic quadrant (Bronfman *et al.* 1988), the FIR mean radial distribution is similar to that of H_2, but the "massive-star" annulus is narrower than the "molecular annulus" (Fig. 1). Examination of this figure suggests a

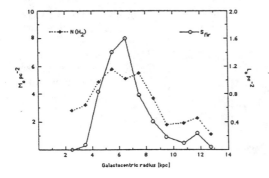

Figure 1.- Face-on FIR surface luminosity associated with embedded massive stars, and H_2 surface density, vs. Galactocentric radius in the fourth Galactic quadrant.

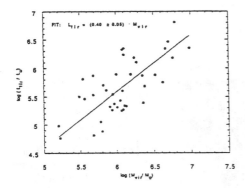

Figure 2.- FIR luminosity associated with embedded massive stars vs. virial mass for molecular clouds in the fourth Galactic quadrant. The line represents the linear relation

$$L_{FIR}/L_\odot = 0.4\, M_{VIR}/M_\odot.$$

power law relation between FIR surface luminosity and H_2 surface density. In fact, for an extended region of the Galaxy, from the 3-kpc arm to the Carina region at $R = 11$ kpc, the FIR surface luminosity is proportional to the second power of the H_2 surface density (Bronfman 1992).

Apparently, giant molecular complexes are responsible for most of the massive star formation in the Galaxy. For clouds undergoing massive star formation there appears to be a fairly linear relation between FIR luminosity associated with embedded point-like sources and cloud virial mass (Fig. 2), with a mean ratio of FIR luminosity to virial mass of 0.4 $L_\odot\, M_\odot^{-1}$. Such a linear relation also holds true, but with a mean ratio of $3.2\, L_\odot\, M_\odot^{-1}$, when the FIR luminosity is computed from IRAS sky-flux images for molecular clouds associated with bright H II regions (Mooney and Solomon 1988). While the FIR luminosity computed here arises from the close neighborhood of embedded massive stars and in that sense is a lower limit for the emission they are responsible for, the FIR luminosity computed by Mooney and Solomon for whole clouds is possibly overestimated owing to line of sight confusion. Thus, a lower limit of about ~13% (0.40/3.2) is suggested for the apparently sustantial contribution of embedded massive stars to the total FIR luminosity of giant molecular clouds associated with bright H II regions.

2. A MASSIVE STAR CLUSTER IN THE MOLECULAR ANNULUS

One of the most luminous IRAS point-like sources ($L_{FIR} = 1.1$ x $10^6\, L_\odot$) of

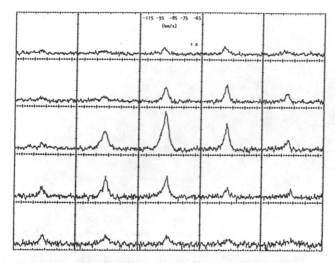

Figure 3.- Array of CS (2-1) spectra centered at the position of the IRAS/CS source G328307. The beam size is 45 "; spectra were taken with a separation of 1 beam.

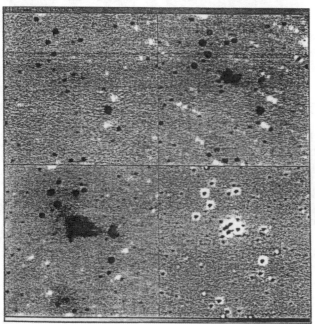

Figure 4.- Near-infrared images of the massive star forming region associated with the IRAS/CS source G328307. All images were taken at the position of the central CS spectrum in Fig. 3. The total field for each image is 40 ", which corresponds to about 1 pc at the kinematic distance of the associated gas. Images are presented as a mosaic: in the upper left the J filter; upper right H filter; lower left K filter; and lower right K filter with a smoothed version of itself subtracted in order to identify individual stellar objects.

our sample is located at ℓ= 328.307, b = 0.432 . We have mapped this source (Fig.3) in the CS (2-1) line with the SEST Telescope at a resolution of 45". The spatial distribution of the CS emission is fairly symmetric, peaking right at the position of the IRAS source. The mean radial velocity of the CS profile, 92 km s^{-1} (VLSR), implies that this source, which we will call G328307, is located near the subcentral point, at a Galactocentric radius of 4.7 kpc and at a distance of 5.6 kpc from the Sun. The distribution of dense gas has an approximate radius of 1 pc (HWHM).

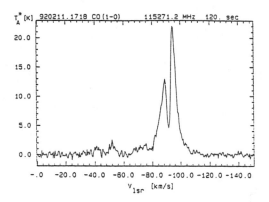

Figure 5.- Spectrum of the CO (1-0) emission in the position of the IRAS/CS source G328307. The dip in the profile is at the same VLSR as the peak emission in the central CS (2-1) spectrum of Fig. 3.

We have obtained near-infrared images of G328307 with the IRCAM-3 detector mounted on the Dupont Telescope at the Carnegie Southern Observatory in Cerro Las Campanas, Chile. The images (Fig. 4) were obtained using J, H, and K filters; the detector array has 256 x 256 pixels, and the total field is of 40". Only a local sky has been subtracted from the J, H, and K images in this preliminar presentation. A conspicuous dust structure can be discerned, using the K filter, at the center of the field. The dust structure is very reddened and almost invisible in the J filter. The fourth image, at the lower right corner, was prepared by smoothing the K image in boxes of 20 pixels side (3 "), and subtracting this smoothed version to the original K observations. The dust structure appears to surround a trapezium-like cluster, most probably of OB stars. This may be one of the farthest stellar fields embedded in molecular clouds in the inner Galaxy that has been imaged in the near-infrared.

A spectrum of the CO (1-0) line obtained at the central position of G328307 shows a very pronounced double peaked profile. We interpretate this profile as self-absorption of the CO line by the external envelopes of the molecular cloud against emission from the dense and hotter gas near the stars. The central velocity of the CS profile (Fig. 3), tracing the dense core gas, falls right between the CO profile peaks.

We acknowledge Dr. Lars-Åke Nyman and Dr. Miguel Roth for help with the CS (2-1) and near-infrared observations, respectively. This research has been supported by FONDECYT, República de Chile, through Grant 92-0944.

REFERENCES

Boulanger, F., and Perault , M. 1988, Ap. J. 330, 964.
Bronfman, L., Alvarez, H., May, J., Cohen, R., and Thaddeus, P. 1988, Ap. J. 324, 248.
Bronfman, L. 1992, in "The Center, Bulge, and Disk of the Milky Way", ed. L. Blitz (Kluwer: Dordrecht) p.131
Mooney, T. J., and Solomon, P. M. 1988, Ap. J. 334, L51.
Wood, D.O.S, and Churchwell, E. 1989, Ap. J. 340, 265.

SUBMILLIMETER IMAGES OF W51

E. F. Ladd, J. R. Deane, J. D. Goldader, D. B. Sanders,
and C. G. Wynn-Williams
Institute for Astronomy, University of Hawaii,
2680 Woodlawn Drive, Honolulu, HI 96822

Email ID
ladd@galileo.ifa.hawaii.edu

ABSTRACT

We report on submillimeter observations of the star forming region W51 on 11pc size scales with 0.7pc resolution. These observations show at least nine emission centers, each of which has enough luminosity to be a potential site for massive star formation. We also detect a region of extended emission surrounding these emission centers. This extended emission component has a steeper submillimeter spectrum, suggesting that it is warmer than the star forming cores.

1. INTRODUCTION

The W51 complex of HII regions and molecular clouds ranks among the most luminous regions of active star formation in the galaxy. It is well-known as a collection of radio continuum sources (e.g. Beiging 1975), many of which have been associated with near and mid infrared counterparts (Wynn-Williams, Becklin, and Neugebauer 1974). The submillimeter emission toward this region has been less well-sampled, with only a large-beam map at 400μm (Jaffe, Becklin, and Hildebrand 1984) and more recently higher spatial resolution coverage at 870μm and 1300μm (Sievers *et al.* 1991). The luminosity from this region can be roughly divided into two spatial components – a region often referred to as W51 MAIN containing the radio continuum sources e1 and e2 (Martin 1972, Scott 1978), H_2O maser activity (Genzel and Downes 1977), and dust emission seen in the millimeter and submillimeter; and a second region around the radio continuum source W51d associated with strong 20μm emission (IRS2; Wynn-Williams *et al.* 1974 , Genzel *et al.* 1982) and H2O maser activity (Genzel and Downes 1977).The low spatial resolution maps in the far infrared and submillimeter suggest that these two activity centers contain the bulk of the region's luminosity (Thronson and Harper 1979, Jaffe *et al.* 1984).

Upon closer examination of the region with higher spatial resolution, it becomes clear that the luminosity is not generated by just two sources, but rather by a collection of 9 or more sources and a diffuse component in emission over 10pc scales. The high spatial resolution maps presented here show that W51 is an even more complex region than previously assumed, with a distribution of sources of various luminosities.

2. OBSERVATIONS

In this paper, we present submillimeter observations taken with the 15-m James Clerk Maxwell Telescope (JCMT) on the summit of Mauna Kea, HI. The observations were conducted during observing runs in May 1991 and May 1992, and used the facility UKT-14 bolometer system wth the standard broadband submillimeter filter set with passbands roughly centered at 350μm, 450μm, 800μm, and 1100μm. We chopped in azimuth with a frequency of 7.813 Hz, and raster scanned in the chop direction to create maps covering $5'$ on the sky. From this chopped, raster scanned dataset, we reconstructed maps using the NOD2 system (Haslam 1974) as modified for use at the JCMT. At all wavelengths but one, we mapped the source multiple times to confirm our results and increase sensitivity. The dates of the observations,chopper parameters, and map sensitivities are listed in the Table 1 below.

Table I — Log of Observations

Date Observed	Wavelength μm	Number of Maps	Beamsize $''$	rms Jy
May 1991	1100	2	19.5	0.1
May 1991	800	3	16	0.5
May 1992	450	2	18	5.8
May 1992	350	1	16	6.5

We obtained calibration for these maps from frequent observations of Uranus and Mars, as well as other sources with reasonably well known submillimeter fluxes. Several maps of Uranus ($r=1.9''$) at all wavelengths were used to determine the FWHM size of the main beam, as well as the contribution to the total flux from the error beam. For the short wavelength observations (350μm and 450μm), we did not use a diffraction limited aperture, largely because the beam shape is not well defined at the diffraction limit, but also because the larger effective beam size allowed us to generate large field of view maps in a reasonable amount of time.

We used the Uranus maps as estimates of the beam response and "cleaned" our maps using an image plane point source extraction technique similar to the CLEAN procedure used in interferometric maps (Hogbom 1974). The resulting maps were then convolved to common resolution equal to the resolution of our largest beam size map ($19.5''$). These maps are presented in Figure 1.

3. RESULTS

From these submillimeter images, we have identified 9 emission centers. The brightest two are associated with W51 MAIN and IRS2, and another (#9) has been identified previously (Sievers *et al.* 1991). The positions of these new sources are listed in Table 2, along with submillimeter fluxes toward each soure in a $19.5''$ beam. The spectral slopes of all of the sources are very similar. The average slope of $\log(F_\nu)$ vs. $\log(\nu)$ for the ensemble is 3.9 ± 0.3, and all of the spectral energy distributions look consistent with emission from relatively warm (T > 40 K) dust with an emissivity law which varies as ν^2 in the submillimeter.

We have calculated the luminosity contained within a 19.5″ beam at each emission center assuming a dust temperature of 50 K and an emissivity law of the type described by Hildebrand (1983). The derived vales vary from 5×10^4 L_o to nearly 2×10^6 L_o, suggesting that each of these emission centers may be a region of vigorous high mass star formation. The 50 K luminosities are listed in the Table 2.

Table 2 — Positions and Fluxes of Emission Centers

ID	Right Ascension 1950.0	Declination 1950.0	19.5″ Flux 1100	800	450	350	L_{50K} L_o
1	19:21:17.5	14:24:30	1.8	8.5	80	161	1.8×10^5
2	19:21:18.6	14:25:15	2.1	13.0	121	256	2.8×10^5
3	19:21:20.0	14:24:27	3.2	16.4	166	313	3.8×10^5
4	19:21:20.7	14:24:41	3.3	17.7	186	356	4.2×10^5
5	19:21:22.3	14:25:15	12.8	50.9	362	655	8.2×10^5
6	19:21:22.9	14:26:43	0.4	1.3	23	33	5×10^4
7	19:21:24.5	14:25:09	5.5	20.0	183	348	4.2×10^5
8	19:21:26.4	14:24:39	20.8	99.0	768	1200	1.8×10^6
9	19:21:28.6	14:23:57	3.4	17.2	179	299	4.1×10^5

In order to compare the spectrum of the extended emission with that from the bright sources in W51, we have plotted the spectrum as a function of 450μm flux level in Figure 2. As more and more flux is included in the composite spectral energy distribution, the spectrum "hardens" – that is, it becomes more like a power law. Note the increase in the slope, particularly between 450μm and 350μm. Such behavior can be explained in the extended emission surrounding the identified sources is warmer that the source condensations themselves.

ACKNOWLEDGEMENTS

E. F. L. has been supported by the James Clerk Maxwell Telescope Fellowship while working on this project.

REFERENCES

Genzel, R., Becklin, E. E., Wynn-Williams, C. G., Moran, J. M., Reid, M. J., Jaffe, D. T., and Downes, D. 1982 ApJ 255:527
Genzel, R., and Downes, D. 1977 A&A Suppl 30:145
Haslam, C. G. T. 1974 A&A Suppl 15:333
Hildebrand, R. H. 1983 QJRAS 24:267
Hogbom, J. 1974 15:417
Jaffe, D. T., Becklin, E. E., and Hildebrand, R. H. 1984 ApJ 279:L54
Martin, A. M., 1972 MNRAS 157:31
Scott, P. F. 1978 MNRAS 183:435
Sievers, A. W., Mezger, P. G., Gordon, M. A., Kreysa, E., Haslam, C. G. T., and Lemke, R. 1991 A&A 251:231
Thronson, H. A., and Harper, D. A. 1979 ApJ 230:133
Wynn-Williams, C. G., Becklin, E. E., and Neugebauer, G. 1974 ApJ 187:473

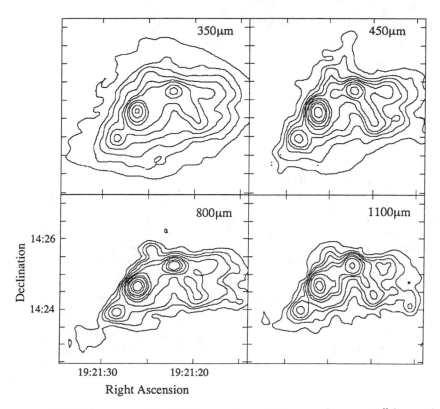

Figure 1: JCMT images of the W51 complex. All images have a 20″ beamsize. Contours are 350μm: 20, 40, 60, 100, 150, 200, 300, 400, 600, and 800 Jy/beam, 450μm: 10, 30, 50, 75, 100, 150, 200, 300, 400, and 600 Jy/beam, 800μm: 2, 4, 6, 10, 15, 20, 30, 40, 60, and 80 Jy/beam, 1100μm: 0.5, 1.0, 1.5, 2, 3, 4, 6, 12, and 18 Jy/beam.

Figure 2: Spectral energy distributions as a function of 450μm flux level. The lowest line includes all pixels whose flux is greater than 50% of the peak map flux. The higher lines include all pixlels whose flux is greater than 30%, 20%, 10%, and 5% of the peak map flux.

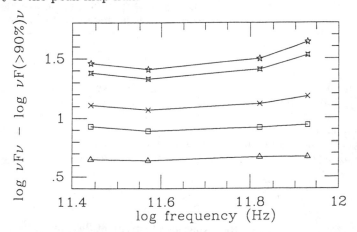

Infrared Emission of Galactic HII Regions

Thomas Kuchar
Phillips Laboratory/GPOB
Geophysics Directorate, Hanscom AFB, MA 01731

Email ID
kuchar@plh.af.mil

ABSTRACT

Infrared emission from HII regions has been extracted from the *IRAS* databases for a sample of Galactic HII regions identified by radio continuum and recombination line surveys. This investigation, based on a sample of radio HII regions, should lessen the bias towards local HII regions which are larger and brighter on average. The infrared luminosity, color and physical properties have been extracted from this data set as a function of Galactocentric distance. The global infrared properties of the HII regions are quite uniform with some very interesting exceptions.

1. INTRODUCTION

The general infrared properties of HII regions in the Galaxy were sought by compiling the IR fluxes listed in the *IRAS* Point Source and Small Scale Structure Catalogs. Also, the detailed properties for a group of $\sim$ 70 HII regions were studied using the *IRAS* Sky Survey Atlas for a 30 square degree field in the Galactic plane ($30° \lesssim \ell \lesssim 60°$). A list of 1000+ HII regions was compiled from the recombination line surveys listed below in the references. Distances (for $R_0 = 8.5$ kpc, and $\Theta_0 = 220$km s^{-1}) are derived from these references. Approximately 730 HII regions were listed in the PSC and 160 in the SSSC.

2. FLUX RATIOS FROM *IRAS* IMAGES AND PSC

Figure 1 shows the *IRAS* 12/100 and 60/100 flux ratios for a sample of HII regions located in the first Galactic quadrant: $30° \lesssim \ell \lesssim 60°, |b| < 1°$. The ratios are derived from the peak fluxes as they appear in images from the Sky Survey Atlas. Thus most of the HII regions in this sample are resolved in the images (*i.e.*, $\theta > 6'$). The histogram shows the mean ratio for the HII regions located in a 0.5 kpc wide bin. The error bars indicate the rms scatter about the mean. Both ratios show peaks near 6 – 6.5 kpc. These HII regions are quite possibly associated with giant molecular clouds and star forming complexes in the molecular ring.

The mean 12/100 ratio for the diffuse interstellar medium and molecular clouds is 0.06±0.02 (Laureijs 1989). Thus it appears that these HII regions are deficient in 12 μm emitters relative to 100 μm when compared with the interstellar medium. The mean 60/100 ratio for molecular clouds is 0.21±0.03. This is a smaller value the mean presented in Figure 1. However, this is not a surprising result, since the dust surrounding the HII regions is being be heated

Figure 1

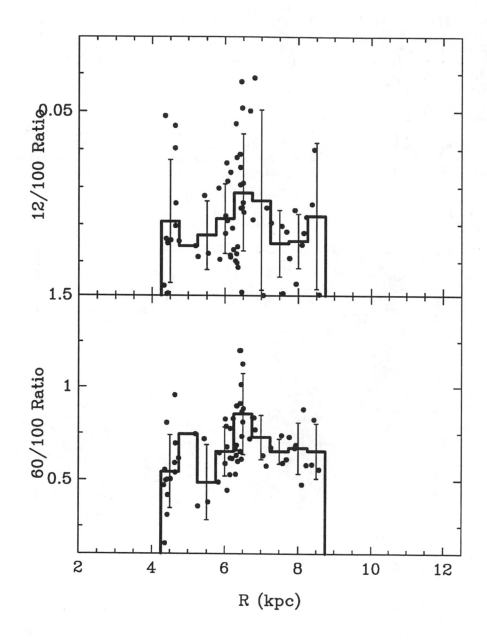

by the OB stars associated with the HII regions.

3. IR LUMINOSITIES

The far infrared color (60/100) temperatures and optical depths ($\tau \propto \lambda^{-2}$) were calculated for 48 HII regions in the Galactc plane for $30° \lesssim \ell \lesssim 60°$. Along with the the kinematic distances (Kuchar 1992), these data were used to determine the FIR luminosities of the HII regions. Figure 2 shows the distribution of luminosities in a face–on view of the Galaxy as seen from the north Galactic pole. The distribution peaks at Galactocentric radius of 6 kpc. The positions of the HII regions are marked by circles the size of which corresponds to the FIR luminosity of the HII region. Fiducial symbols mark the position of the Sun (⊙) and the Galactic center (+). Galactic longitudes are marked as well. The brightest HII regions ($L > L_\odot$) appear to be associated with the densest molecular clouds. This can be seen by comparing Figure 2 with a similar figure in Clemens *et al.* (1988) which shows the surface density of molecular gas.

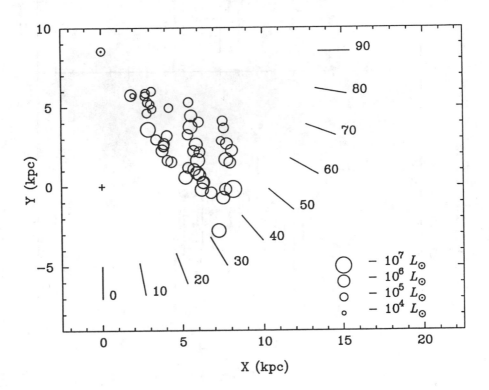

Figure 2

REFERENCES

Caswell, J. L. & R. F. Haynes 1987, A&A, 171, 261

Clemens, D. P., Sanders, D. B., & Scoville, N. Z. 1988 ApJ, 327, 139

Downes, D., Wilson, T. L., Bieging, J., & Wink, J. 1980, A&AS, 40, 379

Fich, M. & Blitz L. 1984, ApJ, 279, 125

Kuchar, T. A. 1992 Ph.D. dissertation, Boston University

Lockman, F. J. 1989, ApJS, 71, 469

Reifenstein, E. C., Wilson, T. L., Burke, B. F., Mezger, P. G., & Altenhoff, W. J. 1970, A&A, 4, 357

Wilson, T. L. 1980, in Radio Recombination Lines, ed. P. A. Shaver (Reidel: Dordrecht) p. 205

Wilson, T. L., Mezger, P. G., Gardner, F. F. & Milne, D. K. 1970, A&A, 6, 364

Wink, J. E., Altenhoff, W. J., & Mezger, P. G. 1982, A&A, 108, 227

INTERSTELLAR
MEDIUM

COOLING OF THE INTERSTELLAR GAS: RESULTS FROM COBE

Charles L. Bennett[1] & Gary Hinshaw[2]
[1]Code 685, NASA Goddard Space Flight Center, Greenbelt, MD 20771.
[2]USRA, Code 685.3, NASA Goddard Space Flight Center, Greenbelt, MD 20771

bennett@stars.gsfc.nasa.gov

ABSTRACT

We report the results of new analyses of far infrared spectral line observations from the FIRAS instrument on the *COBE* satellite with special attention to the spatial distributions of the 158 μm [C II] and the 122 μm and 205 μm [N II] emission lines. These distributions provide new insight into the large-scale distribution and excitation of gas in the Milky Way. Transitions of [C I] at 370 μm and 609 μm and CO J=2-1 through 5-4 are also discussed.

1. INTRODUCTION

An unbiased far infrared survey of the spectral line emission from our Galaxy has been reported from the Cosmic Background Explorer (*COBE*) mission by Wright *et al.* (1991). Unlike other observations that concentrate on specific small regions of the sky at particular wavelengths, the *COBE* FIRAS instrument observed nearly the entire sky with two decades of wavelength coverage. Thus, the *COBE* data show that the large scale diffuse emission from our Galaxy is dominated by a far infrared continuum that cools dust grains, and the 158 μm ground state transition of [C II] that cools the neutral gas in the Galaxy. Two transitions of [N II] trace the large-scale low-density extended ionized component of the galaxy. In addition, the 370 μm and 609 μm lines of [C I] and the J=2-1 through 5-4 lines of molecular CO were detected, and trace the interiors of neutral regions.

Mechanisms for heating interstellar gas include collisions, radiation from stars, shocks, and cosmic rays. Examination of the spectral lines that cool the gas can help determine the dominant excitation mechanisms and conditions. Studies of particular bright regions in our Galaxy and observations of external galaxies have suggested that stellar ultraviolet radiation can ionize vast volumes of a galaxy and that the UV radiation impinging on neutral cloud surfaces is responsible for a large fraction of the far infrared spectral line emission that cools the gas.

Tielens & Hollenbach (1985a) defined *photodissociation regions* (hereafter referred to as PDRs) as neutral "regions where FUV radiation dominates the heating and/or some important aspect of the chemistry. Thus photodissociation regions include most of the atomic gas in a galaxy, both in diffuse clouds and in the denser regions..."

Early conceptual work on PDRs includes Glassgold & Langer (1974, 1975, 1976), Langer (1976), Black & Dalgarno (1977), Clavel, Viala & Bel (1978), and de Jong, Dalgarno & Boland (1980). Tielens & Hollenbach (1985a) presented a detailed and quantitative model for one dimensional PDRs.

The basic photodissociation process can be summarized as follows. Far ultraviolet photons in the energy range 6-13.6 eV, often from O and B stars, impinge on the surfaces of neutral clouds where they will be stopped by impacts with dust grains. The dust grains emit photoelectrons and cool by continuum far infrared emission. The photoelectrons result in the heating of the atoms and molecules in the gas. The UV generally photodissociates the molecules and photoionizes atoms with ionization potentials less than the Lyman limit. Most importantly, photoelectrons heat the gas, resulting in the excitation of C II atoms from their ground state to their first excited state, 91 K above the ground state. C II cools by the emission of 158 μm radiation. The process thus converts far UV radiation to far infrared continuum and spectral line emission. Only far UV radiation can contribute to this process since the typical work function of an interstellar grain is $\sim$6 eV (de Jong 1980). The photoelectric mechanism can typically convert a maximum of $\approx$4 % of the far UV energy into gas heating. This includes the assumption of a $\approx$ 10% probability that an absorbed photon results in an ejected electron (the photoelectric yield). Wright *et al.* (1991) determined, based on *COBE* data, that the 158 μm [C II] line alone emits 0.3% of the bolometric far infrared ($\approx$ UV) lumionosity in our Galaxy.

The UV flux incident on a neutral cloud, G_0, is usually expressed in units of the Habing (1968) flux of 1.6×10^{-3} ergs cm^{-2}s^{-1}. For unusually highly excited regions, such as Orion, the UV flux incident on the neutral PDR region is $G_0 \sim 10^{4-5}$ and the gas density is $n \sim 10^5$ cm^{-3}. The ionization potential of carbon is 11.3 eV, so carbon can be singly ionized where hydrogen is neutral. The ionization potential of oxygen is 13.6 eV, like that of hydrogen, so oxygen is neutral where hydrogen is neutral. In general, for high UV fields ($G_0 > 10^{2.5}$) and/or high densities ($n > 10^{3.5}$ cm^{-3}), [O I] fine structure lines at 63 μm and 145 μm contribute significantly to the cooling of the neutral gas. For lower densities and UV field strengths, the 157.741 μm (Cooksy, Blake, & Saykally 1986) transition of [C II] ($^2P_{3/2} - ^2 P_{1/2}$) is the major cooling line. This is because in O I the $^3P_1$ excited state is 228 K above the $^3P_2$ ground state, and the $^3P_0$ state is 326 K above the ground state. In C II the $^2P_{3/2}$ excited state is only 91 K above the ground state. Dust grains become positively charged, especially near the surfaces of clouds, resulting in a decrease of the photoelectric heating efficiency. An increased electron density results in a higher dust-electron recombination rate, allowing the UV energy to be converted to gas heating more efficiently. Thus, an increase in density acts like an increase in the gas excitation.

The visual extinction into a neutral cloud, A_V, is related to the hydrogen column density, N_H, by $N_H/A_V = 1.8 \times 10^{21}$ cm^{-2} mag^{-1} (cf. Burstein & Heiles 1978; Bohlin, Savage & Drake 1978), where the exact value of the conversion factor depends on the local gas-to-dust ratio. Hydrogen in the PDR is predominantly in the form of H I for $A_V < 2$ and H$_2$ for $A_V > 3$. [C II] dominates the cooling of the outer regions of the cloud ($A_V < 4$) for low-excitation PDRs, as described by Hollenbach, Takahashi, & Tielens (1991). [C I] dominates the cooling for $4 < A_V < 6$ and CO rotational transitions dominate the cooling for $A_V > 6$. The A_V values, above, vary according to the detailed physical properties of the particular PDR (i.e. n, G_0, and geometrical effects). Significant UV shielding (i.e. $A_V \sim 10 - 20$) is required before OH, H$_2$O, and O$_2$ can be sustained. For $A_V > 6$ heating from cosmic rays and gas-grain collisions become significant relative to the UV heating.

The photodissociation model has been successfully applied to Orion, which is a face-on PDR (Tielens & Hollenbach 1985b), and to M17SW, which is an edge-on PDR (Meixner *et al.* 1992). Wolfire, Tielens, & Hollenbach (1990) applied the model to the nuclear region of our Galaxy and to M82 to derive UV field strengths and gas densities. Low density PDR models, which are important for interpreting the average PDR conditions in our Galaxy, were computed by Hollenbach, Takahashi, & Tielens (1991). Wolfire, Hollenbach, & Tielens (1992) conclude that CO 1-0 emission from giant molecular clouds is primarily the result of the PDR physics. Crawford *et al.* (1985) and others have found that PDR models account for the excellent correlation both between 158 μm [C II] emission and CO 1-0 emission, and between [C II] and far IR continuum emission.

2. OBSERVATIONS & PREVIOUS RESULTS

Results are presented from the *COBE* Far InfraRed Absolute Spectrophotometer (FIRAS) instrument. The instrument is a scanning Michelson interferometer that covers the wavelength range from 100 μm to 1 cm. While the primary purpose of the instrument is to make precise measurements of the cosmic microwave background radiation, the instrument design offers many advantages for the large scale study of diffuse emission from our Galaxy. For example, the FIRAS has a 7° beam, is an instrument that makes absolute measurements of intensity with no beam switching, and has observed nearly the entire sky. For these reasons, the FIRAS is well-suited for measurements of diffuse emission. The nominal spectral resolution is 0.7 cm^{-1} (or R $\equiv \lambda/\Delta\lambda \approx$ 30 at 500 μm). Some higher spectral resolution data exist, but are not discussed here.

The first cosmological results from the *COBE* FIRAS instrument were reported by Mather *et al.* (1990), and recent overall summaries of *COBE* results are found in Boggess *et al.* (1992) and Bennett *et al.* (1992a,b).

The first results of the *COBE* FIRAS unbiased far infrared survey of the Milky Way were published by Wright *et al.* (1991). Table 1 in that paper summarizes spectral line detections from *COBE* FIRAS: 158 μm [C II], 370 μm and 609 μm [C I], J=2-1 through J=5-4 CO, and 122 μm and 205 μm [N II]. Since that time, improved data processing algorithms and techniques have allowed for a more detailed examination of the data.

The [C II], [C I], and CO emission lines detected by *COBE* can immediately be recognized as the expected cooling lines from PDRs with low densities and small UV fields (Hollenbach, Takahashi, & Tielens 1991). The result that the galactic-averaged PDR properties are described by lower excitation conditions than objects such as Orion is strengthened by the lack of detection by *COBE* FIRAS of the 145 μm [O I] line.

The two transitions of [N II] are recognized as arising from the ionized medium. As pointed out by Wright *et al.* , this emission can not be coming from only classical, high-density, H II regions. The ratio of the intensities of the $^3P_2$ to $^3P_1$ 122 μm [N II] line to the $^3P_1$ to $^3P_0$ 205 μm ground state transition is a probe of the density of the ionized gas. Rubin (1985) provides a grid of classical H II region models that result in I(122 μm [N II])/I(205 μm [N II]) ranging from 3 (for $n_e \approx 10^2$ cm^{-3}) to 10 (for $n_e \approx 10^5$ cm^{-3}). In the low-density limit, $n_e \ll 100$ cm^{-3}, the expected intensity ratio is 0.7. The measured *COBE* ratio is 1.0 to 1.6, depending on the adopted values for the instrumental resolution at the two wavelengths. Thus the ionized gas measured by *COBE* FIRAS is likely to arise mostly from large, diffuse, regions. Reynolds (1990, and references therein) has extensively studied the warm ionized medium (WIM), concluding that a large fraction of the Galaxy contains $\sim$ 8000 K fully ionized gas with a density of $n_e \approx 0.2$ cm^{-3}. However, the galactic disk is opaque to the Hα and other optical lines used to study this medium so only a local (typically < 2 kpc) view exists. The galactic disk is transparent to microwave continuum free-free emission, but it is difficult to separate the dust and synchrotron components from the overall microwave emission (Bennett et al. 1992c). Radio recombination line observations have not been sensitive enough to observe the WIM.

3. ANALYSIS RESULTS

In the Wright et al. (1991) analysis the dust continuum emission map of the sky was used as a spatial filter for averaging the emission from the spectral lines, effectively assuming that the spectral line intensities follow the continuum spatial distribution. That assumption is not made here. Rather, for each pixel we perform a simultaneous least squares fit of a continuum spectral "baseline" plus a series of Gaussian spectral line profiles centered on the wavelengths of known lines. The result are a set of spectral line intensities for each sky position. We find that while the [C II] emission (color Plate 5, top) follows the continuum closely, and the [N II] emission (color Plate 5, bottom) follows it approximately, the [C I] and CO emission are distinct from the continuum template used by Wright et al. , in that the [C I] and CO flux distributions are strongly concentrated towards the galactic center.

In Figure 1 the longitude profiles of the [C II] and 158 μm far infrared emission are shown for the entire galactic plane. The line and continuum emission are seen to track one another very well, as would be expected for PDR emitting gas. The emission appears to peak towards the molecular ring, the galactic center direction, and in the spiral arms. Only towards the galactic center does the [C II] emission appear to be relatively weak compared with the far infrared continuum. In contrast, the 609 μm [C I] emission, shown in Figure 2, peaks strongly towards the galactic center direction, as does the CO J=5-4 emission. These observed differences in the behavior of the spectral lines towards the galactic center may be explained in a number of ways. They may be indicative of a difference of mean PDR parameters. For example, emission from the galactic center direction may reflect significantly different UV field strength and/or color, gas density, dust properties (size distibution and photoelectric efficiency), gas-to-dust ratio, or the 158 μm [C II] may be saturated towards the galactic center. If the galactic center neutral material were less clumpy or of higher density, relatively more cooling could be done by [C I] and CO than by [C II] because of the higher relative filling factor of high-A_V neutral gas. The [C II] emission towards the galactic center may be self absorbed, but the 1.5 power law scaling with the [N II], discussed below, still holds in this region.

Figure 3 shows the profiles of the two [N II] transitions. Although the 122 μm transition is clearly noisier due to the poorer instrument sensitivity at that wavelength, the transitions clearly track one another, as expected. Figure 4 compares the [C II] and [N II] distributions. The distributions are unmistakably related, but the [N II] is relatively stronger than the continuum while the [C II] follows the continuum. Figure 5 shows that an excellent correlation exists between the fluxes of [C II] and [N II], in the sense that $I(N\ II) \propto I(C\ II)^{1.5}$, as pointed out by Petuchowski & Bennett (1993a,b). They suggest that the 3/2 power law may be the result of a volume to surface area ratio sampling of the interstellar gas. Figure 6 shows that $I(C\ II) \propto I(FIR)^{0.95}$ while $I(N\ II) \propto I(FIR)^{1.55}$.

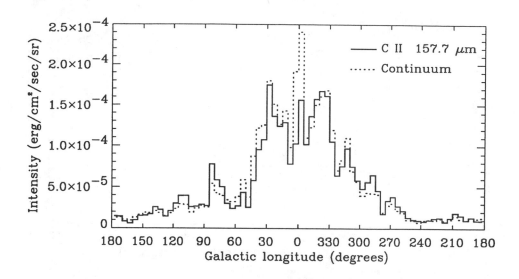

Figure 1: The longitude profile of the [C II] and 158 μm far infrared emission is shown for the entire galactic plane. The flux in each bin represents an average over 5° in galactic longitude and 10 ° in galactic latitude.

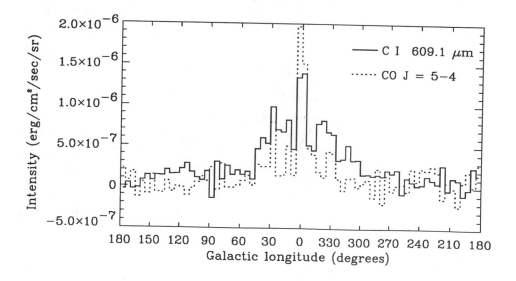

Figure 2: The 609 μm [C I] and CO J=5-4 emission peak strongly towards the galactic center direction.

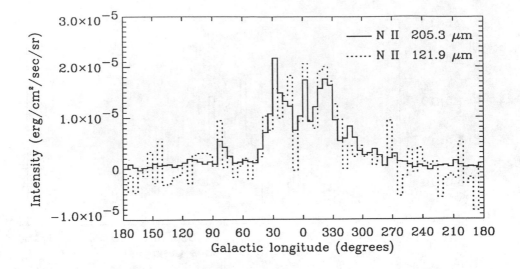

Figure 3: The intensity distributions of the two [N II] transitions.

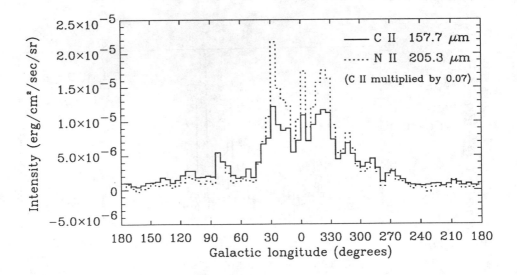

Figure 4: The [C II] and [N II] distributions. While the qualitative distributions are unmistakably related, the [N II] tends to be relatively stronger than the continuum where the [C II] follows the continuum.

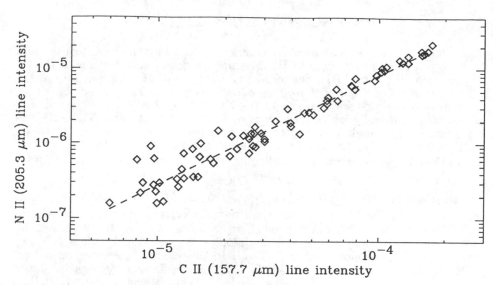

Figure 5: An excellent correlation exists between the intensity of the [C II] and [N II] intensities, in the sense that I(N II) $\propto$ I(C II)$^{1.5}$.

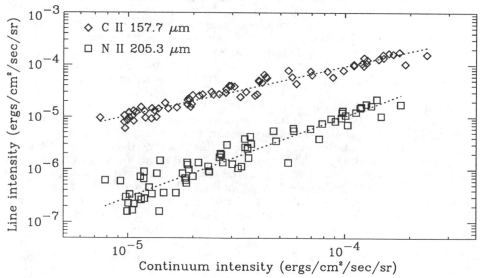

Figure 6: I(C II) $\propto$ I(FIR)$^{0.95}$ while I(N II) $\propto$ I(FIR)$^{1.55}$.

4. DISCUSSION & CONCLUSIONS

The 158 μm [C II] emission is linearly correlated with the far IR continuum. The 205 μm and 122 μm from [N II] also appears closely correlated with the far IR contiuum and the [C II] emission. The flux of the [N II] emission goes as the 3/2 power of the [C II] emission. This power-law may be considered suggestive of a volume to surface area ratio geometric effect. In such a scenario the volumes of very large internally ionized regions give rise to [N II] emission. These regions are partially surrounded by neutral gas with the interface surface traced by [C II] emission. If the emission from each line of sight is dominated by one such region that is characterized by an effective radius R_{eff}, then $I(N~II) \propto R_{eff}^3$ and $I(C~II) \propto R_{eff}^2$, so from one line of sight to another the ratio will be characterized by $I(N~II)/I(C~II) \propto R_{eff}^{1.5}$. For the 3/2 power law to hold each line of sight must be better characterized by an effective radius of an individual region than by the number of regions along each line of sight. If several regions contribute substantially to the emission along each line of sight, then $I(N~II)$ would be linearly proportional to $I(C~II)$. In this picture, the emission, which peaks in the direction of the molecular ring, is dominated by approximately beam-sized bubbles of ionized gas surrounded by neutral gas. For the *COBE* 7° beam, this corresponds to bubbles 1 kpc in diameter. Comparable size bubbles are known to exist in external galaxies (e.g. Bruhweiler, Fitzurka & Gull 1991; Kamphius, Sancisi & van der Hulst 1991). Note that the [N II] emission need not arise from the volume of the bubble, but could be from swept up gas so long as the amount of gas is characterized by the bubble volume.

The above scenario requires large-scale structures. Shibai *et al.* (1991) report results from a 3.4 arc-minute beam size survey of [C II] emission concluding that, "most [C II] emission of the Galactic disk is radiated not from discrete sources but from more extended and diffuse regions." Nakagawa *et al.* (1993), at this conference, have shown [C II] maps made from observations with a 12 arc-minute beam that also shows structures that are much larger than the beam size. Doi *et al.* (1993), at this conference, conclude that discrete sources are significant when looking through the length of the $l=90°$ spiral arm.

Massive and luminous O and B stars tend to form in clusters and then become Type II supernovae that are correlated in space and time. The correlated supernovae create bubbles in the plane of the Galaxy. If the bubble is large enough it can "break through" the thickness of the disk. If the breakthrough is particularly energetic it will blow material into the halo of the Galaxy. These are called "blowout" bubbles. These "blowouts" lead to Galactic fountains. In these models significant amounts of material from the galactic plane are thrown out of the plane and at least some of this material rains back down onto the plane. These processes have been described by Bregman (1980), Houck & Bregman (1990), Mac Low & McCray (1988), Mac Low, McCray, & Norman (1989), Heiles (1990), Norman & Ikeuchi (1989), and Shapiro & Field (1976). Heiles (1990) estimates an area filling factor of breakthough bubbles of 23% in our Galaxy. McKee & Williams (1993a,b) derive the somewhat lower filling factor of superbubles of 10%. We feel that the cycle involving giant molecular clouds, massive clustered O and B stars, and clustered supernovae must be greatly enhanced in the molecular ring of the Galaxy so that the filling factors of bubbles, quoted above, are probably underestimates for the ring.

Petuchowski & Bennett (1993a, 1993b) examined the possible make-up of the *COBE* FIRAS line emission in terms of three morphologies: classical ionization bounded H II regions, boundary ionized clouds, and an extended low density warm ionized medium. They conclude that the classical ionization bounded H II regions contribute very little to the total line emission intensities, largely because of their small volume filling factors. The other two morphologies can occupy large filling factors and contribute significantly to the total signal intensity. Since bubbles, superbubbles, and neutral material around their periphery are examples of large filling factor regions, they are capable of dominating observed signals along lines of sight that also contain several, much smaller, classical H II regions.

Volume-to-surface geometry is only one possible explanation for the 1.5 power scaling law between [C II] and [N II]. It may be that the [C II] scales proportionally with the far infrared continuum for the average PDR conditions in the Galaxy, but that $I([N\ II]) \propto I(FIR)^{\sim 1.5} \propto I(UV)^{\sim 1.5}$ for some other physical reason, for example, having to do with the mean galactic color of the UV radiation field in the Galaxy. It is worth noting that a similar 1.5 power law relation has been reported for different transitions in an external galaxy (Petuchowski & Bennett 1993a, 1993b).

Bock *et al.* (1993) conclude that at a high galactic latitude the [C II] cooling rate is $(2.3 \pm 0.3) \times 10^{-26}$ erg/s per H-atom. Using this value with the estimated total [C II] luminosity from our Galaxy from *COBE* of $10^{7.7}$ solar luminosities (Wright et al. 1991), we derive the 158 μm line from [C II] is cooling $> 10^9$ solar masses of gas. It is not known if the H I gas in the outer Galaxy and halo is dominated by PDR physics, but it is still likely to be the case that 158 μm [C II] emission is the dominant coolant.

Preliminary analysis results from the *COBE* DIRBE experiment, reported at this conference by Hauser, is interpreted by Sodroski *et al.* , who conclude that 90% of the far IR dust emission comes from neutral regions. This conclusion is consistent with the PDR model results suggested here. The fact that H I and C II *appear* to have different spatial distributions is not contradictory with the picture that C II cools H I gas. Except where the emission is self-absorbed, the H I flux is directly proportional to the H I column density. On the other hand, the 158 μm [C II] flux does not, in general, scale directly with the gas density. Since [C II] emission is the result of a collisional process between electrons and carbon ions, one might naively expect [C II] emission to scale as the square of the gas density. In the low excitation PDR models of Hollenbach, Takahashi, & Tielens (1991), the 158 μm [C II] emergent intensity scales somewhat less than the square of the density for low densities and low UV field strengths. Nevertheless, since H I and C II intensities scale differently with density the fact that they have *apparently* different spatial distributions is not surprising.

ACKNOWLEDGEMENTS

COBE is supported by NASA's Astrophysics Division. The Goddard Space Flight Center (GSFC), under the scientific guidance of the *COBE* Science Working Group, is responsible for the development and operation of *COBE*. We are grateful for the contributions of the entire *COBE* Science Team and the support personnel at Goddard's Cosmology Data Analysis Center (CDAC).

REFERENCES

Bennett, C. L. *et al.* 1992a, The Third Teton Summer School: *The Evolution of Galaxies and their Environment*, H. A. Thronson & J. M. Shull eds., in press

Bennett, C. L. *et al.* 1992b, Advances in Space Research, in press

Bennett, C. L. *et al.* 1992c, ApJ, 396, L7

Black, J. H. & Dalgarno, A. 1977, ApJS, 34, 405

Bock, J. *et al.* 1993, this volume

Boggess, N. W. *et al.* 1992, ApJ, 397, 420

Bohlin, R. C., Savage, B. D. & Drake, J. F. 1978, ApJ, 224, 132

Bregman, J. N. 1980, ApJ, 236, 577

Bruhweiler, F. C., Fitzurka, M. A. & Gull, T. R. 1991, ApJ, 370, 551

Burstein, D. & Heiles, C. 1978, ApJ, 225, 40

Clavel, J., Viala, Y. P., & Bel, N. 1978, A&A, 65, 435

Crawford, M. K., Genzel, R., Townes, C. H., & Watson, D. M. 1985, ApJ, 291, 755

de Jong, T. 1980, Highlights of Astronomy, 5, 301

de Jong, T., Dalgarno, A. & Boland, W. 1980, A&A, 91, 68

Doi, Y. *et al.* 1993, this volume

Glassgold, A. E. & Langer, W. D. 1974, ApJ, 193, 73

Glassgold, A. E. & Langer, W. D. 1975, ApJ, 197, 347

Glassgold, A. E. & Langer, W. D. 1976, ApJ, 206, 85

Habing, H. J. 1968, Bull. Astr. Inst. Netherlands, 19, 421

Hauser, M. G. 1993, this volume

Heiles, C. 1990, ApJ, 354, 483

Hollenbach, D., Takahashi, T., & Tielens, A. G. G. M. 1991, ApJ, 377, 192

Houck,J., & Bregman, J. N. 1990, ApJ, 352, 506

Kamphuis, J., Sancisi, R., & van der Hulst, T. 1991, A&A, 244, L29

Langer, W. D. 1976, ApJ, 206, 699

Mac Low, M.-M., & McCray, R. C. 1988, ApJ, 324, 776

Mac Low, M.-M., McCray, R. C., & Norman, M. L. 1989, ApJ, 337, 141

Mather, J. C. *et al.* 1990, ApJ, 354, L37

McKee, C. F. & Williams, J. 1993a, *Star Forming Galaxies and Their Interstellar Medium*, ed. J. J. Franco, in press

McKee, C. F. & Williams, J. 1993b, ApJ, in press

Meixner, M. *et al.* 1992, ApJ, 390, 499

Nakagawa, T. *et al.* 1993, this volume

Norman, C. & Ikeuchi, S. 1989, ApJ, 345, 372

Petuchowski, S. J. & Bennett, C. L. 1993a, ApJ, in press

Petuchowski, S. J. & Bennett, C. L. 1993b, this volume

Reynolds, R. J. 1990, *The Galactic and Extragalactic Background Radiation*, S. Bowyer & C. Leinert eds., p157

Rubin, R. H. 1985, ApJS, 57, 349

Shapiro, P. R. & Field, G. B. 1976, ApJ, 205, 762

Shibai, H. *et al.* 1991, ApJ, 374, 522

Sodroski, T. *et al.* 1993, this volume

Tielens, A. G. G. M. & Hollenbach, D. 1985a, ApJ, 291, 722

Tielens, A. G. G. M. & Hollenbach, D. 1985b, ApJ, 291, 747

Wolfire, M. G., Tielens, A. G. G. M., & Hollenbach, D. 1990, ApJ, 358, 116

Wolfire, M. G., Hollenbach, D., & Tielens, A. G. G. M. 1992, preprint

Wright, E. L. *et al.* 1991, ApJ, 381, 200

THE DISTRIBUTION OF NEUTRAL GAS IN THE MILKY WAY

T. M. Dame
Harvard-Smithsonian Center for Astrophysics
60 Garden Street, Cambridge, Mass. 02138

ABSTRACT

Over the past decade, the uncertainty on the H_2 mass of the inner Galaxy has been reduced from a factor ~4 to less than a factor 2, comparable to the uncertainty which still exists for H I. Gas masses beyond the solar circle remain more uncertain, owing mainly to uncertainties in the rotation curve and CO-to-H_2 mass conversion factor there. H I and H_2 masses are roughly equal within the solar circle, but H I dominates beyond. GMCs are superior to H I clouds for tracing Galactic structure, but kinematic distance errors and severe velocity crowding near the terminal velocity in the inner Galaxy remain formidable obstacles to sorting out the structure of that region.

Large-scale near-infrared surveys of the Galactic plane, particularly those recently obtained by the COBE satellite, hold great promise for determining both distances and masses for large GMCs in the inner Galaxy. An earlier survey by the Infrared Telescope (IRT) has already revealed a remarkably tight anticorrelation between CO and near-infrared emission toward the inner first quadrant, suggesting a possible near-far asymmetry of the gas distribution there, and providing a new calibration of the CO-to-H_2 mass conversion factor on a Galactic scale.

1. INTRODUCTION

A complete and balanced review of subjects as vast as the distributions of atomic and molecular gas in the Galaxy is not possible in the limited space available here, and it is also unnecessary given the excellent reviews on these subjects that have been published recently (e.g., Dickey & Lockman 1990, Combes 1991). Here I will focus on the most basic aspects of the Galactic gas distribution — the radial distribution and implied total gas mass, and non-axisymmetric or spiral features. Also, because this conference is devoted to the discussion of new data and ideas on Galactic structure, I will discuss a promising new technique for determining masses and distances for GMCs in the inner Galaxy based on comparison of CO and near-infrared surveys, and present some preliminary results and speculative comments based on this technique.

2. RADIAL DISTRIBUTION AND TOTAL GAS MASS

In 1984, two research groups, analyzing their own separate CO surveys of the first Galactic quadrant with rather similar axisymmetric models, published results which differed on the total molecular mass of the inner Galaxy by about a factor of four. Sanders, Solomon, & Scoville

(1984; hereafter SSS) used data from the FCRAO 14 m and NRAO 11 m telescopes to derive a molecular mass surface density at the peak of the molecular ring of ~20 $M_\odot$ pc^{-2}, while profiles published by Thaddeus and Dame (1984; hereafter the Columbia group) based on data from the 1.2 m telescope then at Columbia University implied a peak value of ~5 $M_\odot$ pc^{-2}. This disagreement was all the more puzzling in light of the quite similar results obtained by the two groups on the thickness of the molecular gas layer and its displacement from the plane.

Much of the disagreement between the two groups has been resolved and today the uncertainty in the amount of molecular gas in the inner Galaxy is significantly less than the 1984 papers would together imply. With assistance from both groups, Bronfman et al. (1988) showed that nearly all of the discrepancy could be attributed to three factors which each biased the Columbia masses low (by a factor ~0.6 in total) and the SSS masses high (by a factor ~2.4). The smallest factor was instrumental calibration, which was ~11% high for the SSS data (Sanders et al. 1986) and ~6% low for the Columbia data (Bronfman et al.). A second source of discrepancy was inappropriate weighting of the data in fitting axisymmetric models. Lacey (1985) pointed out that neither of the models was entirely self-consistent, since they could not reproduce the observed profile of CO integrated intensity vs. longitude. Bronfman et al. devised an alternate maximum-likelihood method of model fitting which was self-consistent in the sense just described and concluded, as did Rivolo & Solomon (1988), that the SSS fit overestimated CO emissivities in the inner Galaxy by ~40% and the Columbia fit underestimated them by ~30%.

The remaining discrepancy between the SSS and Columbia results can be attributed to the different values adopted for the ratio of H_2 column density to CO velocity-integrated intensity, $X \equiv N(H_2)/W_{CO}$, which throughout this article will be given in units of 10^{20} cm^{-2} K^{-1} km^{-1} s. The value 2.0 adopted by the Columbia group was derived using high-energy gamma rays as a tracer of total gas column density in the first Galactic quadrant (Lebrun et al. 1983), and is just slightly less than the value determined subsequently by a similar analysis over the entire Galaxy (2.3; Strong et al. 1988). Because the cosmic rays which produce high-energy gamma rays through interactions with interstellar gas nuclei are thought to permeate essentially all of the gas (Bertsch et al. 1993), the value of X determined in this way represents a true Galactic average, and is therefore arguably the most appropriate value to use for determining global gas properties. In contrast, the value of 3.6 adopted by SSS is based on star counts and near infrared extinction measurements in 14 small, dense clumps within the Taurus dark cloud complex (Frerking, Langer, & Wilson 1982); because of their much higher densities and narrower linewidths, the CO radiative transfer in these objects is likely to be quite different from that of the large GMCs which contain most of the Galactic molecular mass.

Two other methods of calibrating X on a Galactic scale support a value of about 2. Solomon et al. (1987) determined total masses for 273 molecular clouds throughout the first quadrant on the assumption that they are in virial equilibrium. Although, not surprisingly, they found that X varied with cloud mass, the value applicable to large GMCs and

therefore to the bulk of molecular gas in the Galaxy was found to be 2.2.[*]
As will be shown in § 4, a new method which uses near-infrared
extinction as a gas tracer on a Galactic scale also leads to a mean Galactic X
of order 2. It is the good agreement among these three quite independent
methods, using gamma rays, the virial theorem, and near-infrared
extinction, that gives us some confidence that CO can be used reliably, at
least on the scale of GMCs or larger, to trace molecular gas.

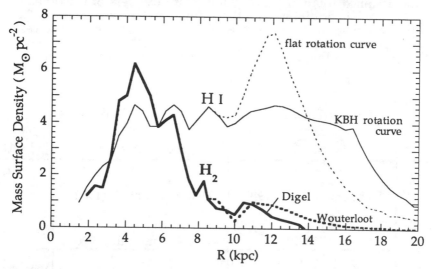

Figure 1. Average H I and H2 mass surface densities in the Milky Way. H2 at R
< R$_\odot$ *(heavy solid):* Bronfman et al. (1988) Table 4, scaled to X = 2.3 and R$_\odot$ = 8.5
kpc. H2 at R$_\odot$: Dame et al. (1987), scaled to X = 2.3. H2 at R > R$_\odot$ *(heavy solid):*
Digel (1991) Table VI-2 with factor 1.36 correction for helium removed. H2 at R >
R$_\odot$ *(heavy dashed):* Wouterloot et al. (1990), renormalized to the H2 mass surface
density at R = R$_\odot$ adopted here. H I at R < R$_\odot$ *(light solid):* derived from
midplane number densities given in Burton & Gordon (1978) and constant H I layer
thickness of 220 pc (FWHM) given by Dickey & Lockman (1990); values have also
been arbitrarily scaled by a factor of 2 to approximately match the value at R =
R$_\odot$ (see Liszt 1992). H I at R = R$_\odot$: Dickey & Lockman (1990) Figure 6. H I at R >
R$_\odot$ *(light solid):* derived by Lockman (1988) using the rotation curve of Kulkarni,
Blitz, & Heiles (1982; KBH). H I at R > R$_\odot$ *(light dashed):* derived by Lockman
(1988) using a flat rotation curve, which differs from the KBH curve by < 14% at
any point.

Figure 1 compares recent determinations of atomic and molecular
mass surface densities in the Galaxy. Liszt (1992) has recently argued that
the amount of atomic gas in the inner Galaxy is considerably less certain
than commonly assumed. He points out that H I gas densities derived by
various workers in 21 cm studies often vary by 50-100%, and that H I mass
surface densities derived on the crude assumption of a constant spin
temperature are commonly scaled — by about a factor of 2 — to match the

---

[*] The value actually reported in the paper, 3.0, was not corrected for helium in the clouds.

value observed at high latitudes in the solar neighborhood. The inner-Galaxy H_2 profile shown in Figure 1 matches on well to the value determined in the solar neighborhood from a simple inventory of clouds (Dame et al. 1987) with no such scaling required. Beyond the solar circle, both atomic and molecular observers are faced with even more formidable uncertainties. The two H I profiles at $R > R_\odot$ in Figure 1 were derived by Lockman (1988) to demonstrate how sensitive mass surface densities there are to the assumed rotation curve. Molecular observers of the outer Galaxy are faced, in addition, with the arduous task of searching for CO lines which are normally both very weak, due to large cloud distances and possibly a larger X ratio (Digel, Bally, & Thaddeus 1990), and distributed over a large latitude range, owing to the warping and flaring of the outer disk.

Table 1: Total H I and H_2 Masses of the Milky Way[*]

	$R < R_\odot$	$R > R_\odot$	Total
H I	$0.7^{(1)}$ Henderson et al. (1982)	$2.8^{(1)}$ Henderson et al. (1982)	3.5
	1.7 Liszt (1992)	5.3 Wouterloot et al. (1990)	7.0
H_2	$0.7^{(1,2)}$ Bronfman et al. (1988)	$0.6^{(2,3)}$ Digel (1991)	1.3
	$1.1^{(1,2)}$ Rivolo & Solomon (1988)	0.6 Wouterloot et al. (1990)	1.7

[*]10^9 $M_\odot$; Galactic center excluded; associated helium not included.

[1] scaled to $R_\odot = 8.5$ kpc.
[2] scaled to $X = 2.3 \times 10^{20}$ cm^{-2} K^{-1} km^{-1} s.
[3] helium correction removed.

Table 1 compares estimates of total H I and H_2 masses in the Galaxy, both within and outside the solar circle. Two recent or widely-quoted estimates are given for each to provide some indication of the uncertainties which still exist. Few, if any, authors have hazarded an explicit estimate of the H I mass within the solar circle. Even the widely-quoted value of Henderson, Jackson, & Kerr (1982) is simply based on midplane densities given in Burton's (1976) review article and a scale height determined by Jackson & Kellman (1974), and the other value given in Table 1 was derived by the author from mass surface densities published by Liszt (1992). The much higher value of Liszt can be attributed to his scaling of H I surface densities to match the value at the solar circle, as discussed above. The discrepancy between the two masses given for H I at $R > R_\odot$ is not surprising, given that kinematic uncertainties *alone* can make surface densities in parts of the outer Galaxy uncertain by a factor of four (Lockman 1988). The relatively good agreement of the values given for H_2 at $R < R_\odot$ reflects the recent convergence of opinions discussed above.[*] Much of the difference between the masses can be attributed to Rivolo & Solomon's attempt to account for what they believe are large-scale optical depth effects for CO near the tangent points, where severe velocity crowding causes "cloud blocking". The agreement of masses for H_2 at $R > R_\odot$ must be somewhat fortuitous, given the small fraction of the outer Galaxy observed in CO to date, and the uncertainties in the rotation curve and X ratio in that region.

In summary, it now appears that the H I and H_2 masses of the inner Galaxy are roughly equal both in value and in uncertainty. The surface density of H_2 may exceed that of H I near the peak of the so-called molecular ring at $\sim R_\odot / 2$, but one cannot claim that the gas is mainly molecular at any radius, except perhaps at the Galactic center. On the other hand, it is certainly true that the gas is mainly atomic beyond the solar circle.

3. NON-AXISYMMETRIC DISTRIBUTION

Efforts to map the spiral structure of the inner Galaxy in H I were largely derailed about 20 years ago by Burton's (1973) clear and convincing demonstration of how the putative spiral features seen in 21 cm surveys could as easily be produced by low-amplitude, large-scale perturbations in the Galactic velocity field as by variations in gas density along the line of sight. The influence of such streaming motions, coupled with the two-fold distance ambiguity which exists in the inner Galaxy, make it seem a nearly impossible task to derive quantities such as the arm-interarm density contrast for H I there. Studies in the outer Galaxy with H I have been somewhat more successful, mainly because no distance ambiguity exists in that region (e.g., Henderson et al. 1982).

Molecular clouds as traced by CO have two significant advantages over H I for sorting out the spiral structure of the inner Galaxy. First, there is now compelling observational evidence, both in the Milky Way (Dame et al. 1986; Grabelsky et al. 1987) and in some nearby external spirals (Casoli 1991), as well as theoretical reasons for believing (Elmegreen 1989) that H_2 is simply more confined to the spiral arms than H I. Second, unlike the bulk of H I, H_2 is mainly concentrated in large, fairly well-defined complexes that can be studied individually to great distances. Using a variety of techniques, it is possible to resolve the kinematic distance ambiguity for many of the large complexes and plot their positions in the Galaxy.

A combination of several such analyses of large GMCs within or near the solar circle, based on CO surveys with the 1.2 m telescope, is shown in Figure 2a. A problem with analyses of this sort is the kinematic distance errors introduced by the clouds' random velocities, which will blur the derived arm pattern in even the most beautiful spirals (see

---

[*] The Rivolo & Solomon value has been scaled to X = 2.3 since one of the authors has subsequently argued, as I do here, that the 3 main methods of calibrating X lead to that value (Solomon & Barrett 1991).

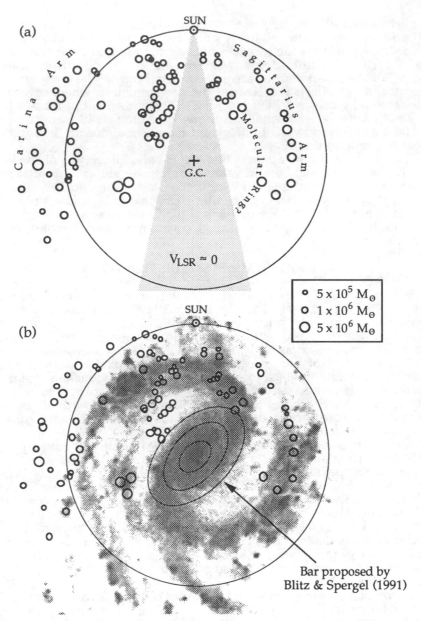

Figure 2. (a) The distribution of large molecular clouds in the inner first quadrant (Dame et al. 1986), the inner fourth quadrant (Bronfman et al. 1993), and the Carina Arm (Grabelsky et al. 1988). (b) Same clouds as in (a), with the bar proposed by Blitz & Spergel (1991) and a faint image of NGC 4535 overlaid. This figure is meant to demonstrate how a bar such as the one proposed by Blitz & Spergel might produce a near-far asymmetry in the molecular ring region of the first quadrant of the sort suggested by the cloud distribution and the tight CO-near infrared anticorrelation (see § 4).

Combes 1991 for a good example). Still, several spiral features are evident in Figure 2*a*, the most obvious by far being the Sagittarius Arm in the first quadrant and its probable extension into the fourth quadrant, the Carina Arm.

An even more serious problem with such analyses is severe velocity crowding and cloud blending near the terminal velocity in the inner Galaxy, which makes the identification of individual clouds difficult: the total mass of all the clouds shown in the first quadrant is just 17% of the total molecular mass inferred to be there by the axisymmetric model of Bronfman et al. (1988). Some of the missing mass is certainly in small clouds, more in unidentified larger clouds near the terminal velocity, and it may also be that the axisymmetric model overestimates the molecular mass in the region (see § 4). The relatively low angular resolution of the 1.2 m telescope is not an issue, since this telescope can detect and resolve an isolated cloud of a few $10^5 \, M_\odot$ anywhere within the solar circle, and according to the molecular cloud mass spectrum (e.g., Dame 1983) most of the Galactic molecular mass ($\geq 80\%$) is contained in clouds of that mass or larger. That higher angular resolution helps little has been demonstrated by Solomon et al. (1987), who carried out a similar analysis in the first quadrant using CO observations which, although seriously undersampled, were at roughly ten times higher angular resolution. They cataloged about ten times as many clouds, but most were small and in total carried only a tiny fraction of the Galactic molecular mass. Their cataloged clouds account for just 18% of the total molecular mass they estimate to be in the region, essentially the same fraction accounted for by the analysis at lower angular resolution.

4. GMCs AS DARK NEBULAE AT 2 MICRONS

Star counting, a fairly direct method of determining masses and distances for molecular clouds, relies on their visibility as dark nebulae, and thus has been limited to relatively small clouds in the solar neighborhood (e.g., Dickman 1978). Here I would like to describe a new technique, involving the intercomparison of large-scale 21 cm, CO, and near-infrared surveys, which can loosely be considered a form of star counting on a Galactic scale. The technique has already provided new information on the X ratio in the inner Galaxy and hints as to the large-scale gas distribution there, and it holds great promise, when applied with the near-infrared data now available from COBE, for sorting out the structure of the inner Galaxy.

The Infrared Telescope (IRT) flown aboard the Space Shuttle as part of the Spacelab-2 mission successfully surveyed the entire first quadrant of the Galactic plane at 2.4 μm with an angular resolution of ~1° (Kent et al. 1992). As Figure 3 shows, there is a remarkably tight anticorrelation along the Galactic equator in the first quadrant between the emission detected by this survey and that of CO. Virtually every peak at 2.4 μm corresponds to a dip in CO and *vice versa*. There can be little doubt that what we are seeing is near-infrared absorption by large GMCs in the inner Galaxy, and there is no clearer or more convincing demonstration of the ability of CO emission to trace dust, and therefore indirectly H_2 mass, than this very tight anticorrelation.

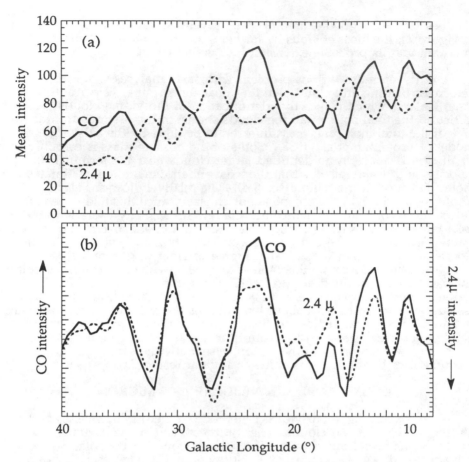

Figure 3. (a) Comparison of CO and 2.4 μm longitude profiles toward the inner Galaxy, averaged over approximately 1° of Galactic latitude. The 2.4 μm profile is from the IRT survey (Kent et al. 1992); units are 0.77×10^{-11} W cm^{-2} sr^{-1} μ^{-1}. The CO profile is from the survey of Dame et al. (1987), integrated over all velocities and smoothed to an angular resolution of 1° to match that of the IRT survey; units are K km s^{-1}. (b) 2.4 μm profile "flipped" to demonstrate the very tight anticorrelation of the two profiles. The slowly varying background level evident in (a) was removed by fitting 4-th order polynomials.

The ability to see GMCs in silhouette at near-infrared wavelengths provides a new method of resolving the kinematic distance ambiguity for these objects. Since 2.4 μm emission is thought to arise mainly from late-type K and M giants (Jones et al. 1981) whose number density drops sharply with Galactic radius but is otherwise relatively smooth, the emission provides an ideal background against which to study GMCs. In most cases, the extent to which a GMC is visible as a dark nebula at 2.4 μm will be radically different depending on whether it lies at the near kinematic distance, in front of most of the disk and bulge stars, or at the far distance and behind.

An intriguing question raised by the comparison in Figure 3 is why the anticorrelation is so tight, since it is only the CO emission from near-side GMCs that should be anticorrelated with the 2.4 μm emission; those on the far-side should add considerable "noise" by contributing CO emission but little 2.4 μm absorption. At least part of the answer is simply that far-side clouds, by virtue of their greater distance, would be expected to contribute only about one-quarter of the emission in a longitude profile integrated over 1° of Galactic latitude. Still, it seems on first sight that the anticorrelation is significantly better than would be expected from a random distribution of clouds, and suggestive of a distribution with relatively few large clouds on the far side in the inner first quadrant.

Just such an asymmetry is evident in the distribution of GMCs shown in Figure 2a. It is very reassuring to see that the only large complex in Figure 2a that lies in the far side of the molecular ring is evidenced in Figure 3 at $l = 23°$ as the only notable breakdown in the anticorrelation! In this direction the amount of near-infrared absorption is small relative to the amount of CO emission — the signature of a large far-side complex. This complex was identified by both Dame et al. (1986; cloud 23,78F) and Solomon et al. (1987; cloud 97) as the most massive in the inner first quadrant, and both groups assigned it to the far kinematic distance. Thus the CO-near infrared comparison in Figure 3 supports the assignment of this large complex to the far distance, and more generally supports the assignment of all the other large complexes to the near distance. The rather abrupt disappearance of the anticorrelation at $l > 40°$ (not shown) also supports the cloud distribution shown in Figure 2, because in that region the CO emission is dominated by GMCs in the far side of the Sagittarius arm.

Even though the evidence for a gross asymmetry in the inner first quadrant must still be considered weak, given the difficulties of identifying GMCs in that region and the relatively low angular resolution of the IRT survey, it is worth considering briefly how such a distribution might come about. As Figure 2b shows, the cluster of near-side clouds lies just off the end of the Galactic center bar proposed by Blitz & Spergel (1991), in the so-called Scutum and 4 kpc spiral arms. In 21 cm (Shane 1972) and CO (Cohen et al. 1980) longitude-velocity diagrams, these arms seem to disappear just past their tangent regions as they spiral into the region of low gas density at R < 3 kpc (see, e.g., the insert to Fig. 2 of Cohen et al.). Also overlaid in Figure 3b is the image of a galaxy with a bar similar to the one proposed by Blitz & Spergel, to demonstrate how a bar might cause the type of asymmetry suggested by the GMC distribution and the CO-near infrared anticorrelation.

As discussed elsewhere in these proceedings (Hauser), the near-infrared survey data from COBE has already provided some direct evidence in support of a bar at the Galactic center, and these data can also be used to test for the sort of spiral pattern that might be driven by a bar. In particular, if the pattern bears any resemblance to that shown in Figure 2b, with the main "molecular ring" arm spiraling closer to the Sun in the fourth quadrant, the CO-near infrared anticorrelation there should both extend further from the Galactic center in longitude than in the first quadrant, and break down more rapidly with increasing angular resolution. The COBE data provides the full longitude coverage necessary

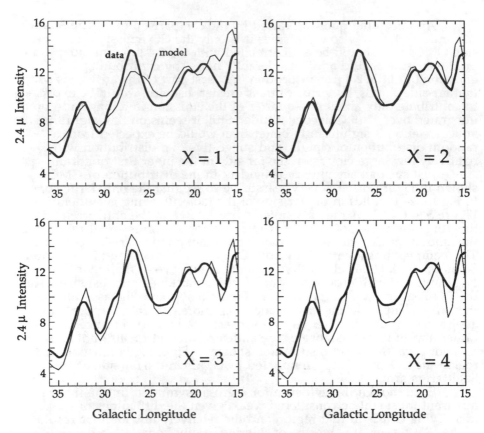

Figure 4. A demonstration of the sensitivity of the Galactic near-infrared model of Kent, Dame, & Fazio (1991) to the CO-to-mass conversion factor X. The model assumes that a smooth, axisymmetric background of near-infrared light is absorbed by dust in atomic and molecular clouds. In deriving the gas distribution, all CO emission more than 15 km s^{-1} from the terminal velocity was assigned to the near kinematic distance, except that here cloud 23,78F from Dame et al. (1986) was placed at the far distance (see § 4). The heavy line is the 2.4 μm intensity from the IRT survey averaged over the inner 1° of Galactic latitude, in units of 5.4 x 10^{-11} W cm^{-2} sr^{-1} μ^{-1}. Note that the intensities are not plotted from zero.

for testing these ideas, as well as a modest but crucial increase in angular resolution over that of IRT, from 1° to 0.7°, which will allow resolution of the distance ambiguity for many more individual GMCs in the inner Galaxy.

Finally, Figure 3 leaves little doubt that the 2.4 μm intensity along the plane is strongly absorbed by molecular clouds, but can the intensity variations be accounted for quantitatively? Using a simple model in which a smooth background of near-infrared light is absorbed by dust in atomic and molecular clouds, Kent, Dame, & Fazio (1991) were able to reproduce in considerable detail the near-infrared intensity variations observed by IRT in the first quadrant. For the reasons just discussed, this model assigned all CO emission displaced from the terminal velocity to the near kinematic distance, and adopted the gamma-ray calibration of X (2.3; Strong et al. 1988). As Figure 4 shows, the model is quite sensitive to the adopted X ratio, and strongly supports a value of ~2 in the inner Galaxy.

REFERENCES

Bertsch, D. L., Dame, T. M., Fichtel, C. E., Hunter, S. D., Sreekumar, P., Stacy, J. G., & Thaddeus, P. 1993, ApJ, in press
Blitz, L., & Spergel, D. N. 1991, ApJ, 379, 631
Bronfman, L., Cohen, R. S., Alvarez, H., May, J., & Thaddeus, P. 1988, ApJ, 324, 248
Bronfman, L., Dame, T. M., May, J., Nyman, L.-Å, & Thaddeus, P. 1993, ApJ, in prep.
Burton, W. B. 1973, PASP, 85, 679
Burton, W. B. 1976, ARA&A, 14, 275
Burton, W. B., & Gordon, M. A. 1978, A&A, 63, 7
Casoli, F. 1991, in Dynamics of Galaxies and Their Molecular Cloud Distributions, ed. F. Combes & F. Casoli (Dordrecht: Kluwer), p. 51
Cohen, R. S., Cong, H., Dame, T. M., & Thaddeus, P. 1980, ApJ, 239, L53
Combes, F. 1991, ARA&A, 29, 195
Dame, T. M. 1983, Ph. D. thesis, Columbia Univ.
Dame, T. M., Elmegreen, B. G., Cohen, R. S., & Thaddeus, P. 1986, ApJ, 305, 892
Dame, T. M. et al. 1987, ApJ, 322, 706
Dickey, J. M., & Lockman, F. J. 1990, ARA&A, 28, 215
Dickman, R. L. 1978, ApJS, 37, 407
Digel, S., Bally, J., & Thaddeus, P. 1990, ApJ, 357, L29
Digel, S. 1991, Ph. D. thesis, Harvard Univ.
Elmegreen, B. G. 1989, ApJ, 344, 306
Frerking, M. A., Langer, W. D., & Wilson, R. W. 1982, ApJ, 262, 590
Grabelsky, D. A., Cohen, R. S., Bronfman, L., & Thaddeus, P. 1987, ApJ, 315, 122
Grabelsky, D. A., Cohen, R. S., Bronfman, L., & Thaddeus, P. 1988, ApJ, 331, 181
Henderson, A. P., Jackson, P. D., & Kerr, F. J. 1982, ApJ, 263, 116
Jackson, P. D., & Kellman, S. A. 1974, ApJ, 190, 53
Jones, T. J., Ashley, M., Hyland, A. R., & Ruelas-Mayorga, A. 1981, MNRAS, 197, 413
Kent, S. M., Dame, T. M., & Fazio, G. 1991, ApJ, 378, 131
Kent, S. M., Mink, D., Fazio, G., Koch, D., Melnick, G., Tardiff, A., & Maxson, C. 1992, ApJS, 78, 403
Kulkarni, S., Blitz, L., & Heiles, C. 1982, ApJ, 259, L63
Lacey, C. 1985, private communication
Lebrun, F. et al. 1983, ApJ, 274, 231
Liszt, H. S. 1992, in The Center, Bulge, and Disk of the Milky Way, ed. L. Blitz (Dordrecht: Kluwer), p. 111
Lockman, F. J. 1988, in The Outer Galaxy, ed. L. Blitz & F. J. Lockman (Berlin: Springer), p. 79

Rivolo, A. R., & Solomon, P. M. 1988, in Molecular Clouds in the Milky
    Way and External Galaxies, ed. R. L. Dickman, R. L. Snell, & J. S.
    Young (Berlin: Springer), p. 42
Sanders, D. B., Clemens, D. P., Scoville, N. Z., & Solomon, P. M. 1986,
    ApJS, 60, 1
Sanders, D. B., Solomon, P. M., & Scoville, N. Z. 1984, ApJ, 276, 182 (SSS)
Shane, W. W. 1972, A&A, 16, 118
Solomon, P. M., & Barrett, J. W. 1991, in Dynamics of Galaxies and Their
    Molecular Cloud Distributions, ed. F. Combes & F. Casoli (Dordrecht:
    Kluwer), p. 235
Solomon, P. M., Rivolo, A. R., Barrett, J. W., & Yahil, A. 1987, ApJ, 319, 730
Strong, A. W. et al. 1988, A&A, 207, 1
Thaddeus, P., & Dame, T. M. 1984, Workshop on Star Formation, in
    Occasional Reports of the Royal Observatory, Edinburgh, ed. R.
    Wolstencroft, 13, 15
Wouterloot, J. G. A., Brand, J., Burton, W. B., & Kwee, K. K. 1990, A&A,
    230, 21

THE LEIDEN/DWINGELOO SURVEY OF GALACTIC HI AT $\delta \geq -30°$

Dap Hartmann and W.B. Burton
Leiden Observatory

The Leiden/Dwingeloo survey of galactic HI, which has occupied the 25–meter telescope of the Netherlands Foundation for Research in Astronomy full–time during the past 4 years, is now essentially complete. The new material provides 21–cm coverage of the entire sky at $\delta \geq -30°$ on a $0°.5$ grid, over a velocity range of 1000 km s^{-1} at 1 km s^{-1} resolution. The Leiden/Dwingeloo data improve the 1974 (Heiles & Habing) Berkeley material by an order of magnitude in velocity coverage and in sensitivity; they improve the 1992 (Stark *et al.*) Bell Labs survey by about as much in spatial and kinematic coverage; and they extend the latitude coverage of the 1985 (Burton) Leiden/Green Bank survey of the northern Milky Way.

Observational parameters of the Leiden/Dwingeloo HI survey:

sky coverage	all sky at $\delta \geq -30°$
number of spectra	$\sim 300,000$
telescope beam size	36 arcminutes
Δb spacing	$0°.5$
Δl spacing	$\leq 0°.5 \cos(b)$
number of channels	1024
Δv per channel	1.03 km s^{-1}
effective vel. coverage	$-450 < v_{\text{LSR}} < +400$ km s^{-1}
system temperature	~ 38 K
integration time	≥ 180 seconds
rms sensitivity	~ 0.06 K

Several queries motivated the Leiden/Dwingeloo HI survey:

1. Many IR cirrus features have anomalous–velocity HI counterparts (Deul & Burton, 1990; Burton *et al.* 1992) which populate the realm of IVC's, *i.e.* with velocities differing from the simply rotating, plane–parallel case, by some 40 to 70 km s^{-1}. Furthermore, IVC HI structures are commonly found correlated with normal–velocity material. The topography of the IVC gas and its relation to the normal–velocity HI are being explored. The link of HI at anomalous velocities with infrared cirri (and with higher–density gas at conventional velocities) suggests the importance of exploring the possibility of the HI having been produced by star formation processes, rather than being precursor to them; the IVC gas might not be material in some only weakly processed state (as has been suggested for the IVC gas and as remains an idea current for the HVC gas), but might have been accelerated away from the galactic plane following star formation. The new data also reveal aspects of the HVC distribution which are evidently related

to those of gas at less extreme velocities. The HVC/IVC distinction was originally introduced rather arbitrarily and largely remains so; an effort is being made to establish if this distinction has a physical basis.

2. The Milky Way serves in many ways as prototype of a warped galactic system. The sensitivity and spectral resolution of the new HI data allow an improved quantitative description of the warp parameters. The velocity coverage at high $|b|$ allows separating information from the warped outer Galaxy from that contributed by the local and $R \geq R_o$ gas and dust, and will allow examining the gas–to–dust trends separately for the outer, local, and inner regions.

3. Several regions of exceptionally low total HI column depth have been identified, in addition to the region of low N(HI) found by Lockman $et\ al.$ (1984). Perhaps even more interesting is the discovery of regions in which the N(HI) at conventional velocities (within ~ 20 km s^{-1} of $v_{\rm LSR} = 0$ km s^{-1}) is exceptionally low ($\leq 7 \times 10^{19}$ cm^{-2}), but where substantial densities do occur at anomalous velocities. One of the reasons for interest in regions of low N(HI) centers on the role of HI in obscuring hard radiation. Indications are being pursued that X–ray shadows are cast by HI features in regions of low total N(HI).

4. Measures of the HI areal as well as volume filling factors may be made. The extension of velocity information beyond the coverage and resolution available earlier shows that the new HI data may be analyzed in terms of the areal filling factor; derivation of the volume filling factor is a more difficult matter, although important constraints may be put on this parameter even using the 0°.5 angular resolution of the new data. The new HI data show many structures with velocity–width dispersion on the order of 1 km s^{-1} or less, and which are clearly isolated in velocity: variations in N(HI) at large $|b|$ are evidently not due to gradients in a generally smooth, diffuse distribution. Determinations of filling factors (and, in general, of characteristic kinematic and structural scalelengths) are important to discussions of interstellar energetics, in particular of turbulence and scaleheight maintenance, of radiation penetration, and of global HI optical depth.

5. The gas–to–dust interrelationships among HI column densities, galaxy counts, and reddening of galactic and extragalactic objects were analyzed by Burstein & Heiles (1978) and led to the establishment of correlations predicting galactic reddening. Their work raised important astrophysical questions which could not be verified with the data then at hand: a zero–point offset was found in the relation between reddening and N(HI) ($i.e.$ regions with essentially no reddening showed nevertheless substantial HI intensities), and the HI gas–to–dust ratio was found to vary widely from region to region, for physical reasons not identified. That work was based on Berkeley HI data confined in velocity such that substantial quantities of gas were not accounted for (in particular the crucial contribution from the warped outer layer), and with spectra of sensitivity lower than now possible. The gas/dust/reddening problem will be reanalyzed, exploiting the qualities of the new HI survey and of the more modern dust data. Particular attention will be paid to the evident breakdown of the correlation between HI and dust emissivities, which is tight in the inner Galaxy and locally, but not in the outer Galaxy, possibly because of changing environmental conditions.

6. Much HI gas is marshalled in the shell and supershell objects, and perhaps also in the so–called worms. The motions of these structures reveal
important aspects of the macroscopic energetics of the ISM. The limited
velocity extent of the $|b| > 10°$ Berkeley data of Habing and Heiles (1974)
has been a constraint in this regard. A few known shells have already
been traced in the Leiden/Dwingeloo data to velocities well outside the
range of the earlier data; this suggests a substantial upward revision of the
currently accepted energetics. If verified as general, this conclusion would
be important to interpretation of the IVC class of objects, as well as to
considerations of the energetics. Kinematic unravelling of the IRAS cirri
is important in this regard as well as to understanding the nature of the
IVC objects; there are firm indications that some dust features are spatially correlated with HI tracing the shell structures at highly anomalous
velocities.

The data shown in the accompanying color plates (XIII–XVI) represent
the preliminary reduction of the Leiden/Dwingeloo survey. The observations
displayed have not yet been calibrated in final form: interference spikes and
other birdies still contaminate individual spectra, some individual positions must
be re–observed, a correction for time–varying gain is yet to be applied, and the
intended correction for stray radiation, being applied in collaboration with P.
Kalberla (see Kalberla *et al.*, 1982) remains to be made. The survey is being
extended over the portion of the sky inaccessible from Dwingeloo by Bajaja
and collaborators at the Argentine Institute for Radioastronomy using the IAR
100–foot telescope and observational parameters similar to those reported here.

The survey utilized the 1024–channel Dwingeloo digital autocorrelator
spectrometer (DAS) designed by A.Bos and constructed at the NFRA. The data–
reduction software was largely home–brewed; it incorporated the DrawSpec
package written by H.S. Liszt at the NRAO. The moment maps shown in the
accompanying color plates were produced from PC-based programs written in
Turbo Pascal to generate .GIF files which were converted to color postscript files
using assorted MS–Windows software and then printed on a Tektronics Phaser
printer.

After the finishing touches are applied to the data and the first intended
scientific goals achieved, the Leiden/Dwingeloo galactic HI survey will be published and subsequently made generally available, most likely on CD–ROM in
FITS format. Requests for the data may be made to Dap Hartmann or W.B.
Burton, Leiden Observatory, P.O. Box 9513, 2300 RA Leiden, The Netherlands.

The Radio Observatory in Dwingeloo is operated by the Netherlands Foundation for Research in Astronomy, under contract with the Netherlands Organization for Scientific Research (NWO). Dap Hartmann has been supported by
a graduate fellowship from the NWO.

REFERENCES

Burstein, D., & Heiles, C. 1978, ApJ, 225, 40
Burton, W.B. 1985, A&AS, 62, 365
Burton, W.B., Bania, T.M., Hartmann, Dap, & Tang, Y. 1992, in Proc. CTS
    Workshop No. 1, Evolution of Interstellar Matter and Dynamics of Galax-

ies, ed. J. Palouš, W.B. Burton, & P.O. Lindblad (Cambridge University Press), 25

Deul, E.R., & Burton, W.B. 1990, A&A, 230, 153

Heiles, C., & Habing, H.J., 1974 A&AS, 14, 1

Kalberla, P.M.W., Mebold, U., & Reif, K. 1982, A&A, 106, 190

Lockman, F.J., Jahoda, K., & McCammon, D. 1986, ApJ, 302, 432

Stark, A.A., Gammie, C.F., Wilson, R.W., Bally, J., Linke, R.A., Heiles, C., & Hurwitz, M. 1992, ApJS, 79, 77

2300 MHz CONTINUUM EMISSION FEATURES IN THE SOUTHERN GALAXY

Justin Jonas

Department of Physics & Electronics, Rhodes University,
Grahamstown 6140, South Africa
phjj@ruchem.ru.ac.za, phjj@hippo.ru.ac.za

ABSTRACT

The Rhodes/HartRAO 2300 MHz radio continuum survey of the southern sky is nearly complete. We present maps of the Galactic radio continuum emission after subtraction of the large-scale structure. These residual maps show structures with scale sizes ranging from the resolution of the data (20 arc-min) up to tens of degrees.

1. INTRODUCTION

As rapporteur at the conclusion of this conference, Carl Heiles emphasized the usefulness of all-sky surveys in the understanding of physical processes in the Galaxy. There has been a paucity of such surveys, with the relatively low frequency 408 MHz survey (Haslam *et al.* 1982) being the vanguard of the radio surveys. The aim of the Rhodes/HartRAO survey is to map the 2300 MHz continuum emission visible from the Hartebeesthoek Radio Astronomy Observatory (HartRAO), which is situated near Johannesburg, South Africa. By the end of 1992 we hope to have completed the survey to a northern limit of $+13°$ declination. Some sections of the survey have been published (Jonas, de Jager & Baart 1985, Mountfort *et al.* 1987), but the completed survey will be published next year.

Combined with the 2720 MHz Stockert survey of the northern sky (Reif 1985), these data provide the highest frequency, highest resolution all-sky survey of the celestial radio continuum emission currently available. These surveys will probably maintain this status for some time because of the expense of operating large diameter radio telescopes in Earth orbit, and the difficulty of making high frequency radio observations through the atmosphere.

2. MAPS OF EMISSION FEATURES

Jonas (1986) has used preliminary 2300 MHz data to identify a number of relatively bright loops and spurs with a wide range of scale sizes and morphologies that extend out of the Galactic plane. The survey now covers three quadrants of the Galaxy, and extends to higher latitudes, so we are able to make a more complete search for these extended Galactic sources. Bearing in mind that the 2300 MHz survey has a factor of 2.5 resolution improvement over the 408 MHz survey we expect to see more detail in these structures, which often have a filamentary morphology. The higher frequency of the survey should allow us to detect thermal emission sources more easily.

In order to detect and emphasize these sources it is necessary to subtract some model of the bright ridge of emission associated with the Galactic disk. For the maps presented in this poster contribution I used a simple two-dimensional running-median filter to attenuate low spatial frequency components in the radio brightness images. The filter employed a rectangular mask measuring 20° in Galactic longitude by 4° in latitude, which resulted in structures larger than 10° (longitude) by 2° (latitude) being removed. This ensures the Galactic disk emission is removed at latitudes higher than $|b| = 2°$, and the non-linear nature of the median filter reduces the occurrence of anomalous regions of negative brightness temperature. Printing restrictions for these proceedings allow only two representative grey-scale maps, which appear in Figures 1 and 2.

Figure 1 shows the residual emission from the region in the first Galactic quadrant where the North Polar Spur (NPS) intersects the Galactic plane. The bright ridge of the NPS is clearly visible, and it appears to fracture as it approaches the plane. Sofue and Reich (1979) made a detailed study of the NPS in this area, but their data did not include the fractured component of the spur near Galactic latitude $\ell = 21°$. There is a very suggestive correlation between this break in the NPS and a high velocity HI cloud mapped in the survey by Hulsbosch and Wakker (1988). This cloud, which has a velocity $v > +100$ kms^{-1}, could have decelerated the expanding SNR shell, causing the fracture. A number of other faint loops and spurs of emission are also evident in this image.

Figure 2 is a radio image of the residual emission from an area surrounding the Galactic centre. Again, a complex network of spurs and loops of emission are evident. There is some suggestion of a faint, filamentary continuation of the NPS below the Galactic plane.

3. CONCLUSION

These maps are clear evidence of emission features that extend out of the Galactic plane, but further investigation is required to determine their nature and size. This analysis will be a test for the existence of "worms", "chimneys" and superbubbles that are hypothesized as energy and matter channels into the Galactic halo.

ACKNOWLEDGEMENTS

The author is supported by a Core Programme grant from the Foundation for Research Development.

REFERENCES

Haslam, C. G. T., Salter, C. J., Stoffel, H., & Wilson, W. E. 1982, A&AS, 47, 1
Hulsbosch, A. N. M., & Wakker, B. P. 1988, A&AS, 75, 191
Jonas, J. L., de Jager, G., & Baart, E. E. 1985, A&AS, 62, 105
Jonas, J. L. 1986, MNRAS, 219, 1
Mountfort, P. I., Jonas, J. L., de Jager, G., & Baart, E. E. 1987, MNRAS, 226, 917
Reif, K. 1985, Bull. Inf. CDS, 28, 65
Sofue, Y., & Reich, W. 1979, A&AS, 38, 251

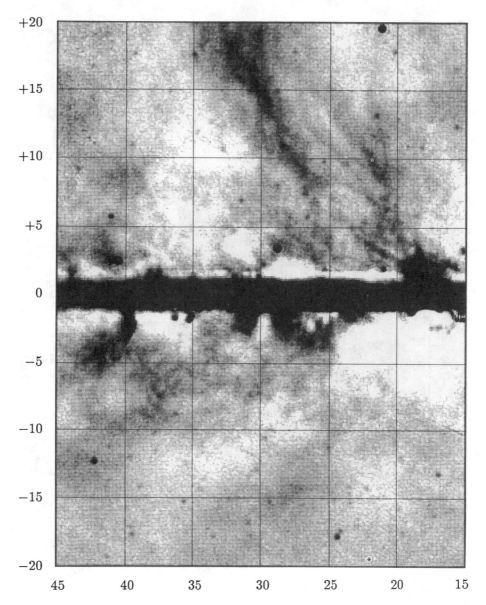

Fig. 1. The residual 2300 MHz continuum emission after high-pass median filtering for the region of sky $45° \geq \ell \geq 15°$, $|b| \leq 20°$. The grey-scale covers a range of 700mK T_b.

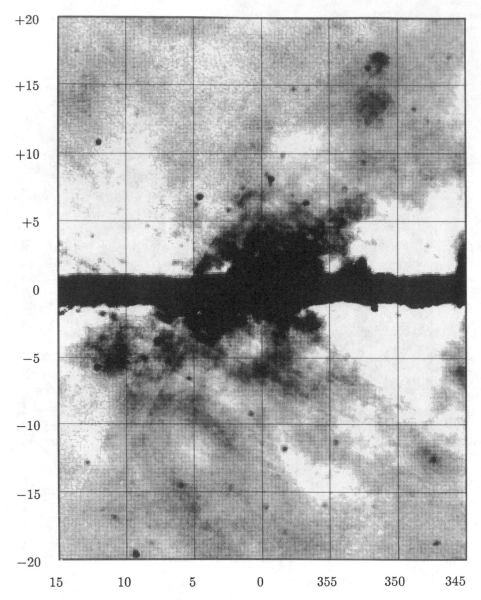

Fig. 2. The residual emission from the Galactic centre region, $15° \geq \ell \geq 345°$, $|b| \leq 20°$. The grey-scale covers a range of 700 mK T_b.

THE I_{CO}/T_{NT} RELATION IN THE GALAXY AND IN OTHER GALAXIES

R. J. Allen
Space Telescope Science Institute
3700 San Martin Drive, Baltimore, MD 21218.

ABSTRACT

The relation between the projected face-on velocity-integrated CO(1-0) brightness I_{CO} and the 20-cm non-thermal radio continuum brightness T_{20} is shown to be virtually constant as a function of radius in the Galactic disk. Averaged in 1 kpc annuli, the ratio I_{CO}/T_{20} has a mean value of 1.51 ± 0.34 km s^{-1} from 2 to 10 kpc. This value is very close to that reported recently for the disks of 8 normal spiral galaxies, where $<I_{CO}/T_{20}> = 1.3 \pm 0.6$ km s^{-1} in spite of the fact that the values of surface brightness in CO and radio continuum in the sample vary by more than a factor of 100.

An explanation of this correlation is given in the framework of the hypothesis that both I_{CO} and T_{NT} have a common origin in the local density of cosmic rays in galaxy disks. The inner disk of M31 is apparently deficient in cosmic rays; the molecular clouds there are likely to be very cold, and may have temperatures close to the cosmic microwave background, rendering them nearly invisible in emission.

1. INTRODUCTION

Adler, Allen, and Lo (1991) have reported a detailed correlation between the CO(1-0) intensity I_{CO} and the 20-cm nonthermal radio continuum brightness T_{20} in the disks of 8 nearby galaxies. The brightnesses refer to values averaged in concentric annular rings about 1 kpc wide. With some small (but probably significant) departures, the ratio of I_{CO}/T_{20} is remarkably constant as a function of radius, and is even numerically *nearly the same value* of 1.3 ± 0.6 km s^{-1} over the sample. The total range of surface brightness covered is more than a factor of 100.

The interpretation offered is that both brightnesses have a common origin in the local density of *cosmic rays*. The dominant source of heating of the diffuse molecular gas far from star-forming regions is thought to be low-energy cosmic ray protons (e.g. Goldsmith and Langer 1978; Black 1987). This heating would therefore control the temperature of the optically-thick CO(1-0) emission, so a higher cosmic ray density leads to brighter CO. The energetic electrons which accompany these protons are responsible for the nonthermal radio emission. The magnetic field is generally monotonic in the density of the cosmic ray component (e.g. by equipartition), so a higher cosmic ray density is generally expected to produce brighter nonthermal radio emission.

2. THE SITUATION IN THE GALAXY

As has been recently discussed in more detail (Allen 1992), the Galaxy also follows this observed I_{CO}/T_{20} correlation in spite of the fact that the data going into the calculation are obtained in a very different way. An axisymmetric model fitted to the all-sky CO(1-0) survey from the Columbia group by Bronfman et.al.(1988) provides the I_{CO} data, and the axisymmetric model fitted by Beuermann et.al.(1985) to the all-sky 408 MHz Galactic background survey of Haslam et.al.(1981) gives the T_{20} information (when suitably translated to 20 cm wavelength). The ratio $I_{CO}/T_{20} = 1.51 \pm 0.34$ km s^{-1} from 2 to 10 kpc in the Galaxy (also averaged in 1 kpc annuli).

The manner in which I_{CO} and T_{20} are obtained for the Galaxy is quite different from the more direct observational determinations possible in nearby galaxies. The fact that the Galaxy also follows the correlation strengthens the generality of the result but, of course, does not necessarily strengthen the interpretation offered in terms of cosmic rays; this latter point will be discussed further below.

3. THE CASE OF M31

Adler et.al. have already shown that, averaged in annular rings 1 kpc thick, the CO and radio continuum brightness data on M31 follow the I_{CO}/T_{NT} correlation quite well. However, the resemblance of the CO distribution to the radio continuum distribution in M31 goes beyond these annular averages and also applies to the maps themselves. As an example of the many radio continuum maps of M31 available in the literature, the upper panel of Figure 1 shows the combined Cambridge-Effelsberg 408 MHz map discussed by Beck and Gräve (1982). The lower panel of Figure 1 shows the CO(1-0) map by Koper et.al. (1991); there is great similarity between the two maps, and the overall correspondence of the CO with the radio continuum is clearly much better than with other components such as the atomic hydrogen or the 100-micron IRAS data (see for example Figure 2 in Koper et.al.).

4. DISCUSSION

Suchkov, Allen, and Heckman (1993) have recently discussed the hypothesis that cosmic rays dominate the energy balance of the molecular ISM both in starburst and in normal galaxies, and have shown that this hypothesis permits a unique determination of the density and temperature of the molecular gas provided the gas pressure and the cosmic ray density are known. The results are in agreement with available data for such diverse cases as M82 and the Galaxy.

The model by Suchkov et.al. leads to an intriguing speculation concerning the physical state of molecular clouds in regions of space where the cosmic ray density is low, of order 1/10 or less of the cosmic ray density in disk of the Galaxy, and where there are no other sources of local heating such as HII regions. The equilibrium kinetic temperatures in these molecular clouds would be close to the cosmic microwave background over a very wide range of density (~ 10 to 10,000 cm^{-3}); these clouds (or parts of clouds) would be virtually invisible in emission, and large amounts of molecular gas could in this way be effectively hidden from view.

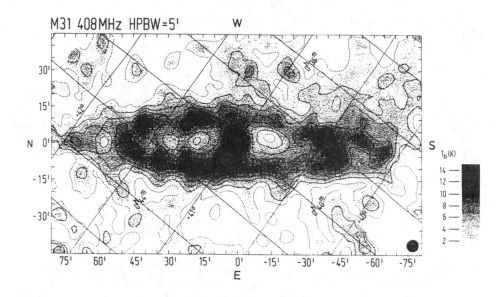

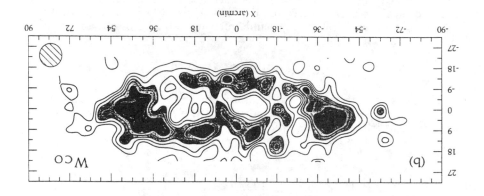

Fig. 1. (Upper panel): M31 at 408 MHz, as analyzed by Beck and Gräve (1982). The first solid contour is about 2.5σ. Copyright ESO; by permission of the Editor, Astronomy and Astrophysics.

(Lower panel): M31 in CO(1-0) emission from Koper et.al. (1991). This panel is intentionally printed upside-down in order to have the same orientation as the upper panel. The first contour is 1σ. Copyright American Astronomical Society, by permission of the author.

The inner disk of M31 (inside the star-forming ring found at 8 - 11 kpc radius) is a place where the cosmic ray density is likely to be considerably lower than in the Galaxy: from the 11 cm data by Berkhuijsen (1977) one can determine that the radio continuum level in the inner disk (radius 1 - 5 kpc, or 5' - 25') is typically 0.08 K at 20 cm averaged over a 1 kpc beam. This is 7 or 8 times fainter than the typical average surface brightness of the Galaxy in this same range of radius (about 0.6 K at 20 cm, from the data in Allen 1992).

Consistent with the picture sketched above is the fact that the inner disk of M31 is indeed nearly devoid of CO emission in the map by Koper et.al., as can be seen in Figure 1. Following the I_{CO}/T_{20} correlation, one can predict that the average CO(1-0) brightness there would be 0.10 ± 0.05 K km s^{-1}. Such faint profiles would be undetectable with current instrumentation unless the emission is clumped on scales less than about 1 kpc. Just such faint CO emission has very recently been detected in several dust clouds in the inner disk of M31 by Allen and Lequeux (1993).

4. CONCLUSIONS

The Galaxy displays the same surface brightness correlation between its CO(1-0) emission and its radio continuum emission as has been seen in the disks of a sample of 8 nearby normal galaxies. Cosmic rays appear to provide the necessary heating to account for this correlation. If molecular clouds exist in regions of space where the cosmic ray density is very low, they would cool rapidly to the temperature of the cosmic microwave background, and would be invisible in emission.

ACKNOWLEDGEMENTS

I thank A. Suchkov and T. Heckman for their suggestions, and all my colleagues at the Space Telescope Science Institute for providing a stimulating atmosphere in which to work.

REFERENCES

Adler, D.S., Allen, R.J., & Lo, K.Y. 1991, ApJ, 382, 475

Allen, R.J. 1992, ApJ, 399, 573

Allen, R.J. & Lequeux, J. 1993, in preparation

Beck, R. and Gräve, R. 1982, A&A, 105, 192

Berkhuijsen, E.M. 1977, A&A, 57, 9

Beuermann, K., Kanbach, G., & Berkhuijsen, E. 1985, A&A, 153, 17

Black, J.H. 1987, in "Interstellar Processes", ed. D.J. Hollenbach and H.A. Thronson (Reidel, Dordrecht), 731

Bronfman, L., Cohen, R.S., Alvarez, H., May, J., & Thaddeus, P. 1988, ApJ, 324, 248

Goldsmith, P.F., & Langer W.D. 1978, ApJ, 222, 881

Haslam, C.G.T., Salter, C.J., Stoffel, H., & Wilson, W.E. 1981, A&ASupp, 47, 1

Koper, E., Dame, T.M., Israel, F.P., and Thaddeus, P. 1991, ApJ, 383, L11

Suchkov, A., Allen, R.J., & Heckman, T.M. 1993, submitted

THE IONIZATION OF DIFFUSE GAS IN GALAXIES:
A Comparison with the Reynolds-layer in the Milky Way

Ralf-Jürgen Dettmar[†]
Radioastronomisches Institut der Universität Bonn, Bonn, FRG
and
Space Telescope Science Institute[‡]
3700 San Martin Dr., Baltimore, MD 21218, USA
e-mail address: dettmar@stsci.edu

ABSTRACT

A thick disk of ionized gas similar to the Reynolds-layer of the Galaxy has recently been detected in several edge-on galaxies. We have measured line ratios of diagnostic emission lines such as [NII]/Hα, [SII]/Hα originating in this extraplanar *diffuse ionized gas* (DIG). In addition, we have compiled from the literature emission line ratios measured for diffuse ionized gas in nearby late-type spiral, early-type disk (S0/Sa), and Magellanic-type irregular galaxies.
From this compilation one trend emerges. For most objects [SII] *and* [NII] lines are both stronger in DIG spectra than in the emission from HII regions. It is only for DIG in Milky Way and irregular galaxies that [SII] lines are strong while [NII] line strengths are comparable to HII regions. The *DIG in the Milky Way is therefore not typical* for physical conditions in the diffuse ionized gas in external galaxies.
The observed spread in excitation conditions allows us to confine possible ionization models. It is concluded that photoionization by a diluted radiation field is most likely, however, additional heating is required.

1. INTRODUCTION

Over the past decade a thick layer of "Diffuse Ionized Gas" (DIG) has been recognized as an important constituent of the Galactic ISM. This so-called *Reynolds-layer* is characterized by a scale height of ~1 kpc and excitation conditions that differ from those in conventional HII regions (Reynolds 1990, 1991, this conference). The favoured model for the ionization of DIG by the diffuse radiation field from OB stars was put forward by Mathis (1986).

Such thick disks of ionized gas have recently been detected in several edge-on galaxies (Rand et al. 1990, Dettmar 1990, Rand et al. 1992, Dettmar 1992) and first spectroscopic observations of diagnostic emission lines are available (Keppel et al. 1991, Dettmar 1992, Dettmar and Schulz 1992).

Here we present emission line ratios for diffuse ionized gas in the disk-halo interface of external galaxies in order to compare them with the observed line ratios for the Reynolds layer. In addition, we have compiled observed emission line ratios of diffuse ionized gas in some nearby galaxies from the literature.

[†] Affiliated with the Astrophysics Division in the Space Science Department of ESA

[‡] Present address

2. OBSERVATIONS

Long-slit spectra of NGC 891 and NGC 4631 were obtained with the Ohio State University CCD spectrograph at the Perkins 1.8 m telescope at Lowell Observatory. The spectra cover a wavelengths range from $\lambda 6150\,\text{Å}-\lambda 6850\,\text{Å}$ with a spectral resolution of $\Delta\lambda = 2\,\text{Å}$. Several spectra parallel to the minor and major axes of the two objects allow us to map out the ionization structure and the excitation conditions in the extraplanar H^+ layer.

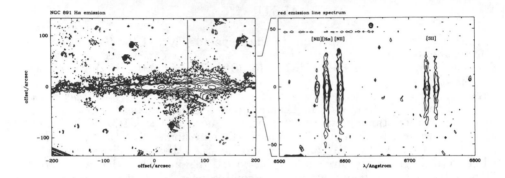

Fig. 1: (a) Distribution of Hα emission in NGC 891. Contour levels correspond to emission measures of $20,40,80,160...\text{cm}^{-6}\text{pc}$. NE is to the left, SW to the right. The slit position for the spectrum presented in the right panel is indicated. **(b)** The observed continuum corrected emission line spectrum near Hα shows the presence of strong [NII] and [SII] lines in the DIG out to 1.5 kpc above the disk plane. The scale along the slit is stretched with respect to (a). The [NII] and [SII] lines become stronger with increasing distance from the mid-plane. Measured line ratios are given in Tab. 1 (from Dettmar and Schulz 1992).

3. RESULTS

In Figures 1 and 2 we present one representative spectrum for each of the two objects. The main findings can be summarized as follows: For NGC 891 the line ratios [NII]/Hα and [SII]/Hα vary with distance from the plane, with the central spectra (near the mid-plane) showing line ratios typical for HII regions. The [NII] and [SII] lines, however, become relatively stronger with increasing distance from the plane. They reach values of [NII]/Hα=1.1 and [SII]/Hα=0.6 at $z=\pm 20''$ (or 900 pc if a distance of 9.5 Mpc is assumed). In NGC 4631, the line ratios at $z=45''$ or 1.5 kpc above the plane change with distance from the centre of the galaxy and reach values of 0.6 for [NII]/Hα and 0.8 for [SII]/Hα.

Values for the measured line ratios and the observed spread in radial or z-direction for NGC 891 and NGC 4631 are given in Tab. 1 in comparison to those for the Reynolds-layer in the Galaxy. In addition, line ratios of DIG in other objects found in the literature are included. As these line ratios are expected to be influenced by variations in metal abundances between the various objects, the line ratios for HII regions within each object are used for reference.

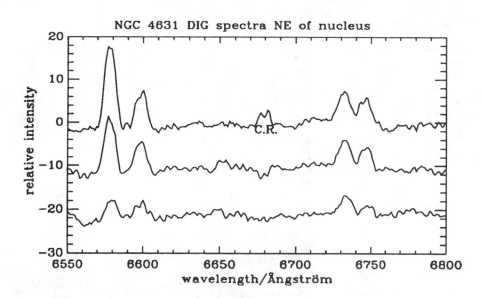

Fig. 2: Spectra in the wavelength range near Hα extracted from a long-slit spectrum parallel to the plane of NGC 4631 in the NE, offset by 45″ (1.5 kpc) from the midplane. If compared to Hα, the emission-line ratios change with position in the sense that [NII] and [SII] lines become stronger compared to Hα with increasing distance (top to bottom) from the nucleus (from Dettmar 1992).

4. DISCUSSION

From the compilation in Tab. 1 one trend emerges. For most objects [SII] *and* [NII] lines both are stronger in DIG spectra if compared to Hα than in the emission from HII regions. This is different from the Reynolds-layer where only the [SII] lines are stronger while [NII] line strengths are comparable to HII regions.

This observed difference in excitation conditions allows us to confine possible ionization models. If we make use of standard diagnostic diagrams (Veilleux and Osterbrok 1987) with the available scarce upper limit for other important lines (e.g. from [OI] and [OIII]) we find that the photoionization models by Mathis (1986) do not allow for the observed emission line ratios. The extragalactic observations rather require either additional heating (Dettmar and Schulz 1992) or a very well tuned photoionization scenario (e.g. Sokolowski 1993). Besides the use of harder photon spectra, additional heating could be provided by various processes such as *cosmic rays* (Dettmar 1992), *photoelectric heating by dust* (Dettmar 1992, Reynolds and Cox 1992), or *magnetic reconnection* (Raymond 1992). A more detailed discussion of the properties of extraplanar DIG and the interstellar disk-halo connection in spiral galaxies can be found in Dettmar (1992).

Acknowledgements. The observations presented in this contribution were obtained at Lowell Observatory, Flagstaff, Arizona. The author's work at the University of Bonn was partly supported by DFG and MPIfR, Bonn.

Table 1: Observed emission line ratios in DIG

		[NII]/Hα	[SII]^a/Hα	[OI]/Hα	[OIII]^b/H$_\beta$
Galaxy	(1)	0.3	0.35	<0.02	<0.2
Orion	(2)	0.17	0.006	-	4.3
NGC 891 reg. 1	(3)	0.6–1.1	0.3–0.6	<0.05	-
NGC 891 reg. 2	(4)	0.4–0.8	0.25–0.35	-	-
NGC 891-HII	(3)	0.3	0.1	<0.05	-
M31	(5)	-	0.29	-	-
M31-HII	(6)	0.09	0.09	~0.02	0.85
NGC 1068-DIM	(7)	1.1	0.23	-	0.3
NGC 1068-HII	(8)	0.49	0.12	<0.05::	0.8
NGC 4594	(9)	~2.5	-	-	-
NGC 4594-HII	(9)	0.3	-	-	-
NGC 4631	(4)	0.4–0.6	0.35–0.8	-	-
NGC 4631-HII	(4)	0.16	0.13	-	-
Irr's	(10,11)	0.15	0.23	<0.05	0.87
Irr's-HII	(10,11)	0.10	0.15	-	2.6
Orion	(11)	0.17	0.006	-	4.3

Notes: ^{a} [SII]λ6717Å,^{b} [OIII]λ5007Å, (1) Reynolds 1990, 1991 (2) Lester et al. 1978 (3) Dettmar and Schulz 1992 (4) Dettmar 1992 (5) Walterbos 1991 (6) Blair et al. 1982 (7) Bland-Hawthorn et al. 1991 (8) Evans and Dopita 1987 (9) Schweizer 1978 (10) Hunter and Gallagher 1990 (11) Hunter 1984

REFERENCES

Blair, W. P., Kirshner, R. P., Chevalier, R. A. 1982, ApJ 254, 50

Bland-Hawthorn, J., Sokolowski, J., Cecil, G. 1991, ApJ 375, 78

Dettmar, R.-J. 1990, A&A 232, L15

Dettmar, R.-J. 1992, Fund. of Cosmic Physics, in press

Dettmar, R.-J., Schulz, H. 1992, A&A 254, L25

Evans, I. N., Dopita, M. A. 1987, ApJ 319, 662

Hunter, D. A. 1984, ApJ 276, L35

Hunter, D. A., Gallagher, J. S. 1990, ApJ 362, 480

Keppel, J. W., Dettmar, R.-J., Gallagher, J. S., Roberts, M. S. 1991, ApJ 374, 507

Lester, D. F., Dinerstein, H. L., Raul, D. M. 1978, ApJ 232, 139

Mathis, J. S. 1986, ApJ 301, 423

Rand, R. J., Kulkarni, S. R., Hester, J. J. 1990, ApJ 352, L1

Rand, R. J., Kulkarni, S. R., Hester, J. J. 1992, ApJ 396, 97

Raymond, J. C. 1992, ApJ 384, 502

Reynolds, R. J. 1990a, in *I.A.U. Symposium No. 139, Galactic and Extragalactic Background Radiation*, eds. S. Bowyer and C. Leinert, Kluwer, p. 157

Reynolds, R, J.: 1991, in *IAU Symp. 144 The interstellar disk-halo-connection in galaxies* ed. H. Bloemen, Kluwer, p. 67

Reynolds, R. J., Cox, D. P. 1992, ApJ 400, L33

Schweizer, F. 1978 ApJ 220, 98

Sokolowski, J. J. 1993, in: *Massive Stars: Their Lives in the Interstellar Medium* eds. J. Cassinelli, E. Churchwell, ASP Conf. Ser., in press

Veilleux, S., Osterbrock, D. E. 1987, ApJS 63, 295

Walterbos, R. A. M. 1991, in *IAU Symp. 144 The interstellar disk-halo-connection in galaxies* ed. H. Bloemen, Kluwer, p. 223

PROBES OF THE WARM IONIZED MEDIUM
IN THE DISK OF THE GALAXY

S. J. Petuchowski & C. L. Bennett
MC 685, NASA Goddard Space Flight Center, Greenbelt, MD 20771.

Email ID
spet@stars.gsfc.nasa.gov

ABSTRACT

Although emission in the [S II] λ6716 forbidden line is commonly assumed to trace ionized gas, we propose state-selective photoionization of sulfur within neutral regions as a mechanism to explain the 1.5 power scaling between optical forbidden lines of N II and S II in M 82 and identify it with the 1.5 power scaling found between the far infrared emission lines of N II and C II associated with large scale structure of the Galaxy. While N II emission can be attributed uniquely to ionized regions, C II emission arises in neutral photodissociation regions as well. Boundaries between ionized and neutral regions are found to contribute significantly to the Galactic luminosity of both species. Forbidden line emission of ionized sulfur, it is suggested, similarly probes neutral photodissociation regions.

1. GALACTIC IMPLICATIONS OF FAR INFRARED LINE RATIOS

The volumetric fraction of warm ($T_e \sim 8000$ K) ionized gas in the Galaxy increases from ~10% in the disk to greater than 40% at a height of 1 kpc (Kulkarni & Heiles 1987). Much remains to be learned about this gas, notably the sources of its ionization far from the disk, its heating, and its distribution throughout the Galaxy. The ionization has been ascribed to far-reaching ultraviolet flux from the young stellar population of the disk, shocks (Sivan *et al.* 1986), Galactic fountain flows (Benjamin & Shapiro 1992), and decaying dark matter (Sciama 1990).

A picture has emerged of two Galactic warm ionized components. The first, comprised of H II regions that have diffused to the point of blending into one another, is characterized by an effective density of ~ 3 cm^{-3} and a scale height of approximately 100 pc (Mezger 1978). The second component extends to a scale height of ~ 900 pc with an effective density of ~ 0.1 cm^{-3} (Reynolds 1992).

The recent 7° all-sky far infrared survey by *COBE*'s Far Infrared Absolute Spectrophotometer (FIRAS) (Wright *et al.* 1991) has added to this picture. For the most recent analysis of FIRAS continuum and spectral line data, see Bennett & Hinshaw (1993). The [N II] 205 μm line is of particular significance to Galactic studies because the 14.5 eV ionization potential of nitrogen exceeds that of hydrogen (13.6 eV), so that emission by singly ionized nitrogen can arise *only* in ionized regions. It therefore traces the Galactic ionized medium and is not subject to dust extinction in the disk.

A number of conclusions are suggested by these recent findings (Petuchowski & Bennett 1993).

1. A ratio of L(122 μm)/L(205 μm) =1.6±0.3 was observed by FIRAS (Wright *et al.* 1991) while the low-density limiting value is 0.7. As a consequence, some component other than a diffuse low-density medium, must contribute significantly to the total Galactic N II luminosity.

2. H II region models indicate that no more than ~ 1% of the luminosity of an illuminating O star will emerge in the N II and C II fine structure lines. By adopting a Galactic total of 2.5×10^4 O stars, derived from Mezger's (1978) census of Galactic H II emission, and allowing that 10% - 20% of these are still embedded in the clouds out of which they were formed and thus surrounded by classical H II regions (Leisawitz & Hauser 1988), we can account for only ~ 1% of the overall luminosity detected in the fine structure lines.

3. We are led to the conclusion that about one half of the observed N II line emission emanates from the boundaries of neutral clouds, *i.e.*, from a morphology characterized by external UV irradiation of a sheath of fully ionized gas, shielding, in turn, a photodissociation region (PDR) of neutral gas in which the [C II] 158 μm radiation arises. The ionized sheath is sufficiently dense (~30 cm^{-3}) for [N II] 122 μm intensity to be enhanced.

One picture consistent with these observations is that of associations of O stars illuminating the interiors of vast superbubbles in the ISM where the concerted occurrence of many type II supernova explosions has blown interstellar material to the fringes of the disk. Such a cavity is filled with the hot (~10^6 K) gas of evaporated clouds within the shell (*cf.* Heiles 1990). However, so long as members of the original association of massive stars continue to burn, and the superbubble remains ionization-bounded, the shell of the cavity must be lined with an ionized sheath protecting a subsequent layer of PDR.

Such a picture might explain a number of observed features such as the extent to which ionizing radiation penetrates above the plane, as has been suggested by Norman (1991). Additional evidence brought to bear by the recent far infrared observations is the 1.5 power scaling of [N II] 205 μm with [C II] 158 μm line intensity discussed by Bennett & Hinshaw (1993) and Petuchowski & Bennett (1993). If each FIRAS pixel were to be characterized by a dominant structure of this sort, then the N II emission, arising from ionized gas swept out of the volume, would scale as the third power of a linear dimension while the C II emission, arising from an extinction-limited neutral shell surrounding the superbubble, would scale as the square of the same characteristic dimension.

2. EVIDENCE IN SPECTRA OF NEARBY GALAXIES

We find further evidence for such a picture in the observed enhancement of S II optical emission, both in 'background' fields in our Galaxy (Reynolds 1988) and away from the midplane of a number of external galaxies. The emission intensity in the 6716 Å forbidden line of ionized sulfur has been termed 'anomalous' in that its ratio to Hα emission exceeds that observed in any identified H II region. This quandary was set out and discussed in detail by Reynolds (1985).

Hunter & Gallagher (1990), in a study of a sample of Magellanic irregular galaxies, observed a similar enhancement within large-scale Hα filamentary structures that they characterize as ionized froth, which lie outside classical H II regions. Dettmar & Schultz (1992) observed the anomaly increasing with height off the plane in NGC 891.

We suggest that the observed enhancement of [S II]/Hα, and, in particular, the 1.5 power scaling of [N II] $\lambda6583$ with [S II] $\lambda6716$, can be understood if S II emission arises within PDRs rather than fully ionized regions as is generally believed. UV penetration of thin layers of denser neutral medium can give rise to S II forbidden line emission due to state-selective photoionization of neutral sulfur within PDR regions surrounding ionized bubbles. In this sense, the S II plays an analogous role to that of the far infrared C II line. The data of Götz *et al.* (1990) are presented in Fig. 1, as a function of displacement along the plane of the nearby starburst galaxy M 82, and plotted logarithmically by 3.''2 pixels $(1'' = 15.2$ pc), and as smoothed to 16''. A 1.5 power scaling law is apparent.

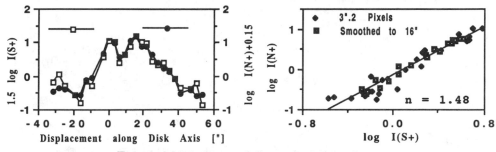

Fig. 1. M 82. Data of Götz *et al.* (1990).

3. STATE-SELECTIVE PHOTOIONIZATION

The partial photoionization cross sections of sulfur to the various product states of S II have been discussed by Altun (1992). We have gone back to readily applicable results of Chapman & Henry (1971), which, while not incorporating resonance effects, qualitatively reflect the relative weights of the photoionization products. By convolving the interstellar radiation field (ISRF) derived by Draine (1978) with the partial cross sections of Chapman & Henry (1971), we find that 24.8% of the product sulfur ions are left in the $^2D^\circ$ states.

An electron collision strength extrapolated from the values calculated by Tayal, Henry, & Nakazaki (1987) imply a collision rate of 6.2×10^{-7} $(100/T(K))^{1/2}$ $(n_e(\mathrm{cm}^{-3}))/1)$ s^{-1}. We assume the temperature-independent rate of 2.2×10^{-9} cm^{-3} s^{-1} (Spitzer 1978) for neutral collisions. The collisional deexcitation is dominated by neutrals since, in a PDR, the electron abundance is, at most, the carbon abundance (Tielens & Hollenbach 1985). The rate of spontaneous emission of 6716 Å photons $(2.60 \times 10^{-4}$ $\mathrm{s}^{-1})$ is comparable to the rate of collisional deexcitation under typical $(n(\mathrm{H\ I}) \sim 10^2$ - 10^5 $\mathrm{cm}^{-3})$ low-density PDR conditions.

Under typical ISRF conditions, every recombined sulfur atom is immediately photoionized, and the [S II] λ 6716 flux is governed by the recombination rate, 1.1×10^{-11} $(T(K)/100)^{-0.593}$ $(n_e(\mathrm{cm}^{-3})/1)$ s^{-1} (Péquignot & Aldrovandi 1986). Since the detected emission arises within an optical depth of the surface, the corresponding column density of S II is 1.6×10^{16} cm^{-2}, assuming a sulfur abundance of 8.5×10^{-6} (Rubin *et al.* 1991) and $N_H/A_v = 1.9 \times 10^{21}$ (Bohlin, Savage, & Drake 1978). The surface brightness corresponding to this column is 4.0×10^3 $(T(K)/100)^{-0.593}$ $(n_e(\mathrm{cm}^{-3}))$ photons cm^{-2} s^{-1} sr^{-1}, or 0.2 Rayleighs at T = 100 K, and $n_{HI} = 10^4$ cm^{-3}.

If we associate [S II] $\lambda6716$ emission with the boundary surface of a spherical ionized volume of radius R, and [N II] $\lambda6583$ emission with the interior volume, then the ratio of luminosities is:

$$\frac{L(N^+ : 6583)}{L(S^+ : 6716)} = \frac{(4\pi/3)R^3 X(N^+)n^2(H^+)\gamma(N^+)}{4\pi R^2(0.15)N(S^+)\gamma_{recomb}f^+(PDR)\ n(PDR)}$$

where $X(N^+)$ is the fractional N II abundance, $\gamma(N^+)$ is the collisional excitation rate of N II, $N(S^+)$ is the S II column density, and $f^+(PDR)$ is the fractional electron density of the PDR,

so that, $R = 1.4\frac{L(N^+)}{L(S^+)}\frac{n(PDR)}{1000cm^{-3}}$ kpc at $T_e = 10^4$ K.

For typical line intensity ratios of unity, the radius implied is of the order of a kiloparsec, comparable to the larger superbubbles evident in H I surveys of the Galaxy (Heiles 1990) and to those identified in external galaxies (Bruhweiler, Fitzurka & Gull 1991; Kamphius, Sancisi & van der Hulst 1991).

REFERENCES

Altun, Z. 1992, JPhysB, 25, 2279

Benjamin, R. A., & Shapiro, P. R. 1992, McDonald Observatory preprint No. 178

Bennett, C. L., & Hinshaw, G. 1993, this volume

Bohlin, R. C., Savage, B. D., & Drake, J. F. 1978, ApJ, 224, 132

Bruhweiler, F. C., Fitzurka, M. A., & Gull, T. R. 1991, ApJ, 370, 551

Chapman, R. D. & Henry, R. J. 1971, ApJ, 168, 169

Dettmar, R.-J., & Schulz, H. 1991, ApJL, 254, L25

Draine, B. T. 1978, ApJS, 36, 595

Götz, M., McKeith, C. D., Downes, D., & Greve, A. 1990, A&A, 240, 52

Heiles, C. 1990, ApJ, 354, 483

Hunter, D. A., & Gallagher, J. S., III 1990, ApJ, 362, 480

Kamphius, J., Sancisi, R., & van der Hulst, T. 1991, A&A, 244, L29

Kulkarni, S. R., & Heiles, C. 1987, in Interstellar Processes, ed. D. J. Hollenbach & H. A. Thronson, Jr. (Dordrecht: Reidel), 87

Leisawitz, D. & Hauser, M. G. 1988, ApJ, 332, 954

Mezger, P. G. 1978, A&A, 70, 565

Norman, C. A. 1991, in IAU Symposium No. 144, The Interstellar Disk-Halo Connection in Galaxies, ed. H. Bloemen (Dordrecht:Kluwer), 336

Péquignot, D., & Aldrovandi, S. M. V. 1986, AA, 161, 169

Petuchowski, S. J. & Bennett, C. L. 1993, ApJ, to be published

Reynolds, R. J. 1985, ApJ, 294, 256

Reynolds, R. J. 1988, ApJ, 333, 341

Reynolds, R. J. 1992, in ASP Symposium, Massive Stars: Their Lives in the ISM, ed. J. Cassinelli & E. Churchwell

Rubin, R. H., Simpson, J. P., Haas, M. R., & Erickson, E. F. 1991, ApJ, 374, 564

Sciama, D. W. 1990, ApJ, 364, 549

Sivan, J.-P., Stasinska, G., & Lequeux, J. 1986, A&A, 158, 279

Spitzer, L., Jr. 1978, Physical Processes in the Interstellar Medium, (NY:Wiley)

Tayal, S. S., Henry, R. J., & Nakazaki, S. 1987, ApJ, 313, 487

Tielens, A. G. G. M. & Hollenbach, D.1985, ApJ, 291, 722

Wright, E. L. et al. 1991, ApJ, 381, 200

OBSERVATION OF [CII] 158μm EMISSION FROM THE DIFFUSE INTERSTELLAR MEDIUM

J. Bock, V. V. Hristov, P. D. Mauskopf, P. L. Richards, and A. E. Lange
Department of Physics, University of California, Berkeley, CA, 94720, USA

M. Kawada, H. Matsuhara, S. Matsuura, M. Tanaka, and T. Matsumoto
Department of Astrophysics, Nagoya University, Nagoya 464, Japan

ABSTRACT

We report the first direct detection of 158μm [CII] line emission from the diffuse interstellar medium at high galactic latitude. We measured the integrated line intensity in a 36' field-of-view along a triangular scan in a 5° x 20° region in Ursa Major using a rocket-borne, liquid helium cooled spectrophotometer. The scans included high latitude infrared cirrus, molecular clouds, a bright external galaxy, M82, and the HI Hole, a region of uniquely low neutral hydrogen column density. Emission from [CII] is observed in all regions, and in the absence of CO emission it is well correlated with neutral hydrogen column density. We deduce a [CII] gas cooling rate of (2.6 ± 0.6) x 10^{-26} ergs s^{-1} H-atom^{-1}, in good agreement with a recent observation of UV absorption lines at high galactic latitude.

I. INTRODUCTION

Fine structure emission from singly ionized carbon ([CII]) at 157.7μm is thought to be the dominant cooling mechanism for the gas component of the diffuse interstellar medium (ISM) over a wide range of temperature and density (Dalgarno and McCray 1972). Observations of [CII] emission can probe the distribution and energetics of many phases of the diffuse ISM, including the cold medium (CM), the warm neutral medium (WNM) and the warm ionized medium (WIM).

Observations of [CII] 158μm line emission using balloon-borne and airborne telescopes have been limited to bright photodissociation regions (PDRs) in the plane of the Galaxy and in bright external galaxies (Shibai et al. 1991, Stacey et al. 1991). Emission from [CII] in the diffuse ISM is too faint to be detected by these techniques; the gas densities and UV photon fields in PDRs are orders of magnitude higher than in the diffuse ISM. The [CII] 158μm intensity from the diffuse ISM has been inferred from the population of the excited $^2P_{3/2}$ state of C$^+$ by ultraviolet absorption measurements (Pottasch et al. 1979, Savage et al. 1992). Recently, the FIRAS experiment aboard COBE produced a full sky map of [CII] 158μm emission in a 7° beam with sensitivity sufficient to detect the line in bright regions near the plane of the Galaxy (Wright et al. 1991). This paper reports the first detection of [CII] 158μm emission from the diffuse ISM at high galactic latitude.

II. INSTRUMENT

A detailed description of the instrument is given by Matsuhara et al. (1992) (also see Matsuhara et al. in these proceedings). The velocity-integrated line flux in a 36' (FWHM) beam is measured using an absolute spectrophotometer at the focus of a 2K, 10cm Cassegrain telescope (Bock et al. 1992). A fixed etalon divides the incident beam into two channels with resolution $\lambda/\Delta\lambda_{FWHM} = 170$: a line channel (LC) at 157.66μm and a continuum channel (CC) at 152.43μm. The detectors are stressed Ge:Ga photoconductors (Haller et al. 1979) cooled to 1.0K by a closed cycle ^{4}He refrigerator and read out with charge integrating amplifiers reset at 2.5s intervals. A cold bi-stable shutter intermittently interrupts the beam at the focal plane to establish the absolute zero flux level. The responsivity of the instrument is monitored in flight by activating a calibration lamp.

For the line-to-continuum ratios we observe, in-band continuum flux contributes on order 50% of the total signal measured in the LC. The measured [CII] line flux, determined by differencing the LC and the CC, depends weakly on the spectral index, s ($I_\nu \propto \nu^s$), of the continuum between 152μm and 158μm. For the sky-averaged ISD spectrum reported by Wright et al. (1991), $I_\nu \propto \nu^{1.5} B_\nu(21.75K)$, we compute s = 0.2. This result is heavily weighted towards the galactic plane. For a colder dust spectrum, $I_\nu \propto \nu^{1.5} B_\nu(15K)$, s = -1.7. In this work we assume s = -1. For the line-to-continuum ratios observed during the flight, the fractional error in the line flux associated with uncertainty in s is small, $\delta I_{[CII]} / I_{[CII]} < 0.13$ (s+1).

III.   OBSERVATIONS

The instrument was launched on an S-520 sounding rocket from Kagoshima Space Center, Japan, at 1:00 JST on Feb. 2, 1992. The telescope was scanned from point A, the HI Hole, to point B2 passing through B1 along the path shown in Fig 1. After acquiring B2, the telescope scanned back to point A for a 120s pointed observation. Both scans were at a constant rate of 36' (1 beam) per second. The data are binned in 45' intervals along the scan. Statistical errors, determined from the pointed observation of the HI hole, are in good agreement with the noise measured in laboratory calibrations. Three regions are distinguished from the rest of the scan: regions in which CO has been detected along the line of sight, M82, and a region with relatively strong (>2x10^{-4} erg s^{-1} cm^{-2} sr^{-1}) continuum intensity but with an anomalously low line-to-continuum ratio.

The neutral hydrogen (N_{HI}) column density in a 36' (FWHM) beam was measured at 18' intervals along the scan path using the 85ft telescope at the Hat Creek Observatory (Heiles 1992) and was binned in the same manner as the infrared data. The Hat Creek data have a minimum value at the HI Hole of N_{HI} = 1.05 x 10^{20} cm^{-2}, which is higher than the level of 5 x 10^{19} cm^{-2} reported by Jahoda et al. (1990) who corrected their data for sidelobe response to bright regions of the Galaxy. Comparing with the Jahoda et al. map we estimate that the level of sidelobe contamination in the Hat Creek data is 6.5 ± 1 x 10^{19} cm^{-2}. Subtracting this level from the Hat Creek data we find agreement between the two maps better than 1.5 x 10^{19} cm^{-2} at every point.

Emission from the J=1 $\rightarrow$ 0 transition of carbon monoxide (CO) is computed from a map obtained by deVries et al. (1987) using the Columbia 1.2m telescope with a 8.7' beam. The data was convolved with the measured [CII] beam before binning.

The observation of M82 provides an in flight calibration check. The result of our observation, assuming the laboratory calibration, gives a line flux of $(2.6 ^{+2.7}_{-0.8})$ x 10^{-10} erg s^{-1} cm^{-2} and a line-to-continuum ratio of $I_{[CII]} / \lambda I_\lambda = (11.2 ^{+4.0}_{-1.6})$ x 10^{-3}. The errors are dominated by the uncertainty in the pointing accuracy (±12') of the telescope as determined from inertial gyroscopes and star sensors. The result is in good agreement with the absolute line flux of (2.8 ± 0.7) x 10^{-10} erg s^{-1} cm^{-2} and the line-to-continuum ratio of (12.9 ± 2.6) x 10^{-3} measured by Stacey (1992).

IV. DISCUSSION

The measured [CII] 158µm flux is well correlated with N_{HI} as shown in Fig 2. Excluding regions with CO emission from the correlation, we obtain a gas cooling rate of $(1.79 \pm 0.45) \times 10^{-26}$ ergs cm^{-2} s^{-1} H-atom^{-1}. Further excluding the region with low line-to-continuum ratio gives a better fit and a gas cooling rate of $(2.56 \pm 0.57) \times 10^{-26}$ ergs cm^{-2} s^{-1} H-atom^{-1}. The observed gas cooling rate agrees well with the rate of 2.1×10^{-26} ergs cm^{-2} s^{-1} H-atom^{-1} reported by Savage et al. (1992) from a measurement of the UV absorption of the excited $^2P_{3/2}$ state of C$^+$ along a line of sight at high galactic latitude. It is much lower than the typical value of 1×10^{-25} ergs cm^{-2} s^{-1} H-atom^{-1} obtained by Pottasch et al. (1979), measured along several lines of sight which sample the local ISM along the plane of the galaxy. Data with appreciable CO emission clearly exceeds the correlation value. We measure a line-to-continuum ratio which varies from $I_{[CII]} / \lambda I_\lambda (\text{cont}) = 2.4 \times 10^{-3}$ to 8.2×10^{-3} in comparison with the all sky average of 8.2×10^{-3} reported by FIRAS, which is heavily weighted towards the Galactic plane.

The correlation with N_{HI} excluding the low line-to-continuum region yields an upper limit on [CII] emission associated with the ionized ISM, $I_{[CII]} < 7 \times 10^{-8}$ erg s^{-1} cm^{-2} sr^{-1} (2σ). The intensity of the Hα line is directly proportional to the emission measure of the WIM. Measurements of Hα by Reynolds (1992) at high galactic latitudes ($45° < b < 60°$) imply that the average [CII] intensity from the WIM should be 3×10^{-7} erg s^{-1} cm^{-2} sr^{-1} assuming a gas temperature of 8000K and a C$^+$ abundance of 3.3×10^{-4}. The level of Hα emission in the HI Hole (Jahoda et al.), however, is anomalously low compared to the average emission at high galactic latitudes and would imply $I_{[CII]} < 3 \times 10^{-8}$ erg s^{-1} cm^{-2} sr^{-1} (2σ), in agreement with the observations. Observations of Hα emission along our scan will allow a more sensitive test for [CII] emission from the WIM.

We thank Carl Heiles for making the 21cm measurements, Jeff Beeman for manufacturing the detectors, and Andreas Heithausen for providing CO data. We acknowledge the help of Gordon Stacey in providing data for M82, H. Okuda and the staff at ISAS for their support at KSC. This work was supported partly by ISAS, the Diako Science Foundation, grant-in-aids from the Ministry of Education, Science and Culture in Japan and by NASA grant NAGW-1352 and GSRP NGT-50771.

Bock, J. et al. 1992, in prep.
Dalgarno, A., and McCray, R. M. 1972, Ann. Rev. Astr. Ap., **10**, 375.
deVries, H. W., Heithausen, A., and Thaddeus, P. 1987, Ap. J., **319**, 723.
Haller, E. E., Hueschen, M. R., and Richards, P. L. 1979, Appl Phys Lett, **34**, 495.
Heiles, C. 1992, priv. comm.
Jahoda, K., Lockman, F. J., and McCammon, D. 1990, Ap. J., **354**, 184.
Kulkarni, S. R. and Heiles, C. 1987 in Interstellar Processes, ed. D. J. Hollenbach
        and H. A. Thronson (Dordrecht : D. Reidel), 87.
Matsuhara, H. 1992, in preparation.
Pottasch, S. R., Wesselius, P. R., and van Duinen, R. J. 1979, Astr. Ap., **74**, L15.
Reynolds, R. 1992, Ap. J. Letters, **392,** L35.
Savage, B. D., Lu, L., Weyman, R. J., Morris, S. L.and Gilliland, R. L. 1992, prept.
Shibai, H. et al. 1991, Ap. J., **374,** 522.
Stacey, G. J., Geis, N., Genzel, R., Lugten, J. B., Poglitsch, A., Sternberg, A., and
        Townes, C. H. 1991, Ap. J., **373,** 423.
Stacey, G. J. 1992, priv. comm.
Wright, E. L. et al. 1991, Ap. J., **381,** 200.

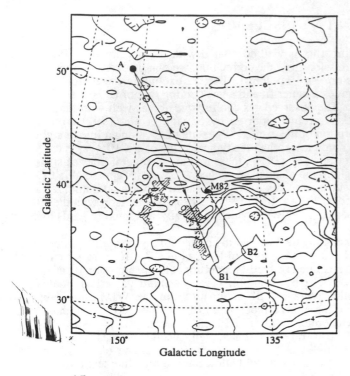

Fig 1: The scan path of the observation. The contours indicate neutral hydrogen column density (Heiles 1975) in units of 1 x 10^{20} cm^{-2}. The shaded contours indicate CO emission from a high latitude molecular cloud complex. The HI map, used for illustration, has not been corrected for sidelobe contamination. The HI data used in the analysis comes from a separate measurement made in 1992. The beam scans over the galaxy M82 (shown as the spiral). The size of the beam is shown as the solid circle at point A, the HI Hole.

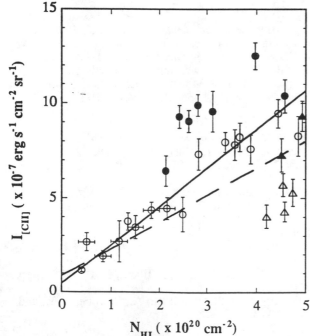

Fig 2: Correlation of [CII] brightness with HI column density. Filled points indicate data with appreciable CO brightness (>0.1 K km s^{-1} averaged over the 36' beam) and triangular points denote a region of low line-to-continuum observed in the scan. The observation of the HI Hole is shown as the point with the lowest [CII] brightness. Points with horizontal error bars represent scanning data binned over intervals of 3.33 x 10^{19} cm^{-2} in N_{HI}. The best fit correlation given by the dashed line excludes points with CO emission. The solid line indicates the correlation if the low line-to-continuum region is also excluded. An offset has been removed from the HI data to account for sidelobe contamination.

FAR-INFRARED [C II] LINE SURVEY OF THE GALAXY

Takao Nakagawa, Yasuo Doi, Kenji Mochizuki, Yukari Y. Yui,
Haruyuki Okuda, Masao Yui, and Hiroshi Shibai
The Institute of Space & Astronautical Science, Sagamihara, Kanagawa, 229, Japan

Tetsuo Nishimura and Frank J. Low
Steward Observatory, University of Arizona, Tucson, AZ 85721

ABSTRACT

We have obtained the first detailed map of the far-infrared [C II] line emission from the Galactic plane using the Balloon-borne Infrared Carbon Explorer (BICE). The map covers wide areas of the Galactic plane ($255° < l < 25°$, $|b| < 4°$) with an effective spatial resolution of 15 arcmin. Many compact sources and bright diffuse components are detected. The scale height of the [C II] emission is similar to those of FIR continuum and mm CO emission. However, the longitude profile of the [C II] emission is quite different from those of other tracers of young populations; the [C II] emission does not show a prominent peak toward the Galactic center region.

We have also obtained a central-velocity distribution map of the [C II] emission, which shows the Galactic differential rotation and the arm structure very clearly.

1. INTRODUCTION

The far-infrared (FIR) [C II] line ($^2P_{3/2} \rightarrow {}^2P_{1/2}$, 157.7409 µm) is thought to be the dominant coolant of interstellar gas (Dalgarno and McCray 1972), and hence observations of this line are essential to determine physical conditions (cooling rate, pressure, etc.) of interstellar clouds and also are very useful for the study of the Galactic structure.

Pioneering observations of the large-scale FIR [C II] emission by Stacey et al. (1983, 1985) and by Shibai et al. (1991) showed that the [C II] emission is very bright and spatially extended. However their observations have very limited spatial coverages. On the other hand, the FIRAS on the COBE satellite made a whole sky map of the [C II] line emission (Wright et al. 1991), but the beam size (7 degree) was too large for detailed study of the Galactic plane.

Here we present preliminary results of balloon-borne observations of the [C II] emission from the Galactic plane. Our observations have much better spatial resolution than the COBE and covers wide areas of the Galactic plane with full sampling.

2. OBSERVATIONS

The observations were carried out with the Balloon-borne Infrared Carbon Explorer (BICE; Nakagawa 1992), which consists of an off-axis telescope and a liquid helium cooled Fabry-Perot spectrometer. The beam size is 12.4 arcmin (FWHM) and the velocity resolution is 175 km s^{-1} (FWHM).

Wide areas of the Galactic plane ($255° < l < 25°$) were observed in two balloon flights. The northern galactic plane ($348° < l < 25°$) was observed in a flight from Palestine, Texas in 1991 and the southern plane ($255° < l < 355°$) was observed in a flight from Alice Springs, Australia in 1992. Observational results were smoothed with a 9 arcmin boxcar, reducing the effective spacial resolution to 15 arcmin (FWHM). The

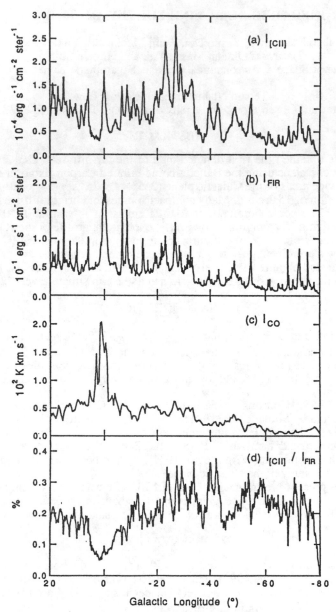

Fig. 2. Longitude profiles of various emission components along the galactic plane averaged over $|b| \leq 2°$: (a) FIR [C II] line intensity (this work), (b) total FIR continuum intensity calculated from IRAS measurements assuming $I_{FIR} = 1.26$ $(\nu f_{\nu}(100\mu m) + \nu f_{\nu}(60\mu m))$ (Cataloged Galaxies in the IRAS Survey 1985), (c) CO emission integrated over all velocities (Dame et al. 1987), and (d) the relative intensity of the [C II] line emission over the FIR continuum.

resulting detection limit is 1×10^{-5} erg cm^{-2} s^{-1} sr^{-1} (3σ).

3. SPATIAL DISTRIBUTION

Fig.1 (Plate #7) shows the spatial distribution of the [C II] intensity. We have detected very strong [C II] emission from most of the regions we observed. We can see many compact sources, most of which are spatially correlated with H II / star-forming regions. Their relative intensities over FIR continuum, however, varies greatly from source to source, which suggests variations of physical conditions.

Fig.1 also shows very strong diffuse components. The scale height of the diffuse [C II] emission is almost the same as those of CO and FIR continuum, and much smaller than that of H I. Hence the [C II] emission has a strong relation with the young population of stars, but the origin of the diffuse components is still controversial.

Moreover, the longitude profile of the [C II] emission (Fig.2*a*) is quite puzzling. Although both the FIR continuum and CO emission (Fig.2*b*, *c*) shows a prominent peak toward the Galactic center region, the [C II] line emission does not show such a peak. This situation is much more clearly illustrated when we make the relative intensity profile of the [C II] emission over FIR continuum (Fig.2*d*); the ratio is almost constant over a wide range of longitudes except for the center region, which shows a clear dip.

The CO emission shows that there is plenty of molecular gas toward the center, and the FIR continuum indicates that the interstellar radiation field is very strong in the center region. Hence we can expect that the [C II] emission should show a prominent peak toward the center, which is totally inconsistent with the observations.

We should mention here that the the [C II] line intensity can be depressed by the self-absorption of the line. However the self-absorption alone is not sufficient to explain the longitude profile, and there must be some other mechanisms depressing the [C II] line emission toward the Galactic center.

The relative intensities of the [C II] line emission over the FIR continuum is 0.2 ~ J.3 % except for the Galactic center region. Typically ~ 0.5 % of the radiation energy is converted to gas heating via photoelectric ejection of electrons from grains (Wolfire, Tielens, & Hollenbach 1990). Hence our observations clearly show that the [C II] line is a dominant coolant of interstellar gas over wide areas along the galactic plane.

4. KINEMATICS

We can determine the central-velocity of the [C II] emission line more accurately than the FWHM resolution of the spectrometer. We fit each measured line profile with a single Lorentzian, assuming a single velocity component at each observed point. Fig.3 shows the results in the form of a longitude vs. central velocity map.

Although our single-component assumption is much too simplified, the velocity profile in Fig.3 clearly shows the patterns expected from the differential Galactic rotation. Fig.3 also shows three jumps (indicated by three arrows) which corresponds to tangential directions of the Galactic spiral arms. Hence we conclude that our [C II] observations are useful also for the study of Galactic kinematics.

In summary, we have obtained the first detailed map of the FIR [C II] line emission along the galactic plane, and the data will serve as one of the basic data sets for the study of the Galactic structure and energy budgets of interstellar clouds.

We thank the staff at the National Scientific Balloon Facility for their excellent flight operations, N. Hiromoto and H. Matsuhara for manufacturing the detectors, and the balloon group at the Institute of Space and Astronautical Science for their support in this project. This work has been supported by the Ministry of Education, Science, and Culture in Japan, and by NSF and by NASA in the United States.

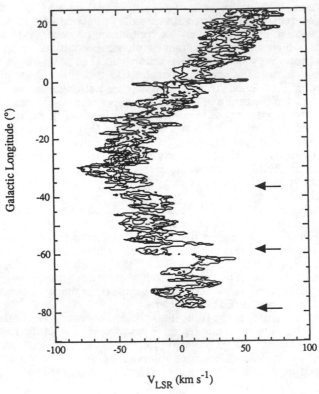

Fig. 3. Longitude vs. central-velocity diagram of the FIR [C II] line emission for $|b| \leq 2°$. Three arrows indicate the tangential directions of the proposed Galactic arms. The central-velocities are determined by Lorentzian fitting, assuming a single velocity component at each observational point.

REFERENCES

Cataloged Galaxies and Quasars Observed in the IRAS Survey, Version 1, 1985, prepared by C.J. Lonsdale, G. Helou, J. Good, and W. Rice (Pasadena: JPL)
Dalgarno, A., & McCray, R.A. 1972, ARA&A, 10, 375
Dame, T.M., et al. 1987, ApJ, 322, 706
Nakagawa, T. 1992, in Astronomical Infrared Spectroscopy: Future Observational Directions, ed. S. Kwok (San Francisco: ASP) in press
Shibai, H. et al. 1991, ApJ, 374, 522
Stacey, G.J., Smyers, S.D., Kurtz, N.T., & Harwit, M. 1983, ApJ, 268, 199
Stacey, G.J., Viscuso, P.J., Fuller, C.E., & Kurtz, N.T. 1985, ApJ, 289, 803
Wolfire, M.G., Tielens, A.G.G.M., & Hollenbach, D. 1990, ApJ, 358, 116
Wright, E.L., et al. 1991, ApJ, 381, 200

CLUMPY STRUCTURE OF INTERSTELLAR CLOUDS IN THE CYGNUS X REGION

Yasuo Doi[1,2], Takao Nakagawa[1], Yukari Y. Yui[1,2], Haruyuki Okuda[1]
Hiroshi Shibai[1], Tetsuo Nishimura[3], and Frank J. Low[3]

[1]The Institute of Space and Astronautical Science,
Sagamihara-shi, Kanagawa 229, JAPAN
[2]Department of Astronomy, Univ. of Tokyo, Bunkyo-ku, Tokyo 113, JAPAN
[3]Steward Observatory, Univ. of Arizona, Tucson, AZ 85721

ABSTRACT

The first detailed survey of the Cygnus X region in the far-infrared [C II] line is presented. Spatial distribution of the [C II] emission correlates well with that of IR continuum, but not with that of CO molecular line emission. The total luminosity of the [C II] line in this region amounts to 0.3% of the total IR luminosity. Under the assumption that atomic gas is the dominant source of the [C II] emission, we find that the total atomic gas mass corresponds to $\sim 30\%$ of the total molecular gas mass. The observed high [C II] cooling efficiency is out of proportion to high dust color temperature, suggesting that the structure of interstellar clouds is better represented by clusters of many optically thin, small cloudlets, rather than a few lumps of large, optically thick clouds.

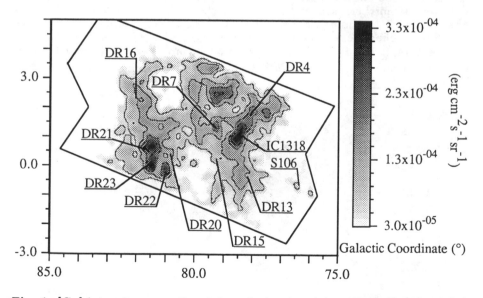

Fig. 1. [C II] intensity map. Spacial resolution is 15′ (FWHM). Contour levels are from 6×10^{-5} to 2.6×10^{-4} erg cm^{-2}s^{-1}sr^{-1} at an interval of 6×10^{-5} erg cm^{-2}s^{-1}sr^{-1}. Main H II regions are also showed.

1. INTRODUCTION

The [C II] 157.7μm $(^2P_{3/2} \to {}^2P_{1/2})$ line is the dominant coolant of warm atomic gas and a good tracer of such a gas phase (Tielens & Hollenbach 1985: hereafter TH). It has been pointed out that an appreciable portion of the gas in star forming regions is in a warm atomic phase (Stacey et al. 1991, Shibai et al. 1991), and comprehensive study of such gas is required. But previous observations have been limited to active star forming regions in dense molecular clouds.

Cygnus X region is a huge complex of various objects, since it falls on the tangential direction of the Local spiral arm. So this region is favorable for the study of general properties of interstellar matter covering a wide range of physical conditions.

Here we show the first detailed, comprehensive survey of the [C II] line in the Cygnus X region and deduce the spatial structure of interstellar atomic gas.

2. OBSERVATIONS AND RESULTS

Details of our telescope system (the Balloon-borne Infrared Carbon Explorer: BICE) are given by Nakagawa (1992). Spatial resolution of BICE is 12.4' (FWHM), which is reduced to 15' (FWHM) in analysis, smoothed with a $\sigma = 5'$ Gaussian. Detection limit of the observation is $\sim 1 \times 10^{-5}$erg cm^{-2}s^{-1}sr^{-1}(3σ).

We detected intense [C II] emission ($I_{[C II]} = 10^{-5}$-10^{-4}erg cm^{-2}s^{-1}sr^{-1}) from most of the observed region (Fig. 1). The [C II] emission is distributed mainly on the northern side of the Galactic plane. This distribution is similar to that of the IR continuum, but not to that of the CO emission (see Fig. 2). In the [C II] line, diffuse components are more prominent than in the IR continuum.

Assuming a distance of 1.5 kpc, we find a total [C II] luminosity $L_{[C II]} = 4.9 \times 10^4 L_\odot$, which corresponds to 0.3 % of the total IR luminosity ($L_{IR} = 1.26 \times$

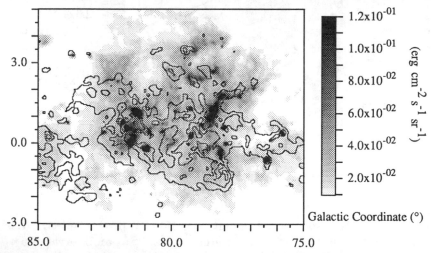

Fig. 2. Gray scale represents IR continuum emission. CO line intensity is shown in contour plot with levels of 6, 12, and 18 K km s^{-1}. CO data are from courtesy of Leung & Thaddeus (1992).

($L_{60} + L_{100}$) found from IRAS 60μm and 100μm observations: Lonsdale *et al.* 1989).

The total atomic gas mass (M_a) can be estimated from the [C II] line intensity, if the [C II] emitting regions map out the warm atomic regions. To estimate the cooling rate of the [C II] line (Wolfire, Tielens, & Hollenbach 1990: hereafter WTH), we assume a gas temperature (T_{gas}) of 100K and assume that the incident UV intensity equals the observed IR intensity. As a result, we get $M_a = 2.8 \times 10^5 M_\odot$. Estimating molecular gas mass (M_m) from the CO observation using a conversion factor $\sigma_{[H_2/I_{CO}]} = 3 \times 10^{+20} \mathrm{cm}^{-2} (\mathrm{K \ km \ s}^{-1})^{-1}$ (Young & Scoville 1991), gives a mass ratio of atomic gas to molecular gas as $M_a/M_m = 33\%$. This result is consistent with previous values derived from observations of the general galactic plane (Shibai *et al.* 1991).

3. CLUMPY STRUCTURE OF INTERSTELLAR CLOUDS

Here we estimate the spatial structure of warm atomic gas using the $I_{[C \ II]}/I_{IR}$ intensity ratio as an indicator of physical conditions.

A C^+ ion is excited collisionally, and hence this excitation state is proportional to T_{gas} and number density (n_0). Atomic gas is heated by electrons which are ejected from dust grains by UV photons (photoelectric heating; TH). Efficiency of photoelectric heating reduces if the dust grains are positively charged due to frequent hitting by UV photons. This implies that the gas to dust temperature ratio decreases as UV flux (G_0) becomes larger. Thus, $I_{[C \ II]}/I_{IR}$ is a function of n_0 and G_0, and large $I_{[C \ II]}/I_{IR}$ means large n_0 and/or small G_0 (WTH).

In the Cygnus X region, we find $I_{[C \ II]}/I_{IR}$ becomes larger at the upper right areas in Fig. 3. So larger n_0 and/or smaller G_0 are required. However, the observed higher color temperature in these regions (T_{col}: $F_{60}/F_{100} \sim 0.4$) is inconsistent with smaller G_0, and it is difficult to expect larger n_0 for these regions because little CO emission is observed.

To explain this discrepancy, we introduce a new model, in which interstellar

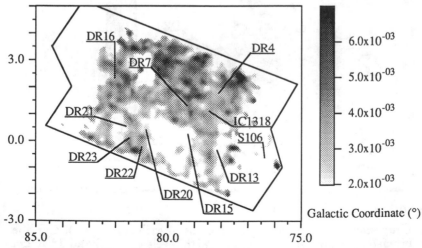

Fig. 3. $I_{[C \ II]}/I_{IR}$ distribution. Spacial resolution is 15' (FWHM).

clouds consist of optically thin, small cloudlets (Cloudlet model; Doi *et al.* in preparation). In the Cloudlet model, the optical thickness of each cloudlet (A_V^{cl}) is an important parameter. If interstellar clouds are clumpy, optically thinner (small A_V^{cl}) clouds should show higher T_{col}, because of easier penetration of UV radiation. Hence, in the Cloudlet model, high T_{col} with small IR intensity (*e.g.* the upper right areas in Fig. 3) is attributed to small A_V^{cl}.

In the Cloudlet model, A_V^{cl} is estimated from T_{col} and I_{IR}, assuming a constant area filling factor (Φ_A). The calculated A_V^{cl} is spatially well correlated with CO intensity (Fig. 4: $\Phi_A = 1$ is assumed). Note that A_V^{cl} is determined only from IRAS 60μm and 100μm observations, and independent of CO data.

As a result of these considerations, we conclude that our observations are well represented by clumpy structure of interstellar clouds.

REFERENCES

Doi Y. *et al.* in preparation
Hollenbach, D. J., Takahashi, T., & Tielens, A. G. G. M. 1991, ApJ, 377, 192
Leung, H. O., & Thaddeus, P. 1992, ApJS, 81, 267
Lonsdale, C. J., Helou, G., Good, J. C., & Rice, W., *Cataloged Galaxies and Quasars Observed in the IRAS Survey, Version 2.* Pasadena: JPL
Nakagawa, T. 1992, in Astronomical Infrared Spectroscopy: Future Observational Directions, ed. S. Kwok (San Francisco: ASP), in press
Shibai, H. *et al.* 1991, ApJ, 374, 522
Stacey, G. J., Geis, N., Genzel, R., Lugten, J. B., Poglitsch, A., Sternberg, A., & Townes, C. H. 1991, ApJ, 373, 423
Tielens, A. G. G. M., & Hollenbach, D. J. 1985, ApJ, 291, 722
Wolfire, M. G., Tielens, A. G. G. M., & Hollenbach, D. 1990, ApJ, 358, 116
Young, J. S., & Scoville, N. 1991, ARA&R, 29, 581

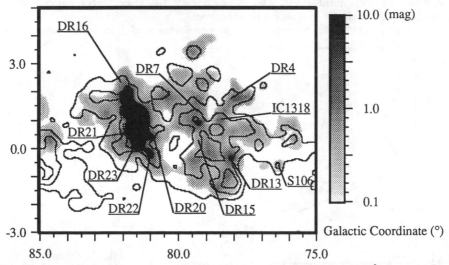

Fig. 4. Gray scale represents optical thickness of each cloud (A_V^{cl}). CO line intensity is shown in contour plot with levels of 6, 12, and 18 K km s^{-1}. Spatial resolution of both data is 15' (FWHM).

A MULTI-TRANSITION CO AND ^{13}CO SURVEY OF THE GALACTIC PLANE

D. B. Sanders
Institute for Astronomy, 2680 Woodlawn Dr., Honolulu, HI 96822.

N. Z. Scoville, R. P. J. Tilanus, Z. Wang
California Institute of Technology, Pasadena, CA 91125.

S. Zhou
Dept. of Astronomy, University of Texas, Austin, TX 78712.

ABSTRACT

A survey of the Galactic Plane in the (2→1) and (3→2) rotational transitions of CO and ^{13}CO has been carried out using the 10m telescope of the Caltech Submillimeter Observatory (CSO). These data, plus previous observations of (1→0) emission from CO and ^{13}CO obtained with at the Five College Radio Astronomy Observatory (FCRAO), have been used to determine the excitation properties of the molecular gas (T_k, $n(H_2)$) throughout the disk of the Milky Way. The bulk of the molecular gas outside the galactic center ($R/R_\odot > 0.05$ kpc) appears to be slightly colder, $T_k = 10\text{-}15$ °K , and denser, $n(H_2) = 10^3\text{-}10^4$ cm^{-3}, than previously assumed. There are no obvious gradients in temperature and cloud H_2 density with galactocentric radius, although the molecular clouds in the galactic center and near the peak of the galactic ring at $R/R_\odot = 0.4\text{-}0.6$ are on average warmer ($T_k \sim 15\text{-}20$ °K) than the molecular material at other radii. The large H_2 volume density derived from the CO data implies that a general internal property of molecular clouds is a low volume filling factor ($\sim 1\text{-}10\%$) for the molecular gas, implying a larger mean-free-path for ionizing photons within molecular clouds than previously assumed. A low volume filling factor would imply that high excitation lines (e.g. C^+) are not simply emitted from the outer cloud surface, but may originate throughout the cloud volume.

1. INTRODUCTION

During the 1970's galactic surveys of CO(1→ 0) emission provided the first accounting of the distribution of molecular gas in the disk of the Milky Way (Scoville & Solomon 1975, Burton et al.1975, Cohen & Thaddeus 1977). A summary of these and more recent, larger CO surveys plus a discussion of the empirical evidence used to convert observed CO integrated intensities into H_2 column densities is given in Scoville & Sanders (1987). Surveys of ^{13}CO provide an important test of the CO-to-H_2 conversion factor, and observations of the higher rotational transitions of both CO and ^{13}CO provide additional information concerning the gas excitation. This report presents preliminary results from a survey of the first galactic quadrant in the (2→1) and (3→2) lines of both CO and ^{13}CO using the 10m CSO telescope.

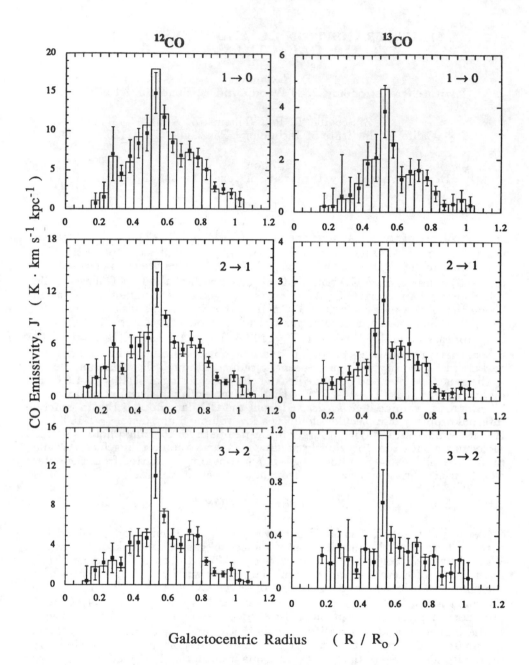

Fig. 1. ^{12}CO and ^{13}CO emissivity at b $= 0°$, J', versus galactocentric radius for the three lowest rotational transitions of CO. Histogram values represent the mean determined from the total CO emission summed over all lines of sight through a given annulus divided by the total pathlength through that annulus. Data points and bars represent the mean of the emissivity values determined for individual lines of sight, and the uncertainty in the mean, $\sqrt{\sigma/N}$, respectively.

2. OBSERVATIONS

A survey of the galactic midplane covering the longitude range 7°- 88° ($\Delta\ell = 1°$) in the (2→1) and (3→2) lines of CO and ^{13}CO was carried out during six separate observing runs during the months of May - August, 1989–91, using facility spectral line receivers of the CSO. System temperatures were typically 500 - 700 °K in the 220-230 GHz band, and 800 - 1500 °K in the 330 - 345 GHz band. Integration times varied from 1-2 min for the CO(2→1) line to 20-40 min for the much weaker ^{13}CO(3→2) line. These CSO data have been combined with data at the same longitudes from a previous survey of CO(1→0) emission (Sanders *et al.* 1985).

3. RESULTS

Figure 1 presents the main results of the new multi-transition survey, showing the radial distribution of CO emissivity, J', in all six lines. The definition and calculation of the emissivity at b = 0°, J', is identical to that defined in Sanders (1981). Extensive (1→0) observations of CO and ^{13}CO for individual clouds throughout the galaxy have in the past been interpreted as indicating that the galactic CO emissivity distribution is a good representation of the distribution of H_2; extensive ℓ, b surveys indicate that the number distribution of molecular clouds mimics the distribution of CO emissivity which can be understood if the majority of molecular clouds have similar internal properties (e.g. $n(H_2)$, T_k), such that the observed CO emission is relatively unaffected by heating and abundance effects (c.f. Scoville & Sanders 1987).

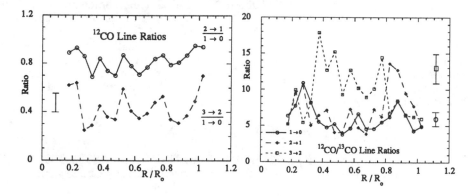

Fig. 2. (left) Mean 12CO line ratios versus galactocentric radius. The error bar represents the typical uncertainty in the mean. (right) Mean ^{12}CO/^{13}CO line ratios versus galactocentric radius. The uncertainties in the mean values for the (3→2) ratio are typically twice as large as for the (2→1) and (1→0) ratios.

Figure 2 shows the radial variation of the CO line ratios and the CO/^{13}CO line ratios with galactocentric radius. There do not appear to be any systematic variations in these ratios with galactocentric radius in the range $R/R_\odot = 0.2$ - 1.0.

4. CONCLUSIONS

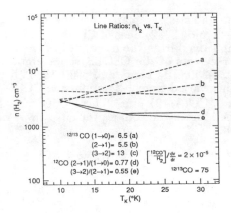

Fig. 3. An LVG analysis of H_2 density vs. gas kinetic temperature corresponding to observed galactic plane mean values for the different CO line ratios (a-e).

Figure 3 shows the results of an analysis of the CO line ratios plotted in Figure 2 using a Large Velocity Gradient Model of radiative transfer, assuming reasonable values for the CO isotope ratio and the quantity CO/H_2/dv/dr. The five curves represent the loci of n(H_2) vs. T_k corresponding to the mean observed ratios from Figure 2. All five line ratios can be understood only if the bulk of the molecular gas in the galaxy is both uniformly cold, $T_k \sim 10°K$, and relatively dense, n(H_2) = 2000–6000 cm^{-3}. This latter, somewhat surprising result, is in contrast to the derived *mean* H_2 densities for molecular clouds of 50–200 cm^{-3} (e.g. Scoville & Sanders 1987), indicating that the volume filling factor of molecular gas in molecular clouds is typically less than 10%. An important implication of such low filling factors would be a longer mean-free-path for ionizing photons than previously assumed, implying that high excitation lines (e.g. C$^+$, N$^+$, CI) may originate throughout the cloud volume rather than being restricted to the outer surfaces of clouds.

ACKNOWLEDGEMENTS

D.B.S. acknowledges patial support from NSF grant AST-8919563. S.Z. was supported in part by the James Clerk Maxwell Telescope Fellowship.

REFERENCES

Burton,W.B., Gordon,M.A., Bania,T.M., & Lockman,F.J. 1975, ApJ, 202, 30
Cohen, R.S., & Thaddeus, P. 1977, ApJ, 239, L53
Sanders, D.B., 1981, Ph. D. Thesis, SUNY at Stony Brook
Sanders,D.B., Clemens, D.P., Scoville, N.Z.,& Solomon, P.M. 1985, ApJS,60,1
Scoville, N.Z., & Sanders, D.B. 1987, in Interstellar Processes, eds. D. J. Hollenbach, & H. A. Thronson (Dordrecht: Reidel) p. 21
Scoville, N.Z., & Solomon, P.M. 1975, ApJ, 199, L105
Solomon, P.M., Scoville, N.Z., & Sanders, D.B. 1979, ApJ, 232, L89

A GRADIENT OF THE CO J=2-1/J=1-0 INTENSITY RATIO OVER THE GALACTIC PLANE

TOSHIHIRO HANDA[1], TETSUO HASEGAWA[1]

MASAHIKO HAYASHI[2], SEIICHI SAKAMOTO[2], TOMOHARU OKA[2]

THOMAS M. DAME[3]

1) Institute of Astronomy, The University of Tokyo, Osawa, Mitaka, Tokyo 181, Japan
2) Department of Astronomy, The University of Tokyo, Hongo, Bunkyo-ku, Tokyo 113, Japan
3) Harvard-Smithsonian Center for Astrophysics, 60 Garden Street, Cambridge, Massachusetts, U.S.A.

ABSTRACT

We report initial results of a survey of the galactic CO J=2-1 line emission being made with the Tokyo-NRO 60-cm survey telescope. From the data taken on the galactic plane, we find a gradient of the intensity ratio between the J=2-1 and J=1-0 transitions of CO that varies as a function of galactocentric radius. The ratio, $R_{2\text{-}1/1\text{-}0}$, is ~0.6 in the solar neighborhood, increases toward the inner region of the galaxy (~0.8 at ~4 kpc from the galactic center) and exceeds 1 near the galactic center. This trend is interpreted in a context of a large scale change of the physical properties of molecular gas in the galactic disk. The higher ratio in the inner galaxy may mean that the molecular gas probed by the CO emission is warmer (T_{gas} ~50 K or higher), denser (n_{H2} ~$10^{3.5}$ cm^{-3} or higher), and less opaque than in the solar neighborhood.

1. INTRODUCTION

Millimeter wave emission lines of carbon monoxide are the best probes of interstellar molecular gas with moderate densities, n_{H2} ~$10^{2\text{-}3}$ cm^{-3}. Large-scale distribution and kinematics of molecular gas in the Galaxy have been revealed by surveys of the CO J=1-0 emission at 115 GHz in the last decade (e.g., Scoville & Solomon 1975; Gordon & Burton 1976; Sanders et al. 1986; Cohen et al. 1986). The physical conditions of the gas have been studied by observations of molecular clouds in the solar vicinity. However, our knowledge about their galaxy-scale variation (if any) is very limited. For external galaxies, on the other hand, multi-transition studies of the CO emission are beginning to reveal the molecular gas properties that varies over a galaxy as well as from a galaxy to another. Interpretation of these results is, however, limited by the spatial resolution of the observations.

We have started a survey of the CO J=2-1 emission at 230 GHz from the galaxy with a dedicated 60-cm telescope. The data reveal the variation of the physical conditions of molecular gas in the galaxy, and also serves as a basis to understand the CO multi-transition observations of other galaxies.

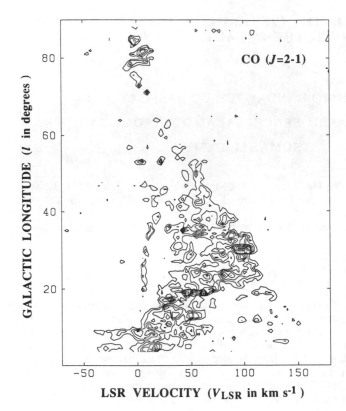

Figure 1
The galactic longitude-
LSR velocity (l-V) plot of
the observed CO J=2-1
emission from the first
quadrant of the galactic
plane. The contours are
drawn at every 0.75 K
from 0.75 K in Tmb, the
antenna temperature
corrected for the main
beam efficiency.

In this paper, we present the initial results of our survey that sampled the first quadrant of the galactic plane. Observations of the galactic center region and the giant molecular clouds in the Orion-Monoceros region are reported separately in this volume (Oka et al. 1992; Sakamoto et al. 1992a).

2.   OBSERVATIONS AND RESULTS

Observations were carried out with the University of Tokyo-Nobeyama Radio Observatory 60-cm survey telescope located at Nobeyama. The telescope has an offset Cassegrain optics and achieves a high beam efficiency of 0.93. The telescope beamsize is 9'±1' at 230 GHz, which is quite close to that of the Columbia 1.2-m telescope at 115 GHz. Because of the matched beamsizes we can make direct comparison of our results with the Columbia survey. The data presented in this paper is a pilot survey sampling the first quadrant of the galactic plane ($b=0°$). We got 85 profiles from $l=4°$ to $88°$ in one degree spacing in l. A more complete description of the data is given in a separate paper (Hasegawa et al. 1992).

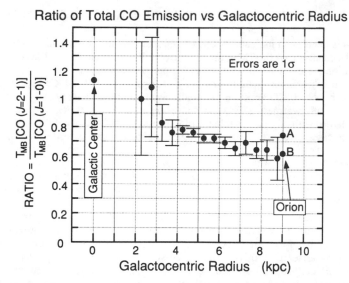

Figure 2 The gradient of $R_{2-1/1-0}$ as a function of galactocentric distance

Figure 1 shows the distribution of the observed CO J=2-1 line intensities in the galactic longitude vs. LSR velocity (l-V) plane. The J=2-1 emission reproduces the characteristic distribution found in the J=1-0 line. We refer to the J=1-0 data of the Columbia survey (Cohen et al. 1986; Leung & Thaddeus 1992; Bitran 1987) and get J=2-1/J=1-0 line intensity ratios, $R_{2-1/1-0}$, for 5 km s^{-1} wide bins after correction for the main beam efficiencies. When we plot them in the l-V plane, we notice that $R_{2-1/1-0}$ scatters within a range 0.5-1.2 from region to region. This scatter is significantly larger than observational errors, and reflects variation of the physical conditions of molecular gas within clouds as we see in the Orion giant molecular clouds (Sakamoto et al. 1992a,b), or their variation between clouds as suggested from an analysis of the J=1-0 line galactic survey (the "warm cloud cores" and "cold cloud cores" classified by Solomon et al. 1985).

We get averaged $R_{2-1/1-0}$ values within a series of annuli 0.5 kpc wide at galactocentric distances 2-9 kpc. Galactocentric distance for a point on the longitude-velocity diagram is estimated with an assumption of circular rotation at a constant velocity Θ =220 km s^{-1} (i.e., a flat rotation curve). Because we add up intensities for all pixels in the l-V plane that fall in an area corresponding to an annulus, accuracy of the resultant averaged ratio is mainly limited by errors in determining the spectral baseline. To minimize the error we weigh all pixels (Δl by ΔV) evenly. On the galactic plane, this sampling underweight areas with severe velocity crowding (e.g., near the subcentral circle and along spiral arms).

Figure 2 shows the average $R_{2-1/1-0}$ values plotted as a function of galactocentric radius. Also shown are ratios derived by integrating over maps of the galactic center molecular cloud complex (NMC, Oka et al. 1992) and Orion A and B giant molecular clouds (Sakamoto et al. 1992a). Fig. 2 reveals, for the first time, a systematic increase of $R_{2-1/1-0}$ toward the inner region of the galaxy. The ratio is ~0.6 in the solar neighborhood and approaches to ~0.8 at ~4 kpc from the galactic center. The molecular gas in the galactic center region exhibit an even higher ratio, which is in accordance with the gradient discovered in the disk.

3.   DISCUSSION: MEANING OF THE $R_{2\text{-}1/1\text{-}0}$ GRADIENT

We interpret this trend as an indication of a galactic scale change of the density, temperature, and the CO fractional abundance of the CO emitting regions. With the aid of the Large Velocity Gradient model calculation, we estimate the density and temperature of the gas from the $R_{2\text{-}1/1\text{-}0}$. Molecular gas in the solar neighborhood is characterized by a temperature of 10-20 K and density of $n_{H2} \sim 10^{2.5}$ cm^{-3}. On the other hand, CO emitting molecular gas in the inner galaxy (~4 kpc from the galactic center) is significantly warmer (>50K) and denser ($n_{H2} > 10^{3.5}$ cm^{-3}). $R_{2\text{-}1/1\text{-}0}$ values larger than ~1 cannot be explained with any reasonable set of temperature and density if the CO lines are optically thick. For these cases, a reduction of the CO fractional abundance is further required as well as the high temprature and density.

As the $J=2$-1 mapping of the Orion clouds has revealed, clouds in the solar neighborhood have low density envelopes which is optically thick in the CO lines (Sakamoto et al. 1992a,b). The envelope show a $R_{2\text{-}1/1\text{-}0}$ value as low as ~0.3, and contribute significantly to the total luminosity of a cloud in the CO emission lines.

Molecular clouds in the inner galaxy may be different from this picture. The high density derived for the CO emitting gas implies that the clouds may have little or no low density envelopes. The denser parts of the clouds are directly exposed to the interstellar radiation field and heated. The CO emission may arise from this warm, dense surface of molecular clouds. In this layer, CO molecules may partially be dissociated, resulting in the lower fractional abundance. The rate of massive star formation in a unit mass of molecular gas is known to increase toward the inner region of the galaxy. This gives a possible mechanism to remove the cloud envelopes in the inner region of the galaxy and form warm and dense dissociation layers on their surfaces. It is also possible that the extended envelopes are stripped off the giant molecular clouds in the inner galaxy due to stronger galactic tidal force.

The survey is being continued to cover the emission within 0.5°-1.0° from the galactic plane. Mapping observations of several important regions are also in progress. These detailed observations will test our present hypothesis outlined above.

This work was financially supported by grants-in-aid from TORAY Science Foundation and those from the Japanese Ministry of Education, Science, and Culture, No. 03249205 and 04233208.

REFERENCES

Bitran, M. E. 1987, PhD thesis, Univ. of Florida
Cohen, R. S., Dame, T. M., & Thaddeus, P. 1986, ApJS, 60, 695
Gordon, M. A., & Burton, W. B. 1976, ApJ, 208, 346
Hasegawa, T., Hayashi, M., Handa, T., Sakamoto, S., Oka, T., & Dame, T. M. 1992, ApJ, to be submitted
Leung, H. O., & Thaddeus, P. 1992, ApJS, 81, 267
Oka, T., Hasegawa, T., Hayashi, M., Handa, T., & Sakamoto, S. 1992, this volume
Sakamoto, S., Hasegawa, T., Hayashi, M., Handa, T., & Oka, T. 1992a, this volume
Sakamoto, S., Hasegawa, T., Hayashi, M., Handa, T., & Oka, T. 1992b, ApJ, to be submitted
Sanders, D. B., Clemens, D., Scoville, N., & Solomon, P. M. 1986, ApJS, 60, 1
Scoville, N. Z., & Solomon, P. M. 1975, ApJ, 199, L105
Solomon, P. M., Sanders, D. B., & Rivolo, A. R. 1985, ApJ, 292, L19

A LARGE AREA CO(J=2–1) MAPPING OF THE ORION GIANT MOLECULAR CLOUDS

Seiichi SAKAMOTO,[1] Tetsuo HASEGAWA,[2] Masahiko HAYASHI,[1] Toshihiro HANDA,[2] and Tomoharu OKA[1]

[1] Department of Astronomy, University of Tokyo, Bunkyo-ku, Tokyo 113, Japan
[2] Institute of Astronomy, University of Tokyo, Mitaka, Tokyo 181, Japan

Email ID
seiichi@sgr.nro.nao.ac.jp

ABSTRACT

A large-area CO(J=2–1) map of the Orion A and B clouds is presented. The J=2–1/J=1–0 intensity ratio of CO varies systematically over the whole extent of these clouds, i.e. the ratio is ~1 in their main ridges and declines to ~0.5 in their peripheries. This variation of the intensity ratio is understood in terms of the variation of the surface gas density of clumps which is $>3\times10^3$ cm^{-3} for those in the ridges and ~1×10^2 cm^{-3} for those in the peripheries. The peripheral regions seen in low-J transitions of ^{12}CO is more surface-filling ($\gtrsim0.7$) than expected.

The J=2–1/J=1–0 luminosity ratio for the Orion A and B clouds is 0.75 and 0.62, respectively. These values are consistent with those observed typically along the Solar circle and are significantly lower than large values often observed in the inner Galaxy and the Galactic center.

1. INTRODUCTION

Large scale mapping observations of the entire extent of GMC complexes play major roles in understanding overall structures and formation mechanisms of molecular clouds. The J=1–0 lines of ^{12}CO and its rarer isotopes have been mainly used to carry out large-scale mapping observations which have revealed clumpy or filamentary nature as well as the spatial extent, mass, and kinematics of molecular gas (e.g. Maddalena et al. 1986). Recent surveys of molecular clouds with higher spatial resolutions showed that clumps or filaments in the regions consist of even smaller subclumps (e.g. Tatematsu et al. 1992), in which star- and planetary system-formation is supposed to be undergoing. We know, however, little about physical conditions of molecular gas and their variation in their parent clouds. Studies of physical conditions of molecular clouds are important to understand the entire structure, formation, evolution, and destruction of the clouds.

2. OBSERVATIONS AND RESULTS

Mapping observations in the ^{12}CO($J=2-1$) emission were made from December 1990 to April 1992. We have taken ~2300 spectra within 35 square degrees on the sky using the Tokyo-NRO 60 cm radio telescope, which has the same angular resolution of 9′ at 230 GHz as that of the Columbia 1.2 m telescopes. The observed area radiates more than 85 % of the CO($J=1-0$) luminosity and covers 74 % of the area observed in CO($J=1-0$) by Maddalena et al. (1986). Achieved noise level is 0.5 K rms with 0.3 MHz frequency resolution.

The spectra were integrated in the LSR velocity range of 0–20 kms^{-1} and were compared with the CO($J=1-0$) data taken with the same angular resolution (Maddalena et al. 1986). As shown in Figure 1, the $J=2-1$ emission exhibits basically similar spatial distribution to that of the $J=1-0$ emission presented in Maddalena et al. (1986), i.e., both maps show two fragments of elongated clouds with their lengths of ~50 pc and widths of ~10 pc, with their strongest emissions occurring at the positions of HII regions NGC 1976 and NGC 2024 for the Orion A and B clouds, respectively.

Although the basic appearance of the two maps are similar, a systematic difference in spatial distribution between the two maps can be recognized when we make a closer comparison of these line intensities. In Figure 2, the $J=2-1/J=1-0$ integrated intensity ratio of the transitions ($=R_{2-1/1-0}$) is shown in a gray-scale representation. The regions with high $R_{2-1/1-0}$ values of around unity or larger are concentrated around two HII regions, reflection nebulosities, and the western edges of the clouds. We can clearly see that $R_{2-1/1-0}$ decreases systematically toward the peripheries of the clouds.

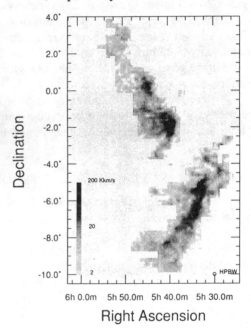

Figure 1._ A gray-scale representation of the CO($J=2-1$) integrated intensity map. The effect of atmospheric attenuation and beam efficiency has been corrected.

3 DISCUSSION

What causes this systematic variation of $R_{2-1/1-0}$? Opaque low-J transitions of ^{12}CO mainly originate from thin dissociative surface of clumps (Gierens et al. 1992). Results of detailed LVG analysis attribute this variation of the intensity ratio to the variation of gas density of clumps rather than that of gas kinetic temperature and chemical abundance. Derived surface gas density is $>3\times10^3$ cm^{-3} for clumps in the ridges and $\sim1\times10^2$ cm^{-3} for clumps in the peripheries.

More careful treatment considering dilution of brightness due to unfilled beam bring interesting conclusion on clumpiness of the clouds. The $R_{2-1/1-0}$ value in the peripheral regions takes a small value of typically 0.5, while the peak J=2–1 intensity is as large as 2 K. If portion f of the beam is filled with clumps, observed brightness temperature of the clumps should be divided by f to derive real brightness temperature. If a surface filling factor in the peripheries is significantly smaller than unity, such a high J=2–1 intensity cannot be explained simultaneously with the observed low intensity ratio. The lower limit of the surface filling factor is ~0.7.

The J=2–1/J=1–0 luminosity ratio integrated over the observed regions of the Orion A and B clouds is 0.77 and 0.66, respectively. Extrapolation to less bright unobserved regions gives 0.75 and 0.62 as plausible luminosity ratios for these clouds. These values are consistent with those observed typically in disks of spiral galaxies (~0.6; Casoli 1991 and references therein) and in the Galactic disk along the Solar circle (Handa et al. 1993), and are significantly lower than large values often observed in active star-forming regions in galaxies (~2; e.g. Eckart et al. 1990) and in the Galactic center (~1.1; Oka et al. 1993). Active star-forming regions in galaxies cannot be explained by an ensemble of Orion-like GMCs.

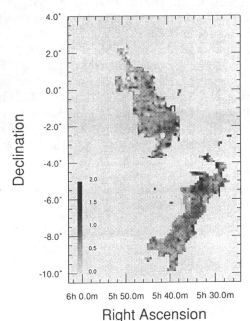

Right Ascension

Figure 2._ A gray-scale representation of the CO(J=2–1)/CO(J=1–0) integrated intensity ratio map. The effect of atmospheric attenuation and beam efficiency has been corrected.

ACKNOWLEDGEMENT

We are deeply indebted to Drs. P. Thaddeus, T. M. Dame and R. J. Maddalena for provision of the CO(J=1–0) data. The radio astronomy group members of the Nagoya University kindly provided us with a graphic display software. S.S. is financially supported by the Japan Society for the Promotion of Science. This research was supported financially in part by the Scientific Research Fund of the Ministry of Education, Science, and Culture, Japan, under Grant No. 01420001 and No. 03249205 and by the Toray Scientific Foundation.

REFERENCES

Casoli, F., 1991, in IAU Symposium 146, Dynamics of Galaxies and Their Molecular Cloud Distributions, ed. F. Combes, & F. Casoli, (Dordrecht : Kluwer), 51

Eckart, A., Downes, D., Genzel, R., Harris, A. I., Jaffe, D. T., & Wild, W., 1990, ApJ, 348, 434

Gierens, K. M., Stutzki, J., & Winnewisser, G., 1992, A&A, 259, 271

Handa, T., Hasegawa, T., Hayashi, M., Sakamoto, S., & Oka, T., 1993, this conference

Maddalena, R. J., Morris, M., Moscowitz, J., & Thaddeus, P., 1986, ApJ, 303, 375

Oka, T., Hasegawa, T., Hayashi, M., Handa, T., & Sakamoto, S., 1993, this conference

Sakamoto, S., Hasegawa, T., Hayashi, M., Handa, T., & Oka, T., 1992, in preparation

Tatematsu, K., Umemoto, T., Kameya, O., Hirano, N., Hasegawa, T., Hayashi, M., Iwata, T., Kaifu, N., Mikami, H., Murata, Y., Nakano, M., Nakano, T., Ohashi, N., Sunada, K., Takaba, H., & Yamamoto, S., 1992, ApJ, submitted

OBSERVATIONS OF EXTENDED CO(7→6) EMISSION IN ORION

J. E. Howe
Department of Astronomy, University of Maryland, College Park, MD 20742

D. T. Jaffe
Department of Astronomy, University of Texas, Austin, TX 78712

E. N. Grossman
NIST, Div. 814.03, 325 Broadway, Boulder, CO 80303

W. F. Wall
NASA Goddard Space Flight Center, Code 685, Greenbelt, MD 20771

J. G. Mangum
Steward Observatory, University of Arizona, Tucson, AZ 85721

G. J. Stacey
Department of Astronomy, Cornell University, Ithaca, NY 14853

ABSTRACT

We mapped 807 GHz CO(7→6) emission in Orion along a strip extending from 0.7 pc west to 1.2 pc east of Ori θ^1C, which lies in front of the Orion A molecular cloud. The similarity of the distribution of the CO(7→6) emission with 158 μm [C II] line emission suggests that over the entire region mapped the narrow CO lines arise in warm photodissociation regions excited by the UV photons from Ori θ^1C. The CO(7→6) lines east of θ^1C split into two velocity components which persist over several arcminutes, indicating that at least one of the components arises from a dense, UV-illuminated clump or filament within the molecular cloud and not at the cloud surface.

1. INTRODUCTION

Recent observations of 158 μm [C II] fine-structure line emission from M17, W3, and NGC 1977 indicate that UV photons from nearby or embedded OB stars can penetrate dense molecular cloud cores over parsec and greater scales, even though the interior regions appear well shielded by large column densities of dust (Stutzki et al. 1988; Matsuhara et al. 1989; Howe et al. 1991). Modeling of the intensity and distribution of the [C II] emission in these sources shows that UV photons penetrate through a relatively transparent interclump medium, excite photodissociation regions (PDRs) on the surfaces of dense clumps or filaments, and produce the extended [C II] emission in PDRs *throughout* the molecular clouds (Stutzki et al. 1988; Howe et al. 1991). Underlying the predominantly atomic region of a PDR where most of the [C II] emission arises is a warm molecular layer which radiates strongly in submillimeter and far-IR CO line emission, primarily from transitions from rotational states J = 6 to 10 (Burton, Hollenbach, & Tielens 1990). The CO(7→6) line is therefore an excellent tracer of the warm molecular component of a PDR.

The extensive distribution of PDRs in the M17, W3, and NGC 1977 molecular clouds suggests that widespread CO(7→6) emission, like [C II], may be common in high-mass star-forming regions. We mapped the CO(7→6) line across a nearly 2 pc region in Orion A to study the extended spatial distribution of high-excitation molecular gas. Nearly all previous observations of CO(7→6) lines in Orion were confined to the Orion KL region or were spatially chopped with an amplitude not much larger than the beamsize, and are therefore not very sensitive to extended emission (Schultz *et al.* 1985; Schmid-Burgk *et al.* 1989). To insure against contamination by emission at the reference position, we position-switched the telescope to a position free of CO(7→6) emission. Our sensitivity to extended low-level line emission is therefore limited only by receiver noise and the opacity of the atmosphere.

2. OBSERVATIONS AND RESULTS

We observed the CO(7→6) line ($\nu = 806.6517$ GHz) using the University of Texas submillimeter laser heterodyne receiver on the 10.4 m telescope of the Caltech Submillimeter Observatory. The receiver uses a liquid-nitrogen cooled Schottky diode as a mixer and an optically pumped submillimeter molecular laser as a local oscillator. T_{Rx} was 3850 K DSB and the zenith atmospheric transmission at 807 GHz was stable at 0.32. The telescope beam efficiency and profile were measured by observing Jupiter and Mars. The 807 GHz beam consists of a Gaussian main beam with a FWHM = 20″ and a broader error beam of FWHM = 90″, with an amplitude ratio of 9.4 to 1. We also observed the J = 2→1 transition of ^{13}CO ($\nu = 220.3987$ GHz) with the CSO 1.3 mm SIS receiver. T_{Rx} was ∼ 210 K DSB and the atmospheric transmission was 0.92 or higher. The beam at 220 GHz is a single Gaussian with FWHM = 33″ and a main beam efficiency $\eta_{MB} = 0.72$.

We observed bright, narrow CO(7→6) lines along a cut in R.A. extending from 300″ west to 525″ east of θ^1C Orionis, the dominant star of the Trapezium cluster. The CO(7→6) and ^{13}CO(2→1) spectra are shown in Figure 1. The vertical scale gives the Rayleigh-Jeans antenna temperature of the CO(7→6) emission corrected for telescope losses and atmospheric attenuation (T_A^*) divided by a 807 GHz beam coupling efficiency factor of 0.4, which is appropriate for sources like Orion with brightness distributions which vary on ∼ 90″ scales.

The CO(7→6) lines are brightest within a few arcminutes of the Trapezium, with intensities ranging from 60 to 140 K and requiring molecular gas temperatures of at least 160 K near the exciting stars. Emission is still detected 525″ (1.2 pc) east and 300″ west of θ^1C, and most likely extends farther in both directions. The line widths decrease from ∼ 7 km s^{-1} at θ^1C to less than 2 km s^{-1} 300″ east of θ^1C. At positions beyond 300″ east of θ^1C, the CO(7→6) line profiles break up into two velocity components with velocity widths ΔV ∼ 1.5 – 2.0 km s^{-1} and line temperatures of 25 – 30 K. Figure 2 compares the ^{13}CO(2→1) and CO(7→6) line profiles for a spectrum synthesized by coadding spectra at 450″ east and 525″ east of θ^1C. While the CO(7→6) velocity components at 6 and 9 km s^{-1} have similar line temperatures (∼ 25 K), the corresponding ^{13}CO(2→1) component line temperatures differ by factor of 3 ($T_{MB} =$ 12 K and 4 K). More complete ^{13}CO(2→1) observations show that the velocity components persist over scales of ∼ 60″ – 120″.

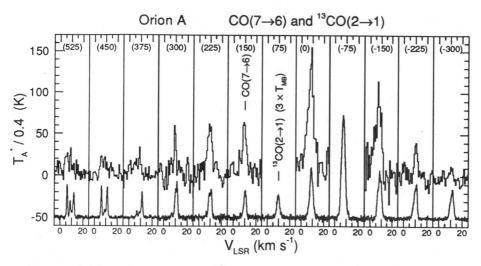

Fig. 1.– CO(7→6) (upper) and ^{13}CO(2→1) (lower) spectra observed along a cut in R.A. in Orion A. The 0″ position is at θ^1C Ori. The R.A. offset from θ^1C in arcseconds is given above each spectrum.

3. DISCUSSION

There is good evidence that the CO(7→6) emission arises from PDRs excited by the far-UV flux from the Trapezium cluster. The local CO(7→6) intensity maximum near θ^1C indicates that the UV photons from the Trapezium stars are the excitation source for the CO(7→6) emission. Near θ^1C, a PDR model successfully reproduces observed far-IR [O I], [C I], and [C II] line intensities (Tielens & Hollenbach 1985). The narrow width of the CO(7→6) lines also makes it unlikely that shocks are important in heating the gas near the Trapezium stars. The large-scale distribution of CO(7→6) emission in Orion, especially east of θ^1C, is similar to the distribution of [C II] emission (Stacey *et al.* 1992) and suggests that the [C II] and CO lines share a common PDR origin across the *entire* region mapped. The CO(7→6) emission in the region ~ 1 pc east of θ^1C probably also arises in PDRs, since [C II] emission from that region indicates the presence of a PDR there. The flux of UV photons in this region is similar to the UV flux in the reflection nebula NGC 2023, where bright CO(7→6) lines are emitted from a PDR (Jaffe *et al.* 1990).

Perhaps the most interesting aspect of the CO(7→6) spectra is the emergence of multiple velocity components in the region 375″ east of θ^1C and beyond. A high-resolution spectrum of the 158 μm [C II] line 4′ east of θ^1C also shows two components, at $V_{LSR} \sim 6$ and 11 km s^{-1} (Boreiko, Betz, & Zmuidzinas 1988). The coexistence of the 6 and 9 km s^{-1} CO(7→6) velocity components over arcminute length scales makes it unlikely that both components are emitted from the surface of the molecular cloud. We interpret the multiple velocity components as emission from two spatially distinct clumps or filaments separated along the line of sight, at least one of which must be embedded in the cloud. The presence of bright [C II] emission from the same region and observations of multiple [C II] line velocity components suggest that the CO(7→6) lines are

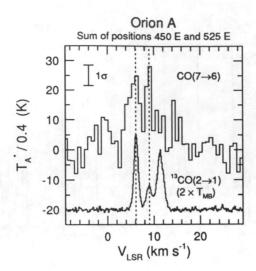

Fig. 2.– CO(7→6) and ^{13}CO(2→1) spectra obtained by coadding positions 450″ and 525″ east of θ^1C Ori. The center velocity of the 6 and 9 km s^{-1} ^{13}CO line components and the 1-σ noise per channel for the CO(7→6) spectrum are indicated.

emitted from PDRs on clump or filament surfaces illuminated by the UV flux from the Trapezium. Like the [C II] lines emitted in M17, W3, and NGC 1977, the CO lines in Orion probably arise in PDRs throughout the interior of the molecular cloud and not only at the surface.

This research was supported by the University of Texas at Austin, NSF grant 88-17544, and a grant from the W. M. Keck Foundation.

REFERENCES

Boreiko, R. T., Betz, A. L., & Zmuidzinas, J. 1988, ApJ, 325 L47
Burton, M. G., Hollenbach, D. J., & Tielens, A. G. G. M. 1990, ApJ, 365, 620
Howe, J. E., Jaffe, D. T., Genzel, R., & Stacey, G. J. 1991, ApJ, 373, 158
Jaffe, D. T., Genzel, R., Harris, A. I., Howe, J. E., Stacey, G. J., & Stutzki, J. 1990, ApJ, 353, 193
Matsuhara, H., et al. 1989, ApJ, 339, L67
Schmid-Burgk, J., et al. 1989, A&A, 215, 150
Schultz, G. V., Durwen, E. J., Röser, H. P., Sherwood, W. A., & Wattenbach, R. 1985, ApJ, 291, L59
Stacey, G. J., Jaffe, D. T., Geis, N., Genzel, R., Harris, A. I., Poglitsch, A., Stutzki, J., & Townes, C. H. 1993, ApJ, 404, 000
Stutzki, J., Stacey, G. J., Genzel, R., Harris, A. I., Jaffe, D. T., & Lugten, J. B. 1988, ApJ, 332, 379
Tielens, A. G. G. M., & Hollenbach, D. 1985, ApJ, 291, 747

Massive Clumps in the W3-Main Cloud Core

J. R. Deane, E. F. Ladd and D. B. Sanders
Institute for Astronomy, University of Hawaii,
2680 Woodlawn Drive, Honolulu, HI 96822

ABSTRACT

We present high-resolution (30″ ; 0.3 pc) ^{12}CO (2-1), ^{13}CO (2-1) and C^{18}O (2-1) observations of the 7′ x 5′ region comprising the W3-Main molecular cloud core. An analysis of velocity channel-maps of ^{13}CO optical depth and column density reveals seven major clumps, ranging in mass from 180 M$_\odot$ to 965 M$_\odot$. The total H$_2$ mass of the mapped region is $\sim$ 6500 M$_\odot$; 58% of the total mass of the mapped region is contained within these seven major clumps, and 26% within just two clumps, one centered on IRS-4 and the other 30″ south of that position. The IRS-4South clump is coincident with a recently discovered peak in submillimeter (350 - 800 μm) continuum emission, but appears not to be associated with previously identified IRS or radio continuum sources. IRS-4South is likely the newest center of massive star formation activity in W3-Main.

1. INTRODUCTION

W3-Main is one of the nearest examples of an extremely massive Giant Molecular Cloud (GMC) core (M $\sim 10^4$ M$_\odot$). It contains a complex of well-studied infrared sources (IRS 1-5) and radio continuum peaks (c.f. Wynn-Williams 1971), presumably sites of embedded massive star formation, whose total luminosity exceeds 10^5 L$_\odot$. Although W3-Main has been extensively observed at millimeter wavelengths, the overall gas kinematics and density structure are still not well understood. Early observations of CO(1→0) were interpreted as a single massive clump in which rotation caused the observed velocity structure (Brackmann and Scoville 1980). Subsequent observations at higher resolution suggested instead two mass concentrations (e.g. Jaffe *et al.* 1983). The identification of velocity components in this core is complicated by substantial self-absorption in the ^{12}CO line by cold, overlying material (Dickel *et al.* 1980). The new observations reported here attempt to minimize the effects of self-absorption by using both the ^{12}CO and ^{13}CO lines to calculate the total column density of the molecular material, and use C^{18}O observations to test the optical thinness assumption for ^{13}CO .

2. OBSERVATIONS

The observations in this paper were made at the 10-m Caltech Submillimeter Observatory (CSO) on the summit of Mauna Kea, HI. The data were taken during observing runs in August and December 1991 using the 220-270 GHz SIS facility receiver (FWHM $\sim$ 30″). Typical system temperatures were $\sim$700 K for the ^{12}CO and ^{13}CO observations and $\sim$900 K for the C^{18}O observations. The C^{18}O data were taken using a raster scanning technique to improve the observation efficiency; all other data were taken in position-switching mode.

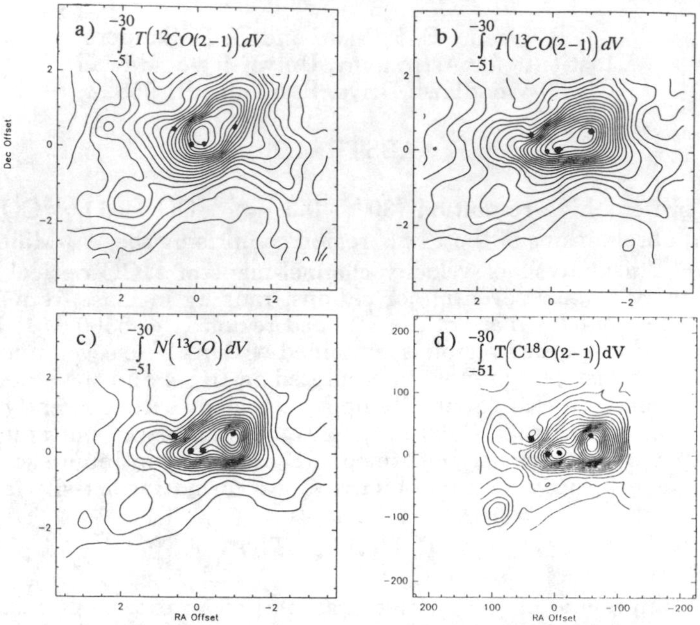

Fig. 1. Maps of the W3-Main core: a,b,c) offsets are arcminutes; d) offsets are arcseconds.

Figures 1a and 1b show integrated brightness temperature maps of the original ^{12}CO and ^{13}CO data. Because the ^{12}CO is optically thick, it is more sensitive to the effects of heating than is the ^{13}CO . The ^{12}CO map peaks near the position of IRS-5 which is known to have a high far-infrared luminosity (Wynn-Williams *et al.* 1972; Werner *et al.* 1980). The maximum in the ^{13}CO map is considerably westward of the IRS-5 position; it is coincident with, or slightly south of, IRS-4.

3. ANALYSIS

In order to remove the effects of heating from the ^{13}CO map, we assumed a solar abundance of ^{12}CO $/^{13}$CO $= 89$, and that T_{ex} was the same for both isotopes at any given position and velocity. The optical depth of the ^{13}CO gas was calculated using the line ratios in each velocity channel, and the excitation temperature of the CO gas was determined from the equation of radiative transfer, and finally the total ^{13}CO column density at each position and velocity was calculated.

The integrated ^{13}CO column density plot is shown in Figure 1c. There is a strong similarity between the ^{13}CO integrated intensity and column density maps which can be understood if the ^{13}CO line is optically thin. However, towards the larger mass concentrations this line is saturated, which accounts for the minor differences in the strength of the peaks in the two maps.

All of our calculations were performed as a function of position *and veloc-*

ity. Using the resulting data cube of calculated column densities, we were able to identify distinct emission maxima which appear to represent the centers of separate clumps. These clumps were traced out to their half-power contours, where their physical parameters (linewidth, mass, etc.) were measured (see Table 1).

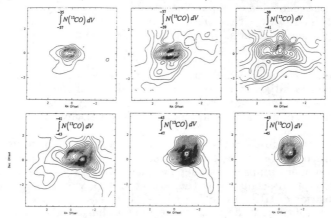

Fig. 2. Channel maps of column density. Offsets are arcminutes.

To confirm that the structure in the calculated column density map is indeed representative of the true structure of the core, and not heavily influenced by either our assumptions or artifacts of our calculations, we observed the same region in the same transition of the $C^{18}O$ molecule. The canonical ^{13}CO /$C^{18}O$ abundance ratio is 5.5, and this value is borne out by the majority of the line ratios in the map, suggesting that the ^{13}CO is indeed optically thin. The structure in the ^{13}CO column density map and the $C^{18}O$ integrated intensity map is nearly identical (allowing for the slightly smaller spatial extent, and lower sensitivity, of the $C^{18}O$ observations). This suggests that our technique of computing the ^{13}CO column density is truly sampling the bulk of the material in the W3-Main core.

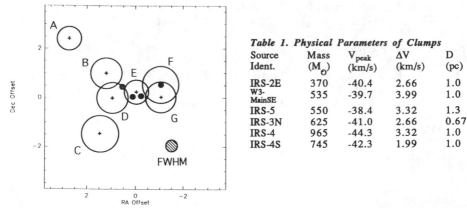

Table 1. Physical Parameters of Clumps

Source Ident.	Mass ($M_\odot$)	V_{peak} (km/s)	ΔV (km/s)	D (pc)
IRS-2E	370	-40.4	2.66	1.0
W3-MainSE	535	-39.7	3.99	1.0
IRS-5	550	-38.4	3.32	1.3
IRS-3N	625	-41.0	2.66	0.67
IRS-4	965	-44.3	3.32	1.0
IRS-4S	745	-42.3	1.99	1.0

Fig. 3. Schematic diagram of clumps in the W3-Main core. Offsets are arcminutes.

4. RESULTS

It is clear that W3-Main is composed of several major mass concentrations rather that being a single centrally-peaked core. Figure 3 shows the positions and sizes of the identified clumps. Nearly 60% of the total mass of the W3-Main core is contained within these seven clumps, with the remaining 40% of the molecular gas being extended over the map. Depending on their star formation efficiencies, each clump is massive enough to form several OB-stars and/or a sub-group of lower mass stars. It is not clear whether the formation of stars in the clumps is coeval or whether massive stars formed in one clump influence star formation in adjacent cores. Given the observed relative velocities of the clumps, it does appear that W3-Main is dynamically bound, i.e. it is neither collapsing nor expanding on the whole. Individual clumps may, however, be undergoing collapse.

Most of the clumps in the W3-Main core are not associated with the currently identified IRS or radio-continuum sources. In the case of IRS-5 and IRS-3, the resolution of our maps cannot separate the two sources within our $30''$ beam. The main exception seems to be the largest clump, which is spatially centered upon the position of IRS-4. IRS-5, IRS-4 and IRS-2 are all prominent sources in submillimeter continuum maps of this region (Richardson, $et\ al.$ 1989). Interestingly, there is also another submillimeter source that is now seen to be coincident with the clump that we identify to the south of the IRS-4 position. Though its mass and size are comparable to the IRS-4 clump, it appears to be somewhat colder and has a CO linewidth only half that of the IRS-4 clump. IRS-4South may be the most recent site of massive star formation in the W3-Main core.

ACKNOWLEDGEMENTS

E. F. L. was supported by the James Clerk Maxwell Telescope Fellowship. D. B. S. was supported in part by NSF grant AST-8919563.

REFERENCES

Brackmann, E., & Scoville, N. 1980, ApJ, 242, 112

Dickel, H. R., Dickel, J. R., Wilson, W. J., & Werner, M. W. 1980, ApJ, 237, 711

Jaffe, D. T., Hildebrand, R. H., Keene, J., & Whitcomb, S. E. 1983, ApJ, 237, L89

Richardson, K. J., Sandell, G., White, G. J., Duncan, W. D., & Krisciunas, K. 1989, A&A, 221, 95

Werner, M. W., Becklin, E. E., Gatley, I., Neugebauer, G., Sellgren, K., Thronson, H. A., Harper, D. A., Loewenstein, R., & Moseley, S. H. 1980, ApJ, 242, 601

Wynn-Williams, C. G. 1971, MNRAS, 151, 397

Wynn-Williams, C. G., Becklin, E. E., & Neugebauer, G. 1972, MNRAS, 160, 1

Balloon Borne Molecular Oxygen Search:
SIS Spectrometer

T.C.Koch[1], P.M.Lubin,[1]

T.B.H.Kuiper[2], M.A.Frerking,[2] K.Chandra,[2]

R.W.Wilson[3]

ABSTRACT

Oxygen is the third most abundant element in our galaxy. The oxygen bearing molecules that have been detected do not account for the expected oxygen abundance in Molecular Clouds. Molecular oxygen (O_2) could be a major reservoir for the missing oxygen. We at UCSB in conjunction with JPL and AT&T have designed, built, and tested a balloon borne radio telescope with the capability to observe O_2 ($118750MHz$) and CO ($115271MHz$). Using an SIS mixer from NRAO[4] and a digital auto-correlator from JPL, the SIS Spectrometer (SISS) has achieved double sideband receiver temperature of $50K$ and a spectral resolution of $1km/s$. Using the 1 meter primary mirror on the UCSB balloon borne gondola, the SISS has a $10'FWHM$ beam.

1. INTRODUCTION

We know that oxygen is the third most abundant element in our galaxy from observations of atomic oxygen in stellar atmospheres and ionized nebulae. Molecular line studies of oxygen bearing molecules, CO, OH, H_2O, etc, cannot account for all the oxygen that is expected in Molecular Clouds. Measurements in highly doppler shifted extra-galactic sources and of the $O^{18}O$ isotope have set limits of $x(O_2) = n(O_2)/n(H_2) \sim 8 \times 10^{-6}$ and $\sim 10^{-5}$ respectively (Lizst 1992). The abundance of H_3O^+, the chief precursor of O_2, implies $x(O_2) \sim 10^{-5} - 10^{-6}$ (Phillips 1992). Using a simple radiative transfer model and an oxygen abundance 10^{-6}, we expect $T_{ant}(O_2) \sim T_{ant}(C^{18}O)$ if the CO abundance is $\sim 4 \times 10^{-5}$, and the isotope ratio in the galactic disk is $n(^{16}O)/n(^{18}O) \sim 675$ (Penzias 1981). Antenna temperatures for our beamsize for $C^{18}O$ are typically 0.1-1K.

2. THE MEASUREMENT

The reason O_2 has not been observed is that the terrestrial O_2 line is opaque and pressure broadened to $> 5GHz\ FWHM$ near the ground. At

---

[1] Department of Physics, University of California, Santa Barbara CA
[2] Jet Propulsion Laboratory, Pasadena CA
[3] Radio Physics Research Department, AT&T Bell Laboratories
[4] A.R.Kerr, S.K.Pan, National Radio Astronomy Observatory, Virginia

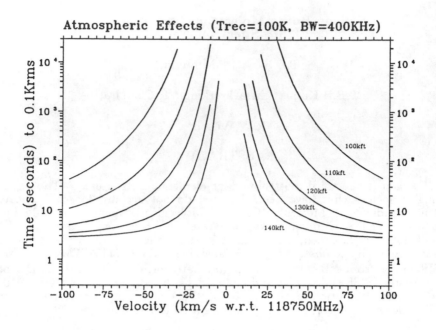

Fig. 1. Integration time to reach $0.1K$rms per channel.

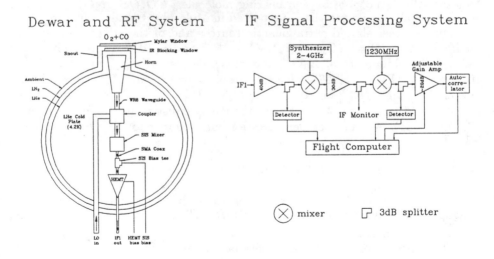

Fig. 2. Schematic Representation of the Dewar and IF Signal Processing.

balloon altitudes the terrestrial O_2 line is much narrower ($\sim 60MHz = 150km/s$), although still opaque at the center. By using the motion of the Earth and choosing sources with high doppler velocities, we can shift the extra-terrestrial lines to reduce the atmospheric emission and attenuation. Figure 1 shows the integration time necessary to reach $0.1K$rms per channel ($400kHz$) outside the atmosphere as function of doppler velocity w.r.t. the center of the terrestrial O_2 line. We have several sources available with total doppler velocities of greater than $50km/s$ that will allow us to fly at $37km$ and reach $0.1K$rms in less than 5 minutes.

3. THE INSTRUMENT

The UCSB 1 meter balloon borne gondola, ACME (Advanced Cosmic Millimeter-wave Explorer), has flown two different Cosmic Microwave Background Radiation (CMBR) detectors on four flights in the last four years. A gyro based inertial guidance system provides pointing information. By using a simple ballscrew for elevation and reaction wheel for azimuth control, this platform has demonstrated arc-minute pointing in flight.

The heart of the SISS is a six junction niobium SIS mixer block from NRAO (Kerr 1990). This brings CO ($115271MHz$) from the lower sideband and O_2 ($118750MHz$) from the upper sideband down to the first IF ($1500\ to\ 2000MHz$) by mixing it with a phase-stabilized Gunn oscillator at $117160MHz$. Mounted on the liquid helium cold plate with the SIS mixer is a low noise, HEMT (High Electron Mobility Transistor) amplifier that boosts the signal by $\sim 30dB$ before it leaves the dewar. Typically, the double sideband noise temperature over the entire first IF is $50K$. Further warm IF signal processing *via* one computer controlled and one fixed local oscillator (Fig. 2) brings a $20MHz$ bandwidth signal down to DC where it is fed into a 52 channel digital auto-correlator built at the Jet Propulsion Laboratory (JPL). This gives us a velocity resolution of $\sim 1km/s$ at $118750MHz$. An on-board 80286-based computer collects and integrates spectra from the auto-correlator as it directs and monitors all other aspects of the experiment: position/frequency chopping, calibration, thermal control, local oscillator lock conditions, *etc.* The spectra and house keeping data are sent down a radio link provided by NSBF (National Scientific Balloon Facility) during the flight as well as saved on board on an $80MB$ hard-disk for later recovery.

4. PRE-FLIGHT TESTS

Beam tests were performed using the a Gunn diode source at 125 meters. With the 1 meter primary mirror on ACME the SISS has a $10' FWHM$ beam. At the same distance, measurements of an LN_2 load show the telescope efficiency to be $\sim 80\%$. We tested system stability by hanging in the lab with the pointing system tracking and the detector looking at the ambient load. We integrated the system temperature of $350K$ DSB and a chopped noise of $1.3K\sqrt{sec}$ per channel, down to theoretical noise levels for more than an hour. The detector and new electronics (without the gondola) have been tested under flight pressure and temperature conditions at JPL. An exaggerated pressure and temperature profile harsher than what is expected in flight was used. One pressure sensitive oscillator was found and a simple pressure box has been constructed.

5. "OBSERVATIONS"

Of course the definitive pre-flight test for our telescope is to use its ability to look at CO in the lower sideband. With the gondola hanging from the awning outside our lab and with system temperature of $\sim 250K$ DSB, dominated by the completely saturated O_2 line which fills the upper sideband, we have been able to measure CO spectra in sources such as NGC7538, DR21, W51, *etc*, within minutes.

6. CONCLUSIONS

In flight we hope to have a system temperature of $100K$ DSB or less and to reach a chopped noise level per channel of less than $10mK$rms on several sources. Because of scheduling conflicts for the use of the ACME gondola, we were unable to fly August 1992. We are building "ACME-II", a nearly duplicate platform, as a backup for our next flight opportunity in June-August 1993. We are refining our observing strategy for this period with $C^{18}O$ maps from the Bell Labs 7m telescope which will allow us to concentrate on lines of sight with the highest column density. JPL is also working on a higher resolution, 128 channel, CMOS auto-correlator with the same $20MHz$ total bandwidth.

7. ACKNOWLEDGMENTS

At JPL, Bill Langer provided help with source selection, Bill Wilson advised on spectrometers and calibration, and Paul Batelaan contructed electronics and supervised the thermal/vacuum test. In Santa Barbara, Mark Lim pilots the gondola and Brad Pendleton helped in mechanical and electronic construction. We are very grateful to Anthony Kerr at the National Radio Astronomy Observatory for loaning us the SIS block and Arthur Lichtenberger of University of Virginia for providing the junctions. This research was performed in part at the Jet Propulsion Laboratory, California Institute of Technology, under contract with the National Aeronautics and Space Administration.

8. REFERENCES

Kerr, A., Pan, S.K. 1990 Inter.J.IR&M.W., 11, 1169
Liszt, H.S. 1992, ApJ, 386, 139
Penzias, A.A. 1981, ApJ, 249, 518
Phillips, T.G., van Dishoeck, E.F., Keene, J. 1992, ApJ, 399, 533

DYNAMICS

THE SHAPE OF THE GALAXY

David N. Spergel
Princeton University Observatory, Princeton, NJ 08544-0010 USA

ABSTRACT

There is growing evidence that our Galaxy is a complex triaxial system. In this article, I will present evidence that the disk of the Galaxy is not axisymmetric and conclude by discussing the implications of its triaxiality for galaxy formation and evolution.

1.Introduction

In this talk, I will review several recent papers that suggest that the Galaxy is not axisymmetric. Since there is not yet a successful synthesis, the goal of this talk is to raise questions rather than to provide a comprehensive synthesis.

I will begin by reviewing how stars and gas respond to non-axisymmetric perturbations. I will then describe how the distribution of the gas in the outer Galaxy suggests that it is distorted and then discuss whether the outer Galaxy has an $m = 1$ or an $m = 2$ distortion. I will consider three different models for the disk of the Galaxy: (1) the disk of the Galaxy is elliptical near the local standard of rest (LSR) and becomes increasingly axisymmetric near $2 R_0$ (Blitz and Spergel 1991); (2) the disk of the Galaxy is axisymmetric near the LSR and is lopsided near $2 R_0$ (Kuijken 1991); (3) the disk of the Galaxy is elliptical with the LSR near the minor axis of the ellipse (Kuijken and Tremaine 1993). I will conclude by discussing the implications of this distortion for our understanding of the formation and evolution of our Galaxy and how new observations can eventually lead to a comprehensive picture of our Galaxy.

In this talk, I will focus on the outer regions of the Galaxy. In particular, I will discuss whether the disk of our Galaxy is lopsided or elliptical. Earlier talks at this meeting (e.g., Blitz 1993, Hauser 1993) describe the growing evidence that the inner regions of the Galaxy are also triaxial.

2.Orbits

Exploring the physical and projected shape of the closed loop orbit in a given potential is a useful place to begin understanding gas and stellar motions. In the disk of the Galaxy, most stars are believed to move along orbits that oscillate around a closed loop orbit. These trajectories are well approximated as epicyclic motion around around a circle, a formulation that Ptolemy would have appreciated. In the disk, the older stars have large amplitude epicycles. Thus, these stars have higher velocity dispersions and asymmetric drifts.

Since gas is collisional, its energy in epicyclic motions is easily dissipated and the gas is expected to settle onto orbits that are very similar to the closed loop orbit. This, however, is only an approximation and is not likely be a very good one near resonances, where the closed loop orbits intersect and shocks are likely to form.

Observations usually measure the position of a star or gas parcel and v_{los}, its line-of-sight velocity relative to the local standard of rest. v_{los} depends both on the motion of the LSR and the motion of the distant star or gas parcel:

$$v_{los} = v_\phi(R,\phi)\sin(\phi+l) - v_\phi(LSR)\sin(l) - v_r(R,\phi)\cos(\phi+l) + v_r(LSR)\cos(l) \tag{1}$$

Here, R is the galactocentric radius and ϕ is the polar angle between the position of the gas parcel and the Sun-center line. If the Galaxy were axisymmetric with a logarithmic potential, then $v_\phi = v_c$ and $v_r = 0$. Thus, equation (1) reduces to the familiar form:

$$v_{los}^{axisym.} = v_c\left(1 - \frac{R_0}{R}\right)\sin(l) \tag{2}$$

where R_0 is the galactocentric radius of the LSR.

Figure 1 shows closed loop orbits in this potential projected onto the longitude-velocity $(l-v)$ plane. There are several important symmetries to note in an axisymmetric Galaxy: (1) the gas in the Galactic anti-center is at rest relative to the LSR; (2) similarly, the gas in the Galactic center should also be at rest; and (3) the $l-v$ diagram is symmetric around $l = 180, v = 0$. As we will see, the gas in the Galaxy does not have these symmetries, an important hint that the disk of the Galaxy is not axisymmetric.

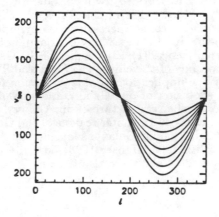

Figure 1. Projection of Circular Orbits into the Longitude-Velocity Diagram

If the Galaxy is not axisymmetric, then its signature may be apparent in the $l-v$ diagram. A useful way of describing these asymmetries in the Galaxy is to expand the potential and the stellar and gas distributions in spherical harmonics. Non-axisymmetric perturbations in the plane can than be expressed as Legendre polynomials, $P_m(\cos\theta)$:

$$\Phi(r,\theta) = \sum_m \Phi_m(r)P_m(\cos\theta)$$

Axisymmetric models have only $m = 0$ terms in the potential. The $m = 1$ term describes lopsided models. Elliptical distortions are described by the $m = 2$ terms.

It is important to distinguish between two possible sources of asymmetry. The gas distribution may be asymmetric either due to the recent accretion of material or due to asymmetries in the background potential. These two physical processes have different observational signatures. While both produce distortions in the shape of the $l - v$ diagram, if the source of the asymmetry is recently accreted gas, then these asymmetries should eventually wind up into a spiral and should appear at different longitudes along each isotherm. If the gas is not rapidly moving inward, then there should not be any asymmetries around $l = 180$ in the gas distribution. On the other hand, if the source of the asymmetry is a lopsided or elliptical potential, then the asymmetries in the $l - v$ diagram should appear at the same location in each isotherm and the gas distribution in the Galactic anti-center need not be centered around $v = 0$. As I will show in Figure 5 and 6, the asymmetries in the outer Galaxy do not appear to wind up and are very striking at $l = 180°$. This suggests that the asymmetries are in the underlying Galactic potential and not in the distribution of gas.

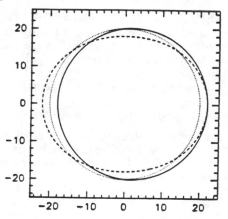

Figure 2. Lopsided (solid) and Elliptical (dashed) Orbits

If the galactic potential is not axisymmetric, then the closed loop orbits will no longer be circular. In a non-rotating elliptical potential, the closed loop orbit is an ellipse whose major axis is at right angles to the principal axis of the potential. In a non-rotating egg-shaped potential, the closed loop orbit is also egg-shaped. However, the orientation of the egg is opposite that of the potential. Because of the 'donkey-like' tendency of the loop orbit family to oppose the orientation of potential, these orbits can not account support self-consistent non-axisymmetric potentials. Binney and Tremaine (1987) describe the response of the closed loop orbit to a weak non-axisymmetric perturbation:

$$v_\phi = v_c - \epsilon v_c \cos[m(\phi - \psi)]$$
$$v_r = -\epsilon v_c \sin[m(\phi - \psi)]$$

$$(3)$$

Here, ϵ is the strength of the perturbation and ψ is the orientation of the bar.
When these non-axisymmetric orbits are projected onto the $l - v$ diagram, they have a distinctive signature. Combining equations (2) and (3) yield

$$v_{los} = v_{los}^{axisym} + v_c \sin\left[(m-1)\phi - l - m\psi\right]$$

$$(4)$$

Blitz and I found that a useful way to look for these asymmetries was to compare the radial velocities of the isotherms on opposite sides of the Galaxy. We quantified this by defining, v_{diff}:

$$v_{diff}(T, l) = v_r(T, l) + v_r(T, 360° - l) \qquad (5)$$

where T is the temperature of the isotherm and $v_r(T, l)$ is the radial velocity of the isotherm at a given longitude.

The position of the gas isotherm in the $l - v$ diagram depends upon the gas surface density, Σ_{gas} and the gas velocity field:

$$T(l, v_{los}) = \frac{\Sigma_{gas}(R, \phi)}{|\frac{dv_{los}}{dR}|_{l, v_{los}}} \qquad (6)$$

Thus, computing the predicted v_{diff} and v'_{diff} curves requires modeling the gas response to non-axisymmetric perturbations. In Blitz and Spergel (1991), we calculated these curves under the assumption that the gas moved on closed orbits. In this limit, the gas isotherms are reasonably well approximated by simply projecting the closed loop orbits into the $l - v$ plane. Thus, we can estimate the signature of non-axisymmetric perturbations by combining equations (4) and (5):

$$v_{diff}(T, l) = 2\left\{v_r(LSR)\cos(l) - \epsilon v_c \sin(m\psi)\cos[(m-1)\phi + l]\right\} \qquad (7)$$

This approximation needs to be checked with detailed numerical simulations. However, it is likely to correctly describe the qualitative effect of motion in a non-rotating axisymmetric model. In the next section, we will compare these predicted curves to the observations.

Another possible source of asymmetry in the $l-v$ diagram is the warping of the disk in the outer Galaxy. This, however, probably has a relatively negligible effect on the *projected* $l - v$ diagram. If we model the gas has moving on a tilted ring of radius, R, with constant circular velocity, v_c:

$$z(\phi) = \epsilon R \cos(\phi - \psi_t)$$
$$r(\phi) = \sqrt{R^2 - z(\phi)^2} \qquad (8)$$
$$\approx R(1 - 0.25\epsilon^2 \sin[2(\phi - \psi_t)])$$

then the effect of the warp is to add an $m = 2$ oscillation to the line-of-sight velocity of order $v_c(z_{warp}/R)^2 \sim 2$ km/s. This effect is much smaller than the observed asymmetries.

In a galaxy with a rotating non-axisymmetric potential, the orbital behavior and its observational signature becomes more complicated. In an barred Galaxy, the prograde short-axis loop orbit family bifurcates into two distinct orbital families (Contopoulos and Mertzanides 1977). Deep inside the bar lie the x_2 orbits, positioned along the major axis of the bar. Further out, these orbits co-exist with the x_1 orbits whose long axis lies perpendicular to the major axis of the density distribution. At large radii, the x_1 orbits are non-intersecting; while at small radii, they are self-intersecting. Binney et al. (1991) discuss the observational signatures of rotating bars in the context of the inner Galaxy.

3.Observations of gas in the outer Galaxy

One of the most striking pieces of evidence for the Galaxy not being ax-isymmetric comes from the observations of gas near the anti-center. If the gas in the outer Galaxy moved on circular orbits and the LSR did not have any radial component in its motion, then the gas at the Galactic anti-center should appear to be at rest.

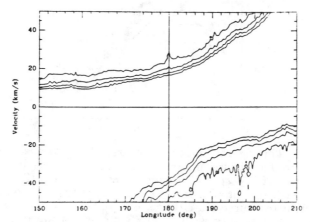

Figure 3. This figure shows the isotherms in the Galactic anti-center re-gion. In an axisymmetric Galaxy in equilibrium, these contours should be cen-tered around $v = 0$.(From Blitz and Spergel 1991)

The offset seen in the Galactic anti-center region is just part of a large scale asymmetry that is most apparent in the v_{diff} diagram shown in Figure 4.

This figure is a useful tool for testing models for the outer Galaxy. The large systematic nature of the asymmetry suggests that the gas in the outer Galaxy is in equilibrium and that this 200 kiloparsec long effect is not due to random motions of the gas. In this figure, gas velocity are compared at two points that are geometrically similar, thus, projection effects are unlikely to explain the effect in the figure. As equation (5) suggests, there are two distinct possible mechanisms that can produce the effect seen in the figure: either the LSR is moving outward at 15 km/s or the gas in the outer Galaxy is moving in a lopsided potential (m=1) with $\epsilon v_c \sin(\psi) = 15$ km/s. The former possibility was explored in Blitz and Spergel (1991), while the later was explored in Kuijken (1991). While dynamically stable models of such a lopsided galaxy have not been constructed, numerous galaxies are observed to have an irregular distribution of gas in their outer regions (Baldwin, Lynden-Bell & Sancisi 1980). We do not know whether these $m = 1$ asymmetries in these external galaxies are due to asymmetries in the distribution of dark matter or are due to the gas having not yet settled into an equilibrium.

The v_{diff} diagram appears to contradict the predictions of the recent model of Kuijken and Tremaine (1993) who suggested that the outer Galaxy was elliptical with $\epsilon \sin(2\psi) = 0$ and $\epsilon \cos(2\psi) = 0.1$ They are forced to assume that by some lucky coincidence, the 30 km/s asymmetries between their predictions and observations are erased by non-equilibrium effects.

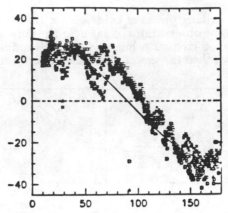

Figure 4. This figure compares the predicted and observed v_{diff} curves. The symbols are the velocity differences between isotherms at a given longitude and colongitude, $(360° - l)$. The open squares are the 0.5 K contours, the solid triangles are the 1 K contours, and the crosses are the 2 K contours. The solid line is the predicted curve for either the lopsided Galaxy model described in Kuijken (1991) or a Galaxy with the LSR moving outward at 15 km/s (Blitz and Spergel 1991). The dashed line is the predicted curve for the elliptical Galaxy described in Kuijken and Tremaine (1993). In an axisymmetric Galaxy, $v_{diff} = 0$, independent of l.

Can we use the gas distribution to discriminate between an $m = 1$ mode and non-circular LSR motion as the source of the irregularities seen the gas distribution? The closed orbits will cause a different spatial distribution of the HI in each case. An $m = 1$ mode will have more gas in the third and fourth Galactic quadrants than in the first and second (these quadrants are specified by the negative velocities of the gas seen in the Galactic anticenter). A local $m = 2$ mode will cause gas to pile up in the second and fourth Galactic quadrants. Thus, it is possible to distinguish between these two cases by looking at the differential column densities of HI measured with respect to the Sun-center line. If one subtracts the column density at a longitude in the third and fourth Galactic quadrant from an equivalent co-longitude in the first and second, one would expect an $m = 1$ to produce an even function, and an $m = 2$ to produce an odd function as a result of the differences in gas distribution.

Figure 5 is such a plot, where N(diff)% is the column density difference (expressed as a percentage of the N(HI) at longitudes > 180°) between the two Galactic hemispheres plotted as a function of longitude. The figure clearly shows that the difference function is in fact odd. However, it is important to be sure that local features do not contaminate the results. To do this, we simply take the differential column densities starting at larger and larger distance from the solar circle. This is equivalent to integrating the HI emission from a given non-zero velocity to infinity, and again plotting N(diff)%; larger values of the velocity correspond to larger distances from the solar circle. The figure shows that the character of the plot does not change even for values of the velocity as large as

30 km s^{-1} , and the oddness of the plot is determined largely by gas at large distances from the Sun.

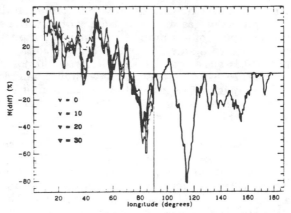

Figure 5. The difference between the column density of HI at $180° < l < 360°$ and $0° < l < 180°$ as a fraction of N(HI) in the range $180° < l < 360°$. The three different curves plot three different velocities from which N(HI) is integrated to infinity. As the velocity increases, we omit gas at increasingly large distance from the solar circle. It is clear that in all cases, the function N(diff)% is an odd function of longitude implying that the HI in the Milky Way is distributed in an $m = 2$ mode. (From Blitz and Spergel 1991)

The scale height of the HI layer as a function of azimuthal position and velocity may provide another test to discriminate between an $m = 1$ and an $m = 2$ mode. Merrifield (1992) demonstrated how the scale height of the HI layer can be used as method for determining the rotation curve of the Milky Way outside the solar circle. Merrifield assumes that the scale height of the gas layer is independent of azimuth and then solves for the layer thickness and distance as a function of velocity. This analysis shows systematic differences between positive and negative longitudes which are perhaps due to the outer Galaxy being either lopsided or triaxial. It also yields a surprisingly small value for the rotation rate of the Galaxy.

The energetics of the gas in the outer Galaxy is relatively poorly understood. Near the solar circle, supernova explosions are thought to be an important source of energy for cloud motions. However, in the outer Galaxy, the supernova rate appears to be significantly lower. In particular, there may be very little energy input in the interarm region and cosmic rays, diffusing outwards in the disk, may be the dominant source of pressure support. It the gas in the outer Galaxy is indeed supported by cosmic ray pressure is Parker unstable at wavelengths larger than the width of the disk (Spitzer 1978). This instability may account for the corrugation seen in the outer Galaxy at roughly this wavelength (Burton 1988.). These effects may weaken the assumption that the scale height of the gas layer is constant at fixed galactocentric distance and complicate the analysis.

4. Observations of Stellar Kinematics

Observations of stellar kinematics provide information that is complementary to the observations of gas motions. Stellar observations have several advantages over observations of gas motions: (1) it is often possible to accurately determine the distance to a star; (2) old stellar populations are almost likely in equilibrium; and (3) viscous and magnetic forces do not play a role in star motions. On the other hand, we do not have stellar samples that probe the extensive volume probed by the gas and interpretation of stellar orbits can be complicated by the need to explore the response of non-closed orbits.

Observations of the local velocity ellipsoid probe the shape of the potential locally. For a well-mixed stellar population in an axisymmetric galaxy, the long axis of the velocity ellipsoid points towards the Galactic center. If however, the Galactic potential is not axisymmetric, the velocity ellipsoid should point roughly normal to the closed loop orbit. Thus, if the local of standard of rest is moving outward then the velocity ellipsoid should point towards negative longitude. This deviation of the velocity ellipsoid is called the vertex deviation. Kuijken and Tremaine (1991) have examined this effect in detail and concluded that the observational data suggests a vertex deviation opposite in sign from that expected if the LSR is moving outward along an elliptical orbit. This conclusion is complicated by the effects of local spiral arms and the younger stars in the local neighborhood not being a well-mixed population.

Observations of stars in the outer Galaxy are another tracer of the Galactic potential. If the Galaxy has a rotating stellar halo with an ILR at $\sim 2R_0$, then closed orbits should become increasingly elliptical as they approach the Inner Linblad Resonance (ILR). As they cross the ILR, the orientation of the ellipse changes by $90°$ implying that the non-circular component of the stellar motion changes sign. Kuijken (1991) argues that this increase in outward motion with radius towards the ILR and the subsequent sign reversal should be apparent in the gas. Gas streamlines cannot cross, thus, rapid variations in gas radial velocity will be damped by dissipation— a result confirmed by hydrodynamical simulations by Bies (1989). On the other hand, stellar systems can move on intersecting orbits since stars are not collisional. The K giant survey of Lewis and Freeman (1989) may show a hint of this effect. They find that the K giants near the anticenter appear to deviate systematically from the LSR motion: at $R = 1.5R_\odot$, $1.75R_\odot$ and $2.1R_\odot$, they find that the K giants are moving away from the LSR at 13 ± 4.2 km/s, 16 ± 4.6 km/s and 5.7 ± 4.5 km/s implying either an inward motion of the LSR or perhaps, an outward motion increasing with distance from the galactic center. The test of this hypothesis would be observations of K giants at $\sim 2.5R_\odot$: if the rotating stellar spheroid model is correct, then these stars should be moving *towards* the LSR at ~ 15 km/s.

A important new sample of the stellar velocity field is the carbon star survey of Schechter and Metzger (1993). They have obtained both distances and radial velocities for a large sample of Carbon stars. They find that the Carbon stars, like the K giants in the Lewis and Freeman sample, are moving away from the LSR. Weinberg (1993, this proceedings) discuss implications of these observations for models of the Galactic halo. The theoretical models for the galactic potential need to be compared to this sample in detail.

Preliminary analysis of the Lewis and Freeman and Metzger and Schechter data, however, imply that the stellar kinematic data is not consistent with the Blitz and Spergel (1991) model. The most surprising result of the stellar kinematic surveys is that the stars and the gas in the outer Galaxy appear to move in opposite directions. This may be a signature of a resonance in the outer Galaxy. Since gas is collisional, the gas velocities should vary smoothly across a resonance, the stellar velocities, on the other hand, can change discontinuously. Thus, as we approach a resonance, the radial motions of the stars should increase and then change sign. The gas velocities, on the other hand, should change smoothly.

5.Testing the Model: Observations of the Galactic Center

If the LSR is moving toward the gas in the outer Galaxy, then it should also be seen to be moving away from the center of the Galaxy. Kuijken and Tremaine (1991) have summarized the stellar data from the inner Galaxy and concluded that the LSR is moving relative to the galactic center with velocity -1 ± 9 km/s, a result that is consistent at the 1.5 σ level with both the standard axisymmetric Galaxy and the Blitz and Spergel model. The molecular gas in the galactic center, on the other hand, appears to be moving away from the LSR at ~ 15 km/s, as predicted in the Blitz and Spergel model. Since the molecular gas in the galactic center displays many kinematic anomalies, this result can not be viewed as definitive.

An improved stellar sample could provide the definitive measure of the LSR motions. Michael Rich, Neil Tyson and I have begun a program of measuring radial velocity of stars along the minor axis of the Galactic Bulge. If the asymmetries in in the outer gas contours is due to LSR motion, then we should be moving away from the stars along the minor axis. If on the other hand, the LSR is moving on a circular orbit and the gas in the outer Galaxy is moving on a lopsided orbit, then there should be no motion with respect to the LSR. If the weather in Chile is good to us, then we may be able to better understand the dynamics of the outer Galaxy by these observations of the Galactic center.

6.Implications and Conclusions

There is mounting evidence that our Galaxy is not axisymmetric, but rather a complex triaxial system. The Milky Way appears to have a central triaxial component that may be rapidly rotating. There is also evidence that either the Galactic spheroid or inner halo is triaxial or the Galaxy is lopsided. These observations of the three dimensional structure of the Milky Way are of more than cartographic interest, for they provide hints about the Galaxy's formation and evolution. (It is not clear whether we should divide the Galaxy's history into these distinct phases since Galaxy formation may be an ongoing process.)

The current observational situation is very confusing. The observations of the kinematics and the distribution of the gas suggest that the LSR is either moving outward on a non-circular orbit. The stellar kinematics also shows evidence for non-axisymmetric motions. However, while the gas in the Galactic anti-center region appears to be moving towards us at about 15 km/s, the stars in the same region appear to moving away from us at 5 km/s. This hints that there may be an orbital resonance in the outer Galaxy, perhaps driven by the rotation of a non-axisymmetric halo.

The shape of Galactic halo is a sensitive probe of the initial conditions that led to galaxy formation. Carlberg and Dubinski (1991) found in their simulations of galaxy formation in the CDM scenario that most galaxy halos were highly triaxial. I suspect that this will be true in any hierarchical scenario in which the halo forms dissipationlessly from Gaussian initial conditions. Blitz and Spergel (1991) and Kuijken (1991) have shown how HI observations can be used to probe not only the monopole term in the potential, but also higher order terms. Observations of the Galaxy's triaxiality combined with observations of the gas motions in the outer regions of other galaxies are a potentially powerful test of galaxy formation scenarios.

If the triaxiality of the bulge and spheroid are driving spiral structure, then this may have important implications for the chemical evolution of the Galactic disk. Gas crossing spiral shocks loses angular momentum. Given the velocity changes observed across the spiral arms of our Galaxy, gas should be driven inwards on roughly 20 dynamical times, a timescale short compared to the age of the Galaxy. This constant infall means that the chemical evolution of various zones of the Galaxy are linked and that the Galactic disk should not be viewed as a closed box. This slow infall of gas will not produce significant X-ray flux— one of the classic objections to accretion models (*e.g.* Binney and Tremaine 1987).

Detailed observations and analysis of the complex structure of our Galaxy affords us the opportunity to learn the physics of galaxy formation and evolution and to unravel our own history.

ACKNOWLEDGEMENTS

I would like to thank Leo Blitz for an enjoyable collaboration upon which this article is based and for his contributions to this article. My research is supported in part by an A.P. Sloan Foundation Fellowship, by NSF grants AST 88-58145 (PYI) and AST 91-17388 and by NASA grant NAGW-2448.

REFERENCES

Baldwin, J.E., Lynden-Bell, D.E. and Sancisi, R. 1980. MNRAS 193, 313.
Bies, W 1989, Princeton University senior thesis.
Binney, J. & Tremaine,S. 1987. *Galactic Dynamics*, Princeton Univ. Press, NJ.
Binney, J., Gerhard, O. E., Stark, A. A., Bally, J. & Uchida, K. I. 1991. MN-RAS252, 210
Blitz, L. & Spergel, D.N. 1991. ApJ 370, 205.
Burton, W.B. 1988 in Galactic & Extragalactic Radio Astronomy (ed. G.L.Verschuuer & K.I. Kellerman (Springer- Verlag: Berlin))
Carlberg, R.A. & Dubinski, K.A. 1991, ApJ 378, 496.
Contopolous, G. and Mertzanides, C. 1977. Å 61, 477.
Kuijken, K 1991. in *Warped Disk & Inclined Rings around Galaxies*, Casertano, S., Sackett, P.D. and Briggs, F.H., Ed. (Cambridge Univ. Press.), p. 159.
Kuijken, K. & Tremaine, S. 1991. in *Dynamics of Disk Galaxies*, ed. B. Sundelius, (Göteborg: Göteborg University Press), 71
Kuijken, K. & Tremaine, S. 1993. CITA preprint 93/4.
Lewis, J.R. & Freeman, K.C. 1990. AJ 97, 139.
Merrifield, M.R 1992, CITA preprint.
Metzger, P. and Schechter, P.L. 1993, MIT preprint.
Spitzer, L., "Physical Process in the ISM", (Wiley Scientific: New York), 1978.

DISTRIBUTION OF STARS IN THE DISK

Martin D. Weinberg
Dept. of Physics and Astronomy
University of Massachusetts, Amherst, MA 01003.

ABSTRACT

Recent developments in determining the large-scale structure of the Galaxy with stellar tracers are discussed. Among the possible approaches, direct mapping using standard candles is emphasized because of its promise to provide the most unbiased and consistent view, especially when kinematic data is available.

Candidate standard candle tracers include HII regions, planetary nebulae, Cepheid variables, and AGB stars (carbon stars, Mira variables, OH/IR stars). Current progress is briefly reviewed in turn. Mapping techniques are discussed, and as an example, the author's experience analyzing the IRAS PSC and detection of a bar is described.

Bars and similar rotating non-axisymmetric features in the Galaxy provide observable stellar kinematic signatures, even if the perturbations are relatively weak (several percent). Therefore, only a modest velocity data set is required to place meaningful constraints. For example, the moderate-strength bar with half-length of 3 kpc described previously will produce a strong jump in radial velocity dispersion at $R \lesssim 6$ kpc which is consistent with Lewis and Freeman's K-giant velocity data. Similarly, this same data and the carbon star velocity data presented by Metzger and Schechter is inconsistent with a rotating triaxial spheroid.

1. INTRODUCTION

Much of our overall view of the Galaxy comes from the observation of its gas whereas most of its luminous mass is in stars. Since the dynamical evolution of the Milky Way (such as the formation of non-axisymmetric structure) is mediated by stars, a fundamental understanding of the underlying structure must begin with an understanding of the stellar distribution.

A comprehensive review of the subject is a tall order given its long history. Sir William Herschel, over 200 years ago, made one of the first attempts to determine the structure of the Galaxy using star counts (*e.g.* Hoskin 1964 or an introductory astronomy text). From the rough symmetry of counts in 683 regions on the sky, Herschel deduced that the Sun was at the center of a somewhat elongated disk. Since Herschel's time, many improvements have been made in star counting techniques, but the two basic problems that led Herschel to wrong conclusions are with us today: 1) interstellar extinction; and 2) intrinsic faintness of stars which limits our ability to probe large-scale structure. Rather than trace the evolution of these research efforts, I will restrict the discussion to recent developments, emphasizing the two-dimensional structure of the Galactic disk, and leave aside the interesting problems of its vertical structure.

Nearly all methods for ascertaining the large-scale structure of the Milky Way fall into three catagories:

- *Morphological comparison* with external galaxies;

- *Parameter estimation* using models; and

- *Direct mapping* using calibrated tracers.

The appearance of external galaxies has framed our expectation of how the Galaxy should look. Frank Kerr gave us some nice examples of *morphological comparison* between (esp.) the observed distribution of neutral hydrogen in our Galaxy and optical images of other galaxies in the opening talk of this conference. Based on these comparisons and quantitative measurements of surface brightness, we have developed a set of model profiles (*e.g.* Hubble models, de Vaucouleurs profiles, exponential disks) that describe galaxies. Finding the best fit between these profiles and observed data constitutes *parameter estimation* as exemplified by the Bahcall & Soneira (1980) model. The third method pioneered by Herschel, *direct mapping*, remains the hardest and has received the least attention, although it also requires the fewest a priori assumptions. However, the rapid pace of detector technology coupled with dramatic improvements in computational power has brought within our reach the possibility of range finding with stars at tens of kpc. In addition, kinematics is dynamically coupled to inferred density map and should provide very strict constraints which are just beginning to be investigated. Discussion of this final method will dominate this talk.

The plan for the remainder of this presentation is as follows. In §2 I will review the tracers that have been used to infer the global structure of the Galaxy to date, or could be in the near future, and outline a direct mapping strategy in §3. I will discuss the kinematic evidence for the observed non-axisymmetric structures in §4 and conclude with §5.

2. GALACTIC TRACERS

Successful tracers must be intrinsically bright, so that they can be seen anywhere in the Galaxy despite dust extinction; bright red stars therefore are better than bright blue stars. Secondly, the tracers must be numerous in order to resolve structural detail. Thirdly, their physical properties must be sufficiently well understood that we have confidence in their identification and inferred luminosity. Finally, tracer populations with both distances *and* velocities will be powerful indicators of disk dynamics, giving us the opportunity to probe the evolutionary history of the Galaxy. Since there is insufficient space for a complete review, I will concentrate on some of the populations that I believe have the promise of meeting these goals. I apologize in advance for omissions.

2.1. HII regions/OB associations

HII regions, although not stars per se, were among the earliest tracers of the disk stars, albeit only the young ones. Georgelin & Georgelin (1976, GG) combined optical data with H α and H 109α radial velocities of HII regions to trace the spiral arms (their Fig. 11). These findings were in agreement with the earlier HI work, such as described by Kerr. In their paper entitled "An Outsider's View of the Galaxy", de Vaucouleurs & Pence (1978) determined the structural parameters of the Milky Way using the work of GG and other available data, and determined its morphology by comparing with external galaxies. I would like to draw your attention to their Fig. 6 which depicts the authors' impression of the Milky Way (an artistic rendering!) inferred from the apparent four-armed spiral pattern. Scaling to the position of the Sun, the well-defined bar whose major axis lies in the first quadrant has a half length of approximately 3 kpc. This prescient inference is in concert with results to be discussed below and a number of other claims for a Milky Way bar.

2.2. Planetary nebulae

Several investigators have recently attempted to use planetary nebulae in the Galaxy as standard candles. Planetary nebulae are representative of an intermediate-age disk population, bright, readily identified, their systemic velocities may be determined. Schneider et al. (1983) compiled a catalog of nebulae which was subsequently analyzed by Schneider & Terzian (1983). They were able to determine the rotation curve between $4\,\mathrm{kpc} \lesssim R \lesssim 11\,\mathrm{kpc}$ and out to nearly $20\,\mathrm{kpc}$ with lower precision. Although the typical expansion velocity for planetary nebula envelopes is $\sim 20\,\mathrm{km\,s}^{-1}$, systemic velocity errors with multiple measurements in the Schneider et al. catalog is typically several $\mathrm{km\,s}^{-1}$. More recently, Jacoby and his collaborators (e.g. Jacoby et al. 1992) have sought to improve the standard calibration. Together with additional systemic velocities, planetary nebulae photometric/kinematic tracers could become an important tool for galactic structure.

2.3. Cepheid variables

Cepheids have played an important role as a distance calibrator since the P-L relation was discovered by Leavitt in 1908. Since Cepheids are supergiants, like planetary nebulae, they trace the spiral arms. With the work of a large number of researchers (see Jacoby et al. 1992 for a review), they now have an well-determined period-luminosity-color (P-L-C) relation which yields accurate photometric distances and their pulsation provides an unambiguous detection signature. In short, Cepheid variable stars are the *standard* standard candles. Unfortunately their space density is low and preferentially distributed in regions of recent star formation rather than being representative of the old disk. Nonetheless, their high quality permits sensitive studies of intermediate distance kinematics. For example, Caldwell & Coulson (1987) used statistical estimation to simultaneously determine the Galactic scale R_o and the rotation curve V_o using a circular kinematic model. With these approaches in mind, Caldwell, Keane, & Schechter (1991) identified Cepheids with $I \leq 14$ using a drift-scan technique. In a follow-up paper, Metzger et al. (1992) determined the velocities of some of these stars with $\sigma \lesssim 1\,\mathrm{km\,s}^{-1}$. In principle, more general uses are possible, and in §4 we will discuss some applications to constraining additional global structure parameters.

2.4. K-giants

Although traditionally important for determining the vertical distribution of stars and "weighing" the Galactic disk, K-giants may be used to investigate the radial structure. Unlike HII regions and classical Cepheids, K giants are representative of the old disk. Lewis & Freeman (1989) observed 600 disk K giants, obtaining both photometry and spectroscopy. They selected longitudes near the Galactic plane in regions of low reddening as calibrated by globular clusters. Fig. 1 shows their data at $l = 0°, 180°$ (from their Table V). This data

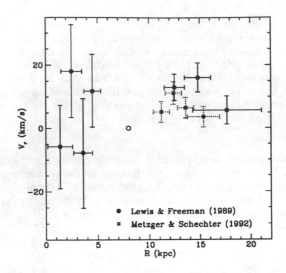

Fig. 1.   Lewis & Freeman's K-giant line-of-sight velocity measurements at
$l = 0°, 180°$ and Metzger & Schechter's carbon star measurements at $l = 180°$.
The open circle shows the position of the Sun.

suggests an interesting feature at $\approx 13\,\text{kpc}$ although the estimated errors make
the significance marginal. We will discuss further implications of this data in
§4.

2.5. AGB stars

The availability of near and mid-infrared data now makes it possible to
detect cool bright low-mass stars in their final stages of evolution, the asymptotic
giant branch (AGB) phase. AGB stars may be detected in the near- and mid-
infrared throughout the Galactic disk, aided by significantly lower extinction.
They therefore hold great promise as a tracer of large-scale Galactic structure.
However, AGB stars have a broad range of properties, so in order to use them
as tracers one must find ways of selecting subclasses whose intrinsic properties
can be ascertained.   Note that these stars are often grouped into physically
overlapping catagories. For example, both carbon stars and OH/IR stars can
be Mira variables, but need not be. A Mira variable may be an oxygen-rich
AGB without maser emission, etc. For the sake of argument, I will define these
subsets by their primary observational signature, with the understanding that
they need not be physically distinct.
Carbon stars may be identified spectroscopically and over the last several
decades have been well studied. Blanco $et\ al.$ (1980) studied carbon stars in the
Magellanic Clouds and found a distinct peak in magnitude ($e.g.$ their Fig. 7)
and Frogel $et\ al.$ (1980) studied their temperatures and colors. This has been
followed up by Cohen $et\ al.$ (1981), Fuenmayor (1981), Claussen $et\ al.$ (1987)
and Jura $et\ al.$ (1989) among others. The absolute K-magnitude dispersion is
≈ 0.6 mag which makes them a quasi-standard candle with a $\sim 30\%$ relative

distance error. In recent survey work (Aaronson *et al.* 1989, 1990), $\sim$ 400 stars with magnitudes $K \lesssim 8$ have been catalogued with typical velocity errors $\sigma \sim 3\text{--}5\,\text{km s}^{-1}$ per measurement in each of a Northern and Southern survey. Metzger & Schechter (1992, this meeting) use similar quality carbon star data to determine the radial velocity as a function of galactocentric radius in the direction of the anticenter. Their data is shown in Fig. 1 (see also §4).

Mira variables are long-period ($\sim$ 200 − 1000 day) pulsators and may be identified by their colors and variability. Infrared-bright Mira stars often have enhanced mass loss, which causes the formation of optically thick circumstellar shells. These dusty shells may absorb much of the optical emission of the central star. Over the last 10 years, extensive infrared photometry has been obtained for Miras in the Magellanic Clouds and towards Galactic center. In particular, precise P-L relations have been determined by Feast *et al.* (1989) for LMC Mira stars. Van der Veen (1989) has found a relation between luminosity, pulsation period and derived envelope mass for Miras observed toward the Galactic center (*e.g.* his Fig. 10) consistent with those for the LMC Miras. These results suggest that the prospects of calibrating infrared standard candles for use in Galactic structure are quite promising. However, unlike carbon stars and other tracers, the velocities of variability-selected AGB stars have not been systematically surveyed. This is a difficult task but would be a *major* contribution to the field.

AGB stars, especially those with dusty shells are bright at $12\,\mu\text{m}$ and could be easily detected by IRAS (IRAS Point Source Catalog 1985, PSC). Habing *et al.* (1985) pointed out that PSC colors—$12, 25, 60\,\mu\text{m}$ flux ratios—may be used to select OH/IR star candidates. Rowan-Robinson & Chester (1987) exploited these color criteria to estimate the scale parameters of the bulge and disk. Habing (1988) used a similar approach to claim evidence for thin/thick disk components. Since these contributions, the Leiden group has been active in following up the candidate lists with OH surveys to determine both the physical properties of the stars and acquire kinematic data (*e.g.* te Lintel-Hekkert 1990). Further clasification of color-selected sources is necessary to understand and improve the efficiency of the color criteria (see Allen *et al.* 1992 for one approach).

These analyses based on IRAS data prove the utility for Galactic astronomy of an infrared survey. Except where confusion limited, the proposed Two-Micron All Sky Survey (2MASS) should discern nearly every AGB star in the Galaxy! Three wavebands—J, H, K—will be observed and the survey will be complete at brightness limit of $K = 14$; however at these faint magnitudes one will observe mostly dwarfs and early-type giants. However, at $K = 10$, 1 out of 10 sources will be an AGB star and I estimate that the survey should be capable detecting hundreds of thousands of AGB stars. The 2MASS database will be useful for Galactic structure studies in its own right and as the basis for future (*e.g.* velocity) surveys.

3. DIRECT MAPPING

Most recent global structural analyses of the Milky Way use parameter estimation based on models of external galaxies similar to de Vaucouleurs & Pence (1978). For example, the Bahcall & Soneira (1980) fit local star counts to an exponential disk and a de Vaucouleurs model for the luminous spheroid. Habing (1988) fits IRAS source counts to exponential disks with exponential scale heights. But one need not be limited to parameter estimation. Given an extinction model and a statistically complete sample of standard candles, we can reconstruct the sample's spatial distribution and map the Galaxy *directly*. This approach has the key advantage of requiring very few assumptions about the "shape" of the Galaxy and therefore one's findings are not biased by one's preconceptions. Here, I will describe my recent attempt to use this approach.

At present, there is no ideal sample. However to begin, I select a sample which allows photometric parallaxes to be obtained at infrared wavelengths and with broad sky coverage. In particular, we have seen the Mira-like variables have a distinct period–luminosity relation and that "dusty" AGBs were copiously detected by IRAS. The IRAS PSC has the advantage of significant sky coverage, low interstellar absorption, and an understood flux-completeness limit. Since the data comprising the PSC had roughly two epochs separated by ~ 200 days, it is likely that the Mira variables which varied between the epochs were preferentially detected at certain periods. Harmon & Gilmore (1988) show evidence for this. If there is a well-defined Galactic P–L relation like that for the LMC Miras, then it is plausible that Miras with IRAS variability represent a narrow range of luminosities.

I found 3170 candidate AGB variables by selecting sources with: (1) $P_{var} \geq 0.98$; (2) $12\,\mu m$ flux $F_{12} \geq 2$ Jy; and (3) galactic latitude $|b| \leq 3°$. The P_{var} statistic quoted in the PSC is the unit-normalized cumulative ranking of the maximum relative flux excursion between all hours-confirmed observations for each source. Thus, $P_{var} \geq 0.98$ means that 98% of the sources have smaller relative flux variation. Near $l = b = 0°$ the source density is high and faint sources will be confused. By plotting the source counts as a function of l at $b = 0°$, the confusion is manifest for $F_{12} < 2$ Jy. The restriction to $|b| \leq 3°$ is for analysis purposes only; I would like only to consider a two-dimensional distribution of stars. I now assume that the sample is truly Mira-like variables with similar period and therefore a narrow range of luminosity. To determine the distances, I assume that all stars have $L_{bol} = 8000\,L_{\odot}$, similar to the neighborhood carbon star variables (Jura *et al.* 1989), and take a uniform extinction model with $A_{12} = 0.16$ mags/kpc based on measurements by Becklin *et al.* (1978).

The resultant list of spatial positions is a very grainy representation of the Milky Way, and we would like to construct a smooth picture. To do this, rather than fit model profiles to the data which precludes finding unanticipated structures, one can represent the point distribution by a complete set of functions, such as a polar moment expansion (analogous to an expansion in spherical harmonics and wavefunctions). One now chooses an expansion origin. The obvious choice is the Galactic center since the Galaxy is probably roughly axisymmetric and centrally concentrated. However, selection effects such as systematic distance errors and contamination by intrinsically faint but nearby stars are more easily treated in the observer's frame. Fortunately, harmonic expansions have translation operators, providing unique well-defined mappings between the coefficients at shifted expansion centers, which allows expansion about the solar position followed by translation to the Galactic center. In the case of a spherical expansion $j_n(r)Y_{lm}(\theta,\phi)$, the shift preserves m and mixes l and n; in the case of the two-dimensional polar expansion, $j_n(r)S_m(\phi)$, both n and m are mixed. The translation technique itself is quite general and may be used for any regular coordinate system (see Weinberg 1992 for details). In an ideal situation where the positions are known precisely, this approach is equivalent to expanding about the translated origin from the start. However, in practice, position on the sky and derived distance are quite different. In the present case in particular, the ability to translate the origin after expanding the data in the observer's frame has the following advantages:

1) Faint stars (*e.g.* M dwarfs) will contribute to the $m = 0$ component about the solar position. This component plays little role in the reconstruction about the Galactic center, and therefore the translation operator effectively filters out any local contamination.
2) Missing longitudes in the IRAS coverage may be straightforwardly treated in the observer coordinate system but would be unwieldy in the translated system.
3) Systematic errors in extinction or standard candle calibration primarily cause scale changes in the distribution as seen by the observer. These may be easily corrected in the observer's frame but would produce non-axisymmetric distortions in a direct expansion about the translated orgin.

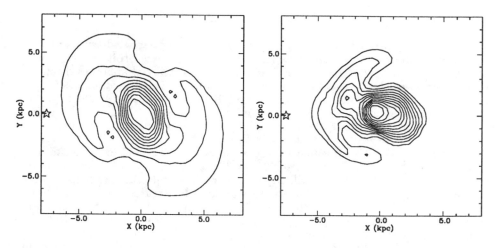

Fig. 2.   Harmonic reconstruction of the source distribution with $F_{12} \geq 2$ Jy including even terms only (left). Twelve evenly spaced levels between 10% and 100% of the peak are shown. The 'star' indicates the solar position.

Fig. 3.   As in Fig. 2 but for $F_{12} \geq 1$ Jy and no restriction on symmetry (right).

There is one additional difficulty. A dusty AGB star at a distance of 9 kpc and small $|l|$ has $F_{12} \approx 2$ Jy. Therefore, the inner disk region is not evenly sampled on both sides of the Galactic center in sample limited to $F_{12} \geq 2$ Jy. To reconstruct the spatial distribution, I first assume that it is inversion symmetric. Fig. 2 clearly gives the impression of a barred galaxy (the bar is in the first quadrant with half-length of ≈ 3 kpc as predicted by de Vaucouleurs & Pence 1978, see *e.g.* Ohta *et al.* 1990 for extragalactic comparisons). But one worries that the appearance of a bar may be imposed by assumption of symmetry. In addition, simulations suggested that the stretch along the y-axis in the inner few kpc may due to the distance cutoff. To address this issue, I repeated the analysis with a lower flux limit of 1 Jy which increased the distance limit to ≈ 13 kpc. Even though the source counts are diminished by confusion in the inner few degrees, I present the reconstruction of this sample in Fig. 3. No assumption of symmetry has been made. Note that the bar distortion occurs at similar position angle on either side of the center with similar length and the y-axis stretch is not apparent. An aside: the sharpness of the profile about the Galactic center suggests the use of variability-selected AGBs is well-motivated!

The direct mapping approach gave a fairly robust signature of a bar even under difficult circumstances. The model-free method presented here does require a rich data set; all features including the background profile must be determined by the data since the a priori constraints are minimal. It could not be used with a low-space density tracer, such as Cepheid variables. However, I firmly believe this approach together with a complete fainter survey, such as the proposed 2MASS, will provide the means of rigorously defining the large-scale structure of our Galaxy.

4. KINEMATIC SIGNATURES

Non-axisymmetric structures in galaxies, such as bars and triaxial spheroids, have dynamical consequences for the stellar and gas orbits which can be observed kinematically. In fact, the existence of the triaxial spheroid discussed by Blitz & Spergel (1991 and this meeting, hereafter BS) is based on this type of kinematic inference. Kuijken & Tremaine (1991 and this meeting) have exploited local kinematic signatures to explore a different scenario for the LSR orbit. Here, I would like to investigate the potential use of *global* kinematics to refine our picture of non-axisymmetric features in the Galaxy.

In particular, the BS spheroid and a Galactic bar are all expected to have rotating patterns. Since it is likely that these perturbations are dominated by their quadrupole terms, I will neglect everything but the quadrupole here. (Other multipole terms, such as the dipole may be treated similarly). For a disk of nearly circular orbits there are 3 dominant resonances; the inner and outer Lindblad resonances (ILR, OLR) and the corotation resonance (CR). The rough location of these resonances in the disk depends on the pattern speed and their existence may have a number of observable consequences. First, the time-averaged density of an orbit at all phases will be an oval which is elongated in a direction parallel to the bar between ILR and CR and outside of OLR and will be elongated perpendicular to the bar elsewhere (*e.g.* Binney & Tremaine 1987). Secondly, near the resonances the distortion of an orbit may be much larger than between the resonances. Therefore, even a relatively weak rotating bar may produce observable consequences. Finally, the effects of the perturbation may depend on the evolutionary history of the Galaxy, not just its current state.

To appreciate these consequences, note that in the absence of any bar, a disk orbit traces out a "rosette"—an ellipse whose apocenter slowly precesses in position angle. Now imagine observing the orbit from a frame rotating with the bar. Exactly at ILR the trajectory appears to close, making two radial oscillations for every azimuthal oscillation. Outside (inside) of the ILR radius, the apocenter of the average orbit will slowly precess backward (forward) in the bar frame. Due to the gravitational potential of the bar, the rate of precession is slower when the apocenter is aligned with the bar (outside ILR) which causes the net oval distortions in the orbit density. As the bar strength grows, the direction of precession may reverse for some orbits near the resonance, and thereafter their apocenters are constrained to a small range of angles either aligned or anti-aligned with the bar. These orbits are said to be "trapped" into libration. Both aligned and anti-aligned librating orbits may occur simultaneously, enhancing the velocity dispersion near the resonance. If the pattern speed is changing, the number of trapped orbits and the fraction of aligned or anti-aligned apocenters also changes. Different fractions of trapped orbits change both the line-of-sight mean velocity and velocity dispersion. In this way, the dynamical response depends on the evolutionary history of the Galaxy. A sequence of non-linear models describing the orbital motion near resonance may be tabulated and used to compute the kinematics throughout the disk. The details will be discussed in a future paper (Weinberg 1993). More general self-consistent models may be computed numerically (*e.g.* Long & Weinberg, this meeting).

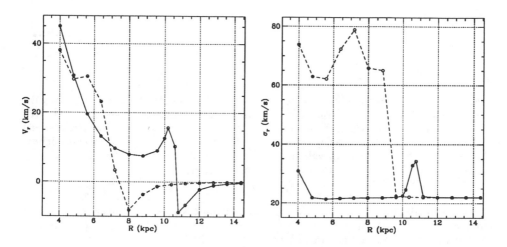

Fig. 4.   Kinematic response to the BS triaxial spheroid model. Line-of-sight velocity (left) and velocity dispersion (right) for an observer at the Galactic center are shown. The solar position is at 8 kpc. The open circles show model computations and the connecting lines are only meant to guide the eye. The solid curve represents a constant pattern speed with ILR at 10.7 kpc. The dashed curve represents a slowly decreasing pattern speed, from 8.7 km s^{-1}/ kpc to 6 km s^{-1}/ kpc, for which the ILR increases from 7.2 to 10.7 kpc.

Let me illustrate observational importance of these effects using the proposed triaxial spheroid and inner bar as examples. BS estimate a spheroid pattern speed of 6 km s^{-1}/ kpc (which puts ILR at $\approx$ 11 kpc), a quadrupole strength of 2% and a pattern which lags the solar position by $\approx$ 45°. I assume the perturbation grows slowly and use the non-linear models described above to compute the response of the disk with radial velocity dispersion of 20 km s^{-1}. Fig. 4 shows the line-of-sight radial velocity and velocity dispersion for an observer at the Galactic center. For example, the LSR (at 8 kpc) will appear to be moving outward at $\approx$ 8 km s^{-1} and therefore we would observe the outer galaxy to be approaching us. This appears inconsistent with the stellar velocity data (Lewis & Freeman 1989, Metzger & Schechter 1992). However, this discrepancy might be resolved if the pattern speed has slowed with time. A model with a

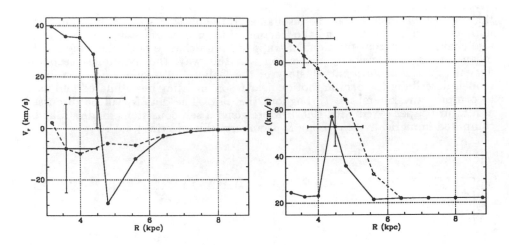

Fig. 5.   Kinematic response to a rotating bar of moderate strength with half-length of 3 kpc. The solid curve represents a constant pattern speed with OLR at 5.1 kpc. The dashed curve represents a slowly decreasing pattern speed for which the OLR increases from 5.1 to 6.1 kpc. The squares with errorbars show the Lewis & Freeman velocity and velocity dispersion data. The velocity dispersion at the LSR is 20 km s^{-1} by construction.

pattern speed which decreases over the age of the Galaxy is also shown in Fig. 4 (dashed curve). Although this is now in the sense observed in Fig. 1 (a match in magnitude can easily be achieved by a change in perturbation strength), this will no longer explain the asymmetries in the HI l–v diagram for which the model was designed! Plots similar to Fig. 4 may be made for an arbitrary line of sight. For example, the ILR model predicts an asymmetry between V_r vs. distance at the $l = 90°, 270°$ directions which could be checked with Cepheid and carbon star data (*cf.* §2.3, Weinberg 1993).

There are a number of additional features in Fig. 4 that deserve comment. For a constant pattern speed (solid curve), the increase in dispersion at $R \approx 10.7$ kpc is largely caused by orbits trapped into libration. Also, V_r increases since the perturbation strength is increasing inward. However, nearly all orbits are trapped into libration for $R \lesssim 4.5$ kpc which gives rise to the jump in dispersion at small R. The evolution of the pattern speed (dashed curve) allows a large fraction of all trajectories with guiding centers $R \lesssim 8$ kpc to be trapped into libration which gives the large jump in dispersion inside of this radius.

Fig. 5 show the effects of the OLR for a bar like that shown in Figs. 2 and 3. The bar leads the LSR by $\approx 45°$ and with CR assumed to be at 3 kpc, the OLR is at 5.1 kpc. The strength is moderate: the quadrupole force is 40% of the background force at maximum. Fig. 5 shows models with both a constant pattern speed (solid) and a decreasing pattern speed (dashed) in which OLR increases from 5.1 to 6.1 kpc. All in all, if there is a moderate bar, these models predict that there should be an increase in radial velocity dispersion (by at least a factor of two near or inside of ≈ 5 kpc. The decreasing pattern speed model and the K-giant data (from Lewis and Freeman 1989, Table V) are consistent, although the error bars are somewhat large. The correspondence between the model and the jump in the velocity dispersion is notable. Further work should be capable of definitively testing and refining this proposed model.

5. CONCLUSIONS

My overall conclusions, in inverse order of presentation, are:

- Rotating bars and spheroids have observable kinematic consequences. We should soon be able to make definitive tests of the scenarios discussed at this meeting. The evolutionary history of the Galaxy may leave tell-tale kinematic signatures and theoretical scenarios should attempt to provide kinematic predictions.

- Joint photometric–spectroscopic tracers are good probes of Galactic structure because, coupled by dynamics, the derived stellar distribution *and* the kinematics strongly constrain the inferred density and evolutionary scenario. As an example, we have seen that a spatial analysis based on a sample IRAS-selected variables led to the inference of a inner-Galaxy bar. The kinematics predicted by the stellar response to this bar are supported by the existing data. Similarly, the predicted signature of the rotating outer spheroid is in conflict with the same data. Additional stellar velocities (only several hundred are in the current samples!) will certainly help answer these questions.

- AGB stars (esp. carbon stars and Mira variables) are promising standard-candle tracers which might be used to attain the above two goals. We need to further investigate the infrared properties of Galactic Miras with the aim of understanding the P-L relation and providing the means of unique identification. A systematic velocity survey of infrared Mira variables would be especially useful.

ACKNOWLEDGMENTS

I thank Tom Arny and Susan Kleinmann for comments on the manuscript and the conference organizers for their support. This work was supported in part by NASA grant NAG 5-1999.

REFERENCES

Aaronson, M., Blanco, V. M., Cook, K. H., Olsewski, E. W., & Schechter, P. L. 1990, ApJS, 73, 841

Aaronson, M., Blanco, V. M., Cook, K. H., & Schechter, P. L. 1990, ApJS, 70, 637

Allen, L. E., Kleinmann, S. G., & Weinberg, M. D. 1992, ApJ, submitted

Bahcall, J. N., & Soneira, R. M., 1980, ApJS, 43, 73

Becklin, E. E., Matthews, K., Neugerbauer, G., & Willner S. P. 1978, ApJ, 220, 831

Binney, J., & Tremaine, S. 1987. Galactic Dynamics (Princeton: Princeton Univ. Press)

Blanco, V. M., McCarthy, M. F., & Blanco, B. M. 1980, ApJ, 242, 938

Blitz, L., & Spergel, D. N. 1991, ApJ, 370, 205 (BS)
Caldwell, J. A. R., Keane, M. J., & Schechter, P. L. 1991, AJ, 101, 1763
Caldwell, J. A. R.,& Coulson, I. M. 1987, AJ, 93, 1090
Claussen, M. J., Kleinmann, S. G., Joyce, R. R., & Jura, M. 1987, ApJS, 65, 385
Cohen, J. G., Frogel, J. A., Persson, S. E., & Elias, J. H. 1981, ApJ, 249, 481
De Vaucouleurs, G., & Pence W. D. 1978, AJ, 83, 1163
Feast, M. W., Glass, I. S., Whitelock, P. A., & Catchpole, R. M. 1989, 241, 375
Frogel, J. A., Persson, S. E., & Cohen, J. G. 1980, ApJ, 239, 495
Fuenmayor, F. J. 1981, Rev. Mexicana Astr. Ap., 6, 83
Georgelin, Y. M. & Georgelin, Y. P. 1976, A&A, 49, 57
Habing, H. J. 1988, A&A, 200, 40
Habing, H. J., Olnon, F. M., Chester, T., Gillett, F., Rowan-Robinson, M., & Neugebauer, G. 1985, A&A, 152, L1
Harmon, R., & Gilmore, G., 1988, MNRAS, 235, 1025
Hoskin, M. 1964. William Herschel and the Construction of the Heavens (New York: Norton)
IRAS Point Source Catalog 1985, Joint IRAS Science Working Group (Washington, DC: GPO) (PSC).
Jacoby, G. H., Branch, D., Ciardullo, R., Davies, R. L., Harris, W. L., Pierce, M. J., Pritchet, C. J., Tonry, J. L, & Welch, D. L. 1992, AJ, 104, 599
Jura, M., Joyce, R. R., & Kleinmann, S. G. 1989, ApJ, 336, 924
Kuijken, K., & Tremaine, S. 1991, in Dynamics of Disk Galaxies, eds. S. Casertano, P. Sackett, & F. Briggs (Cambridge: Cambridge Univ. Press), p.71
Kuijken, K., & Tremaine, S. 1992, this meeting
Lewis, J. R., & Freeman, K. C. 1989, AJ, 97, 139
Long, K., & Weinberg, M. D. 1992, this meeting
Metzger, M. R., Caldwell, J. A. R., & Schechter, P. L. 1992, AJ, 103, 529
Metzger, M. R., & Schechter, P. L. 1992, preprint and this meeting
Ohta, K., Hamabe, M., & Wakamatsu, K. 1990, ApJ, 357, 71
Rowan-Robinson, M., & Chester, T. 1987, ApJ, 313, 413
Schneider, S. E., Terzian, Y., Purgathafer, A., & Perinotto, M. 1983, ApJS, 52, 399
Schneider, S. E., & Terzian, Y. 1983, ApJL, 274, L61
Te Lintel-Hekkert, P. 1990, PhD thesis, University of Leiden
Van der Veen, W. E. C. J., 1989, A&A, 210, 127
Weinberg, M. D. 1992, ApJ, 384, 81
Weinberg, M. D. 1993, in preparation

THE DYNAMICAL EVOLUTION OF THE GALAXY

Rosemary Wyse
Department of Physics and Astronomy, Johns Hopkins University,
Baltimore, MD 21218.

Gerard Gilmore
Institute of Astronomy, Cambridge, England

ABSTRACT

The dynamical evolution of the Galaxy has left its imprint on the older stellar populations, affecting kinematics, chemical abundances and spatial distributions. Efficient star formation requires retention of gas through many generations of stars, despite the momentum and energy injection of massive stars and supernovae and thus can only occur in sufficiently deep potential wells, or in sufficiently massive clouds. Thus the dynamical evolution, which determines the mass of bound structures at a given time, has a great influence on the rate of chemical enrichment. Element ratios can provide more information than simple overall enrichment, and observations of, for example, [O/Fe] in stars in various locations in the Galaxy should constrain the timescale of formation of different components of the Galaxy, such as the Galactic Bulge. Scatter about the mean trends in chemical enrichment is expected in many models, and can be used to set limits on the homogeneity of the star formation history of the Galaxy. Dynamical evolution also affects the structure of the Galaxy, in particular the vertical thickness of the disk. Hierarchical clustering models with dynamically important substructure are perhaps favored by observations, but with caveats.

1. INTRODUCTION

The morphological type of a galaxy is determined in part by its merging and star formation histories (*e.g.* Silk and Wyse 1993). The properties of the stellar populations of the Galaxy are determined by how and when the material that now makes up the Milky Way Galaxy was accumulated, and whether it was gaseous or stellar at that time. The dynamical evolution is related to the star formation history since sustaining a high star formation rate requires a sufficiently deep potential well to balance the energy and momentum output from massive stars and supernovae, allowing recycling of gas through successive generations of stars. An estimate of this minimum potential well depth for a system forming stars on a dynamical time can be found from the condition that supernova remnants cool before they overlap, and is (Wyse and Silk 1985; Dekel and Silk 1986)

$$\sigma_{3D} \gtrsim 100 \text{kms}^{-1}.$$

There is now supporting evidence (Meurer *et al.* 1992; A. Marlowe priv. comm.) that some dwarf galaxies which are actively star-forming (Blue Compact Dwarfs) are ejecting their interstellar medium in the form of a wind (and also that some extreme starburst galaxies are blowing 'superwinds'; Heckman, Armus and Miley 1990). The skewed metallicity distributions of stars in individual dwarf spheroidal satellite galaxies to the Milky Way (Suntzeff 1993) are also suggestive of abrupt truncation of the chemical enrichment process by a catastrophic event.

Star formation in disks will also be inefficient, since fragmentation, via gravitational instability, in a centrifugally-supported disk has a *maximum* unstable wavelength (unlike normal, non-rotating, Jeans' instability), which for a typical disk is well below this threshold. The steep vertical density gradients in disks compound this, leading to eruption through the disk by 'chimneys' etc. (Norman 1991).

Thus if the Galaxy were built by hierarchical clustering of substructure, there will be inefficient star formation until this critical potential well depth is reached, and again in a disk once formed. The connection between chemistry and dynamics suggests that one may be able to constrain the dynamical history of the Milky Way by studying the old stars which are tracers of the early epochs of Galaxy formation (*cf.* Eggen, Lynden-Bell and Sandage 1962; Searle and Zinn 1978). The thick disk stellar population of the Galaxy is considered in some detail here, including the possibility that it is a relic of previous mergers. The effects of dynamical evolution – as manifest by the star formation history – on the chemical evolution, and in particular on elemental abundances and on the scatter about mean trends, are also discussed.

2. THE THICK DISK

The thick disk was first identified by Gilmore and Reid (1983) from star counts in the South Galactic Pole. Their data were best fit by a two component model, consisting of the standard thin disk, plus a component with a scaleheight of $\sim$ 1200pc, a factor of 3–4 larger than that of the thin disk. Their original interpretation was that the thick disk was simply the metal-poor spheroid responding to the locally-dominant potential of the thin disk, but follow-up observations revealed that the thick disk stars do not show the kinematics or chemical abundances that would be expected if this were the case.

The characteristics of the thick disk at the solar neighborhood are reasonably well defined, but the global properties of this stellar population are not. Locally it can be described by a predominantly old population, the same age as globular clusters of the same chemical abundance (Gilmore and Wyse 1987; Schuster and Nissen 1989; Carney *et al.* 1990), which is about one-third that of the Sun, with a significant dispersion in metallicity, of $\sim$ 0.3 dex (Gilmore and Wyse 1985; Sandage and Fouts 1987; Friel 1987; Laird *et al.* 1988; Morrison, Flynn and Freeman 1990). It is a kinematically 'warm' component, with a lag behind the Sun in rotational velocity in the range $35 - 90 \mathrm{kms}^{-1}$ (*e.g.* Wyse and Gilmore 1986; Sandage and Fouts 1987; Ratnatunga and Freeman 1989), perhaps dependent on height above the plane (Murray 1986; Majewski 1992). The most recent determination of the asymmetric drift of the thick disk is from a very large proper motion survey of the NGP (Soubiran 1992) from Schmidt plates, complete to $V = 17$, covering 7 square degrees, with a baseline of 40 years and 4400 proper motions accurate to 2 milliarcsec/yr. The kinematics of the thick disk stars in this survey are most consistent with a constant lag behind circular velocity of amplitude $\sim 50 \mathrm{kms}^{-1}$, out to $z = 2.5 \mathrm{kpc}$. The vertical velocity dispersion of the thick disk is $\sim 50 \mathrm{kms}^{-1}$ (Sandage and Fouts 1987; Ratnatunga and Freeman 1989).

The radial scalelength of the thick disk can also be an useful discriminant for formation scenarios (see below) but is poorly determined due to the small size of published samples far from the Sun. The extant data are consistent with a scalelength equal to that of the thin disk (Ratnatunga and Freeman 1989).

Larger datasets are needed to decide the evolutionary status of the thick disk: current data are such that it is difficult to distinguish, with statistical confidence, between either a discrete thick disk or a continuous sequence of disks (*e.g.* Gilmore, Wyse and Kuijken 1989; their Fig. 6a).

Hierarchical clustering models for the growth of structure in the Universe, such as Cold Dark Matter (CDM) dominated cosmologies, predict that a typical galaxy formed by a significant amount of merging. The effect of the mergers on the stellar population of the resulting final galaxy depends in large part on the gas content of the merging entities, since gas can radiate energy while stars cannot. Thus the star formation rate again plays a crucial role. Should a thin *stellar* disk exist at the epoch of most merging, then with maximal coupling of satellite orbital energy to the random energies of the disk stars, the disk would be heated such that (*cf.* Ostriker 1990)

$$\Delta v^2_{random} \simeq v^2_{orbital} \frac{M_{satellite}}{M_{disk}}.$$

Tóth and Ostriker (1992) used the observed thinness of the stellar disk of our Galaxy to limit the allowable accretion of stellar satellites since the birth of the Sun to only a few percent of the mass of the disk. However, as discussed by Quinn, Hernquist and Fullager (1993), the disk heating effect of an incoming satellite is quite uncertain, due in part to the many parameters – both of the orbit and of the satellite structure. For example, the orbital energy may rather be transferred to the internal degrees of freedom of the satellite, puffing it up and allowing it to be tidally shredded with little damage to the thin disk. Further, any gas present will most likely be simply shock-heated and subsequently radiate away the orbital energy.

It remains that a thick disk could be a tracer of the merging history of a galaxy. Quinn, Hernquist and Fullager (1993) have carried out the most complete set of simulations of *stellar* mergers between a spherical satellite and a disk galaxy. They find that a generic heated disk will indeed have larger velocity dispersions in each direction (vertical, radial and azimuthal), and will have a lower rotational velocity than the initial disk configuration. Not only is the vertical scalelength increased, but the *radial* scalelength is too. It should be noted that the asymmetric drift – the lag behind circular velocity – depends on *both* the scalelength and amplitude of random motions, in such a way that an increase in radial scalength offsets an increase in radial velocity dispersion. Quinn *et al.* offer a 'standard' model which produces a thick disk with properties very similar to those of our Galaxy but with radial scalelength a factor of $\gtrsim 1.5$ greater than that of the thin disk. This standard model is a merger with a satellite of mass 10% of that of the thin disk, initially on a prograde circular orbit, inclined at 30°.

A somewhat related scenario for formation of the thick disk, in that it too involves the relaxation into virial equilibrium of the Galaxy, was proposed by Jones and Wyse (1983). This latter model appealed to star formation in transient 'pancaking' in the early stages of the dissipative collapse of gas in a non-spherical dark halo. This forms a stellar disk which will be thin initially, but fattens somewhat by violent relaxation as the potential equilibrates. In each case, the present thin disk forms after dynamical evolution ends, with the thick disk older and more metal-poor. As demonstrated by White (1980), in the context of forming an elliptical galaxy by the major merger of two disks, and by Quinn *et al.* explicitly for the case of a thick disk, merging and violent relaxation do *not* erase a pre-existing metallicity gradient. Thus the presence of such a gradient in the thick disk does it not imply that it formed by a predominantly dissipative process.

Star formation during the dissipative settling of gas to the plane of a steady potential can also provide a thick disk, perhaps smoothly connected to the thin disk (Larson 1976; Norris 1987). A gradient in asymmetric drift, as found by Murray (1986) and by Majewski (1992), may favor a continuum of properties from thick to thin disk. The metallicity of the thick disk is such that one might expect enhanced cooling and presumably star formation once this enrichment is reached, since then line radiation takes over from bremsstrahlung as the major cooling mechanism from temperatures typical of the virial temperatures of galaxies (noted by Wyse and Gilmore 1988). Indeed, this may be the physics underlying the results of Burkert, Truran and Hensler (1992) who analyse the 'chemodynamics' of disk collapse and find that a short-lived phase of enhanced star formation may occur, driven by early supernovae, leading to a distinct thick disk. Their calculation is one-dimensional and thus the role of angular momentum is ignored.

It is difficult to distinguish a dissipationless heating origin for the thick disk from a dissipational settling origin without a more detailed understanding of the star formation rate and angular momentum transport/conservation.

One can, however, rule out two scenarios for the formation of the thick disk, both involving scattering of thin disk stars. The thin disk stars show a well-defined increase of random motions with age which can (mostly!) be understood as due to scattering by potential fluctuations in the disk – giant molecular clouds and bits of transient spiral structure (*e.g.* reviewed by Lacey 1991). This increase saturates well below the random motions characteristic of the thick disk, essentially due to the fact that the scatterers are confined to the plane, and lose influence once a star spends little time there. Thus the thick disk cannot be due to this mechanism. Rare, close encounters with high-velocity scatterers, such as may be expected if the dark halo consisted of massive black holes, could heat a small fraction, $\sim 1\%$, of the thin disk to a velocity dispersion more typical of the thick disk (Lacey and Ostriker 1985). However, this mechanism cannot provide an explanation for the thick disk, since one would then expect the thick disk to be a random sample of the thin disk, which it manifestly is not, having for example, a very different metallicity distribution.

Infall of unenriched gas to the thin disk has been suggested as a possible solution to the G-dwarf problem, which is the name given to the fact that the thin disk at the solar neighborhood contains few long-lived G stars of low metallicity, and in particular, fewer than expected if the disk evolved as a closed box, from initially zero metallicity. Allowing the mass of the disk to increase with time, with infall, skews the normalisations and hence the expected number of oldest low metallicity stars (*e.g.* Audouze and Tinsley 1976). Infall has also been appealed to to sustain star formation in some external galaxies, where the detected gas is inferred to last only a fraction of a Hubble time (*e.g.* Kennicutt 1983), and to help provide a suitably cold component for long-lived spiral structure (*e.g.* Sellwood and Carlberg 1984). However, there is no observational evidence to support the smooth infall usually assumed, but rather the available observations favor accretion of gas-rich, metal-poor companion galaxies (Sancisi 1990). This would presumably be accompanied by heating of the disk and a comprehensive survey for thick disks in external galaxies is clearly desirable.

The S0 galaxy NGC4550 provides intriguing observational input to the fate of disks in mergers. Rubin *et al.* (1992) discovered the existence of two co-spatial, approximately equal luminosity *counter-rotating* disks, which have the same exponential profile. The galaxy is normal in its overall properties, such as obeying the Tully-Fisher relationship, and the stellar populations in both disks are apparently typical of old disk stars. The current understanding of stellar mergers outlined above implies that this galaxy could not have resulted from a straightforward merger of two equal mass stellar disks, without destruction of the disks. Adiabatic accretion of gas on retrograde orbits is a possibility, but requires a rather contrived star formation rate to produce the apparently co-eval stellar population in each disk (Rix *et al.* 1992). However, this galaxy serves to remind us that disks can apparently somehow survive a merger between comparatively massive systems despite our limited understanding of the process. A less massive counter-rotating disk has been identified in the Sa/Sb galaxy NGC 7217 by Merrifield and Kuijken (1992). These authors suggest that if galaxies form by the merging of many smaller subsystems, the merger may produce both a counter-rotating subsystem and the bulge/stellar halo. Spiral structure is suppressed in counter-rotating disks (Kalnajs 1977), leading to a trend of larger bulge – less obvious spiral structure, as observed. A systematic study of disk galaxies for counter-rotating subsystems is required.

3. SCATTER IN METALLICITY AND ELEMENT RATIOS

The overwhelming majority of published models for the chemical evolution of the Galactic disk are concerned only with fitting the mean trend with time, assuming a one-to-one correlation between time and metallicity. This is due in part to the large observational uncertainties of earlier datasets (see Tinsley 1975 who never-the-less investigated the possible influence of a spread in metallicity on the G-dwarf problem). However, real intrinsic scatter has now been unambiguously detected in at least two independently selected solar neighborhood samples of F/G stars, with scatter of a few Gyr at given metallicity, or a few tenths of a dex at given age (see Nissen 1992 and Freeman 1993). Recent determinations of *stellar* abundances in the Orion association (Cunha and Lambert 1992; see also Gies and Lambert 1992) have shown that the stars have the same low, sub-solar chemical abundance as has long been known for the associated HII region, in particular for elements not depleted onto grains. The scatter thus continues to the present epoch. Indeed, one could plausibly argue that the scatter is as large as any mean trend. This scatter has to be taken into account in models of the evolution of the disk. White and Audouze (1983) did analyse stochastic effects in chemical evolution, but they used the instanteous recycling approximation which is poor for comparison with stellar abundance measurements, which are dominated by iron (as discussed below a large fraction of iron is ejected to the ISM on long timescales). It should be remembered that a typical disk star, of age $\gtrsim$ 3Gyr, has kinematics corresponding to epicyclic radial excursions of $\lesssim$ 3kpc. Thus a mean metallicity gradient of -0.1 dex/kpc (*e.g.* Friel 1993) would produce scatter of a few tenths of a dex. Star formation should be modelled as a local process, within molecular cloud complexes, forming stellar associations with internal enrichment prior to disruption (*cf.* Tinsley 1976).

Element ratios, by contrast, show little intrinsic scatter (e.g. Gratton 1993). Element ratios are very important for understanding galactic evolution, in part since different elements come from different sources, and are ejected into the interstellar medium on different timescales (Tinsley 1979). For example, oxygen is predominantly synthesized in short-lived, massive stars that explode in Type II supernovae, while iron has an additional important source in Type Ia supernovae. The best current model, but by no means generally accepted, for these latter supernovae invokes the eventual merging, driven by gravitational radiation, of a pair of white dwarfs, with total mass greater than the Chandrasekhar mass (Iben and Tutukov 1987). Thus the explosion timescale is set not by intrinsic stellar parameters, but rather by orbital parameters, and can be a Hubble time (note that this is a problem for advocates of white dwarfs as the dark matter in the haloes of galaxies, unless one suppresses the binary fraction of the progenitor stars by several orders of magnitude below that of even the old, metal-poor stars in our Galaxy; Smecker and Wyse 1991). Detailed models suggest that half of the possible progenitors (white dwarf binaries of suitable mass) from a single burst of star formation explode in $\sim 10^9$ yr ($e.g.$ Smecker-Hane and Wyse 1992). Current understanding of supernova rates and nucleosynthesis implies that for a solar neighborhood IMF, Type II supernovae provide a yield of about $\frac{1}{4}$ Fe$_\odot$, while Type Ia provide a yield of $\gtrsim$ Fe$_\odot$ ($cf.$ Rana 1991). Thus a stellar population whose mean iron abundance is above solar metallicity either self-enriched sufficiently slowly that Type Ia supernovae contributed fully, or had an IMF very biased towards massive stars.

The pattern of [O/Fe] $versus$ [Fe/H] for halo stars in our Galaxy – a plateau at low [Fe/H] with a turndown at higher [Fe/H] (reviewed by Wheeler, Sneden and Truran 1989) – is easily understood with the identification of a long-lived source of iron (Tinsley 1979). It is very important to pin down, observationally, the iron abundance at which the element ratios reflect enrichment by Type Ia supernovae, since this allows identification of a timescale, $\sim$ 1Gyr, for the duration of the star formation in the progenitors of the halo stars that are represented in the sample ($e.g.$ Wyse and Gilmore 1988; Truran 1987; Matteucci and François 1992). As discussed by York (this volume), the trend of elemental abundances in the metal-line absorbers along the line of sight to high redshift quasars can be used in a similar way to constrain the timescale of chemical evolution at other locations.

Further information is contained in the plot of [O/Fe] $versus$ [Fe/H]. The lack of intrinsic scatter, especially about the location of the turndown, has important ramifications for the evolution of the halo (Wyse and Gilmore 1988). The evolution of three independent star-forming regions, such as is envisaged in 'chaotic' models for the formation of the halo ($cf.$ Searle and Zinn 1978), is indicated in Figure 1.

One region, indicated by asterisks, forms with a massive star IMF with slope biased towards massive stars compared with that in the solar neighborhood, which results in a higher enhancement of [O/Fe] at early times (oxygen is a steeply increasing function of progenitor mass, whereas the iron from Type II supernovae is more constant; see Wyse and Gilmore 1992). The higher nucleosynthetic yield of this region means a higher [Fe/H] can be reached prior to the onset of Type Ia supernovae, for given star formation rate. Of the remaining two 'blobs', the one indicated by squares has a higher star formation rate and so has more enrichment through successive generations of massive stars prior to the onset of chemically-detectable Type Ia supernovae. It should be clear that to produce the smooth trend seen in halo stars, unless all 'blobs' have identical enrichment histories, the 'blobs' cannot continue internal enrichment beyond the metallicity at which the turndown due to Type I supernovae occurs, and all 'blobs' must have the same massive star IMF, since otherwise there would be scatter, contrary to the observations. The suggestion (Matteucci and François 1992) that the turndown indicating significant enrichment by Type Is occurs at [Fe/H] ∼ −1.7 is then, in the context of a 'chaotic' model, somewhat difficult to reconcile with the observations, given that this is more metal poor than the bulk of the field halo stars.

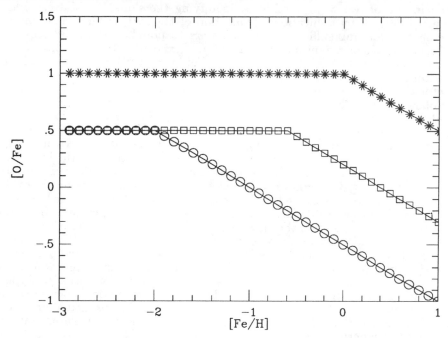

Figure 1.   Schematic evolution of the [O/Fe] ratio as a function of [Fe/H] for three star-forming regions with different star formation rates and/or massive star IMFs. The asterisks indicate the evolution of a region with an IMF dominated by massive stars, compared to the other two, which have a massive star IMF similar to the halo subdwarfs. Of the remaining two regions, the squares indicate the evolution with a faster star formation rate.

Type Ia supernovae can provide significant iron enrichment *without* accompanying star formation (*cf.* Gilmore and Wyse 1991 for a particular application to a galaxy with a star formation rate characterised by bursts with intervening hiatuses). Thus a further requirement of the lack of scatter is that the gas ejected from the 'blobs', assuming they did have different enrichment histories, did not form stars at the solar neighborhood, since otherwise stars could form at low [Fe/H] and low [O/Fe], as in the continuation of the circles in Figure 1. The central regions of the Galaxy provide a natural receptacle for the halo gas, in part due to its low angular momentum (Carney 1990; Wyse and Gilmore 1992), provided it is not permanently ejected in a wind.

Element ratios will also be important discriminants among models of the formation of the nuclear bulge, constraining the accumulation time for material at the center, if the bulge is not primordial, and the star formation timescale. These timescales can be measured relative to the timescale for the enrichment by Type Ia supernovae, since this produces observable signatures in the element ratios. If the proto-bulge material accumulated, and formed stars, more rapidly than this (of order a Gyr), then only Type II supernovae are available for chemical enrichment. In this case, the high mean metallicity of the bulge population, well above solar metallicity (Rich 1988), requires an IMF biased towards massive stars, when compared to that of the solar neighborhood. The IMF can be biased in one of two ways – either by modifying the slope (*cf.* Matteucci and Brocato 1990), or by truncation. Truncation of the IMF is limited by the requirement of forming sufficient low-mass stars to populate the main sequence as observed in color-magnitude diagrams (*cf.* Terndrup 1988). A conservative lower limit is then $0.5\,M_\odot$. As tabulated by Sandage (1986), a variation in lower mass limit from $0.1\,M_\odot$ to $0.5\,M_\odot$ decreases the locked-up fraction by at most a factor of 1.3. This is not sufficient to increase the Type II yield (~ 0.25 of the solar iron abundance for a normal IMF) high enough to provide the iron in bulge stars (mean well above solar). The more plausible way to increase the nucleosynthetic yield is to flatten the slope of the IMF, which provides a clear observational signature by predicting a plateau in [O/Fe] at some value *significantly higher than* that of the halo stars (assuming of course that the halo stars formed with a normal IMF; Wyse and Gilmore 1992). Iron will eventually be synthesised in Type I supernova and ejected into the bulge, releasing about the same amount of iron as ejected by Type II supernovae, for a reasonable range of IMFs and models of Type I supernovae (the Type I supernovae from the halo background population that is also present are merely perturbations to the present model, due to the much larger mass, $\sim \times 10$, of the bulge). This gas must be accounted for in the model. Such iron-rich gas, which is not incorporated in the main bulge population, is produced in any model for the bulge which has sufficiently rapid star formation that it is essentially complete prior to Type I explosions. Should this gas sustain a low level of star formation, then one may expect to find extremely iron-rich stars which have very low oxygen abundances (*cf.* Gilmore and Wyse 1991). The relatively slow, and uncorrelated, rate at which Type I supernovae explode contrasts with Type II supernovae, and it is unlikely that the iron-rich gas will be maintained at a high temperature (and thus will probably not be a detectable X-ray source).

If the star formation were slow, following rapid accumulation, allowing the incorporation of the ejecta from Type Ia supernovae, then the IMF can be normal, producing a pattern of element ratios like that seen in the solar neighborhood, with a downturn at higher iron abundances (the location of the downturn is set by the star formation rate).

If the accumulation of proto-bulge gas were due to the outflow of gas from slow, chaotic halo formation, then the possibility of incorporation of iron from Type Ia supernovae prior to star formation in the bulge can produce stars with [Fe/H] $\gtrsim -1$ but with [O/Fe] ~ 0 (rather than the enhanced oxygen seen in the solar neighborhood at this iron abundance).

ACKNOWLEDGEMENTS

RFGW acknowledges support from the Alfred P. Sloan Foundation, from the NSF and from NASA.

REFERENCES

Audouze, J. and Tinsley, B.M. 1976 ARAA, 14, 43
Burkert, A., Truran, J.W. and Hensler, G. 1992 ApJ, 391, 651
Carney, B.W., Latham, D.W. and Laird, J.B. 1990 AJ, 99, 572
Carney, B.W. 1990 in 'Bulges of Galaxies', eds B. Jarvis and D. Terndrup (ESO, Garching) p26.
Cunha, K. and Lambert, D.L. 1992 ApJ, 399, 586
Dekel, A. and Silk, J. 1986 ApJ, 303, 39
Eggen, O.J., Lynden-Bell, D. and Sandage, A. 1962 ApJ, 136, 748
Friel E. 1993, in 'The Globular Cluster – Galaxy Connection', eds G. Smith and J. Brodie (A.S.P., San Francisco), in press.
Friel, E. 1987 AJ, 93, 1388
Freeman, K.C. 1993, in 'The Globular Cluster – Galaxy Connection', eds G. Smith and J. Brodie (A.S.P., San Francisco), in press.
Gies, D.R. and Lambert, D.L. 1992 ApJ, 387, 673
Gilmore, G. and Reid, I.N. 1983 MNRAS, 202, 1025
Gilmore, G. and Wyse, R.F.G. 1985 AJ, 90, 2015
Gilmore, G. and Wyse, R.F.G. 1987, in 'The Galaxy', eds G. Gilmore and B. Carswell, (Reidel, Dordrecht) p247
Gilmore, G. and Wyse, R.F.G. 1991 ApJ, 367, L55
Gilmore, G., Wyse, R.F.G. and Kuijken, K. 1989 ARAA, 27, 555
Gratton, R. 1993, in 'The Globular Cluster – Galaxy Connection', eds G. Smith and J. Brodie (A.S.P., San Francisco), in press
Heckman, T.M., Armus, L. and Miley, G.K. 1990 ApJS, 74, 833
Iben, I. and Tutukov, A.V. 1987 ApJ, 313, 727
Jones, B.J.T. and Wyse, R.F.G. 1983 AA, 120, 165
Kalnajs, A. 1977 ApJ, 212, 637
Kennicutt, R.C. 1983 ApJ, 272, 54
Lacey, C. 1991, in 'The Dynamics of Disc Galaxies', ed. B. Sundelius (University of Gothenburg Press), p257
Lacey, C.G. and Ostriker, J.P. 1985 ApJ, 299, 633
Laird, J.B., Rupen, M.P., Carney, B.W. and Latham, D.W. 1988 AJ, 96, 1908
Larson, R.B. 1976 MNRAS, 176, 31
Majewski, S.R. 1992 ApJS, 78, 87
Matteucci, F. and Brocato, E. 1990 ApJ, 365, 539
Matteucci, F. and François, P. 1992, preprint
Merrifield, M. and Kuijken, K. 1992, preprint
Meurer, G.A., Freeman, K.C., Dopita, M.A. and Cacciara, C. 1992 AJ, 103, 60
Morrison, H., Flynn, C. and Freeman, K.C. 1990 AJ, 100, 1191
Murray, C.A. 1986 MNRAS, 223, 64
Nissen, P.E. 1992 in 'Elements and the Cosmos' eds R.J. Terlevich et al. (C.U.P., Cambridge) in press
Norman, C.A. 1991 in IAU Symposium 144 'The Interstellar Disk-Halo Connection in Galaxies', ed H. Bloemen, (Reidel, Dordrecht) p337
Norris, J. 1987 in 'The Galaxy', eds G. Gilmore and B. Carswell, (Reidel, Dordrecht) p297
Ostriker, J.P. 1990 in 'Evolution of the Universe of Galaxies', ed R.G. Kron, A.S.P. Conf. Series 10, (A.S.P., San Francisco) p25
Quinn, P.J., Hernquist, L. and Fullager, D.P. 1993 ApJ in press
Rana, N.C. 1991 ARAA, 29, 129
Ratnatunga, K.U. and Freeman, K.C. 1989 ApJ, 339, 126

Rich, R.M. 1988 AJ, 95, 828
Rix, H.-W., Franx, M., Fisher, D. and Illingworth, G. 1992 ApJ, 400, L5
Rubin, V.C., Graham, J.A. and Kenney, J.D.P. 1992 ApJ, 394, L9
Sandage, A. 1986 AA, 161, 89
Sandage, A. and Fouts, G. 1987 AJ, 92, 74
Sancisi, R. 1990 in 'Windows on Galaxies', eds G. Fabbiano, J.S. Gallagher and
        A. Renzini (Kluwer, Dordrecht) p199
Schuster, W.J. and Nissen, P.E. 1989 AA, 222, 69
Searle, L. and Zinn, R. 1978 ApJ, 225, 357
Sellwood, J. and Carlberg, R. 1984 ApJ, 282, 61
Silk, J. and Wyse, R.F.G. 1993 Phys Rep to appear
Smecker-Hane, T.A. and Wyse, R.F.G. 1992 AJ, 103, 1621
Smecker, T.A. and Wyse, R.F.G. 1991 ApJ, 372, 448
Soubiran, C., 1992 PhD Thesis, Observatoire de Paris
Suntzeff, N. 1993, in 'The Globular Cluster – Galaxy Connection', eds G. Smith
        and J. Brodie (ASP : San Francisco), in press
Terndrup, D.M. 1988 AJ, 96, 884
Tinsley, B.M. 1975 ApJ, 197, 159
Tinsley, B.M. 1976 ApJ, 208, 797
Tinsley, B.M. 1979 ApJ, 229, 1046
Tóth, G. and Ostriker, J.P. 1992 ApJ, 389, 5
Truran, J.W. 1987 in proc. 13th Texas Symposium on Relativistic Astrophysics,
        ed M.P. Ulmer (Singapore, World Scientific) p430
Wheeler, J.C., Sneden, C. and Truran, J.W. 1989 ARAA, 27, 279
White, S.D.M 1980 MNRAS, 191, 1P
White, S.D.M and Audouze, J. 1983 MNRAS, 203, 603
Wyse, R.F.G. and Gilmore, G. 1986 AJ 91 855 (Erratum 1986 AJ 92 1215)
Wyse, R.F.G. and Gilmore, G. 1988 AJ, 95, 1404
Wyse, R.F.G. and Gilmore, G. 1992 AJ, 104, 144
Wyse, R.F.G. and Silk, J. 1985 ApJL, 296, L1

GALAXY MODEL PARAMETERS
USING NUMERICAL MAXIMUM LIKELIHOOD ESTIMATION

Kavan U. Ratnatunga
Center for Astrophysical Science, Johns Hopkins University, Baltimore, MD 21218

Stefano Casertano
Astronomy Department, University of Illinois, Urbana, IL 61801

ABSTRACT

We discuss a numerical algorithm based on maximum likelihood for estimating parameters for models of the Galaxy. We use simultaneously all the information available in a catalog of stellar data in a global optimization, to derive unbiased estimates of intrinsic stellar properties, such as luminosity and velocity dispersion. The likelihood function is defined in the observed domain using quantities such as photoelectric photometry, line-of-sight velocity, proper motion, trigonometric parallax, and metallicity. Individual stars included in the statistical analysis can have different amounts of information available. This method includes an explicit treatment of observational errors, can identify outliers objectively and allows use of stellar data with relatively large errors. It can self-consistently detect and correct for systematic deviations in the observations, such as zero point residuals or underestimated errors.

1. INTRODUCTION

The study of stellar statistics has a long and distinguished tradition. Since the days of Hipparchus, a major task of astronomers has been to compile star catalogs as large and as accurate as possible.

The traditional approach of statistical astronomy given in Trumpler & Weaver (1953) has had only few advances over the next four decades of astronomical research. Among them, particularly for the problem of calibrating the luminosity of stars with measured parallaxes, are Jung (1971), who developed a maximum-likelihood scheme to deal with complete, magnitude-limited samples, and Hansen (1973), who developed a method for non-linear optimization of the spectroscopic calibration. Turon Lacarrieu & Crézé (1978) first suggested the use of magnitude constraints to remove the divergence found by Lutz & Kelker (1973) in parallax-based distance estimators. Smith (1988) wrote a series of papers on the calibration problem, specifically discussing the advantages of maximum likelihood in this context and gave an application to synthetic data.

As computation time was the major constraint, all these methods were based on analytical equations derived using clever approximations, and assumptions about the distribution function and observational error. For example, the standard χ^2 test assumes that error distributions are Gaussian. In this case the likelihood function, defined as the sum of the natural logarithm of the probabilities for individual stars, equals $-\frac{1}{2}\chi^2 +$ constant. The χ^2 minimization is therefore equivalent to a special case of maximum likelihood. The principle of maximum likelihood can however be used for arbitrary probability distributions, and can be implemented by evaluating numerically the individual probabilities for each element (star) from a computer model of the distribution function, which can then be as complex as necessary to model the observations.

In many cases, we are mostly interested in global properties of the stellar distribution. Therefore it is not necessary to estimate an individual distance to each star. We represent distance by a probability distribution over 'true' distance, rather than a single mean value and evaluate the likelihood function by numerically integrating the Galaxy model over 'true' distance. We also need not classify each star into a particular stellar population. For example, a low velocity star, which most probably belongs to a population with low velocity dispersion, has a non-negligible probability of belonging to a component with high velocity dispersion. Attempts to bin the stars into stellar populations and analyzing them separately will bias kinematic parameter estimates. We evaluate the conditional probability of membership in each stellar component and sum over the mixture of stellar populations. This numerical approach is only feasible thanks to the general availability of fast computers.

2. GLOBAL MODELS

The most important advantage of using global models for studying the structure of the Galaxy is to avoid the transformation of observed quantities. Consider the estimation of the space velocity for a star. This involves at least three independent measurements: line-of-sight (radial) velocity; proper motion; and a distance estimator, which in turn may depend on one or more observables. Each measurement will contribute its own error, which is transformed in a complicated way into the derived quantity. Unless the errors are very small, their resulting distribution can be very distorted, and is in general poorly behaved. Even worse, in some cases there may not be *any* well-defined transformation. For example, measurement errors can make the observed trigonometric parallax negative. Although a negative parallax cannot be transformed to a real distance, the probability of making this observation for a star at any 'true' distance is well defined. Global modeling always brings the comparison between model and data into the observed domain, where the errors are well-behaved.

Another useful advantage of global modeling is the flexibility it allows as to which data must be available. To continue the above example, if the line-of-sight velocity is not available for a star, the space velocity for that star cannot be computed, and thus any other data available for that star are wasted. On the other hand, with global modeling the comparison can be made separately in each observable, and thus all available data can be used.

Finally, the global approach can detect and compensate for selection effects in the sample. A numerical computation of the expected distributions of the observables can include the various selection criteria applicable to the catalog. A properly constructed model will show incompleteness with respect to any deficiency by direct comparison.

Among the main proponents of the global modeling approach to galactic structure problems have been Bahcall and Soneira (see references in Bahcall 1986). Following in their path, we have produced the IASG model (Ratnatunga, Bahcall & Casertano 1989), which, thanks to improved software, has enabled us to study all-sky catalogs, including kinematics. The IASG model allows for selection limits to be imposed on any observed quantity and for tailoring a model sample of stars on an observed catalog on a star-by-star basis, thus all but eliminating the consequences of incompleteness of the sample. It has been used to study both the luminosity function and the kinematics of intrinsically bright stars, and to derive the kinematic properties of spheroid and old disk from high-proper motion stars.

The global approach described above is not without some disadvantages. The first is inherent in the process of comparison between model and data. Such comparison can only be done effectively in one observable; with more than one, the comparison becomes more difficult, since the data points cover only sparsely a multidimensional domain. The best that can usually be done is a study of the marginal distribution, or a limited comparison involving correlation coefficients or principal components. Any of these possibilities reduces the information actually used in the test.

Another difficulty is parameter optimization. The process of finding the best model must proceed by trial and error, with all the subjectivity that this entails. The adopted model is necessarily guided by parameters derived using the traditional approach, which studies the distribution of those quantities that, while not directly observed, have a more obvious physical meaning. The global model then essentially complements the classical analysis by checking self consistency of independently derived parameters.

3. MAXIMUM LIKELIHOOD

Both difficulties listed above can be overcome by using the method of maximum likelihood. With this approach, the global Galaxy model is used to numerically derive the *probability* of a set of observables with associated errors. To correct for incompleteness in some observables, say for instance photometry and region of sky surveyed, we use *conditional probabilities* to estimate the probability of other observables, for example the kinematic properties, at the *observed* position, apparent magnitude and color for each star individually. This probability is interpreted as a measure of the *likelihood* of the model given that particular observation. The likelihood of the model given the sample is the product of individual likelihoods for each star in the catalog; in practice, we use its natural logarithm, the *likelihood function*. The best model is the one for which the likelihood function is largest. The matrix of second derivatives (Hessian) of the likelihood function at the maximum is the inverse of the covariance matrix of the model parameters.

There is essentially no limit to the complexity of the model, although the quality and quantity of data available will limit of course the ability to constrain many parameters. Several independent components can be defined; each will have its own density law, geometrical shape, luminosity function, color-magnitude relationship and cosmic scatter, mean motion (including overall motion, rotation, and expansion), and velocity dispersion.

The number of free parameters included to model the observed distribution is determined using the likelihood ratio test. The number of parameters is increased until there is no significant increase in the likelihood function. After estimating the Galaxy model parameters the observed distributions is compared with the global model. Outliers and deviations from the functional form of assumed distributions are detected at this stage. A non-Gaussian error distribution would lead to a poor match between model and data. Also, significant differences will arise in the presence of problems in the data, such as a zero-point error and/or underestimate of the observational errors.

With our method, the information content of each star is exploited to the fullest. No binning is used. Error estimates and correlations are derived for each parameter. Since an explicit cost function is computed for each model, the search for the *best* model is objective, not subjective. Good algorithms are available to locate the maximum; several have been tried, and different ones can be use for each problem to ensure optimum performance and to avoid false (local) maxima.

A practical application of the numerical maximum likelihood method to stellar statistics is due to Casertano *et al.* (1990), who determined the properties of the old disk population of the Galaxy from kinematics and photometry only, without any metallicity information and without any preconceived notion of its properties. We have subsequently used variants of this method, with different combinations of photometry and observables: parallaxes (Ratnatunga & Casertano 1991), metallicity and line-of-sight velocity (Ratnatunga & Yoss 1991), and proper motions and line-of-sight velocity (Ratnatunga & Upgren 1992).

4. CONCLUSIONS

With personal workstations becoming as fast as main frame computers of a decade ago, statistical analysis of astronomical data need nolonger be constrained by analytical approximations. Astrometric data has been costly to acquire in terms of human resources, instrumentation, and time and numerical algorithms are useful to extract unbiased parameters from them. We have developed one such maximum likelihood algorithm for the analysis of stellar data and we have tested it on existing ground-based catalogs in preparation for large uniform database from the Hipparcos satellite which will surely give a giant leap for astronomy.

ACKNOWLEDGMENTS

Our work owes its initial impetus to John Bahcall, who initiated both authors to the mysteries of Galactic modeling and instigated the use of maximum likelihood for such problems. We have benefited during the years from discussions with many colleagues, and especially with Bruce Carney, Heinz Eichhorn, Ken Freeman, Ivan King, Mario Lattanzi, Art Upgren, Ken Yoss, and Wayne Warren. We gratefully acknowledge partial NASA support through grant NAGW-2945.

REFERENCES

Bahcall, J. N. 1986 ARAA, 24 577
Casertano S., Ratnatunga K. U., & Bahcall, J. N. 1990, ApJ, 357, 435
Hansen, L. 1973, A&A, 27 355
Jung, J. 1971, A&A, 11, 351
Lutz T. E. & Kelker, D. H. 1973, PASP, 85, 573
Ratnatunga, K. U. Bahcall J. N. & Casertano S. 1989 ApJ 339, 106
Ratnatunga, K. U. & Casertano, S. 1991, AJ, 101, 1075
Ratnatunga, K. U. & Upgren A., 1992, ApJ, submitted
Ratnatunga, K. U. & Yoss, K. M. ApJ, 1991, 377, 442
Smith, H. 1988, A&A, 198, 365
Turon Lacarrieu C. & Crézé M. 1977, A&A, 56, 273
Trumpler, R. J. & Weaver, H. F. 1953, Statistical Astronomy (Berkeley: University of California Press)

THE IMF AT LOW MASSES

A model of stellar evolution of the galactic disc

Misha Haywood

Observatoire de Besançon, BP 1615, 25010 Besançon cedex

EARN::"haywood@frobes51"

1. INTRODUCTION

The difficulty in obtaining reliable ages for field stars makes the search for variations in the star formation rate of our Galaxy an elusive objective. While interesting results have been recently obtained from the white dwarf luminosity function or the chromospheric activity of red dwarfs, constraints remain weak because of the limited size of the samples available for these objects. The work presented here is based on the suggestion that the investigation of the SFR and IMF in the disc of our Galaxy could take advantage of the abundant star-count data that now exist in magnitudes, colours, proper motions or radial velocities, for samples that are usually well defined statistically. These observational parameters are all linked to the intrinsic properties of the stars (mass, age, dynamical evolution). They can therefore be utilized to trace the stellar formation history of the disc of our Galaxy, through a model of stellar evolution of the Galactic Disc, in connection with a model of stellar population synthesis. The present poster briefly describes the model that allow us to compute the theoretical luminosity function $\Phi(M_v, \text{colour})$ for any IMF and SFR, and give its main inputs. The analysis of star count data will be publish in future studies. The model is used here to compute luminosity functions which are compared to the local luminosity function on the visual magnitude range $4 \leq M_v \leq 16$. We discuss the consequences for the IMF of the Wielen Dip feature and the slow decrease in the observational LF at $M_v > 12$. We give an estimate of the slope of the IMF in the corresponding mass range $(0.1 \leq M/M_\odot \leq 1)$.

2. MODEL OF STELLAR EVOLUTION IN THE GALACTIC DISC

The model is based on a code developed by Rocca-Volmerange et al. (1981). While it is a classical model of spectrophotometric evolution, it has been somewhat adapted to allow comparisons with non-integrated quantities, such has magnitudes and colours of field stars. We model the distribution of stars in the theoretical HR diagram making the classical assumption of an IMF $\Phi(m)$ independant of time. The SFR $\Psi(t)$ represents the total mass of stars born per unit time per unit mass of galaxy. The number of stars appearing on the ZAMS at time t on a given mass range is obtained by integrating the following equation

$$d^2N = \Phi(m)\Psi(t)dmdt$$

The stars are binned and evolve in the HR diagram on a finite number of tracks. These tracks have been obtained by interpolation of the following grid of stellar evolutionary tracks.

2.1 The library of evolutionary tracks

The library of stellar evolutionary tracks contains 33 evolutionary sequences with masses between 0.12 and 120 $M_\odot$:

(1) Low mass stellar evolutionary tracks with $0.12 \leq M/M_\odot < 1.00$ are from VandenBerg (private communication) and have been computed with Alexander opacities for Y=0.27 and Z=0.0169.

(2) Tracks with $1 \leq M/M_\odot \leq 120$ come from Schaller et al (1992), with Z=0.02 and Y=0.30. These tracks have been computed with updated input data, including the new OPAL opacities (Rogers & Iglesias, 1992). They allow for moderate overshooting (l/Hp=0.20).

2.2 The conversion of the HR diagram into the CM diagram

The conversion of the theoretical HR diagram to the CM diagram is done using the bolometric corrections (BC) of Flower (1977) for supergiants and main sequence stars with spectral type earlier than K0, Flower (1975) for M giants, together with the temperature scale from Tsuji (1981). The conversion of effective temperature to colours were made using tables from Schmidt-Kaler (1982), through $LogT_{eff}$-Spectral type relation from de Jager & Nieuwenhuijzen (1987).

3. THE SLOPE OF THE IMF AT LOW MASSES ($M < 1M_\odot$)

The IMF for low mass stars has received an increasing interest in the last decade, partly because of the missing mass problem in the Galactic Disc, which has sometimes been thought to exist in the form of brown dwarfs. Most of the discussion has been based on the behavior of the IMF at low masses $M < 1M_\odot$ and its extension to the very low mass end.

3.1 The theoretical Luminosity Function

In this mass range, the shape of LF is almost entirely due to the slope of the IMF, via the mass-luminosity relation. The lack of good bolometric corrections at the low mass end of the main sequence has been stressed by many authors (Malkov, 1987, Reid 1987). In order to overcome this problem, we selected from the recent litterature bolometric luminosities determined by UBVRIJHKL photometry for large parallaxe K and M stars. The bolometric luminosities, together with absolute visual magnitudes from the Gliese and Jahreiss (1992) Catalogue of Nearby Stars provided us with a ($LogL/L_\odot$, M_v) relation (see figure 1). This relation was used to convert theoretical luminosities to absolute visual magnitudes, and to calculate the theoretical $\Phi(M_v)$.

In order to have a LF comparable to the observed LF, the increase with age of the scale height of the stars due to dynamical evolution in the disc has to be taken into account. This was done using the vertical distribution of the stars given by Bienaymé et al. (1987). The resulting number of stars is then counted in 1 magnitude interval.

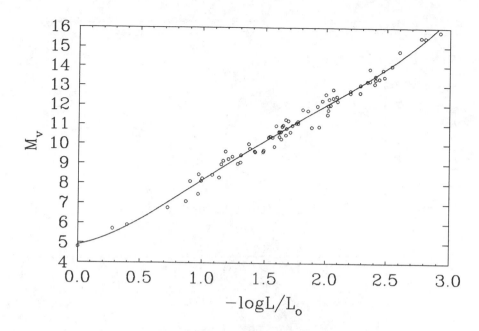

FIG. 1 Plot of bolometric luminosities from the litterature for 75 stars, versus M_v magnitudes from Gliese & Jahreiss (1992).

3.2 Comparison with the observed Luminosity Function

All studies of the IMF from solar neighbourhood stars at low masses must account for a dip at $M_v = 6 - 8.$, and a flattening or a decrease after a maximum reached at $M_v = 12 - 13$. There are some indications that the non-linearity of the mass-luminosity relation in this range of masses could at least partly account for these features (Mazzitelli, 1972, D'Antona & Mazzitelli, 1983).
We plotted on Fig 2 four LFs computed with slopes x=0., x=0.55, x=0.85, x= 1.35 (Salpeter IMF). It clearly illustrates that slopes x=0 or x=1.35 can't give a satisfactory fit on the whole range from 0.1 to 1 $M_\odot$. More reasonnably, best probable values of the slope lies in the range [0.55, 0.85], in good agreement with the value derived by D'Antona & Mazzitelli (1986) (0.68 ± 0.1).

3.3 The dip at $M_v = 6 - 8$

The dip at $M_v = 6 - 8$ has long been proved a real feature in the LF (Upgren & Armandroff, 1981). It has sometimes been taken as indicative of a discontinuous or bimodal IMF, while it could also be ascribed to the appearance of an inflexion point in the mass-luminosity relation caused by the increasing importance of the H^- opacity (Kroupa et al, 1991). It is demonstrated on fig 2 that the dip is well reproduced by the computed LF with continuous power-law IMF, and there is no need to invoke a bimodal IMF to fit the observed LF in this magnitude range. It is worth noting here that, due to the large uncertainties in the LF at $M_v > 12.$, a change of slope at $M < 0.3 M_\odot$ cannot be excluded, and higher values of x would give acceptable fit to the LF.

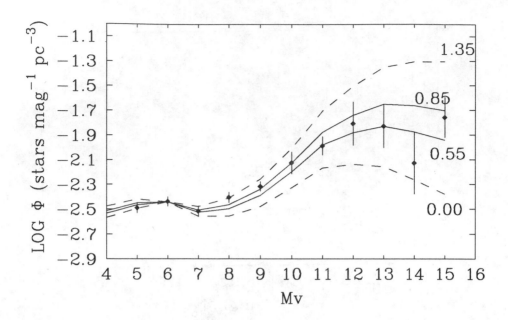

FIG. 2 Computed luminosity functions for four different power-law IMF, with index 1.35, 0.85, 0.55 and 0. Diamonds are the observational LF from Wielen, Kruger and Jahreiss (1983) up to M_v=12., then LF from Dahn, Liebert and Harrington (1986). Error bars are statistical counting uncertainties. All computed LFs are normalized to the observational value at $M_v = 6$.

REFERENCES

Bienaymé O., A.C. Robin, Crézé M., 1987, A&A 180, 94

D'Antona, F., Mazzitelli, I. 1983, A&A 127, 149

D'Antona, F., Mazzitelli, I., 1986, A&A 162, 80

de Jager, C. & Nieuwenhuijzen H. 1987, A&A 177, 217

Flower P.J., 1975, A&A 41,391

Flower P.J., 1977, A&A 54,31

Gliese, W., Jahreiss, H., 1992, The Third Catalogue of Nearby Stars, Preliminary Version, in Selected Astronomical Catalogues, Volume I, CD-ROM eds. L.E. Brotzman & S.E. Gessner

Kroupa P., Tout C., Gilmore G.F., 1991, MNRAS, 251, No. 2, 293

Malkov, O., Y., 1987, Astrophysics 26(3), 288

Mazzitelli, I. 1972, Astrophys. Space Sci. 17, 378

Reid N., 1987, MNRAS 225, 873

Rocca-Volmerange B. Lequeux, J., Maucherat-Joubert, M. 1981 A&A 104, 177

Rogers F.J. & Iglesias C.A., 1992, ApJS 79, 507

Schaller,G., Scharer,D., Meynet, G., Maeder, A., 1992, preprint

Schmidt-Kaler T., Landolt-Bortein, 1982, vol. 2, subvol. 6, Springer-Verlag

Tsuji, T., 1981 A&A 99, 48

Upgren A.R., Armandroff, T.E. 1981, AJ 86, 1898

SHEAR, TIDES, AND CLOUD FORMATION IN SPIRAL ARMS

Charles F. Gammie
Astronomy Dept., University of Virginia
Box 3818, Charlottesville, VA, 22903-0818

This paper considers how dynamical processes that affect cloud formation in the galactic disk depend on the shear rate, which is proportional to Oort's A constant. The low shear rate expected in spiral arms has both stabilizing and destabilizing effects that, on balance, encourage cloud formation.

To show that the effect of shear can be counterintuitive, consider the angular momentum of a small patch of disk about the patch's center of mass (Mestel, 1966). In what follows it is assumed that spiral arms are tightly wound, so that A can be sensibly defined throughout the disk. The fluid velocity at separation $(\delta R, \delta \phi)$ from the patch center of mass, measured in an inertial frame moving with the patch, is

$$\mathbf{V} \approx \mathbf{\Omega} \times \delta \mathbf{R} - 2A\, \delta R\, \hat{\phi}$$
$$= -\Omega (R\delta\phi)\hat{\mathbf{r}} + (\Omega - 2A)\, \delta R\, \hat{\phi} \tag{1}$$

Here $\Omega > 0$ is the rotation frequency of the patch around the galactic center and $A \equiv -(1/2)d\Omega/d\ln R > 0$ is Oort's constant. Then the *spin* angular momentum of the patch is

$$\mathbf{L}_{spin} = \int \delta \mathbf{R} \times \mathbf{V}\, \rho\, d^3 r$$
$$= \int \left[\Omega\, (R\delta\phi)^2 + (\Omega - 2A)\, \delta R^2 \right] \rho\, \hat{\mathbf{z}}\, d^3 r, \tag{2}$$

which implies that

$$\frac{\partial \mathbf{L}_{spin} \cdot \hat{\mathbf{z}}}{\partial A} = -2 \int \delta R^2\, \rho\, d^3 r \quad < 0. \tag{3}$$

The partial derivative signifies that the density and region of integration are fixed. The shear is thus destabilizing in the sense that it decreases the spin angular momentum, which will oppose collapse if the patch is condensed but not torqued.

On the other hand, reduced shear is a symptom of lower tidal stress. The difference in radial acceleration between the inner and outer radial edge of a nonrotating cloud of width δR is

$$\delta R\, T \equiv \delta R\, \frac{\partial}{\partial R}(-\Omega^2 R)$$
$$= \delta R\, \left(4A\Omega - \Omega^2 \right) \tag{4}$$

Here $T > 0$ means the tide tends to pull things apart. To leading order the circular frequency Ω is fixed, but A can change a lot if Ω changes by even a small amount on a small enough scale. Thus

$$\delta T \simeq 4\Omega\, \delta A, \tag{5}$$

i.e., change in tidal stress is proportional to change in shear rate. Lowering tidal stress may encourage the growth of existing clouds (Stark and Blitz, 1978).

Why is the shear rate lower in spiral arms? It is possible to simply impose a wavelike disturbance in the potential and show that at the potential minima the Oort constant A is also a minimum. An entirely independent approach uses the conservation of *potential vorticity* ξ:

$$\frac{D\xi}{Dt} \equiv \frac{D}{Dt}\left(\frac{\nabla \times \mathbf{v} + 2\Omega}{\Sigma}\right) = 0 \tag{6}$$

where $\mathbf{v}$ is the velocity in the rotating frame, Σ is the surface density, and D/Dt is the convective derivative. The conservation of potential vorticity then gives the useful relation

$$\delta A = -(\Omega - A_{av})\frac{\delta\Sigma}{\Sigma_{av}}, \tag{7}$$

where the subscript av implies an average over azimuth. Since $\Omega - A_{av}$ must be positive (so that the disk is stable by the Rayleigh criterion) density maxima corespond to minima in the shear rate.

Strictly speaking, potential vorticity is conserved only in the absence of viscosity, heating and cooling, and magnetic fields. However, large scale magnetic fields and the effective cloud viscosity are probably too weak to cause the potential vorticity to evolve much within a spiral arm crossing time in our galaxy (Gammie, 1992).

A good indicator of gravitational stability in a disk is the Toomre parameter $Q \equiv c\kappa/\pi G\Sigma$ (c is the sound speed and κ is the epicyclic frequency). Recall that $Q > 1$ implies local stability of a thin disk. How does the Q parameter change in spiral arms? Our strategy is to write both κ and Σ as a function of the shear rate. Now $\kappa = \sqrt{4\Omega(\Omega - A)}$, which increases as A decreases. In this sense lower shear rate raises Q and is stabilizing. Conservation of potential vorticity allows us to relate the surface density to the shear rate:

$$\Sigma = \Sigma_{av}\left(\frac{\Omega - A}{\Omega - A_{av}}\right) \tag{8}$$

so that Σ is also increases as A decreases. In this sense lower shear is destabilizing. If the ISM is isothermal (c is constant), then

$$Q = Q_{av}\sqrt{\frac{\Omega - A_{av}}{\Omega - A}}, \tag{9}$$

so that the destabilizing increase in surface density wins over the stabilizing increase in epicyclic frequency.

Finally, one might worry that some mysterious dynamical process contributes to the growth of disturbances in regions of low shear. A way of testing this possibility is to study the evolution of small perturbations in a shearing patch of disk (Goldreich & Lynden-Bell, 1965; Julian & Toomre, 1966; Goldreich & Tremaine, 1978). The disk model used by these authors is not strictly applicable to spiral arms because it does not allow for divergence of the flow (see Balbus, 1988). Nevertheless it can serve as a rough indicator of the responsiveness of the disk. As the shear rate declines, the instantaneous growth rate of the fastest growing disturbance declines as well, if Q is fixed. If Q varies as described in the preceding paragraph, then as A vanishes the maximum growth rate asymptotes to $\kappa\sqrt{Q^{-2} - 1}$, which increases as A decreases.

In sum, the lower shear rate expected in spiral arms has these effects: (1) lower tidal stress, which is conducive to cloud growth; (2) a higher epicyclic frequency, which is stabilizing; (3) an increased surface density, which is destabilizing; (4) a higher growth rate for small amplitude shearing disturbances.

REFERENCES

Balbus, S., 1988, Ap.J., 324, 60.

Gammie, C. F., 1992, Ph.D. Thesis, Princeton Univ.

Goldreich, P., and Lynden-Bell, D., 1965, M.N.R.A.S., 130, 125.

Goldreich, P., and Tremaine, S., 1978, Ap.J., 222, 850.

Julian, W.H., and Toomre, A., 1966, Ap.J., 146, 810.

Mestel, L., 1966, M.N.R.A.S, 131, 307.

Stark, A., and Blitz, L., 1978, Ap.J., 225, L15.

VELOCITY EVOLUTION OF DISK STARS DUE TO GRAVITATIONAL SCATTERING BY GIANT MOLECULAR CLOUDS

Eiichiro Kokubo and Shigeru Ida
Department of Earth Science and Astronomy, College of Arts and Sciences,
University of Tokyo, 3-8-1 Komaba Meguro-ku Tokyo 153 Japan.
Email: kokubo@kyohou.c.u-tokyo.ac.jp

ABSTRACT

We investigate the evolution of velocity dispersion of disk stars due to gravitational scattering by giant molecular clouds (GMCs) through numerical integration of orbits in a differentially rotating disk. We find that the evolution of the velocity dispersion can be divided into two phases. In the first phase where the velocity dispersion is small and the relative velocity to the GMC is determined by the shear velocity of the galactic disk, the radial and tangential components of the velocity dispersion increase with time as $\sigma_R, \sigma_\theta \propto t^{1/2}$, while the vertical component as $\sigma_z \propto \exp(t)$. In the later phase where the velocity dispersion becomes large enough to govern the relative velocity, all the components increase as $\sigma_R, \sigma_\theta, \sigma_z \propto t^{1/4}$ and ratios among them converge to $\sigma_R : \sigma_\theta : \sigma_z \simeq 1 : 0.7 : 0.6$, independent of the initial conditions of velocity dispersion. These behaviors agree well with the observed velocity dispersion of disk stars with various ages.

1. INTRODUCTION

The velocity dispersion of stars in the galactic disk correlates positively with their ages, for example, see Wielen (1977). Spitzer and Schwarzschild (1951) first pointed out that this correlation could be explained in terms of the gravitational scattering by the massive gas clouds. Such massive gas clouds are observed as giant molecular clouds (GMCs) with the mass of $10^{5-6} M\odot$ at present. The random velocities of newly born stars are expected to be as low as that of GMCs ($\simeq 5$ kms^{-1}), since they are formed in GMCs. As disk stars rotate in the galactic disk, their velocity dispersion increases through gravitational interaction with GMCs (gravitational scattering by stars is negligible because a star is much less massive than a GMC).

2. METHOD

We assume the galactic potential to be axisymmetric. We assume that the GMC is on a noninclined circular orbit and adopt Plummer model for the potential of the GMC: $\Phi_{GMC} = -GM/\sqrt{r^2 + c^2}$, where r is the radius from the center of the GMC, G is the gravitational constant, M is the mass of the GMC, and c is the core radius of the GMC. We reduce an encounter between a star and GMCs into two-body problems of a star and a GMC under the galactic potential. We use the three-dimensional epicycle approximation for stellar orbits. We define the local rotating coordinates (x, y, z) centered at the GMC. In this approximation, the equations of motion of the disk star is described

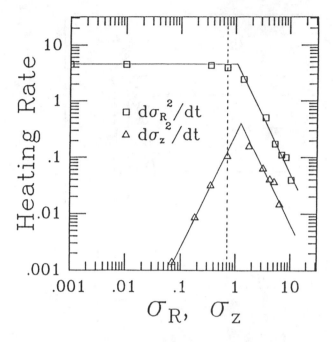

Fig.1 The heating rates along the typical evolutional path are plotted against σ_R and σ_z. We can see two phases in the heating. The dotted line shows the values of σ_R and σ_z where the random motion of stars is comparable to the tidal radius of the GMC.

as follows:

$$
\begin{cases}
\ddot{x} = 2\Omega\dot{y} + (4\Omega^2 - \kappa^2)x - \dfrac{GM}{(r^2 + c^2)^{\frac{3}{2}}}x \\[2mm]
\ddot{y} = -2\Omega\dot{x} - \dfrac{GM}{(r^2 + c^2)^{\frac{3}{2}}}y \\[2mm]
\ddot{z} = -\nu^2 z - \dfrac{GM}{(r^2 + c^2)^{\frac{3}{2}}}z,
\end{cases}
\tag{1}
$$

where Ω is the angular velocity of the GMC and κ and ν are the epicycle and the vertical frequencies, respectively. We calculate the scattering cross-sections of star-cloud encounters with various initial conditions, by the numerical integrations of orbits in the differentially rotating disk. Taking into account the phase space distribution of stars, we average the scattering cross-sections and then obtain the increase rate of velocity dispersion (heating rate) of stars. From the heating rate we follow the evolution of the velocity dispersion of disk stars. For detail description of the method, see Kokubo and Ida (1992).

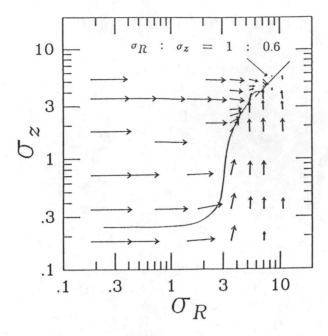

Fig.2 The directions of the heating are plotted on the σ_R vs. σ_z plane. The arrows represent the direction of the heating. The length of arrows is proportional to the heating rate. Solid curve is the typical evolutional path of the velocity dispersion. We can see the ratio σ_z/σ_R converges to $\simeq 0.6$, indipendent of the intial conditions of velocity dispersion.

3. RESULTS

We find that the evolution of the velocity dispersion can be divided into two phases (Fig.1). In the first phase, where the velocity dispersion is small and relative motion is mainly determined by the shear velocity, we find $d\sigma_R^2/dt \simeq$ const. and $d\sigma_z^2/dt \propto \sigma_z^2$, where σ_R and σ_z are the radial and vertical components of velocity dispersion. Then we obtain the velocity dispersion increases with time, t as

$$\sigma_R, \sigma_\theta \propto t^{1/2}, \quad \sigma_z \propto \exp(t). \tag{2}$$

It should be noted that in the epicycle approximation the tangential component of the velocity dispersion, σ_θ, is related to σ_R as $\sigma_\theta \simeq 0.7\sigma_R$ in the solar neighbourhood. In this first phase, $\sigma_R, \sigma_\theta \gg \sigma_z$, since $d\sigma_R^2/dt \gg d\sigma_z^2/dt$ as seen in Fig.1. After the velocity dispersion exceeds a characteristic shear velocity, $r_t\Omega$ (r_t is a tidal radius of the GMC), the second phase starts. In the second phase, where the velocity dispersion mainly dominates the relative velocity, we find $d\sigma_R^2/dt \propto \sigma_R^{-2}$ and $d\sigma_z^2/dt \propto \sigma_z^{-2}$. Then we obtain the velocity dispersion increases with time as

$$\sigma_R, \sigma_\theta, \sigma_z \propto t^{1/4}. \tag{3}$$

This age dependence agrees with Chandrasekhar's two-body relaxation time applied for the disk system (Kokubo and Ida 1992). In the solar neighborhood, the ratios among the velocity dispersion components converge to

$$\sigma_R : \sigma_\theta : \sigma_z \simeq 1 : 0.7 : 0.6, \tag{4}$$

independent of the initial conditions (Fig.2). These ratios agree well with observations. This agreement comes from our realistic treatment for the galactic shear motion and the close encounters. The ratios (4) are also obtained by a semianalytical calculation (Ida, Kokubo, and Makino 1993).

The observed magnitude of the velocity dispersion of nearby stars can be explained except for stars older than 5Gyr if GMCs clusters into complexes of mass $M \sim 10^{6-7} M_\odot$.

REFERENCES

Ida, S., Kokubo, E., & Makino., J. 1993, in preparation
Kokubo, E., & Ida, S. 1992, PASJ, in press
Spitzer, L., & Schwarzschild, M. 1951, ApJ, 114, 385
Wielen, R. 1977, A&A, 60, 263

THE CIRCULAR VELOCITY OF THE THICK DISC

Annie C. Robin
Observatoire de Besançon, France, ANNIE@FROBES51.BITNET

Bing Chen
Observatoire de Strasbourg, France, CDSXB2::CHEN

ABSTRACT

The circular velocity of the thick disc is investigated using data in the 5 dimensional space (V, B-V, U-B, μ_l, μ_b) in a direction at intermediate latitude. We show that using a discriminant analysis we are able to distinguish the thick disc from the disc and the halo and that these data constrain the asymmetric drift of this population. It is found to be 80 ± 20 km/s at a 2 sigmas level.

1. INTRODUCTION

The existence of an intermediate stellar population, call it a thick disc, between the disc and the halo is now well established. Different approaches have been used to detect and to describe it. Observations of distant giants (for example Rose 1985, Ratnatunga & Freeman 1989) as well as star counts (Gilmore & Reid 1983, Wyse & Gilmore 1986, Yoshii et al. 1987, Robin et al. 1989, Majewski, 1992, among others) show the existence of this population with characteristics intermediate between the old disc and the halo on any point of view (metallicity, kinematics, age, density laws). Therefore the difficulty to accurately measure its characteristics lies in detecting possible pollution of any sample of supposed thick disc stars by disc or by halo stars.

Up to now this population may be described by a set of parameters such as a scale height ranging from 600 to 1500 pc, a local density (1% to 12% of the disc), a vertical velocity dispersion (from 30 to 80 km s^{-1}), an asymmetric drift (20 to 100 km s^{-1}), a metallicity between -0.5 and -1.0, and may be down to -1.5 (Morrison et al. , 1990).

We here concentrate on the determination of the circular velocity of the thick disc using a photometric and astrometric sample in a direction close to M5 (l=4, b=47). The data set contain five observational parameters (V, B-V, U-B, μ_l, μ_b) making difficult the extraction of all the information by a standard method. Therefore we use both the synthetic approach and a new method based on the discriminant analysis.

In section 2 we describe the observational material and the method of analysis while comparisons with other investigations and discussions are given in section 3.

2. DATA ANALYSIS

Bienaymé et al. (1992) produced a complete photometric and astrometric sample of stars in a direction close to M5 (l=4, b=47) from Schmidt plates. An area of 1.78 square degrees on each plate was scanned with the MAMA machine in Paris giving rise to an accuracy of 0.2"/cen on a time baseline of 30 years. The photometric accuracy was about 0.08 to 0.10 in B and V to magnitude 17.5 and 0.12 in U at the plate limit.

In order to constrain thick disc parameters from this sample a sub-sample was selected to increase the contrast between disc and thick disc populations. This sub-sample includes stars with V > 16 and B-V < 0.8.

The Besançon model of population synthesis (Robin & Crézé 1986, Bienaymé, Robin & Crézé 1987, Robin & Oblak 1987) was used to simulate several sets of thick disc parameters and to test the sensitivity of this particular sample to the thick disc hypothesis. The choice of the most usefull area of the 5-dimensional space which should be the most sensitive to the thick disc parameters is not straightforward. Therefore we use model simulations of the sample to find the best axis where to project the data in order to separate the intermediate population from the disc and the halo. The discriminant analysis method allows to compute this projection axis. Applied to model simulated catalogues, where we know the population of each star, this analysis gives a first discriminant axis as a linear combination of the five parameters:

$$x = -0.08 \times (B - V) - 0.29 \times (U - B) + 0.14 \times V - 0.40 \times \mu_l + 0.02 \times \mu_b$$

The resulting axis shows that the U-B colour is necessary to make a good discrimination between the three populations because of its sensitivity to the metallicity. On the other hand the μ_l parameter is parallel to the V velocity and discriminates the populations by their asymmetric drift.

Trying to fit the observed distribution with a suitable model for the thick disc we found that these data are not at all sensitive to the scale height and the velocity dispersion used for the thick disc because they are correlated by the potentiel (through the Boltzmann equation) and the proper motions is a function of distance and velocity. Models with a small scale height and a small velocity dispersion are compatible with the data as well as models with large scale height and large velocity dispersion. However these data are very sensitive to the circular velocity of the thick disc as seen in fig. 1.

We use a Kolmogorov-Smirnov test to compare the distribution of the sample on this discriminant axis with a set of model predicted distribution, assuming different circular velocities for the thick disc (figure 2). Table 1 shows the values of the probability of each model to come from the same distribution as the observed sample. It shows that the circular velocity of the thick disc is of the order of 150 km/s (corresponding to a lag of 80 km/s). Lags of 60 km/s or 100 km/s are at 2 sigmas.

Table 1. Probability of each model to come from the same distribution as the data. Models differ by their circular velocity for the thick disc (col. 1). The corresponding lag is given in col. 2.

Vc (km/s)	Lag (km/s)	Probability
190	40	0.020
170	60	0.125
150	80	0.586
130	100	0.134
110	120	0.046

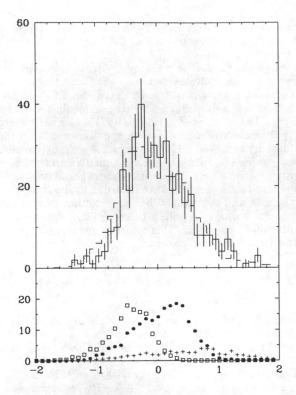

Fig. 1. (a) Distribution over discriminant axis of observed stars (full line, with 1 sigma error bars) and best model predictions (dashed line). (b) Distribution of model predicted stars according to their population. Open squares: Disc; Dots: Thick disc; Crosses: Halo.

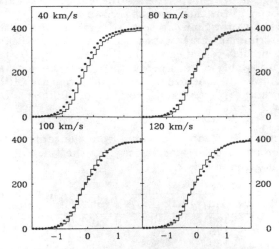

Fig. 2. Cumulated distribution on the discriminant axis of observed stars (full line). Models assuming different asymmetric drift for the thick disc (dots).

3. DISCUSSIONS

Contradictory results for the thick disc asymmetric drift have been published up to now: Yoss *et al.* (1991) found a value of -13 km/s; Norris (1987) found -20 km/s; Carney *et al.* (1989) -30 km/s as well as Ratnatunga & Freeman (1989); Morrison *et al.* (1990) -35 km/s; Sandage & Fouts (1987) -50 km/s; Spaenhauer (1989) -80 km/s at a distance of 2 kpc; Wyse & Gilmore (1986) -100 km/s. Most recently Majewski (1992) obtained an asymmetric drift regularly growing from nearly 0 in the plane to about 120 km/s at 5 kpc above the plane. In our field the mean distance of the thick disc stars is about 1.5 to 2 kpc above the plane. So our result is a priori more consistent with Spaenhauer (1989) result who found a drift of -80 at a distance of z = 2 kpc, than to the Majewski value of -50 $\pm$ 9 at a mean distance of 1.9 kpc. Taking into account the radial gradient of the asymmetric drift in direction of M5 (l=4, b=47) (i.e. about 15 km/s to add to the local drift) the asymmetric drift from Majewski would give about -65 km/s in our sample which is compatible with our value at a 2 sigmas level. We plan to apply the same analysis in other directions in order to confirm this result.

ACKNOWLEDGEMENTS

This research was partially supported by the Indo-French Centre for the Promotion of Advanced Research / Centre Franco- Indien Pour la Promotion de la Recherche Avancée.

REFERENCES

Bienaymé, O., Robin, A. C. & Crézé, M., 1987, A&A 180, 94
Bienaymé, O., Mohan, V., Crézé, M., Considère, S. & Robin, A.C., 1992, A&A 253, 389
Carney, B.W., Latham, D.W. & Laird, J.B., 1989, AJ 97, 423
Gilmore, G. & Reid, N., 1983, MNRAS 202, 1025
Majewski, S., 1992, ApJS 78, 87
Morrison, H.L., Flynn, C. & Freeman, K.C., 1990, AJ 100, 1191
Norris, J., 1987, AJ 93, 616
Ratnatunga, K.U. & Freeman, K.C., 1989, ApJ 339, 126
Robin, A.C. & Crézé, M. 1986, A&A 157, 71
Robin, A.C., Crézé, M. & Bienaymé, O., 1989, *The Gravitational Force Perpendicular to the Galactic Plane*, ed. A.G.D. Philip and P.K. Lu (Schenectady: L. Davis Press), p. 33
Robin, A.C., Crézé, M. & Mohan, V., 1992, A&A 265, 32
Robin, A.C. & Oblak, E., 1987, *Xth European Astronomy Meeting of the IAU*, Vol. 4, p. 323, ed. J. Palous
Rose, J., 1985, AJ 90, 787
Sandage, A. & Fouts, G., 1987, AJ 92, 74
Spaenhauer, A. 1989, *The Gravitational Force Perpendicular to the Galactic Plane*, ed A.G.D. Philip & P.K. Lu (Schenectady: L. Davis Press), p. 45
Wyse, R.F.G. & Gilmore, G., 1986, AJ 91, 855
Yoshii, Y., Ishida, K. & Stobie, R.S., 1987, AJ 92, 323
Yoss, K.M., Bell D.J. & Detweiler, H.L., 1991, AJ 102, 975

RELAXATION AND STABILITY OF THE DISK OF OUR GALAXY

E. Griv

Physics Department, Ben Gurion University of the Negev
Beer-Sheva 84105 Israel

From the observed random velocities of 269 stars of spectral type F the relation for the increase of the velocity dispersion upon age $\sigma \sim t^\alpha$ is obtained. Both the value $\alpha = 0.2 - 0.3$ for star's velocities in the plane of our Galaxy and the value $\alpha = 0.4 - 0.6$ in vertical direction are determined.

Oscillations and stability of stellar-giant molecular clouds disk of the Galaxy with account taken on binary collisions are considered. Kinetic equation with a model Bhatnagar-Gross-Krook collisional integral and the Poisson equation are the basis for our investigation. A dispersion relation is derived to describe the dynamics of small perturbations in the plane of a differentialy rotating disk. It is shown : In parts of a colliding two-component system that are at the boundary of gravitational stability (Jeans' stability) a specific dissipative instability may grow. As a result, this dissipative instability can increase the dispersion of the random velocities of stars, as the observations have shown.

1. INTRODUCTION

There are two general mechanisms which may effect the peculiar (random) velocities of stars of our Galaxy. Spitzer and Schwarzschild (1951) proposed to explain the observed increase of velocity dispersion upon age by gravitational collisions between stars and hypothetical interstellar clouds. Relatively recent radio data have shown that these massive objects (giant molecular clouds) are actually present in the Galaxy. Icke (1982), Lacey (1984) and Villumsen (1985) extended the analysis of Spitzer and Schwarzschild. Their calculations show on an increase of the velocity dispersions of all three components upon age $\sigma \sim t^\alpha$, where $\alpha = 0.25 - 0.31$ while from the observations $\alpha = 0.5$ (Wielen 1977). Therefore Binney and Tremaine (1987) claimed that the gas clouds can not explain of the observed increase of dispersion.

Barbanis and Woltjer (1967), Carlberg and Sellwood (1985) considered the "collisions" stars with unstable spiral density waves. This hypothesis encounters another objection (Grivnev and Fridman 1990). The instability, by increasing the peculiar velocities of the disk particles, brings the waves to the limit of stability. So in the Galaxy the spiral density waves are practically standing, and in interaction with these waves there is no energy exchange between stars and wave.

In this paper the age-dispersion dependence is redetermined by using new observational data. Also I present here the several results from the kinetic theory that characterize the spectra of the natural oscillations of the colliding star-cloud Galaxy's disk. The instability conditions are analyzed below within the framework of the asymptotic theory of Lin *et al.* (1969). Kumar (1960), Lynden-Bell and Pringle (1974), Mishurov *et al.*(1976), Morozov *et al.*(1985) investigated the stability of colliding system in the hydrodynamic approach. In plasma physics the analogous task of drift-dissipative instabilities has been solved by using both the kinetic and hydrodynamic approach (Mikhailovskii 1974).

2. AGE-VELOCITY-DISPERSION RELATION FOR F-STARS

The catalogue of random velocities, coordinates, logarithmic metallicity parameters [Fe/H], effective temperatures, distances from the sun of 1304 stars of spectral type F is given by Shevelev (1986). This catalogue is based on a ubvyβ survey of Philip and Egret (1980) with β-values between 2.61 and 2.72 (F2-F9 stars), and (b-y) - values between 0.19 and 0.45. The stars with distances $r < 80$ pc and a total space peculiar velocity $V < 60$ km/s relative to the Sun are used here. The selected stars were distributed over several subgroups in accordance with their age t using the mean age-metallicity relation of Carlberg *et al.* (1985). Only the stars with $t > 1.5 \times 10^9$ yr were considered in order to exclude the effect of a rapid increase of radial-velocity dispersion during the first billion years of evolution. This increase is a consequence of the differential rotation of the Galaxy's disk (Grivnev and Fridman 1990). The final sample that I treat contains 269 stars.

Than the dispersions σ_u, σ_v and σ_w were calculated. The results of the analysis are presented in Table, where N is the size of each subgroup.

Table. Mean motion V and velocity dispersions σ of F-stars.

N	AGE (10^9 yr)	V_u	V_v (km/s)	V_w	σ_u	σ_v (km/s)	σ_w	σ_{total}
72	2.5	12.6	-11.9	-6.1	19.4	11.1	9.7	24.4
53	4.0	5.9	-10.1	-7.0	19.8	12.5	9.8	25.3
56	5.5	7.8	-10.2	-4.7	21.7	10.0	11.8	26.6
62	7.0	6.6	-10.7	-4.1	25.3	13.9	13.8	32.0
66	9.0	7.6	-10.8	-6.7	21.0	15.4	13.4	29.3
26	11.0	5.4	-6.4	-5.9	22.8	19.6	16.5	34.3

All stars reveal the well-known increase of dispersions with age. The observed increase can be modeled by a diffusion process in velocity space (Wielen 1977)

$$\sigma = \sigma_0 (1 + t/t_E)^\alpha \quad ,$$

σ_0 is the initial velocity dispersion, t_E is the characteristic time. From results presented in Table the values $\alpha = 0.2 - 0.3$ for horizontal and $\alpha = 0.4 - 0.6$ for vertical velocity dispersions are obtained. As it was pointed in INTRODUCTION these results represent approximately the same age-velocity dependence in the plane of the Galaxy as one that predicted by the Spitzer-Schwarzschild mechanism.

These observations favour the view that the star-cloud collisions in the plane of the Galaxy are important for the disk's evolution.

3. THE STABILITY OF COLLIDING DISK

Let us consider the dynamics of the colliding star-cloud disk of our Galaxy. The linearized collisional kinetic equation for the non-equilibrium addition f_1 to the equilibrium distribution function f_0 can be written as

$$\frac{df_1}{dt} = \frac{\partial \Phi_1}{\partial r}\frac{\partial f_0}{\partial E} + \frac{1}{r}\frac{\partial \Phi_1}{\partial \theta}\frac{\partial f_0}{\partial E} + I_{ij} \quad , \qquad (1)$$

where r, θ and z are the cylindrical coordinates, Φ_1 denotes the perturbed gravitational potential, $E = (v_r^2 + v_\theta^2)/2$; v_r and v_θ are the radial and azimuthal random velocities, I_{ij} - linearized collisional integral.

The exact expression for the collisional integral is replaced by the approximation of Bhatnagar *et al.* (1954)

$$I_{ij} = -\sum_j v_{ij}\,\{f_1^{(i)} - \frac{n_1^{(i)}}{n_0^{(i)}}f_0^{(i)}\} \qquad (2)$$

In Eq. (2) v_{ij} is the frequency of collisions of particles of the species i with particles of the species j, $n_1^{(i)}$ is the small perturbation of the surface density $n_0^{(i)}$ of the i-th component of the system. The collisional term of (2) satisfies the laws of conservation of particles and momentum; it corresponds to the "isothermal" approximation by taking the velocity dispersion constant (Mikhailovskii 1974). Further the star-cloud interactional term in (2) retained only; according to the observations $v = 10^{-9}$/yr (Grivnev and Fridman 1990).

Integrating the Eq. (1) along paths and velocities, and allowing for the asymptotic connection between perturbations of the density and potential from Poisson equation, the dispersion relation can be obtained (homogeneous differentially-rotating disk)

$$\frac{\kappa^2\sigma^2}{2\pi Gn_0|\kappa|} = 1 - \omega_* \sum_{s=-\infty}^{\infty} \frac{I_s(x)e^{-x}}{\omega_* + iv - s\chi} + iv\frac{\kappa^2\sigma^2}{2\pi Gn_0|\kappa|}\sum_{p=-\infty}^{\infty}\frac{I_p(x)e^{-x}}{\omega_* + iv - p\chi} \quad , \qquad (3)$$

where $\kappa^2 = \kappa_r^2 + \kappa_\theta^2$; κ_r and κ_θ are the radial and azimuthal wave numbers, σ is the radial-velocity dispersion, n_0 is the local surface density of a stellar disk, $\omega_* = \omega - m\Omega$; ω is the frequency of the oscillations that are excited, m is the number of spiral arms, $\Omega(r)$ is the local angular rotational velocity, $I_s(x)$ is a Bessel function, $x = \kappa_*^2 \sigma^2/\chi^2$; $\kappa_*^2 = \kappa^2\{1 + [(2\Omega/\chi)^2 - 1] \sin^2 \psi\}$; ψ is the pitch angle of spiral structure, $\chi(r)$ is the epicyclic frequency. See Mikhailovskii (1974), Morozov (1981) and Grivnev (1988) for a more detailed derivation of (3). The Eq. (3) without allowence for star-cloud collisions ($v = 0$) and for axisymmetric perturbations ($\sin \psi = 0$) coincides with the well-known Lin-Yuan-Shu equation.

The dispersion relation (3) describes three branches of oscillations of a disk, two ordinary gravitational and one dissipative branch. It can be shown that in the marginally stable gravitationally disk (in the disk lying at the boundary of Jeans stability) for which $\sigma = \sigma_T = 3.36 \ Gn_0/\chi$ and $\kappa = \kappa_T = \chi^2/2\pi Gn_0$ (Morozov 1981; Grivnev 1988), the roots of the (3) will become

$$\omega_{*1} = i \ (v\chi^2)^{1/3} \ ; \qquad \omega_{*2,3} = \pm \frac{\sqrt{3} - i}{2} \ (v\chi^2)^{1/3} \qquad (4)$$

According to (4), if encounters are present the rotating Galaxy's disk would be unstable against non-axisymmetric small perturbations, which would grow aperiodically. This specific instability due to it large $\sim \chi$ increment can increase the dispersion of the random star velocities.

Barbanis, B., and Woltjer, L. 1967, *Astrophys. J.,* **150**, 461.
Bhatnagar, P. L., Gross, E. P., and Krook, M. 1954, *Phys. Rev.,* **94**, 511.
Binney, J., and Tremaine, S. 1987, *Galactic Dynamics,* Princeton University Press, Princeton.
Carlberg, R. G., and Sellwood, J. A. 1985, *Astrophys. J.,* **292**, 79.
Grivnev, E. M. 1988, *Sov. Astron.,* **32**, 139.
Grivnev, E. M., and Fridman, A. M. 1990, *Sov. Astron.,* **34**, 10.
Icke, V. 1982, *Astrophys. J.,* **254**, 517.
Kumar, S. S. 1960, *Publ. Astron. Soc. Japan,* **12**, 552.
Lacey, C. G. 1984, *Mon. Not. R. Astron. Soc.,* **208**, 687.
Lin, C. C., Yuan, C., and Shu, F. 1969, *Astrophys. J.,* **155**, 721.
Lynden-Bell, D., and Pringle, J. E. 1974, *Mon. Not. R. Astron. Soc.,* **168**, 603.
Mikhailovskii, A. B. 1974, *Theory of Plasma Instabilities,* Consultants Bureau, Plenum Press, New York.
Mishurov, Yu. N., Peftiev, V. M., and Suchkov, A. A. 1976, *Sov. Astron.,* **20**,152.
Morozov, A. G. 1981, *Sov. Astron.,* **25**, 421.
Morozov, A. G., Torgashin, Yu. M., and Fridman, A. M. 1985, *Sov. Astron. Lett.,* **11**, 94.
Philip, A. G. D., and Egret, D. 1980, *Astron. Astrophys. Suppl. Ser.,* **40**, 199.
Shevelev, Yu. G. 1986, *Nauchn. Inform. Astrosov. Akad. Nauk SSSR,* Nr. **59**, 64 (in Russian).
Spitzer, L., and Schwarzschild, M. 1951, *Astrophys.J.,***114**,385.
Villumsen, J. V. 1985, *Astrophys. J.,* **290**, 75.
Wielen, R. 1977, *Astron. Astrophys.,* **60**, 263.

SIMULATIONS ON THE GRAVITO-ELECTRODYNAMICAL EVOLUTION OF CHARGED DUST GRAIN PLASMA: A MODEL OF GALAXY FORMATIONS

G. Lee
Center for Astrophysical Sciences
Department of Physics and Astronomy
The Johns Hopkins University
Baltimore, MD 21218

ABSTRACT

We present results of numerical simulations on the dynamical evolution of a two-component plasma system consisting of charged dust grains and oppositely charged low-mass ions. We consider the case that the electrostatic and the gravitational forces are similar in magnitude (charge-to-mass ratio is of order of square root of the gravitational constant). Wollman (1988, Phys. Rev. A37, 3052) showed that such case is possible in some situations and that the gravitational condensation of the system can explain many observed properties of our galaxy. Morphological changes of the plasma system are studied by two-dimensional particle-in-cell (PIC) simulation.

1. INTRODUCTION

Electromagnetic forces are normally neglected in astrophysical N-body simulations. But in the case that the charge-to-mass ratio q/m of the constituent body is of order of square root of the gravitational constant G, the electrostatic and the gravitational forces are similar in magnitude, and thus we must include the two forces in dynamical evolution of an astrophysical system. We consider a two-component plasma system consisting of charged dust grains and oppositely charged low-mass ions. If q/m of the dust grains is of order $\sqrt{G}$, then self-gravitational condensation of the two-component system will lead to large charge separation due to difference in the equilibrium scale height of density distribution between the two components. Since q/m $\approx$ Jeans length l_J $(= \sqrt{kT/4\pi G\rho m}) \approx$ Debye length l_D $(= \sqrt{kT/4\pi n q^2})$, the condensation is an unshielded collection of grains and the clumped grain component coexists with more uniform low-mass ions. Wollman (1988) proposed that the primordial galaxy is the result of the condensation of grains, and that collisions among grains might be the origin of the hot galactic halo gas. We present the morphological changes of such a plasma system.

2. NUMERICAL METHOD

We simulate the dynamical evolution of the two-component plasma system using two-dimensional finite-size particle-in-cell method in which particle dynamics are solved a self-consistently in the gravitational and the electrostatic fields (Tajima 1989; Birdsall & Langdon 1991). Fast Fourier transformation method is used to solve for the gravitational and the electrostatic fields on a computational grid and then subtracted dipole scheme is used to calculate the forces on the particle from the grid fields. The leapfrog method is used to solve the time-dependent particle motions. Following parameters are used in numerical simulation: 1. number of simulation dust grains (ions) = 10000, 2. number of grid points = 256 × 256, 3. unit grid spacing = l_J of grains = l_D of grains = length of the finite-size particles, 4. time step = 0.0002 in units of Jean's time t_J $(= (4\pi G\rho)^{-1/2}$ = plasma oscillation time) of grains, and 5. mass ratio between grain and ion = 20. As an initial condition we use random maxwellian velocity distribution with small velocity dispersion. We use q/m = $\sqrt{G}$ for grains (in one dimensional simulations Gisler and Wollman (1988) showed that the energy stored in the electrostatic field is maximized at this value). We allow particle escape from the system during iteration if the particle velocity exceeds the escape velocity.

3. RESULTS

Figure 1 and 2 show the time evolution of grains and ions respectively. The size of box is 256 × 256 (in units of Jean's length of grains) and the time (in units of Jean's time of grains) is shown in each figures. The time evolutions of each component are similar to the result of pure gravitational N-body simulation (Binney & Tremaine 1987 p. 274): 1. starting from approximately homogeneous 2D-spherical distribution, gravity causes the system to form a tight minimum configuration, 2. core bounce occurs due to increase in kinetic energy, 3. after a series of complex oscillations the system settles to an elongated quasi-steady state; but a large electric polarization produced by charge separation between grains and ions makes nonuniform sublumps of grains. Thus the final steady state is that the clumped grain component coexists with more uniform ions.

4. CONCLUSIONS

The gravito-electrodynamical evolutions of a two-component plasma system consisting of charged dust grains and oppositely charged low-mass ions appear to be interesting in that the system might provide the origin of the hot galactic halo gas by collisions among massive grains; if the grains are composed of heavy elements, gravitationally aggregated hot vapor produced by grain-grain collisions could generate nuclear energy (Wollman 1988). Horanyi and Goertz (1990) discussed possible physical processes which can form such kind of dusty plasmas. But to make physically realistic model it may be necessary to take into account a chemical evolution of the system in three dimensional situation.

The author thanks Prof. Tajima for providing helpful notes on plasma simulations.

REFERENCES

Binney, J. and Tremaine, S. 1987, Galactic Dynamics
        (Princeton : Princeton Univ. Press)
Birdsall, C.K. and Langdon, A.B. 1991, Plasma Physics via Computer
        Simulation (New York : Adam Hilger)
Gisler, G.R. and Wollman, E.R. 1988, Phys. Fluids, 31, 1101
Horanyi, M. and Goertz, C.K. 1990, ApJ, 361, 155
Tajima, T. 1989, Computational Plasma Physics
        (New York : Addison-Wesley)
Wollman, E.R. 1988, Phys. Rev., A37, 3052

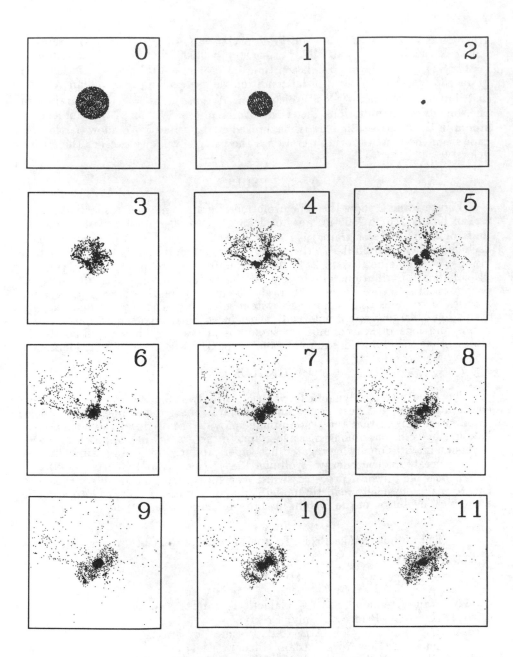

Figure 1. Time evolution of grains. The size of box is 256 × 256 in units of Jean's length of grains and the time is in units of Jean's time of grains.

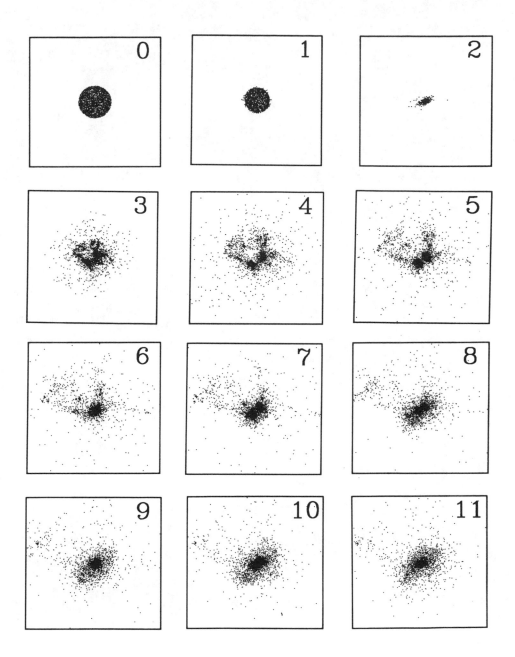

Figure 2. Time evolution of ions. The size of box is 256 × 256 in units of Jean's length of grains and the time is in units of Jean's time of grains.

ENERGETIC

PARTICLES

AND FIELDS

DIFFUSE <1 MEV GALACTIC GAMMA RAYS

Reuven Ramaty and Jeff Skibo
Laboratory for High Energy Astrophysics
NASA Goddard Space Flight Center, Greenbelt, MD 20771.

ABSTRACT

Observations of positron electron annihilation radiation address some fundamental problems in Galactic astrophysics, including nucleosynthesis, supernovae, black holes and the interstellar medium. The 0.511 MeV line resulting from positron annihilation is superposed on a continuum which probably is, in large part, bremsstrahlung of cosmic ray electrons. The power deposited by these electrons could have important implication for the interstellar medium.

1. INTRODUCTION

The diffuse Galactic gamma ray emission below 1 MeV consists of a strong line at 0.511 MeV superposed on a continuum. The positrons responsible for the line probably result from radionuclei produced in various processes of nucleosynthesis. An important goal of current research is the determination of the spatial distribution of the emission and the separation of a time variable component from the diffuse component. This variable emission could result from positrons produced at or near Galactic black holes. The diffuse continuum appears to be in large part bremsstrahlung of low energy electrons. Because of the very large number of electrons involved, this continuum could have very important implications on the energetics of the interstellar medium.

2. POSITRON ANNIHILATION RADIATION

Positron annihilation radiation from the direction of the Galactic center was first observed in 1970 (Johnson, Harnden & Haymes 1972), but it was not until 1977 that the line energy (expected at $m_e c^2 = 511.0$ keV) was accurately determined with a Ge instrument (Leventhal, MacCallum & Stang 1978). The observed line center energy, 510.7±0.5 keV, clearly established that the radiation was due to the annihilation of positrons. In this and all subsequent detections with Ge instruments (Leventhal *et al.* 1980; Riegler *et al.* 1981; Gehrels *et al.* 1991; Chapuis *et al.* 1991; Leventhal *et al.* 1993), the line width was found to be very narrow (full width at half maximum < 3.5 keV) and the line center energy to be at 0.511 MeV within errors less than a keV. Line emission at 0.511 MeV from the direction of the Galactic center has also been observed with lower resolution NaI detectors, most notably a gamma ray spectrometer on the Solar Maximum Mission (SMM, Harris *et al.* 1990) and the OSSE instrument on the COMPTON Gamma Ray Observatory (CGRO, Purcell *et al.* 1992). In these observations the line center energy was also around 0.511 MeV and the line width was narrower than the instrumental resolution.

a. Variable Component

The flux in the narrow 0.511 MeV line was found to vary in time. Evidence for this variability was first provided by the HEAO-3 observations (Riegler *et al.* 1981), which showed that the line flux decreased by about a factor of 3 in less than 0.5 years. Even though the significance of this result was weakened by a different analysis (Mahoney 1988), additional evidence for the variability was provided by observations with similar detectors measuring different fluxes at different times (Leventhal *et al.* 1978; 1980; Paciesas *et al.* 1982; Leventhal *et al.* 1982; 1986). Skibo, Ramaty and Leventhal (1992) have recently attempted to fit a large body of 0.511 MeV data to various assumed diffuse Galactic distributions. They showed that the supposition that all of the observed fluxes result from a time independent distribution can be rejected at the $\sim 3\sigma$ level. The implied time variations suggest that a significant fraction of the emission is produced by a single source in the central region of the Galaxy. The positrons could be produced as a relativistic plasma near the inner region of the accretion disk of a black hole of mass less than several hundred solar masses (Lingenfelter & Ramaty 1982; Liang & Dermer 1988). But because of the lack of any measurable redshift, the positrons must annihilate far from the hole, at a site removed by at least 300 Schwarzschild radii. The narrow width of the line and the time scale of the flux variations imply that the temperature and density of the ambient gas at this site are, respectively, less than about 10^5 K and greater than $\sim 10^5$ cm^{-3} (Lingenfelter & Ramaty 1982).

Recent observations with the imaging gamma ray spectrometer SIGMA on the GRANAT spacecraft revealed two discrete sources of Galactic annihilation radiation: the X-ray source 1E1740.7-2942 located at an angular distance of 0.9° from the Galactic center (Bouchet *et al.* 1991; Sunyaev *et al.* 1991) and Nova Muscae at $l = 64.7°, b = -7.1°$ (Goldwurm *et al.* 1992; Sunyaev *et al.* 1992). Annihilation radiation may also have been observed from the black hole candidate Cygnus X-1 (Ling & Wheaton 1989) and from an object removed by about 15° from the Galactic center (Briggs 1991). The 1E1740.7-2942 source is the most likely candidate to be associated with the variable source. Millimeter observations of this source have shown that it is aligned with a dense molecular cloud, probably located at the distance of the Galactic center (Bally & Leventhal 1991; Mirabel *et al.* 1991). Furthermore, the radio structure of the object is that of a double sided jet emanating from a compact and variable core, possibly due to synchrotron radiation of electrons and positrons in the jet (Mirabel *et al.* 1992). The signature of the annihilation radiation from 1E1740.7-2942 consists of a broad feature centered at 480^{+96}_{-72} keV with a full width at half maximum of 240^{+101}_{-94} keV. This feature is different from the observed 0.511 MeV Galactic line, which is much narrower and centered almost exactly on 0.511 MeV. It has been proposed (Ramaty *et al.* 1992) that the variable 0.511 MeV line emission results from positrons ejected from 1E1740.7-2942 (presumably via the jets) into the molecular cloud which is dense enough to slow down and annihilate the positrons on a time scale of about 1 year.

The observed flux in the broad annihilation line from 1E1740.7-2942 was 1.3×10^{-2} photons cm^{-2} s^{-1} (Bouchet *et al.* 1991). This broad line is probably formed by positrons annihilating in the accretion disk of the black hole. The number of positrons injected into the cloud is then $6.5 \times 10^{-3}(4\pi R_\odot^2)\alpha\eta$ e$^+$ s^{-1}, where α is the ratio of the number of positrons which escape from the hole to the number annihilating in the disk, η is the duty cycle for pair production in

1E1740.7-2942, $R_{\odot} = 7.7$ kpc (Reid 1989), and it is assumed that the positrons in the accretion disk annihilate directly without forming positronium. The 0.511 MeV line flux from the variable point source, although not well known, is probably on the order of a few times 10^{-4} photons cm^{-2} s^{-1} and not higher than 10^{-3} photons cm^{-2} s^{-1} (Share et al. 1990; Ramaty & Lingenfelter 1993). Adopting a value of 4×10^{-4} photons cm^{-2} s^{-1} and assuming a positronium fraction of 0.9 for annihilation in the cloud (Guessoum, Ramaty & Lingenfelter 1991; see also below), yields $\alpha\eta = 0.1$. Thus, if the duty cycle for pair production in 1E1740.7-2942 is less than 10%, the number of positrons escaping into the molecular cloud should exceed the number annihilating in the disk.

b. Diffuse Component

In addition to being variable in time, the 0.511 MeV line flux from the direction of the Galactic center also depends on the fields of view of the detectors (Dunphy, Chupp & Forrest 1983). This first became evident from the comparison of two observations (Leventhal et al. 1978; Gardner et al. 1982) carried out in 1977 with detectors of widely different field of views (15° vs. 100°). Even though the two observations were carried out within 10 days, the flux obtained with the 100° detector exceeded significantly the flux recorded with the 15° instrument. The correlation with detector field of view was also demonstrated by the comparison of observations with the 130° field of view SMM detector (Harris et al. 1990), which revealed large fluxes of 0.511 MeV line emission from 1980 through 1988, with balloon borne observations (Paciesas et al. 1982; Leventhal et al. 1982; 1986) with 15° field of view detectors in 1981 and 1984, which had comparable sensitivity but yielded only upper limits. The disparity in observations between detectors with wide and narrow fields of view suggests that at least a fraction of the observed annihilation radiation from the central region of our Galaxy is spatially distributed. A two component model has been proposed (Ramaty & Lingenfelter 1987; Lingenfelter & Ramaty 1989a) according to which the total observed 0.511 MeV line emission is the superposition of a steady diffuse component and time variable emission from a compact source. The flux in the diffuse component has now been directly measured with the balloon borne Ge detector GRIS (Gehrels et al. 1991) at a Galactic longitude of -25°, and with OSSE (Purcell et al. 1992) at longitudes -21° and 25°.

In a previous paper (Skibo, et al. 1992) we considered various models for the spatial distribution of the diffuse component, in particular a model in which the distribution of positron annihilation followed that of Galactic novae (Higdon & Fowler 1987). Each model yielded an unnormalized longitude and latitude intensity distribution. We normalized the nova distribution by fitting it to all the available off center data and the OSSE Galactic center data (Gehrels et al. 1991; Purcell et al. 1992). The inclusion of the off center data is clearly necessary. We included the OSSE Galactic center measurements because they yielded the lowest finite Galactic center fluxes and hence could be solely due to the diffuse component. It is also possible that the OSSE fluxes include a contribution from the variable point source. Therefore, in the present analysis, we use the same data to determine the normalizations. We consider two cases: (i) the nova distribution alone and (ii) a point source at the location of 1E1740.7-2942 superposed on this nova distribution.

In Figure 1 we show the resultant 0.511 MeV line fluxes as functions of Galactic longitude for $b = 0$ obtained by folding the calculated intensities through the response function of the GRIS detector (N. Gehrels, private communication 1991). For the calculated curves we used the nova distribution from

Higdon & Fowler (1989), which differs from the Higdon & Fowler (1987) model (used by Skibo *et al.* 1992) in the relative normalization of its disk and spheroidal components. The present distribution is given by

$$q(\rho, z) = N_d \exp[-44.5z^2 - 0.297(\rho - R_\odot)],\qquad (1)$$

for the disk component, and

$$q(R, z) = N_{sph} \exp\left[10.093\left(1 - (\frac{R}{R_\odot})^{1/4}\right)\right]$$

$$\times \begin{cases} 1.25(\frac{R}{R_\odot})^{-3/4}, & R \leq 0.03R_\odot \\ (\frac{R}{R_\odot})^{-7/8}\left[1 - 0.08669(\frac{R}{R_\odot})^{-1/4}\right], & R \geq 0.03R_\odot \end{cases} \qquad (2)$$

for the spheroid. Here ρ and z are Galactocentric cylindrical coordinates, R is distance from the Galactic center, $R_\odot = 7.7$ kpc, and $N_{sph}/N_d = 5.57 \times 10^{-3}$ (Higdon & Fowler 1989). This ratio is larger by 23% than the corresponding ratio in the Higdon & Fowler (1987) model.

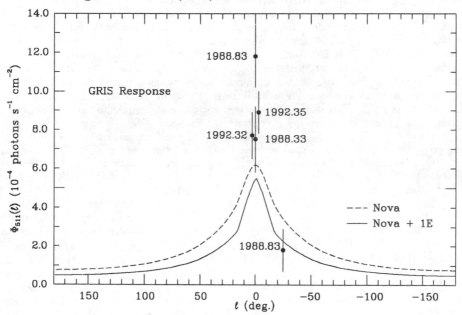

Figure 1.—Fluxes resulting from the convolution of the intensity distributions for the Nova and Nova + 1E models with the GRIS detector response. These intensities are normalized to the parameters of Table 1. The 1988 data are from Gehrels *et al.* (1991), while the 1992 data are from Leventhal *et al.* (1993). The position of these data relative to the curves, in particular the 1988.83 point, suggests the existence of time variable 0.511 MeV line emission.

The fitting parameters for the two models are given in Table 1, where $\phi_{0.511}$ and $Q_{0.511}$ represent fluxes from the central radian of Galactic longitude (integrated over all latitudes), and total Galactic 0.511 MeV luminosities. For the model consisting of the point source superposed on the nova distribution (Nova + 1E), the contributions of the two components are given separately. We see that both models provide acceptable fits to the data. This conclusion, however, should be reexamined after the much larger set of OSSE data, currently under analysis, is released. Considering the central radian fluxes, we first note that for the nova model it is slightly larger (by about 10%) than the corresponding flux in Skibo $et~al.$ (1992). This difference is caused by the different values of N_{sph}/N_d in the two Higdon & Fowler (1987; 1989) models. As for $Q_{0.511}$, the present value for the nova model exceeds that in Skibo $et~al.$ (1992) by a factor of 1.5. This difference is due to a numerical error in our earlier article, the fact that in the present analysis we have integrated the distribution out to infinity rather than out to only $2R_\odot$), as well as the difference in N_{sph}/N_d for the two nova models.

TABLE 1   Fitting Parameters for the off center and OSSE Galactic center data

Model	χ^2	ν	$P(\chi^2)$	$\phi_{0.511}$ $(10^{-4}$ ph cm^{-2} s$^{-1})$	$Q_{0.511}$ $(10^{43}$ ph s$^{-1})$
Nova	5.06	7	0.65	19	1.6
Nova + 1E	1.43	6	0.96	12.4 + 1.5	1.1 + 0.11

Comparing the values of $Q_{0.511}$ for the nova and nova + 1E models, we note that when the point source is included in the analysis, the total diffuse 0.511 MeV luminosity decreases by a factor of 0.7. Assuming a positronium fraction of 0.9 (see below), these luminosities (Table 1) require the production of about $(1.7 - 2.5) \times 10^{43}$ e$^+$ sec^{-1}. The decay of ^{56}Co, ^{44}Ti and ^{26}Al resulting from processes of Galactic nucleosynthesis (Clayton 1973; Ramaty & Lingenfelter 1979; Woosley & Pinto 1988; Signore & Vedrenne 1988; Lingenfelter & Ramaty 1989b) could account for the required production. The relative importance of ^{56}Co depends on the fraction ϵ of positrons that can escape into the interstellar medium from a Type Ia supernova. This escape fraction is uncertain with theoretical estimates (Colgate 1970; Chan, Lingenfelter & Ramaty 1993; Chan & Lingenfelter 1993) ranging as high as 0.1. However, ϵ could also be vanishingly small. The contributions of both ^{56}Co and ^{44}Sc depend on the current rate of Galactic nucleosynthesis, which is also uncertain. The contribution of ^{26}Al, determined from direct observations (Mahoney $et~al.$ 1984; Share $et~al.$ 1985) of the 1.809 MeV line that accompanies the ^{26}Al decay, amounts to about 0.2×10^{43} e$^+$ s^{-1}. The remaining requirement, 1.5 to 2.3 $\times 10^{43}$ e$^+$ s^{-1} could be understood in terms reasonable values for both the current rate of ^{56}Fe nucleosynthesis and the positron escape fraction (Chan $et~al.$ 1993).

However, with the discovery of localized Galactic sources of annihilation radiation, the question arises whether such sources could contribute significantly to Galactic positron production. If the narrow line source near the Galactic center is indeed fed by 1E1740.7-2942 to an average 0.511 MeV flux of 4×10^{-4} photons s^{-1}, then, as we have estimated above, the positron output of 1E1740.7-2942 is about 4.4×10^{42} e$^+$ s^{-1}, using $R_\odot = 7.7$ kpc. To account for the required diffuse Galactic positron production rate of $(1.5 - 2.3) \times 10^{43}$ e$^+$ s^{-1} (after subtracting the contribution of ^{26}Al), about 3 to 5 sources similar to 1E1740.7-2942 are needed. If the diffuse annihilation radiation is produced by such a small number of sources, its spatial distribution would be very inhomogeneous. Whether this is indeed the case will be tested by future observations.

c. Positronium Fraction and Line Shape

The positron annihilation processes have been reviewed (Guessoum *et al.* 1991; Ramaty & Lingenfelter 1991). If the initial energy of the positrons is around 1 MeV, as expected for most positron production mechanisms, then the energy loss is almost exclusively due to Coulomb collisions. The positrons then, having lost the bulk of their energy, either form positronium in flight or thermalize with the electrons in the gas. The thermal positrons can annihilate directly with both free and atomic electrons, they can form positronium by charge exchange with hydrogen atoms and radiative combination with free electrons, and they can interact with dust grains (Zurek 1985). (The effects of dust are negligible for positronium formation in flight.) Direct annihilation produces two line photons at 0.511 MeV. Positronium is formed in either the para or the ortho states. Parapositronium decays into two 0.511 MeV line photons, while orthopositronium decays into a 3-photon continuum below 0.511 MeV.

If the ambient gas density, radiation, and magnetic field are sufficiently low, the ionization, excitation, and spin flip of positronium formed in the gas can be ignored so that once positronium is formed in a given state, it will always annihilate from that state. These conditions are well satisfied in the interstellar medium, but not necessarily in the vicinity of a compact object. Assuming that once formed positronium will decay, the number of line photons per positron is $2 - 1.5f$, where f is the fraction of annihilations occurring via positronium.

Values of f have been calculated (Guessoum *et al.* 1991) for four phases of the interstellar medium (McKee & Ostriker 1977): cold cloud cores, mostly neutral or mostly ionized warm (8000K) gas, and hot (4.5×10^5 K) gas. In the cold clouds the positronium fraction is very nearly 0.9, independent of the amount of dust in the clouds. In the warm gas, f is also near 0.9, except if the amount of dust is excessive, in which case the f can be significantly lower. In the hot gas f is lower than about 0.5. A recent determination based on OSSE data (Purcell *et al.* 1992) yielded $f = 0.9 \pm 0.2$, in good agreement with the expected value for annihilation in the interstellar medium. For $f = 0.9$, there are 0.65 line photons per positron. All of the determinations of f have been obtained by attributing the excess flux above an assumed continuum at energies ≤ 0.511 MeV to orthopositronium annihilation. It was first pointed out by Forrest (1982) that the excess could also result from Compton scattering in the source. While Compton scattering in the vicinity of a compact object can be quite important (Lingenfelter & Hua 1991; Hua & Lingenfelter 1992), for positron annihilation in the interstellar medium it is more likely that the excess will be produced by orthopositronium annihilation.

The annihilation processes also affect the shape of the 0.511 MeV line. Figure 2 (from Ramaty & Lingenfelter 1991) shows spectra for the cold, warm neutral, warm ionized, and hot phases. In the cold phase (panel a) the spectrum shows a narrow line due to direct annihilation with bound electrons on top of a broader line due to positronium annihilation in flight. For the warm cases (panels b and c), the annihilation line is very narrow. For the hot phase (panel d) the line is quite broad if the effects of the grains are ignored, but it becomes narrower if these effects are included. Thus, if the positrons can penetrate the cloud cores, the line would show a broad base which would be absent if the positrons are excluded from the cores. Positrons could be prevented by magnetic fields from penetrating the clouds, but so far there is no data to indicate whether this indeed is happening (see however Wallyn *et al.* 1993).

The above considerations are applicable to steady state situations. It has been shown (Ramaty *et al.* 1992) that in the case of impulsive injection of positrons into an ambient neutral gas (for example positron injection from the 1E1740.7-2942 source into the surrounding molecular cloud), both the positronium fraction and the 0.511 MeV line width will vary with time. Initially, as the positrons form positronium in flight the line is broad (full width at half maximum ~6.4 keV). But subsequently, as the surviving positrons pass through the positronium formation regime and begin to annihilate directly (Drachman 1983), the line will eventually narrow.

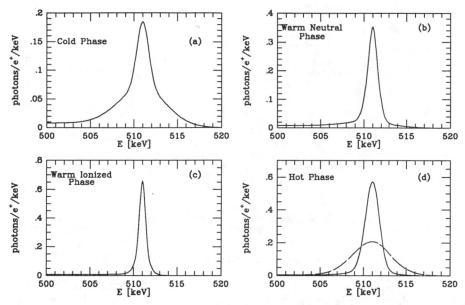

Figure 2.—Calculated 0.511 MeV line profiles for the four phases of the interstellar medium.

3. DIFFUSE LOW ENERGY CONTINUUM

Gamma ray continuum observations from the direction of the Galactic center span a broad range of energies from about 0.1 MeV to over a GeV. Above about 10 MeV a large fraction of the observed emission results from cosmic ray interactions with interstellar gas and radiation (pion production, bremsstrahlung, inverse Compton scattering). At lower energies the origin of the gamma ray continuum is less certain. An analysis of the observations made with GRIS from the direction of the Galactic center and a direction in the plane away from the Galactic center (l=335°) suggests that the emission below 1 MeV is of diffuse origin and has a relatively broad longitude distribution over the central radian of the Galaxy (Gehrels *et al.* 1991). Furthermore, hard X-ray observations from the direction of the Galactic center with detectors of moderate fields of view (15° to 30°) show (Gehrels & Tueller 1992) that, even though the observed fluxes vary in time, they have a lower envelope which coincides with the flux observed with GRIS from l = 335°. However, between about 0.5 and 3 MeV, observations (Riegler *et al.* 1985) with HEAO-3 have indicated that the observed flux may be time variable, suggesting contributions from unidentified discrete sources. In addition, recent measurements made with COMPTEL on CGRO (Bloemen *et al.* 1993; Strong *et al.* 1993) show that the diffuse emission around 1 MeV is somewhat lower than that estimated previously (Gehrels & Tueller 1993).

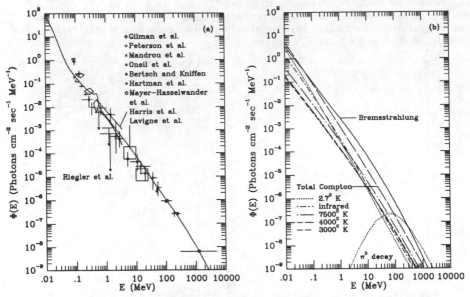

Figure 3.—Diffuse gamma ray continuum (refs. in Skibo & Ramaty 1993).

The diffuse continuum emission observed from the central radian of the Galaxy is shown in Figure 3a. Skibo & Ramaty (1993) calculated the expected continuum taking into account the Galactic distributions of molecular, atomic and ionized gases, and infrared, visible and microwave background radiations. The results are shown in Figure 3b. The solid curve in Figure 3a is the total gamma ray emission obtained by summing the components shown in Figure 3b and adding the expected contributions of orthopositronium and thermal bremsstrahlung from the Galactic hot (10^8 K) plasma. It is evident that the calculated curve fits the data reasonably well over many decades of energy. However, at low energies there are some discrepancies, perhaps due to the presence of hard X-ray sources in the Galactic center region. Also, the calculated continuum runs lower than the continuum measured with SMM by about 1 σ. Preliminary reports (Strong *et al.* 1993) indicate that this is in accordance with the continuum measured with COMPTEL.

A very large power is required to maintain the population of low energy electrons if the low energy gamma ray continuum is of diffuse origin. This situation is similar to that in solar flares, where the observed hard X-ray emission is known to be nonthermal electron bremsstrahlung and where the energy contained in the 0.01 to 0.1 MeV electrons is comparable to the total flare energy (Dennis 1988). For the electron spectrum used to calculate the results of Figure 3, the local power input into the interstellar medium due to electrons of energies greater than 0.05 MeV is 1.0×10^{-26} erg s^{-1} cm^{-3}. The corresponding local ionization rate is $\zeta = 6.0 \times 10^{-16}$ s^{-1} for a density of 0.3 cm^{-3}. This value is not inconsistent with recent estimates (R. J. Reynolds, private communication 1992). Thus the electrons responsible for the bremsstrahlung will not over ionize the interstellar medium.

The electrons responsible for the production of the low energy diffuse gamma ray continuum could provide the ionization and heating required to maintain the warm ($T \sim 10^4$K) ionized component of the interstellar medium (Chi & Wolfendale 1991). The power required to maintain this ionized gas is 7×10^{-27} to 7×10^{-25} erg s^{-1} cm^{-3}, depending on the temperature of the gas (Reynolds 1990). Our calculated local power input of 1.0×10^{-26} erg s^{-1} cm^{-3} falls within this range. We calculate the total power input into the Galaxy. The result, 1.6×10^{41} erg s^{-1}, exceeds earlier estimates of the total power supplied to the nuclear cosmic rays by about an order of magnitude. The mechanism that accelerates the electrons, however, is not known.

ACKNOWLEDGEMENTS

Material from this paper will appear in a dissertation to be submitted to the graduate school, University of Maryland, by J. G. Skibo in partial fulfillment of the requirements for the Ph.D. degree in Physics.

REFERENCES

Bally, J. & Leventhal, M. 1991, Nature, 353, 234

Bouchet, L. *et al.* 1991, ApJ, 383, L45

Briggs, M. 1991, PhD Dissertation, Univ. of Calif. San Diego

Chan, K. W., Lingenfelter, R. E. & Ramaty, R. 1993, Physics Reports, in press

Chan, K. W. & Lingenfelter, R. E. 1993, ApJ, in press

Chapuis, C. G. L. *et al.* 1991, in Gamma-Ray Line Astrophysics, eds. P. Durouchoux & N. Prantzos (New York: AIP), 54

Chi, X. & Wolfendale, A. W. 1991, in The Interstellar Disk-Halo Connection in Galaxies, ed. H. Bloemen (IAU Symp. 144), 197

Clayton, D. D. 1973, Nature Phys. Sci. 244, 137

Colgate, S. A. 1970, Astrophys. & Space Sci., 8, 457

Dennis, B. R. 1988, Solar Phys., 118, 49

Drachman, R. J. 1983, in Positron-Electron Pairs in Astrophysics, eds. M. L. Burns *et al.* (New York: AIP), 242

Dunphy, P. P., Chupp, E. L. & Forrest, D. J. 1983, in Positron-Electron Pairs in Astrophysics, ed. M. L. Burns *et al.* (New York: AIP), 237

Forrest, D. J. 1982, in The Galactic Center, ed. G. R. Riegler and R. D. Blandford (New York: AIP), 160

Gardner, B. M. *et al.* 1982, in The Galactic Center, (New York: AIP) p. 144

Gehrels, N. 1991, in Gamma-Ray Line Astrophysics, ed. P. Durouchoux and N. Prantzos, (New York: AIP), 3

Gehrels, N., Barthelmy, S. D., Teegarden, B. J., Tueller, J., Leventhal, M. & MacCallum, C. J. 1991, ApJ, 375, L13

Gehrels, N. & Tueller, J. 1992, in The Compton Observatory Science Workshop, eds. C. R. Shrader, N. Gehrels & B. Dennis, 446

Gehrels, N. & Tueller, J. 1993, ApJ, in press

Goldwurm, A. *et al.* 1992, ApJ, 389, L79

Guessoum, N., Ramaty, R. & Lingenfelter, R. E. 1991, ApJ, 378, 170

Harris, M. J. *et al.* 1990, ApJ, 362, 135

Higdon, J. C. & Fowler, W. A. 1987, ApJ, 317, 710
Higdon, J. C. & Fowler, W. A. 1989, ApJ, 339, 956
Hua, X.-M. & Lingenfelter, R. E. 1992, ApJ, 397, 591
Johnson, W. N., Harnden, F. R. & Haymes, R. C. 1972, ApJ, 172, L1
Leventhal, M. et al. 1980, ApJ, 240, 338
Leventhal, M. et al. 1982, ApJ, 260, L1
Leventhal, M. et al. 1986, ApJ, 302, 459
Leventhal, M. et al. 1993, ApJ, in press
Leventhal, M., MacCallum, C. J. & Stang, P. D. 1978, ApJ, 225, L11
Liang, E. P. & Dermer, C. D. 1988, ApJ, 325, L39
Ling, J. C. & Wheaton. W. A. 1989, ApJ, 343, L57
Lingenfelter, R. E. & Hua, X.-M. 1991, ApJ, 381, 426
Lingenfelter, R. E. & Ramaty, R. 1982, in The Galactic Center, eds. G. R.
    Riegler & R. D. Blandford (New York: AIP), 148
Lingenfelter, R. E. & Ramaty, R. 1989a, ApJ, 343, 686
Lingenfelter, R. E., & Ramaty, R. 1989b, NucPhysB, (Proc. Supp), 10B, 67
Mahoney, W. A. 1988, in Nuclear Spectroscopy of Astrophysical Sources, eds.
    N. Gehrels and G. H. Share, (New York: AIP), 149
Mahoney, W. A. et al. 1984, ApJ 286, 578
McKee, C. F. & Ostriker, J. P. 1977, ApJ, 218, 148
Mirabel, I. F. et al. 1991, A&A, 251, L43
Mirabel, I. F. et al. 1992, Nature, 358, 215
Paciesas, W. S. et al. 1982, ApJ, 260, L7
Purcell, W. R. et al. 1992, in The Compton Observatory Science Workshop, eds.
    C. Shrader, N. Gehrels & B. Dennis, 431
Ramaty, R., Leventhal, M., Chan, K. W., & Lingenfelter, R. E. 1992, ApJ, 392,
    L63
Ramaty, R. & Lingenfelter, R. E. 1979, Nature, 278, 127
Ramaty, R. & Lingenfelter, R. E. 1987, in The Galactic Center, ed. D. C. Backer
    (New York: AIP), 51
Ramaty, R. & Lingenfelter, R. E. 1991, in Gamma-Ray Line Astrophysics, eds.
    P. Durouchoux & N. Prantzos (New York: AIP), 67
Ramaty, R. & Lingenfelter, R. E. 1993, A&A, in press
Reid, M. J. 1989, in The Center of the Galaxy, ed. M. Morris (Dordrecht:
    Kluwer Academic), 21
Reynolds, R. J. 1990, ApJ, 349, L17
Riegler, G. R. et al. 1981, ApJ, 248, L13
Riegler, G. et al. 1985, ApJ, 294, L13
Share, G. H. et al. 1985, ApJ, 292, L61
Share, G. H. et al. 1990, ApJ, 358, L45
Signore, M. & Vedrenne, G. 1988, A&A, 201, 379
Skibo, J. G. & Ramaty, R. 1993, COMPTON Workshop (St. Louis), in press
Skibo, J. G., Ramaty, R., & Leventhal, M. 1992, ApJ, 397, 135
Strong, A. W. et al. 1993, COMPTON Workshop (St. Louis), in press
Sunyaev, R. et al. 1991, ApJ, 383, L49
Sunyaev, R. et al. 1992, ApJ, 389, L75
Wallyn, P. et al. 1993, ApJ, in press
Woosley, S. E. & Pinto, P. A. 1988, in Nuclear Spectroscopy of Astrophysical
    Sources, eds. N. Gehrels and G. H. Share, (New York: AIP), 98
Zurek, W. H. 1985, ApJ, 289, 603

DIFFUSE GALACTIC GAMMA-RAY EMISSION
ABOVE 1 MEV

Hans Bloemen
SRON-Leiden and Leiden Observatory
P.O. Box 9504, 2300 RA Leiden, The Netherlands

ABSTRACT

The diffuse continuum γ-ray emission from the Galaxy provides a unique tracer of cosmic rays. The emission above $\sim$ 50 MeV is well studied by the SAS-2 and COS-B satellites and can soon be analyzed with better angular resolution and sensitivity now the all-sky survey by the EGRET telescope aboard the Compton Observatory has been completed. Information for the 1-50 MeV range was far more limited prior to the launch of the Compton Observatory. The COMPTEL telescope on this observatory provides for the first time extensive imaging possibilities at MeV energies. This paper presents a global review of the diffuse gamma radiation between about 1 MeV and 10 GeV, with emphasis on implications for the low-energy cosmic-ray electron spectrum and the distribution of cosmic rays in the Galaxy. A complementary review on the diffuse emission below 1 MeV is given by R. Ramaty in this volume.

1. INTRODUCTION

The observed gamma radiation from the Galactic plane is the sum of diffuse emission and point sources. The radiation at energies $E_\gamma \gtrsim 50$ MeV, well studied with the SAS-2 and COS-B satellites, seems largely of diffuse origin. Preliminary results from EGRET confirm this finding. Constraints from high-energy γ-rays on the Galactic cosmic-ray (CR) distribution are summarized in Section 2. Observations in the 1-50 MeV range are difficult. They have mainly been made with non-imaging wide-field instruments and are largely restricted to the inner region of the Galaxy. Thus our knowledge of the Galactic γ-ray emission at these energies is very limited. In particular, the point-source contribution is uncertain. In Section 3, a compilation of measurements is presented and implications for the CR electron spectrum are discussed. COMPTEL provides for the first time good possibilities to map γ-ray emission in the 1-30 MeV range, which also contains nuclear γ-ray lines. Preliminary imaging results from COMPTEL are presented in Section 4, including a first map of the 1.809 MeV line emission from ^{26}Al radioactive decay.

Diffuse continuum γ-ray emission originates from the interaction of cosmic-ray particles with the interstellar gas (bremsstrahlung and π°-decay emission) and the interstellar radiation field (inverse-Compton emission). Beyond $E_\gamma \simeq 50$ MeV, CR protons and α-particles with energies of typically 1-10 GeV play a dominant role through the decay of π°-mesons that originate from nuclear interactions with the interstellar matter. These cosmic rays are of particular interest because they provide most of the CR pressure in the interstellar medium. Each

γ-quantum has an energy of $m_{\pi^\circ}c^2/2 \approx 68$ MeV in the rest frame of the π°-meson, which transforms into a broad energy distribution centered on ~ 68 MeV in the observer's reference system. At γ-ray energies above ~ 1 GeV, the shape of the π°-decay γ-ray spectrum is similar to that of the parent CR spectrum, i.e., for a power-law CR proton spectrum, $I(E_p) = KE_p^{-\Gamma}$, the differential (volume) emissivity converges to a similar power-law spectrum, $Q_\gamma(E_\gamma) \propto nKE_\gamma^{-\Gamma}$, where n is the density of the target gas nuclei.

For $E_\gamma \simeq 1 - 50$ MeV, the diffuse γ-ray emission is almost entirely due to CR electrons through bremsstrahlung and inverse-Compton interactions, although the latter is generally estimated to be of only minor importance for γ-ray studies of the Galactic disk. The bremsstrahlung emission results from CR electrons with low energies of typically 1-100 MeV ($E_e \approx 3E_\gamma$). For an electron spectrum of the form $I(E_e) = KE_e^{-\Gamma}$, the emissivity spectrum is in first-order approximation a similar power-law spectrum, $Q_\gamma(E_\gamma) \propto n[K/(\Gamma - 1)]E_\gamma^{-\Gamma}$, where n is the density of the gas nuclei. Note that γ-ray observations at MeV energies add valuable information to observations of the diffuse radio synchrotron emission, which trace electrons with energies of typically 100 MeV to 10 GeV ($E_e[\text{GeV}] \approx \frac{1}{4}\sqrt{\nu[\text{MHz}]/B[\mu\text{G}]}$). Both for the bremsstrahlung and π°-decay processes, it is useful to define the emissivity per hydrogen atom, $q_\gamma \equiv Q_\gamma/n$, which is a measure of the CR density.

Inverse-Compton emission results from the scattering of electrons with energies (well) above 10 GeV ($E_e \approx m_ec^2\sqrt{E_\gamma/\varepsilon}$) on photons of energy $\varepsilon \ll m_ec^2$ (mainly optical and infrared photons and the 2.7 K background radiation). If the incident electrons have a power-law spectrum, $I(E_e) = KE_e^{-\Gamma}$, then the inverse-Compton volume emissivity has approximately a power-law spectrum of the form $Q_\gamma(E_\gamma) \propto n\langle\epsilon\rangle^{(\Gamma-1)/2}KE_\gamma^{-(\Gamma+1)/2}$, where $\langle\epsilon\rangle$ and n are the average energy and number density, respectively, of the target photons. A good estimate of the local interstellar electron spectrum at these high energies can be obtained from direct CR measurements near Earth. Below 10 GeV, solar modulation effects become increasingly important, flattening the spectrum appreciably.

Clearly, many pieces of information have to be combined to study the interstellar CR electron spectrum. Several attempts have been made, but we will see in Section 3 that the electron spectrum below 100 MeV is still highly uncertain. The importance of low-energy cosmic rays for the ionization and energy balance of the interstellar medium is therefore poorly known.

2. IMPLICATIONS FOR THE GALACTIC COSMIC-RAY DISTRIBUTION

2.1 Radial Gradient and Disk Thickness

Several high-energy γ-ray studies of the Galactic disk, based on SAS-2 and COS-B observations, have been aimed at determining the radial distribution of cosmic rays in the Galaxy, using the γ-ray emissivity as a tracer of the CR intensity (a review is given by Bloemen 1989). This is feasible if independent information is available on the distribution of the target gas particles with which the cosmic rays interact. The result from the latest and most robust analyses (Bloemen et al. 1986; Strong et al. 1988), were obtained from correlation studies of the observed γ-ray emission with various HI surveys and the CO

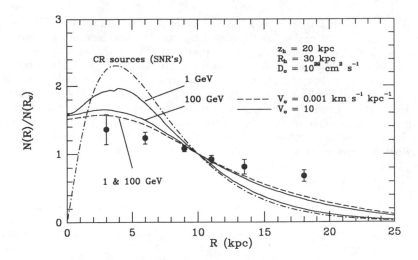

Fig. 1:   Radial γ-ray emissivity profile (points) for the energy range 70 MeV - 5 GeV (Strong et al. 1988) together with examples of Galactocentric distributions of the CR (proton) density for $E_\gamma = 1$ and 100 GeV in two-dimensional convection-diffusion models with $V_o = 10$ and 0.001 km s^{-1} kpc^{-1} (the latter approaches the pure diffusion model). A SNR-like CR source distribution was chosen. All quantities are normalized at the radius of the solar circle.

survey of Dame et al. (1987), using the distance information from the HI and CO line velocities. In addition to $q_\gamma(R)$, the CO-to-H$_2$ calibration factor is a free parameter in these studies, which provides useful constraints on the H$_2$ content of the Galaxy (see e.g. reviews by Wolfendale 1988 and Bloemen 1989).

The resulting q_γ distribution, presented in Fig. 1, shows only a weak Galactocentric gradient. The correspondingly weak gradient for the relevant CR particles sets constraints on the thickness of the CR disk in the framework of a given CR propagation model. Latest results from this method, first applied by Stecker and Jones (1977), are presented by Bloemen et al. (1993a). They considered a convection-diffusion model in which cosmic rays, produced in the Galactic plane, diffuse into a halo and are convected outward in a Galactic wind, assuming a linear increase of the convection velocity with distance from the plane ($V = 3V_o z$) and allowing for an energy-dependent diffusion coefficient ($D = D_o(E[\text{GeV}])^{0.6}$). They first compared their model predictions with direct CR measurements near Earth (the observed grammage traversed by cosmic rays and the abundance of radioactive ^{10}Be), which indicated that $V_o \lesssim 15$ km s^{-1} kpc^{-1} and $D_o \approx (0.5 - 3) \times 10^{28}$ cm^2 s^{-1}, implying that the effective CR scale height in the solar vicinity has to be less than ~ 3 kpc. These findings are in good agreement with the results from a recent study by Webber et al. (1992), using a convection-diffusion model with a constant wind speed. The implied 'small' CR scale height, however, is hard to reconcile with the weak Galactocentric CR gradient deduced from the γ-ray observations, unless the radial distribution of CR sources shows a weak radial fall-off as well. The key point is that a small scale height implies a radial CR distribution similar to the source distribution, because the probability to escape from the Galaxy is large and the CR density at a certain location is mainly determined by the local source density and the characteristics of the ambient interstellar medium. Fig. 1 shows some examples of the modelling by Bloemen et al. The radial distribution of the CR sources was assumed to resemble that of supernova remnants (SNR's). The radial extent of the halo volume, R_h, is 30 kpc and the vertical extent, z_h, is 20 kpc. It can be seen in Fig. 1 that even a very large CR diffusion halo (equivalent to the case $V_0 \approx 0$ for the large R_h and z_h values used) barely explains the observed weak CR density gradient for such a SNR-like source distribution. Also, Fig. 1 shows that including convection produces an even steeper CR gradient, which results from the fact that convection leads to a smaller effective diffusion halo. A source distribution with a radial exponential scale length $\gtrsim 10$ kpc (at least beyond a few kpc from the Galactic centre) seems to be required. In fact, because of strong selection effects, the radial distribution of SNR's is not well known, but the upper limit on the radial scale length appears to be ~ 10 kpc (Li et al., 1991; D. Green, priv. comm.). The same holds for pulsars (Lyne et al. 1985).

2.2 Cosmic-ray – matter coupling

Instead of adopting a CR distribution with a radial variation only, Bertsch et al. (this volume) have made a preliminary analysis of the EGRET observations in which they assume that cosmic rays are preferentially located in regions of high gas density, based on the argument that the weight of the matter ties the magnetic fields and hence the cosmic rays to these regions (Bignami and Fichtel 1974; Fichtel and Kniffen 1984). Their model agrees well with the observations and one can ask whether this implies that the available γ-ray data cannot distinguish between such a *coupling model* and the *gradient model* discussed above. This question was already raised in SAS-2 and COS-B studies, but a detailed comparison of the findings was hampered by the fact that the coupling option was applied to a model of the gas distribution whereas studies of the gradient option made use of detailed HI and CO surveys.

If the density of (GeV) cosmic rays and the matter density are indeed correlated, it is most realistic that this occurs on scales of a few kpc (also used by Bertsch et al.), which is the characteristic CR diffusion path length (Ormes and Protheroe 1983). In this case, the γ-ray intensity distributions from the coupling model and the gradient model can be expected to be very similar, because the radial gradient of the γ-ray emissivity distribution shown in Fig. 1 is similar to that of the total gas surface density when smoothed over a few kpc (at least when the Columbia/CfA CO survey and the radial-unfolding method of Bronfman et al., 1988, are used).

Melisse and Bloemen (1990) have made a comparative study of the two models, using the same HI and CO surveys. Despite the fact that they considered a rather extreme case (CR density proportional to the gas density, $n_{CR} \propto n_{gas}^\alpha$, on a scale of typically 100 pc), the longitude profiles of the γ-ray intensity distributions for both models were found to be not drastically different. The main free parameters in the coupling model are α, $q_\gamma(R_\odot)$, and the CO-to-H_2 calibration factor. A maximum-likelihood test showed, however, that the gradient model fits the data much better. The main reason is the small scale height of the γ-ray emitting disk in the coupling model, imposed by the concentration of

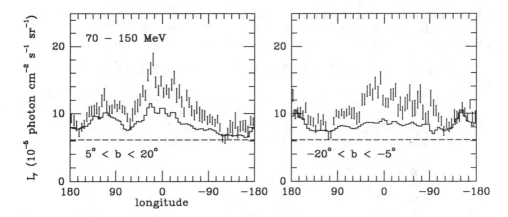

Fig. 2: Longitude distributions of the observed (COS-B, $\pm 1\sigma$ error bars) and modelled γ-ray intensity at medium latitudes. The model prediction includes only CR interactions with atomic and molecular gas and is an extension of the model of the Galactic-disk emission presented by Strong et al. (1988). Also the isotropic background level is from Strong et al. The excess visible toward the general direction of the inner Galaxy is most pronounced at these energies, but clearly visible at higher energies as well (see Bloemen 1989).

cosmic rays in the Galactic disk. Hence, the main conclusion from these findings is that the scale height of the CR distribution is significantly larger than that of the gas distribution. This result is probably the most direct evidence from γ-ray astronomy for the existence of a thick-disk (or halo) distribution of GeV CR protons and confirms similar conclusions from radio synchrotron data for CR electrons (e.g., Baldwin 1976; Beuermann et al. 1985).

2.3 Inverse-Compton Halo and 'Medium-latitude excess'

There is good evidence that almost half of the observed γ-ray emission at medium latitudes ($5° \lesssim |b| \lesssim 20°$) in the general direction of the inner Galaxy cannot be explained by CR-matter interactions if only atomic and molecular gas are involved (Bloemen 1989 and references therein). Fig. 2 shows that the excess extends over essentially the entire first and fourth Galactic quadrants. Inverse-Compton emission can explain only part of the excess if the scale height of high-energy (typically 100 GeV) electrons is of the order of 1 kpc (Shukla and Paul 1976; Stecker 1977; Kniffen and Fichtel 1981; Bloemen 1985). Electrons with such high energies are subject to severe energy losses and fill no more than $\sim 30\%$ of the volume taken up by the proton-nuclear CR component and low-energy ($\lesssim 1$ GeV) electrons. It is most likely that a significant fraction of the excess is due to the fact that ionized gas was not included in the γ-ray modelling (provided that this medium does not extend far beyond the solar circle). If the scale height of this ionized gas is as large as suggested by pulsar dispersion-measure data, which is ~ 1 kpc (Lyne et al. 1985, Reynolds 1991, and references therein), then its column density towards medium latitudes is about half the HI column density. This implies that at least half of the medium-latitude excess should be attributed to CR-matter interactions, with little room for inverse-Compton emission. In fact, the very accurate intensity measurements with EGRET may set a meaningful upper limit on the disk thickness for high-energy CR electrons.

3. GAMMA-RAY SPECTRUM OF THE INNER GALAXY

Fig. 3a shows a compilation of flux estimates for the central radian of the Galactic disk. Most results below ∼ 50 MeV were obtained with non-imaging instruments with fields of view of typically tens of degrees. A detailed comparison is not possible because of differences in the longitude intervals covered by the observations and differing assumptions in calculating the flux from the central radian of the Galaxy. Above ∼ 2 MeV, all measurements are consistent, with the possible exception of the SMM spectrum presented by Harris et al. (1990). Around 1 MeV (between the positron annihilation line at 511 keV and ∼ 2 MeV), however, significant differences are seen, which may be real. Most remarkable is the drastic flux decrease between the two HEAO-3 observations of the Galactic-centre region (Riegler et al. 1985), which cover a field of view as wide as ∼ 35° (the 'low-state' flux is included in Fig. 3a). This suggests a strong contribution from one or two point sources; the high-state spectrum between ∼ 100 keV and several MeV obtained by Riegler et al. is very similar to that of the blackhole candidate Cyg X-1 during its γ_1 phase (Ling et al. 1987). At 100 keV, the observed emission from the central 20° of the Galactic disk is known to be variable — most of the emission can be attributed to two sources, 1E1740.7-2942 and GRS1758-258 (see review by Gehrels and Tueller 1993).

Essentially all studies referred to in Fig. 3 and several others (Fichtel et al. 1978, Lebrun et al. 1982, Lebrun and Paul 1983, Sacher and Schönfelder 1984, Gualandris and Strong 1984, and Strong 1985) have provided estimates of the CR electron spectrum below a few hundred MeV from the observed γ-ray spectrum. In order to illustrate the uncertainties that still exist, we consider two rather extreme scenarios for the bremsstrahlung production.

Fig. 3b shows the π°-decay spectrum derived from the demodulated proton spectrum observed near Earth and the bremsstrahlung spectrum determined from the local electron spectrum given by Webber (1983). The latter is based on measurements near Earth ($E_e \gtrsim 10$ GeV) and radio observations ($E_e \simeq 100$ MeV - 10 GeV), with a power-law extrapolation in the range $E_e^{-2.3}$ to $E_e^{-2.1}$ for $E_e \lesssim 200$ MeV. The sum of these π°-decay and bremsstrahlung spectra (together with the weak inverse-Compton contribution, also shown) was normalized

---

Fig. 3 [next page]: (a) Gamma-ray spectrum of the inner Galaxy, multiplied by E^2. An E^{-2} spectrum was assumed to calculate effective energies. The published fluxes from HEAO-3 and GRIS were converted to flux-per-radian by dividing by the FWHM of the instrument field of view. (b) Modelled γ-ray spectra derived from the local interstellar CR electron and proton spectra (the sum of the individual components is normalized to the observed γ-ray spectrum at ∼ 1 GeV). The upper and lower bound of the bremsstrahlung spectrum correspond to an $E_e^{-2.3}$ and $E_e^{-2.1}$ extrapolation of the electron spectrum for $E_e \lesssim 200$ MeV. (c) The same as figure b, but with a quite different treatment of the bremsstrahlung from molecular clouds, based on a study of cloud penetration and secondary production of low-energy electrons by Morfill (1982).

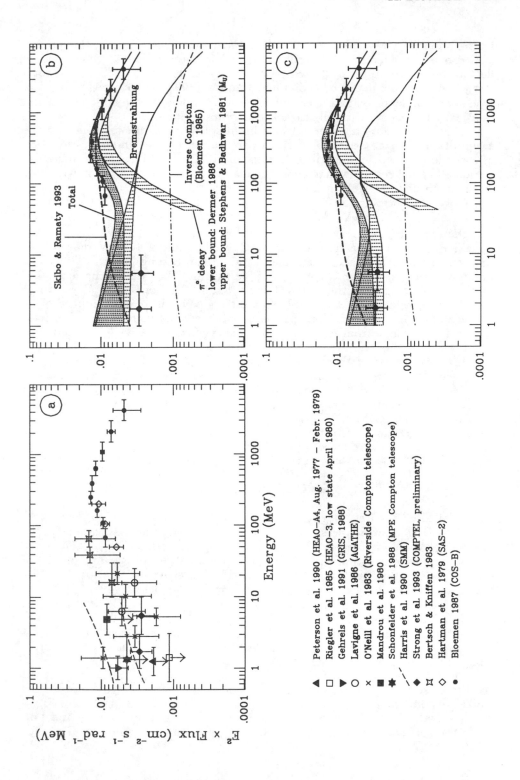

to the observed γ-ray spectrum at ~ 1 GeV. At low γ-ray energies, this prediction can be considered an upperlimit because ionization losses, which tend to flatten the spectrum, are neglected. For comparison, Fig. 3b also shows the total γ-ray spectrum predicted by Skibo and Ramaty (1993), who derived the electron spectrum by propagating an unbroken $E_e^{-2.4}$ power-law injection spectrum in a leaky-box model with all relevant energy losses included.

Fig. 3c presents another extreme bremsstrahlung estimate, based on the assumption that the scenario of cloud penetration and secondary production of low-energy electrons described by Morfill (1982) is applicable to molecular clouds. The net result is that dense clouds produce enhanced bremsstrahlung emission around 100 MeV (due to secondary electrons), but do not contribute to the γ-ray emission below ~ 30 MeV. The latter is due to the fact that low-energy electrons ($E_e \lesssim 100$ MeV) do not penetrate dense clouds in Morfill's scenario and the secondary production rate is very low for $E_e \lesssim 100$ MeV. Evidence for excessive γ-ray emission from the inner Galaxy around 100 MeV may indeed be seen in the COS-B data (Bloemen et al. 1986, Strong et al., 1988). The bremsstrahlung spectrum shown in Fig. 3c assumes that this scenario applies to all molecular gas in the inner Galaxy (almost half of the total gas mass there), whereas it is not applicable to atomic gas which is more diffuse. The $\pi°$-decay and inverse-Compton spectra are the same as in Fig. 3b. The total spectrum is again normalized to the observed spectrum near 1 GeV. The upper and lower bound of the bremsstrahlung spectrum reflect the uncertainties in a simple extrapolation of Webber's electron spectrum, as in Fig. 3b.

Clearly, very accurate flux measurements and imaging information in the 1-30 MeV range are needed in order to distinguish between different models and to identify the contribution from point sources. COMPTEL can provide this information. The COMPTEL data points from Strong et al. (1993) in Fig. 3 have 30% uncertainties (including statistical and systematic errors), but these uncertainties will be reduced when systematic effects are better understood. Note that the 10-30 MeV measurement from COMPTEL is not available yet.

4. FIRST MEV IMAGING RESULTS FROM COMPTEL

COMPTEL's first image of the inner Galaxy in the 1-10 MeV range (Bloemen et al. 1993b), which was obtained by combining several observations of the first half of the all-sky survey, shows a clear ridge of emission (Fig. 4). The same imaging analysis was applied to Monte Carlo simulations of a simple model of the diffuse emission, with an intensity distribution given by $I_\gamma = q_\gamma/4\pi \cdot N_H$, where N_H is the total gas column density (from HI and CO surveys). The value of q_γ was derived first by fitting the model to the observations. Fig. 4 shows that the global appearance of the observations is similar to that of the simulated observations of the diffuse emission, but a likelihood-ratio test revealed significant differences. For the moment, it cannot be excluded that an ensemble of sources along the plane contributes significantly to the observed ridge of emission.

A nice illustration of COMPTEL's imaging capability, although beyond the direct scope of this paper, is the map of the Galactic-centre region in the 1.809 MeV line from the decay of interstellar ^{26}Al ($\tau \simeq 10^6$ yr), shown in Fig. 5 (Diehl et al. 1993). The detection of this line (predicted by Ramaty and Lingenfelter in 1977 and first seen with HEAO-C) provides compelling evidence for

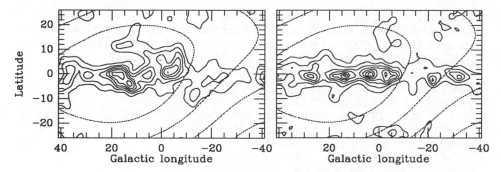

Fig. 4: Left: maximum-entropy image of combined COMPTEL observations of the inner Galaxy (1–10 MeV). Right: similar image of simulated observations of the diffuse emission. Contours of effective exposure, at 20% intervals from the peak value, are superimposed. From Bloemen et al. (1993).

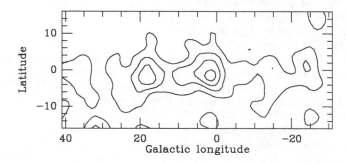

Fig. 5: Maximum-entropy map of the Galactic-centre region in the 1.8 MeV line, obtained from a combination of 4 COMPTEL observations (Diehl et al. 1993).

nucleosynthesis in our Galaxy during the past million years. Various populations have been proposed as sources of ^{26}Al (see e.g. review by Prantzos 1991). Imaging of the 1.8 MeV emission will help to distinguish among the potential candidates.

Although no firm constraints can be obtained yet from the preliminary COMPTEL maps, it is clear that COMPTEL's imaging potential is of great value in disentangling the MeV emission from the inner Galaxy.

REFERENCES

Baldwin, J.E. 1976, in The Structure and Content of the Galaxy and Galactic Gamma Rays, eds. C. Fichtel and F.W. Stecker (Greenbelt: GSFC), p. 206
Bertsch, D.L., Kniffen, D.A. 1983, ApJ 270, 305
Beuermann, K., Kanbach, G., Berkhuijsen, E.M. 1985, A&A 153, 17
Bignami, G.F., Fichtel, C.E. 1974, ApJ Lett. 189, L65
Bloemen, H. 1985, A&A 145, 391
Bloemen, H. et al. 1986, A&A 154, 25
Bloemen, H. 1987, ApJ Lett. 317, L15

Bloemen, H. 1989, Ann. Rev. A&A 27, 469
Bloemen, H., Dogiel, V.A., Dorman, V.L., Ptuskin, V.S. 1993a, A&A, in press
Bloemen, H. et al. 1993b, Proc. Compton Science Symp. (St. Louis), in press
Blumenthal, G.R., Gould, R.J. 1970, Rev. Mod. Phys. 42, 237
Cesarsky, C.J., Paul, J.A., Shukla, P.G. 1978, Ap. Space Sci. 59, 73
Dame, T.M. et al. 1987, ApJ 322, 706
Dermer, C.D. 1986, A&A 157, 223
Diehl, R. et al. 1993, in Gamma Ray Astronomy (COSPAR Washington), ed.
    G.C.C. Palumbo (Oxford: Pergamon Press), in press
Fichtel, C.E. et al. 1976, ApJ 208, 211
Fichtel, C.E., Kniffen, D.A. 1984, A&A 134, 13
Gehrels, N. et al. 1991, ApJ Lett. 375, L13
Gehrels, N., Tueller, J. 1993, ApJ, submitted
Gualandris, F., Strong, A.W. 1984, A&A 140, 357
Harris, M.J. et al. 1990, ApJ 362, 135
Hartman, R.C. et al. 1979, ApJ 230, 597
Kniffen, D.A., Fichtel, C.E. 1981, ApJ 250, 389
Lebrun, F., Paul, J.A. 1983, ApJ 266, 276
Lebrun, F. et al. 1982, A&A 107, 390
Lavigne, J.M. et al. 1986, ApJ 308, 370
Ling et al. 1987, ApJ Lett. 321, L117
Li, Zongwei, Wheeler, J.C., Bash, F.N., Jeffreys, W.H. 1991, ApJ 378, 93
Lyne, A.G., Manchester, R.N., Taylor, J.H. 1985, MNRAS, 213, 613
Mandrou, P., Bui-Van, A., Vedrenne, G., Niel, M. 1980, ApJ 237, 424
Melisse, J., Bloemen, H. 1990, Proc. 21st Int. Cosmic Ray Conf. 1, OG 3.1-18
Morfill, G.E. 1982, ApJ 262, 749
O'Neill, T. et al. 1983, Proc. 18th ICRC 9, 45
Peterson, L.E. et al. 1990, Proc. 21st Int. Cosmic Ray Conf. 1, 44
Prantzos, N. 1991, in Gamma-Ray Line Astrophysics, eds. P. Durouchoux and
    Prantzos (New York: AIP), p. 129
Ramaty, R., Lingenfelter, R.E. 1977, ApJ Lett. 213, L5
Reynolds, R.J. 1991, in The Interstellar Disk-Halo Connection in Galaxies, ed.
    H. Bloemen (Dordrecht: Kluwer), p. 67
Riegler, G.R. et al. 1985, ApJ Lett. 294, L13
Sacher, W., Schönfelder, V. 1984, ApJ 279, 817
Schönfelder, V., von Ballmoos, P., Diehl, R. ApJ 335, 748
Shukla, P.G. and Paul, J.A. 1976, ApJ 208, 893
Skibo, J.G., Ramaty, R. 1993, Compton Science Symp. (St. Louis), in press
Stecker, F.W. 1977, ApJ 212, 60
Stecker, F.W., Jones, F.C. 1977 ApJ 212, 60
Stephens, S.A., Badhwar, G.D. 1981, Ap. Space Sci. 76, 213
Strong, A.W. 1985, Proc. 18th Int. Cosmic Ray Conf. 9, 90
Strong, A.W. et al. 1988, A&A 207, 1
Strong, A.W. et al. 1993, Proc. Compton Science Symp. (St. Louis), in press
Webber, W.R. 1983, in Composition and Origin of Cosmic Rays, ed. M.M.
    Shapiro (Dordrecht: Reidel), p. 83
Webber, W.R., Lee, M.A., Gupta, M. 1992, ApJ 390, 96
Wolfendale, A.W. 1988, in Molecular Clouds in the Milky Way and External
    Galaxies, eds. R. Dickman et al. (Heidelberg: Springer-Verlag), p. 76

Diffuse Galactic Gamma-Rays from Pulsars

Dieter H. Hartmann, Lawrence E. Brown, Neil G. Schnepf

Department of Physics & Astronomy, Clemson University, Clemson, SC 29634

ABSTRACT

A significant fraction of the diffuse galactic γ-ray emission could be due to γ-rays from individually undetected pulsars. Using the polar cap model of pulsar γ-ray emission, the contribution of aging radio pulsars to the diffuse galactic γ-ray glow is estimated. We calculate photon flux maps for a dynamically evolving pulsar population and, using COS-B data, we derive constraints on the Galactic birth rate of pulsars. We emphasize the value of analyzing the collective contribution of pulsars that perhaps individually are too faint to be detected by the Compton Observatory.

1. INTRODUCTION

Observations of COS−B and SAS−2 have long identified the plane of the Galaxy as the predominant source of high energy γ-rays. The origins of the diffuse glow of the plane is attributed to emission processes involving interactions between high energy cosmic rays and interstellar matter (Strong *et al.* 1988). However, a significant fraction of the diffuse γ-ray glow from the galactic plane could be due to unresolved point sources. Pulsars, known to be γ-ray sources, could provide such an unresolved population. The galactic distribution of pulsars is sufficiently well known to accurately estimate their contribution to the diffuse flux if their emission processes were understood. Overall, the trend in recent years has been towards a down-scaling of their contribution to the total detected γ-ray emission. However, because no generally accepted model for γ-ray emission from pulsars exists at present we investigate whether existing data can be used to provide constraints. In particular, we constrain the Pulsar Birth Rate (PBR), averaged over a timescale of order 10^7 yrs, by requiring that nowhere on the sky the pulsar contribution exceeds that observed by the COS-B experiment.

2. PULSAR EVOLUTION

The work of Emmering & Chevalier (1986, 1989), Lyne, Manchester & Taylor (1985), and Narayan (1987) provides a comprehensive review of the overall characteristics and evolution of the Galactic pulsar population. Drawing from these and the polar cap model of γ-ray emission, we extend the studies undertaken by Higdon & Lingenfelter (1976), Harding (1981b), Hartmann *et al.* (1990; 1992), and Bailes & Kniffen (1992: BK) to estimate the pulsar contribution to the diffuse glow of the plane. We perform Monte Carlo simulations to determine the γ-ray glow form the evolving pulsar population. Initial birth velocities were taken to be gaussian with 1-d dispersions of 70 km s^{-1}. Originating positions were selected from a density profile that is exponential in z, with a scale height of 100 pc, and from either one of two fits to the observed radial distribution (see BK). We followed the evolution of 5 10^5 pulsars with lifetimes randomly selected to a maximum age of 2.5 10^7 years (PBR =1/50 yrs). The orbits are integrated in a Miyamoto-Nagai Potential fitted to the Galactic rotation curve. Initial magnetic field strengths and periods were assumed to be log-normal with mean values log B$_0$ = 12.4, log P$_0$ = -0.4, and the magnetic field strength was assumed to decay exponentially on a timescale of τ = 5 10^6 years (Emmering & Chevalier 1989). Evolution of pulsar properties is determined by

$$P^2 = P_0^2 + Q_0^2 \left(1 - \exp\left(-2t/\tau\right)\right) \tag{1}$$

for the period, by

$$\dot{P} = \frac{Q_0^2}{P\tau} \exp(-2t/\tau) \tag{2}$$

for the rate of change of the pulsar period, and by

$$B^2 = \frac{3Ic^3}{8\pi^2 R^6} P\dot{P} = B_0^2 \exp\left(-2t/\tau\right) \tag{3}$$

for the magnetic field. Initial field strength and decay time are related by

$$Q_0^2 = \frac{8\pi^2 R^6}{3Ic^3} B_0^2 \tau . \tag{4}$$

The total γ-ray photon luminosity was calculated using the polar cap predictions of Harding (1981a)

$$L_\gamma(> 100\text{MeV}) = 1.2 \; 10^{35} B_{12}^{0.95} P^{-1.7} \text{ photons s}^{-1} . \tag{5}$$

We assumed that the γ-ray beaming fraction is unity. It is certainly more realistic to assume that the radiation is beamed into solid angles that are significantly smaller than 4π. We approximately compensate for this by neglecting the diffuse γ-ray emission caused by cosmic-ray interactions. Cosmic ray (CR) models (Strong et al. 1988) fit the observed γ-ray glow very well, leaving little space for an unresolved point source contribution. Thus, ignoring the CR contribution is roughly equivalent to assuming a small beaming fraction. The production of γ-rays in the pulsar magnetospheres has been modeled in the context of "polar cap" models (Daugherty & Harding 1982) in which particle acceleration occurs near the pulsar surface, and in "outer gap" models (Cheng, Ho, & Ruderman 1986). Both production scenarios have recently been applied to millisecond pulsars (Chiang & Romani 1992). We do not consider recycled pulsars in this paper. It is not clear at present which of the two scenarios describes γ-ray pulsars with better accuracy, but GRO observations might soon provide guidance when phase-resolved spectra for several new pulsars as well as Crab and Vela are obtained. For the time being, we employ the γ-ray fluxes of the polar cap model (Harding 1981a), but set the γ-ray efficiency (fraction of spin-down energy that is emitted above 100 MeV) to zero after it has reached 100%.

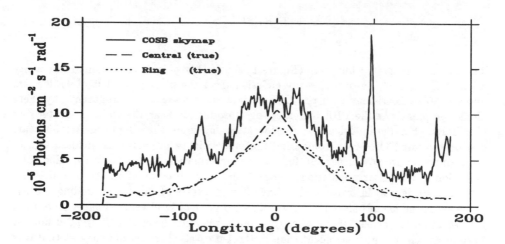

Figure 1: *Latitude integrated flux from indicated models presented against the COS-B skymap. Each of the models were filtered through the COS-B PSF, with summation ranging over -10 to 10 degrees in 1 degree bins.*

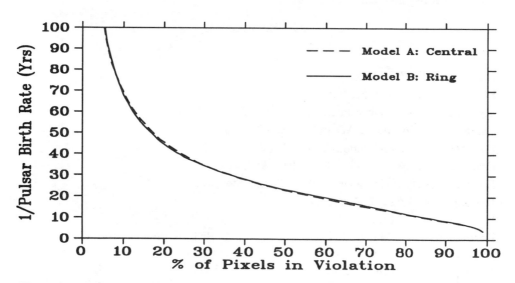

Figure 2: *Each of the curves represents a 2-d comparison of our orbit-integrated models to the COS-B skymap. Percent violation measures the fraction that, when enhanced by some scaling factor, exceed the associated bin of the COS-B data.*

3. RESULTS

The latitude-integrated flux map (Figure 1) shows that pulsars could contribute a significant fraction of the overall γ-ray emission from the plane. A PBR of 1/50 years appears to be consistent with the COS-B data since no excess is apparent anywhere along the plane. As the PBR increases, over-production near the Galactic center sets the limits. For the model under consideration, no more than 1 pulsar per 30 years can be tolerated. However, comparing the models in two dimensions yields a tighter limit since we are now using the full angular information that reflects the kinematic evolution of the pulsar population. Figure 2 shows the fraction of all the bins between -10 and 10 degrees in latitude that have flux in excess of the corresponding COS-B value as a function of pulsar birth rate. The 1-d limit of 1/30 years violates more than a third of all the pixels. If one were to adopt a tolerance level of 10% the 2-d limit is 1/70 years, clearly more stringent than in the 1-d case. These results suggest that it is important to incorporate the 2-d aspects of this kind of study. The EGRET detector aboard Compton Observatory will provide a sky map with unprecedented sensitivity. Combining that information with the knowledge we gain from EGRET data on new γ-ray pulsars (Bertsch et al. 1992; Thompson et al. 1992)) holds great potential for constraining radio pulsar evolution, their γ-ray emission mechanisms, and their birth rate. This study has shown that rigorous data analysis must be based on full 2-d sky maps which require realistic pulsar orbits.

REFERENCES

Bailes, M. & Kniffen, D.A. 1992, ApJ, 391, 659

Bertsch, D. L., et al. 1992, Nature, 357, 306

Cheng, K. S., Ho, C., & Ruderman, M. 1986, ApJ, 300, 522

Chiang, J., & Romani, R. W. 1992, ApJ, 400, 629

Daugherty, J. K., & Harding, A. K. 1982, ApJ, 252, 337

Emmering, R.T. & Chevalier, R.T. 1986, ApJ, 304, 140

Emmering, R.T. & Chevalier, R.T. 1989, ApJ, 345, 931

Harding, A.K. 1981a, ApJ, 245, 267

Harding, A.K. 1981b, ApJ, 247, 639

Hartmann, D. H., Liang, E. P., & Cordes, J. M. 1990, in High Energy Gamma-Ray Astronomy, ed. C. Ackerlof, J. Matthews, and J. van der Velde, AIP, in press.

Hartmann, D. H., Brown, L. E., Clayton, D. D., Schnepf, N., Cordes, J. E., & Harding, A. K. 1992, in Isolated Pulsars, ed. K. A. Van Riper & C. Ho, AIP, in press

Higdon, J. C., & Lingenfelter, R. E. 1976, ApJL, 208, L107

Lyne, A.G., Manchester, R.N. & Taylor, J.H. 1985, MNRAS, 213, 613

Narayan, R. 1987, ApJ, 319, 162

Strong, A.W., et al., 1988, A&A 207, 1

Thompson, D. J., et al. 1992, Nature, 359, 615

HIGH ENERGY GAMMA RAY OBSERVATIONS AND THE COUPLING BETWEEN GALACTIC COSMIC RAYS AND MATTER

S.D. Hunter[1], D.L. Bertsch[1], T. Dame[2], B.L. Dingus[1,8], C.E. Fichtel[1],
R.C. Hartman[1], G. Kanbach[3], D.A. Kniffen[4], P.W. Kwok[1,9], Y.C. Lin[5],
J.R. Mattox[1,10], H.A. Mayer-Hasselwander[3], P.F. Michelson[5],
C. von Montigny[3], P.L. Nolan[5], P. Sreekumar[1,8], P. Thaddeus[2],
E. Schneid[6], G. Stacy[7], D.J. Thompson[1]

[1] NASA/Goddard Space Flight Center, Code 662, Greenbelt, MD 20771, USA
[2] Harvard-Smithsonian Center for Astrophysics, Cambridge, MA 02138, USA
[3] Max-Plank Institut für Extraterrestrische Physik, 8046 Garching bei München, Germany
[4] Hampden-Sydney College, P.O. Box 862, Hampden-Sydney, VA 23943, USA
[5] Hansen Experimental Physics Laboratory, Stanford University, Stanford, CA 94305, USA
[6] Grumman Aerospace Corporation, Mail Stop A01-26, Bethpage, L.I., NY 11714, USA
[7] University of New Hampshire, Durham, NH 03824, USA
[8] Universities Space Research Association
[9] NASA/NRC Research Associate
[10] COMPTON Science Support Center, Ast. Progs., Comp. Sci. Corp.

ABSTRACT

One of the primary goals of the EGRET mission on the Compton Observatory has always been the study of the diffuse gamma ray emission of both galactic and extra-galactic origin. The first 18 months of the mission, following the initial activation period, is being devoted to the all-sky survey with the goal of obtaining a nearly uniform exposure. In anticipation of the improved capabilities of the EGRET instrument relative to the earlier SAS-2 and COS-B missions, a new model of the galactic diffuse gamma ray emission was developed that incorporates recent surveys of the matter distribution and which permits a variety of assumptions regarding galactic cosmic ray distribution to be tested. This paper summarizes the model calculation and gives a preliminary comparison of the model with EGRET observations.

1. MODEL DESCRIPTION

The following is a brief summary of the complete model description given by Bertsch et al. (1993).

The diffuse gamma ray emission in the galaxy is considered to be due to the interaction of cosmic rays and electrons with matter and, to a lesser extent, with photons. The dominant production mechanisms over the EGRET energy range, 50 MeV < E < 30 GeV, which are included in this calculation are: nuclear interactions between cosmic rays and matter (Stecker, 1988), bremsstrahlung interactions between electrons and matter (Koch and Motz, 1959), and inverse Compton scattering of electrons with low energy photons (Ginzburg and Syrovatskii, 1964). Secondary mechanisms which are not included are: synchrotron emission from electrons in magnetic fields, line emission from dust and grains excited by cosmic ray collisions, and unresolved point sources. The first two of these mechanisms provide an insignificant contribution to the total emission. Unresolved point sources are not included so that the predictions of this calculation can be used as a basis for the evaluation of various source distributions.

The intensity of diffuse galactic gamma rays of energy E_γ from galactic longitude l and latitude b is expressed in a general form by

$$j(E_\gamma, l, b) = \frac{1}{4\pi} \int \Big(c_e(\rho, l, b) q_{em}(E_\gamma) + c_n(\rho, l, b) q_{nm}(E_\gamma) \Big) \times$$

$$\Big(n_{HI}(\rho, l, b) + n_{H_2}(\rho, l, b) \Big) d\rho$$

$$+ \frac{1}{4\pi} \sum_i \int c_e(\rho, l, b) q_{pi}(E_\gamma, \rho) u_{pi}(\rho, l, b) \, d\rho \quad \gamma \, \mathrm{cm}^{-2} \, \mathrm{s}^{-1} \, \mathrm{sr}^{-1} \, \mathrm{GeV}^{-1}$$

The integrations are over the line-of-sight distance, denoted by ρ, measured from the solar region in the direction given by l and b. The first integral represents the gamma ray production due to cosmic ray interactions with matter where q_{em} and q_{nm} are the production functions per target atom based on the local cosmic ray density. The functions $c_e(\rho, l, b)$ and $c_n(\rho, l, b)$ are ratios of the electron and nucleon cosmic ray intensities relative to the local intensities. The quantities $n_{HI}(\rho, l, b)$ and $n_{H_2}(\rho, l, b)$ are the atomic and molecular hydrogen densities expressed as atoms per unit volume. The second term describes the contribution from inverse Compton interactions between electrons and photons. The production function $q_{pi}(E_\gamma, r)$, where r is the galacto-centric radius, is based on the local electron intensity. The summation is over six discrete wavelength bands: blackbody, far infra-red, and four stellar components emitting in the range $8 < \lambda < 1000$ μm with photon energy density distributions $u_{pi}(r, l, b)$.

This calculation models the galactic matter distribution using a three dimensional array of cells with dimension, $\Delta l = 0.5°$, $\Delta b = 0.5°$, $|b| \leq 10°$, and $\Delta \rho = 0.5\,\mathrm{kpc}$. The atomic and molecular hydrogen density for each cell is determined by a kinematic deconvolution of the HI and CO radio observations using the galactic rotation curve of Clemens (1985). The atomic hydrogen density is obtained from a combination of the 21 cm northern survey of Weaver and Williams (1974), the Maryland-Parks southern survey (Kerr, et al., 1986) and the Leiden-Green Bank survey (Burton and Liszt, 1983). The $^{1}2CO$, J=1–0 transition, survey of (Dame et al., 1987) is used to deduce the H_2 density.

The cosmic rays are assumed to be in quasi-static equilibrium with the matter, i.e. the expansive pressures of the cosmic rays, interstellar gas and magnetic fields are in dynamic balance with the attractive gravitational pressure. Thus, assuming dynamic balance exists in the Milky Way (Parker 1966, 1969, 1977), the cosmic rays are coupled to the matter distribution. The cosmic ray distribution, coupled to the matter, is modeled by convolving the HI and H_2 distributions with a two dimensional Gaussian characterized by a sigma equal to the assumed cosmic ray-matter coupling scale. The cosmic ray coupling scale, and the X factor $(N(H_2)/W_{CO})$ are the only adjustable parameters in this calculation. The value of these parameters used are 2.3×10^{20} $\mathrm{mol\,cm}^{-2}\,(\mathrm{K\,km\,s}^{-1})^{-1}$ and 2 kpc, respectively.

The photon energy densities are determined in the following manner: the blackbody radiation is taken to be 2.7° K every where with an energy density of 0.25 $\mathrm{eV\,cm}^{-3}$, the far IR radial distribution is modeled on the cold dust emission curve of Cox, et al. (1986), and the near IR, optical and UV are taken from the work of Chi and Wolfendale (1990) who used the four stellar component model of Mathis et al. (1983).

2. RESULTS AND COMPARISON WITH EGRET OBSERVATIONS

The total gamma ray emission above 100 MeV predicted by this calculation, averaged over the latitude range $|b| \leq 10°$, is compared to the COS-B data as a function of longitude in figure 1 without being convolved with the COS-B point spread function. The bright gamma ray sources Vela, Geminga and Crab can be seen in the COS-B data, shown as a dashed line. The tangent points to the Local, Scutum, Norma, Crux, and Carina arms, at $l =$ 80° & 270°, 35°, 345°, 310°, and 280°, respectively, are clearly visible. The contribution to the gamma ray emission from inverse Compton interactions is indicated by the dotted line. The nucleon-nucleon and bremsstrahlung contribution from molecular and atomic hydrogen are shown as dash-dot and dash-dot-dot lines, respectively. The total gamma ray emission is shown as the solid line. The high degree of correlation between the model and the COS-B data indicates that this model is a reasonably good prediction of the diffuse gamma ray emission.

Figure 2 shows a longitude plot of the EGRET data, averaged over the latitude range $|b| \leq 10°$ and the predicted diffuse emission. The vertical scale is given in counts. The modulation apparent in these plots (nearly zero counts in the second quadrant and at $l = 230°$) is due to non-uniformity in the EGRET exposure. The fit of the model to the EGRET data is fairly good along the entire galactic plane except in the fourth quadrant. This over prediction may be due to the use of a symmetric galactic rotation curve.

3. SUMMARY

The model for diffuse emission in the Galaxy presented here incorporates recent data on the distribution of galactic atomic and molecular hydrogen into a 3-dimensional spatial array. Using known interaction cross sections and a cosmic ray density distribution based on dynamic balance, it has been shown that the diffuse gamma ray emission is remarkably consistent in magnitude and structure in the longitudinal variation with existing data.

Unlike most models in the past, no multi-parameter fits are used to adjust the results. The two quantities that are not firmly established have been treated as free, and approximate values based on present knowledge of the galactic diffuse gamma ray emission have been used. Future observations can be expected to further refine their values. The model also serves as a diffuse background model for the EGRET analysis of point sources.

REFERENCES

Bertsch, D.L., Dame, T.M., Fichtel, C.E., Hunter, S.D., Sreekumar, P., Stacy, J.G. and Thaddeus, P., ApJ, 1993, submitted
Burton, W.B., and Liszt, H.S. 1983, A&AS, 52, 63
Chi, X. and Wolfendale, A.W. 1991, J. Phys. G., 17, 987
Clemens, D.P. 1985, ApJ, 295, 422
Cox, P., Krügel, E., and Mezger, P.G. 1986, A&A, 155, 380
Dame, T.M., Ungerechts, H., Cohen, R.S., De Geus, E.J., Grenier, I.A., May, J., Murphy, D.C., Nyman, L.A., and Thaddeus, P. 1987, ApJ, 322, 706
Ginzburg, V.L., and Syrovatskii, S.I. 1964, *The Origin of Cosmic Rays*, Oxford: Pergamon Press
Kerr, F.J., and Lynden-Bell, D. 1986, MNRAS, 221, 1023
Koch, H.W., and Motz, J.W. 1959, Rev. Mod. Phys., 31, 920
Mathis, J.S., Mezger, P.G., and Panagla, N. 1983, A&A, 128, 212

Parker, E.N., 1966, ApJ, 145, 811

Parker, E.N., 1969, Space Sci. Rev., 9, 654

Parker, E.N., 1977, in *The Structure and Content of the Galaxy and Galactic Gamma Rays*, ed. C.E. Fichtel and F.W. Stecker, NASA CP-002, Washington, GPO, 283

Stecker, F.W. 1988 *Cosmic Gamma Rays, Neutrinos and Related Astrophysics* ed., Shapiro, M.M. and Wefel, J.P., (Dordrecht: Reidel) 85

Weaver, H.F., and Williams, D.R.W. 1973, A&AS, 8, 1

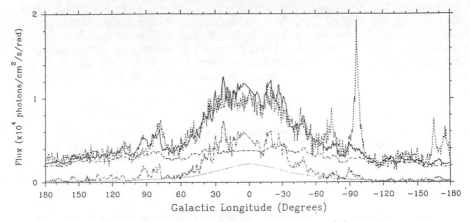

Figure 1. Longitude plot of the COS-B data, averaged over $|b| \leq 10°$, compared with the result of the calculation described in the text. The bright gamma ray sources Vela, Geminga, and Crab, as well as the tangent points to the Local, Scutum, Norma, Crux, and Carina arms are clearly visible.

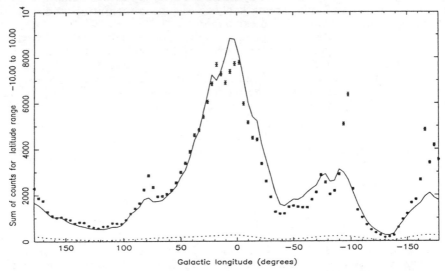

Figure 2. Longitude plot of the EGRET data, averaged over $|b| \leq 10°$, compared with the result of the calculation described in the text. The calculated diffuse gamma ray emission has been convolved with the EGRET point spread function. The sharp peaks in the data are due to the contribution from Vela, Geminga, Crab and sources in the Cygnus region.

WHAT CAN THE MAGELLANIC CLOUDS TELL US ABOUT COSMIC RAYS IN THE MILKY WAY?

P. Sreekumar[1,7], D.L. Bertsch[1], B.L. Dingus[1,7], C.E. Fichtel[1], R.C. Hartman[1],

S.D. Hunter[1], G. Kanbach[2], D.A. Kniffen[5], Y.C. Lin[3], J.R. Mattox[1,6],

H.A. Mayer-Hasselwander[2], P.F. Michelson[3],C. von Montigny[2],

P.L. Nolan[3], E.J. Schneid[4], D.J. Thompson[1]

[1].NASA/Goddard Space Flight Center, Code 662, Greenbelt, MD 20771
[2].Max-Planck Institut fur Extraterrestrische Physik, D/W-8046 Garching
[3].Hansen Experimental Physics Laboratory, Stanford University,CA 94305
[4].Grumman Aerospace Corporation, Mail Stop A01-26, Bethpage,NY 11714
[5].Hampden-Sydney College, P.O.Box 862, Hampden-Sydney, VA 23943
[6].Compton Observatory Science Support Center operated by Astronomy
Programs, CSC, Greenbelt, MD 20771
[7].Universities Space Research Association, Code 610.3 NASA/GSFC

ABSTRACT

A critical question in the study of the dynamics of our own galaxy is whether the cosmic rays are galactic or universal. Diffuse gamma ray emission from the Magellanic Clouds allows us to carry out studies of the cosmic ray distribution in these external galaxies and in turn permit us to answer the question on the origin of cosmic rays in our own galaxy. These gamma rays are believed to be produced primarily through the interaction of cosmic rays with interstellar matter. Hence, combined with a knowledge of the interstellar matter distribution, they can provide a direct measure of the cosmic ray density. Results of a model calculation assuming quasi-stable equilibrium between cosmic rays, magnetic fields and interstellar gas, are presented and compared with recent data from the Energetic Gamma Ray Experiment Telescope (EGRET) aboard the Compton Observatory. On the important question of whether the origin of the bulk of cosmic rays is galactic or extragalactic, a definitive test of the universality of cosmic rays can be accomplished through the measurement of the high energy gamma ray emission from the SMC and the LMC. The results obtained from the EGRET on the Magellanic Clouds clearly resolves this debate in favor of a galactic origin.

1. INTRODUCTION

The interaction of cosmic rays with the interstellar medium and magnetic fields in the Galaxy has been a subject of significant interest over several decades. However, a complication in this study is that, being embedded in the galactic disk with different parts of the disk being at varying distances from us, the interpretation of observations is difficult. The Magellanic Clouds are our closest neighboring galaxies and hence are of great interest in studying many astrophysical aspects of galaxies. The proximity of the Magellanic Clouds allows us to carry out detailed studies of external galaxies without the problems that arise from studying our own galaxy from within. A long-standing, critical question in the study of the dynamics of our own galaxy is whether the origin of the bulk of cosmic rays is galactic or extragalactic. Although there exists evidence indicating that they are primarily galactic in origin, there are counter arguments for an extra-galactic origin, based on contributions from radio galaxies, quasi-stellar objects and galactic winds (For a general discussion, see Bercher & Burbidge 1972). Ginzburg and Ptuskin (1976) have noted that a definitive test of whether the bulk of the cosmic rays is galactic or universal is to compare the level of high energy gamma ray emission from Magellanic Clouds. Radio synchrotron observations provide information on the cosmic ray electron component, although the major energy component (at least in the Milky Way) is the nucleonic one. Using multi-frequency radio measurements, it is possible to construct a spectrum of the non-thermal radiation and to examine its distribution. Fichtel *et al.*(1991) carried out such an analysis for the Large Magellanic Cloud (LMC) in which the electron spectrum and spatial density distribution was calculated. Alternately, using the principle of dynamic balance and containment, these authors calculated the cosmic ray nucleon density distribution, and showed that the derived cosmic ray density distribution is consistent, within uncertainties, to that obtained from the radio synchrotron data. However, Sreekumar and Fichtel (1991) showed in a similar analysis, that the Small Magellanic Cloud (SMC) appears to have a cosmic ray density well below that in our galaxy. This finding is consistent with the concept that the SMC is in a state of irreversible disintegration, in agreement with the independent experimental findings by Mathewson, Ford & Viswanathan (1986,1988) and the tidal interaction model of Murai and Fujimoto (1980). Since high energy gamma rays are created in the interaction of cosmic ray nucleons and electrons with interstellar matter, a study of these photons represents a more direct estimate of the cosmic ray density. Using recent LMC observations by the Energetic Gamma Ray Experiment Telescope (EGRET) on the Compton Observatory, Sreekumar *et al.*(1992) have in fact shown that the high energy gamma ray emission from the LMC is consistent with the calculations assuming a quasi-stable equilibrium scenario. Quasi-stable equilibrium involves a dynamic balance between the attractive gravitational pressure and expansive pressures arising from cosmic rays, interstellar gas and magnetic fields. It is expected to exist if the external pressures are small and cosmic ray sources are adequate. See Parker (1966,1969,1977) for a full discussion. The measured flux is however also consistent with the predicted value for a universally constant cosmic ray distribution model. Thus, a definitive test does not exist from LMC observations.

For the SMC, however, the predictions differ sufficiently for the two cases to permit a meaningful test of the underlying model by comparing with recent observations. For completeness, it should be noted that in principle it should be possible to look for variations of the cosmic ray density in our own galaxy, but this is complicated by the interpretation of the data resulting from our location in the galactic plane as well as uncertainty in estimates of the molecular hydrogen density (Bertsch *et al.*1993; Fichtel & Trombka 1981).

2. DISCUSSION

The high energy gamma ray emission to be expected from the SMC for a quasi-stable equilibrium scenario and a universal or metagalactic case has been calculated by Sreekumar and Fichtel (1991). The results for the flux above 100 MeV are 2.4×10^{-7} photons cm^{-2} s^{-1} for the universal cosmic ray case and 1.2×10^{-7} photons cm^{-2} s^{-1} for the quasi-equilibrium case. Further, based on the synchrotron radiation data, they concluded that the SMC cosmic ray density is not at a level consistent with quasi-stable equilibrium. If the level that they deduced in this way is the one that exists in the SMC, the expected high energy gamma ray flux above 100 MeV is much lower, (2 to 3) x 10^{-8} photons cm^{-2} s^{-1}, with the range in values being associated with the part of the matter that is assumed to be relevant for the disintegrating case. The uncertainty in all of these numbers is estimated to be about 20%. EGRET observations provide an upper limit (95% confidence) for gamma ray emission from the SMC (E > 100 MeV) of 0.5×10^{-7} photons cm^{-2} s^{-1}. From the discussion above, this indicates clearly that the cosmic ray density observed in the local region of out galaxy is not universal. Hence, this long standing question seems to be answered since the conclusion is based only on the amount of matter present, the known cosmic ray density in our galactic neighborhood, and measured nuclear cross sections. The result is consistent with the current theoretical beliefs that there is sufficient energy within a galaxy to fill it with cosmic rays, but that it is much more difficult to envision a means of filling the whole universe or the Local Group with cosmic rays at the energy density level seen in the local region of our galaxy. It also seems that the observed upper limit is not consistent with quasi-stable equilibrium conditions in the SMC, although this conclusion is less compelling because of the statistical limitations of the gamma ray data. Together with other indications, the results suggest that the SMC is not in quasi-stable equilibrium, and is likely to be in a state of disruption. The upper limit is well above the level suggested by Sreekumar and Fichtel for the disrupted state based on synchrotron data and the assumption that the SMC is disrupted. With the result obtained here for the SMC effectively eliminating the universal cosmic ray hypothesis, it is now possible to re-examine the LMC result (Sreekumar *et al.*1992) and answer the questions that were described in the introduction. If the cosmic rays are not universal then the LMC high energy gamma ray result can be interpreted as showing that the LMC is most likely in quasi-stable equilibrium since the gamma radiation is at the level expected for such a condition and would be much different if it were not.

3. IMPLICATIONS FOR THE MILKY WAY

Gamma ray observations of nearby external galaxies have provided important clues on the origin and distribution of cosmic rays. The analysis above clearly demonstrates that the bulk of the cosmic ray energy density is almost certainly not metagalactic nor universal, but is galactic in origin; otherwise the high energy gamma ray flux from the SMC would be much higher than the upper limit obtained from EGRET observations. In addition, with the elimination of the universal cosmic ray concept, the LMC high energy gamma ray data indicate that like the Milky Way, the LMC is most likely in quasi-stable equilibrium, with a cosmic ray energy density near the maximum that can be contained. The Magellanic Cloud analysis presented here thus points to a galactic origin for cosmic rays in the Milky Way, consistent with conclusions derived from high energy diffuse gamma ray observations (Bertsch *et al.*1993).

ACKNOWLEDGEMENTS

The EGRET team gratefully acknowledges support from the following: Bundesministerium fur Forschung und Technologie, grant 50 QV 9095 (MPE); NASA grant NAG5-1742 (HSC); NASA grant NAG5-1605 (SU); and NASA contract NAAS 5-31210 (GAC).

REFERENCES

Bertsch, D. L., et al., 1993, ApJ, (submitted)
Brecher, K., & Burbidge, G. R. 1972, ApJ, 174, 253
Fichtel, C. E., & Trombka, J, I., 1981, Gamma Ray Astrophysics, New Insights
    into the Universe, NASA SP-453 (Washington: GPO)
Fichtel, C. E., Ozel, M., Stone, R., & Sreekumar, P., 1991, ApJ, 374, 134
Ginzburg, V. L., & Ptuskin, V. S. 1976. Rev. Mod. Phys., 48,161
Mathewson, D. S., Fors, V. L., & Viswanathan, N., 1986, ApJ, 301, 664
Mathewson, D. S., Fors, V. L., & Viswanathan, N., 1988, ApJ, 333, 617
Murai, T., & Fujimoto, M., 1980, Publ. Astron. Soc, Jpn., 32,581
Parker, E. N., 1966, ApJ, 145, 811
Parker, E. N., 1969, Space Sci. Rev., 9, 654
Parker, E. N., 1977, in The Structure and Content of the Galaxy and Galactic
    Gamma Rays, edited by C.E. Fichtel and F.W. Stecker, NASA CP-002
    (Washington: GPO), 283
Sreekumar, P., & Fichtel, C. E. 1991, A&A, 251, 447
Sreekumar, P. et al., 1992, ApJL, 400, L67

COSMIC RAY NUCLEOSYNTHESIS IN THE EARLY GALAXY WITH ACCOUNTING FOR ASTRATION AND ACCRETION

B. V. Vayner[1], L. M. Ozernoy[2,3], and Yu. A. Shchekinov[4]

[1] Dept. of Astron., Case Western Reserve Univ., Cleveland, OH 44106
[2] Code 665, NASA/GSFC, Greenbelt, MD 20771
[3] Natl. Res. Counc./Natl. Acad. of Sci. Senior Research Associate
[4] Inst. of Physics, Rostov State Univ., Rostov-on-Don, 344104 Russia

Email: ozernoy@heavax.gsfc.nasa.gov

ABSTRACT

We present an analytical model of light element production by energetic protons of cosmic rays in the active Galactic nucleus early in its evolutionary history. Subsequent change of the element abundances in the course of the galactic chemical evolution (such as deuterium astration in stars and accretion of intergalactic gas) is calculated, also analytically, which enables us to examine the dependencies on the parameters. The model is capable to explain both the observed mean deuterium abundance and its positive radial gradient if during the first billion years the proton power of cosmic rays in the Galactic nucleus was $(3-5) \cdot 10^{42}$ erg/s and the ratio of the accretion rate to star formation rate was higher in the central region of the Galaxy. We present also some results on cosmic ray spallation production of Li, Be, and B by the active galactic nucleus in the early Galaxy.

1. INTRODUCTION

According to a wide-spread point of view, the origin of light isotopes (D, ^{3}He, ^{4}He, and ^{7}Li) is due to the cosmological nucleosynthesis. Naturally, a question arises whether this scenario is compatible with all cosmological data, including the most recent data on dark matter and temperature inhomogeneities in the microwave background radiation. As is known, there is no unambiguous answer to this question yet. Meanwhile some new problems have appeared which need to be solved. They include unexpectedly high abundance of deuterated molecules in the vicinity of the Galactic center (Jacq *et al.* 1990); theoretical explanation of beryllium and boron observations in old (Population II) stars (Gilmore *et al.* 1991; Ryan *et al.* 1992); correlations between lithium abundance and metallicity of halo stars, etc.

All the issues listed above have recently revived interest to exploring a non-cosmological production of at least some of light isotopes. One of the first models of this kind was an attempt to calculate the generation of deuterium and lithium in the active nucleus of the early Galaxy by cosmic rays interacting with low-metallicity gas surrounding the nucleus (Ozernoy & Chernomordik 1975). It has been shown that the model is capable to explain the observed abundances of deuterium and lithium (without overproduction the latter) if the cosmic ray power during the first billion years of the Galaxy life was as high as 10^{43} erg/s.

In the present paper, we describe some results of our work on analytical calculations of light element production by cosmic rays in the active nucleus of the Galaxy early in its history and subsequent chemical evolution of these elements as well. Athough these calculations are, by necessity, not as detailed as numerical computations, an opportunity to analyse analytically the dependencies on input parameters is of an evident interest. The full account of the work is given elsewhere (Vayner *et al.* 1993).

2. LIGHT ELEMENT PRODUCTION

The main processes for generation of D, He and Li are the interactions of fast protons and α-particles with hydrogen and helium in the dense region near the center of the Galaxy: $p + ^4\text{He} \to ^3\text{He} + \text{D}$; $^3\text{He} + 2p$; $^3\text{He} + n + p$; $\text{D} + \text{D} + p$; $\text{D} + n + 2p$ and $\alpha + ^4\text{He} \to ^6\text{Li} + n + p$; $^7\text{Li} + p$. The duration of these processes can not be more than $0.5 \cdot 10^9$ years because the computations of the Galaxy chemical evolution show that the observed metallicity of old (Population II) stars is reached in a time interval like this (Chuvenkov & Vayner 1989). Assuming the spectrum of fast particles to have a form $F_p = F_0(1 + E/E_0)^{-\gamma}$, the total mass of deuterium which can be generated during the active phase is given by $M_\text{D} = 0.5 \cdot 10^6 \, M_\odot \, (W_p/10^{43} \, \text{erg/s}) \, (\Delta t/10^9 \, \text{yr})$ (Ozernoy & Chernomordik 1975). It implies that the abundance of deuterium could be as high as $X_\text{D} \approx 10^{-4}$ by the moment when the age of the Galaxy was $t \approx 10^9$ years. In this model the abundances of other light species are calculated to be $X(^7\text{Li})/X(\text{D}) \approx 2 \cdot 10^{-4}$ and $X(^3\text{He})/X(\text{D}) \approx 2$.

To determine a spatial distribution of element abundances we use a kinetic equation

$$\frac{\partial N_i}{\partial t} + \nabla(N_i v_i) + \frac{\partial}{\partial E}\left[b_i(E)N_i\right] + \sum_j n_j v_j \sigma_{ij} N_i = 0, \qquad (1)$$

where $i = p, \alpha$; $N_i \equiv N_i(E, r, t)$ is the energy spectrum of protons and α-particles in units $\text{cm}^{-2}\text{s}^{-1}\text{MeV}^{-1}$; $b_i(E) = \alpha_i E^{-1} + \beta_i E$ are energy losses, and $v_i \equiv (2E/m_i)^{1/2}$.

The ratio of deuterium to lithium abundances is found to be:

$$\frac{X_\text{D}}{X_{^7\text{Li}}} = \frac{W_{p\,42}}{(\gamma - 1)W_{\alpha\,42}} \left(\frac{E_{\text{cr}\,\alpha}\epsilon_{th\,\text{Li}}}{E_{\text{cr}\,p}\epsilon_{th\,\text{D}}}\right)^{\gamma-1} \left(\frac{f_1}{\epsilon_{th\,\text{Li}}^2}\right)^{(\gamma+1)/2}$$

$$\times \exp\left[\frac{(\tau_\text{H} - \tilde{\tau})(1 - x^{-\mu})}{1 - \mu}\right]\left[\bar{\gamma}\left(\frac{\gamma+1}{2};\ \frac{f_1}{\epsilon_{th\,\text{Li}}^2}\right)\right]^{-1} . \qquad (2)$$

Here $W_{i\,42} \equiv W_i/10^{42}$ erg/s; $E_{\text{cr}\,i} \equiv \alpha_i/\beta_i$; $\epsilon = E/E_{\text{cr}}$; $x = r/R$, R being the radius of the nucleus where the light elements are produced; $n(x) = n_0 x^{-\mu}$, n_0 is number density of gas in that region; $\tau_i = \sigma_i n_0 R$; $\tilde{\tau}$ is optical depth for energetic particles whose interaction with ambient gas results in production of D and ^7Li, respectively; $f_1 = \tau_\alpha (\mu - 1)^{-1} (1 - x^{-\mu+1})$; and $\bar{\gamma}(y, z)$ is the incomplete gamma function.

Almost all the lithium is produced in a spatially constrained region within the nucleus and close to it. For example, $^7\text{Li/D} = 0.027$ at $x = 1$, and $^7\text{Li/D}=8.4 \cdot 10^{-9}$ at $x = 3$. As a result, the problem of lithium overproduction does not arise in this model.

So far the atomic deuterium and deuterated molecules have been only observed in the interstellar medium. That is why it is necessary, before comparing the yield of deuterium with observations, to consider the evolution of deuterium abundance in the interstellar gas with incorporating its astration as well as accretion of an external gas.

3. CHEMICAL EVOLUTION

The equation for abundance of i-th element has a form:

$$\frac{\partial(X_iG)}{\partial t} + \nabla j_i = -X_i\Psi(r,t) + \int E_i(m)\varphi(m)\Psi(r,t-\tau_m)dm, \qquad (3)$$

where G is gas density; $\Psi(r,t)$ is star formation rate; $E_i(m)$ is the mass of i-th element ejected by a star with initial mass m; $\varphi(m)$ is the initial mass function; τ_m is the life time of a star with initial mass m. Obviously, $\int \varphi(m)mdm = 1$; $\sum_i X_i = 1$.

a) Deuterium abundance

The flow of the i-th element, j_i, consists of two terms: radial one, j_r, and that perpendicular to the galactic plane, $A(r,t)$. If we neglect, in the zero's approximation, the radial flow and suppose the accretion rate not to depend on time, we obtain the deuterium abundance as a function of gas density to be

$$X(G) = G^{-1}\left(X_0G_0 - \frac{AX_A}{g}\right)\left(\frac{G - A/g(1-\Re)}{G_0 - A/g(1-\Re)}\right)^{\frac{1}{1-\Re}} + \frac{AX_A}{gG}, \qquad (4)$$

where $\Re = \int E(m)\varphi(m)dm$; $g = \Psi/G$, and X_A is deuterium abundance in the accreting gas. Below, we consider important particular solutions given by Eq. (4):

- (I) $A(r,t) = 0$ (Closed model of the Galaxy, no accretion of intergalactic gas).

In this case, $X(G) = X_0(G/G_0)^{\frac{\Re}{1-\Re}}$. If $\Re = 0.5$, then we get $X = X_0(G/G_0)$. Ostriker & Caldwell (1979) estimate the present gas density $G \approx 0.1\,G_0$. Therefore, the astration factor is estimated to be $X_0/X \approx 10$.

- (II) $A(r,t) \neq 0$ (Accretion of intergalactic gas is essential).

Two particular cases are of interest here:

(i) $X_0 = X_A$, *i.e.* the Galactic production of deuterium is much less than the primordial one. The deuterium abundance has, in this variant, an asymptotic value $X_\infty = X_0(1 - \Re)$.

(ii) $A \neq 0, X_A \neq X_0$. In this variant, we get $X_\infty = X_A(1-\Re)$. If $\Re = 0.5$, then $X_\infty = 0.5\,X_A$. Thus, the Galactic production of deuterium is asymtotically "washed out" by deuterium abundance in the accretion flow.

Our model enables us to account for a positive radial gradient of deuterium abundance in the Galaxy. The case when $X_A \ll X_0$ seems to be the most appropriate for this. We have:

$$X(G) = X_0\left(\frac{G}{G_0}\right)^{\frac{\Re}{1-\Re}}\left(1 - \frac{A}{gG(1-\Re)}\right)^{\frac{1}{1-\Re}}. \qquad (5)$$

Since $\Re = 0.4$ in the solar vicinity and $(A/gG)_\odot = 0.1$, we get the astration factor $(X_{0\,\odot}/X_\odot) \approx 9$. If one supposes that near the Galactic center $(A/gG)_c \approx 0.5$, then $X_c/X_\odot \approx 6.5 \cdot 10^{-2}(X_{0\ c}/X_{0\ \odot})$. In the reality, the observed gradient in deuterium abundance might be affected by the two factors: $\nabla\Re < 0$ and $\nabla(A/gG) < 0$.

b) Berillium and Boron abundances

The isotopes of Be and B are generated by spallation of CNO nuclei in the gas surrounding a source of energetic protons (*e.g.* Walker *et al.* 1992, Prantzos *et al.* 1992). We deal with the active galactic nucleus early in the history of the Galaxy. The rate of beryllium production is given by $dN_{Be}/dt = \sum_{i=C,N,O} \alpha_i N_i$. The Be abundance in the Population II stars having a low metallicity ([Fe/H]$<$ -2) can be calculated as $N_{Be}(t_*) = \sum_{i=C,N,O} \int_0^{t_*} \alpha_i N_i(t)dt$, where t_* is the time interval between the onset of the activity of the nucleus and the starburst epoch. According to calculations by Chuvenkov & Vayner (1989), $t_* < 0.5 \cdot 10^9$ yr and therefore $N(t) \propto t$. As a result, we find a quadratic dependence of the beryllium abundance on metallicity: $N_{Be}(t_*) \propto t_*^2 \propto [Fe/H]^2$.

It is easy to show that the cosmic ray power $W_p \approx 10^{43}$ erg/s is just enough for the production of the observed amount of beryllium. However, if a few CNO elements were synthesized in the pregalactic epoch (*i.e.* in Pop.III stars) then the Be abundance is $N_{Be}(t_*) \propto t_* \propto [Fe/H]$. To make a choice among the above two variants (or adopt some combination of them) more observational data is required.

4. CONCLUSIONS

The model presented above is consistent with available data on deuterium and lithium abundances although it requires a tremendous output of $(3-5)\cdot10^{42}$ erg/s in cosmic rays early in the early history of the Galactic nucleus. If the ratio of the accretion rate to star formation rate was higher in the central region of the Galaxy than in the solar neighbourhood, the model would be consistent with available data on deuteriun gradient. More data and detailed numerical computations will show the directions in which this model needs further development. Our model can also be helpful in exploring of how the interactions of energetic particles produced in active galactic nuclei and starburst galaxies would change the light element abundancies while interacting with the ambient media.

REFERENCES

Chuvenkov, V.V. & Vayner, B.V. 1989, Astrophys. Space Sci., 154, 287
Gilmore, G., Edvardsson, B., & Nissen, P.E. 1991, ApJ, 378, 17
Jacq, T. *et al.* 1990, A&A, 228, 447
Ostriker, J. & Caldwell, J.A.R. 1979, in The Large-Scale Characteristics of the Galaxy, Dordrecht, p.441
Ozernoy, L.M. & Chernomordik, V.V. 1975, Sov.Astron., 19, 693
Prantzos, N., Cassé, M., & Vangioni-Flam, E. 1992, Preprint No. 382 - MAI
Ryan, S.G. *et al.* 1992, ApJ, 388, 184
Vayner, B.V., Ozernoy, L. M., & Shchekinov, Yu. A. 1993, ApJ (to be submitted)
Walker, T.P., Steigman, G., Schramm, D.N., Olive, K.A., & Fields, B. 1992, Preprint OSU-TA-2/92

OUTER
GALAXY

ROTATION, SCALEHEIGHT AND MASS

Michael R. Merrifield

Department of Physics, University of Southampton, Highfield, S09 5NH, UK.

ABSTRACT

The observational interplay between the rotational motion and scaleheight of the galactic atomic hydrogen (HI) distribution is described. The implications of these observations for the large scale structure of the Galaxy are discussed. A picture emerges of the Milky Way as a normal spiral system in which the sun lies at a radius of $R_0 = 8$ kpc, two scalelengths out in the disk, where the circular speed is $\Theta_0 = 200$ km s^{-1}. The observed thickness of the HI layer can only be sustained against the gravitational pull of the disk if it is supported by significant magnetic field and cosmic ray pressures in addition to its own turbulent motion. The combination of these constraints results in a model in which the mass of the Galaxy is dominated by its dark halo at all radii, and it argues against the "maximal disk" hypothesis.

1. INTRODUCTION

Atomic hydrogen (HI) provides an ideal tracer for investigating the overall structure of the Milky Way. Its ubiquitous presence out to large radii in the Galaxy means that it probes this structure on the largest scales, and the 21cm emission line means that its kinematics can be directly measured from radio observations. This paper concentrates on two of the most basic properties of the galactic HI distribution: the thickness of the HI layer perpendicular to the galactic plane (defined by some characteristic scale such as its full width at half maximum), and the speed at which this material orbits the Galaxy. Because the geometry of the Milky Way is more complicated than is the case for external galaxies, even these gross properties are difficult to disentangle from one another. Nonetheless, the effort required to extract this information is well worthwhile, as it provides us with critical constraints on the overall distribution of mass in the Galaxy.

The close interplay between rotation and scaleheight arises because the angular scaleheight of any component of the Milky Way can only be converted to an absolute scale once its distance from us is known. We can only generally measure such distances by assuming a rotation curve and using the observed kinematics of the material to solve for its location. Section 2 describes how this analysis has been applied to measure the variation in scaleheight of the HI distribution with position in the Milky Way. This method can also be turned on its head by using the observed angular scaleheight of HI to provide a new measure of the kinematics of the outer Galaxy. The resulting rotation curve is also presented in Section 2. The HI scaleheight is dictated by the distribution of mass close to the plane of the Galaxy, as it is the balance between gravity's pull toward the plane and the various buoyancy forces acting on the HI that sets the equilibrium thickness of the layer. There has been some recent debate as to the size and nature of the buoyancy forces; since this issue has a direct bearing on the amount of mass in the plane of the Galaxy, it is addressed in

Section 3, and the corresponding constraints on the mass of the galactic disk are presented. Finally, the density distribution derived for the galactic plane can be combined with the total mass constraints obtained from the rotation curve, and the resulting models for the mass distribution of the Milky Way are discussed in Section 4.

2. ROTATION AND SCALEHEIGHT

By obtaining spectra in the 21cm band, the distribution of HI in the Galaxy has been extensively mapped as a function of galactic longitude and latitude, $\{l, b\}$, and the line-of-sight component of its velocity, v_{los} (Kwee, Muller & Westerhout 1954; Weaver & Williams 1974; Kerr $et\ al.$ 1986; Stark $et\ al.$ 1992; and many in between). Translating these maps into the spatial distribution of HI in the Milky Way is complicated for two reasons: firstly, unlike external galaxies, the Milky Way does not all lie at the same distance from us, and so its geometry is difficult to disentangle; and secondly, the relation between v_{los} and the third spatial dimension is only indirect, and it requires that we assume something about the kinematics of the HI. The simplest kinematic assumption is that the material follows circular orbits with a known rotation curve $\Theta(R)$. Even this simple assumption will not allow us to make a complete map of the HI distribution, since there are two points in the Galaxy at radii less than the solar radius, R_0, which map to the same point in the observable $\{l, b, v_{los}\}$ coordinates. The distribution of HI away from the galactic plane in the inner Galaxy has only been studied at the small locus of points where there is no such ambiguity — the tangent points — and these studies show that the scaleheight of HI in the inner Galaxy is constant (Schmidt 1957; Dickey & Lockman 1990), or slowly rising (Celnick, Rohlfs & Braunsfurth 1979; Knapp 1987) over the inner ~ 6 kpc.

For $R > R_0$, there is a one-to-one relation between the observable and intrinsic coordinate systems for any reasonable assumption about $\Theta(R)$, and we can therefore explore the full three-dimensional structure of the outer Galaxy. Unfortunately, the absence of tangent points in the data from this region prevents the simplest approach for calculating the rotation curve (Kwee $et\ al.$ 1954, and many thereafter) from being extended beyond the solar radius. Instead, less accurate methods using standard candles to obtain distances must be adopted: the results of such analyses have recently been summarized by Fich & Tremaine (1991), but new data sets are continually refining the determination of $\Theta(R)$ ($e.g.$ Metzger & Schechter 1993; Turbide & Moffat 1993). Since the rotation curve is still not well determined beyond R_0, there is a significant degree of uncertainty in the mapping from $\{l, b, v_{los}\}$ to true spatial coordinates.

A detailed investigation of the HI distribution in the whole of the outer Galaxy was performed by Henderson, Jackson & Kerr (1982), who adopted a flat rotation curve of $\Theta(R) = 250$ km s^{-1} in order to calculate kinematic distances. As had been known prior to this complete analysis ($e.g.$ McGee & Milton 1964), these authors demonstrated that the HI layer flares dramatically outside the solar radius at all azimuthal angles in the Galaxy, with a scaleheight that doubles between R_0 and $\sim 1.8 R_0$. Somewhat disturbingly, Henderson $et\ al.$ also found that, of all the points at the solar radius in the Galaxy, the sun seems to lie in the region where the HI scaleheight is smallest. There also appears to be a systematic gradient to greater scaleheights as one moves around the solar circle toward the far side of the Galaxy, and this gradient is present in data from both $l > 180°$ and $l < 180°$. This heliocentric phenomenon is clearly spurious, and the simplest explanation for it is that the assumed rotation curve is incorrect,

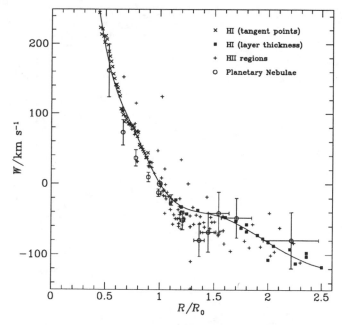

Fig. 1. The function $W(R) = (R_0/R)\Theta(R) - \Theta_0$ as derived from HI data (tangent points in the inner Galaxy, layer thickness in the outer Galaxy) and from standard candle data. The planetary nebula data are from Schneider & Terzian (1983), and the HII region data are from Fich, Blitz & Stark (1989). The solid line represents a spline fit to the HI data.

so that material has been placed at the wrong kinematic distance. If the true scaleheight of the galaxy were constant around the solar circle, then such an error would produce apparent scaleheights that vary with galactic azimuthal angle with just the heliocentric appearance that was observed.

The above explanation suggests a new possibility: the true rotation curve for the outer Galaxy might be calculated by finding the form for $\Theta(R)$ which produces a map of the variation in thickness of the HI layer with position that shows no heliocentric distortion. It turns out that this constraint allows one to solve simultaneously for $W(R) = (R_0/R)\Theta(R) - \Theta_0$, [where $\Theta_0 = \Theta(R_0)$], and the variation in the thickness of the HI layer with radius, $h_z(R/R_0)$ (Merrifield 1992). The functional form obtained by this method for $W(R)$ is illustrated in Fig. 1 together with the same quantity derived from standard candle analyses. The general agreement between the two methods is most heartening, as each is subject to a variety of potential systematic errors which reduce confidence in any one of the rotation curves obtained. One interesting source of such systematic error arises from the assumption that the material in the Galaxy follows circular orbits: if the galactic disk is actually elliptical, then both methods will produce incorrect rotation curves, but with errors that scale differently with the degree of ellipticity. This possible handle on the shape of the outer Galaxy has recently been investigated by Kuijken & Tremaine (1993).

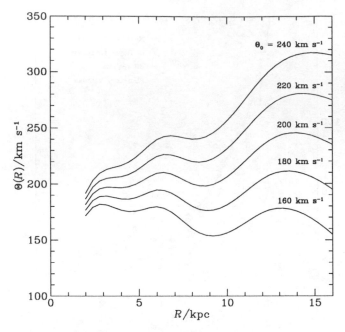

Fig. 2. The rotation curve of the Galaxy as derived from the spline fit to $W(R)$ in Fig. 1 for various values of Θ_0.

In order to convert $W(R)$ into a rotation curve, we must ascribe values to the galactic parameters Θ_0 and R_0. Much effort has been directed toward measuring these quantities: in 1986, Kerr & Lynden-Bell reviewed the available constraints on the local galactic constants for the I.A.U. and came up with best estimates of $R_0 = 8.5\pm1.1$ kpc and $\Theta_0 = 220\pm20$ km s^{-1}, although several more recent analyses have suggested that both these values are on the high side of the truth (Rohlfs & Kreitschmann 1988; Reid 1989). As one more piece of evidence that the value of Θ_0 may be lower than its I.A.U.-sanctioned value, consider the rotation curves derived from $W(R)$ for various values of Θ_0 that are plotted in Fig. 2. For larger values of Θ_0, the rotation curve rises rapidly with radius — more rapidly, in fact, than is observed in any normal spiral galaxies. An analysis of the 29 Sb and Sc galaxies from Rubin $et\ al.$ (1985) yields rotation curves that rise by $5 \pm 10\%$ between 8 and 16 kpc. Adopting the Bayesian stance that the Milky Way should be drawn from this parent population, we find the probable range for Θ_0 is 180 ± 20 km s^{-1}. Combining this constraint with the various previous measurements produces a consistent best estimate of $\Theta_0 = 200$ km s^{-1}. This value and $R_0 = 8$ kpc are adopted as the galactic parameters for the remainder of this paper.

3. SCALEHEIGHT AND MASS

The observed increase in HI scaleheight with radius essentially results from the decreasing surface density of the galactic disk, although the exact variation also depends on the size of the restoring forces which maintain the equilibrium thickness of the layer. Detailed three-dimensional models of this equilibrium can be found in Kundić, Hernquist & Gunn (1993), but, to a very good approximation, the HI layer at any radius can be treated as a one-dimensional system with properties that vary only with distance from the galactic plane. The appropriate component of the tensor virial theorem then implies that its equilibrium thickness is given by:

$$h_z(R) = \left(\frac{fQ^2}{G\rho_0}\right)^{1/2},$$

(1)

where ρ_0 is the mass density in the galactic plane; $Q^2 = \sigma^2 + B^2/8\pi\rho_{HI} + P_{cr}/\rho_{HI}$ describes the pressure support due to turbulent motions, magnetic fields, and cosmic rays, respectively; and all the details of the structure of the HI layer, variations in the different pressure terms with distance from the galactic plane, etc., are hidden in the dimensionless factor f (Kellman 1972, van der Kruit 1988).

In order to use Eq. (1) to obtain the large scale distribution of mass in the galactic plane, $\rho_0(R)$, we must specify the variation in Q^2 with radius. Unfortunately, neither the absolute nor the relative sizes of the terms that make up Q^2 are well understood, and possible new contributions to the buoyancy force such as radiation pressure (Ferrara 1993) are still being suggested. Boulares & Cox (1990) have examined the observational constraints on the value of Q^2 in the solar neighborhood, and they conclude that turbulent motion, magnetic fields, and cosmic rays all contribute approximately equally to the total pressure support. On the other hand, Lockman & Gehman (1991) have constructed a collisionless, purely hydrodynamic model of the HI layer which is consistent with both high latitude observations of the local HI kinematics and the spatial distribution of HI in the inner Galaxy — this model implies that Q^2 is dominated by the turbulent term. If such a hydrodynamic model is appropriate, then it is possible to go further and combine kinematic and spatial observations of the HI distribution to solve directly for the gravitational potential of the galactic disk in the solar neighborhood. The application of this analysis results in a potential that implies an unphysical negative mass density above ~ 250 pc from the galactic plane (Merrifield 1993), and hence it rules out support purely by turbulent motions. The failure of this collisionless hydrodynamic model should come as no particular surprise: the total energy of the HI layer is re-supplied by supernovae and stellar winds on a timescale of only 2×10^6 years, an order of magnitude less than the period of oscillation of a single HI cloud in the galactic potential (Lockman & Gehman 1991). Significant dissipation of energy must hence be occurring in order for equilibrium to be maintained: since this dissipation is likely to take the form of inelastic cloud–cloud collisions, a collisionless model is not likely to provide an accurate description of the HI layer dynamics.

Even with all the uncertainties, we do have some physical constraints on the value of Q^2. Firstly, observations of face-on external galaxies all yield values for the HI velocity dispersion σ of ~ 10 km s^{-1}, independent of radius (van der Kruit 1981). The reason for this apparently universal value is probably because collisions between HI clouds with higher relative velocities are highly dissipative due to the enormous rise in the H–H collision cross-section at this energy (Dalgarno & McCray 1972). To test this idea, a simple model can be constructed which balances the energy input to the HI layer kinetic motions

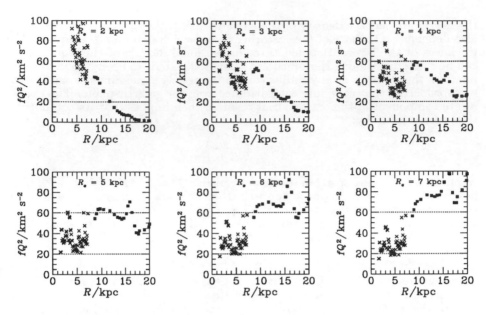

Fig. 3. The value of fQ^2 as a function of radius derived from the HI scale-height (crosses indicate tangent point measurements, and squares show the scaleheight as derived in Section 2). Mass models for the galactic disk with various e-folding lengthscales, R_e, have been assumed. The dashed lines indicate the physically acceptable range for fQ^2.

from supernovae and stellar winds, η, against cooling by cloud–cloud collisions. The value of η is somewhat uncertain, and it probably varies with galactic radius (Abbott 1982). However, the cooling efficiency rises so steeply with the kinetic energy of the clouds that a heating rate anywhere within *five orders of magnitude* of the best estimate of $\eta = 1 \times 10^{38}$ erg s^{-1} kpc^{-2} will result in an equilibrium velocity dispersion for the gas of approximately 10 km s^{-1}! A second constraint on Q^2 is that the magnetic field and cosmic ray terms are unlikely to significantly exceed the turbulent term. The interstellar gas is somewhat ionized (Falgarone & Lequeux 1973) and hence it is coupled to the magnetic field. The turbulent motion of the gas will therefore tangle the galactic magnetic field lines and increase the magnetic energy density. Once the magnetic contribution to Q^2 becomes comparable to the turbulent term, however, the magnetic field will start to suppress the random motions of the gas, and further enhancement of the magnetic energy density will be halted. Similarly, cosmic rays will be confined to the Milky Way by its magnetic field, and their contribution to Q^2 will increase until it is approximately equal to the magnetic energy term, at which point the cosmic rays have sufficient energy to break free from the Galaxy. Conservatively, therefore, the plausible range for Q^2 is 100 – 300 km^2 s^{-2}. The details of the HI equilibrium configuration embodied in the factor f also introduces a degree of uncertainty, but physically reasonable model of the HI layer imply that it lies in the relatively narrow range $f = 0.2 \pm 0.05$ (Merrifield 1992).

The possible values for f and Q^2 can now be combined with the observed variation in HI scaleheight with radius in Eq. (1) in order to measure the constraints that they impose on the distribution of mass in the galactic plane. The simplest model for $\rho_0(R)$ is one in which the mass in the plane follows the distribution of stellar light which, presumably, obeys the same simple exponential law that is observed in external galaxies. Figure 3 shows the variation in fQ^2 with radius, as derived from Eq. (1) and the observed $h_z(R)$, and assuming various values for the radial e-folding length of the disk mass distribution. The disk mass has been normalized to give a density in the solar neighborhood of $0.1M_\odot\mathrm{pc}^{-3}$, as measured by Kuijken & Gilmore (1989a) [but see also Bahcall (1984) where the local disk density was found to be almost twice this size]. It is apparent from this figure that because the HI scaleheight has been measured over several e-foldings of the galactic disk density, even the rather weak physical limits on fQ^2 translate into quite tight constraints on the possible distribution of disk mass — only models with R_e in the range 4 – 5 kpc produce physically acceptable values for fQ^2 at all radii.

This result is consistent with the range of properties observed in other spiral galaxies: from the compilation of Casertano & van Gorkom (1991), galaxies with maximum rotation velocities in excess of 200 km s^{-1} can have scalelengths of anywhere between 2 and 9 kpc. The derived value for R_e also agrees very well with existing measurements of the Milky Way's disk scalelength: Kent, Dame & Fazio (1991) have collated previous estimates of R_e and find that they range from 1.8 to 6 kpc, with a mean estimate of ~ 4 kpc. Interestingly, almost all of these measurements were based on the photometric properties of the disk, whereas the HI scaleheight analysis provides us with a direct measure of the mass in the disk. It is therefore apparent that the disk of the Milky Way can be satisfactorily modeled as a constant mass-to-light ratio system at all radii. Since the local density of the disk can be explained by the sum of its known constituents (Kuijken & Gilmore 1989a), the global properties of the disk can also be described by a similar mix of constituents, and there is no need for recourse to dark matter of unknown origin anywhere in the disk. The higher local density advocated by Bahcall (1984) is harder to reconcile with the observed HI scaleheight, since the values of fQ^2 derived from this normalization would be a factor of two greater than those shown in Fig. 3, and so they could not lie between the physically acceptable bounds at all radii for any disk scalelength.

It is also interesting to note that the simple constant mass-to-light ratio model with $R_e = 4$ kpc implies a value for fQ^2 that does not vary much with radius and that contains a significant contribution from the non-turbulent terms out to at least ~ 17 kpc (see Fig. 3). These observations are consistent with a variety of independent measures of the magnetic field and cosmic ray energy densities [summarized by Boulares & Cox (1990)], and they suggest that the mechanisms outlined above which drive the different buoyancy forces toward equipartition are, indeed, operating.

4. ROTATION, SCALEHEIGHT AND MASS

Finally, the distribution of mass in the galactic plane obtained in the last section can be combined with the more global measure of the mass in the Milky Way provided by the rotation curve analysis of Section 2 in order to construct a complete mass model for the Galaxy. Figure 4 presents the results of such an exercise, where the Milky Way has been decomposed into the usual spheroid, disk and halo components. The mass distribution for the spheroid was taken from

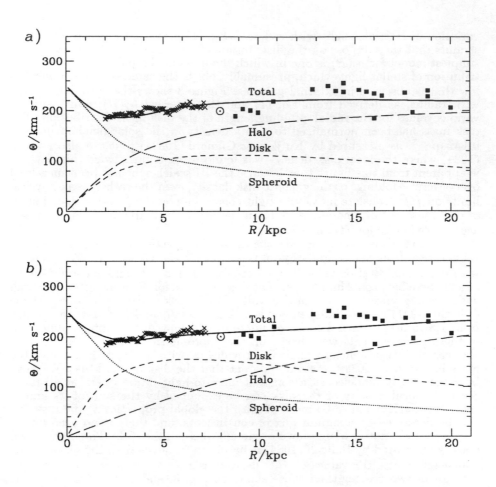

Fig. 4. Three component mass model fits to the HI rotation curve of the Milky Way: (*a*) a fit where the disk component has been normalized to agree with local density measurements; (*b*) a "maximal disk" fit.

Schmidt (1985), the disk was assumed to be thin and exponential in radius, and the halo was modeled using the pseudo-isothermal sphere density distribution, $\rho \propto 1/[1+(r/a)^2]$. In Fig. 4*a*, the disk was taken to have a scalelength of $R_e = 4$ kpc, as implied by the best fit to the HI layer thickness. Its mass was normalized to give a local surface density of $\Sigma(R_0) = 50 M_\odot \text{ pc}^{-2}$, consistent with Kuijken & Gilmore's (1989b) measurement which was made from stellar kinematics. A least-squares fit to the halo parameters then yields values of $a = 3$ kpc, and an asymptotic rotation speed of 220 km s^{-1}.

This model is dominated by the contribution from the dark halo at all radii outside the central few kiloparsecs, but it has been strongly argued that the majority of external galaxies may have "maximal disk" rotation curves where most of the mass of each system within its optical radius comes from its luminous components (van Albada & Sancisi 1986). Such models have also been constructed for the Milky Way (Sellwood & Sanders 1988, Salucci & Frenk 1989), and Fig. 3b shows a similar fit to the current HI rotation curve. For this model, the same spheroid distribution was employed, then the disk model was fitted to the data from $R < R_0$, and finally the halo was matched to any remaining discrepancy in the outer Galaxy. The parameters of the halo are only poorly determined by this process, as it is the dominant contributor to the rotation curve for only a very small range in radii. The more massive disk in this model implies that its local surface density does not agree with Kuijken & Gilmore's measurement of $\Sigma(R_0) = 46 \pm 9 M_\odot$ pc^{-2}. In order to minimize this discrepancy, the disk scalelength has been fixed at the low value of $R_e = 3$ kpc, which is already marginally inconsistent with the analysis of Section 2, but the derived local surface density is still $70 M_\odot$ pc^{-2}, outside the expected error range on $\Sigma(R_0)$. The maximal disk model does not seem to be able to self-consistently reproduce both the rotation curve of the Galaxy and the local value for $\Sigma(R_0)$.

The only way in which the maximal disk model might be salvaged is by adopting a higher value for $\Sigma(R_0)$. Indeed, earlier stellar kinematic measurements of this quantity have suggested that it lies in the range $70 - 80 M_\odot$ pc^{-2} (Oort 1960, Bahcall 1984). The HI data presented in this paper cannot be modeled in the detail required to provide a direct measure of $\Sigma(R_0)$ because of the uncertain contribution to the HI layer's structure from magnetic fields, cosmic rays, *etc.*. However, the HI analysis does provide one important clue: there is general agreement between all the stellar kinematic analyses as to the scale-height of the matter which contributes to the disk mass, and the discrepancy in measurements of $\Sigma(R_0)$ arises from the normalization of this density distribution [see Kuijken & Gilmore (1984b) Fig. 14]. A value of $\Sigma(R_0) = 70 M_\odot$ pc^{-2} would require a mass density in the plane of the Galaxy of almost $0.2 M_\odot$pc^{-3}, and, as was demonstrated in Section 2, such a high local density is not consistent with the observed HI distribution. It is noteworthy that the Milky Way provides us with a system where the disk-halo decomposition can be based on more than just the rotation curve of the galaxy, and that the resulting model does not seem to be consistent with the maximum disk hypothesis. Although maximal disk fits to external galaxies' rotation curve frequently look highly convincing, it might pay to ask whether halo-dominated models are really ruled out for any of these systems.

REFERENCES

Abbott, D.C. 1982, ApJ, 263, 723

Bahcall, J.N. 1984, ApJ, 276, 169

Boulares, A. & Cox, D.P. 1990, ApJ, 365, 544

Casertano, S. & van Gorkom, J.H. 1991, AJ, 101, 1231

Celnick, W., Rohlfs, K. & Braunsfurth, E. 1979, A&A, 76, 24

Dalgarno, A. & McCray, R.A. 1972, ARA&A, 10, 375

Dickey, J.M. & Lockman, F.J. 1990, ARA&A, 28, 215

Falgarone, E. & Lequeux, J. 1973, A&A, 25, 253

Ferrara, A. 1993, this volume

Fich, M., Blitz, L. & Stark, A.A. 1989, ApJ, 342, 272

Fich, M. & Tremaine, S.D. 1991, ARA&A, 29, 409

Henderson, A.P., Jackson, P.D. & Kerr, F.J. 1982, ApJ, 263, 116

Kellman, S.A. 1972, ApJ, 175, 353

Kent, S.M., Dame, T.M. & Fazio, G. 1991, ApJ, 378, 131

Kerr, F.J., Bowers, P.F., Jackson, P.D. & Kerr, M. 1986, A&AS, 66, 373
Kerr, F.J. & Lynden-Bell, D. 1986, MNRAS, 221, 1023
Knapp, G.R. 1987, PASP, 99, 1134
Kuijken, K. & Gilmore, G. 1989a, MNRAS, 239, 651
Kuijken, K. & Gilmore, G. 1989b, MNRAS, 239, 571
Kuijken, K. & Tremaine, S.D. 1993, this volume
Kundić, T., Hernquist, L. & Gunn, J.E 1993, this volume
Kwee, K.K., Muller, C.A. & Westerhout, G. 1954, Bull. Astron. Inst. Netherlands, 12, 211
Lockman, F.J. & Gehman, C.S. 1991, ApJ, 382, 182
McGee, R.X. & Milton, J.A. 1964, Australian J. Phys., 17, 128
Merrifield, M.R. 1992, AJ, 103, 1552
Merrifield, M.R. 1993, MNRAS, in press
Metzger, M. & Schechter, P. 1993, this volume
Oort, J.H. 1960, Bull. Astron. Inst. Netherlands, 15, 45
Rohlfs, K. & Kreitschmann, J. 1988, A&A, 201, 51
Reid, M.J. 1989, in The Center of the Galaxy, IAU Symposium 136, ed. M. Morris (Kluwer, Dordrecht)
Rubin, V.C., Burstein, D., Ford, W.K. & Thonnard, N. 1985, ApJ, 289, 81
Salucci, P. & Frenk, C.S. 1989, MNRAS, 237, 247
Schmidt, M. 1957, Bull. Astron. Inst. Netherlands, 13, 247
Schmidt, M. 1985, in The Milky Way Galaxy, IAU Symposium 106, ed. H. van Woerden, R.J. Allen & W.B. Burton (Reidel, Dordrecht)
Schneider, S.E. & Terzian, Y. 1983, ApJL, 274, L61
Sellwood, J.A. & Sanders, R.H. 1988, MNRAS, 233, 611
Stark, A.A., Gamie, C.F., Wilson, R.W., Bally, J., Linke, R.A., Heiles, C. & Hurwitz, M. 1992, ApJS, 79, 77
Turbide, L. & Moffat, A. 1993, this volume
van Albada, T.S. & Sancisi, R. 1986, Phil. Trans. R. Soc. Lond. A, 320, 447
van der Kruit, P.C. 1981, A&A, 99, 298
van der Kruit, P.C. 1988, A&A, 192, 117
Weaver, H. & Williams, D.R.W. 1974, A&AS, 17, 1

THE WARP OF THE MILKY WAY

Linda S. Sparke
Washburn Observatory, 475 N. Charter Street, Madison, WI 53706
sparke@madraf.astro.wisc.edu

ABSTRACT

The outer parts of the Galaxy disk are observed to bend away from the plane of the inner material. The warp is most obvious in neutral hydrogen, but also involves molecular gas and stars. Models for the bending of a self-gravitating disk are discussed with reference to the Galactic warp.

1. THE WARP IN THE GAS LAYER

The warp of the Milky Way was apparent already in the earliest $21\,cm$ radio surveys (Kerr, this volume); in the outer Galaxy, the layer of atomic hydrogen was seen to curve systematically away from the plane of the Galactic equator, bending in opposite senses on northern and southern sides. Since then, the HI distribution has been surveyed within $10°$ of the Galactic plane over the whole range of longitude (Weaver & Williams 1973, Kerr *et al.* 1986); Burton & te Lintel Hekkert (1986) have combined these and other surveys to reconstruct the form of the entire outer HI disk. Surveys over the complete Northern sky have been made from Bell Labs (Stark *et al.* 1992) and from Dwingeloo (Hartmann, this volume).

The observations yield intensities of $21\,cm$ emission in each (l, b) direction at each velocity v; mapping out the warp requires a conversion to Galactocentric coordinates (R, θ, z). In the outer Galaxy, where the warp is observed, this transformation is single-valued, and depends on the distance $R_\circ$ of the Galactic center, on the rotation curve $\Theta(R)$, and on any noncircular motion; Figure 7.8 of Burton (1988) illustrates how surfaces of constant velocity appear in Galactic coordinates. The rotation curve of the outer Galaxy is now believed to be flat or gently rising outwards, so that the distance scale and form of the warp depend mainly on $R_\circ$ and $\Theta(R_\circ)$. Discussions of the HI warp (*e.g.* Henderson *et al.* 1982, Burton & te Lintel Hekkert 1986, Diplas & Savage 1991) have used $R_\circ = 10\,kpc$ and $\Theta_\circ = 250\,km/s$.

Burton (1988) gives a excellent summary of the overall properties of the warp as inferred from the compilation of Burton & te Lintel Hekkert (1986). The gas layer can be traced to about $25\,kpc$ Galactocentric radius. It bends 'up' towards $z > 0$ on the northern side of the galaxy and 'down' on the southern side; this is the 'integral-sign' shape common for warps in external galaxies. Bending sets in close to the solar circle, at about $R = 12\,kpc$, and the peak height grows approximately linearly with radius out to about $R = 17\,kpc$. Within this region, the mean height of the warp varies approximately sinusoidally with Galactocentric azimuth θ, as expected if the gas follows circular orbits tilted with respect to the Galactic plane. Beyond $R = 17\,kpc$, on the northern side the warp amplitude continues to grow; at $R = 24\,kpc$ the mean plane of the HI reaches $4\,kpc$ above the plane $b = 0°$, equivalent to a tilt of $10°$ away from the plane of the central galaxy. On the southern side, the maximum excursion is reached at about $18kpc$, beyond which the hydrogen layer curves back towards the equatorial plane. The azimuth of the line of nodes, at which the warped disk crosses the plane $b = 0°$, appears to twist by about $60°$ between the solar radius and $r = 13kpc$, in a trailing sense relative to the Galactic rotation; but this angle is rather uncertain because the warp is so slight here. Beyond $13kpc$, the line of nodes is fairly straight; the warp reaches furthest from the Galactic plane at $l = 75°$ and $l = 255°$ respectively.

The Milky Way is not unusual in having a warped HI disk. Most (12 out of 20) of the edge-on spiral galaxies for which high-resolution $21cm$ maps are available show some degree of warping in the gas layer (Bosma 1991). Since the bending could only be seen from a favorable orientation, the true fraction of warped systems must be even higher. The basic form of the warp is that of the 'integral sign', but some asymmetry is seen in about half of the warped disks. For example, the HI layer in both NGC4565 and NGC5907 bends back towards the plane of the inner disk on one side (Sancisi 1983), in a similar way to the southern side of the Milky Way. These edge-on warps cannot have a pronounced spiral form, or we would observe a thickening of the disk instead of a warp. When the galactic disk is at intermediate inclination, the form of the warp may be determined by fitting a 'tilted-ring' model to the velocity field, assuming the gas to follow inclined concentric circular orbits (*e.g.* Bosma 1981). Briggs (1990) examined tilted-ring fits for a number of well-mapped galaxies, deprojecting the ring angles to refer to the plane of the inner gas disk. He found that in this reference frame the warp showed a straight line of nodes approximately out to the Holmberg radius; beyond this the warps tended to assume a leading spiral form. As Briggs pointed out, the warp of the Milky Way is in this respect not quite typical, since the line of nodes is straight in the outer part.

The outer HI layer not only warps, but also flares to greater thickness at larger radii; Burton (1988) estimates that the scale height of the warped layer increases by an order of magnitude between R_o and $R = 25kpc$. Diplas & Savage (1991) have analysed a $21cm$ survey taken with the Bell Labs 20ft horn reflector (Stark *et al.* 1992); although this survey has only $2°$ resolution, it covers the entire sky above declination $-40°$. Also, the horn has extremely low sidelobes, reducing stray radiation by an order of magnitude over normal antennas. Diplas & Savage find that their contours of HI density agree well with those of Burton & te Lintel Hekkert (1986) down to $n_H = 10^{-2}cm^{-3}$, but are 'better behaved' below that level; they trace a distinct, flaring, warped layer down to $n_H = 10^{-3}cm^{-3}$. The HI disk extends out to $R = 30kpc$ in some directions ($\theta \sim 130°$), where it can be traced over a region $10kpc$ thick ($z_{rms} \approx 3 - 4kpc$); gas in the warped disk is found at latitudes as high as $b = 25°$, showing the value of a survey beyond $|b| = 10°$. Plots of n_H versus z at fixed (R, θ) show that the HI disk consists of two components: a 'confined' component with density $n_H \geq 10^{-2}cm^{-3}$, which persists only out as far as $R \sim 20kpc$, and an 'extended' component with considerably lower density and larger scale height, persisting out to $R \sim 25 - 30kpc$. Flaring has been seen in the HI disks of external galaxies (Bosma 1991), including M31 (Brinks & Burton 1984); the disk of NGC891 appears to have a two-component structure similar to the Milky Way (see Rand 1992).

The Galactic warp may be traced also in molecular and ionized gas, although both HII regions and molecular clouds are sparse in the outer Galaxy. Fich & Blitz (1984) showed that HII regions beyond a Galactocentric radius of

$15 kpc$ are found preferentially above the Galactic plane in the northern sector $(0° < l < 180°)$, and below that plane in $(180° < l < 360°)$. A CO line survey of molecular clouds in the third Galactic quadrant (May et al. 1988) showed these clouds lying predominantly below the plane $b = 0°$.

Using the IRAS Point Source Catalog to select objects with infrared colors typical of molecular clouds with embedded star-forming regions, Wouterloot et al. (1988) then observed these in the CO line to provide radial velocity measurements for 1077 clouds in the range $(85° < l < 280°)$. The mean plane of this sample shows a clear tilt in the sense of the HI warp. Calculating distances on the basis of a slightly rising rotation curve with $R_o = 8.5 kpc$ and $\Theta_o = 220 km/s$, they trace the molecular clouds out to $20 kpc$ from the Galactic center. Since the IRAS sources are intrinsically very luminous, this should represent the true limits of the source population. A quantitative comparison with the HI observations (using the same rotation curve) shows that the molecular cloud layer follows the same warped shape over the sampled range $(120° < \theta < 240°$ at $R = 2R_o)$, crossing the plane $b = 0°$ at approximately the same azimuth $(165° \pm 10°)$ at all radii. The molecular cloud layer also flares with increasing Galactocentric distance, approximately doubling in thickness between $10 kpc$ and $17 kpc$, though remaining systematically thinner than the HI disk.

In addition to the large-scale warp, with azimuthal wavenumber $m = 1$, the outer edge of the HI disk is 'scalloped', with an up-and-down variation of about $1 kpc$, corresponding to a wavenumber $m \approx 10$. Scalloping is also seen within the solar circle, both in HI (Quiroga 1974) and in CO observations (Sanders et al. 1984), with a similar azimuthal wavenumber but much smaller amplitude. The relation of this to the large-scale warp is unclear.

A puzzling feature of the inner galaxy is that within $2 - 3 kpc$ of the center, the gas disk is tilted in the same sense as the 'S' curve of the outer warp: the gas lies below the plane $b = 0°$ at positive longitudes and below it at negative longitudes. The tilt is seen in HI (Kerr 1967, Burton & Liszt 1978) and in CO (Liszt & Burton 1980, Sanders et al. 1984, Bally et al. 1988), although estimates of its amplitude vary from almost 30° (Liszt & Burton 1980) to only about 4° (Bally et al. 1988). The analysis is further complicated by the presence of noncircular motions perhaps indicative of a rapidly tumbling central bar (Liszt & Burton 1980, Binney, this volume). Blitz & Spergel (1991), analyzing a $2.4 \mu m$ balloon survey of the Galactic center region, suggested that the bulge was tilted by a few degrees in the same sense as the gas disk. However, the more recent COBE data (Hauser, this volume) shows no evidence for a tilted bulge; this would imply that in the central regions of the galaxy, the gas follows a different mean plane from the stars.

2. THE WARP IN THE STARS

The Galactic warp has also been detected in the stellar disk, although this is much harder than measuring the gas warp: it is difficult to select distant stars in crowded fields close to the plane, and the surface density of the stellar disk drops off much faster with Galactocentric radius than that of the HI disk. Miyamoto et al. (1988) examined a sample of OB stars within $3 kpc$ of the sun, obtaining distances by spectroscopic parallax; they found the average plane defined by these stars to be tilted in the sense corresponding to the gas warp. Djorgovski & Sosin (1989) selected from the IRAS Point Source Catalog 89886 objects with colors typical of evolved stars with dust shells. These sources should correspond to Mira-type variables or luminous AGB stars, with ages in the range $4-14Gyr$; such stars must have made many orbits around the Galaxy since their formation, and their presence in the warp would suggest that it is not a sporadic or transient feature. They should also be extremely luminous, and detectable to at least $20kpc$ from the sun. A fit to the mean plane defined by these sources shows a roughly sinusoidal variation with Galactic longitude in the same sense as the gas warp, but with a much smaller amplitude than expected for sources following the HI warp at distances of 10kpc or more; this may indicate that the sample is contaminated by intrinsically fainter objects.

An interesting project now in progress is that of Seitzer and Carney (this volume). They have taken multicolor CCD frames in four directions which intersect the warped plane defined by the HI gas. By comparison with four control fields at the same Galactic latitudes, they find a significant excess of stars; a statistical 'cleaning' of field stars reveals a clear main sequence in each of the warp directions. Spectroscopy of the 'warp' stars should yield distances, establishing the scale and form of the warp independently of assumptions about Galactic rotation.

Recent infrared maps produced from the COBE/DIRBE experiments (refs this volume) also show a warped Galactic plane; the bending amplitude increases from the shorter to the longer wavelengths, as diminishing obscuration makes it possible to see further. The shortest wavelength bands sample mainly stars, while the longest are sensitive to cool dust. Thus the COBE measurements show that both stars and dust in the Galactic disk partake in the warp.

It is difficult to detect warping in the stellar disks of external galaxies, because the bending of the HI layer becomes strong at around 3 – 5 times the exponential scale length in the stellar disk (Sancisi 1983), which is close to the point where the stars appear to end (van der Kruit & Searle 1982). However, warps in the stellar disk are definitely seen, for example in M31 (Innanen *et al.* 1982), and in the edge-on systems NGC4565 (van der Kruit & Searle 1981) and NGC5907 (Sasaki 1987). The 'superthin' galaxy UGC7170 (Shaw 1986) shows a particularly clear stellar warp.

The presence of a warp in the stellar disk has important dynamical implications. If warps are caused by gravitational forces, these should affect stars and gas in the same way; both must share in the warp. Other proposed warping mechanisms, such as the ram pressure of intergalactic gas (Kahn & Woltjer 1959) or large-scale magnetic fields (Battaner *et al.* 1990) would imply much greater forces on the gas; stellar and gaseous disks should separate. Warps observed in edge-on galaxies are the best way to test this point, though in making the comparison one must take account of the fact that the surface density of the stellar disk drops much faster with radius than that of the gas. Detailed analysis for two edge-on galaxies, NGC5907 (Sasaki 1987) and NGC4565 (Casertano & Sancisi, private communication), shows that the gaseous and stellar disks indeed appear to follow the same warped plane.

3. THEORY

The Galactic warp has represented a puzzle for theorists since its discovery. The line of nodes of the warp is straight over most of its extent, whereas differential precession in the gravitational field of the inner disk is rapid (*e.g.* Binney 1992) and should wind the warp into a tight leading spiral. This 'winding dilemma' suggests that the warp either is stabilized in some way to prevent the development of a spiral, or is continuously regenerated, perhaps as an instability of the disk. Binney (1992) presents a general review of warps; some specific developments are discussed below.

One potential mechanism for exciting the warp is via a Galactic bar. Infrared photometry (Blitz & Spergel 1991, Hauser, this volume) and analysis of gas motions (Liszt & Burton 1980, Binney et al. 1991, Binney, this volume) indicate a prolate structure with dimensions of $1 - 2kpc$ in the central region of the Galaxy, tumbling rapidly so that corotation with the orbital period occurs just beyond the ends of the bar. Binney (1978, 1981) has shown that the orbits of particles in a barred potential, the figure of which rotates with pattern speed Ω_p, are unstable to developing vertical oscillations if

$$(\Omega - \Omega_p)^2 = \mu^2/n^2, \; n = 1, 2, 3 \ldots \tag{1}$$

where Ω is the orbital frequency, and μ the frequency of small vertical oscillations, in the azimuthally-averaged potential; the strength of the resonance decreases with increasing n. Since in an oblate potential $\mu > \Omega$, the strongest resonance for a bar which tumbles in the same sense as galactic rotation is at $(\Omega - \Omega_p) = -\mu/2$. Assuming all frequencies to decrease inversely with radius, this is likely to be somewhere around two or three times the bar corotation radius; the model of Binney et al. (1991) has corotation at $R = 2.4kpc$, placing the resonance well within the solar circle. If the disk is self-gravitating, vertical motion at any given radius gives rise to bending waves, which transport energy away from the resonant region. Sparke (1984) examined this process and found that an amplification cycle can be set up, allowing the warp to grow – but only for very low disk masses, below about 10% of the mass of the bar. That calculation assumed that the gas follows nearly circular orbits; in fact the orbits will be somewhat elliptical, which strengthens the forcing (Lubow 1992), but by less than a factor of two (Binney 1981 eq. 18), which is still insufficient to produce the warp. Another argument against this possibility is that the warp is observed to start beyond the solar radius, whereas the resonant region is significantly further in.

It is unlikely that the tidal pull of the Magellanic Clouds is responsible for the warp. This possibility was investigated by Hunter & Toomre (1969); even making the optimistic assumption that the Clouds are currently near apogalacticon, they, concluded that their mass was insufficient to cause the warp, by a factor of two or three. Since then, modelling of the orbits of the Clouds has shown that they are in fact close to their perigalactic distance (Murai & Fujimoto 1980, Lin & Lynden-Bell 1982); the mass discrepancy is then hopelessly large.

Lynden-Bell (1965) suggested that perhaps the warp is not growing, but is held together and prevented from winding by the self-gravity of the disk. If the Galactic disk had a discrete mode of vertical oscillation, external disturbances could excite the mode, and once a warp had been set up it would persist for many rotation periods. Hunter & Toomre (1969) showed that $m = 1$ 'integral-sign' bending disturbances are neutrally stable, neither growing nor decaying. As a consequence, a bending mode would have no spiral form, but would exhibit a straight line of nodes, precessing steadily without change of shape until dissipation became important. However, when these authors investigated the warping modes of a sequence of isolated thin self-gravitating disks, they found that those disks for which the surface density tapered smoothly to zero had no discrete warping modes, but admitted a continuum of dispersive bending waves. When such a disk was initially forced to warp, corrugation waves were set up; these carried the energy of warping outwards, so that the warp became increasingly spiral and confined to the very outer edge of the disk. The only discrete mode was the trivial tilt of the whole disk.

If the disk is surrounded by a dark halo which is somewhat oblate rather than spherical, the trivial tilt is no longer a mode of the system: a discrete warped mode may appear instead. Dekel & Shlosman (1983) and Toomre (1983) suggested that warps might be caused by a misalignment between the plane of the inner disk and the symmetry plane of the halo; the latter paper showed that a discrete bending mode did exist when the halo was idealized as a distant massive fixed ring. Sparke & Casertano (1988: hereafter SC) showed that this modified tilt mode can exist in more realistic galactic models, consisting of a thin exponential disk and a flattened fixed 'pseudo-isothermal' halo. They showed that the shape of the warped disk depended mainly on the core radius of the halo, and constructed models which successfully reproduced the main features of the warps observed in the galaxies NGC4565 and NGC4013. The required halo flattening is modest; the models of that paper featured halos with an oblateness corresponding to an E2 shape, less flattened than the 'best-fit' oblateness indicated for the dark halo of the polar ring galaxy NGC4650A (Sackett & Sparke 1990).

When the halo mass and core radius are chosen in such a way that the rotation curve remains approximately flat, the warped modes start to show a pronounced bending around $3 - 5$ scale lengths of the disk (SC Fig 7). The warp of the Milky Way becomes strong around $11 - 12kpc$. The scale length h has been estimated to be in the range $1.8 - 6kpc$ (Kent, Dame & Fazio 1991); it would be difficult to interpret the warp as a modified tilt mode if h is near the high end of this range. In a model with a relatively small halo core, less than $1 - 2$ disk scale lengths, the warped mode bends upwards, away from the halo equator: SC call these Type I warps. If the halo core is larger than about four times the disk scale length, the warp is of Type II, curving down towards the halo equator with increasing radius. Mass models for the Galaxy (e.g. Caldwell & Ostriker 1981, Bahcall, Schmidt & Soneira 1983, Merrifield 1992) imply core radii in the range $2 - 8kpc$; this would mean that the warp should be of Type I, bending away from the equatorial plane of the unseen halo.

Kuijken (1991) has extended the analysis of SC into the nonlinear regime, and finds that the shapes of downward-bending Type II warps remain fairly close to the predictions of linear theory, while the sharpest upward-curving Type I modes cannot persist more than a few degrees from the halo equatorial plane. As noted above, the relatively small core of the Milky Way halo implies that the warp should be of Type I; but the bending in the symmetric part of the warp is limited to $\approx 5°$, so there is no immediate cause for alarm.

Because the warping modes do not grow, they must be excited in some way: the disk angular momentum may initially be misaligned with the halo, gas could be accreted onto the disk from out-of-plane orbits, the orientation of the halo may change as dark matter is accreted (Ostriker & Binney 1989), or the galaxy may suffer tidal interaction. To examine behavior subsequent to an initial perturbation, Hofner & Sparke (1991, and in preparation) followed the time-evolution, using the same dynamical system as SC. They found that, if the disk has a discrete normal mode of bending, then starting from an initial rigid tilt, it will settle towards the mode from the center outwards. The settled inner warp has a straight line of nodes, while the outer disk warp eventually twists into a leading spiral; thus the warp assumes the typical form found by Briggs (1990). The dynamical model explicitly excludes dissipational effects; settling occurs because bending waves transport outwards the energy associated with the transient response (Toomre 1983). The rate of transport can be estimated using the short-wavelength WKB approximation; the group velocity is proportional to the surface density $\sigma(R)$ of the disk (Hunter 1969), so that waves propagate ever more slowly towards the edge. The time required for the disk to settle out to a given radius depends on the disk mass and scale length, and on the amplitude of the rotation curve; it is largely independent of the halo properties. Typically the disk requires a few billion years to settle out to 6 or 7 scale lengths; the outermost parts of observed HI disks should not settle within a Hubble time.

But what happens to a disk which has *no* discrete warping mode? Hofner & Sparke ran simulations for a model in which there is a discrete mode when the disk ends at 6 scale lengths, but not when it is extended to 9 scale lengths. The settling of the extended disk appeared very similar to the case in which a mode exists: the line of nodes is straight in an inner, settled, region and twists in a leading sense in the outer disk. The settled portion is close to the warped mode for the smaller disk; it is as though the inner part of the disk does not know that the outer part exists. The root of this ignorance can be found in the WKB wave transport equations. Unless the integral $\int dR/\sigma(R)$ converges, wave energy never reaches the edge of the disk; the inner regions of the galaxy cannot 'know' the total extent of the disk, or even whether there is a mode of the whole disk. Thus *even if no normal mode of bending exists*, the inner portion of the warped disk will still show a coherent behavior, with a straight line of nodes. The shape of the inner disk resembles a Type I warp if the halo core is small, and a Type II warp if it is large.

But the Milky Way does not consist only of disk and halo; there is also a substantial bulge component. The bulge is concentrated to the center of the Galaxy, and contains perhaps 10-15% as much mass as the disk (van der Kruit 1986). Both kinematic considerations (van der Kruit 1989) and infrared observations (*e.g.* Hauser, this volume) show the bulge to be flattened, with axis ratio c/a around $0.6 - 0.7$. A precessing warped disk will exert a torque on the flattened bulge, causing the bulge to take part in the overall warp. Jo Pitesky and I (Pitesky 1991, Pitesky & Sparke in preparation) have extended the SC warp model to include a 'wooden' bulge, with a rigid shape but free to tilt about its fixed center.

In this system, we find in general *two* discrete warping modes. The modified tilt mode continues to exist, with the bulge and disk tipped in the same sense relative to the halo midplane. The bulge tips at a different angle from the inner disk, because it is more nearly spherical, so the torque from the halo is less, which reduces the precession rate. But the bulge is likely to have less angular momentum per unit mass than the disk, which works in the opposite direction; the balance of these two factors determines whether the bulge tips more or less than the inner disk. In this mode, the angle between bulge and inner disk is generally smaller than that between disk and halo; the bulge will not show a pronounced tilt away from the disk. In the other, 'bulge', mode, the bulge and disk tip away from the halo midplane in opposite senses. This mode persists and remains warped when the halo is spherical or even absent; in that case, as the bulge is made more massive or more rapidly rotating (and thus harder to tip), the mode passes over into the modified tilt mode that the disk would have if the bulge were fixed.

Fig 1 shows the 'bulge' mode for two mass models based on fits to the rotation curve of the galaxy M31. The first (top left) is based on the fit by Kent (1987); the disk has scale length $h = 5.24kpc$, extends to $30kpc$ (the outer limit of the HI data), and has mass $M_D = 1.5 \times 10^{11} M_\odot$. We modelled the bulge with an $r^{1/4}$ law spheroid, with $r_e = 1.9kpc$ and $M_B = 3.6 \times 10^{10} M_\odot$ chosen to fit the rotation curve. Kent specified a halo of constant density; we used a 'pseudo-isothermal' form with the same central density and a large scale length ($r_c = 6$ disk scale lengths, and $V_\infty = 250km/s$). We then chose, arbitrarily, a modest flattening $c/a = 0.7$ for the halo. The bulge flattening was 0.63, and the angular momentum was 0.35 in units where $G = M_D = h = 1$.

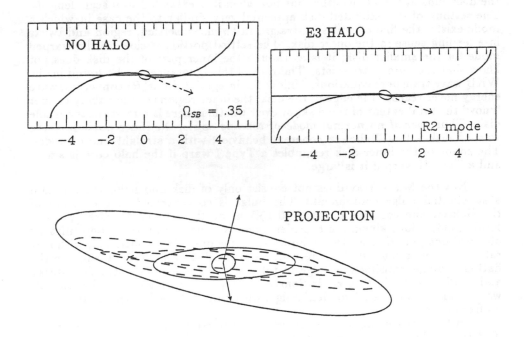

Fig. 1: Warp models with a tilted bulge

The mode in this system (top right) is almost indistinguishable from that in a model with no halo (top left), based on the rotation-curve fit of Braun (1991). Here $h = 5.8kpc$, $M_D = 1.2 \times 10^{11} M_\odot$; $r_e = 2kpc$ and $M_B = 7.8 \times 10^{10} M_\odot$, and there is a small point mass of $10^8 M_\odot$ at the nucleus. The bulge flattening and angular momentum were as in the previous model. The bottom panel shows this no-halo model, projected to the viewing angles of M31; the warp amplitude has been set by taking the angle between bulge and inner disk ($R = 2h$) to be 20°. It has long been known that the bulge of M31 cannot be both axisymmetric and aligned with the plane of the inner disk; difficulties with the triaxial hypothesis (Gerhard 1986) suggest re-interpretation in terms of a tilted bulge.

The mode analysis neglects the huge number of degrees of freedom represented by the orbits of individual bulge and disk stars, and halo particles, assuming that these will not drain away the energy present in a warping mode. It is not currently possible to check this point by N-body simulation, since the problem is intrinsically three-dimensional, and even a million particles, the maximum currently practical, give only an average of 10 per phase space dimension. There is some cause for hope, however, from recent investigations showing that one-dimensional stellar systems can have discrete oscillatory modes (Gerhard & Louis 1988, Mathur 1990, Weinberg 1991). Weinberg (1991) extended his calculations of vertical modes in a one-dimensional disk to two dimensions, in order to approach the problem of warps. He found that pure oscillations can exist if the particle distribution function is truncated at high velocities, as it would be in a finite disk. But even if the Galactic disk does not have undamped modes, warps may still be effectively long-lived. Louis (1992) finds that modes which have crossed into the continuum region (and are thus subject to damping) lose their energy only very slowly; since galaxies are dynamically quite young (the Sun has made no more than 100 orbits around the Milky Way), a warp which is slowly damped over tens of orbital times will be, for practical purposes, quite as good as a truly permanant one.

REFERENCES

Bahcall, J. N., Schmidt, M. S. & Soneira, R. M. 1983, ApJ 265, 730

Bally, J., Stark, A.A., Wilson, R.W. & Henkel, C. 1988, ApJ 324, 223

Battaner, E., Florido, E., & Sanchez-Saavedra, M. L. 1990 A&A 236, 1

Binney, J. 1978, MNRAS 183, 779

Binney, J. 1981, MNRAS 196, 455

Binney, J., Gerhard, O. E., Stark, A. A., Bally, J. & Uchida, K. I. 1991, MNRAS 252 210

Binney, J. J. 1992 Ann Rev A&A 30, 51

Blitz, L. & Spergel, D. M. 1991, ApJ 379, 631

Bosma, A. 1981, AJ 86, 1825

Bosma A. 1991, in Warped Disks and Inclined Rings Around Galaxies, eds. S. Casertano *et al.* (Cambridge: CUP), p181

Braun, R. 1991 ApJ 372, 54

Briggs, F.H. 1990, ApJ 352, 15

Brinks, E. & Burton, W. B. 1984, A&A 141, 195

Burton, W. B. 1988, in Galactic and Extragalactic Radio Astronomy, 2nd Edition, eds. G. L. Verschuur & K. I. Kellerman (Berlin: Springer), p295

Burton, W. B. & te Lintel Hekkert, P. 1986, A&A Suppl, 65, 472

Burton, W. B. & Liszt, H. S. 1978, ApJ 225, 815

Caldwell, J. A. R. & Ostriker, J. P. ApJ 251, 61

Dekel, A. & Shlosman, I. 1983, in Internal Kinematics and Dynamics of Galaxies, IAU Symposium 100, ed. E. Athanassoula (Dordrecht: Reidel), p187

Diplas, A. & Savage, B. D. 1991, ApJ 377, 126

Djorgovski, S. & Sosin, C. 1989, ApJL 341, L13

Fich, M. & Blitz, L. 1984, ApJ 279, 125

Henderson, A. P., Jackson, P. D. & Kerr, F. J. 1982, ApJ 263, 116

Hofner. P., & Sparke, L. S. 1991, in Warped Disks and Inclined Rings Around Galaxies, eds. S. Casertano et al. (Cambridge: CUP), p225

Gerhard, O. E. 1986, MNRAS 219, 373

Hunter, C. 1969, Stud. Appl. Math 48, 55

Hunter, C. & Toomre, A. 1969, ApJ 157, 183

Innanen, K. A., Kamper, K. W., Papp, K. A. & van den Bergh, S. 1982, ApJ 254, 515

Kahn, F. D. & Woltjer, L. 1959, ApJ 130, 705

Kent, S. M. 1987, AJ 93, 816

Kent, S. M., Dame, T. M., & Fazio, G. 1991, ApJ 378, 131

Kerr, F. J. 1967, in Radio Astronomy and the Galactic System, ed. H. van Woerden (London: Academic Press), p575

Kerr, F. J., Bowers, P. F., Jackson, P. D., & Kerr, M. 1986, A&AS 66, 373

Kuijken, K. 1991, ApJ 376, 467

Lin, D. N. C. & Lynden-Bell, D. 1982, MNRAS 198, 707

Liszt, H. S. & Burton, W. B. 1980, ApJ 236, 779

Louis, P. D. & Gerhard, O. E. 1988, MNRAS 233, 337

Louis, P. D. 1992, MNRAS 258, 552

Lubow, S. 1992, ApJ 398, 525

Lynden-Bell, D. 1965, MNRAS 129, 299

Mathur, S. D. 1990, MNRAS 243, 529

Merrifield, M. 1992, AJ 103, 1552

Miyamoto, M., Yoshizawa, M. & Susuki, S. 1988, A&A 194, 107

Murai, T. & Fujimoto, M. 1980, PASJ 32, 581

Ostriker, E. C. & Binney, J. 1989, MNRAS 237, 785

Pitesky, J. 1991, in Warped Disks and Inclined Rings Around Galaxies, eds. S. Casertano et al. (Cambridge: CUP), p215

Quiroga, R. J. 1974, Ap Sp Sci 27, 323

Rand, R. J. 1992, in Star Forming Galaxies and their Interstellar Medium, eds. J. Franco & F. Ferrini (Cambridge: CUP)

Sancisi, R. 1983, in Internal Kinematics and Dynamics of Galaxies, IAU Symposium 100, ed. E. Athanassoula (Dordrecht: Reidel), p55

Sanders, D. B., Solomon, P. M., & Scoville, N. Z. 1984, ApJ 276, 182

Sackett, P. D. & Sparke, L. S. 1990, ApJ 361, 408

Sasaki, T. 1987, PASJ 39, 849

Shaw, M. A. 1986, unpublished PhD thesis, University of Edinburgh

Sparke, L. S. 1984, MNRAS 211, 911

Sparke, L. S. & Casertano, S. 1988, MNRAS 234, 873

Stark, A. A. et al. 1992, ApJS 79, 77

Toomre, A. 1983, in Internal Kinematics and Dynamics of Galaxies, IAU Symposium 100, ed. E. Athanassoula (Dordrecht: Reidel), p177

van der Kruit, P. C. & Searle, L. 1981, A&A 95, 105

van der Kruit, P. C. & Searle, L. 1982, A&A 110, 61

van der Kruit, P. C. 1986, A&A 157, 230

van der Kruit, P. C. 1989, in The Milky Way as a Galaxy, G. Gilmore, P. C. van der Kruit & I. R. King (SAAS-FEE), §5

Weaver, H. F. & Williams, D. R. W. 1973, ApJS 8, 1

Weinberg, M. 1991, ApJ 373, 391

Wouterloot, J. G. A., Brand, J., Burton, W. B. & Kwee, K. K. 1990, A&A 230, 21

A KINEMATIC STUDY OF GALACTIC HII REGIONS.
INDICATION OF A CONTRACTION TOWARDS THE CENTER

P. PİŞMİŞ & E. MORENO

Instituto de Astronomía, Apartado postal 70-264

C.P. 04510, México, D. F.

ABSTRACT. We have considered radial velocities of galactic H II regions with known distances to analyze the velocity field of these regions along the galactic plane. A radial velocity is taken as the average of optical and radio data existing in the literature. A limiting value of 14 km kpc^{-1} s^{-1} for Oort's constant A is attained with a least squares solutions, as we incorporate regions with ever larger distances from the sun. The data also suggest an overall radial contraction of the H II regions towards the galactic center of about 6 km s^{-1}.

DATA ANALYSIS

We have compiled information about radial velocities and photometric distances of 213 H II regions in our Galaxy. Optical data have mainly been obtained from Georgelin & Georgelin (1970a) and Georgelin *et al.* (1973). Radio data are taken mainly from Blitz *et al.* (1982), Brand *et al.* (1987), Wilson *et al.* (1970), Wilson (1972), Whiteoak & Gardner (1974), Gardner & Whiteoak (1984), and Reifeinstein *et al.* (1970). In the analysis we take a straight average of radio and optical radial velocities in a given region. An extended version of the data analysis will be presented in a forthcoming paper (Pişmiş & Moreno 1992). H II regions are analyzed in different circular bands centered at the solar circle, with $R_0 = 8.5$ kpc. A preliminary analysis of the data in the northern and southern hemispheres is shown in Fig. 1, for Oort's constant A, solving by least squares the equation (Kraft & Schmidt 1963):

$$U_\odot \cos b \cos \ell + V_\odot \cos b \sin \ell + 2A(R - R_0) \cos b \sin \ell + 2\alpha(R - R_0)^2 \cos b \sin \ell =$$
$$- (V_r)_\odot - W_\odot \sin b \qquad (1)$$

with $W_\odot = 7.5$ km s^{-1}.

Fig. 1 reproduces, through the variation of A, essentially the trends of the Hα rotation curve presented by Georgelin & Georgelin (1976) in their figure 2; namely, a decreasing northern rotation curve and a nearly flat southern rotation curve. Both trends representing mainly the nearer $R > R_0$ distances, where the majority of our H II regions is distributed.

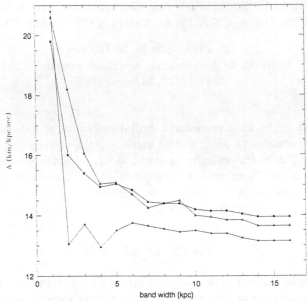

Fig. 1   Least squares solutions for Oort's constant A using eq. (1). Open triangles: northern hemisphere, filled triangles: southern hemisphere, filled squares: both hemispheres.

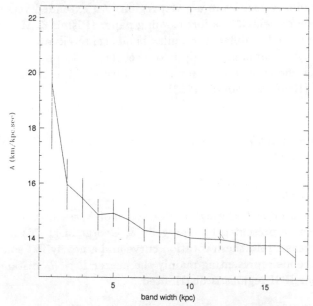

Fig. 2   Least squares solutions for Oort's constant A using the modification of eq. (1). Both hemispheres are considered. Vertical lines show standard deviation 2σ.

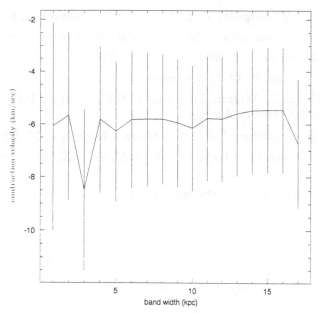

Fig. 3 Same as Fig. 2, but now for the velocity π, being a contraction towards the galactic center.

The analysis of the $U_\odot$ solution emerging from eq. (1) suggests that we must allow for a K–term. Our best fit, with a correlation coefficient about 0.95, has been obtained by adding $\frac{\pi}{R} \{(R - R_0)\cos\ell + r\cos b\}\cos b$ to the left side of eq. (1), with π a contraction or expansion velocity of the H II regions in the R direction. Figs 2 and 3 show the solutions for A and π with this modification of eq. (1).

From overall data a slowly decreasing local rotation curve is obtained, represented by the limiting value $A \sim 14$ km kpc$^{-1}s^{-1}$. Also, a mean contraction towards the galactic center of ~ 6 km$^{-1}s^{-1}$ is found for the assembly of H II regions. This same value is also obtained by Georgelin & Georgelin (1970b) from optical data.

References

Blitz, L., Fich, M., & Stark, A.A. 1982, *Ap. J. Suppl. Ser.*, 49, 183

Brand, J., Blitz, L., Wouterloot, J.G.A., & Kerr, F.J. 1987, *Astron. Astrophys. Suppl. Ser.*, 68, 1.

Gardner, F.F., & Whiteoak, J.B. 1984, *M.N.R.A.S.*, 210, 23.

Georgelin, Y.M., Georgelin, Y.P., & Roux, S. 1973, *Astron. Astrophys.*, 25, 337.

Georgelin, Y.M., & Georgelin, Y.P. 1976, *Astron. Astrophys.*, 49, 57.

Georgelin, Y.P., & Georgelin, Y.M. 1970a, *Astron. Astrophys.*, 6, 349.

Georgelin, Y.P., & Georgelin, Y.M. 1970b, *Astron. Astrophys.*, 8, 117.

Kraft, R.P., & Schmidt, M. 1963, *Ap. J.*, 137, 249.

Pişmiş, P. & Moreno, E. 1992, in preparation.

Reifenstein III, E.C., Wilson, T.L., Burke, B.F., Mezger, P.G., & Altenhoff, W.J. 1970, *Astron. Astrophys.*, 4, 357.

Whiteoak, J.B.,& Gardner, F.F. 1974, *Astron. Astrophys.*, 37, 389.

Wilson, T.L., Mezger, P.G., Gardner, F.F., & Milne, D.K. 1970, *Astron. Astrophys.*, 6, 364.

Wilson, T.L. 1972, *Astron. Astrophys.*, 19, 354.

EXAMINING MOTION OF THE LSR
USING ANTICENTER CARBON STARS

Mark R. Metzger and Paul L. Schechter
Room 6-204, Massachusetts Institute of Technology, Cambridge, MA 02174.

metzger@alioth.mit.edu
schech@achernar.mit.edu

ABSTRACT

We summarize results of a measurement of the motion of the LSR with respect to anticenter carbon stars. New radial velocities were combined with existing K-band photometry to examine the velocity mean and dispersion as a function of distance. The mean velocity of the carbon stars relative to the LSR is 6 km s^{-1} with a dispersion of 23 km s^{-1}, and no trends with distance are evident. We conclude that the carbon star data are not consistent with the model proposed by Blitz and Spergel (1991), and tend to agree with the model of Kuijken (1991) more closely.

1. INTRODUCTION

Observations of gas in our galaxy have yielded several results that are apparently inconsistent with an axisymmetric disk model. Blitz and Spergel (1991, hereafter BS) described several such discrepancies, and proposed a non-axisymmetric disk model driven by a rotating triaxial spheroid. In particular, their model attempts to account for both the apparent 14 km s^{-1} inward motion of the anticenter H I gas and the asymmetry of the outer galaxy H I, as measured in surveys by Weaver and Williams (1974) and Kerr *et al.* (1986). In the BS model, the outer galaxy gas follows roughly circular orbits, and the asymmetry is produced by non-circular orbits near the Sun giving a mean motion to the LSR relative to the outer gas.

Another model to explain the outer galaxy H I asymmetry was put forth by Kuijken (1990), who invoked a large-scale $m = 1$ distortion of the outer galaxy H I. Stars do not participate in the asymmetry of Kuijken's model to the degree they do in the BS model. Thus to put additional constraints on these models one would like to compare the stellar kinematical predictions of the two models with observations. Data on local stellar kinematics were collected by Kuijken and Tremaine (1991) from many sources and compared with model predictions. One of their conclusions was to show a discrepancy between the measured tilt of the velocity ellipsoid and the prediction from the BS model.

To constrain the models on larger scales, we have measured the radial motion of the LSR with respect to anticenter carbon stars. Using radial velocities and distances derived from K-band photometry, the radial profile of the velocity mean (with respect to the LSR) and dispersion of the carbon stars can be used to constrain rotation curve models. This work is described in detail by Metzger and Schechter (1993, hereafter MS), and we summarize the results at this conference.

2. OBSERVATIONS

We obtained spectra of 178 carbon stars from Fuenmayor's (1981) survey of the galactic anticenter region using the 2.1m telecope at Kitt Peak in November, 1990. Radial velocities were obtained from the CN molecular absorption bands near 8000 Å by comparison with a template spectrum, with a precision of 3.5 km s^{-1}. The zero point was aligned with that of Aaronson et $al.$ (1990) using velocities of 62 stars in common. Details of the data reduction and individual velocities for the stars are reported by MS.

3. ANALYSIS

Distances to the stars were computed using K–band photometry from Jura et $al.$ (1989). The carbon stars were assumed to have a Gaussian distribution with average absolute magnitude $M_K = -8.1$ (Claussen et $al.$ 1987) and $\sigma = 0.6$ mag (Schechter et $al.$ 1991). An average extinction law of 0.05 K mag kpc^{-1} was adopted based upon H I column density and average $(I - K)$ colors (see MS for details). A correction for Malmquist's effect was applied using a smooth fit to the observed distribution.

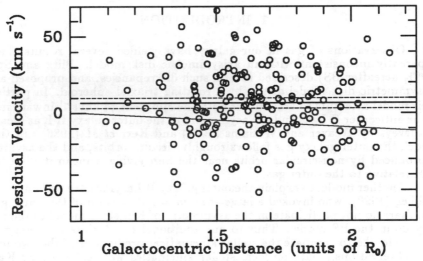

Fig. 1. Residual velocities of anticenter carbon stars plotted vs. galactocentric distance. The solid line is the mean velocity of the sample, and the short-dashed lines show 2 standard errors of the mean. The long-dashed line show the BS model prediction for mean velocities along gas orbits.

Residual velocities were computed from an axisymmetric rotation curve model with $\Theta_0 = 220$ km s^{-1} and $R_0 = 8.5$ kpc (Kerr and Lynden-Bell 1986), and corrected for the Sun's motion peculiar to the LSR from Delhaye (1965). The resulting velocities and distances are shown in Figure 1. There is no significant trend in velocity with distance, and individual quartiles of distance have mean velocities of 5.2 ± 3.3, 11.2 ± 3.6, 6.5 ± 3.3, and 3.6 ± 3.3 km s^{-1} at median distances of 3.4, 4.7, 6.0, and 7.7 kpc, respectively. The velocity dispersion for the sample is 23.1 km s^{-1}, and has no trend with distance.

We can immediately see a difference between the motion of the anticenter gas and the stars. While the outer H I is moving toward us at 14 km s^{-1}, the carbon stars are moving *away* at 6 km s^{-1}. This is consistent with Lewis and Freeman's (1989) K giant measurements, which show a similar average velocity. The BS prediction for the gas velocity is shown in Figure 1 as the slanted line, and if the stars and gas follow similar orbits, the BS model is ruled out at the 3-σ level.

At this conference, D. Spergel proposed that the motion of the carbon stars can be explained by an inner Lindblad resonance from the BS rotating potential which occurs just outside the solar orbit. A resonance large enough to produce a velocity difference of 20 km s^{-1} between the LSR and the outer galaxy should produce a measurable signature in our carbon star data, the most distant of which are not far from the stellar cutoff in the disk. If the mean motion of the stars changes abruptly across the resonance, as proposed, we should see evidence of this in the data near the resonance. The signature of a large shift in mean velocity in the presence of distance errors should appear as an increase in the velocity dispersion near the resonance, but no such increase is seen in our data. The BS model would need to be modified to move the resonance to a larger radius where we would not detect it by an increased dispersion, while remaining consistent with the mean velocity. See M. Weinberg's paper in these proceedings, which analyzes stellar orbits near a resonance in detail.

The new data also raise some old questions about the distribution of carbon stars. In contrast to Lewis and Freeman's (1989) K giant sample, the velocity dispersion of the carbon stars does not decrease with distance. If one adopts the assumption that σ_z/σ_R is constant in the old disk, the scale height of carbon stars would increase with radius as the total column density decreases. One could in principle look for this by comparing densities as a function of both distance and galactic latitude, however there are not enough carbon stars in this sample (even if it were complete) to perform such an analysis over the 300+ sqare degree area surveyed. We also find that we can reproduce the observed star counts with either a radially exponential disk or with a constant radial density as suggested by Jura *et al.* (1989). We need only change the assumption of the effective cutoff of the survey in K magnitude: this dominates the cutoff in the tail of the distribution where most of the information on the radial density gradient is present. A survey for carbon stars in the near-infrared going $\simeq 2 - 3$ mag fainter should be adequate to provide an answer to this question.

4. CONCLUSIONS

The BS model appears to have difficulty reproducing the observed carbon star distribution, at least without careful fine-tuning. Kuijken's (1991) model of an $m = 1$ distortion seems to more readily explain the lack of features in the carbon stars. However, it does not by itself acocunt for the 6 km s^{-1} mean velocity of the carbon stars. To address these models further, information from additional tracers is needed, particularly those with a lower intrinsic velocity dispersion. This will be an observational challenge, as most such potential tracers are fainter and more difficult to identify.

ACKNOWLEDGEMENTS

We thank L. Blitz, D. Spergel, M. Weinberg, and K. Kuijken for informative discussions. We also thank the staff at Kitt Peak National Observatory for their skilled assistance during our observing run. This work was supported by NSF grants AST 89-96139 and AST 90-15920.

REFERENCES

Aaronson, M., Blanco, V. M., Cook, K. H., Olszewski, E. W., & Schechter, P. L. 1990, ApJS, 73, 841

Blitz, L. & Spergel, D. N. 1991, ApJ, 370, 205 (BS)

Claussen, M. J., Kleinmann, S. G., Joyce, R. R., and Jura, M. 1987, ApJS, 204, 242

Delhaye, J. 1965, in Galactic Structure, eds. A. Blaauw and M. Schmidt (Chicago: U. of Chicago Press), p. 61

Fuenmayor, F. J. 1981, Rev. Mexicana Astron. Astrof., 6, 83

Jura, M., Joyce, R. R., & Kleinmann, S. G. 1989, ApJ, 336, 924

Kerr, F. J., Bowers, P. F., Jackson, P. D., & Kerr, M. 1986, A&AS, 66, 373

Kerr, F. J., & Lynden-Bell, D. 1986, MNRAS, 221, 1023

Kuijken, K. 1991, in Warped Disks and Inclined Rings around Galaxies, eds. S. Casertano, P. Sackett, and F. Briggs (Cambridge: Cambridge Univ. Press), p. 159

Kuijken, K., & Tremaine, S. 1991, in Dynamics of Disk Galaxies, ed. B. Sundelius (Göteburg, Sweden: Göteburg Univ.), p. 71

Lewis, J. R., & Freeman, K. C. 1989, AJ, 97, 139

Metzger, M. R., & Schechter, P. L. 1993, ApJ, submitted (MS)

Schechter, P. L., Aaronson, M., Blanco, V. M., & Cook, K. H. 1991, AJ, 101, 1756

Weaver, H. F., & Williams, D. R. W. 1973, A&AS, 8, 1

REMOTE YOUNG STELLAR GROUPS AND THE ROTATION OF THE OUTER GALACTIC DISK

Luc Turbide[1] and Anthony F.J. Moffat
Département de physique, Université de Montréal, and Observatoire du
Mont Mégantic, C.P.6128, Succ "A", Montréal, QC, Canada, H3C 3J7

Email ID
turbide@astro.umontreal.ca and moffat@astro.umontreal.ca

ABSTRACT

We have obtained UBV CCD-magnitudes for stars in eight distant, young groups in the third Galactic quadrant. Comparison with theoretical isochrones yields ages and distances of significantly improved precision compared to previous studies. One of the best observed groups (in the faint HII region S 289) is located twice as far from the Galactic center as the Sun, at the outer limit of active star formation in the Galactic disk. Together with published radial velocities and allowing for a metallicity gradient, the new distances harden the evidence for a flat Galactic rotation curve out to $R \simeq 16$ kpc.

1. A PROBLEM OF DISTANCES

Is the rotation curve $\Theta(R)$ of the Galactic disk flat beyond the central few kpc, as one tends to find in external spiral galaxies (*e.g.* Rubin *et al.* 1977, 1978; Rubin 1979)?

The answer to this question - especially for the most remote parts of the disk - is clearly important in the content of dark matter (DM), since it may ultimately be that only our own Galaxy will turn out to be close enough to reveal the true physical nature of DM.

Although there is already a fair amount of evidence in favor of a flat Galactic rotation curve beyond $R \simeq 2$ kpc (Georgelin & Georgelin 1976; Jackson *et al.* 1979; Blitz 1979; Brand *et al.* 1988), the data are generally noisy and $\Theta(R)$ is poorly determined for $R > R_o$. The reason for this lies almost entirely with the errors in the **distances** of the **stars** necessary to map out unambiguously the rotation curve beyond the solar circle (although: see Merrifield's (1992) new method based on 21 cm: also in these proceedings). Radial velocities (RV; *e.g.* from narrow-line spectra of stars or associated nebulosity) are generally not a problem, since they can usually be determined to relatively high precision (*e.g.* Moffat 1988).

2. POSSIBLE REMEDY

In an attempt to improve the quality of the exterior rotation curve of the Galaxy, we have obtained more accurate distances of distant young stellar groups of luminous blue stars, using PSF-fitting CCD photometry and isochrone

---

Visiting astronomer, Univ. of Toronto Southern Observatory (UTSO), Las Campanas, Chile

matching in color-magnitude diagrams (CMDs). Such groups offer two clear advantages over single stars:

- isochrone fitting avoids systematic errors due to variations in age, which are difficult to detect for single stars, even when the spectrum is known.
- even without the above, a larger number of stars improves the overall error in the distance of the group.

A disadvantage is of course that more work is involved for each distance candle. Other useful objects include: Cepheid variables (Welch 1988), Planetary Nebulae nuclei (Schneider & Terzian 1983) and Carbon stars (Aaronson *et al.* 1990) but only on an individual basis. While Cepheids are quite useful for RVs and distances, they are relatively rare (*cf.* their progenitor B stars). The last two are considerably older than OB stars and Cepheids and are therefore subject to Galactic drift. We chose to study groups of luminous OB stars, that are easy to locate from their blue colors and especially if they excite visible HII or reflection nebulae on survey charts. The third Galactic quadrant ($180° \leq l \leq 270°$) is ideal for deriving $\Theta(R)$ beyond the solar orbit, because interstellar extinction is known to be relatively low out to large distances in that direction (*cf.* Moffat *et al.* 1979).

3. OBSERVATIONS

We have obtained UBV CCD photometry at UTSO for 8 of the most distant known anti-center groups of OB stars (mostly in HII regions), taken from previous works based on photoelectric diaphragm photometry (Moffat *et al.* 1975, 1979). A deeper search by Brand (1986) revealed no new objects that are known to be farther from the Galactic center.

Fig. 1 shows an example of the results for a typical region, S 283. Distance and age were determined for this and other regions by minimizing the χ^2 deviation between the observed group member stars in the CMDs and the solar metallicity ($Z_\odot$) isochrones of Maeder (1990, 1991), as a function of age, reddening (E_{B-V} and E_{U-B}), and distance modulus. Values of the ratio E_{U-B}/E_{B-V} so obtained were checked for consistancy from a few spectral types available in the literature for a few stars in 6 of the 8 regions: agreement was satisfactory, given the small number of spectra and the large cosmic scatter in M_V from individual spectral-luminosity classes.

4. ROTATION CURVE

Fig 2. shows the rotation curve for the whole Galactic disk beyond $R \approx 2$ kpc. For $R \leq R_o$, the tangential point HI data are from Burton & Gordon (1978); for $R \geq 2/3R_o$, the data are for OB stars mostly in HII regions (Fich *et al.* 1989). We show this for $Z = Z_\odot$ (for previous works only and our new 8 values replacing previous ones) and for $Z = Z(R)$ based on a nominal gradient for spiral galaxies (Belley & Roy 1992): $d \log Z/dR \approx -0.08$ kpc^{-1}.

5. CONCLUSIONS

Mainly on the basis of Fig. 2, we draw the following conclusions:

- CCD photometry allows significant improvement of distances (smaller error bars)
- The Galactic rotation curve remains flat (or slightly rising) at $\Theta_o \sim 220$

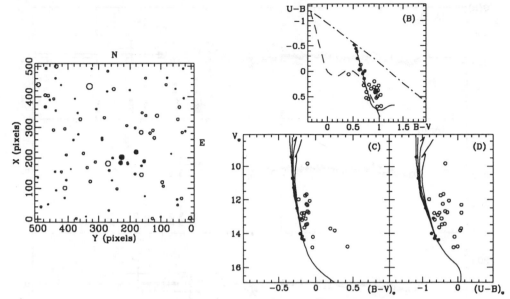

Fig. 1. Example for one of the eight regions, S 283. On the chart, 1 pixel= 0.45". Filled symbols refer to assigned members, open symbols to field stars. The four isochrones (Maeder 1990) refer to ages of 0,4,6 and 10 Myr.

km s^{-1} for $R > R_o$
- The outer limit of active star formation in the disk is at $R_l \sim 17$ kpc (for $R_o = 8.5$kpc, $Z = Z_\odot$)
- Allowing for $Z(R) = Z_\odot 10^{-0.08(R-R_o)}$ makes $\Theta(R)$ slightly flatter and $R_l \sim 16$kpc
- All 8 OB groups are young ($\leq 6 \times 10^6$ yrs mostly) but contain no very massive stars (early-mid O type), probably due to small number statistics, not a change in IMF slope.

ACKNOWLEDGMENTS

AFJM is grateful to NSERC (Canada) and FCAR (Québec) for financial assistance.

REFERENCES

Aaronson, M., Blanco, V. M., Cook, K. H., Olszewski, E. W., & Schechter, P. L. 1990, ApJS, 73, 841
Belley, J., & Roy, J-P. 1992, ApJS, 78, 61
Blitz, L. 1979, ApJ, 231, L115
Brand, J. 1986, The Velocity Field of the Outer Galaxy, Ph.D. Thesis, University of Leiden.
Brand, J., Blitz, L., & Wouteroot, J. 1988, in The Outer Galaxy, Symposium in Honor of F. Kerr, eds. L. Blitz and F.J. Lockman (New York, Springer),p.40

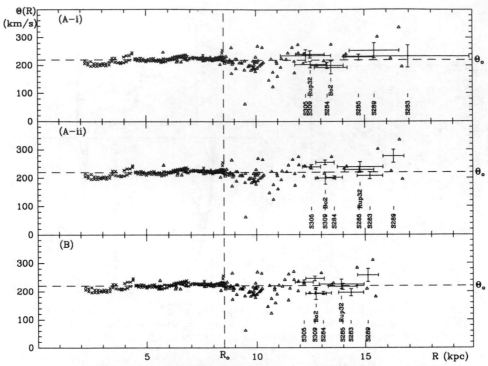

Fig. 2. Galactic rotation curve using HI data (×) for $R \leq R_o$ and OB groups ($\triangle$ or +) for $R \geq 2/3R_o$. (A) for $Z = Z_\odot$ (i) from previously published OB group data and (ii) the new 8 values replacing previous ones. (B) for $Z = Z_\odot 10^{-0.08(R-R_o)}$ for all OB groups, including new data.

Burton, W.B., Gordon, M.A. 1978, A&A, 63, 7

Fich, M., Blitz, L., Stark, A.A. 1989, ApJ, 342, 272

Georgelin, Y. M., & Georgelin Y. P. 1976, A&A, 25, 337

Jackson, P. D., FitzGerald, M. P., & Moffat, A. F. J. 1979, in The Large-Scale Characteristics of the Galaxy, IAU Symposium 84, ed. W.B. Burton (Dordrecht: Reidelt)p.221

Maeder, A. 1990, A&AS, 84, 139

Maeder, A. 1991, A&A, 242, 93

Merrifield, M.R. 1992, AJ, 103, 1552

Moffat, A. F. J., & Vogt, N. 1975, A&AS, 20, 85

Moffat, A. F. J., FitzGerald, M. P., & Jackson, P. D. 1979, A&AS, 38, 197

Moffat, A. F. J. 1988, in The Outer Galaxy, Symposium in Honor of F. Kerr, eds. L. Blitz and F.J. Lockman (New York: Springer)p.47

Rubin, V. C., Ford, W. K., & Thonnard, N. 1977, ApJ, 217, L1

Rubin, V. C., Ford, W. K., & Thonnard, N. 1978, ApJ, 225, L107

Rubin, V. C. 1979, in The Large-Scale Characteristics of the Galaxy, IAU Symposium 84, ed. W.B. Burton (Dordrecht: Reidelt),p.211

Schneider, S. E., & Terzian, Y. 1983, ApJ, 274, L61

Welch, D.W. 1988, The Mass of the Galaxy, ed. M. Fich,(Toronto: CITA),p.29

CHEMICAL HOMOGENEITY WITHIN THE
OLD OPEN CLUSTER M 67

Beth Hufnagel and Graeme H. Smith
University of California Observatories/Lick Observatory
Board of Studies in Astronomy and Astrophysics
University of California, Santa Cruz, CA 95064
beth@lick.ucsc.edu and graeme@lick.ucsc.edu

1. INTRODUCTION

Some models (Larson 1976; Burkert, Truran, and Hensler 1992) of galaxy disk evolution indicate that such disks form radially from the "inside-out". These models predict that, for a given age, stars in the outer disk will have a lower mean abundance than those in the inner disk. Open clusters of stars are valuable objects for testing such models because they can be observed both near the solar circle and in the outer disk. In addition, their overall chemical abundance can be determined quite accurately using photometric methods (Geisler 1987). Hence open clusters can be used as probes of the past chemical evolution of the interstellar medium (ISM) at different Galactocentric radii *if* they accurately reflect the chemical abundance of the ISM from which they were made. One way of testing if this condition holds is to investigate the degree of chemical homogeneity of the older open clusters.

M 67 is a well-studied (Janes 1984) Galactic open cluster that has a near-solar chemical abundance, is massive for an open cluster, and is about the same age as the Sun (Hobbs and Thorburn 1991; Demarque et al. 1992). If unevolved main sequence non-binary stars in M 67 show chemical homogeneity, then it can be inferred that the cluster was initially chemically homogeneous and did not undergo processes that could change its initial chemical abundance from that of its parent ISM. Janes and Smith (1984), however, have found that several red giants in M 67 show enhanced 4216 Å CN bands in their spectra. Interpretation of these enhancements is complicated by the observation by Brown (1987) that the surface carbon abundances of the M 67 giants have been altered by a combination of interior nucleosynthesis and mixing processes. These findings suggest that it is important to investigate the homogeneity of CN and CH bands among the M 67 dwarfs, whose atmospheric compositions are less likely to have been altered by internal evolutionary processes.

2. OBSERVATIONS AND RESULTS

We obtained spectra of five main sequence M 67 dwarfs having colors in the range $0.58 \leq (B - V)_0 \leq 0.68$ using the UV Schmidt cassegrain spectrograph on the Lick Shane 3m telescope in January 1992 during a single night. In addition, we observed thirteen main-sequence dwarfs in the Pleiades cluster. These spectra have a resolution of 3 Å and cover the wavelength range 3600 Å to 4600 Å.

We present measurements made from these spectra of three indices defined following Norris and Freeman (1979) and Janes and Smith (1984) that quantify the strengths of the 3883 Å violet CN, the 4216 Å blue CN, and the 4300 Å

CH (G) bands. These indices, denoted S(3839), S(4142) and S_{CH} respectively, are defined such that an enhancement in molecular band strength produces an increase in the value of the associated index.

The reduced (wavelength and flux calibrated) spectra are presented in Figure 1. Figure 2 shows plots of all three indices versus $(B - V)_0$ for both the M 67 and Pleiades cluster stars. The $(B - V)$ colors for the M 67 stars were taken from Gilliland et al. (1991); the reddening from Nissen et al. (1987). The $(B - V)$ colors and reddenings for the Pleiades dwarfs were obtained from Klemola (1993).

The spectra of the observed M 67 dwarfs exhibit very similar CN and CH band absorption, and their 4216 Å CN and 4300 Å CH indices are of similar value to within ~ 0.04. The M 67 dwarfs have values of these two indices that are nearly identical to those of Pleiades dwarfs of comparable color. The five M 67 dwarfs observed show similar S(3839) values, but their mean S(3839) value is greater than the mean value for Pleiades dwarfs of comparable color. Our data give no evidence that the stars in our M 67 sample are not homogeneous with regard to CN and CH. However, in view of our small sample size, it is premature to draw any firm conclusions as to the chemical homogeneity of M 67 as a whole. Consequently, we would like to increase the sample of M 67 dwarfs observed to ~ 20 to 25 stars and quantify the uncertainty in the M 67 index values from further observations.

3. CONCLUSIONS

Additional evidence for the chemical homogeneity of unevolved stars in open clusters is provided by the [Fe/H] and [C/H] measurements for F dwarfs in Coma, Praesepe, and M 67 by Friel and Boesgaard (1992), which show no evidence for intrinsic intracluster dispersion. The available data provide encouragement that open clusters can be used to study the chemical evolution of the Galactic disk and that their chemical abundances reflect that of the ISM at the time of their formation. It is well to bear in mind, however, that the study of the chemical homogeneity of main sequence stars in old open clusters is still in its infancy.

REFERENCES

Brown, J. A. 1987, ApJ, 317, 701
Burkert, A., Truran, J. W., & Hensler, G. 1992, ApJ, 391, 651
Demarque, P., Green, E. M., & Guenther, D. B. 1992, AJ, 103, 151
Friel, E. D., & Boesgaard, A. M. 1992, ApJ, 387, 170
Geisler, D. 1987, AJ, 94, 84
Gilliland, R. L., et al. 1991, AJ, 101, 541
Hobbs, L. M., & Thorburn, J. A. 1991, AJ, 102, 1070
Janes, K. A. 1984, in Calibration of Fundamental Stellar Quantities, IAU Symposium No. 111, edited by D. S. Hayes, L. E. Pasinetti, and A. G. Davis Philip (Reidel, Dordrecht), 361
Janes, K. A., & Smith, G. H. 1984, AJ, 89, 487
Klemola, A. 1993, in preparation
Larson, R. B. 1976, MNRAS, 176, 31
Nissen, P.E., Twarog, B. A., & Crawford, D. L. 1987, AJ, 93, 634
Norris, J., & Freeman, K. C. 1979, ApJ, 230, L179

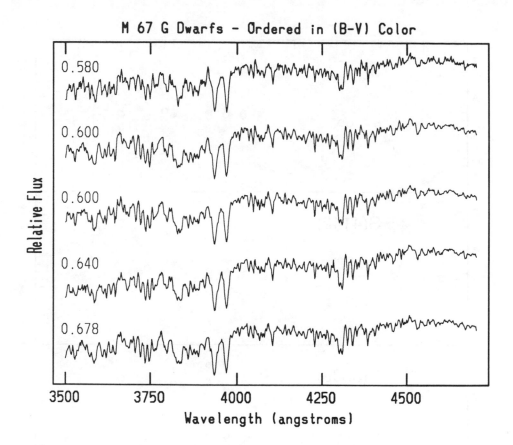

Fig. 1. Spectra of five M 67 G dwarfs, labelled with the $(B - V)_0$ color of each star. The spectra show very similar CN and CH band absorption strengths.

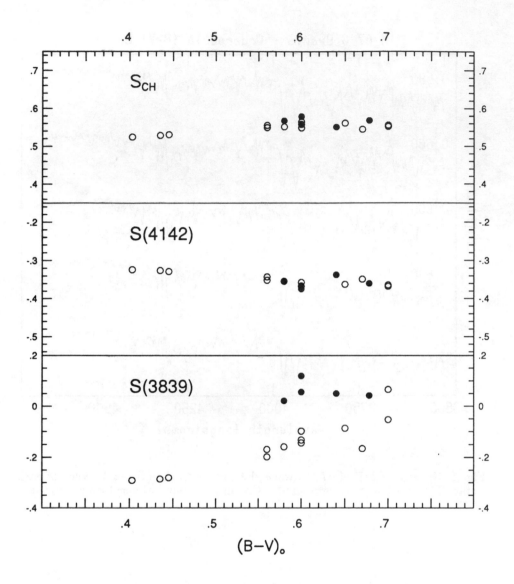

Fig. 2. Plots of CN and CH indices versus $(B - V)_0$ for five M 67 G dwarfs
(filled circles), and thirteen comparison Pleiades dwarfs (open circles).

A SENSITIVE SURVEY IN ^{12}CO J = 1 → 0
OF ARM AND INTERARM MOLECULAR CLOUDS
IN THE OUTER GALAXY

Sean J. Carey, Marc L. Kutner
Physics Department, Rensselaer Polytechnic Institute, Troy, NY 12180.

Kathryn N. Mead
Physics Department, Union College, Schenectady, NY 12308.

ABSTRACT

The effects of environment on the star formation process are currently not well understood. By studying the cloud populations in two different environments, spiral arms and the interarm regions, we hope to gain insight on the effects of environment in the star formation process. To this end, we are currently conducting a sensitive, 60" sampled survey of a one square degree region of the outer Galaxy in the J = 1 → 0 transition of ^{12}CO. Our survey region spans galactocentric radii between 9 and 16 kpc. Two populations of molecular clouds, arm and interarm, have been identified. The spiral arm population consists of GMC's, with masses of 10^4 to $10^5 M_\odot$, and smaller clouds, clustered at velocities indicative of the 13 kpc spiral arm. The interarm clouds are dispersed throughout the remainder of the survey volume. The interarm population is comprised of small clouds with masses less than $10^4 M_\odot$.

1. INTRODUCTION

The outer Galaxy is a fantastic laboratory for studying spiral structure. The lack of a distance ambiguity permits the assignment of unique kinematic distances to molecular material. The surface density of molecular material is lower in the outer Galaxy; consequently, there is less velocity blending permitting the identification of individual clouds. Also, spiral arms are less tightly wound for large galactocentric radii making identification of arm and interarm regions using molecular tracers easier.

Understanding the nature and environment of molecular clouds is a central issue in studying star formation. Cloud-cloud collisions and interactions with spiral density waves are two examples of environmental conditions that impact on the star formation process. Examining the nature of cloud populations in arm and interarm regions provides information on the effects of cloud environment on the star formation process.

Previous work in the outer Galaxy can be divided into two broad categories: searches for individual clouds and surveys of large regions. Typically, searches for individual molecular clouds are biased toward large star forming clouds [Mead et al.(1988), Wouterloot & Brandt (1990)]. Previous surveys [Sanders et al.(1986), Cohen & Thaddeus (1977) and Digel et al.(1992)] are unbiased, but either have poor angular resolution or are undersampled. Small clouds will not be detected because they are either significantly beam diluted (low angular resolution) or fall between beams (undersampling). Both the biased and unbiased observations underestimate the number of small clouds.

Our current work is a sensitive survey of a velocity slice in the outer Galaxy. We are able to identify individual medium ($10^4 M_\odot$) and small ($\sim 10^3 M_\odot$) clouds as well as examining the arm/interarm populations of the survey region.

2. OBSERVATIONS

Our survey consists of a $1° \times 1°$ region centered around $l = 76.5°$, $b = 1.5°$. This region was selected as it contains the most continuous HI "outer arm" feature in the first quadrant [Henderson *et al.*(1982)], and is likely to have the highest contrast between arm and interarm regions. We are using the NRAO 12m telescope* to observe the $J = 1 \rightarrow 0$ of ^{12}CO. We are conducting position switched observations with the new 3mm SIS receiver and a backend of two 250 kHz filterbanks operating in parallel. Our temperature sensitivity (1σ rms) for each position is .1 K on the ΔT_R^* scale. The velocity coverage is from -80 to -20 km/s. Assuming a flat rotation curve, this velocity range translates to galactocentric radii between 9 and 16 kpc. The sample spacing is 60" which corresponds to a linear resolution of 2-5 parsecs for the range of kinematic distances covered. The spectral resolution is .65 km/s, permitting observations of narrow line sources. All spectra were manually processed and any anomalous spectra were discarded. Linear baselines were subtracted from all but a handful of positions. After three observing runs (June 1991, Dec 1991, June 1992), 60% of the survey region has been mapped. The preliminary results from these runs are discussed in the next section.

3. RESULTS

To date, 173 clouds have been detected. A cloud consists of one or more adjacent detections in $l,b,$v space. A signal, 3σ or better in two adjacent channels and above zero in four adjacent channels, constitutes a detection. Formally, a detection is equivalent to a 4.5σ line. For the approximately 2000 positions observed, we expect fewer than 5 false detections. The weakest detections will be confirmed with additional observations.

Virial masses were calculated for the 27 clouds consisting of 5 or more contiguous positions ($N > 4$). A typical mass for the clouds in this sample is $10^4 M_\odot$. For smaller clouds, virial masses cannot be reliably determined as the radii of objects that only fill a beam or two are highly uncertain. A histogram of cloud masses is shown in Figure 1. The cloud masses are typical of giant molecular clouds (GMC's) in the outer Galaxy [Mead (1988)]. The total mass in the survey region is dominated by the most massive clouds.

The distribution of clouds with galactocentric radius (R) is shown in Figure 2. The distribution of large clouds ($N > 4$) is strongly peaked at 12.5 kpc. We associate this peak with the 13 kpc or "outer" arm. The width of the arm is ~ 2 kpc. It is interesting that there are few large clouds in the interarm region. Despite the small sample size, the peak in the large cloud distribution is statistically significant (6σ deviation from a random distribution). It is unlikely that this peak is an artifact created by errors in distance determination. Systematic uncertainties in the kinematic distances of the clouds would shift the centroid of the peak but not change the nature of the distribution. Random uncertainties in the cloud distances would broaden the peak and make the arm distribution less "armlike".

The distribution of small clouds can be described by three components: local clouds, the 13 kpc arm and an interarm background. The number density of clouds in the arm is 1.3 times the average number density for the galactocentric radii covered. The interarm population is a diffuse scattering of small clouds throughout the entire survey volume. The local peak is probably a selection effect due to increased sensitivity to nearby material. In addition, many probable detections $(2 - 3\sigma)$ exist over the entire survey region. These results suggest a background of small and/or cold molecular clouds. More sensitive observations are needed to determine the extent of the very small and/or weak emission clouds.

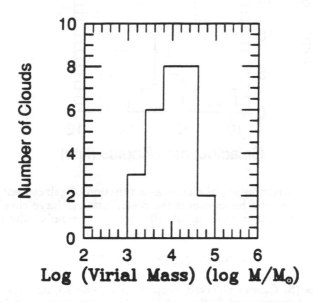

Figure 1. Virial Mass distribution for the large clouds. The bin size is $.4 \times \log(M/M_\odot)$.

4. SUMMARY AND FUTURE WORK

We are currently conducting a sensitive survey of a square degree slice through the outer Galaxy. We have identified two clouds populations, arm and interarm (diffuse). The interarm population consists of small clouds (radii < 5 pc). The arm component is composed of large clouds (typical GMC's) and an enhancement of 30% in the number of small clouds. The molecular arm is centered at $R \sim 12.5$ kpc. Most of the mass in the survey volume is contained in the most massive clouds.

In the future, we will complete the survey (expected completion date November 1992). With improved statistics, we will further explore the nature of the cloud populations. In addition, fully sampled observations will be made to determine the properties of small clouds.

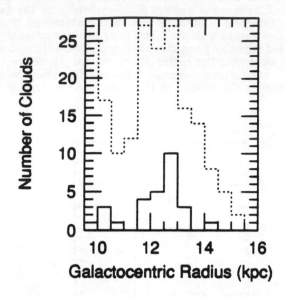

Figure 2. Distribution of clouds as a function of galactocentric radius. The solid histogram is the distribution of large clouds. The dotted histogram is the distribution of all clouds. The bin size is .5 kpc.

ACKNOWLEDGEMENTS

One of us (S. Carey) received support from the student travel support program of the Astronomical Society of New York and was aided by a Grant-in-Aid of Research from the National Academy of Sciences, through Sigma Xi, The Scientific Research Society.

REFERENCES

Cohen, R. S., & Thaddeus, P. 1977, ApJ, 217, L155
Digel, S., Bally, J., & Thaddeus, P. 1992, ApJ, 357, L29
Henderson A. P., Jackson P. D., & Kerr, F. J. 1982, ApJ, 263, 116
Mead, K. N., & Kutner, M. L. 1988, ApJ, 330, 399
Sanders, D. B., Clemens, D. P., Scoville, N. Z., & Solomon, P. M. 1986, ApJS, 60, 1
Wouterloot, J. G. A., & Brandt, J. 1980, A&AS, 80, 149

PROPERTIES OF MOLECULAR CLOUDS IN THE EXTREME OUTER GALAXY

Seth Digel, Patrick Thaddeus

Center for Astrophysics, MS 72, 60 Garden St., Cambridge, MA 02138

Eugène de Geus

University of Maryland, Laboratory for Millimeter Wave Astronomy, College Park, MD 20742

ABSTRACT

We present a study of two molecular clouds with kinematic Galactocentric distances of 22 and 28 kpc ($R_\odot$ = 8.5 kpc). These are several kiloparsecs further from the Galactic center than any previously known and apparently lie beyond the outermost massive star-forming regions. We compare the physical properties of the clouds with molecular clouds of similar size near the solar circle, e.g., those in Taurus and Orion. Observations with the CfA 1.2 m and NRAO 12 m telescopes indicate that these distant clouds have sizes of 40–70 pc, molecular masses of 4–7 x 10^4 $M_\odot$, and velocity widths 2–3 km s^{-1}. They contain dense molecular clumps of sizes 7–10 pc and kinetic temperatures 15–25 K. The distant clouds are similar to molecular clouds near the solar circle in all the properties we have determined. Their primary difference is an apparent lack of embedded *IRAS* infrared sources.

I. INTRODUCTION

This poster presents observations of two newly-discovered molecular clouds that are apparently beyond the star-forming disk of the Galaxy. They are the most distant among seven found with the CfA 1.2 m telescope in a CO $J = 1 \rightarrow 0$ survey of H I complexes at large Galactocentric distances (Digel, de Geus, & Thaddeus 1992). Follow-up observations of both clouds at the NRAO 12 m have been obtained.

The most distant clouds found in the CfA survey, Clouds 1 and 2 have kinematic Galactocentric distances R = 22 kpc and 28 kpc, respectively. They are several kiloparsecs beyond the most remote star-forming regions known (Wouterloot & Brand 1989) and much further from the Galactic center than the observed extent of the stellar disk (Chromey 1978, Robin, Creze, & Mohan 1992). At such distances the total surface density of gas, the metallicity, the intensity of the interstellar radiation field, and the density of cosmic rays are all likely to be significantly less than those near the solar circle. The properties of molecular clouds at the unfamiliar edge of our system are clearly of interest and may bear on the study of molecular clouds in external galaxies.

II. OBSERVATIONS

The association of Clouds 1 and 2 with atomic gas is evident in the maps from the CfA 1.2 m (Fig. 1). Both clouds lie along peaks or ridges of H I column density. The linear resolution of the CfA data is only 40–50 pc, and that of the H I poorer still, so little about the detailed correspondence of the molecular and atomic gas can be inferred.

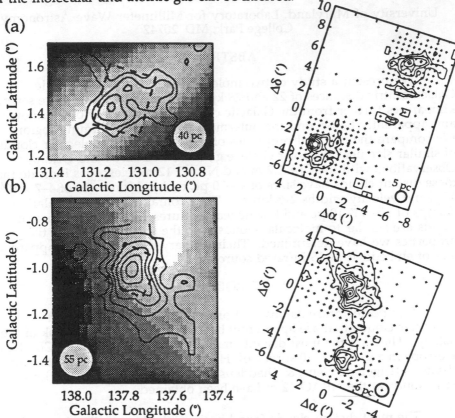

Figure 1. Contours of CO emission from the distant clouds observed with the CfA 1.2 m telescope overlaid on gray scale representation of the H I column density at that distance. Also shown are the higher resolution CO maps from the NRAO 12 m telescope of the central parts of the clouds. For both clouds the CO data are integrated over the velocity range $v_{lsr} = -107$ to -98 km s^{-1}; the angular resolutions are indicated. The H I data, from the Maryland–Green Bank survey (Westerhout & Wendlant 1982), are integrated from -112 to -93 km s^{-1}. (*a*) Cloud 1; the gray scale corresponds to the range $N(H\ I) = 3.0$–4.7×10^{20} cm^{-2} (blackest–whitest), the CfA data are contoured at 0.12 K km s^{-1} intervals from 0.12 K km s^{-1} (approximately 2 σ steps from 2 σ), and the contours for the NRAO data are 2x1 K km s^{-1}. (*b*) Cloud 2; the gray scale corresponds to $N(H\ I) = 2.2$–6.3×10^{20} cm^{-2}, the CfA contours are 0.24x0.24 K km s^{-1} (approximately 2 σ steps from 2 σ), and the NRAO contours are 4x2 K km s^{-1}.

Higher resolution CO observations at the NRAO 12 m telescope resolve the clouds into multiple clumps of size comparable to local molecular cloud complexes such as those in Orion and Taurus (Fig. 1 and Table 1). The main distinction in the table between the distant and nearby clouds is molecular mass. The masses of Clouds 1 and 2 may be less than those of Taurus and Orion, but the mass calibration in the distant outer Galaxy is essentially unknown and the CO luminosity per unit molecular mass is likely to be less there than here (Mead & Kutner 1988, Digel, Bally, & Thaddeus 1990, Sodroski 1991). The inferred kinetic temperatures of the molecular gas in both Clouds 1 and 2 indicate warm clouds, similar in temperature to the molecular clouds in Orion and Taurus and warmer than the ~8 K temperatures reported by Mead & Kutner (1988) for clouds on $R = 11$–14 kpc. Because the balance of the heating and cooling rates depends on many unknown quantities, little more can be said about the gas kinetic temperatures of Clouds 1 and 2, although their temperatures may be taken as some indication that at least Cloud 2 has an internal source of energy.

TABLE 1

Position[a] (l,b)	v_{lsr}[b] (km s^{-1})	R[c] (kpc)	D[d] (kpc)	r[e] (pc)	Δv[b] (km s^{-1})	M_{CO}[f] ($10^3\,M_\odot$)	T_k[g] (K)
Cloud 1 131°.05, 1°.45	-101.8	22 ± 2	16 ± 3	36 ± 6	2.8	7.5	13 ± 5
Cloud 2 137.75, -1.00	-102.4	28 ± 4	21 ± 4	20 ± 4	2.2	37	22 ± 7
Orion (A&B)			0.5	~35	7	~200	~25
Taurus			0.1	~25	4	~50	~10

Notes: Data for Orion are from Maddalena & Thaddeus (1985) and for Taurus from Ungerechts & Thaddeus (1987).

[a] Emission-weighted mean.

[b] From a Gaussian fit to the cloud's composite spectrum; Δv is the FWHM, corrected for channel width.

[c] Kinematic Galactocentric distance, derived on the assumption of a flat rotation curve with $R_\odot = 8.5$ kpc, $V_\odot = 220$ km s^{-1}. The uncertainties are estimated from the effect of a ± 7 km s^{-1} non-circular velocity component.

[d] Distance.

[e] $(A/\pi)^{1/2}$, where A is the area at a distance D *within the half-maximum contour of* W_{CO} *for the velocity range of the cloud's CO emission*, corrected for the telescope beam.

[f] Cloud mass derived from CO luminosity on the assumption $N(H_2)/W_{CO}$ $= 2.3 \times 10^{20}$ cm^{-2} (K km s^{-1})$^{-1}$ (Strong et al. 1988), with an additional factor of 1.4 for helium.

[g] Kinetic temperature of the molecular gas. For Clouds 1 and 2, T_k is derived from an LVG model of CO and CO (2–1) intensities using a CO column density consistent with ^{13}CO observations at the NRAO 12 m telescope. For Orion and Taurus, T_k is derived from the peak intensity in the CO surveys.

The properties of the distant clouds are comparable to those of well-known molecular cloud complexes near the solar circle. The CO-derived masses in Table 1 show that Cloud 1 is markedly fainter than the molecular clouds complexes in either Taurus or Orion, while Cloud 2 is approximately as luminous as Taurus but still much fainter than Orion. The narrower CO line profiles of Clouds 1 and 2 are consistent with somewhat lower masses if the clouds are gravitationally bound.

No sources are apparent in the *IRAS* data for either Cloud 1 or 2, although they are so remote that only massive stars embedded in the clouds would be detected. In the *IRAS* point source catalog, the sensitivity limit for a good detection in the 100 μm band, approximately 3 Jy, corresponds to a luminosity of ~700 $L_\odot$ at the 16 kpc distance of Cloud 1, or a single star of spectral type B4 (Panagia 1973), on the assumption of a color temperature of 25 K. By way of comparison, none of the embedded sources in Taurus would be seen by *IRAS* from 16 kpc distance and only the one or two most luminous in Orion would be detected.

III. CONCLUSIONS

The observation of molecular clouds at R = 22–28 kpc extends by several kiloparsecs the known extent of molecular gas in the Milky Way. The clouds at these extreme Galactocentric distances are apparently not greatly different in size, mass, and temperature from those near the solar circle. There is no evidence for star formation in either cloud, although the sensitivity limits of the *IRAS* data exclude only fairly early type stars. The absence of early star formation in molecular clouds at great distances from the Galactic center is apparently not due to a gross difference between these objects and star-forming molecular clouds near the solar circle.

REFERENCES

Chromey, F. R. 1978, AJ, 83, 162.
Digel, S., Bally, J., & Thaddeus, P. 1990, ApJ, 357, L29.
Digel, S., de Geus, E. J., & Thaddeus P. 1992, BAAS, 24, 817.
Maddalena, R. J., & Thaddeus P. 1985, ApJ, 294, 231.
Mead, K. N., & Kutner, M. L. 1988, ApJ, 330, 399.
Panagia, N. 1973, AJ, 78, 929.
Robin, A. C., Creze, M., & Mohan, V. 1992, ApJ, 400, L25.
Sodroski, T. J. 1991, ApJ, 366, 95.
Strong, A. W., et al. 1988, A&A, 207, 1.
Ungerechts, H., & Thaddeus, P. 1987, ApJS, 63, 645.
Westerhout, G., & Wendlant, H.-U. 1982, A&AS, 49, 143.
Wouterloot, J. G. A., Brand, J. 1989, ApJS, 80, 149.

FORMATION OF MASSIVE STARS
AT THE EDGE OF OUR GALAXY

Eugène J. de Geus
Astronomy Department, University of Maryland, College Park, MD 20742

ABSTRACT

Preliminary results are presented of a high-resolution radio-continuum and molecular line study of high-mass star forming regions in the outer parts of our Galaxy.

1. INTRODUCTION

Extensive studies have made it clear that all star formation in our Galaxy takes place in molecular clouds, with the most massive (O and B-type) stars being formed in Giant Molecular Clouds (GMCs), which have typical masses $\sim 2 \times 10^5$ $M_\odot$, sizes $\sim$50 pc, and average densities of ~ 100 cm^{-3} (Blitz 1978; Sanders, Solomon, and Scoville 1984). Most of the observational work on GMCs and massive star formation has concentrated on regions in the solar neighborhood and in the inner Galaxy. However, to obtain a complete picture of molecular clouds and star formation in the Milky Way, and to be able to use our locally gathered knowledge for interpretation of observations in external galaxies, we need to understand cloud- and star formation in a physical and chemical environment which is different from that in the inner parts of the disk. The ideal locations for such a study are the massive-star forming clouds at the edge of our own Galaxy. Their galactocentric radius is large enough for the physical environment to be significantly different, e.g. molecular clouds are more sparsely distributed (Wouterloot et al. 1990), the diffuse galactic interstellar radiation field is weaker, the metallicity is lower (Shaver et al. 1983, Fich and Silkey 1991), and the cosmic-ray flux is down (Bloemen et al. 1984), and yet these objects are near enough to resolve structures as small as 0.2 pc diameter cores.

The goal of this study is to establish the presence of high-mass star formation at large galactocentric radii, to determine the properties of the star forming molecular clouds down to the scales of the star forming cores and compare these to nearby molecular clouds (e.g. Orion), and to use the HII regions to obtain measurements of the metallicity out to much larger galactocentric radii than possible so far.

2. SOURCE SELECTION AND OBSERVATIONS

We used the Wouterloot and Brand (1989, WB) catalog of CO observations in the second and third Galactic quadrants toward IRAS point sources with colors indicating the presence of H_2O masers and dense molecular cloud "cores"

(Wouterloot and Walmsley 1986) and hence star forming regions. Distances quoted in that catalog (and used in our study) are based on a flat rotation curve, $R_0 = 8.5$ kpc, and $\Theta_0 = 220$ km s^{-1}. We selected sources which have $L_{FIR} > 10^4 L_\odot$ (indicating the presence of stars with spectral type B1 or earlier), galactocentric radii $R_{G.C.} > 15$ kpc, and $|l - 180°| > 15°$. This results in 31 sources, of which 23 are easily observable with existing mm-interferometers (i.e. they are in northern hemisphere).

Preliminary results on the comparison between VLA 6-cm continuum data and IRAS data are presented in the next sextion. For the interpretation of the VLA data we used the model by Mezger and Henderson (1967) with an electron temperature of 10^4 K, which is likely to be appropriate for these objects at large $R_{G.C.}$ (Shaver *et al.* 1983, Fich and Silkey 1991). BIMA interferometer observations of CO and CS (2-1) for two objects are discussed in section 4.

3. COMPARISON OF IR AND RADIO CONTINUUM LUMINOSITIES

Eleven sources were observed at 6-cm, of which 7 objects were detected and 4 were resolved. From the flux densities we have derived N_L, the number of ionizing photons required (Matsakis *et al.* 1976). Using the tables of Panagia (1973), we then derived the spectral type of the most massive single star required to produce this photon flux resulting in a range of O7.5 to B2 (main sequence). Figure 1 shows the far-IR luminosity (from WB) plotted against N_L. In the limit that one star produces all the photons necessary for the ionization of the gas ($\Rightarrow$ the continuum flux) and the heating of the dust ($\Rightarrow$ the far-IR luminosity, L_{FIR}), a simple model can be calculated for the relation between N_L and L_{FIR} as a function of spectral type of the star (see also Mead, Kutner and Evans 1990): $N_L \propto L_{Ly\alpha,star}$, and $L_{FIR} \propto L_{tot,star} - L_{Ly\alpha,star}$. The dashed line in figure 1 shows this model in the limit that the proportional sign is an equal sign in both equations. This is likely to be realistic for the ionizing flux, but less so for the part of the radiation heating the dust: a fraction of the radiation is expected to heat the dust, the rest will escape the region. A more realistic scenario would tend to move the model line to the left. In the case that more than one star is responsible for the heating and ionization, the curve moves up and toward the right. The dotted line at the model point for spectral type B2 indicates the extent of the shift in the case that two B2 stars would be present. The resolved continuum sources lie very close to the model, indicating that a single star could be sufficient to explain the observations. The unresolved continuum sources (indicated by the upward arrows) are consistent with the presence of one or two early-type stars. The sources that were not detected at 6-cm (downward arrows) can only be explained by the presence of a cluster of stars with spectral types later than B2.

4. INTERFEROMETER OBSERVATIONS

S127 is an optical HII region with $R_{G.C.} = 15$ kpc and $R_{hel} = 11.5$ kpc. The IRAS far infrared luminosity of this source is $10^5 L_\odot$, which indicates the

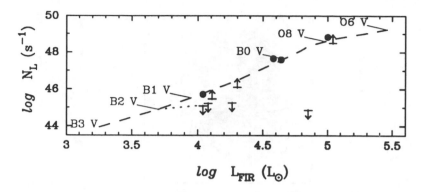

Figure 1. Plot of the number of ionizing photons calculated from the VLA 6-cm continuum flux (Matsakis *et al.* 1976) as a function of the far-IR luminosity. Sources detected and resolved at 6 cm are indicated by filled dots (•), sources detected but unresolved are shown by a bar with an upward arrow. Sources undetected at 6-cm are plotted at the N_L corresponding to the 1 σ rms noise level (with a downward arrow). The dashed line indicates the simplified relation between N_L and L_{FIR} for main sequence stars between B3 and O6.

presence of the equivalent of an O7 main sequence star. The two strongest peaks in the 6-cm observations indicate Ly-α photon production rates of 1.5 and 5 $\times 10^{48}$ s^{-1}, which implies the presence of one O9 V and one O7.5 V star (cf. Felli and Harten 1981 from Westerbork observations). The radio continuum and far-IR luminosities are therefore in good agreement.

A large scale CO map of the molecular cloud associated with this HII region has been made with the Kitt Peak 12-m. From the single dish observations we derive a CO luminosity of $3.3\,10^4$ K km s^{-1} pc^2, which, using the standard CO-to-H$_2$ conversion factor, gives a molecular mass for S 127 of $1.5\,10^4$ $M_\odot$, implying a small GMC. Figure 2 shows the HII region with an overlay of the BIMA interferometer CO map. The mm-interferometer data have been combined with the Kitt Peak 12-m data in order to obtain the zero-spacing information normally missing from interferometer maps. The necessity of this procedure is clear from the fact that the flux seen in the interferometer maps only was about 30% of the total. The two main peaks in the continuum are evidently associated with peaks in the CO distribution. A third peak in CO is devoid of continuum radiation, and might be a good candidate to look for evidence of new signs of star formation. The interferometer observations show clump radii between 0.12 (beam size) and 0.75 pc. We are currently undertaking a full analysis of the clump properties such as sizes, linewidths and CO luminosities (de Geus et al. in prep.).

WB 380 (#380 in Wouterloot and Brand 1989), at a distance of 10.7 kpc from the Sun (R$_{G.C.}$ = 17 kpc) has a far-IR luminosity of $1.1\,10^5$ $L_\odot$, which indicates the presence of an O7 V star. The source does not have a known optical counterpart, which may be due to extinction in the foreground and local to the source. The continuum source is unresolved in right ascension, and marginally

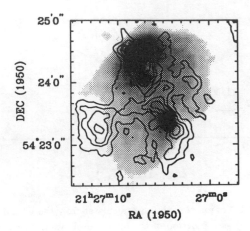

Figure 2. High-resolution image of the S 127 region. 6-cm continuum radiation shown as grey scale, peak flux is 21 Jy/beam. The contour overlay shows the BIMA interferometer CO observations, contour levels: 10,15,...,65 Jy/beam.

resolved in declination. From this, we can derive an upper limit to the source diameter of 0.2 pc (compact or ultracompact) and a lower limit to the Ly-α photon production rate of $3.2\,10^{48}$ s^{-1}, which is equivalent with the presence of an O8 V star or earlier. Given the uncertainty in the size of the 6-cm source, the continuum and far-IR luminosities are in good agreement. The continuum source is well correlated in position with peaks in the interferometer maps of CO and CS. This suggests that this object is in a very early stage of its evolution. The high-resolution interferometer observations show the presence of sub-structure ("clumps") with radii of 0.8 pc down to about 0.1 pc. A full analysis of the clump properties is being carried out.

EdG acknowledges financial support from NSF grant AST-8918912 and travel support for the observations through a research grant funded by the Margaret Cullinan Wray Charitable Lead Annuity Trust.

REFERENCES

Blitz, L., 1978. Ph.D. Dissertation Columbia University.

Bloemen, J.B.G.M., et al., 1984. A& A, 135, 12.

Felli, M., & Harten, R.H., 1981. A& A, 100, 28.

Fich, M., & Silkey, M., 1991. Ap. J. , 366, 107.

Lada, E.A., 1990. Ph.D. Dissertation University of Texas at Austin.

Mezger, P.G., & Henderson, A.P., 1967. Ap. J., 147, 471.

Sanders, D.B., Solomon, P.M., & Scoville, N.Z., 1984. Ap.J., 276, 182.

Shaver, P., McGee, R.X., Newton, L.M., Danks, A.C.,& Pottasch, S.R., 1983. MNRAS, 204, 53.

Wouterloot, J.G.A., Brand, J., Burton, W.B., & Kwee, K.K., 1990. A& A, 230, 21.

Wouterloot, J.G.A., & Brand, J., 1989. A& A Suppl., 80, 149.

Wouterloot, J.G.A., & Walmsley, C.M. 1986. A& A, 168, 237.

DIRBE EVIDENCE FOR A WARP IN THE GALAXY

H.T. Freudenreich[1], T.J. Sodroski[2], G.B. Berriman[3], E. Dwek[4]

B.A. Franz[2], M.G. Hauser[5], T. Kelsall[4], S.H. Moseley[4]

N.P. Odegard[3], R.F. Silverberg[4], G.N. Toller[3], and J.L. Weiland[3]

[1]Hughes STX, Code 685.9, NASA Goddard Space Flight Center
[2]Applied Research Corp., Code 685.3, NASA Goddard Space Flight Center
[3]General Sciences Corp., Code 685.3, NASA Goddard Space Flight Center
[4]Code 685, NASA Goddard Space Flight Center
[5]Code 680, NASA Goddard Space Flight Center, Greenbelt MD 20771

ABSTRACT

A large-scale warp in the surface brightness distribution of the Galactic disk has been found in the data of the Diffuse Infrared Background Experiment (DIRBE) of the Cosmic Background Explorer. In the far-infrared it is seen in seen in thermal emission by dust. In the near-infrared it is seen in starlight. The far-infrared warp matches that of the velocity-integrated HI. The near-infrared warp has approximately the same orientation, but has a smaller amplitude, which which increases with increasing wavelength. This is consistent with the stellar disk of the Galaxy being bent in the same way as the HI layer.

1. INTRODUCTION

While more attention is usually paid to the center of the Galaxy, the fact that the gravitational potential of the outer Galaxy is dominated by dark matter makes its structure of great interest (Binney 1992). It has long been known that the HI layer of the outer Galaxy, like that of many other galaxies, is warped (Burke 1957, Kerr 1957). It rises above the nominal plane in the direction of $l \sim 90°$ and falls below it at $l \sim 270°$ (Burton & Deul 1987). Molecular clouds and cool ($\sim 20K$) dust appear to exhibit the same behavior (Wouterloot et al. 1990, Sodroski et al. 1987), but the distribution of stars in the outer Galaxy is obscured by extinction. A warping of the gas and dust layer does not necessarily require a warping of the stellar disk. The warp may be a young phenomenon, confined to recently accreted gas and dust, or it may commence beyond the edge of the stellar disk.

Evidence for a warping of the population I stellar component has been obtained through studies of Cepheid and OB-type stars, and studies of WR stars, open clusters and SNR's (e.g. Miyamoto et al. 1991, Pandey et al. 1990). However, these objects, bright enough to be seen at great distances, are young enough to have formed within an already-warped gas layer.

Djorgovsky and Sosin (1989) have found relatively old stellar tracers in the form of (mostly) AGB and post-AGB stars of the IRAS Point Source Catalog at 12 and 25 μm, wavelengths at which extinction is negligible and these dust-shrouded stars are highly luminous. Djorgovsky and Sosin find a warp similar to that of the HI layer, though of lesser amplitude, a fact they tentatively ascribe to radial truncation of the stellar disk.

The DIRBE is a photometer with ten infrared spectral bands, covering the wavelength range of 1-240 μm (Hauser *et al.* 1990). The far-infrared bands are sensitive to interstellar dust, while the near-infrared bands, particularly those at 3.5 and 4.9 μm, provide complete, low-extinction views of the stellar component of the Galaxy.

2. METHOD

Latitudinal asymmetries in surface brightness have been calculated from equal-area cylindrical projection maps of the Galactic plane, $|b| < 7°$, at 1.25, 2.2, 3.5, 4.9, 100 and 240 μm. The first four bands are near-infrared (NIR) and their maps are dotted with bright stars, most of which lie relatively nearby. The bright resolved stars are removed using an iterative filtering algorithm. All but the 240 μm band have a significant foreground due to interplanetary dust. Strongest near the ecliptic, the interplanetary dust foreground tends to produce a distortion similar to a warp, but weaker and roughly opposite in sign to that of the HI layer. An empirical model (Hauser 1993) is used to remove this foreground. The latitude centroid of each column of pixels is calculated from the corrected maps, then robustly averaged over 12° bins of longitude.

The warp detected in the DIRBE surface brightness maps is not a subtle effect. Except for the 1.25 μm band, the signature of the warp is clear (though somewhat noisier) even before the removal of foreground and bright stars. It can be discerned by inspection of the maps themselves.

For this preliminary investigation, the primary purpose of which is to establish the presence of a warp in the DIRBE data, the centroid has been taken as the mean latitude, weighted by surface brightness. Unfortunately, a mean is not resistant to outliers, which in this case are usually nearby sources. Limiting the range to $|b| < 7°$ avoids the most prominent local sources, the clouds in the Taurus, Orion and Ophiuchus regions, but also causes the systematic underestimation of true large-scale asymmetries. The latitude limits are imposed symmetrically with respect to $b = 0°$. If the latitude profile is shifted with respect to $b = 0°$, however, the tail of the profile farthest from $b = 0°$ is truncated more than the other, introducing a bias against large displacements. Other means of determining the centroid are being explored.

A more basic problem in using surface photometry is that it provides no distance information. The asymmetries detected are asymmetries integrated over distance. An analysis of point-sources, like that performed by Djorgovsky and Sosin, could be attempted, but the DIRBE's large (.7°-square) beam makes it difficult to find distant sources uncontaminated by less-distant emission.

3. THE FAR-INFRARED: HI AND DUST

At 100 and 240 μm the DIRBE is sensitive to the thermal emission of cool dust. Fig. 1 shows the latitude centroids vs l at 240 μm. The centroids of a velocity-integrated HI map (Weaver & Williams 1973) are included, showing a strong correlation between gas and dust. The warp at 100 μm is very similar.

The same analysis was applied to a map of the optical depth at 240 μm (Sodroski *et al.* 1993), with similar results, demonstrating that the warp seen at 240 μm is not due to variations in dust emissivity. The results of Sodroski *et al.* (1987), who measured the latitude distribution of the IRAS 100 μm data for $|b| < 4°$, have also been verified.

The clumpy distribution of nearby dust leads to irregular structure in surface brightness maps. The relative smoothness and clearly sinusoidal shape of the warp in Fig. 1, and the fact that there is virtually no extinction to limit range in the outer Galaxy, argue that local emission is unlikely to play a major role in our results. Emission from Gould's belt could form a warp-like distortion of the surface brightness, but Gould's belt has a line of nodes at $l \sim 265°$, while that of Fig. 1 is at $l \sim 160°$. The "big dent," a two-kpc vertical depression in the interstellar medium delineated by Alfaro *et al.* (1991), does roughly coincide with the trough in Fig. 1, but they describe no feature matching the peak, and it does not seem likely that this feature would stand out in the velocity-integrated HI map. The sign and line of nodes of the warp shown in Fig. 1 are consistent with those of the distant HI layer. The fact that the two extrema are less than 180° apart is also consistent with the geometry of the HI warp. The approximate north/south symmetry of the warp is to be expected if most emission comes from within ~ 17 kpc of Galactic center, a distance beyond which the southerly displaced edge of the HI layer begins to curve back toward the nominal plane (Burton & Deul 1987).

4. THE NEAR-INFRARED: STARS

NIR photometry is dominated by the older stars of the disk, in particular the G and K giants. Within one kpc of the Sun these stars seem to be more or less uniformly distributed (McCuskey 1965). Older stars in general show little correlation with small- and medium-scale structure of the ISM.

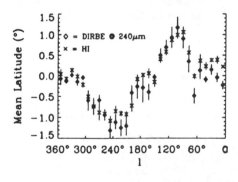

Fig. 1. HI and DIRBE FIR

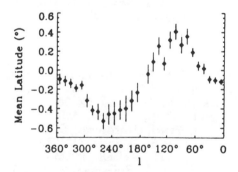

Fig. 2. DIRBE at 3.5 μm

The signature of the warp is visible in all NIR bands. The peak-to-peak amplitude increases with increasing wavelength: $.35°$, $.65°$, $.9°$, $1.0°$, going from 1.25 to 4.9 μm. This is consistent with a stellar disk that curves away from $b = 0°$ beyond the solar circle. The longer the wavelength, the smaller the extinction, and the greater the contribution of distant–and more highly displaced–stars to the surface brightness. Fig. 2 shows the latitude centroids at 3.5 μm. The mean plane of the inner ($330° < l < 30°$) Galaxy seems to be at $b \approx -.1°$. Others (Djorgovsky and Sosin 1989, Pandey et al. 1990, Weiland et al. 1993) find the same effect, plausibly due to the Sun's z height above the Galactic plane.

The DIRBE results at 4.9 μm and $|b| < 7°$ have been compared with those of Djorgovsky and Sosin (1987), who use the latitude range $|b| < 10°$. Good agreement with their "basic" sample is found except for $270° < l < 220°$, where their centroids are $\approx .2°$ below ours. Local sources, perhaps young, highly reddened stars, may contaminate their sample in this region; or using $|b| < 7°$ may exclude more of the warped disk. The DIRBE results agree qualitatively with those of Pandey et al. (1990), who studied the locations of open clusters, WR stars and SNR's. They find a larger amplitude for the warp than is shown in Fig. 2, but this may be due to the bias of our centroid calculations.

ACKNOWLEDGEMENTS

The authors gratefully acknowledge the efforts of the DIRBE data processing and validation teams in producing the high-quality datasets used in this investigation. We also thank the COBE Science Working Group for helpful comments on this manuscript.

COBE is supported by NASA's Astrophysics Division. Goddard Space Flight Center (GSFC), under the scientific guidance of the COBE Science Working Group, is responsible for the development and operation of COBE.

REFERENCES

Alfaro, E.J., Cabrera-Cano, J., & Delgado, A.J., 1991, ApJ, 378, 106
Binney, J., 1992, Ann. Rev. A&A, 30, 51
Burke, B.F., 1957, AJ, 62, 90
Burton, W.B., & Deul, A.R., 1987, in The Galaxy, eds. G. Gilmore & R. Carswell (Dordrecht: D. Reidel) p.141
Djorgovsky, S., & Sosin, C. 1989, ApJ, 341, L13
Hauser, M.G., Kelsall, T., Moseley, S.H., Jr., Silverberg, R.F., Murdock, T., Toller, G., Spiesman, W. & Weiland, J., in After the First Three Minutes, eds. S. Holt, C Bennett & V. Trimble, AIP 222, p 161
Hauser, M.G., 1993, this volume
Kerr, F.J., 1957, AJ, 62, 93
McCuskey, S.W., 1965, in Galactic Structure, eds. A. Blaauw & M. Schmidt, (Chicago: Univ. of Chicago Press) p 1
Miyamoto, M., Yoshizawa, M., & Suzuki, S., 1991, Ast. & Sp. Sci., 177, 399
Pandey, A.K., Bhatt, B.C., & Mahra, H.S., 1990, A&A, 234, 128
Sodroski, T.J., Dwek, E., Hauser, M.G., Kerr, F.J., 1987, ApJ, 322, 101
Sodroski, T.J., et al., 1993, this volume
Weaver, H.F., & Williams, D.R.W, 1973, A&A Suppl., 8, 1
Weiland, J.L., et al., 1993, this volume
Wouterloot, J.G., Brand, J., Burton, W.B. & Kwee, K.K., 1990, A&A, 230, 21

THE RADIAL STRUCTURE OF THE GALACTIC DISC

Annie C. Robin
Observatoire de Besançon, France, ANNIE@FROBES51.BITNET

Michel Crézé
Observatoire de Strasbourg, France, CDSXB2::CREZE

Vijay Mohan
U.P. State Observatory,Manora Peak, Nainital, 263129 India

ABSTRACT

As part of a stellar population sampling program, a series of photometric probes at various field sizes and depths have been obtained in a low extinction window in the galactic anticentre direction. Such data set strong constraints on the radial structure of the disc. These new data, used in combination with lower magnitude photographic data in a wider field, give a strong evidence that the galactic density scale length is rather short (2.5 kpc) and drops abruptly beyond 6 kpc. Over the whole effective magnitude range (12 to 25), all contributions in the statistics which should be expected from old disc stars beyond 6kpc vanish, although such stars dominate by far at distances less than 5 kpc. This is the signature of a sharp cut-off in the star density: the edge of the galactic disc between 5.5 and 6 kpc. As a consequence, the galactic radius does not exceed 14 kpc (assuming Ro=8.5). Colours of elliptical galaxies measured in the field rule out the risk of being misled by undetected extinction.

1. DATA SET

The studied field is located towards the anticenter close to the plane (l=179.7, b=2.8). It is known as a field of low extinction (Kapteyn Special Area 23). Two types of observations have been performed in this field:

Schmidt plates data (Mohan et al., 1988) have been obtained in a field of 2 square degrees in the UBV system with a photometric accuracy of about 0.1 in B and V and 0.12 in U. The completeness limit is reached at V=16.5.

CCD data (Robin, Crézé, Mohan, 1992) have been obtained at the 3.6 meter Canada-France-Hawaii Telescope in 4 CCD fields adding up to 29 arcmin². The photometric accuracy ranges from 0.01 at V=20 to 0.08 at V=25. The completeness limits are: 25 in V, 22.5 to 24 in B, 21 to 23 in U.

2. ANALYSIS

The analysis has been performed in several steps. First we determine the extinction using (U-B, B-V diagrams) and elliptical galaxies in the field (for the CCD data). Then we model the stellar distribution using the Besançon model of population synthesis. Finally we apply a maximum likelihood technics to determine the best fit parameters.

The extinction has been measured using three ways. First (U-B, B-V) diagrams in different apparent magnitude range allows to estimate a total extinction of 1.2 to 1.4 at 4 kpc. Second colours of elliptical galaxies have been measured in two of the four CCD frames. It gives a visual extinction of about 1.6. Finally the HI maps allow to estimate the integrated extinction, giving a value of 1.4 at 4 kpc. We have adopted a value of 1.4 in agreement with the three estimations.

The Besançon model of population synthesis (Robin & Crézé 1986, Bienaymé et al. 1987) has been used to make the star counts simulations assuming different values for both the disc scale length and the cutoff. Although the model uses a typical stellar and galactic evolution model to detail the stellar content of the disc, including the age distributions, it is of little importance here, since star counts are mainly sensitive to the global luminosity function (which is suppose to be the same as in the solar neighbourhood) and the radial density law.

We use a maximum likelihood technics to compare model predictions with Schmidt plate star counts and CCD data in the (V, B-V) plane in the range 12 < V < 25. U-B colours have not been used because their are not complete for V > 21. The details of the analysis are given in Robin, Crézé & Mohan, 1992. The data have been shown to be sensitive to two parameters : the disc scale length and the disc cut-off (the point where the distribution stops to be exponential and drops to nearly zero). We explore a grid of these parameters from 2.0 to 4.5 kpc for the scale length and from 3.0 kpc to ∞ for the cutoff distance.

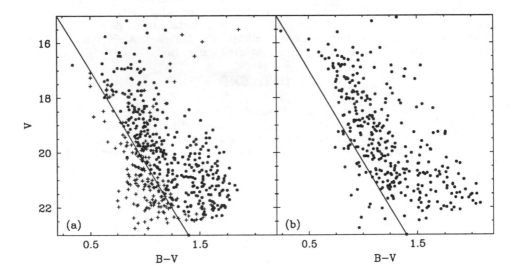

Fig. 1. (V, B-V) diagram of anticentre stars. (a) Model predicted distribution. Dots are stars closer than 5.5 kpc while crosses are stars beyond 5.5 kpc. (b) Observed distribution. The solid line is a guide to identify the zone where most stars are beyond 5.5 kpc.

Figure 1 shows the (V,B-V) diagram predicted and observed in the 4 CCD fields. In the predicted diagram crosses are the stars with distances larger than 5.5 kpc, while dots are closer stars. Nearly all the stars further away than 5.5

kpc have vanished in the observed diagram, indicating a sharp cutoff in the radial distribution in the disc.

3. RESULTS AND DISCUSSION

In fig. 2 the apparent magnitude distribution is compared with a series of model predictions with scale lengths ranging from 2.2 to 4.5 kpc. In each figure the model is plotted without cutoff (dashed line) and with the cutoff value resulting from the maximum likelihood estimation (solid line).

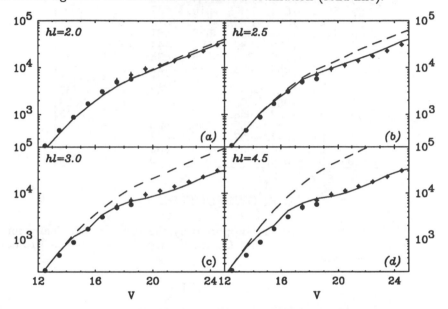

Fig. 2. Distribution in V. Schmidt data (circles), CCD data (diamonds), models with various scale lengths : hl=2.0 (a), 2.5 (b), 3.0 (c), 4.5 (d). Solid line: model density laws with best fit cutoff (9.5, 5.5, 4.0, 3.0 kpc resp.); dashed line: model density laws assuming no cutoff.

In table 1 values of the likelihood for the grid of models are given. The maximum is found when the assumed scale length is 2.5 ± 0.3 kpc and the cutoff is at 5.5 to 6 kpc from the sun. The data are incompatible with scale length larger than 3.0 kpc.

This value is compatible with the scale length measured by the asymmetric drift observations (Mayor 1974, Oblak & Mayor 1987) as well as with IR observations (Kent *et al.* 1991). However it is incompatible with Pioneer result (van der Kruit 1986, see discussion in Robin *et al.* 1992a).

The detection of the cutoff in these data concerns mainly the old disc. The value of 5.5 to 6 kpc is about the distance were the farthest open clusters are found. Young stars have been observed in farther regions (Digel *et al.* 1992, Turbide & Moffat 1992). We emphasize that this sharp edge could be related to a Lindblad resonnance (according to the model of Blitz & Spergel 1991) and/or to the evolution of star formation in the outer part of the Galaxy in a Larson scenario for the formation of the disc.

Table 1. likelihood of data under various assumed model disc scale lengths hl and cutoffs r_{max}.

r_{max} hl	1.5	1.8	2.0	2.2	2.5	3.0	3.5
1.00	-8678.	-8168.	-7237.	-7493.	-6958.	- 5848.	-5931.
1.50	-6844.	-5577.	-4992.	-5220.	-4446.	- 3672.	-3616.
2.00	-4968.	-3808.	-3679.	-3412.	-3022.	- 2387.	-2162.
2.50	-3907.	-3015.	-2541.	-2322.	-2004.	- 1516.	-1598.
3.00	-3387.	-2368.	-1941.	-1730.	-1231.	- 987.	-1127.
3.50	-2847.	-1870.	-1427.	-1110.	-790.	- 690.	-840.
4.00	-2605.	-1624.	-1182.	-910.	-607.	- 611.	-851.
4.50	-2444.	-1453.	-1018.	-748.	-515.	- 656.	-975.
5.00	-2264.	-1264.	-831.	-636.	-461.	- 764.	-1324.
5.50	-2235.	-1212.	-786.	-592.	-444.	- 816.	-1459.
6.00	-2210.	-1139.	-705.	-529.	-438.	- 952.	-1797.
6.50	-2192.	-1086.	-683.	-513.	-492.	- 1131.	-2156.
7.00	-2164.	-1076.	-671.	-514.	-521.	- 1208.	-2321.
7.50	-2138.	-1061.	-663.	-524.	-557.	- 1328.	-2567.
8.00	-2129.	-1056.	-660.	-531.	-571.	- 1384.	-2658.
8.50	-2111.	-1052.	-658.	-536.	-583.	- 1435.	-2753.
9.00	-2107.	-1051.	-659.	-540.	-596.	- 1476.	-2828.
9.50	-2101.	-1047.	-657.	-547.	-608.	- 1514.	-2907.
10.00	-2096.	-1044.	-657.	-555.	-633.	- 1562.	-3033.
10.50	-2093.	-1040.	-658.	-556.	-640.	- 1589.	-3095.

ACKNOWLEDGEMENTS

This research was partially supported by the Indo-French Centre for the Promotion of Advanced Research / Centre Franco- Indien Pour la Promotion de la Recherche Avancée.

REFERENCES

Bienaymé O., Robin A. C., Crézé M., 1987, A&A 180, 94

Bienaymé, O., Mohan, V., Crézé, M., Considère, S., Robin, A.C., 1992, A&A 253, 389

Blitz, L., Spergel, D., 1991, ApJ 370,205

Digel, S., de Geus, E., Thaddeus, P., 1992, this conference

Kent, S.M., Dame, T.M., Fazio, G., 1991, ApJ 378, 131

Larson, R. B. 1976, MNRAS 176, 31

Mayor, M., 1974, A&A 32, 321

Mohan, V., Bijaoui, A., Crézé, M., Robin, A.C., 1988, A&AS 73, 85

Oblak, E., Mayor, M. 1988, Xth European IAU Assembly, Vol 4., p. 263, Ed. J. Palous

Robin, A.C., Crézé, M. 1986, A&A 157, 71

Robin, A.C., Crézé, M. & Bienaymé, O., 1989, *The Gravitational Force Perpendicular to the Galactic Plane*, p. 33, A.G.D. Philip and P.K. Lu (eds.), L . Davis Press

Robin A.C., Crézé M. & Mohan V., 1992, A&A 265, 32

Robin, A.C., Oblak, E., 1987, *Xth European Astronomy Meeting of the IAU*, Vol. 4, p. 323, Ed. J. Palous

Turbide, L., Moffat, A.F.J., 1992, this conference

Van der Kruit, P.C., 1986, A&A 157, 230

"BACK TO THE FUTURE," MAJOR DYNAMICAL EVENTS IN THE LOCAL GROUP 15 BILLION B.C. TO A.D. 15 BILLION

G. Byrd(U. of Alabama), M. Valtonen(Tuorla Obs., Finland), M. McCall and K. Innanen(York Univ.)

Email Byrd@okra.astr.ua.edu

ABSTRACT

Kahn and Woltjer(1959) assumed simple two-body motion of M31 and our galaxy with initial recession and now approach of the two galaxies to estimate the total mass. However, recent measurents of recession speeds and distances of IC342 and Maffei 1 (Buta and McCall 1983, McCall 1989) indicate that each of these galaxies must once have been so close to M31 that there was a gravitational interaction stronger than with our galaxy, violating the two body assumption. We show with computer simulations and a generalized Kahn-Woltjer calculation that a few-body ejection/merger event involving M31 about $5 \cdot 10^9$ years ago can explain the present-day distances and velocities of these galaxies. The smaller mass IC342 and Maffei were dynamically ejected rapidly away and the greater mass M31 slowly toward us in the opposite direction. In contrast, we find our galaxy to have been a "by-stander" to these events. However, besides IC342 and Maffei 1, other much smaller mass objects undoubtedly were ejected, some toward our galaxy. One or more of these captured ejecta may confuse estimates of the mass of our galaxy using satellites. Looking toward the future, in about $10 \cdot 10^9$ years, our simulations predict a wide encounter with M31. Finally, although we find the initial assumption of Kahn-Woltjer's mass calculation of $4 \cdot 10^{12} M_\odot$ for the Local Group invalid, we get a similar result with our generalized Kahn-Woltjer method.

1. INTRODUCTION

The Kahn-Woltjer(1959) Local Group mass estimate of $\sim 4 \cdot 10^{12} M_\odot$ with today's distance, radial velocity and age of universe (Fich and Tremaine 1991) is a very robust result even if the calculation is made more sophisticated (Peebles, Melott, Holms and Jiang 1989). The assumption of two-body motion of our galaxy and M31 is crucial to this method. As can be seen in Fig. 1, the back-extrapolated positions and masses of (Maffei 1 + IC342)/M31 ($\sim 1/2.4$)and our galaxy/M31 ($\sim 1/2.5$) imply the two-body assumption was violated. This violation should be true regardless of whether more sophisticated extrapolations are done or of one's beliefs about masses or extents of halos. A "few-body ejection/merger interaction" (Valtonen and Mikkola 1991 review) at this past time seems the most physically reasonable way to explain the present day Maffei 1 and IC342 radial velocities and distances. In this interaction and our simulations, Newtonian dynamics of multiple bodies occurs with galaxy halo gravitational drag resulting in a merger should two galaxies pass close to one another. A complete description of our simulations will appear in Valtonen, Byrd, McCall and Innanen (1993).

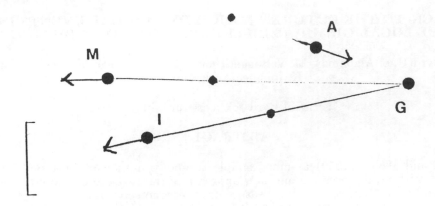

Fig. 1. Present positions (large dots) of Maffei 1 (M), IC342 (I), M31 (A) and our galaxy (G) in the supergalactic plane (scale bar is 500 kpc). Small dots are positions $5 \cdot 10^9$ yr ago via simple extrapolation of constant radial velocities.

2. PHYSICAL FEASIBILITY AND GENERALIZED METHOD

Requiring that IC342, Maffei 1 and M31 reach $\sim$ their present-day distances, directions and radial velocities, we numerically searched thousands of cases rejecting those that did not result in the observed configuration. The most fruitful initial situation was two successive encounters of binary galaxies with M31 (see Fig. 2 next page). The more massive member of two binaries each merges with M31, the less massive (Maffei 1 or IC342) is ejected. The massive M31 slowly recoils toward our galaxy (a by-stander to all this excitment). The six successful examples we found serve as a simple demonstration of the physical feasibility of the hypothesis. Other sequences may be possible e.g. two binaries encountering M31 at the same time or a triple galaxy/M31 encounter, but there were too many free parameters for us to search these encounters. The successful examples require total masses $\geq 3 \cdot 10^{12} M_\odot$ for sufficient ejection speeds. We find that masses $> 5 \cdot 10^{12} M_\odot$ are not possible because only mergers result with no ejections at all!

We now use our simulation mass range to constrain the age of the universe since the major error in the original Kahn-Woltjer mass estimate results from uncertainty in this age ($6 \cdot 10^{12} M_\odot$ if $10 \cdot 10^9$ yr and $3 \cdot 10^{12} M_\odot$ if $20 \cdot 10^9$ yr). Because the two body assumption of this method is probably invalid, we create a "Generalized Kahn-Woltjer Method" with some reasonable approximations to estimate the Local Group mass as a function of the age of the universe. Our method numerically follows the motion of the Maffei 1/IC342 center of mass and M31 back to the merger/ejection point according to Newton's third law constrained by their mass ratio, radial velocities and distances. We assume

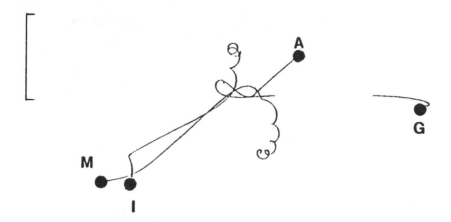

Fig. 2. An interaction which produces today's configuration. Only one member of each binary is plotted with today's positions indicated by large dots.

different transverse velocities for M31 for a given total mass. For one M31 transverse velocity, Maffei 1/IC342 and M31 reach the merger/ejection point and we continue to see when the center of mass of the three comes together with our galaxy at the origin of the universe. By carrying out the above procedure for different total masses, we can determine a plot of age of universe versus total Local Group mass. If the universe is younger, the mass is larger. The total mass $< 3 \cdot 10^{12} M_{\odot}$ if the age is $> 25 \cdot 10^9$ yr old. This is the maximum age for the universe since we found insufficient ejection speeds below this mass. A $15 \cdot 10^9$ yr age results in $3.6 \cdot 10^{12} M_{\odot}$ ($\sim$ the Kahn-Woltjer result but totally different dynamically). Ages $< 12 \cdot 10^9$ yr give masses $> 5 \cdot 10^{12} M_{\odot}$, where we expect only merging. We thus find $12 \cdot 10^9 <$ the age of the universe $< 25 \cdot 10^9$ yr. The bounds of this range are affected by uncertainties in the mass ratios of the Local Group members, the Maffei 1–IC342 distances and the disk orbital speed of the sun in our galaxy among other parameters.

3. PRESENT DAY CONSEQUENCES AND THE FUTURE

Our simulations indicate that low mass companions of Maffei 1, IC342 or of M31 should be preferentially ejected in the encounter. The observation that M31's companions are predominently redshifted relative to their primary can thus be explained (Byrd and Valtonen 1985). Ejecta could well be thrown toward our galaxy to be visible today as companions with unusual orbits. Contamination of the satellite system of our galaxy by a few of these companions could result in an error in estimating our galaxy's mass. We suggest Leo I with its abnormally large radial velocity for its distance (Table 2, Fich and Tremaine) and the Magellanic Clouds with their unexpectedly large angular momentum (Shuter 1992) as prime candidates.

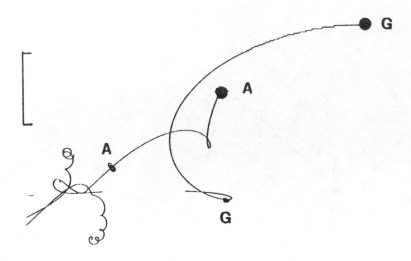

Fig. 3. Future of the Local Group during the next $20 \cdot 10^9$ yr for Fig. 2.

Going from the past back to the future, M31 and our galaxy will swing by one another in a wide encounter $\sim 10 \cdot 10^9$ yr from now (Fig. 3). Observation of this encounter should serve as a proof of our hypothesis some billions of years hence. Measurement of the proper motion of M31 (the transverse component of the recoil) should suffice for verification in the more immediate future. From our simulations, this transverse recoil should be $\approx 2 \cdot 10^{-3}$ arc sec/100 yr away from IC342 and Maffei 1.

ACKNOWLEDGEMENTS

This work was supported by the Finnish Academy and grants NSF EPSCoR RII8996152 and AST 9014137 (GB).

REFERENCES

Buta, R.J. and McCall, M.L. 1983 MNRAS, 205, 131.
Byrd, G.G. and Valtonen, M.J. 1985 ApJ 289, 535.
Fich, M. and Tremaine, S. 1991 Ann. Rev. Astron. Astrophys. 29, 409.
McCall, M.L. 1986 J. Roy. Ast. Soc. Canada, 80, 271.
McCall, M.L. 1989 AJ 97, 1341.
Peebles, P.J.E., Melott, A.L., Holmes, M.R., Jiang, L.R. 1989 ApJ 345, 108.
Shuter, W.L.H. 1992 ApJ 386, 101.
Valtonen, M., Byrd, G., McCall, M. and Innanen, K. 1993 AJ March issue
Valtonen, M. and Mikkola, S. 1991 Ann. Rev. Astron. Astrophys. 29, 9.

DISK/HALO

FOUNTAINS, BUBBLES, AND HOT GAS

Christopher F. McKee
Departments of Physics and Astronomy, University of California, Berkeley CA 94720

Email ID
mckee@bkyast.berkeley.edu

ABSTRACT

Gas is observed to extend far from the disk of the Galaxy, with a scale height that increases with the ionization state of the gas. Recent observations have shown that a substantial amount of the halo gas is hot. This gas could be either heated internally, by halo supernovae for example, and/or it could be energized by gas venting from superbubbles in the Galactic disk—a Galactic fountain. Calculation of the UV emission and absorption spectrum of a Galactic fountain shows good agreement with observation; in the absence of conduction, agreement with X-ray observations requires fine tuning of the initial temperature of the fountain. The superbubbles that provide the mass and energy of the gaseous halo in fountain models are created by OB associations in the disk. A recent calculation of the luminosity distribution of Galactic OB associations is described. The filling factor of superbubbles in the disk is estimated to be of order 0.1; combined with observations of H I and C IV, this suggests a mass flow of several $M_{\odot}$ yr^{-1} between the halo and the disk.

1. OVERVIEW OF OBSERVATIONS OF THE GASEOUS HALO

The initial suggestion of hot gas in the Galactic halo—a Galactic corona—was put forth by Spitzer (1956). One of his main arguments for such a corona was the existence of cool gas clouds high above the plane, which he reasoned must be confined by a hotter ambient medium. We therefore begin our survey of the gaseous halo by considering observations of the neutral gas.

Recent observations have shown that neutral hydrogen extends well into the lower halo ($|z| \lesssim 1$ kpc). Lockman (1984) inferred the existence of three components of H I in the inner Galaxy, two Gaussians with scale lengths of 106 pc and 254 pc, and an exponential with a scale height of 480 pc. Kulkarni & Heiles (1987) identified the lowest component with H I clouds, and the other two with intercloud H I. In a recent review, Dickey & Lockman (1990) obtained results for the high–z H I similar to those of Lockman (1984). UV observations suggest that the H I has an even greater extent: Danly et al (1992) find a density $n(z) = 0.15 \exp(-|z|/510 \text{ pc})$ cm^{-3}, which has several times more H I above a kiloparsec than Lockman (1984) found. Merrifield (1993) has used observations of H I on the far side of the Galaxy to estimate the scale height. He too finds a somewhat larger value than Lockman, although his estimate could be contaminated by foreground H I. Thus, there is general agreement that H I extends far from the plane, and it is a major challenge for theory to account for this observation.

The kinematics of the gas is important in attempting to infer the dynamics of the coupling between gas in the disk and that in the halo. Kulkarni & Fich (1985) identified three velocity components in the H I toward the Galactic poles, with velocity dispersions of 6.1, 15, and 35 km s^{-1} (the properties of the last component, the "fast H I", were only estimated). Based on these observations, McKee (1990) inferred that a substantial fraction of the pressure in the ISM is due to turbulent gas motions (1.8×10^{-12} dyne cm^{-2} out of a total pressure of 3.0×10^{-12} dyne cm^{-2}). Boulares & Cox (1990), on the other hand, concluded that turbulent pressure, magnetic pressure, and cosmic ray pressure are closer to equipartition; they found a total midplane pressure of $(3.9 \pm 0.6) \times 10^{-12}$ dyne cm^{-2}. Direct modeling of the line profiles in the Galactic gravitational potential led Lockman & Gehman (1991) to conclude that turbulent pressure is indeed dominant; Merrifield (1993) comes to the opposite conclusion, based on the greater value of the scale height he infers. However, Ferrara (1993) has argued that the inclusion of the effects of radiation pressure on the dust in the H I clouds reduces the pressure in the magnetic field and cosmic rays needed to support the halo gas. In any case, the substantial level of turbulent motions observed in the H I suggests, but does not prove, that there is reasonably rapid exchange of material between the halo and the disk.

In addition to the turbulent motions in the H I, however, there is a systematic infall to the plane (Weaver 1974) that provides direct evidence for such mass exchange. Lockman & Gehman (1991) find that the mean velocity of this gas is $v_{in} = 5$ km s^{-1}. If this inflow extends over a fraction f of the Galactic disk, then the resulting inflow rate to both sides of the disk is

$$\dot{M}(\text{H I}) = 2\pi f R^2 \rho(0) v_{in} = 42 f R_{10}^2 \quad M_\odot \text{ yr}^{-1}, \tag{1}$$

where R_{10} is the radius of the Galactic disk in units of 10 kpc. The well–known infall of gas above the plane at intermediate velocities ($|v| < 70$ km s^{-1}) has a much smaller mass inflow rate of about $4 f R_{10}^2$ $M_\odot$ yr^{-1}, based on Danly's (1989) estimate of about 0.5 kpc for the distance to this gas. The rate of mass inflow by high velocity clouds is quite uncertain because the distance to the clouds is unknown. Danly (1989) estimated a minimum distance of 1.7 kpc, and they are often taken to be at a distance of several kpc. If the high velocity clouds are at such a large distance, the local infall rate should be a fair sample of the Galactic infall rate, and we can set $f \sim 1$. Using Giovanelli's (1980) survey, which has a low velocity cutoff of 50 km s^{-1} at high Galactic latitudes, Kaelble, de Boer, & Grewing (1985) inferred an infall rate of about $1 \times R_{10}^2$ $M_\odot$ yr^{-1}. For $|v| > 100$ km s^{-1}, Wakker's (1991) analysis indicates that only about 10% of the sky is covered by gas with a column density $N > 10^{18}$ cm^{-2}, provided the Magellanic Stream and the Outer Arm are excluded. If this gas is at a distance of several kpc and is falling in at a velocity of order 100 km s^{-1}, its mass inflow rate is only about 0.04 $M_\odot$ yr^{-1}, comparable to the "drizzle" of low column density gas Danly (1989) inferred from UV observations.

The vertical distribution of the ionized hydrogen is more certain, since it is more uniformly distributed. Most of this H II is warm ionized medium (WIM) at $T \sim 10^4$ K. Observations of pulsar dispersion measures show that the column density of H II toward the pole is 7×10^{19} cm^{-2}, with a scale height of about 1 kpc and a velocity dispersion of about 20 km s^{-1} (Reynolds 1993).

UV observations of ions such as C II and Si II trace both the H I and H II. Observations of the line of sight to 3C 273, for example, indicate that this gas has a velocity distribution with a FWHM of about 150 km s^{-1}; the mean velocity is -4 km s^{-1}, consistent with the H I results cited above (Savage et al

1992). Observations of halo stars indicate that the gas has both positive and negative velocities for $|z| \lesssim 1$ kpc, whereas gas above 1 kpc has predominantly negative velocity (Danly et al 1992). The range of velocities appears to be larger at smaller column densities: Meyer and Roth (1992) have observed Ca II absorption over a 350 km s^{-1} velocity interval in gas that they estimate has a column density $\lesssim 10^{18}$ cm^{-2}.

The presence of highly ionized gas in the halo provides important clues on the energetics of the halo gas (see the review by Spitzer 1990). Table 1 summarizes absorption line observations of C IV, Si IV, N V, and O VI. The results for the halo stars, which have $|z| \lesssim 3$ kpc, have been extrapolated through the halo based on the inferred scale height for each ion (Sembach & Savage 1992). The typical scale height of the highly ionized halo gas is about 3 kpc. A lower limit on the column density of O VI toward 3C 273 has been obtained by Davidsen et al (1992); however, this line of sight passes near two large radio continuum loops, and may not be typical (Burks et al 1991). Data on the remaining ions for 3C 273, the LMC, and the SMC have been taken from Savage et al (1992). There is considerable variation in the values of the column densities from one line of sight to another, with $\log N$(C IV)$\sin b$ ranging from 13.89 for the line of sight toward SN 1987A in the LMC to 14.57 for the line of sight to 3C 273; the variation in the line ratios is much less. Models of this gas (§2) suggest that large mass fluxes, comparable to those in equation (1), are involved.

The highly ionized gas has also been observed in emission, which demonstrates that the gas is hot ($T \sim 10^5$ K) and collisionally ionized rather than warm ($T \sim 10^4$ K) and photoionized (Martin & Bowyer 1990; hereafter MB). Using the Berkeley EUV/FUV Shuttle Telescope, they detected emission in the C IV λ 1550 doublet in 6 of 8 directions, and emission in the O III] λ 1663 line in 2 of 8 directions. They estimate that the emitted intensity of the C IV lines is about 5000 photons cm^{-2} s^{-1} sr^{-1}. The observed intensity is often less because of attenuation by dust. If the emitted intensity extends over a fraction f of the Galactic disk, the luminosity would be L(C IV) $= 5 \times 10^{39} f R_{10}^2$ erg s^{-1}. Generally, the C IV lines constitute less than 20 % of the total cooling of the plasma, indicating that the total luminosity of the C IV emitting gas exceeds $2.5 \times 10^{40} f R_{10}^2$ erg s^{-1}. MB infer an emission measure ~ 0.01 cm^{-6} pc, a temperature $\sim 10^5$ K, and volume filling factor of about 0.05 for the emitting gas.

Upper limits have been placed on the intensity of O VI $\lambda\lambda 1032$, 1037 emission lines from the halo. Dixon et al (1993) used the Hopkins Ultraviolet Telescope to search for O VI emission in 6 directions; in the two directions in which their measurements are most sensitive, the limit is about 6×10^4 photons cm^{-2} s^{-1} sr^{-1}. In a rocket experiment, Edelstein & Bowyer (1992) set a much more stringent limit on this emission in one direction. If the intensity ratio $\lambda 1037/\lambda 1032$ has the standard value of 0.5, their upper limit on the O VI emission is 1.5×10^4 photons cm^{-2} s^{-1} sr^{-1}.

Hotter gas is revealed by its soft X–ray emission. Since the observations of this emission are expertly reviewed elsewhere in this volume (McCammon 1993), I shall not discuss them in detail here. Recent ROSAT observations have shown that foreground gas can cast an X–ray shadow, thereby establishing that some regions of the halo, at least, are soft X–ray emitters (Snowden et al 1991; Burrows & Mendenhall 1991). The latter authors infer that the halo gas behind the Draco molecular cloud has an emission measure $EM \sin b = 0.004$ cm^{-6} pc

TABLE 1

Column Densities of Highly Ionized Gas in the Halo

	Halo stars[a]	Extragalactic[b]	Fountain model[c]
$\log N(\text{CIV}) \sin b$	14.20	14.32	14.00
Si IV/C IV	0.22	0.18	0.21
N V/C IV	0.16	0.14	0.14
O VI/C IV	–	>0.7	2.2

[a] Sembach and Savage (1992), Table 9.
[b] Average of 3C 273, LMC, and SMC; Savage et al (1992).
[c] Average of models with parameter $n_{-2}D_0 = 135$ pc and 250 pc; Shapiro and Benjamin (1992).

at a temperature of 1.3×10^6 K.

The salient features of the halo gas revealed by observations to date are thus: (1) the halo extends well away from the plane (0.5 kpc for the H I, 1 kpc for the WIM, and several kpc for the hot ionized medium [HIM]); (2) the weight of this extended mass distribution leads to a substantial midplane pressure of $(3-4) \times 10^{-12}$ dyne cm^{-2}; (3) substantial mass exchange occurs between the disk and the halo, as revealed by the large turbulent velocities that are observed and by the local infall of H I at a rate of $42fR_{10}^2\ M_\odot$ yr^{-1}; and (4) the C IV luminosity and the X–ray luminosity observed from the halo imply that substantial energy is injected into the halo.

2. MODELS OF THE GASEOUS HALO

2.1 Source of the Thermal Energy of the Halo Gas

Models of the gaseous halo can be characterized by the heating mechanism invoked: external ultraviolet radiation, internal energy sources, or energy injection from the disk. External radiation produces a photoionized halo at a temperature of about 10^4 K with reasonably strong absorption lines of C IV and Si IV (Chevalier & Fransson 1984; Hartquist, Pettini, & Tallant 1984; Bregman & Harrington 1986). Such a model can now be ruled out because it cannot account for the N V and O VI absorption, the C IV emission, the X-ray emission, or the large line widths observed in the halo.

Heating by supernova remnants (SNRs) in the halo was considered by Chevalier & Oegerle (1979), but little detailed modeling of this possibility has been done since. If the filling factor of hot gas in the halo is not too close to unity, it is approximately given by

$$Q = S \int V dt \propto \frac{S}{n_0^{0.11} P_0^{1.36}} \qquad (2)$$

(McKee 1990), where S is the supernova rate per unit volume and V is the volume occupied by a remnant of age t. Since the dependence on the ambient density is so weak, we approximate $n_0 \propto P_0$, so that $Q \propto S/P_0^{1.5}$. Thus the filling factor of hot gas in the halo will exceed that in the disk if the supernova scale height exceeds 2/3 of the pressure scale height. Neither scale height is well determined at present, though. Note that halo supernovae include Type II SN due to runaway OB stars as well as Type Ia SN due to low mass progenitors. Narayan & Ostriker (1990) have argued that a substantial fraction of pulsars are created with a scale height $\gtrsim 500$ pc, so the runaway contribution to the SN rate could be substantial.

An alternative internal heating mechanism has been discussed by Raymond (1990): microflares. By analogy with the solar corona, he envisions conversion of the energy in Galactic differential rotation into magnetic energy, which in turn is dissipated by reconnection. Unfortunately, the physics of reconnection is not well understood, so it is difficult to check this model quantitatively. Hartquist & Morfill (1986) have pointed out that the damping of waves generated by cosmic rays as they stream out of the halo can heat the gas to high temperatures provided the density is sufficiently low ($n_e \lesssim 5 \times 10^{-4}$ cm^{-3}), but such a density is too low to account for the emission observations of the halo gas.

By far the most popular model for the energy source for the halo is a "Galactic fountain", in which the energy from supernovae in the disk is vented into the halo (Shapiro & Field 1976). At that time, it was generally assumed that supernovae occur randomly in space and time. A fountain based on random supernovae fails, however: McKee & Ostriker (1977) showed that disk SNRs cannot break out of the disk, even if much of the disk volume is filled with hot, low–density gas that favors SNR expansion. Cox (1981) reached a similar conclusion by demonstrating that any Galactic fountain must be weak.

A significant fraction of Galactic supernovae are correlated in OB associations, however, and these correlated supernovae create "superbubbles", large volumes of hot gas which can extend out of the disk (Bruhweiler et al 1980; Cowie et al 1981; McCray & Kafatos 1987; Ikeuchi 1987, 1988; Norman & Ikeuchi 1989; Li & Ikeuchi 1990; see review by Tenorio–Tagle & Bodenheimer 1988). In principle, these superbubbles can break through the disk into the halo, and Heiles (1984) has identified H I structures in the inner Galaxy ("worms") which he attributes to such breakthrough. More recently, observations have shown that there is substantial gas in the lower halo (§1), so that one must distinguish between superbubbles that break through the cold gas disk into the extended gas disk in the lower halo and those that are sufficiently energetic to break out of the extended gas disk entirely (Heiles 1990). "Breakthrough" bubbles require about 50 supernovae, whereas "breakout" bubbles require at least 800 (Koo & McKee 1992). The latter estimate is a lower limit because it did not explicitly allow for the inhibiting effects of a magnetic field in the halo, nor did it allow for radiative losses. Insofar as the "chimneys" discussed by Ikeuchi and coworkers correspond to breakout bubbles, they must constitute a small fraction of all superbubbles.

2.2 Cooling Flow Models

Hot gas injected into the halo from the disk will cool, producing a characteristic spectrum of emission and absorption lines. The cooling is thermally unstable, so that clouds are produced in the cooling flow. Models of the halo have generally focussed either on the dynamics of these clouds, or on the spectrum of the gas. A detailed calculation of the cloud dynamics in a Galactic fountain was carried out by Bregman (1980). The fountains he considered were "weak" in the sense that only a small fraction ($\sim 20\%$) of the supernova energy in the disk was vented into the halo. This model is reasonably successful in accounting for the kinematics of the high velocity clouds. However, he made no allowance for the turbulence, magnetic fields, and cosmic rays that together dominate the pressure in the lower halo (see §1). Low temperature fountains (Kahn 1981; Houck & Bregman 1990) are more consistent with the spatial extent and velocity structure of the gas in the lower halo, but they suffer from the same problems: the pressures and mass fluxes are too small (for example, Houck & Bregman [1990] conclude that a model with $\dot{M} \simeq 0.4R_{10}^2 \ M_\odot \ \mathrm{yr}^{-1}$ best fits the kinematic data).

Models which attempt to account for the UV spectrum of the halo gas are highly idealized: The gas is assumed to cool from an initial temperature of order 10^6 K, where it is taken to be in ionization equilibrium; the cooling occurs at constant density (Shapiro & Field 1976) or at constant pressure when it is hot and at constant density when it has cooled somewhat (Edgar & Chevalier 1986; Shapiro & Benjamin 1992). Thermal conduction is assumed to be unimportant. The intensity of the emission lines is proportional to the mass flow rate into the disk from the halo. A fit to the intensity of the C IV emission measured by MB gives a mass flow onto both sides of the disk of

$$\dot{M}(\mathrm{C\ IV}) = 21fR_{10}^2 \quad M_\odot \ \mathrm{yr}^{-1} \qquad (3)$$

based on the calculations of Edgar & Chevalier (1986) or Shapiro & Benjamin (1992). This value is within a factor 2 of the mass flux inferred from H I observations (§1). In fact, the C IV value could be an underestimate, since recent calculations have increased the dielectronic recombination rate of C IV (Romanik 1988). This increase in the recombination rate requires a larger mass flux to give the same amount of C IV.

Measurement of the intensity of the O VI emission lines from the halo would provide a powerful diagnostic for fountain models. The models of Shapiro & Benjamin (1992) predict a photon intensity for O VI that is at least 4.8 times that of C IV. Extinction by interstellar dust reduces the intensity of each line, with O VI being more strongly attenuated because of its shorter wavelength. For lines of sight with $N(\mathrm{H\ I}) > 3 \times 10^{20}$ cm^{-2}, MB observed a C IV intensity $\lesssim 3000$ photons cm^{-2} s^{-1} sr^{-1}. The line of sight for Edelstein & Bowyer's (1992) O VI observation has an H I column density exceeding 3×10^{20} cm^{-2}, so the ratio of their upper limit on the O VI intensity (15000 photons cm^{-2} s^{-1} sr^{-1}) to the expected C IV intensity ($\lesssim 3000$ photons cm^{-2} s^{-1} sr^{-1}) is $\gtrsim 5$, which would be consistent with the Shapiro & Benjamin model. It would be extremely valuable to get an actual measurement of the O VI intensity in a direction which has a low value of $N(\mathrm{H\ I})$ and a known value of the C IV intensity.

The absorption line column densities N_i depend on the flow velocity v and on the temperature at which the flow switches from being isobaric to isochoric. For a given value of v, N_i is independent of the mass flow rate $\dot{M}$, since $N_i \propto n_i v t_{\mathrm{rec}} \propto v/\alpha$, where t_{rec} is the recombination time and α is the recombination coefficient. As the gas cools, it emits ionizing radiation which can alter the column densities of ions such as Si IV. By including this effect, Shapiro & Benjamin (1992) obtained the results listed in Table 1, which are in remarkably good agreement with UV observations of highly ionized halo gas. However, this agreement cannot be taken as conclusive evidence for the cooling flow model, since several other models with non–equilibrium ionization may be equally successful: Models of old SNRs (Slavin and Cox 1992) and magnetized conduction

fronts (Borkowski, Balbus, & Fristrom 1990) calculated without photoionization each have UV absorption spectra similar to those calculated by Edgar &
Chevalier (1986) for cooling flows, and it seems possible that these spectra could
be brought into agreement with the observations if photoionization were self–
consistently included. Furthermore, the model of turbulent mixing layers developed by Slavin, Shull, & Begelman (1993) includes photoionization, and it also
can produce UV absorption spectra similar to those observed.

    It remains to be demonstrated that the cooling flow models which are so
successful in accounting for the UV spectrum can also account for the soft X–ray
emission from the halo. If the initial temperature is too low ($\lesssim 10^{5.8}$ K according
to Shapiro & Benjamin, private communication) the models underproduce X-
rays even in the absence of absorption. On the other hand, models with initial
temperatures above 10^6 K tend to overproduce X–rays. To see why this is so,
we proceed as follows. Let c_A be the specific heat per unit mass when the
thermodynamic variable A is held constant; in particular, if μ is the mean mass
per particle, then $c_V = (3/2)k/\mu$ and $c_P = (5/2)k/\mu$. Let $dM(T)$ be the mass
in the cooling flow in the temperature range T to $T + dT$. The energy emitted
by this gas as it cools through a temperature drop dT at constant A is

$$c_A dT \cdot dM(T) = dL(T)dt, \tag{4}$$

where $dL(T)$ is the luminosity of this gas and dt is the time interval over which
the energy is emitted. In a steady flow, the mass flow rate is

$$\dot{M} = \frac{dM(T)}{dT}\frac{dT}{dt}, \tag{5}$$

so that the luminosity of the gas at temperature T is

$$\frac{dL(T)}{d\ln T} = c_A \dot{M} T \propto T. \tag{6}$$

This equation makes the obvious point that in a cooling flow more energy must
be radiated to cool a hot gas than a warm one: the hottest region is the
most luminous. The observations appear to contradict this, however: MB infer
$EM \sin b \sim 0.01$ cm^{-6} pc at $T \sim 10^5$ K, whereas the X–ray observations toward
Draco give a lower value at $T \sim 10^6$ K, $EM \sin b \sim 0.004$ cm^{-6} pc. Shapiro and
Benjamin (1992 and personal communication) have shown that the discrepancy
can be removed by absorption: a cooling flow with an initial temperature of 10^6
K is consistent with the observations of soft X–ray emission if it is attenuated
by absorption through a column density of $(2-3) \times 10^{20}$ cm^{-2}, comparable to
that observed at high Galactic latitudes (e.g., Danly et al 1992). However, the
model is extremely sensitive to this initial temperature since, as just mentioned,
the hottest part of the cooling flow is also the most luminous. For example,
in regions in which the absorption column density is less than 2×10^{20} cm^{-2},
the initial temperature must be less than 10^6 K but greater than $10^{5.8}$ K to
be consistent with observation. Thus, although the Shapiro & Benjamin (1992)
fountain model is the most successful one to date in accounting for the observations of emission and absorption by the halo gas, the lack of a mechanism
for tuning the initial temperature so precisely must be considered a significant
weakness.

Thermal conduction provides a possible mechanism for reducing the X–ray emission from a fountain, and thereby decreasing the sensitivity to the initial temperature. When thermal conduction is effective, heat is conducted from the hot gas to cooler gas, where it can be radiated more effectively. If the C IV emission arises in the conductive interfaces between a hot halo and cooler clouds, a large number of interfaces ($\sim 10^2$) would be required to account for the intensity (MB). Slavin et al (1993) have pointed out that such a number may not be unreasonable: if there are 10-12 clouds per kpc in the halo as there are in the disk (an important point to check observationally), and if each cloud has two interfaces along the line of sight, then one would expect $\gtrsim 60$ interfaces over a typical 3 kpc line of sight through the halo. As remarked above, it is possible that the UV absorption spectra of magnetized conduction fronts calculated by Borkowski et al (1990) would agree with observation if photoionization were included, as Shapiro & Benjamin (1992) have done for cooling flows.

"Turbulent mixing layers" have been invoked by Slavin et al (1993) to account for the UV observations of the halo. They assume that the instabilities associated with the motion of clouds through hot halo gas result in mixing on such a fine scale that the electrons in the two gases instantaneously come to a single temperature; the mixture is then convected away from the unstable region, and it cools by radiation. For example, a 10^6 K gas might mix with a 10^4 K gas to produce a mixture with an electron temperature of 10^5 K, but with an ionization characteristic of the two initial gases. Since the velocities in the halo are very small compared to the electron thermal velocity, it is difficult to see how thermal conduction between the 10^5 K mixture and the 10^6 K gas can be suppressed. Even though the model is physically unrealistic, it may approximate the spectrum of a conductive interface with non–equilibrium ionization. The UV absorption spectrum produced in their model depends on the initial electron temperature of the mixed gas and on the velocity dispersion. Their favored model has a mean temperature of $10^{5.3}$ K, a turbulent velocity of about 75 km s^{-1}, and depleted abundances. This model has $N(\text{Si IV})/N(\text{C IV})$ and $N(\text{N V})/N(\text{C IV})$ smaller than the observed values by a factor ~ 2, which they suggest could be remedied by an adjustment in the abundances or the depletions. If this discrepancy were removed by a reduction in the C IV column density in an individual layer, then over 200 such layers would be required to account for the observed column density, which seems excessive. Slavin et al (1993) have not attempted to account for the halo X–ray emission.

2.3 What is f?

We have expressed the mass flux into the disk implied by cooling flow models in terms of the fraction of the disk, f, covered by such a flow. Similarly, the H I mass flux observed in the solar neighborhood was translated into a global mass flow rate by including a (possibly different) factor f (eq. 1). If the halo is dominated by the injection of mass and energy from superbubbles, f is of order the filling factor of superbubbles in the disk. The simplest assumption is that the rest of the Galaxy is like the solar neighborhood; in that case, $f \simeq 1$.

There are two arguments, however, that suggest $f \ll 1$. First, the rate at which SNRs heat gas to a temperature of at least 10^6 K is

$$\dot{M}(\text{SNR}) = 20 \left(\frac{\text{SN rate}}{2.2 \text{ century}^{-1}} \right) \frac{(E/10^{51} \text{ erg})}{(T/10^6 \text{ K})} \quad M_\odot \text{ yr}^{-1} \tag{7}$$

(Chevalier & Oegerle 1979); the estimate of the Galactic supernova rate is taken from McKee (1990). Recall that the mass flow implied by the C IV observations is $21 f R_{10}^2\ M_\odot$ yr^{-1}; thus, f is about equal to the fraction of the supernova energy going into the halo. McKee & Ostriker (1977) have argued that this fraction is small because most of the supernova energy is radiated in the disk; from a completely different viewpoint, Cox (1988) has argued that this fraction is small because SNRs occupy only a small fraction of the volume of the disk, so they cannot vent into the halo. (Since Don and I rarely agree on any point of interstellar astrophysics, one is tempted to conclude that such points of agreement must be correct!)

The second argument for a small value of f comes from considering the value of the thermal pressure in the injection region implied by the mass flow rate inferred from the C IV observations. In fountain models, gas is injected into the halo from the disk at a temperature $T \sim 10^6$ K and a velocity $v \sim 10^2$ km s^{-1}. In a steady state, the mass flux observed in the C IV emitting gas at a temperature of order 10^5 K (eq. 3) is the same as that injected at $T \sim 10^6$ K. The thermal pressure in the injection region is then

$$P = \frac{\rho kT}{\mu} = \left[\frac{\dot{M}(\text{C IV})}{2\pi R^2 f v}\right] \frac{kT}{\mu} = 3.0 \times 10^{-12} \left[\frac{(T/10^6 \text{ K})}{(v/10^2 \text{ km s}^{-1})}\right] \text{ dyne cm}^{-2},$$

(8)

where μ is the mean mass per particle. This is comparable to the total pressure in the disk (§1), so the injected gas can indeed expand out of the disk into the halo. However, this value for the thermal pressure in the injection region is almost an order of magnitude larger than the typical thermal pressure in the disk, 5.0×10^{-13} dyne cm^{-2} (Jura 1975). Thus, the source of the hot gas for the Galactic fountain must occupy a small fraction of the disk. (Note that we have glossed over any possible difference between the value of f for the upwelling hot gas and the downflowing cooler gas.) We turn now to a direct estimate of f.

3. THE LUMINOSITY DISTRIBUTION OF GALACTIC OB ASSOCIATIONS

OB associations ionize the surrounding ISM and then create large cavities, "superbubbles", through the combined action of their winds and supernovae. The size of the superbubbles depends on the number of stars in the association that will explode as supernovae, which in turn is directly related to the luminosity of the association. The determination of the filling factor of superbubbles in the disk thus requires a knowledge of the luminosity function of the OB associations. A brief account of preliminary results on this problem (McKee & Williams 1992) is given here.

Galactic OB associations can be identified throughout the Galaxy by radio observations of their associated H II regions, but the complex structure of these regions complicates the determination of their luminosity distribution. Radio surveys of H II regions are sensitive to the dense gas in the immediate vicinity of the association, but there is a good deal of diffuse thermal emission from "extreme low density" H II regions (Smith, Biermann, & Mezger 1978) and low density "envelopes" of H II regions (Anantharamaiah 1985). Some ionizing photons escape the H II regions entirely and ionize gas in the diffuse ISM (the WIM).

The data on Galactic H II regions are inadequate to deduce an accurate luminosity function, so we adopt a simple analytic form that is consistent with the overall properties of the distribution. Observations of extragalactic H II regions (Kennicutt, Edgar, & Hodge 1989) appear to be consistent with a power–law with an upper cutoff. In other words, if $\mathcal{N}_a(S)$ is the number of associations emitting more than S ionizing photons per second, then $d\mathcal{N}_a(S)$ is proportional to $S^{-\alpha}$ and $\mathcal{N}_a(S)$ vanishes above an upper limit S_u:

$$\mathcal{N}_a(S) = \frac{\mathcal{N}_{au}}{\alpha}\left[\left(\frac{S_u}{S}\right)^\alpha - 1\right]. \tag{9}$$

The quantity $\mathcal{N}_{au}$ is approximately equal to the number of associations which have S within a factor two of the upper limit. If $\mathcal{N}_{au} \sim 1$, then it is likely that this limit is set by chance; on the other hand, if $\mathcal{N}_{au} \gg 1$, then the limit is likely to be set by some physical mechanism, since otherwise there would be no reason for some associations not to have $S > S_u$. The data of Kennicutt et al (1989) suggest that normal spiral galaxies have $\alpha \sim 1$ and $\mathcal{N}_{au} \simeq 4 - 5$.

To determine the luminosity function, we must distinguish the ionizing photon luminosity emitted by the stars in an association, S, from the ionizing photon luminosity absorbed by the gas, S'; the difference is absorbed by dust or escapes from the Galaxy entirely. Furthermore, since only a fraction of the ionizing photons are absorbed in the dense gas near the association, where they create a radio H II region, we let S'_R be the ionizing photon luminosity inferred from radio observations of the H II region. Note that optical observations of extragalactic H II regions are not limited to regions of very high emission measure as are the radio observations of Galactic H II regions, so the optically determined ionizing photon luminosity is about equal to S' (Kennicutt et al 1989).

The survey of giant radio H II regions by Smith et al (1978) is consistent with the distribution

$$\mathcal{N}_a(S'_R) = 5.5\left(\frac{88}{S'_{R49}} - 1\right), \tag{10}$$

where we have measured S'_R in units of 10^{49} photons s^{-1} (for example, an O9 star emits 0.2×10^{49} ionizing photons s^{-1}). The correction for dust absorption is uncertain because the dust in H II regions appears to differ from normal inter-stellar dust, being deficient in small grains (Baldwin et al 1991). For simplicity, we shall adopt the correction of Smith et al, which yields an average reduction of a factor 2.2 in the ionizing photon luminosity in radio H II regions due to absorption by dust ($S'_R \simeq S_R/2.2$). The corresponding luminosity distribution is

$$\mathcal{N}_a(S_R) = 5.5\left(\frac{164}{S_{R49}} - 1\right). \tag{11}$$

This gives a total ionizing photon luminosity due to radio H II regions in the entire Galaxy of

$$S_{RT} = \int S_R d\mathcal{N}_a = \mathcal{N}_{au}S_{Ru}\ln\left(\frac{S_{Ru}}{S_{R\ell}}\right) \simeq 6 \times 10^{52} \quad \text{photons s}^{-1}, \tag{12}$$

where $S_{R\ell} \sim 0.1 \times 10^{49}$ photons s^{-1} is the lower limit on the ionizing photon luminosity of a radio H II region.

The actual value of the total ionizing luminosity of the Galaxy can be de-termined in two ways: First, observations of the radio free–free emission indicate

$S'_T = 1.45 \times 10^{53}$ photons s^{-1}, and $S_T = 2.0 \times 10^{53}$ photons s^{-1} after allowing for dust absorption (Smith et al 1978; Güsten & Mezger 1982; the correction for dust is less than the factor 2.2 cited for the radio H II regions because much of the Galactic emission is from low density gas where the dust absorption is only about 20%). More recently, observations of the Galactic N II λ 122 μm line (Wright et al 1991) give $S'_T = 2.45 \times 10^{53}$ photons s^{-1}, based on the assumption that the nitrogen abundance throughout the Galaxy has the local value; because the abundance in the inner Galaxy is higher, this is an upper limit to the true value. We conclude that radio H II regions fail to account for the observed Galactic ionizing photon luminosity by a factor of about 3.

Since OB stars are the dominant source of ionization (as shown in other galaxies by Kennicutt et al 1989), there are two main approaches to account for the discrepancy:

Model A: More Associations. In this model, only about 1/3 of the associations are observable as radio H II regions, with the remainder being "extreme low density" H II regions (Smith et al 1978). As a result, this distribution is characterized by $\mathcal{N}_{au} \simeq 3 \times 5.5 = 16.5$. A possible problem with this model is that the lifetime of the massive stars responsible for most of the ionization is only 4×10^6 yr (e.g., McKee 1989), and it is difficult to destroy the giant molecular cloud from which the association was born in such a short time. A more serious problem is that the Galaxy would be anomalous in this model, since the data of Kennicutt et al (1989) show that normal spiral galaxies have $\mathcal{N}_{au} \simeq 4 - 5$.

Model B: Bigger Associations. In this model, each association emits about 3 times more ionizing photons that inferred from the radio H II region, so that $S \simeq 3S_R$; in particular, the upper cutoff to the distribution is $S_{Ru49} = 475$. The H II regions are partially density bounded, just as in the case of champagne flows. Observed examples include the Rosette Nebula (Cox, Deharveng, & Leene 1990) and the Orion Nebula (e.g., Baldwin et al 1991), each of which has $S \simeq 2S'_R$, and W43, for which Heiles & Koo (1992) infer an ionizing luminosity corresponding to about 3 times the dust–corrected value of Smith et al (1978). Support for this model comes from the work of Anantharamaiah (1985), who has shown that diffuse radio recombination line emission arises in localized regions and that its velocity is correlated with the velocity of the H II regions along the line of sight; he therefore ascribes this emission to low density envelopes around H II regions. The observations of extragalactic H II regions by Kennicutt et al (1989) show that the optical diameters are typically at least 100-200 pc, significantly larger than the radio diameters of the Galactic H II regions catalogued by Smith et al (1978)—the radio observations of Galactic H II regions are sensitive only to the brightest parts and therefore miss much of the ionizing luminosity. Finally, as remarked above, normal spiral galaxies typically have $\mathcal{N}_{au} \simeq 4.5$, comparable to the value for the Galaxy in this model. We therefore conclude that the distribution of OB associations in the Galaxy is given by

$$\mathcal{N}_a(S) = 5.5 \left(\frac{475}{S_{49}} - 1 \right). \tag{13}$$

This implies that the most luminous association in the Galaxy has more than 5 times the luminosity directly measured for the brightest radio H II region. The fact that $\mathcal{N}_{au} = 5.5$ is significantly greater than unity suggests that the maximum luminosity of Galactic H II regions is set by a physical mechanism, not by chance.

How many supernovae are associated with these associations? Let $\mathcal{N}_{*h}$ be the number of high–mass stars in the association that will eventually explode as supernovae. Setting the threshold for supernovae at 8 $M_\odot$ and adopting the Scalo (1986) IMF, we find that the number of stars is related to the ionizing luminosity by $\mathcal{N}_{*h} = 2.6S_{49}$. The effective lifetime of these stars is 4×10^6 yr. Each association has a number of subassociations, and the lifetime of the association as a whole is typically about 2×10^7 yr (Blaauw 1964, 1991). The number of massive stars produced over the entire lifetime of an association is then $\mathcal{N}_{*h} = 13S_{49}$. For the largest association in the Galaxy, which has $S_{49} = 475$, the number of supernova progenitors is 6200. Heiles (1990) inferred a similar number (8880) by comparison with the largest H II regions in the Sb galaxies observed by Kennicutt et al (1989). The distribution of associations with respect to the number of supernova progenitors produced over the lifetime of the association is then

$$\mathcal{N}_a(\mathcal{N}_{*h}) = 5.5 \left(\frac{6200}{\mathcal{N}_{*h}} - 1 \right). \tag{14}$$

These associations are all luminous, and have ages less than 2×10^7 yr. The supernovae in an association explode over a time interval of about 6×10^7 yr (Heiles 1990), so the number of active superbubbles is about three times the value in equation (14). Koo & McKee (1992) gave a lower limit of 800 supernovae for a superbubble to break out of the disk entirely; if the actual number required for breakout is close to this, then equation (14) suggests that there are a number of breakout bubbles, or "chimneys", in the Galaxy.

The filling factor of these associations can be calculated by a generalization of the method outlined by Heiles (1990). A description of this calculation will be given elsewhere (McKee & Williams, in preparation). We find that the fraction of the disk area in the solar neighborhood that is covered by super-bubbles that have broken through the thin, cold gas disk is $Q_{2D} \sim 0.1$. There are several uncertainties in this estimate, but it is consistent with Heiles' (1980) observational estimate.

For cooling flow models, this implies that the covering factor $f \sim 0.1$ as well. The total mass flow rates from the halo onto the disk implied by H I (eq. 1) and C IV (eq. 3) observations are then several $M_\odot$ yr^{-1}. The X–ray emission from the halo should be patchy, as observed (McCammon 1993). We conclude that the superbubble model for energizing the halo appears consistent with observation, but further observational and theoretical work is needed to determine if it is indeed correct.

4. CONCLUSIONS

Gas is observed to extend far from the plane of the Galaxy, with a scale height that tends to increase with the degree of ionization. Much of this high–z gas has a large velocity dispersion, implying a rapid exchange of mass with the disk. The great altitude of the gas means that a large pressure is required in the disk to keep it there; of the total pressure $(3 - 4 \times 10^{-12}$ dyne cm^{-2}, or $\sim 2.5 \times 10^4$ cm^{-3} K), only a relatively small portion (~ 3600 cm^{-3} K) is thermal.

Recent UV observations of halo gas, both in emission and in absorption, are consistent with models in which gas is cooling from a temperature well above 10^5 K, at a sufficiently high rate that it is out of ionization equilibrium.

Cooling flow models in which the gas initially undergoes isobaric cooling and then switches to isochoric cooling work quite well (Table 1). Less work has been done on conduction fronts, turbulent mixing layers, or old supernova remnants, but they may also provide reasonable agreement with the data. In principle, observations of O VI emission lines and of X–rays can discriminate among these models: cooling flow models without conduction are very sensitive to the initial temperature because the hottest gas is the most luminous. As a result, they produce relatively more O VI and X-ray emission than models with conduction.

A variety of heating mechanisms for the halo gas have been proposed. Photoionization models have been ruled out by observation. Heating by halo SNRs remains a viable possibility, but more work is needed to evaluate this model quantitatively. Assessment of Raymond's (1990) microflare model must await a deeper understanding of the physics of magnetic field reconnection. Energization by supernovae in the disk—a Galactic fountain—appears viable, provided the fountain is driven by supernovae in OB associations (which produce superbubbles), rather than by isolated supernovae.

Analysis of the distribution of OB associations in the Galaxy indicates that about 2/3 of the ionizing photons emitted by the associations escape from the radio H II regions excited by the associations, and ionize lower density gas in the vicinity. The distribution of associations in terms of the number of massive stars that will eventually undergo a supernova explosion (taken to be all stars more massive than 8 $M_\odot$) is given by equation (14). The largest associations in the Galaxy are truly enormous, with over 6000 supernova progenitors. The superbubbles produced by such large associations can most likely break out of the gas disk of the Galaxy entirely. The fraction of the disk of the Galaxy covered by superbubbles is about 10%, which indicates that the mass flow between the disk and the halo is several $M_\odot$ yr^{-1}.

ACKNOWLEDGMENTS

The work described in §3 would not have been possible without the contributions of Jonathan Williams. Comments by Carl Heiles, John Holliman, Rob Kennicutt, and Mike Shull, and an extensive correspondence with Paul Shapiro and Robert Benjamin, are gratefully acknowledged. My research is supported by NSF grant AST89-18573.

REFERENCES

Anantharamaiah, K.R. 1985, J. Astr. Ap., 6, 203
Baldwin, J.A., et al 1991, ApJ, 374, 580.
Blaauw, A. 1964, ARA&A, 2, 213.
Blaauw, A. 1991, in *The Physics of Star Formation and Early Stellar Evolution*, ed. C. Lada & N. Kylafis (Dordrecht: Kluwer), p. 125.
Borkowski, K., Balbus, S.A., & Fristrom, C.C. 1990, ApJ, 355, 501
Boulares, A., & Cox, D.P. 1990, ApJ, 365, 544
Bregman, J.N. 1980, ApJ, 236, 577
Bregman, J.N., & Harrington, J.P. 1986, ApJ, 309, 833
Bruhweiler, F.C., Gull, T.R., Kafatos, M., & Sofia, S. 1980, ApJ, 238, L27
Burks, G.S., York, D.G., Blades, J.C., Bohlin, R.C., & Wamsteker, W. 1991, ApJ, 381, 55
Burrows, D.N., & Mendenhall, J.A. 1991, Nature, 351, 629

Chevalier, R.A., & Fransson, C. 1984, Ap. J. (Letters), 274, L43
Chevalier, R.A., & Oegerle, W.R. 1979, ApJ, 227, 398
Cowie, L.L., Hu, E.M., Taylor, W., & York, D.G. 1981, ApJ, 250, L25
Cox, D.P. 1981, ApJ, 245, 534
Cox, D.P. 1988, in *Supernova Remnants and the Interstellar Medium*, ed. R.S. Roger
    and T.L. Landecker (Cambridge: Cambridge University Press), 73
Cox, P., Deharveng, L., & Leene, A. 1990, A&A, 230, 181.
Danly, L. 1989, ApJ, 342, 785
Danly, L., Lockman, F.J., Meade, M.R., & Savage, B.D. 1992, ApJS, 81, 125
Davidsen, A.F., Bowers, C.W., Kruk, J.W., Ferguson, H.C., Kriss, G.A., Blair, W.P.,
    & Long, K.S. 1992, ApJ, 000, 000
Dickey, J.M., & Lockman, F.J. 1990, ARA&A, 28, 215
Dixon, W.V., Davidsen, A.F., Bowers, C.W., Kriss, G.A., Kruk, J.W., & Ferguson,
    H.C. 1993, this volume
Edelstein, J., & Bowyer, S. 1992, Adv. Sp. Res., 000, 000
Edgar, R.J., & Chevalier, R.A. 1986, ApJ, 310, L27
Ferrara, A. 1993, ApJ, in press
Giovanelli, R. 1980, AJ, 85, 1155
Güsten, R., & Mezger, P.G. 1982, Vistas in Astronomy, 26, 159
Hartquist, T.W., & Morfill, G.E. 1986, ApJ, 311, 518
Hartquist, T.W., Pettini, M., & Tallant, A. 1984, ApJ, 276, 519
Heiles, C. 1980, Ap. J, 235, 833
Heiles, C. 1984, ApJS, 55, 585
Heiles, C. 1990, ApJ, 354, 483
Heiles, C., & Koo, B.-C. 1992, in preparation
Houck, J.C., & Bregman, J.N. 1990, ApJ, 352, 506
Ikeuchi, S. 1987, in *Starbursts and Galaxy Evolution*, ed. T.X. Thuan & T. Montmerle
    (Gif sur Yvette: Editions Frontieres), 27
Ikeuchi, S. 1988, Fund. Cosmic Phys., 12, 255
Jura, M. 1975, Ap. J., 197, 581
Kaelble, A., de Boer, K.S., & Grewing, M. 1985, A&A, 143, 408
Kahn, F.D. 1981, in *Investigating the Universe*, ed. F.D. Kahn (Dordrecht: Reidel), 1
Kennicutt, R.C., Edgar, B.K., & Hodge, P.W. 1989, ApJ, 337, 761.
Koo, B.-C., & McKee, C.F. 1992, ApJ, 388, 93
Kulkarni, S.R., & Fich, M. 1985, Ap. J, 289, 792
Kulkarni, S.R., & Heiles, C. 1987, in *Interstellar Processes*, ed. D. Hollenbach & H.
    Thronson (Dordrecht: Reidel), 87
Li, F., & Ikeuchi, S. 1990, ApJS, 73, 401
Lockman, F.J. 1984, Ap. J., 283, 90
Lockman, F.J., & Gehman, C.S. 1991, ApJ, 382, 182
Martin, C., & Bowyer, C.S. 1990, ApJ, 350, 242 (MB)
McCammon, D. 1993, this volume
McCray, R., & Kafatos, M. 1987, ApJ, 317, 190
McKee, C.F. 1989, ApJ, 345, 782
McKee, C.F. 1990, in *The Evolution of the Interstellar Medium*, ed. L. Blitz (San
    Francisco: Astronomical Society of the Pacific), 3
McKee, C.F., & Ostriker, J.P. 1977, ApJ, 218, 148
McKee, C.F., & Williams, J. 1992, in *Star–Forming Galaxies and Their Interstellar
    Media*, ed. J.J. Franco (Cambridge: Cambridge University Press), in press
Merrifield, M.R. 1993, this volume
Meyer, D.M., & Roth, K.C. 1992, ApJ, 000, 000
Narayan, R., & Ostriker, J.P. 1990, ApJ, 352, 222

Norman, C., & Ikeuchi, S. 1989, ApJ, 345, 372
Raymond, J.C. 1990, ApJ, 365, 387
Reynolds, R.J. 1993, this volume
Romanik, C.J. 1988, ApJ, 330, 1022
Savage, B.D., Lu, L., Weymann, R.J., Morris, S.L., & Gilliland, R.L. 1992, ApJ, 000, 000
Scalo, J.S. 1986, Fund. Cos. Phys., 11, 1
Sembach, K.R., & Savage, B.D. 1992, ApJS, 83, 147
Shapiro, P.R., & Benjamin, R.A. 1992, in *Star-Forming Galaxies and Their Interstellar Media*, ed. J.J. Franco (Cambridge: Cambridge University Press), in press
Shapiro, P.R., & Field, G.B. 1976, ApJ, 205, 762
Slavin, J.D., & Cox, D.P. 1992, ApJ, 392, 131
Slavin, J.D., Shull, J.M., & Begelman, M.C. 1993, ApJ, 407, 000
Smith, L.F., Biermann, P., & Mezger, P.G. 1978, A&A, 66, 65
Snowden, S.L., Mebold, U., Hirth, W., Herbstmeier, U. & Schmitt, J.H.H.M. 1991, Science, 252, 1529
Spitzer, L. 1956, Ap.J., 124, 20
Spitzer, L. 1990, ARA&A, 28, 71
Tenorio-Tagle, G., & Bodenheimer, P. 1988, ARA&A, 26, 145
Wakker, B.P. 1991, A&A, 250, 499
Weaver, H. 1974, Highlights Astr., 3, 423
Wright, E.L., et al 1991, ApJ, 381, 200

GLOBULAR CLUSTERS AND PULSARS

R. N. Manchester

Australia Telescope National Facility, CSIRO, Epping, NSW 2121, Australia
and
Physics Department, Princeton University, Princeton, NJ 08544.

ABSTRACT

Globular clusters are the oldest known stellar systems in our Galaxy. It is therefore surprising that they are found to contain relatively large numbers of millisecond pulsars, since these pulsars have a limited active lifetime. Globular cluster pulsars are preferentially found in clusters with massive and dense cores and are often binary with another star. These properties suggest that they are neutron stars which have been captured by a cluster star or binary system and 'recycled', that is, spun up to their present short periods by accretion of mass from the companion. Based on the observed sample of 30 or so pulsars, it is estimated that the total number of pulsars in globular clusters is about 1000. This is about two orders of magnitude greater than the observed number of low-mass X-ray binary systems, which places limits on the lifetime of these systems if they are the progenitors of millisecond pulsars. Some pulsars lying close to the core of dense clusters are observed to be accelerated in the cluster gravitational field, allowing lower limits to be placed on the core mass-to-light ratio and density.

1. INTRODUCTION

There are now about 550 pulsars known. The vast majority of these are located in the disk of our Galaxy and have periods between 0.2 and 2.0 seconds. It is now universally accepted that pulsars are rotating neutron stars, spinning once per pulse period. Ages, or at least upper limits to ages, of pulsars can be readily determined from the observed rate of increase of the pulse period. For a dipolar magnetic field, the characteristic age of a pulsar is given by $\tau_c = P/(2\dot{P})$, where $\dot{P}$ is the first time derivative of the period. For most pulsars, observed period derivatives are $\sim 10^{-15}$, so characteristic ages are typically one to ten million years. A few pulsars have much larger period derivatives indicating smaller ages and most of these are associated with known supernova remnants. The strength of the magnetic field at the neutron-star surface can also be estimated from the observed period derivative, and is proportional to $(P\dot{P})^{1/2}$. For most pulsars the surface field strength is between 10^{11} and 10^{12} G.

A relatively small, but very important, subset of the known pulsars has much shorter periods, typically a few milliseconds. These *millisecond* pulsars have extremely stable periods, with period derivatives three to six orders of magnitude less than 'normal' pulsars. Their characteristic ages are therefore very large, in many cases more than 10^9 years, and their surface magnetic field strengths are low, 10^9 or 10^{10} G.

Although they are very old, it is obvious that these pulsars cannot have been formed by simple aging of normal pulsars. Some additional mechanism must be invoked to explain their very short periods. A strong clue to the nature of this mechanism is given by the fact that about half of the millisecond pulsars are binary, in orbit with another star. In most cases the companion star is a white

© 1993 American Institute of Physics

dwarf, a star at the end of its evolutionary path. At an earlier evolutionary stage, material from this star may have overflowed its Roche lobe and formed an accretion disk around the neutron star. At the Alfvén surface, where the energy density of the corotating neutron-star magnetic field equals the kinetic energy density of the disk material, the gas is trapped by the field and falls down to the magnetic polar cap. If the field corotation velocity is less than the orbital velocity at the Alfvén surface, then angular momentum is transferred to the neutron star, spinning it up. The equilibrium period, when the two velocities are equal, is given by

$$P_{eq} \simeq 1.9\, B_9^{6/7} \left(\dot{M}/\dot{M}_{Edd} \right)^{-3/7} \text{ ms}, \tag{1}$$

where B_9 is the magnetic field at the surface of the neutron star in units of 10^9 G, $\dot{M}$ is the mass-accretion rate and $\dot{M}_{Edd}$ is the Eddington-limited mass-accretion rate. So, to spin up to millisecond periods, the neutron-star magnetic field must be relatively weak. It is not clear whether neutron-star fields decay spontaneously or if the decay is induced by the accretion process itself, but decay they must. Even for accretion at the Eddington-limited rate, it takes a long time, $\gtrsim 10^8$ years, for a neutron star to be spun up to millisecond periods. Therefore, most millisecond pulsars are believed to have been formed in binary systems with a companion of relatively low mass, say one to two solar masses, which evolves sufficiently slowly. The subject of millisecond pulsar formation and evolution has been extensively reviewed by Bhattacharya & van den Heuvel (1991).

During the accretion phase, these systems become powerful sources of X-rays and many have been detected as such. In some, the X-ray emission is modulated by the neutron-star rotation and the spin-up can be directly observed. X-ray surveys showed that Low Mass X-ray Binary systems (LMXBs) are very common in globular clusters (Katz 1975). Although these clusters contain only 0.1% of the stars in the Galaxy, they contain about 10% of the known LMXBs. Therefore, if the formation scheme outlined above is correct, globular clusters should be a good place to look for millisecond pulsars.

2. GLOBULAR CLUSTER PULSARS

The Jodrell Bank group was the first to have success in finding a millisecond pulsar in a globular cluster when, in 1987, Lyne et al. announced the discovery of a pulsar with a 3 ms period lying within the core of the dense cluster M28. Searches since then have been so successful that more than three-quarters of the millisecond pulsars now known lie in globular clusters. Table 1 lists the globular clusters known to contain pulsars. All but a few of these pulsars have periods in the millisecond range (conventionally defined to be less than 25 ms) and more than a third of them are binary. This is a much higher proportion of binaries than is found for ordinary galactic pulsars, lending support to the idea that millisecond pulsars are 'recycled'.

Two clusters in this list stand out: M15 with eight pulsars and 47 Tucanae with eleven. Both of these are massive clusters with very dense cores. Figure 1 gives the location of the M15 pulsars with respect to the cluster, showing that they are strongly concentrated in the core of the cluster, with only PSR B2127+11C lying well outside the core. This pulsar is interesting in its own right as it is an almost carbon copy of the famous relativistic binary pulsar PSR B1913+16 (Taylor & Weisberg 1989); it is in an eccentric orbit around another neutron star with an orbital period of 8.05 hours (Anderson et al. 1990b). PSR B1718−19 is

Table 1. Globular cluster pulsars

Cluster	–	PSR B	Nr of Pulsars	Nr of Binaries	Psr Period Range (ms)	Refs
M28	–	1821−24	1	–	3.05	1
M4	–	1620−26	1	1	11.07	2
M15	–	2127+11	8	1	4.0 – 110	3,4,5,6
M13	–	1639+36	2	1	3.5,10	7,8
M53	–	1310+18	1	1	33	7
M5	–	1516+02	2	1	5.5,7.9	9
47 Tuc	–	0021−72	11	4+	2.0 – 5.8	10,11,12
NGC 6440	–	1745−20	1	–	288	13
Terzan 5	–	1744−24	1	1	11.5	14
NGC 6624	–	1820−30	2	–	5.4 – 378	15
NGC 6539	–	1802−07	1	1	23.1	16
NGC 6760	–	1908+00	1	1	3.6	17
NGC 6342	–	1718−19	1	1	1004	18
			33	13+		

Refs: 1. Lyne et al. (1987) 2. Lyne et al. (1988) 3. Wolszczan et al. (1989b) 4. Anderson et al. (1990b) 5. Prince et al. (1991) 6. Anderson (1992) 7. Kulkarni et al. (1991) 8. Anderson et al. (1991) 9. Wolszczan et al. (1989a) 10. Manchester et al. (1990) 11. Manchester et al. (1991) 12. Robinson et al. (1993) 13. Manchester et al. (1989) 14. Lyne et al. (1990) 15. Biggs et al. (1990) 16. D'Amico et al. (1992) 17. Anderson et al. (1990a) 18. Lyne et al. (1992)

an unusual system in which the relatively long-period pulsar is eclipsed by a wind from the companion. It was found in a search directed toward NGC 6342 and is probably associated with this cluster. The Terzan 5 pulsar is another which is eclipsed by a wind from its companion. In this system the eclipses are extremely variable, sometimes extending over the entire orbital period.

Table 2 lists the known pulsars in the southern cluster 47 Tucanae, with binary period and minimum companion mass where appropriate. Four and maybe six of these eleven pulsars are binary, in contrast to M15, where only one of the eight pulsars is binary. Selection effects have contributed to the low proportion of binaries in M15, since the last five to be discovered (D – H) were found by methods which discriminated against a wide range of binary parameters (Anderson 1992). Another notable difference between the pulsars in these two clusters is that all of the pulsars in 47 Tuc have very short periods, whereas four of the pulsars in M15 (A, B, C and G) have periods greater than 30 ms.

Several letters in the alphabetical sequence are missing from this table. A and B are the two pulsars reported to lie in 47 Tuc by Ables et al. (1988) and Ables et al. (1989). Despite several searches for these pulsars in our Parkes data, we have been unable to confirm them. K was listed by Manchester et al. (1991) as a 1.79 ms pulsar, but it has since been realized that it is spurious; the frequency detected was actually the third harmonic of that for PSR B0021-72D. To compensate for that we have recently discovered PSR B0021-72N (Robinson et al. 1993), so the number of known pulsars in 47 Tuc remains at eleven. PSR B0021-72J is in a very short-period binary system and is eclipsed for about one quarter of the orbital period (Robinson et al. 1993).

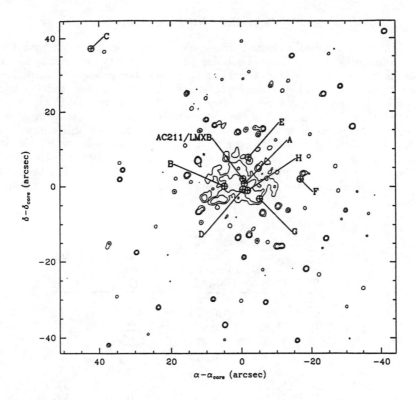

Figure 1: Positions of pulsars and the LMXB AC211 in the cluster M15. The contour lines are B-band photometry. (After Anderson 1992)

Table 2. Pulsars in 47 Tucanae

PSR B0021−72	Pulsar Period (ms)	Binary Period (d)	M_2 ($M_\odot$)
C	5.75	−	−
D	5.35	−	−
E	3.53	2.2	∼ 0.2
F	2.62	−	−
G	4.04	−	−
H	3.12	Yes	?
I	3.48	0.23	∼ 0.02
J	2.01	0.12	∼ 0.02
L	4.34	?	?
M	3.67	?	?
N	3.05	−	−

3. FORMATION AND EVOLUTION

It is clear that globular clusters are fertile breeding grounds for both LMXB systems and millisecond pulsars. Globular clusters are very old, much older than the lifetime of LMXBs. Main-sequence stars massive enough to have formed neutron stars at the end of their evolution would have evolved quickly and no such stars are likely to have formed since. Neutron stars formed in this way are primordial, that is, comparable in age to the cluster. Because of their larger mass, they will be concentrated in the core of a cluster in thermal equilibrium. Table 3 lists the characteristic ages of those globular-cluster pulsars for which long-term timing solutions are available. Since these ages are an upper limit to the time since the pulsar was born, or, for a recycled pulsar, since the recycling terminated, most of the globular cluster millisecond pulsars must have been formed relatively recently.

Table 3. Characteristic ages of cluster pulsars

PSR B	Pulse Period (ms)	Period Derivative $(\times 10^{-18})$	Characteristic Age (Gyr)	Ref.
0021−72C	5.75	−0.05	–	1
0021−72D	5.35	< 0.05	> 3.5	1
1620−26	11.08	0.82	0.21	2
1744−24A	11.56	−0.019	–	3
1821−24	3.05	1.6	0.030	4
2127+11A	110.66	−21.07	–	5,6
2127+11B	56.13	9.56	0.093	5,6
2127+11C	30.53	4.99	0.098	6
2127+11D	4.80	−10.75	–	6
2127+11E	4.65	0.18	0.42	6
2127+11F	4.03	0.032	2.0	6
2127+11G	37.66	2.0	0.29	6
2127+11H	6.74	0.024	4.5	6

References: 1. Robinson et al. (1993) 2. McKenna & Lyne (1988) 3. Nice & Thorsett (1992) 4. Foster et al. (1988) 5. Anderson et al. (1990b) 6. Anderson (1992)

Figure 2 shows that there is a strong preference for pulsars to be found in clusters with dense cores and a somewhat weaker preference for clusters with high escape velocity, which is approximately equivalent to large total mass. Since neutron stars apparently receive a substantial kick velocity at birth, a high escape velocity is necessary to ensure that they are not lost to the cluster.

Analysis of the observed population to determine the pulsar luminosity function and the dependence of pulsar formation on cluster parameters is difficult because of the low numbers. Selection effects also complicate matters. Millisecond pulsars are weak and all surveys are sensitivity limited, so the luminosity function is weakly constrained, especially at the low-luminosity end. Searches also discriminate against very short-period binary systems, although continuum observations (Fruchter & Goss 1990) show that such systems do not dominate the population.

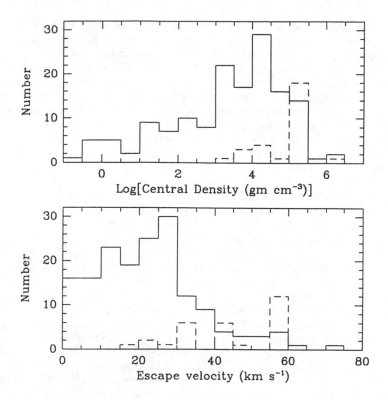

Figure 2: Distribution of central density and escape velocity for 138 galactic globular clusters from the data of Webbink (1985). The dashed line gives the distribution of pulsars sorted by the parameters of their home cluster.

By pooling the data for all clusters, Johnston, Kulkarni & Phinney (1992) find that the cluster pulsars are best fitted by the same luminosity function as field pulsars, $dN \propto L^{-2}dL$, and that the number of pulsars in a cluster is approximately proportional to $M_T \rho_c^{0.5}$, where M_T is the total mass of the cluster and ρ_c is its central density. The total number of pulsars in the cluster system depends on assumptions about the lower end of the luminosity function and is at least $500/f$, where f is the beaming fraction, that is, the fraction of the celestial sphere covered by the beam of a typical pulsar. For millisecond pulsars f is probably about 0.5. A typical lifetime of 10^9 years therefore gives a birthrate for globular cluster pulsars of about 10^{-6} yr^{-1}.

Capture of a neutron star by another star depends on tidal dissipation of a sufficient proportion of the interaction energy and requires a near collision, within a few stellar radii, of the two stars. The rate of tidal capture can be expressed in the form:

$$r_2 \sim 7 \times 10^{-13} \, \rho_V \, \sigma_{10}^{-1.1} \, N_{ns} \text{ yr}^{-1} \qquad (2)$$

where ρ_V is the V-band luminosity density in the core of the cluster in units of $10^4 L_\odot \text{pc}^{-3}$, σ_{10} is the central velocity dispersion in units of 10 km s^{-1}, and N_{ns}

is the number of neutron stars in the cluster (Hut et al. 1992). For clusters such as M15 and 47 Tucanae, which are both massive and have very dense cores, tidal captures of primordial neutron stars by cluster stars could account for the observed millisecond pulsar population. However, millisecond pulsars are found in several clusters with much smaller core densities, for example, M13 and M53. In clusters such as these, the rate of tidal captures is much too small.

Fortunately, other formation routes are possible. In the past few years, considerable evidence for the existence of primordial binary systems in globular clusters has accumulated, and it is now believed that a third or more of the stars in globular clusters may be binary. The main evidence for this comes from the direct detection of eclipsing systems (Mateo et al. 1990; Hodder et al. 1992) and spectroscopic binaries (Pryor et al. 1989) and detection of large numbers of 'blue stragglers' in the cores of several clusters (Aurière, Ortolani & Lauzeral 1990; Paresce et al. 1991). These unusually blue systems are now believed to be binary stars. The cross-section for a three-body interaction is $\sim 5(a/R)^{0.3}$ times the tidal capture cross-section, where a is the binary semi-major axis and R is the stellar radius (Hut et al. 1992). If the binary fraction is as large as 0.3, these interactions will dominate and could produce the millisecond pulsar progenitor systems, even in low-density clusters.

LMXB systems are relatively luminous and, if accreting at the Eddington-limited rate, they can be detected in globular clusters across most of the Galaxy. The total number known in globular clusters is ~ 12, that is, about one percent of the estimated number of active pulsars. Therefore, if LMXBs radiate more-or-less isotropically and are the progenitor systems for millisecond pulsars, their lifetime must be $\sim 10^7$ years. But standard LMXB models suggest that the duration of the accretion phase is $\sim 10^9$ years (Bhattacharya & van den Heuvel 1991). Therefore, either LMXBs are not the progenitors of millisecond pulsars in globular clusters, or their X-ray lifetime is much shorter than 10^9 years.

Neutron-star formation by accretion-induced collapse of a massive white dwarf has been suggested by Michel (1987) and Bailyn & Grindlay (1990). This would provide an alternative formation route in which the LMXB stage would be absent or short-lived. X-ray sources with a luminosity $\sim 10^4$ times lower than the Eddington-limited value for a neutron star ($\sim 10^{38}$ erg s^{-1}) have been detected in several globular clusters (e.g. Hertz & Grindlay 1983), and in these systems, the compact star could be a white dwarf.

There is, however, evidence that the lifetime of LMXBs is much less than 10^9 years. Timescales of the order of 10^7 years for orbital evolution are observed in several X-ray binary systems (Parmar 1992) and the eclipsing millisecond pulsar, PSR 1957+21 (Ryba & Taylor 1991). Theoretical arguments suggest that energetic pulsar winds may accelerate the evolution of the companion star, giving X-ray lifetimes of the order of 10^7 years (Tavani 1991; Podsiadlowski 1991).

These mechanisms may resolve the birthrate problem, but some aspects of the observed population remain difficult to explain. Why, for example, do all of the known pulsars in 47 Tucanae have very short periods? Were they all formed in the relatively recent past? Maybe pulsar formation is accelerated at certain phases of globular cluster evolution, for example, during the lead-up to core collapse.

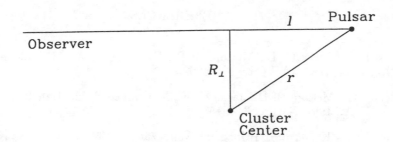

Figure 3: A pulsar is located at radial distance r from the cluster center and a projected distance l behind the cluster center. The pulsar is gravitationally accelerated toward the cluster center.

4. CLUSTER KINEMATICS

Given that pulsars exist in globular clusters, one can exploit their precise periodicity to investigate cluster properties. The energy of pulsars is stored in their rotation, so as pulsars evolve they slow down. Most pulsars do have positive period derivatives but, as Table 3 shows, some globular cluster pulsars have negative observed period derivatives. These pulsars lie close to the core of dense clusters; both PSRs 2127+11A and D in M15 lie within the core radius of 2.2 arcsec (Figure 1). The observed negative derivative is attributed to acceleration of the pulsar in the cluster gravitational field (Figure 3):

$$\frac{\dot{P}}{P} = \frac{\dot{P}_0}{P_0} - \frac{1}{c}\frac{GM(<r)}{r^2}\frac{l}{r} \tag{3}$$

where P_0 and $\dot{P}_0$ are the intrinsic pulsar period and its derivative and $M(<r)$ is the mass within radius r (Phinney 1992). If the pulsar is behind the cluster ($l > 0$), the observed period derivative is made more negative by the cluster acceleration.

Since the intrinsic period derivative is unknown, an observed negative period derivative allows one to place a lower limit on the central density:

$$\frac{\Sigma(<R_\perp)}{10^6\,M_\odot\mathrm{pc}^{-2}} > 2 \times 10^{15}\left|\frac{\dot{P}}{P}\right| \tag{4}$$

where $\Sigma(<R_\perp)$ is the surface mass density within a radius of $R_\perp$ of the cluster center. For M15, the observed negative derivatives show that within the core the mass to U-band light ratio exceeds 1.8 times the solar value and that the central density exceeds $2 \times 10^6\ M_\odot\mathrm{pc}^{-3}$ (Phinney 1992).

The discovery of pulsars in globular clusters has provided new insights into stellar, binary system and cluster evolution. Further study will elucidate the relationship of recycled pulsars to binary X-ray sources and will provide new information on the structure and dynamics of globular clusters.

REFERENCES

Ables J. G., Jacka C. E., McConnell D., Hamilton P. A., McCulloch P. M., Hall P. J., 1988. IAU Circ. No. 4602

Ables J. G., McConnell D., Jacka C. E., McCulloch P. M., Hall P. J., Hamilton P. A., 1989, Nature, 342, 158

Anderson S., Kulkarni S., Prince T., Wolszczan A., 1990a. IAU Circ. No. 5013

Anderson S. B., Gorham P. W., Kulkarni S. R., Prince T. A., Wolszczan A., 1990b, Nature, 346, 42

Anderson S., Kulkarni S., Prince T., Wolszczan A. 1991. Unpublished

Anderson S. J., 1992, PhD thesis, Caltech

Aurière M., Ortolani S., Lauzeral C., 1990, Nature, 344, 638

Bailyn C. D., Grindlay J. E., 1990, ApJ, 353, 159

Bhattacharya D., van den Heuvel E. P. J., 1991, Phys. Rep., 203, 1

Biggs J. D., Lyne A. G., Manchester R. N., Ashworth M., 1990. IAU Circ. No. 4988

D'Amico N. Bailes M., Lyne A. G., Manchester R. N., Johnston S., Fruchter A. S., Goss W. M., 1992, MNRAS, In press

Foster R. S., Backer D. C., Taylor J. H., Goss W. M., 1988, ApJ, 326, L13

Fruchter A. S., Goss W. M., 1990, ApJ, 365, L63

Hertz P., Grindlay J. E., 1983, ApJ, 275, 105

Hodder P. J. C., Nemec J. M., Richer H. B., Fahlman G. G., 1992, AJ, 103, 460

Hut P. et al., 1992, Pub. Astron. Soc. Pacific., In press

Johnston H. M., Kulkarni S. R., Phinney E. S., 1992, in van den Heuvel E. P. J., Rappaport S. A., eds, X-ray Binaries and Recycled Pulsars. Kluwer, Dordrecht, p. 349

Katz J. I., 1975, Nature, 253, 698

Kulkarni S. R., Anderson S. B., Prince T. A., Wolszczan A., 1991, Nature, 349, 47

Lyne A. G., Brinklow A., Middleditch J., Kulkarni S. R., Backer D. C., Clifton T. R., 1987, Nature, 328, 399

Lyne A. G., Biggs J. D., Brinklow A., Ashworth M., McKenna J., 1988, Nature, 332, 45

Lyne A. G. et al., 1990, Nature, 347, 650

Lyne A. G., Biggs J. D., Harrison P. A., Bailes M., 1992, Nature, Submitted

Manchester R. N., Lyne A. G., Johnston S., D'Amico N., Lim J., Kniffen D. A., Fruchter A. S., Goss W. M., 1989. IAU Circ. No. 4905

Manchester R. N., Lyne A. G., D'Amico N., Johnston S., Lim J., Kniffen D. A., 1990, Nature, 345, 598

Manchester R. N., Lyne A. G., Robinson C., D'Amico N. D., Bailes M., Lim J., 1991, Nature, 352, 219

Mateo M., Harris H. C., Nemec J. M., Olszewski E. W., 1990, AJ, 100, 469

McKenna J., Lyne A. G., 1988, Nature, 336, 226, Erratum ibid., 336, 698

Michel F. C., 1987, Nature, 329, 310

Nice D. J., Thorsett S. E., 1992, ApJ, 397, 249

Paresce F., Shara M., Meylan G., Baxter D., Greenfield P., 1991, Nature, 352, 297

Parmar A. N., 1992, in van den Heuvel E. P. J., Rappaport S. A., eds, X-ray Binaries and Recycled Pulsars. Kluwer, Dordrecht, p. 5

Phinney E. S., 1992, Phil. Trans. Roy. Soc. A, 341, 39

Podsiadlowski P., 1991, Nature, 350, 136

Prince T. A., Anderson S. B., Kulkarni S. R., Wolszczan W., 1991, apj, 374, L41

Pryor C., McClure R. D., Fletcher J. M., Hesser J., 1989, in Merrit D., ed, Dynamics of Dense Stellar Systems. Cambridge U.P., Cambridge, p. 175

Robinson C. R., Lyne A. G., Bailes M., Manchester R. N., Johnston S., D'Amico N., 1993, In preparation

Ryba M. F., Taylor J. H., 1991, ApJ, 380, 557

Tavani M., 1991, ApJ, 366, L27

Taylor J. H., Weisberg J. M., 1989, ApJ, 345, 434

Webbink R. F., 1985, in Goodman J., Hut P., eds, Dynamics of Star Clusters, IAU Symposium No. 113. Reidel, Dordrecht, p. 541

Wolszczan A., Anderson S., Kulkarni S., Prince T., 1989a. IAU Circ. No. 4880

Wolszczan A., Kulkarni S. R., Middleditch J., Backer D. C., Fruchter A. S., Dewey R. J., 1989b, Nature, 337, 531

Dissipational Halo Collapse and the Globular Cluster Metallicity Gradient

Dieter Hartmann & Neil Miller
Department of Physics & Astronomy, Clemson University, SC 29634

Grant Mathews
Department of Physics, LLNL, Livermore, CA 94551

Guy Malinie
Commissariat á l'Energie Atomique, 94195 Villeneuve-St.Georges, France

ABSTRACT

Globular clusters provide information on the early evolution of the Galaxy. Correlations between kinematics and composition might allow us to probe the most active period of Galactic Chemo-Dynamics (GCD). If the sub-class of low metallicity globulars traces the star formation history of the halo, we can use them to investigate the chemical evolution of the collapsing proto-galaxy. The absence of an observed radial metallicity gradient presents a constraint on collapse models. Following cluster orbits during collapse, when the potential changes rapidly, and in a static potential thereafter, we find only a small present-day metallicity gradient. Therefore, the absence of a pronounced metallicity gradient is not a strong argument against ELS scenarios of halo formation.

1. INTRODUCTION

Globular cluster systems provide the oldest known galactic fossils, and are therefore uniquely suited to study early phases of galaxy formation. Analysis of cluster kinematics and their metal abundances reveals two distinct subsystems in our Galaxy; halo- and disk clusters (Zinn 1985, 1990). Here, we consider the slowly rotating system of halo clusters, which separate chemically from the disk clusters at around [Fe/H] = −0.8 (Zinn 1985). The abundance distribution function (ADF) of halo clusters is assumed to result from cluster formation in the chemically evolving gas of the collapsing halo.

The observed ADF of halo clusters can be reproduced with solar yields for metal-poor stars, if one considers the collapse of a centrally condensed proto-Galaxy (Malinie, Hartmann, & Mathews 1991:MHM). A consequence of such an inhomogeneous collapse is the generation of a pronounced abundance gradient. However, the present-day abundance gradient observed in the halo globulars is very shallow, if present at all. Thus the question arises as to whether the subsequent dynamic evolution can wash out the gradient established during the collapse.

2. INHOMOGENEOUS HALO COLLAPSE

Consider a centrally condensed proto-galaxy with an initial gas density profile

$$\rho_g(r) = \rho_0 \exp\left[-\left(\frac{r}{R_n}\right)^n\right] , \tag{1a}$$

where the scaling length is chosen such that the total mass of the galaxy, M_g, is independent of the shape index n, i.e.,

$$R_n = \left(\frac{n\, M_g}{4\pi\rho_0\, \Gamma(3/n)}\right)^{1/3} . \tag{1b}$$

In this study we employ $n = 3$, and $\rho_0 = 6\ 10^{-26}$ g cm^{-3}. The stellar density, ρ_*, increases due to star formation described by a Schmidt law

$$\partial_t\rho_* = C_{\mathrm{SFR}}\, \rho_g^s . \tag{2}$$

We use $s = 2$, and a coefficient $C_{\mathrm{SFR}} = 2.5\ 10^8$ g^{-1} cm^3 s^{-1}. Stellar evolution will increase the gas metallicity due to stellar winds and supernova explosions, but also heat the gas and thus increase its velocity dispersion. This "pressure support" leads to an increase in the random velocities of stars at birth, which are super-imposed on the bulk flow of the collapsing cloud. This effect is important for the problem under consideration, because the more metal rich clusters, born later and thus deeper in the gravitational well, will receive larger birth velocities. Thus, these clusters will spread over larger distances, reducing the metallicity gradient. The local change in the gas velocity dispersion, σ_g, is calculated from

$$\partial_t\left(\frac{1}{2}\rho_g\sigma_g^2\right) = \lambda\, \Psi , \tag{3}$$

where Ψ is the star formation rate, and λ is the specific heating rate due to supernovae

$$\lambda = \int \zeta(\mathrm{m})\, \lambda_{\mathrm{II}}\, dm \sim 2\ 10^{15}\ \mathrm{ergs\ g^{-1}} , \tag{4}$$

where we have assumed a Salpeter IMF, $\zeta = 0.13$ m$^{-2.35}$, and a standard energy release of $\lambda_{\mathrm{II}} = 10^{51}$ ergs per supernova event.

The chemo-dynamical evolution of gas and stars is followed with separate numerical techniques. We assume spherical symmetry and include finite stellar lifetimes. Following Larson (1976) and Burkert & Hensler (1988), the gaseous component is represented by a cloud-like system. Inelastic cloud-cloud collisions dissipate kinetic energy and produce an isotropic velocity dispersion. We use Eulerian hydrodynamics to follow the evolution of the gas. The dynamics of stars and their interaction with the gas is simulated with a particle scheme. At every hydrodynamic timestep a "popstar" is generated in each cell of the radial grid. The subsequent orbit evolution of popstars does not depend on their mass, but varying amounts of metal enriched gas as well as momentum and energy are returned to the gas as the stellar population ages. Note, that the stellar feedback mechanism is non-instantaneous and non-local. The coupled equations for this gas-star system are evolved until the stellar metallicity has reached $\langle Z \rangle = 0.08\ Z_\odot$.

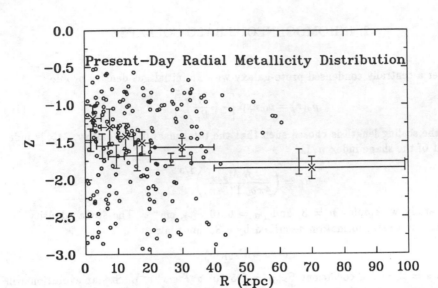

Figure 1: – Metallicities and present-day radii for 210 clusters evolved through inhomogeneous halo collapse. The metallicity gradient is small. Also shown are the observed averages derived by Pilachowski (1984; cross and bar) and Zinn (1985; bars only). Note, that the metallicity spread is large and roughly constant, as observed.

3. METALLICITY GRADIENTS

Based on low-resolution spectroscopy of 177 red giants in 19 globular clusters, Searle & Zinn (1978) concluded that there is no radial abundance gradient in the cluster system of the outer halo. From this lack of a gradient, the authors concluded that the halo globulars did not form in a pressure-supported collapse, but that Galactic clusters more metal-poor than $0.1\ Z_\odot$ were perhaps formed during a more rapid free-fall phase, as proposed by Eggen, Lynden-Bell & Sandage (1962). As an alternative interpretation, Searle & Zinn suggested that cluster formation might have occured in proto-galactic fragments that subsequently merged to form the present halo. (see also Mathews & Schramm 1992).

Pilachowski (1984) concluded from high quality data for 92 clusters that a gradient indeed exists in the halo, extending to ~ 100 kpc from the galactic center. Her results imply a relationship of the form $Z/Z_\odot = 0.10\ R^{-0.43\pm0.13}$. Although the gradient appears to be shallow, its existence implies that the clusters did participate in the general collapse of the Galaxy and did not form by a separate process. Based on data for 85 halo clusters (within R = 33 kpc and [Fe/H] less than −0.8), Zinn (1985) confirmed the presence of a metallicity gradient for clusters in the inner halo (R less than $\sim R_\odot$), but concluded that there is no compelling evidence for a gradient in the outer halo. Zinn (1985) emphasizes that the dispersion in [Fe/H] is roughly independent of distance, supporting the chaotic collapse picture suggested by Searle & Zinn (1978). We show, that a similar behavior of the metallicity dispersion is expected in the inhomogeneous halo collapse picture.

Popstar birth locations show a strong radial metallicity gradient. More stars are born deep inside the gravitational well, where the metallicity is higher, than in the outskirts of the proto-galaxy. An approximately linear correlation between stellar specific energy, E, and metallicity, Z, develops (MHM). As the collapse progresses, the gravitational potential changes rapidly. Thus, stellar energy per unit mass changes with time, details depending on the initial position and velocity. This violent relaxation still leaves a gradient (van Albada 1982), but the velocity dispersion of the more metal-rich stars is larger than that of more metal-poor stars because of supernova heating (equation 3). Thus, the initial correlation between metallicity and binding energy will be reduced by the heating mechanism. MHM find that the velocity dispersion is roughly proportional to the product λZ, which implies that sufficiently strong heating could offset the E-Z correlation (MHM). We followed cluster orbits during the collapse phase, when the potential changes rapidly, until the limiting mean metallicity is reached. After this, cluster orbits are integrated in a static potential. The snapshot of present-day globulars (Figure 1) shows no significant metallicity gradient and a metallicity dispersion that is roughly independent of radius.

4. CONCLUSIONS

We have calculated the dynamical and chemical evolution of the Galactic globular cluster system in the context of inhomogeneous halo collapse. The metallicity gradient is washed out by violent relaxation and the effects of supernova heating. Calculating cluster orbits from their birth to the present, we show that the present-day abundance gradient is small, consistent with the observations. These chemo-dynamical results emphasize that a dissipational halo collapse is not in conflict with the absence of a metallicty gradient in the present-day system of globular clusters.

REFERENCES

Burkert, A. & Hensler, G. 1988, A&A, 199, 131

Eggen, O. J., Lynden-Bell, D. & Sandage, A. R. 1962, ApJ, 136, 748

Larson, R. B. 1976, MNRAS, 176, 31

Malinie, G., Hartmann, D. H. & Mathews, G. J. 1991, ApJ, 376, 520; MHM

Mathews, G. J. & Schramm, D. N. 1992, ApJ, in press

Pilachowski, C. A. 1984, ApJ, 281, 614

Searle, L. & Zinn, R. 1978, ApJ, 225, 357

van Albada, T. S. 1982, MNRAS, 201, 939

Zinn, R. 1985, ApJ, 293, 424

Zinn, R. 1990, J. Roy. Astr. Soc. Can., 84, 89

LIMITS ON O VI EMISSION FROM THE GALACTIC CORONA WITH THE HOPKINS ULTRAVIOLET TELESCOPE

W. V. Dixon, A. F. Davidsen, C. W. Bowers, G. A. Kriss, J. W. Kruk
Center for Astrophysical Sciences, Department of Physics and Astronomy
The Johns Hopkins University, Homewood Campus, Baltimore, MD 21218

H. C. Ferguson
Institute of Astronomy, University of Cambridge
Madingley Road, Cambridge CB3 0HA, UK

Email ID
wvd@pha.jhu.edu

ABSTRACT

We have searched for O VI emission from Galactic coronal gas using long-exposure spectra obtained during orbital night through the $9'' \times 116''$ and $17'' \times 116''$ slits of the Hopkins Ultraviolet Telescope (HUT). In observations along six lines of sight with $\mid b \mid$ ranging from 13 to 57, we detect no O VI emission. Current models of the Galactic halo, when scaled by the reported flux of C IV $\lambda\lambda1548, 1551$, predict O VI $\lambda\lambda1032, 1037$ fluxes below our upper limits.

1. INTRODUCTION

A number of authors (*e.g.*, Edgar & Chevalier 1986, Shapiro & Benjamin 1992) have calculated the ionization balance of low-density gas cooling from a high temperature in various models of the Galactic corona. In this scenario, gas originating in the disk is shock heated by supernova explosions to temperatures above 10^6 K. The hot gas flows into the Galactic corona where it radiatively cools and recombines, loses buoyancy, and falls back toward the disk. These models predict emission from O VI $\lambda\lambda1032, 1037$ with a luminosity nearly an order of magnitude greater than that of C IV $\lambda\lambda1548, 1551$, which has already been observed in the diffuse UV background (Martin & Bowyer 1990). O VI has recently been observed in absorption with a column density through the Galactic halo that is consistent with these coronal models (Davidsen et al. 1992a).

2. THEORY·

Shapiro & Benjamin (1991, 1992) and Benjamin & Shapiro (1992) have modeled the ionization of low-density gas cooling from 10^6 K to 10^4 K, assuming that the cooling gas is isochoric (constant density), isobaric (constant pressure), or initially isobaric with a transition to isochoric. The type of cooling depends upon the assumed size D_0 of the cooling region. By including the effects of an external photoionization field, they were able to match the observed Galactic halo column densities of C IV, Si IV, and N V and UV emission from C IV and O III] in the isochoric limit. They found that self-ionization by the cooling gas is capable of producing the observed column density ratios in cooling regions as small as ~ 15 pc.

Martin & Bowyer (1990) measured emission from C IV $\lambda\lambda 1548, 1551$ along a line of sight $\sim 32°$ from the Galactic pole with an intensity of 5000 ± 800 ph cm^{-2}s^{-1}sr^{-1}. Scaling by this value, Shapiro & Benjamin (1992) predict an O VI flux of 6.14×10^{-7} ergs s^{-1}cm^{-2}sr^{-1}, or 32,000 ph cm^{-2}s^{-1}sr^{-1}, for a cooling region with $D_0 = 20$pc, a case intermediate between the isochoric and isobaric limits.

This model does not consider reddening by Galactic dust. If we model the dust as a plane parallel slab with E(B-V) = 0.03 toward the Galactic poles and apply the Seaton (1979) extinction curve (assuming it is valid to the Lyman edge), we find that the predicted O VI to C IV flux ratio must be scaled by the factor $10^{-0.06 \csc |b|}$. For lines of sight along the galactic pole, this factor is 0.87; for the Martin & Bowyer (1990) observation at $b = 58$, it becomes 0.85, reducing the O VI flux predicted by Shapiro & Benjamin (1992) for targets at this Galactic latitude to $\sim 27,000$ ph cm^{-2}s^{-1}sr^{-1}.

3. OBSERVATIONS

Our observations were carried out with HUT on the Astro-1 mission of the space shuttle *Columbia* in December, 1990. HUT consists of a 0.9-m mirror feeding a prime-focus spectrograph with a microchannel-plate intensifier and photo-diode array detector. First-order sensitivity extends from 830 Å to 1850 Å at 0.5 Å pixel^{-1} with a resolution of ~ 3 Å through the $9'' \times 116''$ slit and ~ 6 Å through the $17'' \times 116''$ slit. Because HUT is a photon-counting detector with an extremely low dark-count rate, we are able to obtain essentially photon-noise limited spectra even of faint sources. The spectrograph and telescope are described in detail by Davidsen et al. (1992b).

Of the 77 objects observed during the mission, spectra of six were analyzed for the presence of emission due to O VI in the Galactic halo. We selected spectra that had been obtained through the large slits of the HUT spectrograph in order to maximize the flux from the diffuse O VI emission, as smaller slits require enormous integration times to achieve comparable signal-to-noise ratios. Targets observed through the smaller slits are generally point sources with considerable continuum emission, making it difficult to isolate potential weak emission lines. Our choice of large slits restricted us to targets observed during the night portion of the shuttle orbit, because the scattered light from geocoronal Ly α greatly increases the background level during the day. Use of night observations also minimizes the strength of the nearby Ly β line and eliminates the O I airglow features at 1026 Å and 1040 Å, which are prominent during the day. For each of the observations considered in this analysis, Table 1 lists the Galactic coordinates, integration time, and a brief description of the object.

To estimate upper limits to the O VI emission-line fluxes, we used the nonlinear curve-fitting package "specfit," written by G.A.K., to fit a Gaussian to the Ly β line profile and a linear continuum to the background (121 pixels centered on Ly β). Uncertainties were taken to be Poisson due to counting statistics alone. Two Gaussian emission features, each with the same width as the Ly β line and at fixed wavelengths from it (corresponding to 1031.9 Å and 1037.6 Å, respectively), were then added to the model. Their fluxes, fixed at a ratio of two to one, were raised, and the values of the other parameters re-optimized, until

Table 1

Limits to O VI Emission from the Galactic Halo

Object	l	b	Time[a]	χ_ν^2 [b]	2σ Limit[c]	Target Description
Observations through the 9″×116″ Slit:						
M32-N206	120.9	−22.0	1150	1.1	130	Between M 32 and NGC 206
NGC 1316	240.1	−56.7	1318	1.1	140	Nucleus of peculiar S0 galaxy
NGC 891	140.4	−17.4	1568	1.1	130	Halo of edge-on spiral
Perseus	150.6	−13.3	1482	0.86	84	Seyfert galaxy w/ cooling flow
Observations through the 17″×116″ Slit:						
Abell 665	149.8	+34.7	1828	1.4	67	Cluster of galaxies
M 74	138.6	−45.7	892	0.98	51	Spiral arm and H II regions

[a] Integration time in seconds

[b] χ_ν^2 of best-fitting model with no O VI emission; $\nu = 116$

[c] 2σ upper limit to O VI flux in 10^3 ph cm^{-2} s^{-1} sr^{-1}

$\Delta\chi^2 = 4.00$ (95.4% confidence for one interesting parameter; see Avni 1976). This set our 2σ upper limit on the combined flux of the O VI lines.

The results of these analyses are presented in Table 1, which lists for each observation χ_ν^2 for the best-fitting model without O VI emission ($\nu = 116$ for all models) and the 2σ upper limit to the O VI flux as determined by our spectral-fitting program. None of our observations shows significant emission from O VI $\lambda\lambda 1032, 1037$. The 2σ upper limits derived from observations made through the 9″×116″ slit are significantly greater than those derived from the 17″×116″ slit observations, reflecting the factor of two difference in solid angle between the two apertures. Our upper limits are consistent with the model predictions of Shapiro & Benjamin (1992).

The HUT spectrum of M32-N206, a field mid-way between the galaxies M 32 and NGC 206, is presented in Figure 1. The data are shown as a histogram of raw counts per 0.5-Å bin and are overplotted by models with no O VI emission (dashed line) and with O VI at the level for which $\Delta\chi^2 = 4.00$ (solid line). Only those regions used in the fit (995 Å to 1057 Å) are shown.

4. CONCLUSIONS

We have searched the HUT spectra of six extragalactic objects for O VI $\lambda\lambda 1032, 1037$ emission from Galactic coronal gas. Observations along six lines of sight yield only upper limits. Current models of the Galactic halo, when scaled by the reported flux of C IV $\lambda\lambda 1548, 1551$, predict O VI $\lambda\lambda 1032, 1037$ fluxes which are below our upper limits. We hope to reduce these limits with further HUT observations on Astro-2.

ACKNOWLEDGEMENTS

We wish to acknowledge the efforts of our colleagues on the HUT team as well as the NASA support personnel who helped make the Astro-1 mission successful.

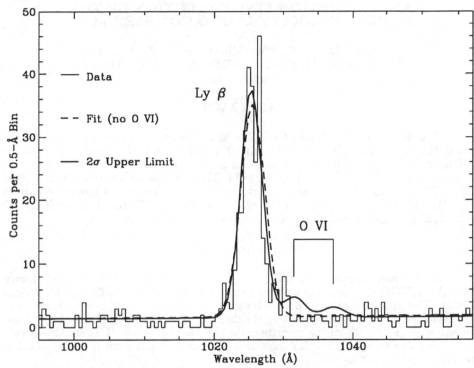

Fig. 1. HUT spectrum of M32-N206, a field mid-way between the galaxies M 32 and NGC 206. The data are shown as a histogram and are overplotted by models with no O VI emission (dashed line) and with O VI at the level for which $\Delta\chi^2 = 4.00$ (solid line). Only those regions used in the fit are shown.

The Hopkins Ultraviolet Telescope Project is supported by NASA contract NAS 5-27000 to The Johns Hopkins University.

REFERENCES

Avni, Y. 1976, ApJ, 210, 642
Benjamin, R. A., & Shapiro, P. R. 1992, in Proceedings of The Tenth International Colloquium on UV and X-Ray Spectroscopy of Astrophysical and Laboratory Plasmas (Cambridge: Cambridge Univ. Press), in press
Davidsen, A. F., Bowers, C. W., Kruk, J. W., Ferguson, H. C., Kriss, G. A., Blair, W. P., & Long, K. S. 1992a, ApJ, submitted
Davidsen, A. F., et al. 1992b, ApJ, 392, 264
Edgar, R. J., & Chevalier, R. A. 1986, ApJ, 310, L27
Martin, C., & Bowyer, S. 1990, ApJ, 350, 242
Seaton, M. J. 1979, MNRAS, 187, 75P
Shapiro, P. R., & Benjamin, R. A. 1991, PASP, 103, 923
Shapiro, P. R., & Benjamin, R. A. 1992, in Star Forming Galaxies and Their Interstellar Media, ed. J. J. Franco (New York: Cambridge Univ. Press), in press

AN ABSORPTION LINE DETECTION OF THE HIGH VELOCITY CLOUD COMPLEX M

Laura Danly, C. Elise Albert, and K. D. Kuntz
3700 San Martin Drive, Baltimore, MD 21218
danly@stsci.edu

ABSTRACT

The High Velocity Cloud Complex M has been detected in absorption toward the halo star BD +38 2182 with the *International Ultraviolet Explorer (IUE)* Echelle spectrograph. Absorption is seen out to -120 km/s in the saturated resonance lines of C II, Si II, and O I. The absorption line detection toward BD +38 2182, along with the non-detection at v < -100 km/s toward the halo star HD 93521 which also lies in the direction of M II (27 arcmin away from BD +38 2182), places the distance to Complex M at 1.5 < z < 4.4 kpc.

1. Introduction

Since their discovery in 1963, the nature and origin of the High Velocity Clouds (HVCs) has remained mysterious. (Muller et al. 1963; see also Oort 1966). The most recent and thorough survey for HVCs carried out by Hulsbosch and Wakker (1988) and analyzed by Wakker (1989) defines a High Velocity Cloud as one with $|V| > 100$ km/s. Intermediate Velocity Clouds (IVCs) are those with velocities less extreme than HVCs, but which are still not readily understood in terms of galactic rotation.

A significant impediment to our understanding of HVCs has been the lack of measurements of their distances. Absorption line measurements of HVCs in the spectra of *extragalactic* sources have shown that metals are present in observable ionization stages (West *et al.*1983, Savage *et al.*1993, Bowen and Blades 1993; see review by Van Woerden, Schwarz and Wakker 1989). Intermediate Velocity Clouds (IVC) seen in the direction of many HVCs have been detected in absorption against galactic sources, (Munch and Zirin 1961, Habing 1969, Songaila *et al.*, Danly 1989) while the HVCs have not, thus only lower limits can be inferred. The claim by Songaila et al. (1988) of a detection at -135 km/s from Complex C has been disputed by several authors, and has never been confirmed (Lillienthal et al. 1990, see review by Danly 1991).

The High Velocity Cloud Complex M has the highest galactic latitude of the HVC Complexes identified by Hulsbosch (1968). It is located in the general region $140 < l < 190$ and $50 < b < 70$, and ranges over velocities of $-120 < v < -85$ km/s. Complex M is made up of several distinct clouds known as MI and MII, as well as other clouds without assigned names. Their distances, like other HVCs, have remained unknown.

We report the results of the interstellar absorption line spectra of the very closely aligned pair of stars, HD 93521 and BD +38 2182, which lie 27 arcmin apart on the sky in the direction of Complex M. The results confirm the detection of Complex M which was reported by Danly (1989; see her Table 6). A full description of the results reported here can be found in Danly, Albert and Kuntz (1993).

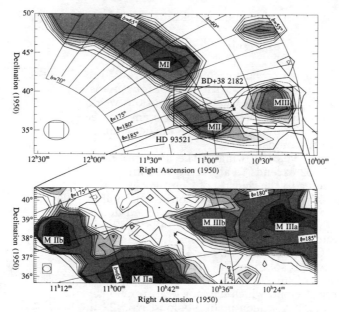

Figure 1. HI emission towards Complex M in the velocity range -120<v<-85 km/s. The contour levels are $4 \cdot 10^{18}, 5 \cdot 10^{18}, 6 \cdot 10^{18}, 7 \cdot 10^{18}, 8 \cdot 10^{18}, 9 \cdot 10^{18}, 10^{19}, 2 \cdot 10^{19}, 3 \cdot 10^{19}, 4 \cdot 10^{19}$, and $5 \cdot 10^{19}$. See text for fuller description.

2. Observations

Figure 1 illustrates the 21-cm HI distribution of the High Velocity (HV) gas toward the HVC Complex M. Both Figures 1a and 1b show emission over the range -120<v<-85 km/s. The positions of the galactic halo stars HD 93521 (z=1.5 kpc, l=183.1, b=62.2) and BD +38 2182 (z=4.1 kpc, l=182.2, b=62.2) are plotted on the contour diagrams. The stars have an angular separation of 27 arcmin, corresponding to roughly 13 pc at the distance of HD 93521.

The map of the H I emission shown in Figure 1a was created from data taken with the Bell Laboratories HI survey (Stark et al. 1992); the beam size is roughly 2.5 x 2.5 degrees, and the sampling was a quarter-beamwidth spacing. In order to study the spatial distribution of H I toward M II in greater detail, we obtained new 21-cm data in August, 1991 using the 140' telescope at NRAO/Greenbank, which is shown in Figure 1b. The beamsize is 21 arcmin, the sampling was a half-beamwidth spacing, and the data were re-binned to a 20'x30' grid.

The top three panels of Figure 2 present 21-cm H I emission spectra centered on the positions of the two halo stars. The data were also taken with the 140' telescope at NRAO/Greenbank, and were "cleaned" of stray radiation according to the prescription of Lockman et al. (1986). The bottom panel of Figure 2 shows absorption line profiles in the resonance transitions of CII (λ1334), Si II (λ1260 and λ1526), and O I (λ1302) toward the two halo stars. The data were taken with the Echelle spectrograph aboard the *International Ultraviolet Explorer (IUE)*. Data for the more distant BD +38 2182 are shown by the dotted lines. The sensitive ultraviolet lines are capable of detecting an H I column density of less than 10^{17} cm^{-2}, assuming solar abundances and that the observed line is the dominant species in neutral Hydrogen gas.

3. Results

The H I spectra toward each star (Figure 2) are nearly identical, in agreement with the contour maps. Emission in the High Velocity Cloud is present but weak: the feature has a column density of 2×10^{18} cm^{-2} over the velocity range of about -120<v<-85 km/s, again in agreement with the maps.

The negative velocity wing of the absorption toward BD +38 2182 extends to about v=-120 km/s, exactly corresponding to the velocity of Complex M in this direction. In contrast, the absorption toward HD 93521 extends only to about -85 km/s. The combined detection of the high velocity gas toward BD +38 2182 and the non-detection toward HD 93521 implies a distance to this part of Complex M of 1.5<z<4.1 kpc.

In the lower resolution Crawford Hill map (Fig 1a), the positions of the clouds M I and M II are shown. A third prominent cloud which has not been previously named is also seen; it is identified as Cloud No. 22 in the catalogue of Wakker and van Woerden (1991). We will call it M III in keeping with the naming convention. HD 93521 and BD +38 2182 lie right at the edge of a low column denstiy bridge between M II and M III. As the lowest contour in the map represents a 2σ detection of 4×10^{18} cm^{-2}, it is highly likely that the region just off the bridge is also filled with lower column density gas which is below the 2σ limit. The fact that the gas is indeed observed in absorption toward BD +38 2182 supports this supposition.

There does not appear to be any significant structure on scales on the order of the beamsize in the vicinity of the two stars which would account for the differences in the absorption line spectra. We therefore conclude that the high velocity absorption observed only toward the distant star BD +38 2182 does indeed represent a cloud located at a z-distance between 1.5 and 4.1 kpc above the galactic plane.

4. Discussion

With a distance between 1.5 and 4.1 kpc above the disk, several of the original models offered by Oort (1966) for the origin of the High Velocity Clouds can be ruled out. It is unlikely at this large distance above the plane that the HVCs are part of a supernova shell. The distance is also too small for the HVC to be gas associated with small satellites or galaxies in the local group.

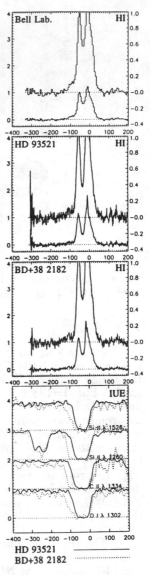

Figure 2. HI emission spectra and IUE ultraviolet absorption line spectra toward HD 93521 and BD+38 2182.

Another proposed origin for the High Velocity Clouds is the returning flow of a possible galactic fountain (Shapiro and Field 1976; Bregman 1980). Additional study of other ionic species are required to determine if indeed the material is cooling from a temperature greater than 10^6 K. At present, the IUE data are too noisy to determine the presence or absence of C IV or N V which might indicate cooling hot gas. Additional data will soon be obtained with the GHRS spectrograph aboard HST, which will place tighter constraints on the likelihood of associated hot gas.

The distance determination for Complex M also aids in the interpretation of the x-ray emission seen toward Complex M which was recently reported by Mebold et al. (1992). Unlike other observations of cool clouds at high latitude by ROSAT, Complex M probably does not create a shadow, but shows an enhancement in emission instead. A more detailed study incorporating all the available data will be required to understand the nature of the High- and Intermediate-Velocity gas that is seen throughout this region of the sky.

ACKNOWLEDGEMENTS

This research was supported by NASA under the Hubble Fellowship grant HF-1004900A to L. Danly.

REFERENCES

Bowen, D.V. & Blades, J.C. 1993, ApJ Letters, in press

Bregman, J.N. 1980, ApJ, 236, 577

Danly, L., 1989, ApJ 342, 78

Danly, L., 1991, in Proc. IAU Symp. No. 144, "The Disk Halo Connection in Galaxies", Bloemen, H., ed. (Dordrecht:Klewer), p. 53

Danly, L., Albert, C.E., & Kuntz, K.D. 1993, ApJ submitted

Habing, H.J. 1969, BAN, 20, 177

Hulsbosch, A.N.M. 1968, BAN, 20, 33

Hulsbosch, A.N.M. & Wakker, B.P. 1988, A&AS 75, 191

Lillienthal, D., Meyerdierks, H., & de Boer, K.S. 1990, A&A, 240, 487

Lockman F.J., Jahoda, K., & McCammon D., 1986, ApJ, 302, 432

Mebold, U., Kerp, J., Herbstmeier, U., Moritz, P., & Westphalen, G. 1992, in Recent Results in X-Ray and EUV Astronomy, proc. COSPAR/IAF World Space Congress, Truemper, J. et al., Washington, DC,

Muller C.A., Oort J.H., & Raimond E., 1963, C.R. Acad. Sci. Paris, 257, 1661

Munch, G.B. and Zirin, H. 1961, ApJ, 133, 11

Oort, J.H., 1966, Bull Astr. Inst. Neth, 18, 421

Savage, B.D., et al. 1992, ApJ, submitted

Shapiro, P.R. & Field, G.B., 1976, ApJ, 205, 762

Songaila, A., York, D.G., Cowie, L.L., & Blades, J.C. 1985, ApJ, 293, L15

Stark, A., 1992, ApJS, 79, 77

van Woerden, H., Schwarz, U.J., and Wakker, B.P. 1989, in Proc. IAU Colloq. 120, The Structure and Dynamics of the Interstellar Medium, Tenorio-Tagle, G., et al., eds. (Berlin: Springer-Verlag), p.389

Wakker, B.P. 1989, PhD Thesis, Groningen, The Netherlands

Wakker B.P. & van Woerden H. 1991, A&A 250, 509

West, K.A., Pettini, M., Penston, M.V., Blades, J.C., & Morton, D.C., 1985, MN 215, 482

The Formation of High-Ionization Species Associated with Galactic, Intermediate Velocity H I

James Sokolowski and Laura Danly

Space Telescope Science Inst., 3700 San Martin Dr., Baltimore, MD 21218

ABSTRACT

We report first results on the photoproduction of the species H I, Al III, Si IV and C IV kinematically associated with the Galactic intermediate-velocity (IV) arch. We find that photoionization of solar abundance gas *cannot* reproduce the observed column densities of these species, while models using stellar radiation fields and the observed x-ray background can reproduce the observed columns in gas having depleted metal abundances. The pattern of metal depletions required is consistent with their incorporation into standard interstellar grains. Limits on the intensity of the halo radiation field indicate that the material of the IV arch is smoothly distributed, having a local density $\sim 10^{-2}$ cm^{-3} and filling factor of order unity. These results suggest that photoionization plays a key role in the production of at least *some* of the high-ionization species Si IV and C IV in the Galactic halo.

1. INTRODUCTION

Interstellar absorption due to the species H I, Al III, Si IV, C IV and N V are observed in the spectra of many high latitude sources (Sembach and Savage 1992). It is believed that the ubiquity of these species indicates that high temperature gas occupies a significant fraction of the volume of the Galactic Halo (Savage 1990; Spitzer 1990) and models of cooling gas can reproduce the observed C IV/Si IV and C IV/N V column density ratios (Shapiro and Benjamin 1991; Slavin, Shull and Begelman 1992). However, due to the extremely short cooling time of this gas, excessive mass fluxes (50 M$_\odot$ to 100 M$_\odot$ per year!) are required to generate the *magnitudes* of these columns. The source of such large quantities of hot gas is uncertain as supernovae are estimated to inject ~ 1 M$_\odot$ yr^{-1} into the Galactic Halo (Norman 1990) and alternate sources are equally insufficient. Models of halo photoionization have difficulty producing the observed N V columns (Bregman and Harrington 1986), because the spectra of common stars contain little flux beyond the ionization edge of He$^+$ (54.4 eV). However, the ionization equilibrium of the Galactic 'Reynolds layer' and a similar Hα emitting component in NGC 891, suggests that the stellar ionizing photon flux in galactic halos must be significant (Reynolds 1990; Sokolowski 1992); starting at $\sim 10^6$ cm^{-2} sec^{-1} at the midplane and decreasing with height. This is in accord with the detection of Hα emission from the surfaces of high velocity clouds (Kutyrev and Reynolds 1989; Songaila, Bryant and Cowie 1989), which places an upper limit on the high-$|z|$, ionizing photon flux of order 10^5 cm^{-2} sec^{-1}. Recent results show that this radiation field dominates the formation of the low-ionization species N II and S II in galactic halos (Sokolowski and Bland-Hawthorn 1992), but its role in the formation and excitation of higher ionization species is uncertain.

The majority of intermediate velocity (IV) H I is contained in one spatially and kinematically coherent structure, the 'intermediate velocity arch' (Danly 1992). The IV arch contains $\sim 10^5$ M$_\odot$ of gas, spans $\simeq 90°$ by $\simeq 20°$ across the sky and has a velocity of -70 km sec^{-1} with respect to the local standard of rest (Kuntz and Danly 1992). The IUE spectra of four stars which lie on sight lines through the IV arch, contain multicomponent absorption at 1550 Å, due to C IV (Danly and Kuntz 1992). Resolution of the velocity structure of these profiles shows one component having a velocity centroid of $\simeq$-70 km sec^{-1}, which is therefore associated with the IV arch itself. While none of these spectra show absorption due to N V, there is clear evidence for an IV absorption component due to Al III and Si IV in the spectrum of HD 121800 (summarized in Table 1). Here, we examine a model for the formation of the observed IV species toward HD 121800, in which a largely neutral IV arch is photoionized on its surface by the combined stellar radiation fields and x-ray background.

Intermediate-Velocity Columns Toward HD 121800

$N_{H\,I}$ (cm^{-2})	$N_{Al\,III}$ (cm^{-2})	$N_{Si\,IV}$ (cm^{-2})	$N_{C\,IV}$ (cm^{-2})	$N_{N\,V}$ (cm^{-2})
9×10^{19}	9.0×10^{12}	1.3×10^{13}	2.8×10^{13}	$\lesssim 1.5 \times 10^{13}$

2. PHOTOIONIZATION MODELING

We use the photoionization code CLOUDY (Ferland 1991) to calculate the ionization and thermal structure of the IV arch, assuming it to have a local density of 0.1 cm^{-3} and either cosmic (Grevesse and Anders 1989) or depleted (Jenkins 1987) abundances appropriate for an average density of 0.1 cm^{-3}. All models have plane parallel geometry and are terminated when the H I column reaches 9×10^{19} cm^{-2}, as observed toward HD 121800.

The Ionizing Radiation Field

The radiation field used throughout our models is a superposition of those produced by O stars, the central stars of planetary nebulae (CPN), white dwarfs (WD) and the observed x-ray background. The radiation field from O stars is formed by a superposition of individual stellar atmospheres models (Kurucz 1979) weighted and summed such that they simulate a Salpeter IMF over the mass range 15 M$_\odot$ $\lesssim$ M$_*$ $\lesssim$ 80 M$_\odot$. The radiation fields from CPN and WD are represented by blackbody emitters having luminosity and temperature appropriate to match the evolutionary tracks of Schönberner (1981) and Schmidt-Voigt and Köppen (1987) for CPN and WD respectively. In addition, the superposed CPN contribution has been modified to account for absorption within the planetary nebulae shells themselves, where, on average, half of all He$^+$ and H^0 ionizing photons are absorbed (Sokolowski 1991). The observed x-ray background is represented by a 3 component power law, normalized to match the observations reported in McCammon and Sanders (1990). O stars, CPN, WDs and the x-ray background contribute 90%, 5%, 5% and 0.1% of the ionizing photons to the composite continuum respectively (Sokolowski 1991). This composite radiation field successfully reproduces the emission properties of the low-ionization, 'Reynolds Layer' in the halos of NGC 891 and the Galaxy (Sokolowski 1992).

Model Results

We present in Figure 1 the results of our photoionization modeling. All columns of interest increase monotonically with ionization parameter and the observed C IV column is reproduced at U $\simeq 10^{-3}$ for both cosmic and depleted abundances. However, cosmic abundance models yield C IV/Al III and C IV/Si IV column density ratios of $\simeq$1:3 and $\simeq$1:1 respectively. For cosmic abundances therefore, the species Al III and Si IV are overabundant with respect to C IV by an order of magnitude and a factor of two respectively, which are incompatible with observations. Because of the relatively strong depletions of Al and Si in the interstellar medium, the depleted abundance models do successfully reproduce the observed Al III, Si IV and C IVcolumns, where at U $\simeq 10^{-3}$, the ratios C IV/Al III and C IV/Si IV are $\simeq$3:1 and $\simeq$2:1 respectively. The non-detection of N V 1238 Å absorption in the spectra of any of the stars behind the IV arch is in accord with these model results, as at the ionization parameters of interest, the N V columns generated ($\lesssim 10^{10}$ cm^{-2}) would be unobservable.

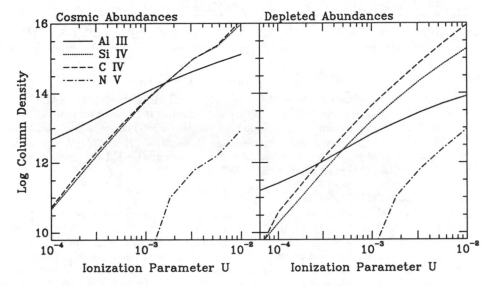

Figure 1: The log of the column densities of the species Al III, Si IV, C IV and N V vs ionization parameter for models having cosmic (left) and depleted (right) abundances.

The Physical Properties of the IV Arch

The low-ionization, halo photoionization models of Sokolowski (1992) and the observed Hα fluxes from high velocity clouds (Kutyrev and Reynolds 1989; Songaila, Bryant and Cowie 1989) suggest that the halo, ionizing photon flux is $\sim 5 \times 10^5$ cm^{-2} sec^{-1}. In order to produce an ionization parameter of 10^{-3} this flux must impinge upon a surface of local density $\sim 10^{-2}$ cm^{-3}. The size and mass of the IV arch limit its particle density to be greater than 10^{-2} cm^{-3}, thus a filling factor of unity is suggested.

3. CONCLUSIONS

The primary conclusions of this work are;

(1) Photoionization by the radiation field from the stellar populations of the Galaxy *can* reproduce the intermediate velocity columns of H I, Al III, Si IV and C IV observed toward HD 121800.

(2) An ionization parameter of $\simeq 10^{-3}$ and metal depletions due to the presence of dust grains within the IV arch are necessary to produce the observed column densities along this line of sight.

(3) Estimation of the ionizing photon flux combined with these model results indicates that the matter of the IV arch is smoothly distributed with filling factor of order unity and local density $\simeq 10^{-2}$ cm^{-3}.

(4) We find that halo photoionization produces unobservable columns of N V, in agreement with the data and previous models.

REFERENCES

Bregman, J.N. & Harrington, J.P. 1986, ApJ, 309, 833

Danly, L. 1992, PASP, in press

Danly, L. & Kuntz, K.D. 1992, in Star Forming Galaxies and Their Interstellar Medium, in press

Ferland, G.J. 1991, OSU Internal Report 91-01

Grevesse, N., & Anders, E. 1989, in Cosmic Abundances of Matter, ed. C.J. Waddington

Jenkins, E.B. 1987, in Interstellar Processes, eds. D.J. Hollenbach and H.A. Thronson

Kuntz, K.D. & Danly, L. 1992, in Star Forming Galaxies and Their Interstellar Medium, in press

Kurucz, R.L. 1979, ApJS, 40, 1

Kutyrev, A.S. & Reynolds, R.J. 1989, ApJ, 344, L9

McCammon, D. & Sanders, W.T. 1990, ARA&A, 28, 657

Norman, C.A., 1990, in The Interstellar Disk Halo Connection in Galaxies, ed. H. Bloemen, (Dordrecht:Kluwer)

Reynolds, R.J. 1990, ApJ, 349, L17

Savage, B.D., 1990, in The Interstellar Disk Halo Connection in Galaxies, ed. H. Bloemen, (Dordrecht:Kluwer)

Schmidt-Voit, M. & Köppen, J. 1987, A&A, 174, 211

Schönberner, D. 1981, A&A, 103, 119

Semback, K. & Savage, B.D. 1992, ApJS, 83, 147

Shapiro, P.R. & Benjamin, R.A. 1991, PASP, 103, 923

Slavin, J.D., Shull, J.M. & Begelman, M. 1992, ApJ, in press

Sokolowski, J. 1991, PhD Thesis, Rice University

Sokolowski, J. 1992, in Massive Stars: Their Lives in the Interstellar Medium, in press

Sokolowski, J. & Bland-Hawthorn, J. 1992, in The Evolution of Galaxies and Their Environment, in press

Songaila, A., Bryant, W. & Cowie, L.L. 1989, ApJ, 345 L71

Spitzer, L. 1990, ARA&A, 28, 71

THE RADIATIVE SUPPORT OF GALACTIC HI

Andrea Ferrara[1,2,3]
[1] STScI, 3700 San Martin Drive, Baltimore, MD 21218, U.S.A.
[2] Osservatorio Astrofisico di Arcetri, Largo E. Fermi 5, 50125 Firenze, Italy
[3] Affiliated with the Space Science Department, ESA

Email ID
ferrara@stsci.edu

1. INTRODUCTION

The best fit curve to the HI z-distribution in the solar neighborhood has a FWHM of about 230 pc, and it extends appreciably beyond 500 pc (Dickey & Lockman 1990). In this paper I want to address the following two questions: a) can the support of the Galactic HI layer in the Galactic potential may be provided by bulk motions of the gas ? b) Why is the HI scale height almost constant over the large range of galactocentric radii $\omega = 4 - 8$ kpc ?

The values of the velocity dispersion of a fast population of clouds, $\sigma_f = 15 - 35$ km s^{-1} suggested by different authors (Anantharamaiah et al.1984; Kulkarni & Fich 1985) are certainly marginally sufficient to account for the observed HI scale height and distribution at $\varpi = \varpi_\odot$, but they are still quite controversial. It is evident that a range of values for σ_f is allowed which is not sufficiently constrained by current observations and suffers many uncertainties, as, for example, the determination of the infalling "high-velocity" and "intermediate-velocity" gas.

In this paper I would like to stress the role of the Galactic radiation field and show how some of the problems related to the HI distribution in the Galaxy may be influenced by it. More details can be found in Ferrara (1993). The above questions are addressed by this paper through the "photolevitation" process (Ferrara et al.1991, Franco et al.1991. Ferrara 1993) in which radiation pressure on dust grains can raise diffuse. dusty clouds to considerable heights above the main gaseous disk. *The support provided by radiation pressure is expected to result in a lower "effective" gravitational potential. thus relaxing the need for high values of σ_f.*

Spiral galaxy disks are characterized by surface mass density and luminosity distributions falling exponentially as a function of ϖ with similar scaleheight. This encourages the analysis of the aspects of the Galactic mass-radiation field relationship relevant to the large scale vertical support of the HI.

2. METHOD AND RESULTS

According to the photolevitation process, when the photon field is anisotropic, a dusty cloud receives a net acceleration which depends on the ratio $\gamma(r)$ between the average energy flux at z directed towards the disk, F_{down}, and the one directed upwards, F_{up}

$$\gamma(r) = \frac{F_{down}}{F_{up}} = \frac{\int_z^\infty F(z)\,dz}{\int_{-\infty}^z F(z)\,dz} = \frac{1 - W(r)}{b + W(r)}, \tag{1}$$

where $r = z/H_*$, the constant b depends on the average extinction of the disk, and $W(r)$ is a function describing the bulk population of disk stars. The contribution to the radiation pressure on the cloud due to stellar clusters, P_{cl}, is also included. I would like to explore the vertical equilibrium of HI at different ϖ. In order to do this it is then necessary to evaluate the galactocentric dependence of the relevant quantities below:

Average radiation field As for the average interstellar radiation field, $J_\lambda(\varpi)$, the adopted one is given by Mathis, Mezger & Panagia (1983). The total intensity $J(\varpi)$ fits well to the the exponential law of the form $J = J_0 e^{-\varpi/\lambda}$, A particularly good fit to the Mathis *et al.*data is obtained when $\lambda \sim 6$ kpc for $5 \leq \varpi \leq 10$ kpc. In addition, the spectral distribution of $J_\lambda(\varpi)$ is remarkably similar at various galactocentric radii.

Cluster distribution Kennicut, Edgar & Hodge (1989) have measured Hα fluxes from the detected HII regions in 30 nearby spirals and Irr galaxies. From their data of Tab. 1, corrected for the faint HII regions according to their Tab. 3, the total average HII luminosity for the 7 Sb galaxies (the Hubble type of the Milky Way) in the sample comes out to be $\langle L_{Sb} \rangle = 9.91 \times 10^6 L_\odot$. As for the galactocentric distribution, I use the results given by Gordon (1988), who has studied the radial distribution $g(\varpi)$ of the HII regions in the Galaxy. I have normalized Gordon's distribution $g(\varpi)$ to the total average HII luminosity $\langle L_{Sb} \rangle$ given by Kennicut et al.(1989). In addition, given the small observed scale height of HII regions I will assume that they are located at $z = 0$.

Metallicity gradient An additional effect may be responsible of a variation in the radiation pressure on the cloud: the dependence of the dust cross section A_d on the fraction of gas mass in heavy elements, Z. In fact, following Franco & Cox (1986) the dust cross section A_d is proportional to the metallicity $A_d \sim A_0(Z/Z_\odot)$, where $A_0 = 1.8 \times 10^{-21}$ cm^2, for a grain radius $a = 0.1\mu m$. In this paper an exponential law with $\alpha = -0.15$ kpc^{-1} will be used for the metallicity, in agreement with Matteucci (1989).

Gravitational field As for the gravitational acceleration perpendicular to the plane, $K_z(z)$, I adopt Oort's (1965) determination. The extrapolation of $K_z(z)$ can be obtained under the hypotheses of constant vertical and exponential radial stellar density distribution. The explicit expression for $K_z(z)$ is

$$K_z(\varpi, z) = \frac{\sigma_g^2(\varpi)}{z_0}\left[\tanh\left(\frac{z}{z_0}\right) + \epsilon(\varpi)\left(\frac{z}{z_0}\right)\right]. \tag{5}$$

where $z_0 = 250$ pc, $\sigma_g^2(\varpi) = (15.4\text{ km s}^{-1})^2 \exp(\varpi_\odot - \varpi)/0.44\varpi_\odot$, and $\epsilon = 0.04, 0.07, 0.14$ for $\varpi = 5, 10, 15$ kpc, respectively. A quadratic interpolation to ϵ has been used whenever required. The radiative force adds to the gravitational one acting on the clouds, resulting in an "effective" potential ϕ_e:

$$\phi_e(\varpi, z) = \int_0^z K_z(\varpi, z)dz - \int_0^z \frac{P_{rad}(\varpi, z)}{N\mu}dz = \int_0^z K_z(\varpi, z)dz$$

$$-\frac{(1-e^{-\tau_e})}{cN\mu}\int\limits_{0}^{z}\left\{[Q_pF]_{\odot}e^{(\varpi_{\odot}-\varpi)/\lambda}\left[\frac{1-\gamma(r)}{1+\gamma(r)}\right]+\frac{\langle Q_p\rangle\langle L_{Sb}\rangle}{4\pi z^2}g(\varpi)\right\}dz, \qquad (6)$$

where N is the cloud total column density along the z-axis and μ is the mean mass per gas particle. Finally, the density distribution $n(z)$ can be obtained by solving the hydrostatic equilibrium equation

$$\nabla p = -\rho\nabla\phi_e. \qquad (8)$$

If one admits that the pressure is provided solely by gas motions, then $p = \rho\sigma^2$; supposing also that the HI can be approximated as the sum of the "fast" and "slow" isothermal components, eq. (8) has the solution

$$\rho(z) = \sum_{j=f,s} \rho_{0,j}\exp\left[-\phi_e(z)/\sigma_j^2\right], \qquad (9)$$

where the indexes f and s stand for fast and slow, respectively, and $\rho_{0,j}$ is the density at midplane.

Fig. 1 shows the fit to the observed vertical density profile for the solar neighborhood ($\varpi = 8.5$ kpc). The theoretical curves represent the solutions of eq. (9) when two cloud populations with different values of the velocity dispersion are considered; the individual distributions of the fast and slow clouds are also reported. A quite good agreement between the observed and the calculated curves is obtained assuming $(\sigma_j, n_{0,j})$ equal to $(\sigma_f = 13.0$ km s^{-1}, $n_{0,f} = 0.145$ cm^{-3}) and $(\sigma_s = 5.0$ km s^{-1}, $n_{0,s} = 0.447$ cm^{-3}).

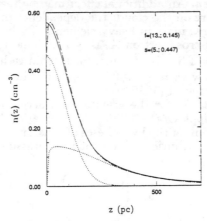

Fig. 1 HI vertical density distribution obtained from the model (dot-dashed line), compared with the observed one given by Dickey & Lockman (1990) (solid) at the solar radius. Also shown are the individual fast and slow components (dotted lines), along with their $(\sigma_j, n_{0,j})$.

At different galactocentric radii, remarkable good fits can be obtained with moderate to low values of the dispersion velocity for the two populations. At small ϖ, an increase of the velocity dispersion either for the fast and slow clouds is required in order to support the gas in the larger gravitational field present towards the Galactic center. However, only a factor less than 2 in the values of the velocity dispersion is necessary in order to account for the observed distribution. This is naturally explained, on a qualitative base, by the distribution of the HII regions $g(\varpi)$, which exhibits a fractional increase of about 5 times passing from the solar circle to $\varpi = 4$ kpc. Therefore, the radiative support is enhanced in regions closer to the Galactic center.

3. DISCUSSION

I have shown that the vertical support of the galactic HI may be completely provided by turbulence- $e.g.$bulk motions of the gas- without requiring any additional pressure due to magnetic fields and cosmic rays when the average Galactic and cluster radiation fields acting on the clouds are included. The photolevitation mechanism tends to create a lower "effective" gravitational potential, as described above. High accuracy fits of the observed vertical density profile of HI are obtained considering two populations of fast and slow clouds with moderate-to-low dispersion velocities. In addition, the observed constancy of the HI scale height with ϖ is naturally explained by the model in terms of the similar average luminosity and mass distribution found in spiral galaxies and, in particular, in the Milky Way. The increase in the velocity dispersion of the fast clouds implied by the fit parameters (Tab. 1) towards the Galactic center closely resembles the global distribution of the HII regions in the disk.

The radiative support alleviates the problem of the presence of a class of very high-velocity ($\sigma \geq 25$ km s^{-1}) dispersion clouds necessary if the support is demanded solely to turbulence without photolevitation.

REFERENCES

Anantharamaiah, K.R., Radnakrishnan. V., & Shaver. P.A. 1984. A&A, 138, 131
Dickey & Lockman 1990, ARA&A. 28. 215
Ferrara. A., Ferrini, F., Franco. J.. & Barsella. B. 1991. ApJ. 381. 137
Ferrara. A. 1993, ApJ, (Apr 10)
Franco. J., & Cox, D.P. 1991. PASP. 98. 1076
Franco. J., Ferrini, F., Ferrara, A.. & Barsella, B. 1991, ApJ, 366, 433
Gordon, M.A. 1988 in Galactic and Extragalactic Radio Astronomy, eds. G.L. Vershuur & K.I. Kellerman, (Berlin:Springer-Verlag), 37
Kennicut, R.C., Edgar, B.K. & Hodge, P.W. 1989, ApJ, 337. 761
Kulkarni. S.R., & Fich, M. 1985. ApJ. 289, 792
Mathis. J.S., Mezger, P.G., & Panagia. N. 1983. A&A. 128. 212
Matteucci, F. 1989, in Evolutionary Phenomena in Galaxies. ed. J. Beckman & B.E.J. Pagel (La Laguna), 297
Oort, J.H. 1965, in Stars and Stellar Systems. Vol. 5. (Chicago: Univ. Press), 455

WORMS OR FROTH?
FINE-SCALE STRUCTURE IN
THE FAR-INFRARED MILKY WAY

William H. Waller
Code 681, NASA Goddard Space Flight Center, Greenbelt, MD 20771.

Email ID
waller@stars.gsfc.nasa.gov

Francois Boulanger
Institut d'Astrophysique Spatiale, B.P. 10, Route des Gatines,
91371 Verrières le Buisson Cedex, France.

Email ID
boulange@friap51.bitnet

ABSTRACT

Mosaics of the inner Galaxy have been constructed from selected IRAS data products. Through median-filtration techniques, we have been able to eliminate the strong gradient in brightness towards the Galactic midplane thus revealing peaks and valleys in the *residual* FIR emission that span several degrees. Some of the negative residuals are coincident with known "windows" in the otherwise opaque screen of dust near the Galactic plane. The processing also reveals diaphanous fine structure in the FIR emission with size scales as small as the 4′ × 4′ resolution of the images. This cirrus-like structure brightens towards the Galactic midplane just like the smoothly varying background, thus indicating its ubiquity throughout the inner Galaxy. Although we had expected to find morphological evidence for supernova-driven "chimneys" or "worms" rooted in the Galactic plane, the far-infrared fine structure appears more complex (e.g. less coherent and less rooted) as viewed in projection. This may be due to the varying distances and sizes of the emitting features. Evidence for a froth-like superposition of nearby and distant filaments and shells is presented.

1. MOTIVATING QUESTIONS

Our current picture of the diffuse interstellar medium is a confusing melange of "phases," each one having a characteristic temperature, ionization state, density, and scale height. The organization, ecology, and evolution of this dynamic mix continues to elicit controversy. Some of the major unanswered questions include...

1. **Morphology:** How are the various phases structured? Are the denser phases mostly in the form of spheroidal clouds, sinuous filaments, warped sheets, or hollow shells? Are the dominant structures embedded within the thick HI and HII disks, or are they rooted to the thin Pop I disk (as in the "worm" scenario)? Do any of these structures penetrate into the halo (as in the "chimney" paradigm)?

Do the *clumps* or the *voids* better define the overall structure? Should we be thinking of Irish stew or Swiss cheese? Or do thick-skinned voids (aka beer froth) provide the best approximation to the spatial distribution of the various phases? How does this distribution change with position in the disk?

2. **Volume Filling:** Can the preferred morphologies and clump/void balance explain Baade's window and other optically transparent regions near the Galactic plane? Are there hitherto unexplored "windows?" These questions depend on the volume filling fractions of the densest components of the diffuse ISM — e.g. the cool neutral medium (CNM), warm neutral medium (WNM), and warm ionized medium (WIM) — as well as the distribution of cold dense molecular clouds along the line of sight.

3. **Dynamics:** What dynamical processes are most responsible for driving the ISM to its present morphological state? Possible candidates include stellar winds, stellar radiation pressure, supernova explosions, supernovae-driven "fountains," hydro-magnetic waves, and hydro-thermal condensations.

These questions of morphology, volume filling, and dynamics have provided the grist for several recent review articles (cf. reviews by McCammon, Reynolds, McKee and Heiles in these proceedings). Here, we present new observational results which address the morphology of the diffuse ISM as traced by the FIR emission from the associated dust.

2. ANALYTIC STRATEGIES

• By mapping the FIR emission from dust that has been warmed by the interstellar radiation field (ISRF), one can trace both the cool and warm phases of the diffuse ISM. These two phases represent most of the mass in the diffuse ISM.

• The recent data products produced by IPAC from the IRAS mission database provide the best resolved and most complete mapping of the Galactic FIR emission. At 100 μm, the fully-sampled resolution is 4 arcminutes.

• Of the 4 IRAS bandpasses, the 100 μm bandpass is most sensitive to the cool Galactic dust and least sensitive to the warm Zodiacal dust — a troublesome contaminant in some regions of the Galactic plane.

• Through median-filtration techniques, the strong gradient in brightness towards the Galactic midplane can be eliminated thereby revealing the fine-scale FIR structure throughout the inner Galaxy.

3. MOSAICS OF THE FIR EMISSION

We have accessed 60° × 20° mosaics of the inner Galaxy which are based on the BIGMAP data product released by IPAC. These mosaics were constructed by Walter Rice and are available at IPAC. In addition, we have constructed a 60° × 60° mosaic covering the Galactic center, bulge, and molecular ring regions. This mosaic is based on the more recent Infrared Sky Survey Atlas (ISSA) data product, portions of which have yet to be released. In all these images, the Zodiacal light contribution has been modeled and removed, with residual contamination (at 100 microns) at negligible levels. These mosaics are dominated by the strong gradient in brightness towards the Galactic midplane.

4. MEDIAN SMOOTHING AND FILTERING

Median smoothing of the images with a 15° × 0.05° window yields excellent maps of the smooth background emission while preserving the steep brightness gradient towards the Galactic midplane. This sort of smoothing window also ensures that structures with significant longitudinal extent are preserved in the subsequent filtering. Subtraction of the median smoothed images yields *median filtered* images, where the smoothly varying background has been effectively eliminated (see Figure 1). The residual emission structure contains both large-scale and fine-scale components, as well as a general brightening towards the Galactic midplane which mimics the *cosecant b* behavior of the smoothly varying background. If the residual structure only came from nearby "cirrus" clouds or from a single superbubble enveloping the Sun, it would be more uniformly distributed in latitude. Therefore, the residual emission (like the smoothly varying background) comes from dust at a variety of distances from the Sun.

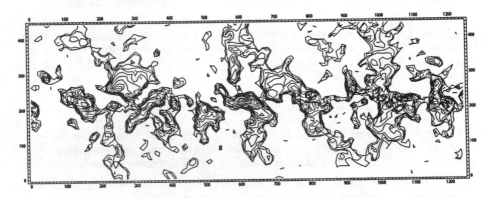

Fig. 1. Fine-scale residual 100 μm emission from the central 60° × 20° after having subtracted off a median-smoothed version of the original mosaic. Emission is contoured logarithmically in increments of 0.25 dex beginning at $10^{0.5}$ MJy/Sr. The Galactic center is located at the central emission peak.

5. WINDOWS AND WINDOW CANDIDATES

Figure 2 shows the depletions in the FIR emission compared to the smoothly varying background for the Galactic bulge region. The "hole" immediately below the Galactic center includes the Sgr I, Sgr II, and Baade *windows* — well-known optically transparent regions in the otherwise opaque screen of dust that obscures the Galactic bulge. The nearby "hole" at negative longitude may qualify as another "window" candidate. This region and many of the other depleted regions near the Galactic midplane show correspondingly depleted CO emission — one of the key tracers of dense interstellar gas and associated dust. Further comparison with HI and optical data will be necessary to determine whether these joint FIR and CO depletions represent hitherto unexplored "windows."

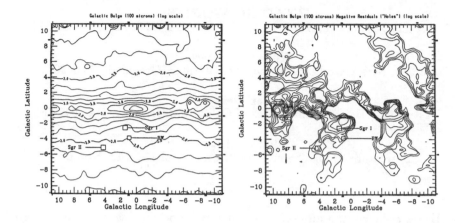

Fig. 2. The central 20° × 20° covering the Galactic bulge before and after median filtering. Only the negative residuals ("holes") in the 100 μm emission are shown in the median filtered image. Same logarithmic contouring as Fig. 1.

6. FINE-SCALE STRUCTURE — RELATIVE RESIDUALS

Images of the residual emission *relative to the ambient emission* were produced by dividing the original images by their median-smoothed counterparts. Figure 3 (Plate 8) shows the relative residuals for the 60° × 60° region covering the Galactic center, bulge, and molecular ring. This particular representation of the fine-scale FIR emission eliminates all of the *cosecant b* brightening towards the Galactic midplane while highlighting the structure above and below the plane. The worm-like features evident in the images of *absolute residuals* now become parts of a larger, more complex scene. Instead of "worms" rooted in the Galactic plane, filamentary and shell-like features are evident at all latitudes. Above and below the Galactic center, the nearby Ophiucus and R Corona Australis clouds stand out in stunning contrast and detail (see Boulanger and Waller, these proceedings). Within 5° of the Galactic midplane the *relative residuals* are probably tracing a mix of nearby and more distant structures.

7. CONCLUSIONS

Different image processing techniques can lead to different conclusions regarding the prevalence of "worms" or "froth" in the diffuse ISM. Our analysis of the diffuse FIR emission from the inner Galaxy leads us to suggest that many of the so-called "worms" are parts of larger more complex structures. Decomposing this "froth" of superposed filaments, shells, and voids will require a more careful comparison of the diffuse FIR, HI, Hα, and CO emission structures.

ACKNOWLEDGEMENTS

This research was funded in part by a NASA/ADP grant (NAG 5-1424) to StarStuff Inc. We also thank the folks at IPAC for their invaluable support.

GALACTIC WORMS:
INFRARED AND RADIO MAPS OF THE FIRST QUADRANT

William T. Reach
Code 685, NASA Goddard Space Flight Center, Greenbelt, MD 20771.

Carl Heiles
Astronomy Department, Univ. of California, Berkeley, CA, 94720

Bon-Chul Koo
Astronomy Dept., Seoul Nat. Univ., Seoul 151-742, Korea

ABSTRACT

Using radio continuum and infrared surface brightness maps we surveyed galactic worms—structures perpendicular to the galactic plane. For the first time, the structures were resolved and distances determined. We find two main classes based on radio spectral index, and suggest an evolutionary sequence.

1. INTRODUCTION

The term 'worm' was originally used by to describe structures "crawling out of the Galactic plane" (Heiles 1984) in spatially-filtered 0.6°-resolution H I maps of the Galaxy. Since then, a worm candidates were defined as any contiguous set of > 5 pixels with brightness above a certain absolute threshold in spatially-filtered H I and infrared maps with 0.5° pixels (Koo, Heiles, & Reach 1992; KHR). In this work, we use higher-resolution (5' pixels) infrared and radio continuum maps of most of the first Galactic quadrant ($68° > l > 7°$) to study the worm candidates of KHR and to identify smaller worms.

Heiles (1984) first suggested that worms may be the walls of cavities evacuated from the Galactic plane by multiple supernovae. Estimates of the effects of large clusters containing many early-type stars revealed that they could indeed punch holes through the interstellar gas layer (Heiles 1990). A distinction is made between clusters that 'break through' the thinner gaseuos disk ($H \simeq 170$ pc) and those that 'blow out' of the thicker gaseous disk ($H \simeq 500$ pc). Clusters with more than 50 stars earlier than B3 can break through the thin disk (Koo and McKee 1992); based on the luminosity function of H II regions (Kennicutt, Edgar, & Hodge 1989), there should be ~ 20 such clusters in the Galaxy. Clusters with more than 800 early stars can blow out the entire disk (including the ionized layer); there should be only only of order 1 such object in the Galaxy.

2. OBSERVATIONS

The infrared maps were created by mosaicing adjacent Infrared Astronomical Satellite (IRAS) galactic-plane images and coadding the images from the 3 separate IRAS coverages of the fields. The maps are binned to 5' pixel size, which is about the resolution of the 100μm detectors. The radio maps were generated from the Effelsberg galactic-plane surveys at 2695 MHz (11 cm) and

1420 MHz (21 cm) (Reich *et al.* 1990a,b). The beamsize of the 100-m telescope at these frequencies is 4' and 9', respectively. Maps were created with the exact same grid as the infrared data—full-beam at 2695 MHz and half-beam at 1420 MHz.

To remove the diffuse background, each pixel was replaced by its value minus the median of a 3.5° square centered on the pixel. This is the same filter used by Heiles (1984) and KHR. The resulting images are shown in Color Plates 11–12. (For comparison, see Color Plate 8, an infrared map produced with a rectangular filter.) The galactic plane itself is noticeable only in the inner galaxy, where it is substantially narrower than the filter. 'Worms' are evident as vertical protrusions from the inner galaxy. A list of worms was created using contour maps of the 2695 MHz and 100μm images. An example of the contour maps, for longitudes 27–37°, is shown in Figure 2.

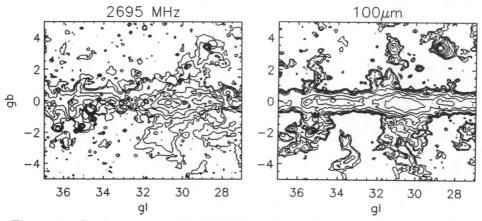

Figure 2. Contour maps of 2695 MHz radio continuum and 100μm infrared surface brightness for part of the first quadrant. The lowest contour levels are 0.1 K and 10 MJy Sr^{-1}, and the spacing is by factors of 2.154. The worms GW31.0-2.1 (width 1.2°) and GW35.3+0.9 (width 0.3°) are evident in these maps. The bright point source at $l = 30.8°$, $b = 0°$ is W43, which we associate with GW41.0-2.1 in position and velocity. Emission north of the plane at $l = 30°-28°$ is part of the North Polar Spur.

3. PROPERTIES OF WORMS

The locations and sizes of the most prominent first-quadrant worms are listed in Table 1. Locations correspond to the portion of the worm on which photometry was performed. Distances were determined from an ongoing radio recombination line (H166α) survey, being undertaken with the 85-foot telescope at Hat Creek Radio Observatory; the recombination line velocities were converted into kinematic distances. Infrared and radio surface brightnesses were determined by integrating over regions 0.3 to 1° in diameter (depending on worm thickness), in portions of the worm with no obvious point sources or H II regions. The background was determined by integrating over identical regions, offset in ± galactic longitude by the worm thickness. The 60/100μm colors and radio spectral indices were determined using the 'chopped' surface brightness.

We define the spectral index such that the brightness temperature is proportional to $\nu^{-\beta}$; for a thermal source $\beta = 2.1$, for supernova remnants the median $\beta = 2.45$, and the background at typical worm latitudes has $\beta = 3.0$. Worm spectral indices were classified as 'thermal' if consistent with 2.1 with uncertainty less than 0.2, and 'nonthermal' if clearly greater than 2.1. Less certain spectral indices are indicated by a question mark. Some of the apparently 'nonthermal' worms have radio recombination lines, and so they are at least partially thermal.

Using the radio surface brightness and angular size, the total 2695 MHz radio flux was determined. The Lyα optical depths of the worms are estimated to be larger than unity, so we assume that the ionized regions in the worms are in 'case B' recombination. Since the dust optical depth to Lyman continuum photons is substantial for some worms, the ionization parameter U was determined using approximations for a dusty H II region. In terms of the production rate of Lyc photons, $N_u = 4 \times 10^{43} U^3 \, \mathrm{s}^{-1}$; for comparison, a single O5 star has $U \simeq 100$.

TABLE 1. PROPERTIES OF WORMS

name	Dist (kpc)	Width (pc)	$|z_{max}|$ (pc)	radio spectrum	U_{ion} (pc cm^{-2})	H II region association
GW 10.9 −2.4	3.0	63	260	nonthermal?	100	
GW 15.1 +3.7	2.1	45	180	thermal	70	S46
GW 15.9 −2.0	2.9	76	210	nonthermal?	120	M17??
GW 18.5 +3.0	2.3	60	220	thermal	190	S54, Ser OB2
GW 25.5 +1.5	1.9:	17:	170:	nonthermal?	30?	
GW 31.0 −2.1	5.6	150	600	nonthermal	350	W43
GW 35.3 +0.9	5.0	35	160	thermal?	60	
GW 43.9 +1.6	...	...	...	nonthermal?	...	
GW 53.4 −1.7	...	...	...	thermal?	...	

4. IDENTIFICATION OF WORMS WITH OTHER OBJECTS

Some worms can be easily associated with bright infrared and radio objects by their positional coincidence and the smooth decline of worm surface brightness with distance from the bright object. In two cases, the bright objects are known optical H II regions (S46 and S54). The spectroscopic distances of the exciting stars, the kinematic distance of the associated molecular gas, and the kinematic distance of the diffuse radio recombination-line-emitting gas agree. There may be a worm associated with S49 (=M16), but further diffuse radio recombination line observations are needed.

In order to determine whether all large clusters have worms, we compiled a list of giant radio H II regions (ionization parameter > 200) from the 4874 MHz survey of Downes et al. (1980). M 17 (2.2 kpc) is possibly associated with GW15.9-2.0, but the worm and H II region velocities differ by 12 km s^{-1} so the association is doubtful. 3C385 (11.4 kpc) has a small, radio worm poorly defined in the infrared. W43 (5.6 kpc) is associated with GW31.0-2.1; radio recombination line velocities of H II region and worm agree. W49A (11.8 kpc) is near the base of GW43.9+1.6; it lacks a diffuse radio recombination line observation. W51 ($\sim$ 5 kpc) has a small, radio worm with confused infrared emission; it also lacks a recombination line observation.

5. CONCLUSIONS: AN EVOLUTIONARY SEQUENCE

During the early phases of its lifetime, a large star cluster contains several O stars that produce H II regions of various size. The most massive stars easily ionize their surroundings and produce wind-blown bubbles. Ionizing photons from adjacent O stars located in rarefied ISM will together form an extended H II region that surrounds the cluster. At this stage, the radio morphology of the cluster is a group of compact sources superposed on a plateau of emission.

If the cluster contains enough O stars in rarefied regions, it can produce an H II region with radius comparable to the scale height of the ISM. Then it will expand much more rapidly perpendicular to the galactic plane, and it will produce a thermal worm. Examples of such structures are associated with S54 and S46.

When the first generation of O and early B stars explode as supernovae, the pressure in the cluster increases dramatically and a large plume of gas can be levitated above the galactic plane. Since supernovae are nonthermal radio sources, it might be suspected that supernova-driven worms have nonthermal radio spectra. However, subsequent generations of early-type stars can ionize the interiors of the worm walls and produce composite spectra. GW31.0-2.1, associated with W43, was almost certainly produced by supernovae (because of its large size and apparently nonthermal radio spectrum), but the walls must be currently ionized (because radio recombination lines were detected). Further, W43 itself is a thermal source superposed on an extended plateau; the worm was likely produced by an earlier generation.

In its final stages, a worm will contain no more early-type stars, and its walls will completely recombine. By this time, it may have ejected cold fragments into the halo, and the walls will form filamentary clouds. Such structures would appear in H I and infrared maps only, and would eventually be inseparable from background and foreground clouds.

ACKNOWLEDGEMENTS

This work was done while WTR held a National Reach Council-NASA/GSFC Research Associateship. We thank Dr. Reich for providing the radio continuum maps. The radio recombination line observations were supported in part by NSF Award No. 91-23362 to Carl Heiles.

REFERENCES

Downes, D., Wilson, T. J., Bieging, J., & Wink, J. 1980, A&AS 40, 379
Heiles, C. 1984, ApJS 55, 585
Heiles, C. 1990, ApJ 354, 483
Kennicutt, R. C., Jr., Edgar, B. K., & Hodge, P. W. 1989, ApJ 337, 761
Koo, B.-C., Heiles, C., & Reach, W. T. 1992, ApJ 390, 108
Koo, B.-C., & McKee, C. F. 1992, ApJ 388, 93
Reich, W., Fürst, E., Reich, P., & Reif, K. 1990a, A&AS 85, 633
Reich, W., Reich, P., & Fürst, E. 1990b, A&AS 83, 539

MAGNETIC SHAPING OF SNRs AND THEIR BUBBLES[*]

Michael L. Norman
Department of Astronomy and
National Center for Supercomputing Applications
405 N. Mathews Street
Urbana, Illinois 61801

Abstract

We have simulated the evolution of a supernova remnant embedded in a galactic magnetic field using 2D numerical MHD. The calculation generalizes the spherically symmetric results of Slavin and Cox (1992) to 2D axisymmetric geometry. We follow the evolution of the remnant for 5×10^6 yr--sufficiently long for the effects of magnetic shaping and radiative cooling of the hot bubble to become important. We find that: (1) the shock wave becomes increasingly oblate and the hot bubble becomes increasingly prolate with respect to the magnetic axis with time; (2) the radiative shell remains thick except at the magnetic poles due to magnetic pressure support; (3) near the magnetic poles a dense polar cap of HI forms at the bubble apex; (4) a complex shock system including a shock "dimple" forms near the magnetic axis when the shock speed has dropped to slightly greater than the ambient Alfven speed; (5) after 5 Myr the hot bubble collapses to a slender, 400 pc-long cylinder with $T \approx 10^7$ K.

1. Introduction

Hydrodynamic models of supernova remnants identify four evolutionary phases based on the shock speed (e.g., Woltjer 1972). During Phase I, the supernova ejecta undergoes free expansion and the shock velocity is essentially constant. Phase II begins once the shock has swept up a mass of ambient material equal to the ejecta mass; thereafter the remnant evolves approximately as described by the Sedov-Taylor self-similar solutions in which the shock velocity scales as $t^{-3/5}$. Phase III is characterized by radiative losses first in the swept up shell, and later in the hot bubble interior. During this phase, the shell conserves momentum and the shock radius obeys an offset powerlaw $R \alpha (t-t_{offset})^{3/10}$ (Cioffi, McKee and Bertschinger 1988). Finally in Phase IV, the remnant merges with the interstellar medium once the shock speed drops below the mean velocity dispersion of the ISM. Rough timescales for the transitions between these various phases are I-II: 100 yr; II-III: 25,000 yr; III-IV: 750,000 (Woltjer 1972).

Ambient magnetic fields are expected to alter this picture somewhat. First, magnetic fields mediate the coupling of the ejecta to the ambient plasma. Recently, Chevalier, Blondin and Emmering (1992) have performed 2D hydrodynamic simulations of the Phase I-II transition and shown that the ejecta/ambient interface is Rayleigh-Taylor unstable. The swept up ambient magnetic field, although dynamically unimportant initially, could be amplified in the turbulent mixing layer via dynamo action. Second, Ferriere and Zweibel (1991), in a simplified 2D treatment using the thin shell approximation, found that magnetic pressure prevents the collapse of the radiative shell except near the magnetic axis during Phase III. Finally, the remnant should become highly nonspherical when the shock has swept up an amount of magnetic energy equal to the explosion energy. Dimensional analysis yields $t_m = E_o^{1/3} \rho_o^{1/2} B_o^{-5/3}$ where E_o is the explosion energy, ρ_o is the ambient density, and B_o is the ambient magnetic field strength. Using typical parameters $t_m = 2.2(E_o/10^{51}$

*to appear in "Back to the Galaxy", proceedings of the 3rd Annual October Astrophysics Conference in Maryland, Oct. 12-14, 1992.

erg)$^{1/3}(n_0/0.1$ cm$^{-3})^{1/2}(B_0/3\mu G)^{-5/3}$ Myr--well beyond Phase IV. This is an upper limit, however, since radiative losses have been ignored. Slavin and Cox (1992; hereafter SC) have modeled the evolution of a spherically symmetric SNR to very late times (6 Myr) including radiative losses, heat conduction and magnetic fields. Since their calculation is 1D, the magnetic field is assumed to be everywhere tangential to the shock front and thus shaping could not be studied directly. However, they find that 50% of the explosion energy is radiated away in 2×10^5 yr, implying that t_m may be of this order. They also found that bubble collapse (Phase V) occurs with or without a magnetic field on timescales of 5-6 Myr and that an ambient magnetic field reduces the size and hastens the collapse of the hot bubble as compared to non-magnetic models.

2. The Calculation

In this paper we present a 2D numerical MHD simulation of SNR evolution covering Phases II-V. The results of this study will have implications to several topics of interest reviewed by Spitzer (1990) concerning the nature of the diffuse ISM: (1) hydromagnetic stability of SNRs; (2) magnetic shock geometries near the magnetic axis; and (3) evolution of hot gas in the ISM. Here we touch on topics 2 and 3. The initial conditions chosen were identical to those used by SC to enable comparison. We thus generalize their results to 2D. The initial conditions are as follows: We consider a rectangular domain in RZ axisymmetric geometry 0<R<200pc, 0<Z<200pc with a uniform mesh spacing of 1 pc. The grid is initially filled with a uniform ambient medium with $n_0=0.2$ cm^{-3}, $T_0=10^4$K, and $\mathbf{B}_0=(0,0,B_{z0})=5\mu G$. The gas is assumed to obey a $\gamma=5/3$ ideal equation of state. To this is added a 5 pc radius hot bubble centered on the origin in which 2 M_{solar} of ejecta mass and 5×10^{50} erg of thermal energy is uniformly distributed. We calculate the subsequent evolution using the ZEUS-2D code (Stone and Norman 1992a,b). ZEUS-2D is a two-dimensional, time-explicit, Eulerian code which integrates the time-dependent, nonlinear equations of ideal MHD via finite differences. Shock waves are handled using artifical viscosity. The magnetic field algorithms satisfy the divergence-free constraint to roundoff, and have proven stable, accurate and robust against an extensive battery of test problems (Stone et al. 1992). Radiative cooling is incorporated as described in MacLow, McCray and Norman (1989), wherein the radiation is assumed to be optically thin, and the cooling rate is determined from the cooling function of MacDonald and Bailey (1981). We ignore the effects of heat conduction for the moment, which are only important at very late times (SC). The R=0 and Z=0 boundaries are taken to be reflection symmetry boundaries, while both outer boundaries are taken to be open (transmitting) boundaries.

3. Results

Fig. 1 shows the evolution of the remnant and the hot bubble out to 5 Myr. At 0.5 Myr (Fig. 1a), the shell is already radiative and magnetic stresses have introduced asphericity. The density isocontours reveal that the shock wave (R≈75 pc), is slightly oblate at this time, whereas the bubble boundary is decidedly prolate (Z_{bubble}≈70pc; R_{bubble}≈50 pc). The former is a simple consequence of the fact that MHD waves propagate faster across magnetic field lines than along them. Notice also that the shell, defined as the region between the shock wave and the bubble boundary, is much thicker at the magnetic equator than at the pole. This is because magnetic pressure dominates thermal pressure in the equatorial plane ($\beta \equiv P_{gas}/P_{mag}$≈0.1) and thus prevents the shell from collapsing. At the magnetic pole, however, magnetic pressure is ineffective and the shell collapses due to radiative losses. The shell is roughly isothermal at T≈6300 K.

At 1 Myr, the shell has developed a curious configuration near the magnetic axis. The shock normal becomes parallel to the ambient magnetic field at (R,Z)=(25,90) pc and dips in toward the symmetry axis as it looses compression. Detailed analysis of this configuration, seen also in simulations of solar coronal transients (Steinolfson and Hundhausen 1990), reveals that it is composed of several MHD shock waves types, including fast, slow, intermediate and switch-on shocks as diagrammed in Fig. 2. This configuration only exists so long as the shock speed exceeds the ambient Alfven speed by a small factor and the ambient plasma beta is less than unity-- conditions which are met here. At the axis, the intermediate shock merges with a slow MHD shock wave shaped like a bow shock. Behind the slow shock is a dense, cool shell of swept-up ambient material. Also, the hot bubble has begun to cool radiatively at its extremities and is becoming increasingly prolate.

At longer times the total pressure interior to the shell drops below the ambient pressure due to inertial expansion and radiative cooling losses, and the bubble collapses. Figs. 1c,d show that the bubble collapses to a slender cylinder by 5 Myr. The shock wave dissipates and magnetic field lines straighten, except at the bubble apex where magnetic tension forces are largest. The bubble apex advances sub-Alfvenically as it "chases" a low pressure zone created by a fast magnetosonic rarefaction wave which precedes it. A dense ($n \approx 1$ cm^{-3}), cool (T≈5000K) polar cap of material advances along with it. The low shock speed and long timescales imply this material would be in the form of HI.

4. Discussion

We now compare our results with SC. Given their assumptions, it is only meanginful to compare the radial evolution of our 2D remnant in the equatorial plane to their spherically symmetric evolution. Our calculation confirms many of their key results both qualitatively and quantitatively: formation of a thick shell due to magnetic pressure support; evolution of R_{shock} and R_{bubble} versus time; onset of bubble collapse at t≈2 Myr. However, our bubble does not achieve zero equatorial radius at late times as theirs does, but settles into R≈15pc. This is likely due to our neglect of heat conduction, which cools the highest entropy material to low enough temperatures so that it can radiate (SC).

Not unsurprisingly, the evolution of the remnant parallel to the magnetic axis is very different from SC. It is also very different from SC's unmagnetized result, which one might naively think applies exactly on axis. In fact, the 2D evolution of the remnant becomes very unlike any 1D solution once the shock speed drops to of order the ambient Alfven speed. In our calculation this occurs at roughly 5×10^5 yr--well into the radiative phase. Thereafter the bubble elongates along the field at roughly the Alfven speed, driving ahead of it a complex MHD wave system. This advance is maintained by a lateral squeezing of the bubble by magnetic and thermal pressure.

The author would like to acknowledge the continued interest and patience of Don Cox, who introduced me to this problem, and also useful discussions with Mordecai-Mark MacLow. The computation was performed using the CRAY YMP at the National Center for Supercomputing Applications, University of Illinois.

References

Chevalier, R. A., Blondin, J. M., & Emmering, R. T. 1992, ApJ, 392, 118
Cioffi, D. F., McKee, C. F., & Bertschinger, E. 1988, ApJ, 334, 252
Ferriere, K. M., & Zweibel, E. G. 1991, ApJ, 375, 239

MacDonald, J., & Bailey, M. E. 1981, MNRAS, 197, 995
MacLow, M.-M., McCray, R. M., & Norman, M. L. 1989, ApJ, 337, 141
Slavin, J. D., & Cox, D. P. 1992, ApJ, 392, 131
Spitzer, L. 1990, Ann. Rev. Astron. Ap., 28, 71
Steinolfson, R. S., & Hundhausen, A. J. 1990, J. Geophys. Res., 95, No. A5, 6389
Stone, J. M., Hawley, J. F., Evans, C. R., & Norman, M. L. 1992, ApJ, 388, 415
Stone, J. M., & Norman, M. L. 1992a, ApJS, 80, 753
_____. 1992b, ApJS, 80,791
Woltjer, L. 1972, Ann. Rev. Astron. Ap., 10, 129

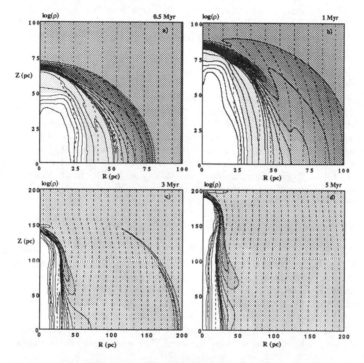

Fig. 1: Evolution of a magnetically-shaped SNR over 5 million years. Solid contours show the logarithm of density; dashed contours show representative magnetic field lines. Note the change of scale between a,b) and c,d).

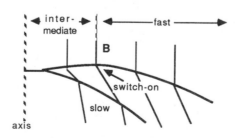

Fig. 2: Diagram of MHD shock wave system corresponding to Fig. 1b.

An X-RAY and OPTICAL study of the SUPERNOVA REMNANT W 44

J.-H. Rho[1], R. Petre, and E. M. Schlegel[2]
Code 666, NASA/Goddard Space Flight Center
J. Hester
Dept. of Physics & Astronomy, Arizona State University
[1]Astronomy Dept.University of Maryland,[2]Universities Space Research Association

ABSTRACT. We report on a 8,000 sec observation of the supernova remnant W44 using the ROSAT PSPC, and mosaiced H α and [S II] images of W44 with 1.2 arcsec resolution. The image has a centrally peaked morphology and contrasts with the shell-like radio. The temperature distribution obtained from X-ray spectra is largely uniform over the remnant, but column density variations are found. An evaporation model with two-phase interstellar medium structure of clumps and interclump gas explains the X-ray centrally peaked morphology of W 44. The mosaiced images reveal the first discovery of optical filaments.

1. INTRODUCTION

W44 (G34.7-0.4) is classed as a "composite" supernova remnant (SNR) whose X-ray emission is centrally peaked, although its radio morphology is limb-brightened. Other members of this class are W 28, 3C 400.2 and W49B. A 267 ms pulsar, PSR 1853 + 01, in this supernova remnant, was discovered by Wolszczan et al.(1991). The pulsar is located 9 arcmin south of the center of W44, within its radio shell and at the southern edge of X-ray emission region. Despite the presence of a radio pulsar and the center-filled X-ray morphology, the X-ray emission from W44 is thought to be thermal rather than the synchrotron radiation. Nearby dense molecular clouds near W44 are interacting with this SNR (Wootten, 1977).

2. OBSERVATION

W44 was observed with the ROSAT X-ray telescope with the position sensitive proportional counter (PSPC). The field of view of PSPC is 2 degree diameter, the angular resolution is 10 arcsec (FWHM, at 1 keV) and the energy resolution is 40% at 1 keV. The PSPC covers the energy range between 0.1 keV and 2.2 keV. Each of four H α, [S II] and red continuum images of W44 were taken using the Wide Field PFUEI on the Palomar 1.5-m telescope with 1.2 arcsec resolution. The wavelengths are 6564Å, 6730Å and 6450Å, respectively, with 15Å filter. Each image has 2,000 sec exposure time. The continuum subtracted H α and [S II] were mosaiced.

3. X-RAY AND OPTICAL IMAGES OF W44

The X-ray image is centrally peaked and shows interior structure as shown in Fig. 1. There is no compact X-ray source coincident with a pulsar, PSR 1853 + 01 (9' to south from the center) in W44. The ROSAT upper limit on a point source luminosity is $1.3 \times 10^{32} ergs^{-1}$, assuming a Crab-like spectrum power law index of 2.2. The ROSAT image superposed on 408 MHz radio continuum is shown in Fig 2. The radio continuum shell is very well aligned with the edge of the X-ray emission. But the eastern X-ray emission does not extend as far east as the radio shell, where a large, dense molecular cloud exists. Spectral analysis also shows the absorption is larger at the east of the remnant. The lack of X-ray emission at the east within the radio continuum shell is due to absorption. The H α image is shown in Fig. 3, superposed with the contour of ROSAT X-ray image. The mosaiced images reveal the presence of optical filaments, in both H α and [S II]. A semicircular filament appears in the northwestern part of SNR, and rather clumpy filaments in the southeastern portion. The optical filaments are mainly confined within the X-ray emitting region. The X-ray image shows locally brighter emission and clumps along the optical filaments, suggesting both are produced by the interaction between the supernova shock front and regions of enhanced ambient density. There is no evidence of a point source on the optical image– even around the pulsar PSR 1853 + 01 in W44. Only clumpy filaments appear around this pulsar position. The eastern optical missing shell is likely due to extinction, because even though this part of optical shell is missing, the radio shell appears as shown in Fig.2. This phenomenon contrasts the case of CTB 109, where half of the shell dissipates in radio and X-ray due to strong interaction with molecular clouds (Tatematsu et al, 1987). The extinction is consistent with the higher absorption measured in X-ray spectra.

4. SPECTRAL OBSERVATIONS

The spectra from the ROSAT PSPC and *Einstein* IPC, EXOSAT ME and *Einstein* SSS, have been jointly analyzed. The spectral analysis of the central part of W44, combining EXOSAT ME and *Einstein* SSS, shows that a two-temperature Raymond model is preferred over a single-temperature Raymond model. This implies that the shocked plasma has not yet reached ionization equilibrium. The column density for the central region is $1.84 \pm 0.05 \times 10^{22}$ cm^{-2} and the temperatures are 0.49 ± 0.02keV and >4keV. These numbers are consistent with those derived from the central ROSAT and IPC data. Assuming the two-temperature Raymond model is appropriate for the rest of the remnant, the variation of temperature and the column density was obtained region by region using the PSPC and *Einstein* IPC. The absorption is high in the east and center of the remnant, where very dense molecular clouds are located (Wootten, 1977). The temperature is largely uniform. The northern peak and southeast edge of X-ray emission show somewhat lower temperature, consistent with the hardness ratio map of IPC and ROSAT. The hardness ratio map of IPC (0.4-4keV) to ROSAT (0.1-2.2keV) images shows two peaks near the remnant center and another weaker peak near a northern peak of the ROSAT intensity map.

5. DISCUSSION

It has been thought that the center-filled X-ray morphology of W 44 is due to high absorption. Smith et al.(1985) suggested the existence of a radially-dependent temperature gradient and that the X-rays created by the cool ($\sim 10^6$K) shell are absorbed by the material along the line of sight. But this model cannot produce the centrally peaked X-ray morphology (White and Long, 1991, WL hereafter). The evaporation model with a two-phase interstellar medium with clump and interclump gas is able to explain the centrally peaked X-rays (WL). The clumps remaining behind a SN shock provide the reservoir of material, and evaporate to increase the density of the interior of a SNR. These clumps gradually evaporate from the hot, postshock gas (Cowie and McKee, 1977). The electron conduction is usually saturated in SNRs until they reach the age of 2x10^4 yr. Only when the evaporation time over SNR age is 20 - 30, the X-ray morphology is centrally peaked. The uniform temperature distribution of W44 strongly supports the predictions of this model. Based on the evaporation model, the interclump gas density is $n_{gas}=0.04$ cm^{-3}, the age of W 44 is 3,500 yrs and the X-ray emitting mass is 38 $M_\odot$. The model assumes "the filling factor is small, the clumps are numerous and the mass ratio between the clumps and interclumps are large"(WL). Using a mass ratio of 100 and $n_{gas}=0.04$ cm^{-3}, the clump density is 500 cm^{-3} with a 1% filling factor This kind of ISM is found in Rosette nebula by interferometric ^{13}CO observations. The density of the clumps is 10^3cm^{-3} and the interclump gas density is very small. The filling factor for the Rosette clumps is about 3% (Blitz, 1992). This also supports the evaporation model by showing the assumed environment exists. W 44 as well as other members W28, 3C400.2 and W49B, exploded in a dense cloud and have evolved rapidly to the radiative stage, being distorted by dense clouds which they encountered. But the clouds cannot have been noticeably compressed by the SNR because the clouds are far from the center of expansion, and are dense to require for compression more energy than the expanding shock can supply.

ACKNOWLEDGE. I thank Dr. Szymkowiak for assistance of non-equilibrium model computation and Dr. Reach for helpful discussions.

REFERENCES

Blitz, L., 1992, in private communication

Cowie, L.L. and Mckee, C.F. 1977, Ap. J., 211, 135.

Smith, A., Jones, L.R., Watson, M.G. and Willingale, R. 1985, MNRS, 217, 99.

Tatematsu, K., Fukui, Y., Nakano, M., Kogure,T., Ogawa, H., and Kawabata, K. 1987, A.& A., 184, 279.

Weiler, K. W. 1983, Observatory, 103, 85.

White, R. L. and Long, K. S. 1991, Ap. J., 373, 543.

Wolszczan, A., Cordes, J.M. and Dewey, R.J. 1991, Ap. J., 372, L102.

Wootten, H. A. 1977, Ap. J., 216, 440.

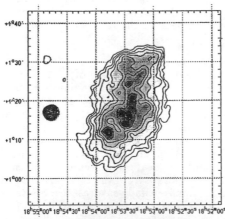

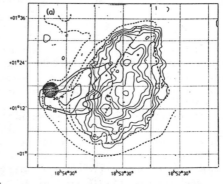

Fig. 1. ROSAT image of W44. Contours are from 1 to 18 photons/15"pixel.

Fig. 2. X-ray image superposed on ^{13}CO and 408MHz radio maps (dashed)

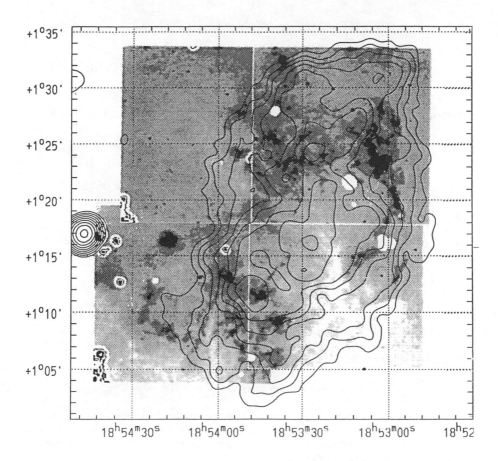

Fig. 3. The Hα image is superposed on X-ray contours. The optical filements are confined within X-ray emission.

IS THE MILKY WAY ELLIPTICAL?

Konrad Kuijken[♠]
Center for Astrophysics, 60 Garden Street, Cambridge MA 02138, USA

Scott Tremaine
Canadian Institute for Theoretical Astrophysics, University of Toronto,
60 St. George St., Toronto M5S 1A7, Canada

ABSTRACT

We show that two significant anomalies in the kinematics of the Galaxy,
viz. the low value of the axis ratio of the velocity ellipsoid of old stars in the solar
neighbourhood and the difference in apparent rotation curves between distant
stars and HI gas, can be explained if the gravitational potential in the Galactic
plane is elliptical. In the best-fit model, the isopotential curves have an axis ratio
of 0.92, roughly constant with radius, and the mean circular speed is $200\,\mathrm{km\,s^{-1}}$.

1. INTRODUCTION

Galaxy halos are unlikely to be round if they are formed by dissipationless
collapse (Dubinski & Carlberg 1991). Galactic disks embedded in such halos
will experience a non-axisymmetric gravitational field Ψ, which to first order we
can characterise as an elliptical perturbation on a circular potential

$$\Psi(R,\phi) = \Psi_0(R) + \tfrac{1}{2}\epsilon(R)v_c(R)^2 \cos 2(\phi - \phi_b). \qquad (1)$$

We might hope to detect signatures of this perturbation in the shapes and
kinematics of galactic disks. Here we focus on available evidence for our own
Galaxy; for results concerning possible ellipticity in other galaxies, see Franx &
de Zeeuw (1992) and references therein.

We will consider a simple 'standard' model, in which the axisymmetric
component has constant circular speed v_c, $\Psi_0 = v_c^2 \ln R$, and the ellipticity of
the equipotentials $\epsilon > 0$ is also constant. The angle ϕ_b is the Galactocentric
azimuth of the minor axis of the potential. It is convenient to work with the
combinations

$$c_\Psi = \epsilon \cos 2\phi_b \quad \text{and} \quad s_\Psi = \epsilon \sin 2\phi_b. \qquad (2)$$

Local constraints on c_Ψ and s_Ψ have been discussed elsewhere (Kuijken
and Tremaine 1991, Kuijken 1992). By way of summary, we present the local
results, and predictions for the 'standard' model, in Table 1. They imply

$$s_\Psi = 0.01 \pm 0.04, \qquad c_\Psi = 0.12 \pm 0.04, \qquad (3)$$

i.e. the Sun lies near the minor axis of the equipotentials, which have an axis
ratio of 0.88 ± 0.04. This number is sensitive to radial gradients in v_c or ϵ, which
are assumed zero in the standard model.

♠ Hubble Fellow

	Observed	$\pm$	Predicted
Oort constant A	$14.4 \, \mathrm{km \, s^{-1} kpc^{-1}}$	1.2	$\frac{1}{2}(1 + c_\Psi)v_c/R_0$
Oort constant B	$-12.0 \, \mathrm{km \, s^{-1} kpc^{-1}}$	2.9	$-\frac{1}{2}(1 + c_\Psi)v_c/R_0$
Oort constant C	$0.6 \, \mathrm{km \, s^{-1} kpc^{-1}}$	1.1	$\frac{1}{2}s_\Psi v_c/R_0$
Oort constant K	$-0.35 \, \mathrm{km \, s^{-1} kpc^{-1}}$	0.5	$-\frac{1}{2}s_\Psi v_c/R_0$
Radial LSR motion $\overline{v_R}$	$-1 \, \mathrm{km \, s^{-1}}$	9	$s_\Psi v_c$
Azimuthal LSR motion $\overline{v_\phi}$	—	—	$(1 - c_\Psi)v_c$
Vertex deviation	$5.5°$	4.2	$-2s_\Psi$
Velocity ellipsoid σ_ϕ^2/σ_R^2	0.42	0.06	$(1 - 3c_\Psi)(0.66 \pm 0.06)$

Table 1. Local constraints on ellipticity.

For the remainder of this paper, we will set $s_\Psi \equiv 0$ and focus on constraints derived from distant tracers of Galactic rotation. In particular, we will re-interpret published determinations of the Galactic rotation curve (derived with the assumption of axisymmetry) from the viewpoint that the Galaxy may not be axisymmetric.

2. DEEP STELLAR RADIAL VELOCITY SURVEYS

An elliptical distortion affects classical methods for measuring the Galactic rotation curve. In an axisymmetric velocity field, the radial velocity with respect to the local standard of rest of 'tracer' stars at distance D towards Galactic longitude ℓ is given by

$$v_{\mathrm{rad}} = \left[\left(\frac{R_0}{R} \right) v_c(R) - v_c(R_0) \right] \sin \ell, \qquad (4a)$$

$$\text{with } R = \left(R_0^2 + D^2 - 2DR_0 \cos \ell \right)^{\frac{1}{2}}. \qquad (4b)$$

In practice, the distance to the Galactic centre R_0 is only known with about 15% precision and the zero point of the distance scale of the tracer stars may be poorly calibrated, so it is best to leave R_0 as a free parameter in fits to $(D, v_{\mathrm{rad}}, \ell)$ data. A fit to v_c = constant models is usually found to be adequate—but remember that with radial velocity data alone it is always possible to add an arbitrary term $v_c \propto R$ to the rotation curve. In the case of a linear rotation curve, constant-v_c fits return the value of $2AR_0$.

We have simulated two recent surveys of distant tracers ($D \lesssim R_0$): the Cepheid survey of Caldwell & Coulson (1989) and the carbon star data of Schechter *et al.* (1989; this volume). We assumed that the Galaxy is of 'standard' elliptical form, with the Sun on a symmetry axis. For a range of ellipticities c_Ψ, we simulated radial velocity surveys sampled in a manner similar to these real stellar surveys, and fitted an axisymmetric model of the form of eq. (4) to these data, assuming constant v_c. The results for the two data sets are plotted in Figure 1; crudely, an ellipticity of $c_\Psi = 0.1$ leads to an overestimate of the circular speed of $\sim 25\%$. However, while these surveys are *sensitive* to ellipticity, they are not good *detectors* of it: the rms deviation of the best-fit axisymmetric

Method	v_c	Reference
HI tangent points: $\ell = 30°-90°$	$220\,\mathrm{km\,s^{-1}}$	Gunn *et al.* (1979)
HI tangent points: $\ell = 30°-90°$	$214\,\mathrm{km\,s^{-1}}$	Rohlfs *et al.* (1986)
HI tangent points: $\ell = 53°-90°$	$284\,\mathrm{km\,s^{-1}}$	Rohlfs *et al.* (1986)
HI tangent points: $\ell = 53°-90°$	$260\,\mathrm{km\,s^{-1}}$	Merrifield (1992)
HI scale heights: $R = R_0-2R_0$	$165\,\mathrm{km\,s^{-1}}$	Merrifield (1992)

Table 2. Determinations of v_c from gas kinematics.

model to radial-velocity data drawn from a Galaxy with intrinsic ellipticity as high as 0.2 is less than 2% of the circular speed!

3. ROTATION OF THE HI GAS

The other traditional tracer of large-scale Galactic structure is the neutral hydrogen 21cm line. This is diffuse emission, and so distance determinations are somewhat problematical, and necessarily less direct than they are for stars: somehow the kinematic information must be used to yield both a distance and a rotation curve simultaneously. Two methods have been used with success: the classical tangent-point method (Kwee *et al.* 1954) for the inner Galaxy ($R < R_0$), and the HI scale-height method (Merrifield 1992) outside the solar circle. Table 2 summarizes recent results from such analyses. Once again, we have simulated the effects of ellipticity on these data, and plotted the results in Figure 1. The most striking result is the low value of v_c derived by Merrifield; also, his value is sensitive to ellipticity in the *opposite* sense to the other results. This is a consequence of the radically different kinematic distance determination he uses.

4. ELLIPTICITY?

We have seen that determinations of the rotation curve of the Milky Way outside the solar circle (i) differ significantly, and (ii) are affected in different ways by any ellipticity present. It is therefore natural to ask if there is a value for the ellipticity with which these analyses can be reconciled. The answer is provided in Figure 1, where the determinations of the circular speed discussed above are shown corrected for a range of possible ellipticities. We see that, if the rotation curve is flat and the ellipticity of the potential constant with radius, an ellipticity of 0.08 is able to account for all of the discrepancy. This is the same value as was derived *independently* from the anomaly in the solar neighbourhood kinematics. A least-squares fit for the ellipticity and circular speed from the data shown in Table 1 and Figure 1 (assuming errors of $20\,\mathrm{km\,s^{-1}}$ for the latter) gives

$$v_c = 197 \pm 10\,\mathrm{km\,s^{-1}}, \qquad c_\Psi = 0.080 \pm 0.013, \qquad \chi^2 = 3.6 \ (6 \ \mathrm{d.o.f.}). \qquad (5)$$

When we relax the assumptions of a flat rotation curve and constant ellipticity, replacing both with power laws ($v_c \propto R^\alpha$ and $\epsilon v_c^2 \propto R^p$), we find

$$v_c = 184^{+19}_{-14}\,\mathrm{km\,s^{-1}}, \quad c_\Psi = 0.08^{+0.04}_{-0.03}, \quad \alpha = -0.1^{+0.1}_{-0.1}, \quad p = 0.1^{+0.3}_{-0.6}, \qquad (6)$$

with $\chi^2 = 2.0$ (4 d.o.f.): the best-fit ellipticity is reasonably constant with

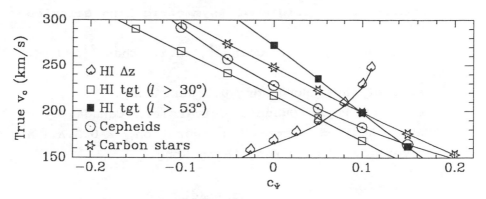

Figure 1. Determinations of the Galactic circular speed, corrected for different possible ellipticities c_Ψ.

radius. The best-fit axisymmetric model has $\chi^2 = 9$, with the best fits requiring, implausibly, that $\alpha \simeq -0.45$, much steeper than in other galaxies.

5. SUMMARY

We have shown that discrepancies between different measurements of the Galactic rotation curve outside the solar circle can be understood as the effect of an elliptical distortion in the gravitational potential. The same distortion also explains a different anomaly, namely the low value of the axis ratio of the velocity ellipsoid in the solar neighbourhood.

ACKNOWLEDGEMENTS

This work was supported by a NASA Hubble Fellowship awarded by the Space Telescope Science Institute (KK), and by a research grant from NSERC (ST).

REFERENCES

Caldwell, J.A.R. & Coulson, I.M. 1989, in The Outer Galaxy, eds. L. Blitz & F.J. Lockman (Springer: Berlin), 68
Dubinski, J. & Carlberg, R.G. 1991, ApJ, 378, 496
Franx, M. & de Zeeuw, P.T. 1992, ApJ, 392, L47
Gunn, J.E., Knapp, G.R. and Tremaine, S.D. 1979, AJ, 84, 1181
Kuijken, K. 1992, PASP, 104, 809
Kuijken, K. & Tremaine, S. 1991, in Dynamics of Disk Galaxies, ed. B. Sundelius (Göteborg: Göteborg University Press), 71
Kwee, K.K., Muller, C.A. & Westerhout, G. 1954, BAN, 12, 211
Merrifield, M.R. 1992, AJ, 103, 1552
Rohlfs, K., Chini, R., Wink, J.E. and Böhme, R. 1986, A&A, 158, 181
Schechter, P.L., Aaronson, M., Cook, K.H. & Blanco, V.M. 1989, in The Outer Galaxy, eds. L. Blitz & F.J. Lockman (Springer: Berlin), 31

Formation and Origin of Asymmetric Ring Galaxies

S.M. Leung[1], K.S. Cheng[1], W.Y. Chau[2] and K.L. Chan[3]

[1] Dept. of Phys. Hong Kong U., Hong Kong.
[2] Dept. of Phys. Queen's University, Canada.
[3] NASA, Appiled Research Corporation, U.S.A. and University of Hong Kong, Hong Kong.

Abstract

We use a modified particle-mesh N-body code, which can enhance the resolution of certain desirable regions to study the interaction between dark halo and its enclosed galaxy. We find that a transient rings with the nuclei displaced toward the edge of the ring can easily form if the center of the halo and the center of galaxy are separated by a distance d and move with a relative velocity υ. More intriguingly, these asymmetric rings can last up to a few billion years for some narrow parametric regimes of d and υ. An estimation of the probability of forming relatively stable asymmetric ring galaxies from our model are consistent with the observed results.

(I) Introduction

There are a number of ring galaxies where their nuclei are found apparently displaced towards the edges of the rings. They are called asymmetric ring galaxies. For examples AM 0552-324, AM 1452-234, AM 1724-622, AM 0058-311, and AM 2021-724 (c.f. Chap. 6 and 10 of Arp and Madore, 1987). We address their origin by the approach of cold dark matter. Although it is believed that the centers of the (dark halo and the visible galaxy) are conincident to each other since they are formed, it is not unreasonable to

assume that the galaxy is originally displaced away
from the dark halo center d kpc and moves around it
with a relative velocity v_{rel} or a relative angular
velocity Ω_{rel}. Intuitively, such initial conditions
should result in some ring structure in galaxies.

(II) The Choice of Initial Conditions

Many observations suggest that galaxies are
embedded into a dark halo. The general form which
fits both the flat rotation curves of many disk
galaxies (Kent 1987) and the isotropic gamma ray burst
distribution (Brainerd 1992) is

$$\rho(r) = \frac{\rho_0 \cdot a^2}{(a^2 + r^2)}$$

where a and ρ_0 are the core radius and density
respectively and r is the isodensity radius. For the
initial distribution of stars in the galaxy, we choose
that all particles are randomly distributed in a
sphere of 10 kpc to produce a uniform mass
distributed. The galactic center initially departs
from the halo center away d kpc. The random
velocities of the particles are chosen at random from
an isotropic Gaussian distribution with a standard
deviation. In addition each particle is assigned a
specific velocity distribution around the center of
the dark halo. There are two velocity distributions
used: constant angular velocity (Ω_{rel}) and constant
velocity (V_{rel})

The figure for Ω_{rel} show a series of morphology
of our model galaxy with d = 11.0kpc and k = 1. Each
picture of these series is the morphology at the
critical time of revolutions. The time of each
revolution is 3.4×10^8 years. It reveals that a tail
first appears. As the core of the galaxy is being
attracted toward the halo center, more particles leave
it and amend the head of tail while the original tail
particles are orbiting around the halo center. In the

meantime, the evaporation of the core particles join the returning end of the tail. Thus a ring galaxy with displaced nucleus is formed.

(III) **Probability of Forming Stable Asymmetric Rings**

Our initial condition of placing all visible particles inside a compact sphere makes the whole galaxy act as it a single object inside the dark halo. The mean velocity of this object will be the Keplerian velocity (V_k) which produce a centrifugal force just balance the gravitational force of the halo. If we assume that the velocity distribution of the particles of the halo mimics the velocity distribution of a huge elliptical galaxy which has a typical velocity dispersion σ_v ~ 300km/sec (ref. Peletier al et 1990), then we can calculate the probability of finding a particle with velocity between V_{min} and V_{max}, where they equals to 0.7 and 1.2 times of the Keplerian velocities of the galaxy rotating around the halo, and it is given by

$$P_v \approx \frac{\int_{V_{min}}^{V_{max}} \exp\left(-\frac{(V-V_k)^2}{2\sigma_v^2}\right)\cdot d^3v}{\int_{-\infty}^{\infty} \exp\left(\cdot\frac{(V-V_k)^2}{2\sigma_v^2}\right)\cdot d^3v}$$

P_v equal 0.068 for both constant Ω_{rel} and constant V_{rel} respectively. Secondly, let's assume that the galaxy is randomly placed inside the halo initially, then the probability of finding the galaxy between 8.0kpc and 11.0 is given by

$$P_d \sim \frac{(11.0 - 8.5\ )kpc}{100kpc} \sim 0.025$$

where 8.5kpc is the lower limit of d that a clear ring can be obtained. With a smaller value, the evaporation of particles from the galaxy will fill up the empty space and ring structure losses. The upper limit concerns the tidal force from the nearest halo (about 200kpc away from centers) which will distort the ring shape. Finally the denominator of 100kpc represents the halo size. The combined probability is 1.7×10^{-3}. However observation shows that about 2/3 of galaxies are spirals or lenticulars. Together with the recent results which support that a certain amount of elliptical galaxies are formed from the merging of spiral galaxies, we believe that our initial conditions about the galaxy itself happened only in about 10% of the total galaxies. Thus our approximated probability is 1.7×10^{-4} . (The details have been presented elsewhere).

References:

1. H.C.Arp and Madore, Barry F., "A Catalogue of Southern Peculiar Galaxies and Association" Cambridge Univ. Press, 1987.

2. J.J.Brainerd, 1992 Nature 355,522.

3. R.F.Peletier, R.L.Davies, G.D.Illingworth, L.E.Davis & M.C.Cawson, 1990 AJ., 100, 1099

4. S.M.Kent, 1987 AJ., 93, 816

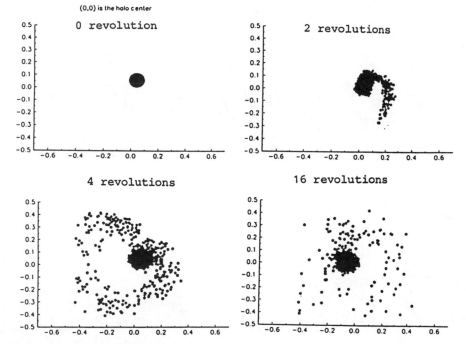

DARK

MATTER

EVIDENCE FOR DARK MATTER

Paul L. Schechter
Physics 6-206, Massachusetts Institute of Technology, Cambridge, MA 02139.

Email ID
schech@achernar.mit.edu

ABSTRACT

By measuring the dimensionless product $(\sigma_z/v_c)^2 R_0/h$ one can estimate the ratio of disk mass to total mass interior to the Sun's orbit, where σ_z is velocity dispersion perpendicular to the disk appropriate to some tracer population, h is the scale height of that tracer, and v_c is the circular velocity at the the Sun's distance from the center of the Milky Way, R_0. While there are model dependent factors of order unity which render this ratio uncertain, it is difficult to avoid the conclusion that there is significant dark matter interior to the Sun's orbit. Beyond the solar circle radial velocity data for disk stars are consistent with a flat rotation curve out to $2R_0$. Beyond that the interpretation of radial velocity data for blue horizontal branch stars, globular clusters and satellite galaxies requires assumptions about their transverse velocities. Proper motion measurements less uncertain than 0.5 mas/yr might resolve this ambiguity.

1. INTRODUCTION

There may be as many as four distinct kinds of dark matter – disk dark matter, galaxy dark matter, cluster dark matter and cosmological dark matter. While economy of hypothesis would argue that two or more of these are the same, one must then ask why this dark stuff finds itself in condensations as different from each other as galaxy disks, galaxy halos and clusters of galaxies, as well as in an (as yet) uncondensed cosmological background.

Only two of these species of dark matter lie within the scope of the present conference. In restricting ourselves to dark matter in galaxies we should nonetheless bear in mind that this represents only a tenth of the dark matter observed in clusters. This, in turn, represents only a fifth of the dark matter needed to provide the universe with a critical mass density. Milky Way dark matter may be interesting, but in the grander scheme of things it's definitely small potatoes. Perhaps the best question one can ask about the dark matter in galaxies is not whether it exists but rather where it ends and where cluster dark matter begins.

It is considerably more difficult to obtain evidence for dark matter in the Milky Way than in an external galaxy, if for no other reason than that a single pointing of a telescope can suffice for an external galaxy, while studies of our own galaxy demand considerably more effort. The Milky Way nonetheless offers unique opportunities arising from the fact that we can, at least in principle, measure all six phase space components for tracers of the Milky Way's potential.

Arguments regarding the presence or absence of dark matter are made doubly difficult by the fact that one must compare a global mass estimate with a census of constituents for which masses are likewise uncertain. In the next section we make two passes at the dark matter question, first using an overly simple model which has the advantage of exhibiting explicitly the interplay of the four principal measured quantities and then using a more refined model which gives a more reliable if less transparent estimate.

Both models rely on tracers in the disk of the Milky Way, which are therefore on nearly closed orbits. Beyond the disk we must use halo tracers whose orbits are far from closed, and whose distribution of shapes is at best poorly known. Dark matter estimates based on these halo tracers are therefore considerably more uncertain than those obtained from disk tracers. But since they extend to considerably greater distances, they permit one to trace the dark matter further from the center of the galaxy.

2. CLOSED ORBITS

a. Toy Model

The Milky Way is widely thought to be an Sbc or Sc galaxy much like NGC 4565 or NGC 891. The census of such a galaxy can be carried out on the fingers of one hand. There is a disk, a bulge, and perhaps, a "thick disk" (Burstein 1979) and a nucleus. The thick disk of the Milky Way contributes of order 10% to the mass of the disk (Gilmore and Reid 1983), and may not be distinguishable from the conventional disk (Norris 1987). The nucleus likewise contributes negligibly. That leaves us with only the bulge and the disk.

Kent (1986, 1987, 1988) has written a series of papers in which he obtains photometric and dynamical decompositions of spiral galaxies into disk and bulge components. Assuming that the mass-to-light ratios for these are independent of position, he finds that the bulges have M/L values which are roughly a factor of 2 greater than galaxy disks.

In an important paper which is only beginning to receive the attention it deserves, Kent et $al.$ (1991) have used $2.4\mu m$ data obtained with the Infrared Telescope flown aboard the Spacelab-2 mission to obtain photometric parameters for the bulge and disk of the Milky Way. Among the many useful quantities presented in that paper is an infrared bulge-to-disk ratio for the Milky Way: $B/D = 1/5$. Unless the mass-to-light ratio for the bulge is very different from that seen in other galaxies, the disk contribution to the Milky Way's mass is very much greater than that of the bulge.

Our census of the Milky Way has produced a single major contributor to its mass – the disk. The evidence for dark matter in the Milky Way can now be presented in cartoon form by assuming that the rotation curve for the Milky Way is constant and that the disk is solely responsible for the potential which produces that rotation curve. A uniform disk with surface density varying as $1/R$ gives a constant rotation curve (Mestel 1963). The surface density at any point in that disk is given by $\mu_{Mestel} = v_c^2/RG$. But the local surface density of the Milky Way can be determined from measurements of the velocity dispersion perpendicular to the plane, σ_z, and the scale height h perpendicular to the plane, with $\mu_{local} = \sigma_z^2/hG$ for the idealized case of a massless isothermal tracer in the field of an infinitely thin disk. The ratio of the local surface density to the density needed to produce the observed circular velocity using our Mestel disk model, would therefore be

$$\frac{\mu_{local}}{\mu_{global}} = \frac{\sigma_z^2}{v_c^2}\frac{R_0}{h} \quad . \tag{1}$$

Taking $R_0 = 8.5$ kpc (Kerr and Lynden-Bell 1986), $h = 300$ pc (c.f. Bahcall and Soneira 1980), $v_c = 220$ km/s (Kerr and Lynden-Bell 1986) and $\sigma_z = 20$ km/s (Hartkopf and Yoss 1982) we find that the local surface density is only one quarter of that required by the Mestel disk model. Since the bulge makes only a small contribution to the total mass, one is driven to the conclusion that the major contribution to the circular velocity at the Sun's position lies in an unseen component – a dark "halo" or "corona."

b. Freeman Disk

A more realistic model would depend upon the four observables in the same way as in our toy model, but with different dimensionless factors which might reverse our conclusion that the disk contributes little of the mass interior to the solar circle. The surface brightness profiles of disks in external galaxies look more like exponentials (Freeman 1970) than Mestel disks. Assuming that the surface mass density in the disk has an exponential scale length R_d, we can use the locally measured surface density to derive a mass for the disk. A dimensionless function, $f(R_0/R_d)$ gives us the fraction of the mass interior to the Sun's orbit. This can then be compared to the total mass interior to the solar circle (again assuming a spherical mass distribution), giving

$$\frac{M_{disk}(< R_0)}{M_{total}(< R_0)} = \frac{2\pi f(R_0/R_d)G\mu(R_0)R_d{}^2 exp(R_0/R_d)}{v_c{}^2 R_0} \quad . \tag{2}$$

Several different methods have been used to estimate an exponential scale length for the light in the Milky Way's disk. Kent et al. (1991) find $R_d = 3.0$ kpc, with model induced uncertainties of 0.3 kpc. Van der Kruit (1986), using Pioneer 10 data, finds find $R_d = 5.5 \pm 1$ kpc. For the sake of argument we adopt Gould's (1990) value local surface density, $\mu = 54 \pm 8$ $M_\odot$, which he derives from the data of Kuijken and Gilmore (1989). For the Kent et al. (1991) value of R_d we have $f = 0.77$, putting a large fraction of the disk mass interior to the Sun. We find that the disk contributes only 42% of the mass inside the solar circle.

There has been a spirited ongoing debate about the presence or absence of dark matter within the disk of the Milky Way (see the volume edited by Philip 1989, and the recent papers by Kuijken 1991 and Bahcall et al. 1991), with estimated ratios of dark matter to luminous matter in the ranging from zero to unity. The differences involve different censuses of luminous components and different methods of analysis using differently chosen tracers of the disk potential. Gould's surface density estimate puts relatively little dark matter in the disk. Were we to use a larger value, we would find less dark matter in the halo, at the expense of more dark matter in the disk. A fundamental difference between the two is that disk dark matter must be dissipative, while halo dark matter need not be. Since many non-baryonic species of dark matter would be non-dissipative, they would be ruled out as candidates for disk dark matter.

Had we used van der Kruit's value for the disk scale length we would have found the disk contributing a yet smaller fraction to the total mass interior to the Sun's orbit, but with a larger fraction of the disk's mass beyond the solar circle. In either case the ratio of disk mass to total mass decreases at radii larger than R_0 as long as v_c remains roughly constant. Since the rotation curves of external galaxies are found to be nearly constant out to their last observable point, it behooves us to determine the rotation of the Milky Way as far as possible from its center.

c. Rotation of the Milky Way to $2R_0$

It is unfortunate that the tangent point method for measuring rotation using neutral hydrogen data, which works so well inside the solar circle, cannot be used beyond it. One needs a tracer of the galactic potential for which accurate distances can be determined, and which is sufficiently abundant that it can still be found 2 or three scale lengths beyond the Sun. Mike Merrifield (1992) has invented an ingenious method for using neutral hydrogen (what could be more abundant?) which relies, however, on the assumption that the scale height of the hydrogen at a given radius is independent of galactocentric azimuth. More traditonal investigators have used individual stars and OB associations (Blitz, Fich and Stark 1989).

There are a number of difficulties associated with such determinations; these can only be touched upon here. One must be sure that the estimated distance to the tracer is on the same *scale* as the distance to the galactic center – otherwise one will make a systematic error in computing galactocentric radius and in deprojecting the observed radial velocity. One must correct for heavy absorption by intervening dust. One must allow for the possibility that the intrinsic luminosity of the chosen tracer might vary with metallicity, and hence galactocentric radius. One must allow for Malmquist's effect, which depends upon $A(m)$, the observed apparent magnitude distribution for one's tracer. This is particularly difficult in the presence of patchy galactic obscuration which varies with both latitude and longitude. And one must allow for the possibility of an elliptical potential.

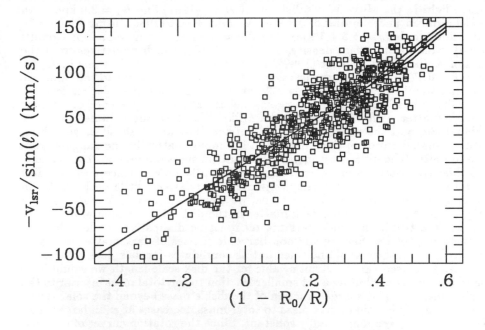

Fig. 1. Radial velocities v_{lsr} and galactocentric radii R for 651 carbon stars. The ordinate and abscissa have been chosen so that a linear rotation curve would yield a straight line with slope $2AR_0$.

My colleagues Marc Aaronson, Victor Blanco, Kem Cook, Ed Olszewski and I (Aaronson *et al.* 1989, 1990) undertook to use carbon stars to trace the potential out to to $2R_0$. Previously unknown carbon stars were identified using the Curtis and Burrell Schmidt telescopes operated by NOAO. Radial velocities and JHK photometry were then obtained at Las Campanas, Cerro Tololo and Mount Palomar. We observed carbon stars *interior* to the solar circle to calibrate their intrinsic luminosities, guaranteeing that all our distances are on a consistent scale. We plot the quantity $-v_{lsr}/\sin\ell$ against an abscissa chosen so that a linear rotation curve produces a linear distribution of points, with slope $2AR_0$, where A is Oort's constant. For the special case of a flat rotation curve the slope is equal to v_c. Results for a subset of our data are shown in Figure 1. One sees first that there is substantial scatter, consistent with the fact that carbon stars have a rms spread in absolute magnitude of roughly 0.60 mag. Second, one sees that we run out of carbon stars at roughly $2R_0$ – we have found the edge of the disk. Third, while we can determine the slope quite accurately, it is difficult to tell whether the points deviate from linearity.

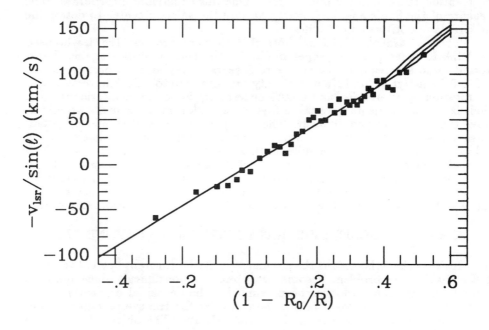

Fig. 2. The same data as in Fig. 1. but with each filled symbol representing an average of 16 carbon stars. The solid lines show the effect on a flat rotation curve of introducing a Keplerian falloff at $1.6R_0$, $1.8R_0$ and $2.0R_0$.

In Figure 2 we have binned the points from Figure 1 in groups of 16. The rotation curve now appears to be very nearly linear. For the sake of comparison we show several model rotation curves which are constant out to some radius and then fall off in Keplerian fashion. A Keplerian onset at $1.8R_0$ is not ruled out, but is at $1.6R_0$. This approach suggests a useful way of parameterizing mass (see, e.g., Fich and Tremaine 1991), which we shall adopt for the remainder of our discussion: we assume a flat rotation curve out to some limiting radius R_{max}, expressed in units of R_0.

There are several embarrassments associated with Figures 1 and 2 to which I must confess. The first is that at distances greater than $2R_0$ more than half of the circular velocity is projected perpendicular to the line of sight. Radial velocities are increasingly inefficient at larger distances – proper motions would give a better rotation curve if they could be measured. The second is that the uncertainty in the overall slope is dominated by the details of the adopted Malmquist correction – carbon stars are simply too poor a standard candle. The points shown have been corrected using one of several alternative schemes for estimating Malmquist's effect. The resulting slope is $2AR_0 = 227$ km/s.

The third embarrassment is that our measurements of velocities relative to the Local Standard of Rest give us only the Milky Way's *differential* rotation. Our stars might have some constant angular velocity with respect to an inertial frame which would not show up in our radial velocities. The data tell us that the rotation curve is linear but not flat. One must therefore incorporate some measure of the Sun's angular velocity around the galactic center to determine the circular velocity at any radius.

The best available data are those of Backer and Sramek (1987), who used the VLA to determine the proper motion of the galactic center against background radio sources over the course of 5 years. They found a proper motion of 6 ± 0.6 mas/yr. This 10% uncertainty propagates into a 20% uncertainty in the rotation speed at $2R_0$, and a 40% uncertainty in the mass interior to $2R_0$. Their last measurements were taken 6 years ago. New measurements would improve the uncertainty enormously. Their proper motion is, it should be noted, consistent with a flat rotation curve. Cognoscenti will recognize that Oort's A and B constants also give a handle on this quantity. But the road to Oort's B is a rocky one and I prefer the simplicity of Backer and Sramek's approach.

For the record, it should be noted that if we use van der Kruit's value for R_d, which maximizes the disk mass out to $2R_0$, we find that the mass in the disk is only one quarter of the mass out to $2R_0$. The rest is presumably dark.

3. RANDOM ORBITS: VELOCITY DISPERSIONS BEYOND $2R_0$

How much further might the circular velocity be constant beyond the edge of the disk? The problem becomes very much more difficult since material beyond the disk can no longer be presumed to be on closed or nearly closed orbits. For objects on closed orbits one can infer the transverse velocity (and hence the circular velocity) from the radial velocity. The objects beyond the disk – globular clusters and halo stars – are on plunging orbits for which the ratio of radial to transverse velocity is a matter of some debate.

The degree to which these orbits plunge is not widely appreciated. One can construct an illustrative spherical model for an elliptical galaxy with a logarithmic potential using Schwarzschild's (1979) method. Only one orbit is needed to produce an r^{-3} density profile and an isotropic velocity dispersion tensor. It has a ratio of apogalacticon to perigalacticon of 3.73.

Several different methods have been developed to measure the mass of the halo beyond $2R_0$, all of which rely first, on measurements of the radial velocity dispersion of a tracer population, and second, on assumptions about the shape of the typical orbit. The most straightforward of these is that used by Hartwick and Sargent (1978), who used the radial Jean's equation for the case of spherical symmetry:

$$v_c(r)^2 = \sigma_r{}^2 \left(-\frac{d\ln\nu}{d\ln r} - \frac{d\ln\sigma_r{}^2}{d\ln r} - 2\beta(r) \right) \quad . \tag{3}$$

Here σ_r is the radial velocity dispersion, ν is the number density of the tracer and $\beta = 1 - \sigma_\theta{}^2/\sigma_r{}^2$, the "anisotropy parameter," is 0 for isotropic orbits, 1 for radial orbits and $-\infty$ for circular orbits. The tracers used for such studies – globular clusters, satellite galaxies and blue horizontal branch (BHB) stars, have number densities which typically fall off as r^{-3} or r^{-4}.

Since $v_c(r)^2$ is inferred from the measured radial velocity dispersion, one needs large numbers of tracers to get a modest fractional uncertainty. Worse yet, for the case of an r^{-3} density profile, isotropic orbits give masses which are larger than those for radial orbits by a factor of 3! In the absence of strong arguments constraining the anisotropy parameter one must measure proper motions to eliminate this ambiguity.

Little and Tremaine (1987) have concocted a scheme for estimating either masses (under the assumption of a Keplerian potential) or $v_c{}^2$ (under the assumption of a logarithmic potential) using a small sample of objects (assumed to be either complete or a fair sample thereof) spanning a wide range of radii. Their method is subject to the same uncertainty regarding the assumed anisotropy, and an added uncertainty arising from the need for some assumption about the relative likelihood of different values of $v_c(r)$.

Norris and Hawkins (1991) have newly identified ten faint blue horizontal branch stars at galactocentric distances ranging from 40 to 65 kpc. The observed radial velocity dispersion in this sample is 111 ± 25 km/s. They estimate the logarithmic derivative of the number density to be -4. Their data are therefore consistent with a flat rotation curve to $6R_0$ under the assumption of isotropic orbits. A major concern in their analysis is possible contamination by foreground blue stragglers, which they attempt to eliminate using Balmer absorption line strengths. One of the great strengths of their method is that there are many more BHB stars at the same radius waiting to be found – their sample was drawn from just two Schmidt plates.

Hartwick and Sargent (1978), Little and Tremaine (1987), Zaritsky et al. (1989) and Kulessa and Lynden-Bell (1992) have used globular clusters and satellite galaxies to work at even greater galactocentric distances. The results depend critically upon whether one includes or excludes the Leo I dwarf, at a radius of 230 kpc with an observed velocity of 177 km/s (Zaritsky et al. 1989). Excluding Leo I and Leo II (at roughly the same radius) one finds that the 12 satellites and globulars in the range $51 < r < 140$ kpc have a line of sight velocity dispersion of 86 km/s.

A simple escape energy argument shows why Leo I so dominates the calculation. If one adopts the a "minimal halo" model (Fich and Tremaine 1991), with a flat rotation curve out to a radius $r_{maz} < r_{Leo}$ and a Keplerian falloff thereafter, and assumes that Leo I is bound to the Milky way, one has

$$\frac{r_{maz}}{r_{Leo}} > \frac{1}{2} \left[\frac{v_{Leo}}{v_c(R_0)} \right]^2 \quad , \tag{4}$$

implying a flat rotation curve out to at least $8R_0$. If Leo I had any transverse velocity, the curve would extend even further.

These same escape arguments may, in principle, be applied to any sample of stars (e.g. Leonard and Tremaine 1989). One of the Norris-Hawkins stars, an RR Lyrae at $r = 61$ kpc, has a radial velocity of 227 km/s.

4. WHAT IS IT?

Calculations of big bang nucleosynthesis (Walker *et al.* 1991) argue for a low present day baryon density, perhaps so low that it rules out baryonic matter not only for closing the universe but even for providing the dark matter within clusters of galaxies. Non-baryonic particles are a natural constituent of many favored approaches to unifying the strong interaction with the weak and the electromagnetic (e.g. Turner 1987). My theorist friends take it for granted that the dark matter is non-baryonic. A number of experimenters have taken such arguments so seriously that they have begun laboratory searches. Experiments are underway in Berkeley, Munich and elsewhere to detect various candidate dark matter particles directly. I am not competent to judge whether any of these is likely to produce interesting or useful constraints.

There is a second kind of dark matter, baryonic objects with substellar masses, which may be somewhat more familiar. Several groups are looking for MACHOs, massive astrophysical compact halo objects, which if present, would produce microlensing of the stars in the Magellanic Clouds and the galactic bulge (Udalski *et al.* 1992, Bennett *et al.* 1992). The groups carrying out these searches involve curious marriages of convenience. On the one hand there are particle physicists who believe that the dark matter is non-baryonic and whose principle interest is in ruling out MACHOs. On the other are classically inclined astronomers whose enthusiasms run toward the Cepheids, RR Lyraes, Algols and SX Phoenicis stars that the are the inevitable byproduct of such a search. The Warsaw/Campanas effort, "OGLE", has already obtained photometry for more than a million bulge stars on each of 45 nights (Udalski *et al.* 1992).

5. BOK TO THE FUTURE?

The future belongs to the astrometrists. Those of us who measure radial velocities have been stymied by our ignorance of the transverse velocities of tracers of the halo potential. We measure only one of the three available velocity components. The expected proper motions for stars at $5R_0$ are of order 0.5 mas/yr. Why not start a proper motion program now? The natural fear is that a superior technology would upstage such a program just as it was time to obtain second epoch data. The expectation of even modest gains leads to procrastination. But wouldn't two sets of results be better than none?

This review was written with the support of NSF grant AST90-15920 to the Massachusetts Institute of Technology.

REFERENCES

Aaronson, M., Blanco, V. M., Cook, K. H., & Schechter, P. L. 1989, ApJS, 70, 637

Aaronson, M., Blanco, V. M., Cook, K. H., Olszewski, E, & Schechter, P. L. 1990, ApJS, 73, 841

Backer, D. C. & Sramek, R. A. 1987, in The Galactic Center, ed. D. C. Backer (New York: American Institute of Physics) p.163

Bahcall, J. N., Flynn, C. & Gould, A. 1992, ApJ 389, 234

Bahcall, J. N. & Soneira, R. M. 1980 ApJS 44,73

Blitz, L., Fich, M. and Stark, A. A. 1982, ApJS, 49, 183

Burstein, D. 1979, ApJ, 234, 829

Fich, M. & Tremaine, S. 1991, ARAA 29, 409

Freeman, K. C. 1970, ApJ, 160, 811

Gilmore, G. & Reid, N. 1983, MNRAS, 202, 1025

Gould, A. 1990, MNRAS, 244, 25

Hartkopf, W. I. & Yoss, K. M. 1982, AJ 87, 1679

Hartwick, F. D. A. & Sargent, W. L. W. 1978, ApJ, 221, 512

Kerr, F. J. & Lynden-Bell, D. 1986, MNRAS, 221, 1023

Kent, S. M. 1986, AJ, 91, 1301

Kent, S. M. 1987, AJ, 93, 816

Kent, S. M. 1988, AJ, 96, 514

Kent, S. M., Dame, T. M., & Fazio, G. 1991, ApJ, 378, 131

Udalski, A., Szymański, M. S., Kaluzny, J., Kubiak, M. & Mateo, M. 1992 Acta Astron., 42, 253

Kuijken, K. 1991, ApJ, 372,125

Kuijken, K. & Gilmore, G. 1989, MNRAS, 239,605

Kulessa, A. & Lynden-Bell, D. 1992, MNRAS, 255, 105

Leonard, P. J. T. & Tremaine, S. D. 1989, ApJ 353, 486

Little, B. & Tremaine, S. D. 1987, ApJ 320, 493

Merrifield, M. R. 1992, AJ, 103, 1552

Mestel, L. 1963, MNRAS, 126, 553.

Norris, J. 1987, ApJL, 314, L39

Norris, J. E. & Hawkins, M. R. S. 1991, ApJ 380, 104

Philip, A. G. D. 1989, The Gravitational Force Perpendicular to the Galactic Plane, (Schenectady: L. Davis Press)

Schwarzschild, M. 1979, ApJ, 232, 236.

Turner, M. S. 1987, in Dark Matter in the Universe, eds. J. Kormendy & G. R. Knapp (Dordrecht: Reidel) p. 445

van der Kruit, P. C. 1986, A&A, 157, 230

Walker, T. P., Steigman, G., Schramm, D. N., Olive, K. A., and Kang, H.-S. 1991, ApJ 376, 51

Zaritsky, D, Olszewski, E. W., Schommer, R. A., Peterson, R. C., & Aaronson, M. 1989, ApJ, 345, 759

KINEMATICAL CONSTRAINTS ON THE DYNAMICALLY DETERMINED LOCAL MASS DENSITY OF THE GALAXY

Burkhard Fuchs and Roland Wielen
Astronomisches Rechen-Institut Heidelberg
Mönchhofstr. 12-14, D-6900 Heidelberg, Germany

1. INTRODUCTION

Since the classical study of Oort (1932) the dynamical determinations of the local mass density of the Galaxy have led to controversial results. Mass densities have been determined using various samples of tracers of population I stars in the – thin – galactic disk. To each of the observed vertical spatial distributions and velocity dispersions of the tracer stars, the hydrostatic equilibrium condition has been applied in order to derive the galactic K_z-force law, which gives via the Poisson equation the total mass density of all gravitating matter in the disk. By selecting stars near to the galactic midplane, one may determine in this way the mass density at the midplane. Recently Bahcall (1984a, b, c) has analyzed samples of F and K stars, which led him to postulate midplane densities as high as $0.2 M_\odot pc^{-3}$, indicating a large amount of dark matter, comparable to the observed amount of matter in the form of stars and interstellar gas of about $0.1 M_\odot pc^{-3}$. A second approach is due to Kuijken and Gilmore (1989a, b, c) who studied a sample of K stars far away from the midplane. They derive a surface density of the galactic disk of about $50 M_\odot pc^{-2}$ which, deconvolved by the observed vertical scale heights of the stellar distributions, gives a dynamical mass volume density estimate in the range of the directly observed mass density in the solar neighbourhood (Kuijken, 1991a).

Most recently a further carefully selected sample of K giants has become available, which has been used to determine the local mass density (Bahcall, Flynn, and Gould, 1992). Following the method of Bahcall (1984a) these authors derive values for the local dynamical mass density, which are higher than the actually observed mass density, so that there is still a controversy (see also Bienaymé, Robin, and Crézé, 1987).

Both methods are subject to substantial systematic and statistical uncertainties as reviewed by Gilmore (1990), and it might be thus useful to set further constraints on the dynamically determined mass densities.

2. CONSTRAINTS ON THE LOCAL MASS DENSITY

It has been already pointed out by von Hoerner (1960) that the distribution function $f(w_0)$ of – vertical – w_0 velocities of stars at the galactic midplane represents in fact the distribution of stars in energy space as far as the vertical oscillations are concerned, if the galactic gravitational potential separates in coordinates perpendicular and parallel to the plane. The energy distribution may be formally integrated over velocities leading to the vertical spatial distribution as a function of the gravitational potential Φ,

$$\rho(\Phi) = 2 \int_{\sqrt{2\Phi}}^{\infty} f(w_0) \frac{w_0 \, dw_0}{\sqrt{w_0^2 - 2\Phi}}. \tag{1}$$

Assuming that the adopted velocity distribution represents the velocity distribution of all gravitating matter and using the Poisson equation,

$$\frac{d^2 \Phi}{dz^2} = 4\pi G \rho, \tag{2}$$

in order to establish self-consistency, one may convert the formal spatial distribution function $\rho(\Phi)$ into a spatial distribution as a function of height above the midplane. Equations (1) and (2) form a complete set of integro-differential equations, which we have solved numerically in order to derive the functions $\Phi(z)$ and $\rho(z)$.

We have analyzed in this way the sample of McCormick K and M dwarfs in the catalogue of nearby stars (Gliese, 1969), which is the most suitable sample of nearby stars for kinematical studies, because it is free of kinematical selection effects. The velocity distribution of the McCormick stars has been corrected for the presence of bright main sequence stars and giants in the actual stellar velocity distribution as well as for the interstellar gas, which we described by a gaussian distribution with a velocity dispersion of 5 km/s and a central mass density equal to the mass density of the stars. The resulting velocity distribution is shown in Fig. 1. Furthermore we have taken into account a halo and dark corona component with a homogeneous density of $\rho_h = 0.01 M_\odot pc^{-3}$ (Bahcall and Soneira, 1980), which leads to a second term, $4\pi G \rho_h$, on the rhs of equation (2).

3. RESULTS AND DISCUSSION

Results are presented for four illustrative models. In the first 'realistic' model we have adopted a midplane density of the disk of $0.092 M_\odot pc^{-3}$, which is twice the observed local mass density of stars (Wielen, 1982), and a halo density of $0.01 M_\odot pc^{-3}$. The K_z-force law and the resulting spatial distribution of the late type stellar component calculated from equation (1) are shown in Figs. 2 and 3, respectively. The spatial distribution has an exponential shape with an exponential scale height of 340 pc for distances greater than 100 pc from the midplane, which compares well with the observed distributions of late type stars. The surface density of the disk according to this model is $45 M_\odot pc^{-2}$, in good agreement with the value found by Kuijken and Gilmore (1989b).

In models 2, 3 and 4 we have assumed additional dark matter in the galactic disk with a midplane density equal to the observed density of stars and interstellar gas at the galactic midplane, so that the total midplane density is $0.184 M_\odot pc^{-3}$, as suggested by Bahcall (1984b). In model 2 the dark matter is distributed like the overall spatial distribution of the stars and interstellar gas. In this case the resulting spatial distribution of late type stars is narrower, but the surface density of all gravitating matter rises to $68 M_\odot pc^{-2}$. In model 3 we have assumed that the dark matter is distributed like the late type stellar component, which leads to an even higher surface density of $86\mathcal{M}_\odot pc^{-2}$. Finally, in model 4 we have assumed that the dark matter is distributed like the interstellar gas. Due to the narrow distribution of the dark matter, the surface density of this model is $49\mathcal{M}_\odot pc^{-2}$, close to that of the 'realistic' model 1.

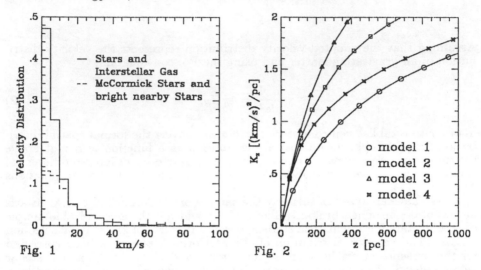

Fig. 1   Fig. 2

In Fig. 3 we compare the vertical spatial distributions of the late type stellar components of the various models with the observed spatial distribution of K dwarfs (Kuijken and Gilmore, 1989b). Because model 1 fits the data best, the spatial distribution of the late type stars as well as the derived surface density seem to indicate that there is either no or only a small amount of dark matter present in the disk.

As a consistency check we have compared the velocity distribution of K dwarfs at distances of 300 to 500 pc from the midplane (Kuijken, 1991b) with velocity distributions, which have been calculated by projecting the velocity distribution of the McCormick K and M dwarfs from the midplane up to 300 to 500 pc above the midplane, using the equation of continuity $f(z, w = \sqrt{w_0^2 - 2\Phi}) = f(0, w_0)$. The velocity distibutions, which depend only weakly on the model actually adopted, fit well to the observed velocity distribution, indicating that the K dwarfs observed by Kuijken and Gilmore (1989a) represent a fair sample of late type stars.

Most recently a further carefully selected sample of K giants has been made available, which is suitable as a tracer population of stars to determine the galactic K_z-force law (Bahcall, Flynn, and Gould, 1992). The spatial distribution of these tracer stars is derived from star counts in a cone oriented towards the south galactic pole (Flynn and Freeman, 1992). Because of the conical shape of the counting volume the spatial distribution thus derived cannot be traced to the galactic midplane. Flynn and Fuchs (1992) have derived the midplane density for this kind of tracer stars by selecting K giants from the Bright Star Catalogue (Hoffleit and Jaschek, 1982) according to the same absolute magnitude, colour and metallicity criteria, and converting their number into a spatial density. The velocity distribution of the K giants from the Bright Star Catalogue has been checked carefully by vertical projection as described above, against the velocity distribution of the K giants from the sample of Bahcall, Flynn, and Gould

(1992), in order to ensure that both samples represent the same population of tracer stars of the galactic gravitational potential. The observed vertical spatial distribution of K giants is shown in Fig. 4 in comparison with density laws constructed by Flynn and Fuchs (1992) from the velocity distribution of K giants at the midplane using equation (1). The potentials for the models shown in Fig. 4 are those of the 'realistic' model 1 and of model 2, respectively. Although statistical errors are obviously large, we conclude also from this sample that there is no need to assume a large amount of dark matter in the galactic disk, because there is no significant discrepancy between the observed spatial distribution and that predicted by using the potential of model 1.

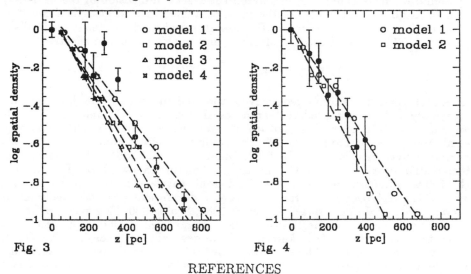

Fig. 3   Fig. 4

REFERENCES

Bahcall, J.N. 1984a ApJ 276 156
Bahcall, J.N. 1984b ApJ 276 169
Bahcall, J.N. 1984c ApJ 287 926
Bahcall, J.N., & Soneira, R.M. 1980 ApJ Suppl. 44 73
Bahcall, J.N., Flynn, C., & Gould, A. 1992 ApJ 389 234
Bienaymé, O., Robin, A., & Crézé,M. 1987 A&A 180 94
Flynn, C., & Freeman, K.C. 1992 A&A Suppl., in press
Flynn, C., & Fuchs, B. 1992, in preparation
Gilmore, G. 1990, in Baryonic Dark Matter, eds. D.Lynden-Bell & G.Gilmore
    (Kluwer) p.137
Gliese, W. 1969 Veröff.Astron.Rechen-Inst. Nr.22
Hoffleit, D., & Jaschek, C. 1982 The Bright Star Catalogue, 4th. ed.(New Haven:
    Yale Univ. Observatory)
Kuijken, K. 1991a ApJ 372 125
Kuijken, K. 1991b, private communication
Kuijken, K., & Gilmore, G. 1989a MNRAS 239 571
Kuijken, K., & Gilmore, G. 1989b MNRAS 239 605
Kuijken, K., & Gilmore, G. 1989c MNRAS 239 651
Oort, J.H. 1932 BAN 6 249
von Hoerner, S. 1960 Fortschritte der Physik 8 191
Wielen, R. 1982, in Landolt-Börnstein New Series Vol. VI, 2c (Springer) p.208

FRACTAL GEOMETRY OF INTERSTELLAR GAS AND DARK MATTER IN HI DISKS

Daniel Pfenniger
Geneva Observatory, CH-1290 Sauverny, Switzerland

ABSTRACT

The projection properties of computer generated fractal clouds are examined. Systematic effects depending mainly on the fractal dimension of the ISM may lead to large underestimates of the mass contained in cold gas. The constant ratio of dark mass to *estimated* HI mass observed in the outer galactic disks (Bosma 1981) might be attributed principally to this geometric effect.

1. FRACTALS IN THE COLD ISM

In recent years the fractal geometry of the cold interstellar gas has been increasingly well documented (Falgarone et al. 1992). Nearby molecular clouds appear self-similar over a range of scales and densities of at least 10^4 (Falgarone et al. 1991), but perhaps up to 10^6, according to different kinds of observations of fine structures (Fiedler et al. 1987, Diamond et al. 1989). The smallest detected structures have a few AU, the largest structures a few 100 pc. Remarkably the fractal dimension of isophotes at different wavelengths is almost constant, around 1.3–1.4, over a wide range of scales and *independent of the molecular or atomic state of the gas* (Bazell & Désert 1988, Scalo 1990, Falgarone 1992).

The projection properties of fractals are not well studied. The *section* of a fractal of dimension D is known to have generally a dimension $D_s = D - 1$. However *projection* is a different operation that does not follow the same rule. A simple counterexample is a line ($D=1$): generally the section of a line is a point ($D_s=0$) but the projection of a line is a line ($D_p=1$). The section and projection of smooth density distributions ($D=3$) have generally $D_s=D_p=2$. For fractal distributions with $D < 3$ an intermediate behavior is expected; thus the projection of a $D=2$ fractal could have a dimension $1 \leq D_p \leq 2$. Mandelbrot (1982, p. 91) argues that "typically" for $D < 2$, $D_p = D$, while for $D > 2$, $D_p = 2$. But, contrary to simple fractal models which are either opaque or transparent, interstellar gas has a continuous distribution of optical depths. Therefore semi-transparency complicates the relation between D and D_p. In any case the general fractal isophotes of nearby molecular clouds over a large range of scales and physical properties strongly suggest that cold HI in the outer disks may also be better described by a fractal model than by "standard" smooth clouds. Virial equilibrium arguments applied to hierarchical clouds ($M \sim r^D$, $v^2 \sim r^{D-1}$, Chièze 1987, Larson 1981, 1992) and recent observational data indicate a fractal dimension around $D=2$. In summary in the cold ISM probably $1.4 < D \lesssim 2$.

Modeling the physics of fractals is presently far from trivial, as one of the most basic traditional assumption, differentiability, is lost. Unless the smallest scale of a hierarchical structure is accessible (to observations, respectively to calculations), where differentiability holds, in principle differential equations, such as the radiation transfer equation, are no longer justified. Although the general evidence is that ISM cold gas is structured down to the smallest accessible scales, lacking better tools, observers apply the equation of transfer and assume homogeneity at subresolution scales to estimate the amount of interstellar gas, which in turn provides the amount of baryons in the outer HI disks of spirals.

2. BUILDING HIERARCHICAL MASS DISTRIBUTIONS

Here, as a first step, simple static hierarchical cloud models are built by Monte-Carlo simulations. The geometric properties of the cloud projections are studied as the fractal dimension, and other parameters are varied.

A *hierarchical* cloud of level $L > 0$ is made of N independent subunits which are themselves hierarchical clouds of level $L-1$. When $L=0$ the recursion stops, a $L=0$ cloud represents the most basic mass unit ("atom", or cloudlet) of the whole hierarchy. The centers of these subunits are distributed randomly according to a spherical density law ρ, which is kept the same at all levels, except that all lengths between two levels are in a ratio $\alpha = r_{L-1}/r_L$. If M is the mass within a radius r, then the *uniform scaling relation* $\underline{M \sim r^D}$ defines a fractal model of dimension D (Mandelbrot 1982). Since the mass M_L of a clump at level L consists of N subclumps of mass M_{L-1} at level $L-1$, it follows $\underline{\alpha = N^{-1/D}}$. The examined density laws ρ are physically motivated by the fact that an isothermal collapsing mass tends rapidly toward an r^{-2} density law, which quickly fragments into a small number N of isothermal subunits, because too singular isothermal distributions are unstable. Fragmentation can then repeat at lower scales as long as the gas remains isothermal. Since the mass of an isothermal distribution lies mostly at large distances, the subclumps must be truncated by collisions, i.e., the density law must decrease faster than r^{-2} in the outer parts. Accordingly, the following density laws have been considered:

$$\text{Density A:} \quad \rho \sim r^{-2} \quad \text{for} \quad r < R, \quad \rho = 0 \quad \text{for} \quad r \geq R,$$
$$\text{Density B:} \quad \rho \sim r^{-2} \left(1 + (r/R)\right)^{-5}, \tag{1}$$

which are abruptly (A) and mildly (B) truncated singular isothermal spheres. In density B about 6% of the mass is exterior to R.

Hierarchical clouds are generated most easily with *recursive programming*. Here Fortran-77 is used with the recursion extension. The calculation of a cloud with about $N^L \gtrsim 10^9$ cloudlets needs a few hours on a Sparc-2 workstation. The number N of fragments per level has typically been varied between 5 and 20, and the number L of levels between 7 and 13. Recently Houlahan & Scalo (1992) built similar models with up to three levels. A "standard" cloud model has only one level and is made of a very large number N of smoothly distributed "atoms".

Simultaneously to our cloud calculation, the 3D mass distribution is projected along one direction on a square array of fixed resolution (up to 2048^2 pixels). The resulting image is made of a superposition of cloudlets the size of which is smaller than or similar to the pixel size. If the smallest cloudlet can be measured in isolation, its own optical thickness must be small. The mass of a single cloudlet averaged over one pixel defines the smallest possible surface density present in the cloud. The cloud can be characterized by a distribution of pixel averaged surface densities Σ. By comparing the different Σ distributions for different D, we can estimate when a large fraction of the cloud mass is contained in large surface density clumps covering a small fraction of the sky, a combination of factors that are likely to induce systematic errors in the usual gas mass determination. Indeed even if a single cloudlet is optically thin, the superposition along the line of sight of up to millions of cloudlets may be com-

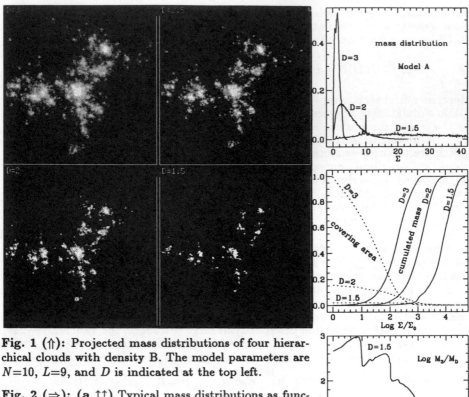

Fig. 1 (⇑): Projected mass distributions of four hierarchical clouds with density B. The model parameters are $N=10$, $L=9$, and D is indicated at the top left.

Fig. 2 (⇒): (a ↑↑) Typical mass distributions as functions of Σ (model A, $N=10$, $L=9$). Clearly lower D models contain increasingly more mass at high Σ.
(b ↗) Cumulated mass at lower Σ and covering area at higher Σ as functions of Σ (model B, $N=10$, $L=9$).
(c →) Ratios of the $D=3$ to the $D=2$ and to the $D=1.5$ cumulated mass as functions of Σ (model B, $N=10$, $L=9$).

pletely opaque to a given wavelength. A cloud with $D<3$ tends to correlate the spatial and projected positions of the individual cloudlets and also tends to have a low covering factor; in such a case most of the radiation coming from behind can peer through, ignoring the large amount of mass hidden in opaque clumps.

Hundreds of different fractal clouds have been generated. The results have been found to depend weakly on N, or L, but strongly on D and also on ρ. A few typical cloud models are shown in Fig. 1. With $D=3$ the hierarchical distribution looks "natural", but the area filling factor of a single cloud is large. Models resemble interstellar or terrestrial clouds, between approximately $D=2$ and $D=3$. When $D\lesssim1.5$, the area filling factor of a single cloud is very low.

The mass distribution as a function of Σ is mainly dependent on D. For example the distribution for a *smooth* singular isothermal sphere $(L=1)$ is similar to a *hierarchical* $D=3$ model one $(L>1)$. Fig. 2a shows the typical Poisson like distributions. The amount of mass at high Σ increases strongly as D decreases. In a real cloud this would correspond to an increasingly larger spread of optical depths as D decreases. Since observations are biased towards low Σ, we show in Fig. 2b the cumulated mass distributions as a function of Σ. Half of the mass is reached for Σ's about $3-7$ times higher with $D=2$ than with $D=3$. The relative area filling factor is also shown in Fig. 2b; it decreases rapidly with decreasing D. The ratios of the cumulated mass with $D=3$ and lower D are roughly *constant* at low surface densities (Fig. 2c). For $D=2$ we get $7-12$, and much more at lower D (> 200 for $D=1.5$). *This means that if mass at higher density is not observed due to e.g. too large optical depths, we underestimate the mass by a factor $7-12$ if $D=2$ when assuming $D=3$, and even more if in fact $D<2$.*

As the number of pixels is decreased, the mass distributions with $D \leq 3$ as functions of Σ (Fig. 2) tend toward the $D=3$ distribution. This means a limited resolution is mimicking a $D=3$ distribution.

3. HOW MUCH MASS HIDDEN IN SMALL SCALE STRUCTURES?

The above simple hierarchical clouds allow to examine geometry effects associated with fractals. In a real interstellar cloud other physical effects may induce systematic blindness to high surface densities. For example since the average density of a clump of size r scales as $\langle \rho \rangle \sim r^{D-3}$, when $D<3$ the clump average density increases indefinitely at small scales. Then in addition to large optical depths, both the formation of H_2 and a low temperature close to $3\,K$ may considerably decrease the detection of the 21 cm emission.

But already projection effects in clouds with a fractal dimension lower than 3 indicate a possible significant underestimate of the gas mass in HI disks. This error would be due to not taking fractal geometry into account, to the limited range of column densities over which the 21 cm line is optically thin, and to the low covering factor of high column density clumps. An underestimate by a factor of order 10 is typical in the models when $D \approx 2$. A factor 10 of HI mass underestimate is enough to remove the need of exotic matter in disk galaxies, because outside the optical disks the ratio of dark mass to HI mass, *assumed smooth and optically thin,* is nearly constant, around $8-10$ (Bosma 1981, Freeman 1993), suggesting a tight relation between HI and dark matter.

A more extensive discussion about the relation between cold gas and dark matter in spirals is in preparation (Pfenniger, Combes and Martinet).

REFERENCES

Bosma, A. 1981, AJ, 86, 1971
Chièze, J.P. 1987, A&A, 171, 225
Diamond, P.J., et al. 1989, ApJ, 347, 302
Falgarone, E. 1992, in "Astrochemistry of Cosmic Phenomena", ed. P.D. Singh (Kluwer: Dordrecht) p. 159
Falgarone, E., Phillips, T.G., & Walker, C.K. 1991, ApJ, 378, 186
Falgarone, E., Puget, J.-L., & Pérault, M. 1992, A&A, 257, 715
Fiedler, R.L., et al. 1987, Nature, 326, 675
Freeman, K. 1993, in "Physics of Nearby Galaxies, Nature or Nurture?", XIIth Moriond Astrophysics Meeting, eds. T. X. Thuan, C. Balkowski, J. T. T. Van (Editions Frontières: Gif sur Yvette) in press
Houlahan, P., & Scalo, J. 1992, ApJ, 393, 172
Larson, R.B. 1981, MNRAS, 194, 809
Larson, R.B. 1992, MNRAS, 256, 641
Mandelbrot, B. 1982, "The Fractal Geometry of Nature", (Freeman)
Scalo, J. 1990, in "Physical Processes in Fragmentation and Star Formation", eds. R. Capuzzo-Dolcetta et al. (Kluwer: Dordrecht) p. 151

The Cosmological Constant and Dark Matter In the Galaxy

Thomas L. Wilson

National Aeronautics and Space Administration,
Johnson Space Center, Houston, Texas 77058

Abstract

Flat galactic rotation curves represent strong evidence that a substantial fraction of the total galactic gravitational mass is not visible or that gravitational dynamics on the galactic scale is not understood. Particle physics models do no offer any simple explanation as to why the dark matter candidates assume the requisite shape (e.g., a coronal halo), because the basic problem of galactic confinement is never addressed. A similar problem has existed in hadron physics until the advent of quantum chromodynamics and soliton bag theory. A new interpretation of the cosmological constant λ in general relativity as a confinement mechanism for the so-called MIT bag (the author's proposed tensor-soliton theory of gravitation) is used to demonstrate that all hadronic matter in the Universe may be comprised of a hidden mass component due to λ.

I. Introduction

The possible existence of dark matter in the Universe has inspired a great deal of current research (Turner 1991; Kolb and Turner 1990; Srednicki 1990; Trimble 1987; Tremaine 1992; Binney and Tremaine 1987). Of particular importance is how this relates to our own Galaxy and our solar neighborhood (Bahcall 1985, 1984, 1986). We want to focus our attention here on several of the gravitational aspects of the dark matter enigma, particularly since the conventional assumption of Kepler's third law determines a dynamical mass M = M(r) of a galaxy as $GM = v^2r$. The "flat" rotation curves (van Albada *et al.* 1986, 1985; Kent 1986, 1987; Rubin *et al.* 1983, 1985) for orbital velocity v then follow, provided G, M, or GM vary linearly with r. A number of *ad hoc* models of non-Newtonian, non-Einsteinian gravity have been proposed to accomplish this (Liboff 1992; Kuhn and Kruglyak 1987).

In a tensor-soliton theory of gravitation (Wilson 1992b), a scalar σ-field is introduced (reminiscent of Brans-Dicke theory), where the soliton potential U(σ) has a broken symmetry induced by quantum vacuum fluctuations. A false vacuum appears at σ = 0, for which $B = \kappa^{-1}\lambda_{Bag}$ is an MIT bag constant. Inside the soliton, $B \neq 0$ and outside $B = 0$ as in Figure 1. Tensor-solition theory will be discussed here, to show that we may be unable to resolve the dark matter question until we understand the origin of mass in the Universe.

II. Classical General Relativity

This reinterpretation of Einstein's cosmological constant (Einstein 1917, 1919) reverses the conventional Kottler-Schwarzschild (1918) problem with a cosmological constant. That is, $\Lambda = 0$ in the exterior solution and $\lambda_{Bag} = \kappa B \neq 0$

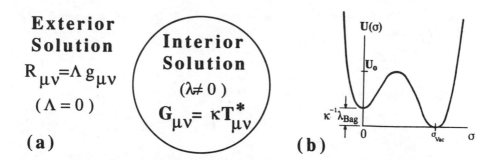

Figure 1. *(a) The tensor-soliton bag, reversing the roles of Λ and λ. (b) The cosmological constant breaks vacuum symmetry in the soliton potential.*

in the interior of the soliton bag (Figure 1).   We will retain Λ and then set it equal to zero, since both Λ and λ are not permissible at zero temperature.

General relativity has long been known to predict a non-Newtonian force law (Lemaitre 1949, Schlüter 1955, Jackson and Forman, 1970) in the presence of Λ.   The tensor-soliton case, however, in Figure 1 has never been done with a soliton bag ($\lambda \neq 0$) embedded in an Schwarzschild exterior background ($\Lambda = 0$ or $\Lambda \neq 0$).   We will do it here to make a point about soliton bags and bubbles in our Galaxy.   The interior equations are:

$$G_{\mu\nu} = \kappa T^*{}_{\mu\nu} \qquad\qquad (\kappa = 8\pi G) \qquad\qquad\qquad (1)$$

$$T^*{}_{\mu\nu} = T_{\mu\nu} - U(\sigma)g_{\mu\nu} \qquad\qquad\qquad\qquad\qquad (2)$$

$$\lambda_{Bag} = \kappa B \quad . \qquad \text{(Cosmological Bag Constant)} \qquad\qquad (3)$$

For this discussion we will simplify $U(\sigma)$ to $U(\sigma) = B$, a variation of the so-called mean field approximation, and ignore the wave equation (Wilson 1992b) for σ. (1)-(3) have the interior metric ($T^*{}_{\mu\nu} \neq 0$ and $B \neq 0$)

$$ds^2 = e^{2\nu} dt^2 - e^{-2\delta} dr^2 - r^2 d\Omega^2 \qquad\qquad\qquad (4)$$

where (Moller 1952, Kottler and Schwarzschild 1918)

$$e^{-\delta} = (1 - r^2/R^2)^{-1/2} \ , \qquad\qquad e^\nu = (A - Be^\delta) \qquad , \qquad\qquad (5)$$

$$R^2 = 3/(\lambda + \kappa\rho) \qquad ; \qquad A = (1 - r_0{}^2/R^2)^{-1/2} \ ; \qquad B = \frac{1}{2} \ . \qquad (6)$$

The exterior metric ($T^*{}_{\mu\nu} = 0$ and $\Lambda \neq 0$) is

$$e^{2\nu} = e^{-2\delta} = (1 - \frac{2m}{r} - \frac{1}{3}\Lambda r^2) \qquad\qquad\qquad (7)$$

for the Schwarzschild mass $m = GM$, while ρ is a uniform mass density.

Classically, Eq. (7) represents in the Newtonian weak-field limit a potential g_{oo} = $e^{2v} \approx 1 + 2\phi$, or

$$\phi = -\frac{GM}{r} - \frac{1}{6}\Lambda r^2 \qquad . \qquad (8)$$

A specific force law ($a = -\nabla\phi$) can be derived from (8) as

$$a = -\frac{GM}{r^2} + \frac{1}{3}\Lambda r \qquad ; \text{ or } \quad a = \left[-\frac{4}{3}\pi\rho + \frac{1}{3}\Lambda \right] r \qquad . \qquad (9a,b)$$

(9a) represents an asymptotic force, increasing with r and without limit, which has been speculated to relate to the stabilization of galaxies (Jackson and Forman 1970). Cosmological limits on Λ, however, make the effect small. Noerdlinger (1970) has repudiated Λ as irrelevant, claiming that (9a) would explode or implode the Universe which, he implies, means Λ must be zero.

The interior solution is different. If one considers the Galaxy as a simplified solitonic bag (Figure 1), the roles of Λ and λ are reversed (Wilson 1992a), and the bag does not have to explode or implode. It can reach an equilibrium state. We now want the interior metric (5), carefully noting that the Schwarzschild parameter is no longer constant since $M=M(r)$, and (5) gives

$$e^{2v} = (A - Be^\delta)^2 = A^2 - 2ABe^\delta - B^2 e^{2\delta} \qquad (10)$$

which in the same weak-field limit $g_{oo} = e^{2v} \approx 1 + 2\Phi$, determines

$$\Phi = \frac{1}{2} (A^2 - \frac{3}{4}) - \frac{1}{2}A\sqrt{1 - r^2/R^2} - \frac{1}{8} r^2/R^2 \qquad , \qquad (11)$$

yielding a specific force ($a = -\nabla\Phi$), at least as tenable as *ad hoc* models, of

$$a = \frac{1}{R^2}\left[\frac{1}{4} - \frac{A}{\sqrt{1 - r^2/R^2}} \right] r \qquad (12)$$

inside the soliton. This vanishes only when $4A=e^\delta$ (derivable from 15a below) or r=0. [Schwarzschild assumed m=GM=0 in arriving at the interior solution (5) to prevent a singularity at r=0. This is not necessary since $M=0 \Rightarrow$ m=0 at r=0. Wyman (1949) and Volkoff appear to have appreciated or noticed this.]

We cannot introduce a universal constant twice in the classical theory (Λ and λ). Therefore we set $\Lambda = 0$ in (9) and, if the Galaxy is a soliton, then (12) clearly predicts an asymptotic acceleration inside the bag ($r < r_0$).

III.   The Galaxy and Nontopological Solitons (Bags)

The exterior solution (7) with $\Lambda = 0$ must coincide with the interior solution (5)-(6) with $\lambda \neq 0$ at the bag radius $r=r_0$, and with zero pressure at the surface of the sphere. These conditions give the equations

$$1 - \frac{2m}{r_0} = 1 - r_0^2/R^2 = [A - B \sqrt{1 - r_0^2/R^2}] \qquad (13)$$

$$3B \sqrt{1 - r_0^2/R^2} - A = 0 ; \qquad R^{-2} = \frac{1}{3}(\lambda + \kappa\rho) , \qquad (14a,b)$$

having the solutions

$$A = \frac{3}{2}\sqrt{1 - r_0^2/R^2} \ ; \quad B = \frac{1}{2} \ ; \quad m = \frac{1}{2}\frac{r_0^3}{R^2} = \frac{1}{6}(\lambda + \kappa\rho) \, r_0^3 \ . \qquad (15a,b,c)$$

Since $M = 8\pi\kappa^{-1}m$, noting that $M=M(r)$ is a function of r in the bag, (15c) gives

$$M = M(r) = (\frac{1}{6}\lambda + \frac{4}{3}\pi\rho) \, r^3 \ ; \qquad M_0 = (\frac{1}{6}\lambda + \frac{4}{3}\pi\rho) \, r_0^3 \ . \qquad (16a,b)$$

For the apparent asymptotic mass at $r=r_0$, $M_0 \neq \frac{4}{3}\pi r_0^3 \rho$. Beyond $r=r_0$ in Figure 1 (outside the bag), $\lambda=0$ and $M_0 = \frac{4}{3}\pi r^3 \rho$. There is a "hidden" mass density, then, in (16) of magnitude

$$\rho_{Hidden} = \lambda/8\pi = GB \qquad . \qquad (17)$$

IV.   Finite-Temperature Phase Transitions

We have discussed $\lambda_{Bag} = \kappa B$ as constant in classical field theory. Finite temperature (T) quantum field theory [FTQFT] treats the full soliton dynamics of $U(\sigma)$ with a chemical potential μ, resulting in a bag parameter $B = B(\mu,T) = B - 3T^4 - 46T^2$ (Song 1992). At hadronic T, $B=60$ MeV fm^{-3} gives $\lambda = 2\times10^{-13}$ cm^{-2}, significantly different from $\Lambda = 0$ (or $\sim10^{-54}$ cm^{-2}). Because space-time has become temperature-dependent in FTQFT, a Galactic soliton is a dynamic object where phase transitions can vary $\lambda = \lambda(\mu,T)$, creating bubbles within the bag.

References

Bahcall, J.N. 1985, In: *Dark Matter in the Universe*, ed. J. Kormendy and G.R. Knapp (Reidel), 17; 1986, *Phil. Trans. Roy. Soc.* A320, 111; 1984, *Ap. J.* 276, 169.
Binney, J., and Tremaine, S. 1987, *Galactic Dynamics*, Ch. 10 (Princeton).
Einstein, A. 1917, Sitsungsber. Preuss. Akad. Wiss., 142. [Paper on Λ].
Einstein, A. 1919, Sitzungsber. Preuss. Akad. Wiss., 349. [Paper on λ].
Forman, W.R. 1970, Ap. J. 159, 719.
Jackson, J.C. 1970, Mon. Not. Roy. astr. Soc. 148, 249.
Kent, S.M. 1986, Astron. J. 91, 1301; *ibid.* 93, 816.
Kolb, E.W., and Turner, M.S. 1990, *The Early Universe* (New York: Addison-Wesley).
Kottler, F. 1918, Ann. d. Phys. 56, 401.
Kuhn, J.R., and Kruglyak, L. 1987, *Ap. J.* 313, 1.
Lemaitre, G. 1949, in *Albert Einstein: Philosopher-Scientist*, Vol. II, 439 (Harper).
Liboff, R.L. 1992, *Ap. J.* 397, L71.
Moller, C. 1952, *The Theory of Relativity*, 331 (Oxford).
Noerdlinger, P.D. 1970, Nature 228, 845.
Ostriker, J.P., Peebles, P.J.E., and Yahil, A. 1974, *Ap. J.* 193, L1.
Rubin, V. 1983, Science 220, 1339.
Rubin, V. *et al.* 1985, *Ap. J.* 289, 81.
Schwarzschild, K. 1918, Sitzungsber. Preuss. Akad. Wiss., 424.
Schlüter, A. 1955, Astron. J. 60, 141.
Song, G., Enke, W., and Jiarong, L. 1992, Phys. Rev. D46, 3211.
Srednicki, M. 1990, *Particle Physics and Cosmology: Dark Matter* (North-Holland).
Tremaine, S. 1992, *Phys. Today* 45, No. 2, 28.
Trimble, V. 1987, *Ann. Rev. Astron. Astrophys.* 25, 425.
van Albada, T.S., and Sancisi, R. 1986, *Phil. Trans. Roy. Soc. Lond.* A320, 447.
van Albada, T.S., Bahcall, J.N., Begeman, K., and Sancisi, R. 1985, *Ap. J.* 295, 305.
Wilson, T.L. 1992a, in *Testing the AGN Paradigm*, AIP Conf. Proc. 254, 109 (American Institute of Physics).
Wilson, T.L. 1992b, "A New Interpretation of Einstein's Cosmological Constant," to be published.
Wyman, M. 1949, Phys. Rev. 75, 1930. $K \neq 0$ is the critical parameter in this reference.

FLARING OF GAS IN GALACTIC DISKS

Tomislav Kundić
Princeton University Observatory, Peyton Hall, Princeton, NJ 08544

Lars Hernquist
Board of Studies in Astronomy and Astrophysics, Univ. of California, Santa Cruz

James E. Gunn
Princeton University Observatory, Peyton Hall, Princeton, NJ 08544

ABSTRACT

The vertical distribution of gas in disk galaxies is studied using analytical and numerical methods. If the sound speed of gas or the velocity dispersion of clouds is only weakly dependent on galactic radius, then the gas flares outward, perpendicular to the galactic plane. The rate of this flaring is determined by the galactic potential and constrains the mass distribution differently from rotation curves. In principle, therefore, high resolution measurements of the widths of HI layers in external edge–on galaxies could be used in conjunction with measured rotation curves to infer the shapes and masses of dark matter halos. This paper investigates the dependence of flaring on the flattening of the halo, ratio of halo–to–disk mass, and extent of the halo. We conclude that flaring can in principle be used to constrain the vertical scale-height of the non–gaseous disk material, as well as the axis ratios of dark matter halos.

1. INTRODUCTION

In has been observed in the spiral galaxies that the vertical scale–height of the gas is not constant and rises outwards from the center (*e.g.* Burton & Gordon 1978; Gunn 1980). If the effective sound speed of the gas is roughly constant across the disk then this flaring can be interpreted as arising from the decreased restoring force to the disk plane as the stellar density drops outward.

As we describe below, the vertical flaring of gas in disks is sensitive to the detailed form of the galactic mass distribution. Since the rate of flaring at any given radius is sensitive to both the local surface density of matter and the enclosed mass, it provides a different diagnostic of the galactic potential than does the rotation curve. Thus, the flaring rate can, in principle, be used together with the rotation curve to infer the dark mass in galaxies and the shapes of galactic halos.

Unfortunately, differences in the flaring rate between various model potentials are noticeable primarily at rather large radii, requiring observations with high sensitivity. In addition, flaring is influenced by other physical effects which are not unambiguously constrained and hence complicate the analysis. Included among these are radial variations in the sound speed or velocity dispersion of the gas, the phase structure of the gas, non–thermal effects arising from magnetic fields and cosmic rays, and warps.

2. GALAXY MODEL

We consider two–component systems consisting of a disk galaxy and a dark matter halo. The disk mass distribution is represented by a density profile which is exponential in cylindrical radius, R, and has a vertical structure corresponding to an isothermal sheet (Spitzer 1942). That is,

$$\rho_{disk}(R, z) = \frac{M_d}{4\pi z_0 h^2} \exp(-R/h) \operatorname{sech}^2(z/z_0), \tag{1}$$

where M_d is the mass of the disk, h is the radial scale–length, and z_0 is the vertical scale–length. We choose to base our calculations on the Bahcall-Soneira model of the Galaxy and take $h = 3.5$ kpc and $M_d = 5.6 \times 10^{10} M_\odot$.

For the halo we consider a non–singular isothermal model with equidensity surfaces stratified on oblate spheroids (de Zeeuw & Pfenniger 1988). That is,

$$\rho(m) = \frac{\rho_0}{1 + m^2}, \tag{2}$$

where

$$m^2 = \frac{R^2}{a^2} + \frac{z^2}{c^2}. \tag{3}$$

For oblate models, $a \geq c$. The potential corresponding to this density profile can be reduced to a quadrature, and gradients of the potential yield elementary functions. Instead of ρ_0 we use the rotation curve in the $z = 0$ plane at large radii to characterize the halos:

$$v_\infty^2 = 4\pi G \rho_0 a^2 c \, \frac{\tan^{-1}(\sqrt{a^2 - c^2}/c)}{\sqrt{a^2 - c^2}}. \tag{4}$$

Finally, we define the vertical scale–height of the gas in terms of the sound speed of the continuum gas (or velocity dispersion of the gas clouds) c_s and the second derivative of the potential with respect to z:

$$z_{0,gas} = \left[\frac{2c_{gas}^2}{\partial^2 \Phi / \partial z^2|_{z=0}} \right]^{1/2}. \tag{5}$$

In the limit where gas contributes negligibly to the potential, the right hand side is independent of $z_{0,gas}$. This approximation is certainly valid for present–day disks in which the mass distribution is strongly dominated by the halo and stellar disk. Note that neglect of the potential of the gas is valid even at large radii where the local surface density of stars may be small relative to the gas, since Φ is then determined mainly by the *enclosed* mass. We confirmed the validity of this approximation using dynamical models. We further assume that $c_s \equiv 10$ km/sec, but our results can easily be scaled to other values of the sound speed, as long as it is independent of z.

3. RESULTS

i) Influence of Stellar Distribution

First we examined the dependence of the scale–height of the gas on the scale–height of the stars $z_{0,s}$. We computed this dependence numerically for several values of $z_{0,s}$ and halo parameters $a = c = 3.5$ kpc and $v_\infty = 220$ km/sec. At sufficiently large R, we found $z_{0,gas} \propto R$, as expected from analytical estimates. Provided that M_d is fixed, the scale–height of the gas becomes almost independent of the $z_{0,s}$, since the disk contribution to the flaring is determined by the enclosed mass, rather than the local surface density of the disk. The maximum shift in $z_{0,gas}$ is a factor of $\approx$ 2-3 at small R if $z_{0,s}$ is varied by a factor of 8 around the preferred value of 0.7 kpc (Bahcall & Soneira 1980). This is a fairly weak dependence in the sense that such a large uncertainty in z_0 is unlikely to arise in practice unless a significant fraction of the disk mass is contained in dark matter which is distributed similarly to the stars in radius but with a very different scale–height. Nevertheless, a detailed comparison of models with various $z_{0,s}$ may constrain the local distribution of dark matter in our disk (e.g. Rupen 1989).

We also examined the rotation curve variation with $z_{0,s}$ and found that a factor of 8 change in $z_{0,s}$ produces a change in circular velocity of less than 10%. Although by today's standards this change is measurable, we find that flaring is affected more significantly and thus provides a better diagnostic of the stellar scale–height.

From now on we adopt $z_{0,s} = 0.7$ kpc for the scale–height of the disk stars to fully specify the stellar distribution,

ii) Influence of Halo Structure

We next studied the effect of varying the core radius of the halo in models with halo density profiles as in equation (2) with $v_\infty = 220$ km/sec. We computed three models with spherical halos in which a and c were varied from 1 to 10 kpc. The effect of this variation on the flaring rate turned out to be smaller than on that of the rotation curve. A factor of 10 change in a and c results in only a 20% variation in $z_{0,gas}$ and is confined almost entirely to large radii. For the same set of parameters, the combined rotation curve varies by more than 40%. If one assumes that the rotation curve is known with relatively high precision, then rotation curves will essentially limit a and c to a narrow range and little additional information could be gleaned from a measurement of flaring of gas.

Of perhaps greater interest is the dependence of the flaring on the axis ratio of the halo. The figure shows the vertical scale–height of the gas for a halo model as described above with $a = 3.5$ kpc, $v_\infty = 220$ km/sec, and various values of c. This figure demonstrates that the flaring is rather sensitive to c/a. At the same time, the rotation curve is virtually unaffected by the change in c/a ratio. This implies that, unlike the rotation curve, the vertical scale–height of the gas provides a diagnostic of the shape of the halo. In principle, a comparison between results like those in the figure and observations of the vertical distribution of gas in galaxies could be used to constrain the shapes of halos. In particular, the fact that gas flares at all strongly suggests that the dark matter cannot be distributed in a "sheet–like" manner and must be rather more extended than the disk itself.

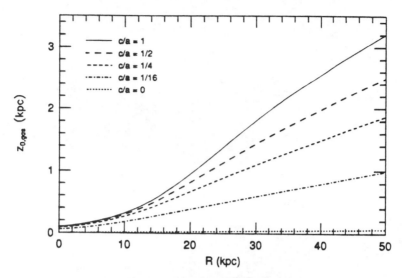

Dependence of the scale–height of the gas on the
galactocentric radius for models with different halo axis ratios.

iii) Dynamical Models

The analysis presented above assumes that the gas is in vertical hydrostatic
equilibrium in the potential well of the disk and halo and that the self–gravity
of the gas is negligible. To test these assumptions, we have run several dynamical models of the evolution of gas and stars in isolated disks using a hybrid N–
body/hydrodynamics code which includes the full self–gravity of all components
(Hernquist & Katz 1989). Typically, these disks develop large–scale spiral structure in both the stars and gas within roughly a dynamical time (*e.g.* Hernquist
1990). Perpendicular to the disk plane, the gas settles into a quasi–equilibrium
state in approximately a vertical sound–crossing time.

The agreement between the analytical and numerical calculations is quite
good. Since the dynamical model includes the self–gravity of the gas, this provides justification for the approximations made in the calculations and indicates
that the assumption of vertical hydrostatic equilibrium is reasonable. Moreover,
the agreement implies that the numerical method used to model the gas in the
simulation (smoothed particle hydrodynamics) can faithfully represent at least
some aspects of the the the dynamics of galactic gas.

REFERENCES

Bahcall, J.N., & Soneira, R.M. 1980, ApJ, 44, S73
Burton, W.B. & Gordon, M.A. 1978, A&A, 63, 7
de Zeeuw, T., & Pfenniger, D. 1988, MNRAS, 235, 949
Gunn, J.E. 1980, Phil. Trans. R. Soc. Lond. A, 296, 313
Hernquist, L. 1990, in Dynamics and Interactions of Galaxies, ed. R. Wielen,
    (Berlin: Springer–Verlag) p.108
Hernquist, L., & Katz, N. 1989, ApJ, 70, S419
Rupen, M. 1989, Ph.D. thesis, Princeton University
Spitzer, L. 1942, ApJ, 95, 329

SUMMARY
OVERVIEWS

A PESSIMIST'S VIEW OF GALACTIC STRUCTURE

Scott Tremaine

Canadian Institute for Theoretical Astrophysics, University of Toronto,
60 St. George St., Toronto M5S 1A7, Canada
tremaine@cita.utoronto.ca

Our conference title is taken from the movie "Back to the Future", which describes the adventures of a time-traveller who is trapped in 1955. To set the conference in historical context, I would like to review briefly the progress we have made since that time towards understanding the major components and fundamental parameters of the Galaxy. The year 1955 is particularly appropriate because the first 21-cm neutral hydrogen surveys of the Galaxy had just been completed. As my title implies, I will focus on areas in which progress has been slow, since these areas are slighted by most speakers.

First consider the distance to the Galactic center, R_0, and the speed of the Local Standard of Rest, v_{LSR}. The first 21-cm surveys used $R_0 = 8.2$ kpc, $v_{LSR} = 216 \, \mathrm{km \, s^{-1}}$, following van de Hulst et al. (1954). In 1964, following recommendations by Schmidt (1965), the IAU adopted the standard values $R_0 = 10$ kpc, $v_{LSR} = 250 \, \mathrm{km \, s^{-1}}$. Most subsequent determinations of both quantities have been smaller than Schmidt's values, and a review by Kerr and Lynden-Bell (1986) led to revised IAU standards, $R_0 = 8.5$ kpc and $v_{LSR} = 220 \, \mathrm{km \, s^{-1}}$. In a more recent review, Reid (1989) concluded that $R_0 = 7.7 \pm 0.7$ kpc (an analysis of globular cluster distances by Racine and Harris 1989 yields almost the same answer); recent estimates of v_{LSR} range from $248 \pm 16 \, \mathrm{km \, s^{-1}}$ (Schechter et al. 1988) to $200 \pm 10 \, \mathrm{km \, s^{-1}}$ (Merrifield, 1992 and this volume).

Thus neither R_0 nor v_{LSR} is known to better than $\pm 20\%$ at the 2σ level (and the uncertainty is even larger if—as is usually the case—there are unrecognized systematic errors); moreover, the best estimates and likely uncertainty of both parameters have not substantially changed since 1955.

Another fundamental parameter is the total mass density in the solar neighborhood, ρ_0 (the Oort limit), which is determined from the dynamics of disk stars in the direction normal to the Galactic plane (Kapteyn 1922). Comparison of ρ_0 to the total density of known stars and gas (about $0.10 M_\odot \, \mathrm{pc^{-3}}$) could reveal the presence of dark matter in the Galactic disk. The first reliable estimate was made by Oort (1932), who found $\rho_0 = 0.09 M_\odot \, \mathrm{pc^{-3}}$; later, Oort (1960) revised his estimate to $\rho_0 = 0.15 M_\odot \, \mathrm{pc^{-3}}$ with an uncertainty of $\pm 10\%$. For comparison, recent estimates include $0.09 \pm 0.03 M_\odot \, \mathrm{pc^{-3}}$ (Kuijken 1991a; see also Fuchs & Wielen, this volume) and $0.26^{+0.19}_{-0.12} M_\odot \, \mathrm{pc^{-3}}$ (Bahcall et al.

1992). Clearly we have not made much progress in converging to an accurate estimate of ρ_0 in the past generation.

The nature and origin of spiral structure were not understood in 1955. Since then, there has been considerable theoretical progress on a related issue, the stability and response of self-gravitating gaseous and stellar disks. Analytical studies, mostly using the WKB approximation (Lin & Shu 1964) or Hill's equations (Goldreich & Lynden-Bell 1965, Julian & Toomre 1966), have been complemented by N-body simulations and numerical calculations of linearized normal modes, to yield a secure understanding of many of the principal features of disk dynamics. However, the relation between disk dynamics and spiral structure remains obscure (see Toomre 1977 for a review), and as Kerr has stressed in his introductory talk, many basic questions about the spiral pattern in our own Galaxy remain unanswered: Is the pattern global or local, i.e. is there a well-defined "grand design" pattern or just a chaotic superposition of spiral patches? Is there a well-defined pattern speed—that is, will the same pattern persist for several Galactic years—and if so what is it? Is the pattern trailing or is there a leading component? What drives the spiral pattern—a central bar? a recent encounter with a satellite galaxy? local gravitational instability? or an unstable normal mode?

The most profound change in our picture of the Galaxy since 1955 has been the introduction of massive halos by Ostriker *et al.* (1974), who argued that an unseen dark halo may extend to many times the radius of the visible disk, with mass increasing roughly in proportion to radius, so that the halo contains virtually the whole mass of the Galaxy. The visible stars and gas constitute a small cesspool at the center, containing only a few percent of the total mass. The evidence cited for the halo included the Local Group timing argument, the dynamics of satellite galaxies, and the rotation curve; in addition, there are strong reasons to believe that such halos form naturally in a range of cosmological models (Gunn 1977). Almost twenty years later, the observational and theoretical evidence for dark halos surrounding our own and other galaxies is far stronger, but our understanding of the distribution of mass in our halo is not greatly improved over that of Ostriker *et al.*

The preceding paragraphs offer a bleak picture of progress in understanding Galactic structure over the last few decades. Obviously, this picture is far from correct. There has in fact been tremendous progress, particularly through surveys at wavelengths that were inaccessible in 1955 (such as the infrared surveys from COBE described here by Hauser). Nevertheless, it is striking that estimates of many of the basic parameters of the Galaxy—the solar radius, the rotation speed of the Local Standard of Rest, the Oort limit, the properties of the spiral structure—are not much better now than they were in 1955. Given the dramatic improvements in the quality and quantity of the data, and the many new types of data available, why can we not do better? *Why is understanding Galactic structure so hard?*

I think this is an interesting question, and the answer could help to direct observational and theoretical research efforts in the future. In the rest of this summary, I will try to identify some of the reasons why accurate models of the Galaxy have been so slow to arrive.

The Galaxy is very responsive   Hot stellar systems, like the spheroid or halo of the Galaxy, do not support sound waves: small-scale disturbances are strongly Landau damped. This well-known result led many of us to believe that all normal modes in hot stellar systems were Landau damped, so that such systems would settle rapidly to a steady state once galaxy formation was complete. This expectation was shown to be false by Mathur (1990), who demonstrated that self-gravitating spherical stellar systems can support discrete large-scale oscillations that do not Landau damp. Mathur's work provides theoretical support for observations of long-lived oscillations in a number of N-body simulations of stellar systems (Hénon 1968, Miller *et al.* 1982, Smith & Miller, this volume).

The Galactic disk is even more responsive than the spheroid. Leading spiral disturbances can be amplified by factors of 10 or more as they shear into trailing disturbances, even in disks that are safely stable to axisymmetric perturbations in the sense that Toomre's thermometer Q exceeds the critical value for stability by 50% or so (Goldreich & Lynden-Bell 1965, Julian & Toomre 1966). One consequence of this "swing amplification" process (Toomre 1981) is that concentrated lumps in the disk, such as giant molecular clouds, are surrounded by a trailing wake of disk stars, whose mass may exceed the mass of the original lump by an order of magnitude or more. A second is that large parts of the combined star-gas disk may quiver on the threshold of instability: gas cooling continually reduces Q and turns up the gain on the swing amplifier, until the heating induced by the amplification of small disturbances and resulting star formation stirs up the gas, increases Q, and quenches the amplifier.

Not only is the disk sensitive to gravitational disturbances, but the disk environment is a noisy one, containing almost 10^4 giant molecular clouds with masses $10^{5.5} M_\odot$ or larger (plus their associated wakes). Note that the ratio of the mass of one such cloud to the total mass of the disk is about the same as the ratio of the mass of the Earth to the mass of the Sun; thus the Galactic disk is as dynamically noisy as a solar system containing 10^4 planets like the Earth!

Non-axisymmetric distortions   Ever since the work of Shapley, Lindblad and Oort in the 1910s and 1920s, astronomers have understood that the Galaxy is approximately axisymmetric. Nevertheless, non-axisymmetric distortions—warps, bars, and spiral structure, to name three—are common in other galaxies and likely to affect our own.

Speakers at this meeting described several independent lines of evidence that yield a similar picture of a central bar in our Galaxy: (i) Blitz argued that 2.4μm balloon observations imply that the stars in the central kpc are arranged in a bar whose near side is at positive Galactic longitude (see also

Blitz & Spergel 1991b). (ii) This result is confirmed by the COBE/DIRBE near-infrared maps of the Galactic bulge described by Hauser and by Spiesman *et al.* (iii) As Weinberg reported, the distribution of variables in the IRAS point-source catalog is consistent with a bar with semi-major axis of about 5 kpc (see also Weinberg 1992). The position angle of the long axis of Weinberg's bar is $\theta = 36° \pm 10°$, where position angle is relative to the Galactic center-Sun axis and positive in the direction of Galactic rotation, so that $0 < \theta < 90°$ implies that the near side is at positive longitude. (iv) Rohlfs and Kampmann's paper shows that HI terminal velocities indicate the presence of a bar with $\theta \simeq 45°$, and semi-major axis 2–3 kpc; (v) Binney argued (following Binney *et al.* 1991) that the CO kinematics in the central kpc indicate the presence of a bar with $\theta \simeq 16° \pm 2°$, pattern speed $63 \,\mathrm{km\,s^{-1}\,kpc^{-1}}$, and corotation radius 2.4 kpc. Sellwood confirmed that such a bar arises naturally in N-body simulations of an unstable disk.

The impressive agreement of all five groups on the quadrant containing the long axis suggests that most or all have identified the same real structure, although the parameters of the bar (length, strength, pattern speed, position angle) are still quite uncertain.

Several authors have invoked other plausible $m = 2$ distortions to explain kinematic features in the solar neighborhood or the outer Galaxy: (i) Blitz and Spergel (1991a; see also Spergel's review) argue that the longitude-velocity distribution of distant HI implies that the Local Standard of Rest is moving outward at $14 \,\mathrm{km\,s^{-1}}$; this outward motion is induced by an $m = 2$ distortion from a triaxial spheroid with position angle $\theta = -45° \pm 20°$ and pattern speed $6 \,\mathrm{km\,s^{-1}\,kpc^{-1}}$. (ii) Kalnajs (1991) analyzes the distribution of stellar velocities in the solar neighborhood and finds evidence for two distinct star streams, which he associates with crossing orbits near an outer Lindblad resonance; he deduces that the Galaxy contains an $m = 2$ disturbance with pattern speed $46 \,\mathrm{km\,s^{-1}\,kpc^{-1}}$ and position angle $\theta = 56°$, possibly due to a bar (though the parameters appear incompatible with those deduced by Binney *et al.*). (iii) Kuijken and I argued at this meeting that the axis ratio of the local velocity ellipsoid and apparent inconsistencies between outer rotation curves measured by different methods can be explained if the Sun is near the minor axis of an $m = 2$ distortion with near zero pattern speed, possibly arising from a triaxial dark halo. (iv) Most prominent "grand design" spiral patterns in other galaxies have two arms, and many authors have sought to explain features of local kinematics and the HI terminal velocity curve in terms of $m = 2$ spiral distortions.

Lopsided ($m = 1$) distortions are also common in many galaxies (Sellwood & Wilkinson 1992). One-armed spirals are seen in M31 and other galaxies (García-Gómez & Athanassoula 1991), although they are not as common as two-armed spirals; the HI distributions in the outer parts of galaxies are often lopsided (Baldwin *et al.* 1980); Kuijken (1991b) has suggested that a similar $m = 1$ distortion of the outer Galaxy could explain the asymmetries in the HI distribution described by Blitz and Spergel (1991a); in Magellanic irregular galaxies the photometric center is offset from the kinematic center (de Vaucouleurs & Freeman 1970).

The distribution of CO near the Galactic center is lopsided, with roughly three times as much gas in the north as the south. Binney *et al.* (1991) interpret this displacement as an effect of the perspective from which we view an $m = 2$ barlike distortion; however, an $m = 1$ distortion is an alternative explanation. Several of the other lines of evidence presented at this meeting for a central bar, including asymmetries in the infrared photometry and the distribution of IRAS variables, could equally well be evidence for an $m = 1$ distortion. A possible weakness of this interpretation is that a rotating $m = 1$ distortion in the inner Galaxy would normally induce a non-zero radial velocity between the Galactic center and the Local Standard of Rest, but observations show that this velocity is very near zero (see Blitz & Spergel 1991a and Kuijken & Tremaine 1991 for two estimates).

There are also theoretical reasons to be interested in lopsided distortions. The most unstable mode in stellar Mestel disks is lopsided (Zang 1976), as is the most unstable mode in models of the Galaxy that have no massive halo (Sellwood 1985); some lopsided modes are only weakly damped in spherical stellar systems (Weinberg 1991a); a lopsided mode is the dominant instability in massive gas disks orbiting a central point mass (Adams *et al.* 1989, Shu *et al.* 1990); and lopsided distortions appear in N-body simulations (see, for example, Smith and Miller, this volume). Finally, I note that as our understanding of stellar dynamics has increased, models of stellar systems have lost their artificial symmetries: from the spherical models investigated by Eddington and Jeans early in this century, we progressed to axisymmetric models in the 1960s and early 1970s (Lynden-Bell 1962, Gott 1973), then to triaxial models in the 1980s (Schwarzschild 1979, de Zeeuw 1985). However, the triaxial models still contain symmetries: they are invariant under reflection in the three principal planes (the D_{2h} point group, Landau & Lifshitz 1977). These symmetries are less artificial than spherical symmetry or axisymmetry, since they appear to be spontaneously generated in violent relaxation: N-body experiments suggest that initial states with no symmetries collapse to form equilibrium systems with D_{2h} symmetry. Nevertheless, it seems worthwhile to investigate models of stellar systems that contain fewer symmetries.

There can also be distortions normal to the Galactic plane; examples include the well-known warp in the outer Galaxy, and discrete vertical oscillation modes of the disk (Weinberg 1991b).

Non-axisymmetric distortions complicate attempts to model the structure of the Galaxy. There are at least five known or likely sources of distortion (central bar, triaxial spheroid, triaxial halo, spiral structure, warp)—not to mention possible $m = 1$ distortions or local inhomogeneities—and each is described by at least three parameters (amplitude, pattern speed, phase)—not to mention the parameters needed to describe the radial dependence. Unless one or two of the distortions dominate, it may prove difficult to disentangle their effects.

A particularly pernicious sort of distortion is one whose symmetry axis coincides with the line between the Sun and the Galactic center. Such distortions are symmetric under reflection of positive and negative Galactic longitude, and hence are difficult to detect, but Kuijken and I have argued here that they can have a drastic effect on measurements of the rotation curve, the speed of the Local Standard of Rest, and other kinematic parameters.

Incomplete mixing   The Galaxy is usually assumed to be in a steady state but this state is approached only gradually, by phase mixing. When mixing is not complete, even large samples of stars may not provide representative samples for analyzing Galactic structure.

The presence of moving groups of stars in the solar neighborhood (Eggen 1965) implies that the disk is not well-mixed, at least for A-type and earlier stars, a conclusion consistent with simple estimates of the clumpiness of star formation and the mixing rate (e.g. Kuijken & Tremaine 1991).

Next consider mixing in the spheroid. We assume that the spheroid is composed of stars from N lumps (protogalaxies? globular clusters?), each of mass m and radius Δr, that were tidally disrupted at the time of spheroid formation, $t = 0$. The lumps would be disrupted at orbital radius $r \approx \Delta r (M/m)^{1/3}$, where M is the mass of the Galaxy inside r. The stars from each disrupted lump spread out into a tidal stream, which we assume has typical radius comparable to the tidal disruption radius r. After time t the length of the stream is $L \approx r(d\Omega/dr)\Delta r\, t \approx \Delta r \Omega t$, where Ω is the orbital angular speed of the stars in the stream, while its cross-sectional area would be $A \approx \Delta r^2$. The filling factor— the fractional volume containing one or more tidal streams—is then $1 - \exp(-f_3)$, where

$$f_3 \approx N\frac{AL}{r^3} \approx \frac{Nm}{M}\Omega t = \frac{M_s}{M}\Omega t,$$

and M_s is the mass in the spheroid. A well-mixed system should have $f_3 \gg 1$; note that f_3 is independent of the lump mass m. If some lumps are disrupted later than $t = 0$ (because of late infall or dynamical friction), then f_3 would be even smaller. A related statistic is the fraction of the sky containing one or more streams seen in projection, $1 - \exp(-f_2)$ where

$$f_2 \approx N\frac{L\Delta r}{r^2} \approx \frac{M_s}{M}\left(\frac{M}{m}\right)^{1/3}\Omega t \approx f_3 \left(\frac{M}{m}\right)^{1/3}.$$

For the spheroid at the solar radius, we may take $M_s/M \approx 0.01$, $\Omega t \approx 100$, so $f_3 \approx 1$ independent of the lump mass m; in other words, at a given spatial location the spheroid is likely to contain stars from at most a few lumps. Given this result it is not surprising that the phase-space distribution of spheroid stars in the solar neighborhood appears clumpy (Doinidis & Beers 1989, Norris & Ryan

1989), or that there are variations between the kinematical properties of various samples of spheroid stars. Oort (1965) already suggested that Eggen's moving groups of spheroid stars (which however are of doubtful statistical significance) might arise from disrupted globular clusters.

We may also apply this argument to stars in the distant halo, say at $r > 50\,\mathrm{kpc}$. Here $\Omega t \approx 20$, and M_s/M may be taken to be the ratio of the mass in stars to the total mass, which is poorly known—since few stars are seen at this distance—but certainly small. Thus $f_3 \ll 1$, and it is likely that $f_2 \ll 1$ as well.

Let us then imagine how the Galaxy would look if we had a special telescope that was only sensitive to stars at $r > 50$ kpc. We would not see a smooth distribution of stars randomly dotted around the celestial sphere; rather, the stars would be concentrated in randomly oriented streaks of various lengths, as well as a few isolated galaxies such as the Magellanic Clouds that have not (yet) disrupted. The Magellanic Stream of HI is presumably one such streak. At some locations in the sky the streaks might fold back on themselves as seen from our perspective near the Galactic center; in such cases there would be a strong local enhancement of the density, which might be mistaken by us for an isolated low surface-brightness galaxy. It is possible that the Draco and Ursa Minor dwarf "galaxies" are a mirage of this sort; this radical hypothesis eliminates the need for dark matter in these systems and is consistent with the clumpy distribution of stars in Ursa Minor, which would not be expected in an equilibrium stellar system (Olszewski & Aaronson 1985).

Insufficient statistics   Even if the Galaxy were well-mixed, fluctuations due to Poisson statistics in samples of limited size can obscure the effects we seek. Suppose, for example, we want to measure some velocity parameter from the globular cluster system (velocity of the LSR, mean rotation velocity of the clusters, etc.). There are about 115 clusters with known radial velocities and the system has a dispersion of $110\,\mathrm{km\,s^{-1}}$ (Thomas 1989). Thus the 2σ limits on any mean velocity will at best be $\pm 20\,\mathrm{km\,s^{-1}}$, and smaller if the sample is restricted to metal-rich or metal-poor clusters (the two types have different kinematics), or if some clusters contribute less weight to the answer (for example, clusters near the Galactic poles do not contribute to estimates of rotation speed). Typical samples of nearby metal-weak stars yield comparable statistics.

The Galaxy is not a relaxed system   Are galaxies important archaeological sites, which retain a record of their formation and history? The answer is not obvious: in a different context, the study of isolated stars tells us very little about star formation, because of the Vogt-Russell theorem (the structure of a non-rotating star is uniquely determined by its mass and chemical composition). Similarly, there may be relaxation processes that operate or have operated in galaxies to erase all memory of the initial conditions and the formation history; such processes might include violent relaxation in halos, late infall of gas, or angular momentum transport by bars and spiral structure.

The weight of evidence now suggests, however, that galaxies are not relaxed, so that many of their properties depend strongly (and stochastically) on their individual histories. For example, dark halos are not spherical, but instead have a range of triaxial shapes determined by the primordial fluctuation spectrum (Dubinski & Carlberg 1991); warps may reflect the excitation of a particular normal mode early in the history of the galaxy (see Sparke's review) or reorientation of the spin axis of the disk in response to continuing infall (E. Ostriker & Binney 1989, Binney 1992); the distribution of gas near the center and nuclear activity may be strongly time-variable, and so on. A particularly important stochastic process is merging, which has affected the star-formation history and morphology of many nearby galaxies (Schweizer 1990). Tóth and J. Ostriker (1992; see also J. Ostriker, this volume) have argued that the present thickness of the disk implies that no more than a few percent of the mass inside the solar radius can have been accreted in recent mergers. This merger rate is substantially smaller than expected in standard (cold dark matter, $\Omega = 1$) cosmological models (Carlberg & Couchman 1989), but an alternative possibility is that much of the vertical energy imparted to the disk in the merger is swept out by bending waves.

Unrelaxed galaxies are expected to be more diverse and harder to understand than relaxed galaxies would be, but the study of such systems offers insights into details of cosmology and galaxy formation that would not otherwise be accessible. Another consequence of the dependence of present structure on past history is that large segments of galactic structure, as a field distinct from galaxy formation, are likely to wither away. More and more, it has become impossible to address issues of galactic structure except in the context of galaxy formation.

The importance of astrometry   Much of our understanding of galactic structure rests, directly or indirectly, on distances and proper motions of nearby stars and clusters. This is a shaky foundation: there are only about 1000 stars with trigonometric parallaxes accurate to within 15% (Gliese *et al.* 1986), while Oort's B-constant, which measures the mean proper motion of nearby stars in an inertial frame, is only known to within 25% at the 1-σ level (Kerr & Lynden-Bell 1986). This situation will improve dramatically with the release of data (in about 1996) from the European astrometric satellite Hipparcos, launched in 1989 (Perryman 1989). Hipparcos is expected to improve the number of stars with useful parallaxes by more than an order of magnitude, in addition to improving the parallax accuracy and greatly reducing systematic errors. A short list of potential applications of Hipparcos data includes studying fine structure in the HR diagram, Galactic kinematics using proper motions of Cepheids, the distance to the Hyades, the bright end of the HR diagram, internal kinematics of clusters and associations, stellar masses, proper motions of globular clusters, structure in the phase-space distribution of spheroid stars, the Oort constants,

the Oort limit, and the velocity ellipsoid and its spatial variation. Even more accurate results could be obtained from a possible Hipparcos II mission, for which
the present mission would provide first-epoch positions. Ground-based optical
and radio astrometric techniques are also advancing rapidly (Monet 1988); a
future milestone for radio astrometry will be measurement of the distance to
the Galactic center from the trigonometric parallax of Sgr A*.

I think that the improved accuracy of the Hipparcos survey of the solar
neighborhood will "trickle down" to larger scales and thereby resolve many of the
puzzles and inconsistencies in our current understanding of Galactic structure.

This is an incomplete listing of some of the particular features of the
Galaxy that make it difficult to study. Psychiatrists tell us that admitting our
problems is the first step to solving them, and, for similar reasons, thinking further about the fundamental obstacles to progress in studying Galactic structure
is likely to be a therapeutic and useful exercise.

REFERENCES

Adams, F. C., Ruden, S. P., & Shu, F. H. 1989, ApJ, 347, 959

Bahcall, J. N., Flynn, C., & Gould, A. 1992, ApJ, 389, 234

Baldwin, J. E., Lynden-Bell, D., & Sancisi, R. 1980, MNRAS, 193, 313

Binney, J. J. 1992, ARAA, 30, 51

Binney, J. J., Gerhard, O. E., Stark, A. A., Bally, J., & Uchida, K. I. 1991,
    MNRAS, 252, 210

Blitz, L., & Spergel, D. N. 1991a, ApJ, 370, 205

Blitz, L., & Spergel, D. N. 1991b, ApJ, 379, 631

Carlberg, R. G., & Couchman, H.M.P. 1989, ApJ, 340, 47

de Vaucouleurs, G., & Freeman, K. C. 1970, Vistas Astr., 14, 163

de Zeeuw, T. 1985, MNRAS, 216, 273

Doinidis, S. P., & Beers, T. C. 1989, ApJ, 340, L57

Dubinski, J., & Carlberg, R. G. 1991, ApJ, 378, 496

Eggen, O. 1965, in Galactic Structure, eds. A. Blaauw & M. Schmidt (Chicago:
    Univ. of Chicago Press), 111

García-Gómez, C., & Athanassoula, L. 1991, in Dynamics of Disc Galaxies,
    ed. B. Sundelius (Göteborg: Göteborg Univ. Press), 365

Gliese, W., Jahreiss, H., & Upgren, A. R. 1986, in The Galaxy and the Solar
    System, eds. R. Smoluchowski, J. N. Bahcall, and M. S. Matthews (Tucson:
    Univ. of Arizona Press), 13

Goldreich, P., & Lynden-Bell, D. 1965, MNRAS, 130, 125

Gott, J. R. 1973, ApJ, 186, 481

Gunn, J. E. 1977, ApJ, 218, 592

Hénon, M. 1968, Bull. Astron., 3, 241

Julian, W. H., & Toomre, A. 1966, ApJ, 146, 810

Kalnajs, A. J. 1991, in Dynamics of Disc Galaxies, ed. B. Sundelius (Göteborg: Göteborg Univ. Press), 323

Kapteyn, J. C. 1922, ApJ, 55, 302

Kerr, F. J., & Lynden-Bell, D. 1986, MNRAS, 221, 1023

Kuijken, K. 1991a, ApJ, 372, 125

Kuijken, K. 1991b, in Warped Disks and Inclined Rings around Galaxies, eds. S. Casertano, P. Sackett and F. Briggs (Cambridge: Cambridge Univ. Press), 159

Kuijken, K., & Tremaine, S. 1991, in Dynamics of Disc Galaxies, ed. B. Sundelius (Göteborg: Göteborg Univ. Press), 71

Landau, L. D., & Lifshitz, E. M. 1977, Quantum Mechanics, 3rd ed. (Oxford: Pergamon)

Lin, C. C., & Shu, F. H. 1964, ApJ, 140, 646

Lynden-Bell, D. 1962, MNRAS, 123, 447

Mathur, S. D. 1990, MNRAS, 243, 529

Merrifield, M. R. 1992, AJ, 103, 1552

Miller, R. H., Vandervoort, P. O., Welty, D. E., & Smith, B. F. 1982, ApJ, 259, 559

Monet, D. G. 1988, ARAA, 26, 413

Norris, J. E., & Ryan, S. G. 1989, ApJ, 336, L17

Olszewski, E. W., & Aaronson, M. 1985, AJ, 90, 2221

Oort, J. H. 1932, BAN, 6, 249

Oort, J. H. 1960, BAN, 15, 45

Oort, J. H. 1965, in Galactic Structure, eds. A. Blaauw and M. Schmidt (Chicago: Univ. of Chicago Press), 455

Ostriker, E. C., & Binney, J. J. 1989, MNRAS, 237, 785

Ostriker, J. P., Peebles, P.J.E., & Yahil, A. 1974, ApJ, 193, L1

Perryman, M.A.C. 1989, Nature, 340, 111

Racine, R., & Harris, W. E. 1989, AJ, 98, 1609

Reid, M. J. 1989, in The Center of the Galaxy, ed. M. Morris (Dordrecht: Kluwer), 37

Schechter, P. L., Aaronson, M., Cook, K. H., & Blanco, V. M. 1988, in The Outer Galaxy, eds. L. Blitz and F. Lockman (Berlin: Springer), 31

Schmidt, M. 1965, in Galactic Structure, eds. A. Blaauw and M. Schmidt (Chicago: Univ. of Chicago Press), 513

Schwarzschild, M. 1979, ApJ, 232, 236

Schweizer, F. 1990, in Dynamics and Interactions of Galaxies, ed. R. Wielen (Berlin: Springer), 60

Sellwood, J. A. 1985, MNRAS, 217, 127

Sellwood, J. A., & Wilkinson, A. 1992, Rep. Prog. Phys., in press

Shu, F. H., Tremaine, S., Adams, F. C., & Ruden, S. P. 1990, ApJ, 358, 495

Thomas, P. 1989, MNRAS, 238, 1319

Toomre, A. 1977, ARAA, 15, 437

Toomre, A. 1981, in The Structure and Evolution of Normal Galaxies, eds.
    S. M. Fall and D. Lynden-Bell (Cambridge: Cambridge Univ. Press), 111

Tóth, G., & Ostriker, J. P. 1992, ApJ, 389, 5

van de Hulst, H. C., Muller, C. A., & Oort, J. H. 1954, BAN, 12, 117

Weinberg, M. 1991a, ApJ, 368, 66

Weinberg, M. 1991b, ApJ, 373, 391

Weinberg, M. 1992, ApJ, 384, 81

Zang, T. A. 1976, Ph. D. thesis, Massachusetts Institute of Technology

THIS MEETING: A BIASED OBSERVER'S VIEW

Carl Heiles
Astronomy Department, University of California, Berkeley CA 94720

ABSTRACT

Letting yourself be nominated for a conference summary talk is considered by some to be a big mistake.* It eliminates the possibility of making up the sleep lost at night, while partying, during the day, while sitting in the talks. It even forces you to look at all the poster papers.

But at a meeting like this, with the wealth of observational data, it is definitely not a mistake: it was even worth missing some of the parties! My problem was to devise a way to be sufficiently selective so as to provide a reasonably coherent summary. I chose to emphasize the multitude of large-scale maps presented at the meeting. Many are relevant to the 'worm paradigm' (§2), and the recent γ-ray and ROSAT results are relevant to the Hot Ionized Medium (§3). And finally, I was impressed by a number of well-crafted smaller-scale observations, which elucidate particular aspects of the interstellar medium (§4).

1. INTRODUCTION

With 124 poster papers and 24 verbally presented papers this has been a stimulating meeting, and we have had to work hard. I can't review all of the papers and have been searching for ways to reduce my task. My original charge was to review the 'structure and energetics'. Most of the energy in the Galaxy is in the stars, and I know nothing about stars! Furthermore, I am going to bypass the fascinating papers on the Galactic center—even though the center is the most energetic region of the Galaxy. The Galactic center is a huge topic, one which has had whole symposia devoted to it, and a quick review could not possibly do it justice.

I will concentrate on interstellar matter, but even this restriction is not enough because the observational developments presented at this meeting have been so plentiful and exciting! So I am going to present a highly personal view, and apologize in advance to people whose work I am not reviewing.

The most powerful impression I have of this meeting is maps of the sky. We have been presented both with new maps at familiar wavelengths and with maps at completely unfamiliar wavelengths. The new maps at familiar wavelength are characterized by superb data quality, by which I mean consistent calibration and zero level and high dynamic range. These features are important because they enable image processing techniques to extract important weak structural details that are overshadowed by large-scale or brighter features. As evidenced by this

---

* I am indebted to the hard-working conference editor organizer, Dr. Fran Verter, for unknowingly providing the original intriguing version of this sentence.

meeting, such image processing has become as important as the data themselves in understanding what the images mean. Much of my review will deal with these maps and what I think they mean for the morphology, physics, and evolution of the interstellar medium (ISM).

In addition to maps, I have been struck by a number of what I regard as particularly well-chosen or well-designed smaller observations that provide specific insights or generate annoying questions, and these will occupy the rest of my review in §4.

Before I begin, I wish to mention perhaps the most energetic interaction that interstellar matter undergoes: the interaction with the gravitational field produced by the stars (and dark matter too!). The most obvious examples of such interaction are external galaxies that suffer mergers and near collisions, and sometimes galaxies that have asymmetric mass concentrations such as bars. The interaction produces bursts of star formation. Frequently these bursts occur in localized regions of a galaxy and produce many young, hot stars in a short time. Occasionally they occur globally over wide regions of a galaxy and produce the 'starburst galaxies', which aside from quasars are the brightest known objects in the Universe. These dynamically-induced bursts of star formation do not seem to be occurring in the Milky Way at the moment, even though we have the requisite conditions: we have the Magellanic clouds as neighbors and we seem to have the 'stellar bar' at the Galactic center, which has been so prominent a topic of this meeting. We may well have had bursts of star formation in the past.

2. MAPS and WORMS

2.1. The worm paradigm.

This section discusses maps that show 'worms'. The current paradigm for producing worms involves multiple supernovae as outlined by McKee (verbal paper) this morning.

If a molecular cloud, or group of clouds, produces O stars in quantity over several generations, O stars in the early generations will have exploded as supernovae. Usually the stars are formed near the edge of the molecular cloud, so their ionizing photons, winds, and supernova shocks carve out a much larger volume in the direction away from the molecular cloud. Furthermore, multiple supernovae produce huge cavities, and these extend much further in the z-direction than in the Galactic plane because the interstellar gas decreases in density with z. This is a 'worm'.

Earlier this morning, Dick McCray bemoaned the terminology. I sympathize with his dismay. One unsatisfactory alternative word is often used: the 'chimney' (Norman and Ikeuchi 1989). A chimney is a large hole in the Galactic disk that extends all the way through the disk to the gaseous halo. The term is apt: the chimney contains hot gas, heated by the cluster of supernovae; the hot gas is polluted with the heavy elements produced by the nuclear burning in the supernovae, much as the smoke in a terrestrial chimney is polluted by the products of combustion; the chimney has a cold dense wall, consisting of the original interstellar material that was pushed aside by the shock front caused by the supernova explosions.

But a worm is not necessarily a chimney: a chimney connects directly to the halo, but a worm may not. And as McKee discussed, it takes a much larger cluster of supernovae to make a chimney than a worm. Thus there should be many more worms than chimneys. I must admit to being the one who coined the term 'worm'. At the time, I had no idea that these structures would turn out to be so popular, and simply assigned a name that was based on nothing more than their appearance on my pictorial data sets (Heiles 1984). Little did I realize that the topics of the Galaxy, diffuse interstellar matter, and the interaction of stars and gas would become so prominent in later years. I agree with McCray that the term, 'worm', is not the best.

Perhaps we should ask the IAU to invent an official name for these objects. On the other hand, I am somewhat reluctant to take this irretrievable step. Despite the seemingly iron-clad observational and theoretical evidence, some beautiful examples of which we have seen at this meeting, I still—in my heart of hearts—have a lingering doubt concerning their reality, both in observations and theory. Although this is simply the internal, illogical voice of frail human subjectivity, it nevertheless speaks to me—ever more softly with the increasing years—whenever I think of these topics.

So we continue with the unfortunate terminology, 'worm'. Except for the molecular cloud, which is not easily destroyed by supernova explosions, the interior of the worm should be devoid of all interstellar gas except for the Hot Ionized Medium (HIM), which was heated by the explosive energy and cannot interact with the ionizing photons. Photons from the O stars ionize part or all of the molecular cloud, producing a 'blister' type H II region. Such regions are ionization-bounded towards the molecular cloud, but are density-bounded in the other direction where the ionized gas streams away supersonically. In this direction, most of the ionizing photons have a direct, unimpeded line of travel to the worm wall. If the worm is open at the top, then it forms a true chimney to the Galactic halo. The ionized gas produces thermal radio emission, and the supernova shocks produce nonthermal emission, so the worm should be easily visible in the radio continuum. However, if O-star production has ceased, then there should be no ionizing photons and no thermal radio continuum.

The Orion nebula/Eridanus shell region is an excellent example of this. Generations of young, massive stars in the Orion region, located at $(\ell, b) \approx 209°, -19°)$, have blown a large worm. Towards lower Galactic longitudes, the worm is bounded by a prominent wall of H I which is easily visible in maps of 21-cm line emission (Hartmann, poster paper (§2.2); or, in more accessible published maps, Colomb, Pöppel, and Heiles 1980). Towards higher longitudes, there is very little H I at negative Galactic latitudes. The Eridanus shell, which is a $\sim 40°$-diameter shell-like structure of expanding H I enclosing hot gas observed in soft X-ray emission, lies further from the Galactic plane and closes the 'top' of the worm. Part of the Eridanus shell is observed in Hα radiation and has been ionized by the O stars in Orion (Reynolds and Ogden 1979); this material can be considered to be the diffusely-distributed warm ionized medium (WIM). Reach (verbal paper) discussed other examples of worms.

2.2. A new 21-cm line survey.

The 21-cm line is the basic tracer of neutral atomic gas. A new survey performed at Dwingeloo by Hartmann (poster paper) covers the entire northern sky with excellent sensitivity, velocity coverage, and completeness. Plans exist to extend it to the southern sky and to correct for the 'stray radiation'. When these important tasks have been completed, this will become the standard reference and will be a very significant improvement on the existing surveys.

2.3. Cm-wavelength radio continuum maps.

Very high-quality radio continuum maps of the northern Galactic plane at about 11-cm wavelength have been made by the Bonn group (Reich et al. 1990). A spatially filtered version of this map was presented by Reach (verbal paper). In addition, Jonas (poster paper) presented an equivalent map of most of the southern sky (all latitudes) obtained at the Rhodes/Hart Radio Astronomy Observatory in South Africa. These are impressive data sets indeed, because even after spatial filtering there are no instrumental artifacts in the maps. After removing the smooth background near the Galactic plane one is left with a narrow-b component.

Both maps show worms very prominently and unambiguously. In fact— aside from the Galaxy itself, individual supernova remnants, and H II regions— worms are the most prominent features on these maps.Nevertheless, some worms observed in the H I line and in the IRAS maps do not appear on these maps. These basic observational statements imply that some—but not all—worms exhibit prominent radio emission. Further studies show that much of this emission is thermal.

One important fact that remains to be established is whether the radio-detected worms are different from the radio-undetected worms, or whether the detection statistics simply reflect an intrinsic distribution of radio brightness. The 'worm paradigm' would predict that most worms are not radio emitters, at least not thermal radio emitters: thermal emission requires ionization, and only some worms are early enough in their evolutionary state to still have formation of new generations of hot stars.

2.4. IR maps: IRAS and FIRAS maps.

The new IRAS data products are not significantly contaminated by Zodiacal dust emission and provide superb, greatly improved maps of far-IR emission. The COBE/DIRBE maps are more accurately calibrated and span a large wavelength range with lower angular resolution than the IRAS maps; they were shown here for the first time. The full analyses of both of these data sets will take time. I predict that the fruits of these analyses will be richly rewarding, even in ways that we cannot foresee at the present time.

We do not yet understand the full ramifications of spatial filtering. Spatial filtering of the IRAS maps with a square filter (Reach, verbal paper) shows the worms very clearly. But a highly elongated filter $[(\ell \times b) = (15° \times 0.05°)]$ (Waller, verbal paper) eliminates the worms and shows a residual, weaker 'froth' which looks much more random. Waller interprets this in terms of a number of independent edge-brightened shells lying along the line of sight.

It is not clear—to me, at any rate—why these different filters produce such radically different-looking results. Perhaps the different types of filters elucidate different, but physically significant, structural aspects of the ISM. Test studies of different types of filters would be most instructive. The filters should also be applied to the H I and CO data sets because they include velocity, which provides the crucial distance information.

2.5. C II 157 μm line maps.

I cannot overemphasize the importance of this line: as discussed in more detail below (§4.5), it is the major coolant for the cold neutral atomic gas (CNM). For the warm neutral medium (WNM), Fe and Si would dominate C II if it were not for their depletion onto dust grains. For the WNM, cooling from collisional excitation of Lyα dominates C II cooling if the temperature is high enough, say $\gtrsim$ 7000 K; indeed, the temperature *must* be this high if the WNM is thermally stable. However, some WNM is observed at somewhat lower temperatures, ~ 2000 K (Verschuur and Schmelz 1989; Heiles, 1989); at these temperatures, Lyα cooling is unimportant and C II cooling dominates.

The C II line provides the total cooling of the CNM—and thus measures the heating. The major heating source for the gas is thought to be photoelectric emission from grains, which under most interstellar conditions should provide a constant heating rate per H atom. However, there may well be additional heating in specialized regions, for example in colliding clouds, where shock heating should dominate, and near supernova remnants, where cosmic rays and soft X-rays might predominate. An increased C II line brightness per H-atom would be an unambiguous indicator of such processes.

In the past, C II observations have been restricted to small, very dense regions, where it is possible to cancel out the terrestrial atmospheric emission by beam switching. But this meeting has provided the first maps of *diffuse* C II from the widely-distributed interstellar gas. Nakagawa (verbal paper) showed 12'-resolution C II maps of the Galactic plane in the Galactic interior, and Bennett (verbal paper; also Petuchowski and Bennett 1993) showed 7°-resolution maps of C II and the 205 μm N II fine structure line. While C II primarily traces the CNM, N II traces the WIM.

Near the Galactic plane we see the usual disk component in these lines. But we also see the worms! In addition to these morphological matters, the quantitative interpretive analysis of the thermal energy balance revealed by the C II and N II lines will be most interesting.

3. MAPS: THE HOT IONIZED MEDIUM

3.1. The highly-ionized Fe 6.7 Kev line.

Koyama (verbal paper; also see Koyama *et al.* 1989) presented maps of the Fe XXVI 6.7 Kev line and the associated free-free continuum made with the GINGA satellite. I've been told that the previous interpretation of these data was that the emission arises from a large number of point sources, which made it look like diffuse emission. However, the current interpretation differs markedly: the intensity ratio of the Fe line and free-free continuum matches that expected from a diffusely-distributed gas with $T \sim 10^8$ K. If this is correct, then this is a *qualitatively* different type of Hot Ionized Medium (HIM) than we have encountered before.

Regarding the HIM, we are running into difficulties with nomenclature! First, the HIM is philosophically different from the other phases (the CNM, the WNM, and the WIM). These other phases are thermally stable, and therefore truly deserve* to be designated as 'phases' of the ISM. But the HIM is thermally unstable**, and exists only because its thermal time constant is very long. Nevertheless, we follow the astronomical convention and refer to it as a 'phase'. And now we must extend this unseemly nomenclature because we now identify at least three subphases: the $\sim 3 \times 10^5$ K subphase revealed by O VI absorption; the $\sim 3 \times 10^6$ K revealed by soft X-ray emission; and this ultrahot, $\sim 10^8$ K subphase.

For this ultrahot subphase, there are three spatially distinct components: a ridge component, with $\Delta\ell \sim \pm 40°, \Delta b \sim \pm 2°$; a bulge component, roughly coincident with the stellar bulge and having $\Delta\ell \approx \delta b \sim 16°$, and a center component with angular diameter $\sim 1.5°$. For the first two components, $n_e \sim 3 \times 10^{-3}$ cm^{-3}; for the last, spatially-confined center component, n_e is about ten times larger.

For the two more extended components, the gas has a very large pressure, $p/k \sim 3 \times 10^5$ cm^{-3} K. This is about 100 times the typical thermal gas pressure in interstellar space and some twenty times the pressure derived from the weight of the overlying layers in the Solar vicinity (Boulares and Cox 1990). This ultrahot gas, which seems to be distributed over large volumes in the interior of the Galaxy, should explode into the halo. What confines it?

For the disk component, the weight of the overlying layers is totally insufficient, by at least one order of magnitude. The apparent coexistence of the ultrahot phase and the ordinary gas phases in the disk is exceedingly difficult to understand: the ultrahot gas should just explode into the halo with utter disregard for the observed layered structure of the cold gaseous disk. But we *do* observe this layered structure! This implies that the explosion hasn't happened! The simplest way out of this quandary, perhaps, is that the original interpretation is correct and the Fe-line emission does, indeed, arise from a large number of hot 'point' sources in which the ultrahot gas is clumped at very high densities.

Similarly, it seems that there is nothing to contain the bulge component, so it should expand freely into the halo; indeed, this may be occurring before our very eyes (§3.2.2).

For any gas phase, even if it is not in thermal equilibrium, there must be a heating and cooling mechanism. For this ultrahot phase, the cooling mechanism is easy: we observe it! The heating mechanism might conceivably be the interaction of cosmic rays with a fully ionized gas, although this would require a much-enhanced low-energy cosmic ray density in these regions which should be ruled out by γ-ray observations (Blitz *et al.* 1985; Strong *et al.* 1988). Supernovae are another possibility, of course, but the required rate is high; for example, the characteristics of the center component can be reproduced with an energy input of $\sim 10^{54}$ erg over the past $\sim 10^4$ yr, equivalent to 1 supernova every 10 years within < 100 pc of the Galactic center (Ozernoy, Titarchuk, and Ramaty 1993). This is a lot. The heating mechanism for any of the three components of this ultrahot phase is not obvious.

3.2. Soft X-ray maps from ROSAT.

---

    * in the sense of Physical Chemistry.

    ** Evaporative cooling from imbedded clouds can make the HIM thermally stable (Begelman and McKee 1990).

McCammon (verbal paper) showed some of the spectacularly beautiful new soft X-ray ROSAT maps, which highlight gas with $T \sim 3 \times 10^6$ K gas. Older X-ray maps at several energies were made by the Wisconsin and MIT groups, but with poor angular resolution (several degrees or worse). Mapping at several energies is important because it provides hints on the location of the gas. Soft X-rays are absorbed by photoelectric ionization of interstellar gas to a degree that depends very sensitively on energy. For the observed energy ranges, the typical path length of the photons ranges from tens to thousand of pc.

3.2.1. The high-latitude sky.

One of the most prominent features of the older maps was a pronounced anticorrelation of the observed X-ray intensity I_x with H I column density N_{HI}. However, the energy dependence showed that it was not caused by simple absorption. In the 'displacement model', the anticorrelation is explained by having all of the X-ray emission arising from hot gas located within the 'local bubble' (Paresce 1984; Snowden et al. 1990); the H I layer lies beyond. With an H I layer of constant thickness, in directions where the local bubble extends higher (with consequently higher I_x) the H I column density is lower. While this model explains the anticorrelation, it is unsatisfying because it is basically ad hoc and makes no further predictions.

The new ROSAT images change all of this. For the first time, the angular resolution is good enough to see absorption produced by individual clouds. Distances to some of these clouds are known from other considerations. The results force the conclusion that some of the hot gas lies within the local bubble, but some lies at high z. Tentatively, we can call the latter a 'halo component' because the high temperature should lead to a large z-scale height. Thus, reality has some of the features of the model treated by Jakobsen and Kahn (1986), which has cold clouds interspersed in the HIM; the two components have different scale heights.

An important point emerges from these data: the 'halo component' of the HIM is patchy, varying widely in intensity from one place to another. This patchiness is not simply a manifestation of the patchy cold clouds, but instead is true structure in the HIM. The current analyses of ROSAT data are restricted to only a few hundred square degrees of high-latitude sky. Clearly, we will not really understand the structure of the HIM until very large sections of sky—basically, the whole sky—is analyzed.

3.2.2. The extended X-ray emission towards the Galactic center.

In the two steradians of sky towards the Galactic interior, one sees prominent X-ray emission from the edge of the North Polar Spur, which is a nearby, 120°-diameter circle in the sky centered near the Sco/Oph star association. We believe that the Spur was produced by multiple supernovae and stellar winds of previous generations of stars in this association. In addition, one sees a large $\sim 50°$ diameter patch of X-ray emission centered near $(\ell, b) = (0°, 0°)$. Most X-ray astronomers consider this large patch to be produced by hot gas within the North Polar Spur.

But the new ROSAT data show clearly (to me, at any rate!) that this large patch is absorbed by the Galactic plane gas, as well as by numerous dust filaments extending up from the plane. In fact, the X-ray image of this region bears a striking resemblance to the wide-angle optical photographs of the Galaxy, which show the disk and dust filaments clearly absorbing the background diffuse light from unresolved distant stars. If this is correct, then the large patch cannot be associated with the North Polar Spur; instead, it would most likely be associated with the Galactic central regions.

One should not revise the current conception regarding this large patch of emission without more detailed analysis and consideration. But nevertheless, we speculate that it is produced by hot gas centered on the Galactic bulge. If so, we can speculate further and relate it to the bulge component of the ultrahot $T \sim 10^8$ K gas mentioned above in §3.1. This gas is highly overpressured and should expand. If it expands adiabatically, $T \propto \rho^{2/3}$, or $T \propto R^{-2}$. The X-ray emission from the large patch centered on the Galactic center is roughly four times the size of the 6.7-Kev Fe line emission and is roughly 16 times colder, so this is consistent with the speculation. Would this speculation hold up under scrutiny of the experts?

4. PARTICULARLY STRIKING smaller-scale OBSERVATIONS

4.1. Extensive CO 2-1 and 1-0 maps and their analysis.

These maps, presented by Sanders (verbal paper), Handa et al. (poster paper), and Sakamoto (poster paper), are valuable because they offer the possibility of deriving H_2 densities and temperatures using an excitation analysis of the two transitions. This has, of course, been done in the past. But these more extensive maps allow such analyses to be applied to a large number of clouds.

The result is that all molecular clouds are highly clumped. This is consistent with several previous observational results on individual clouds. The widespread existence of clumping, by itself, would not be particularly important. However, it has ramifications for interstellar chemistry and heating. The clumping means that the surface-to-volume ratio is much larger than previously thought. In particular, it can be easy for UV radiation, which dissociates molecules, and cosmic rays, which initiate ion-molecule reactions, to reach the innards of a porous, highly clumped cloud. Many current models of the chemical structure within dense interstellar clouds do not include such clumping.

4.2. High-z CO.

Dame (verbal paper) showed high-sensitivity b-scans of CO 1-0 emission at $\ell = 25°, 30°, 40°, 50°$. These show, first, the standard molecular layer with scale height ~ 90 pc. But in addition, they show that weaker CO emission extends up to several hundred pc, comparable to the height of denser parts of the Lockman H I layer. Perhaps these are the Galactic-interior analog of the MBM (Magnani, Blitz, and Mundy 1985) local high-latitude molecular clouds. These high-latitude clouds are neither optically thick in CO nor gravitationally bound, so one cannot use the standard ratio of CO/H_2 to determine the H_2 abundance.

It is curious that two of these scans, at $\ell = 30°$ and $40°$, happen to lie on prominent worms! On the one hand, this makes Dame's current results less than generally applicable—but on the other, it offers a chance to study the molecular structure of the worms.

4.3. Tiny-scale H I structure.

Frail *et al.* (poster paper) used pulsars, and others (Dieter, Welch, and Romney 1976; Diamond *et al.* 1989) have used VLBI, to probe small-scale H I structure. Frail finds changes in H I column density $\Delta N_{HI} \sim 10^{20}$ cm^{-2} over transverse length scales of about 100 Astronomical Units (A.U.). If the length of the cloud along the line of sight is a factor f times the transverse distance, then $n_{HI} \approx 3 \times 10^4/f$ cm^{-3}. This implies a high pressure, and one wonders how it might be attained.

In principle, it isn't terribly difficult to produce such a high density and pressure. A 25 km s^{-1} radiative (roughly isothermal)—but nonmagnetic—shock hitting a 'standard interstellar cloud' having $n_{HI} \sim 30$ cm^{-3} will do (with $f = 1$; larger values, which are likely, make it easier). Such shocks are common: they are routinely observed in H I expanding shells, and are easily produced by individual supernovae. But if the shock is magnetic—as most radiative shocks are—most of the post-shock pressure is magnetic, which might make it difficult to build up such a large gradient in column density. (I say 'might' because we usually observe highly radiative shocks to be highly clumped, probably because of thermal instabilities during the cooling).

A different type of phenomenon, the 'extreme scattering event', is caused by refractive lensing by ionized blobs (Fiedler *et al.* 1987; Fiedler *et al* 1988); these also require large column densities and pressures. The properties of the lenses depend on distance. For 1 kpc distance and a lens size of 2 A.U., the electron column density $N_e \sim 5 \times 10^{17}$ cm^{-2}. This can be produced by a nonradiative magnetic shock with velocity $\lesssim 100$ km s^{-1} hitting a cloud with $n \sim 10$ cm^{-3} (Clegg, Chernoff, and Cordes 1988) or by a lesser nonmagnetic shock. Again, such shocks are plentiful in the interstellar medium because of supernovae. Tiny-scale H I structure has about the same gradient of column density, and perhaps it is simply the cooled version of the tiny-scale ionized gas.

4.4. Faraday Rotation in the outer Galaxy.

Faraday Rotation in the Galaxy was not discussed at the meeting, but the topic came up in a comment by Clegg in the discussion of a paper—and I was struck by the incongruity of the data and our expectations. Clegg *et al.* (1992) have shown that the *mean* large-scale Rotation Measure (RM) through the outer Galaxy near $\ell = 90°$ is large, about -800 rad m^{-2}. This means that the product $\langle n_e BL \rangle \approx -1000$ cm^{-3} μG pc. Near the Sun, $\langle n_e \rangle \approx 0.03$ cm^{-3} (the value of Lyne, Manchester, and Taylor 1985, modified for a revised Solar Galactocentric distance of ~ 8 kpc) and $\langle B \rangle \approx +1.6$ μG (Rand and Kulkarni 1989; by convention, the negative RM corresponds to a positive B). To build up the required $\langle n_e BL \rangle$ product requires $L \sim 21$ kpc—*if* $\langle n_e \rangle$ and $\langle B \rangle$ do not decrease significantly outside the Solar circle, which is unlikely, at least for $\langle n_e \rangle$. The synchrotron emissivity remains high out to large Galactocentric radii, which implies that the magnetic field does not decrease much with radius. But the thermal electrons are produced—presumably—by photons from young, massive stars, and these stars are much less populous at large radii. Either Clegg *et al*'s derived RM's are not representative—which is hard to believe, because they are averages over many randomly-placed extragalactic sources; or the electrons and magnetic field are highly correlated, thus making $\langle n_e B \rangle > \langle n_e \rangle \langle B \rangle$; or our understanding of the interstellar medium is deficient.

4.5. Finally, a point for specialists: C II cooling and photoelectric heating.

Two companion poster papers examine the correlation between, for one, C II 157 μm line emission (Bock *et al.*, poster paper) and, for the other, dust IR emission (Matsuhara *et al.*, poster paper) versus H I column density for a particular isolated high-latitude cloud. Observing an isolated cloud is useful because it removes ambiguity concerning foreground or background material. However, the cloud's high-latitude location may mean that it is not representative of most interstellar gas.

4.5.1. The C II cooling per H atom.

Bock *et al.* determined that the C II line emits 2.3×10^{-26} erg s^{-1} H-atom^{-1}. Because the C II line is essentially the only coolant, this is also a measure of the total heating rate. This value is a full factor of *six* smaller than the traditional value of 1.4×10^{-25} erg s^{-1} H-atom^{-1}, which was obtained many years ago from UV absorption data using the *Copernicus* satellite (Pottasch, Wesselius, and van Duinen 1979, hereafter PWvD). A result similar to that of Bock *et al.* has been obtained for a thin high-latitude cloud towards 3C273 from UV absorption measured with Space Telescope by Savage *et al.* (1993).

In the verbal version of this paper, I assumed that the six-times-smaller cooling per H-atom of Bock *et al.* was a general result, and that it resulted from the uncertainties in the slightly-less-direct UV determinations. However, upon reflection I believe that this assumption is probably unfair both to PWvD and to the people who were associated with the *Copernicus* satellite, which produced generally excellent data. Rather, let us assume that the difference is real, and inquire what it means.

First, some basic physics (Hollenbach and McKee 1989). Consider the C II line intensity from the CNM and WNM, which we assume to have temperatures of 100 K and 7000 K, respectively. For equal column densities of CNM and WNM, the ratio of line intensities $I_{WNM}/I_{CNM} \approx 3.3 n_{WNM}/n_{CNM}$. Invoking pressure equilibrium ($nT = const$), which defines $n_{WNM}/n_{CNM} = 1/70$, we obtain $I_{WNM}/I_{CNM} \approx 0.05$. Because the total mass of WNM is comparable to that of CNM, we conclude that the WNM contributes only negligibly to the observed C II line emission.

Next, consider the C II line intensity from the CNM and the WIM. The ratio of electron to atomic collisional rates $C_e/C_{HI} \approx 350 T_2^{-0.57} n_e/n_{HI}$, where T_2 is the temperature in units of 100 K. For the CNM, $T_2 \sim 1$ and electrons come from ionization of Carbon so that $n_e/n_{HI} \lesssim 4 \times 10^{-4}$; thus H atoms totally dominate. But for the WIM, which is nearly fully ionized with $T_2 \sim 70$, electrons dominate the C II excitation. For equal column densities of CNM and WIM, the line intensity ratio $I_{WIM}/I_{CNM} \approx 77 n_e/n_{HI}$, or for pressure equilibrium $I_{WIM}/I_{CNM} \approx 0.55$.

This result for I_{WIM}/I_{CNM} is interesting because when we look vertically out of the Galaxy the column densities of WIM and CNM are roughly equal. Thus a significant fraction of the total Galactic C II line luminosity comes from the WIM—but the fraction should be smaller than the above-derived value of 0.55 because the scale height of the WIM is about ten times that of the CNM, and the pressure (and thus n_e) decreases with z. The difference in scale height also means that the average density of the WIM at $z = 0$ is much smaller than that of the CNM. Thus, when we look along any one line of sight near the Galactic plane, the column density of the CNM is much larger than that of the

WIM and most of the C II line comes from the CNM.

Thus the C II sampled by both PWvD and Bock *et al.* should have resided predominantly in the CNM. To resolve the discrepancy between PWvD and Bock *et al.*, we must look for differences between the CNM sampled by them. On a per-atom basis there are two parameters that determine the C II line intensity: the volume density n and the temperature T. If we invoke our usual argument of pressure equilibrium, then these two variables are inversely related. The cooling increases with temperature for any thermally stable phase (Field 1965), which the CNM is. Thus Bock *et al.*'s lower C II cooling rate must mean that the temperature in the cloud he observed is lower, and the density correspondingly higher, than in those sampled by PWvD.

This conclusion, that the C II cooling rate per H-atom is inversely correlated with temperature, can be gleaned directly from the data of PWvD—if one blindly forges ahead and assumes, first, that the UV data are absolutely accurate and, second, that a statistical correlation observed by radio astronomers applies individually to these few samples of PWvD. First, examination of PWvD's results reveals that the observed cooling rates are larger for stars with smaller total column density. In particular, the four stars having $N_{HI} < 1.8 \times 10^{20}$ cm^{-2} have above-average cooling rates, while the five having $N_{HI} > 3.7 \times 10^{20}$ cm^{-2} have below-average rates. Second, the column densities for the four low-density stars all lie in the range where the $T - \tau$ relation, derived from 21-cm line data, shows that the cloud temperatures increase with decreasing N_{HI} (see review by Kulkarni and Heiles 1987). So this rather thin argument points toward reassuring result that the C II cooling rate decreases with decreasing CNM temperature.

The current observations, then, suggest that the C II cooling rate is measurably affected by the conditions in the cloud: the cooling decreases with decreasing temperature. We need more confirmatory data!

4.5.2. The efficiency of photoelectric heating by grains.

The very same grains that heat the gas by absorbing photons and producing photoelectrons also emit the infrared radiation observed by Matsuhara *et al.*, by turning most of the photon energy into heat. Thus, the ratio of the C II line intensity to the frequency-integrated IR intensity is a direct determination of the efficiency of photoelectric heating by grains. From the data of Bock *et al.* and Matsuhara *et al.*, the observed efficiency turns out to be a few tenths of a per cent. I believe that this experiment is the best direct determination of this because the cloud is isolated and there is no foreground/background ambiguity; however, these very conditions may make the cloud unusual, and it is important to have better statistics. Happily, a similar number for efficiency was derived by Bennett (verbal paper, §2.5) from large-scale maps, and this provides statistical accuracy.

We compare this observationally determined efficiency with the theoretically estimated efficiencies, which have been recently reviewed by Hollenbach (1989). Theorists needed high efficiencies, in the range of about 3%, to reproduce the PWvD result. However, independent estimates based on laboratory results produced lower efficiencies, less than 1% (Draine 1978). Heating by PAH's increases the total efficiency to $> 1\%$. While this isolated cloud is in line with easier efficiencies, it seems that many clouds require the higher efficiencies.

4.5.3. Prospects...

Under current theory, the C II cooling rates per H-atom, the cloud temperature, and the PAH content are the three essential parameters that describe the heating and cooling in the CNM. They should show interesting correlations. But it is not easy to obtain the data for an interestingly large sample of clouds. C II cooling can be measured both directly, from the 157 μm line, and indirectly, from UV absorption; the latter is easy with Space Telescope, but is restricted to directions where suitable background stars exist. The cloud temperature can be measured from 21-cm line emission/absorption observations; this is also easy, but is restricted to directions where suitable background radio sources exist. The PAH content can be measured from the $\sim$ 12 μm excess in IR emission; again, this is easy for clouds that produce sufficiently strong IRAS signatures. The intersection of these 'easy' cases may be too small, and we may need to pursue more difficult approaches to obtain a sufficiently large sample of clouds.

5. CONCLUSION

The stellar dynamicists at this meeting have remarked that the basic requirement for observational data has not changed for two decades: accurate astrometry provides proper motions and parallaxes. Further observational and interpretive progress hinges on obtaining vast quantities of such data, and some should soon become available because of the *Hipparchus* satellite.

The situation is completely different for the interstellar medium. At this meeting we have been presented with new maps and new data sets at wavelengths spanning the radio to the gamma ray, and each different data set elucidates a different aspect of the interstellar medium. We cannot understand the interstellar medium without understanding all of its constituents, because they all mutually interact. My summary here, which only scratches the surface, illustrates this interdependence.

At this meeting we have gotten our first glimmer of some important new data sets which cover only pieces of the sky, for example the C II 157 μm line. These provide important new information. However, it is not enough to observe just small pieces of the sky, and our appetite is whetted for more! We need maps of large sections of sky because different phases, morphological structures, and types of interaction occur in different places. This meeting has brought us 'back to the Galaxy'; now it is time for us astronomers to get 'back to our telescopes and laboratories'!

ACKNOWLEDGEMENTS

This work was supported in part by NSF grant AST91-23362. I would like to thank Chris McKee for his comments on early draft of this manuscript. It is a pleasure to thank the conference organizers for travel support and, most especially, for a truly excellent meeting.

REFERENCES

Begelman, M.C., & McKee, C.F. 1990, ApJ, 358, 375

Blitz, L., Bloemen, J.B.G.M., Hermsen, W., & Bania, T.M. 1985, A&A, 143, 267

Boulares, A., & Cox, D.P. 1990, ApJ, 365, 544

Clegg, A.W., Chernoff, D.F., & Cordes, J.M. 1988, in *Radio Wave Scattering in the Interstellar Medium* (AIP Conference Proceedings 174), ed. J.M. Codes, B.J. Rickett, and D.C. Backer, p. 174

Clegg, A.W., Cordes, J.M., Simonetti, J.H., & Kulkarni, S.R. 1992, ApJ, 386, 143

Colomb, F.R., Pöppel, W.G.L., & Heiles, C. 1980, A&A Suppl., 40, 47

Diamond, P.J., Goss, W.M., Romney, J.D., Booth, R.S., Kalberla, P.M.W., & Mebold, U. 1989, ApJ, 347, 302

Dieter, N.H., Welch, W.J., & Romney, J.D. 1976, ApJ, 206, L113

Draine, B.D. 1978, ApJ Suppl., 36, 595

Fiedler, R., Simon, R., Johnston, K., Dennison, B., & Hewish, A. 1988, in *Radio Wave Scattering in the Interstellar Medium* (AIP Conference Proceedings 174), ed. J.M. Codes, B.J. Rickett, and D.C. Backer, p. 150

Fiedler, R.L., Dennison B., Johnston, K.H., & Hewish, A.. 1987, Nature, 326, 675

Field, G.B. 1965, ApJ, 142, 531

Heiles, C. 1984, ApJ Suppl., 55, 585

Heiles, C. 1989, ApJ, 336, 808

Hollenbach, D.J. 1989, in *Interstellar Dust* (IAU Symposium No. 135), ed. L.J. Allamandola and A.G.G.M. Tielens, p. 227

Hollenbach. D, & McKee, C.F. 1989, ApJ, 342, 306

Jakobsen, P. and Kahn, S.M. 1986, ApJ, 309, 682

Koyama, K., *et al.* 1989, Nature, 339, 603

Kulkarni, S.R., & Heiles, C. 1987, in *Interstellar Processes*, ed. D.J. Hollenbach and H.A. Thronson, Jr., p. 87

Lyne, A.G., Manchester, R.N., & Taylor, J.H. 1985, MNRAS, 213, 613

Magnani, L., Blitz, L., & Mundy, L. 1985, ApJ, 295, 402

McKee, C.F., & Ostriker, J.P. 1977, ApJ, 1996, 565

Norman, C.A. & Ikeuchi, S. 189, ApJ, 345, 372

Ozernoy, L., Titarchuk, L., & Ramaty, R. 1993, preprint

Paresce, F. 1984, AJ, 89, 1022

Petuchowski, S.J., & Bennett, C.L. 1993, ApJ, 10 March 1993

Pottasch, S.R., Wesselius, P.R., & van Duinen, R.J. 1979, A&A, 74, L15

Rand, R.J., and Kulkarni, S.R. 1989, ApJ, 343, 760

Reich, W., Fürst, E., Reich, P., & Reif, K. 1990, A&A Suppl., 85, 633

Reynolds, R.. & Ogden, P.M. 1979, ApJ, 229, 942

Savage, G.D., Lu, L., Weymann, R.J., Morris, S.L., & Gilliland, R.L. 1993, preprint

Snowdon, S.L., Cox, D.P., McCammon, D., & Sanders, W.T. 1990, ApJ, 354, 211

Strong, A.W., *et al.* 1988, A&A, 207, 1
Verschuur, G.L., & Schmelz, J.T. 1989, AJ, 98, 267

Appendix A

Conference Program

MONDAY, OCTOBER 12, 1992

0845   **1.**   **Introductory Session**   **S. Holt**, chair

F. Kerr   The Milky Way: A Typical Spiral Galaxy?   45 min

J. Ostriker   Current Issues in the Study of the Large Scale Properties
of the Milky Way   45 min

1000   **Coffee**

1030   **2.**   **Galactic Center**   **C. Townes**, chair

M. Morris   The Central Engine and Activity at the Galactic Cente   30 min

M. Rieke   Stellar Distributions Near the Galactic Center   30 min

J. Marr   The Distribution and Excitation of Molecular Gas in the Galactic
Center: CO Observations from -170 to -250 km/s   5 min

1230   **Lunch**

1400   **3.**   **Bulge/Bar: Stars, Dynamics and ISM**   **V. Rubin**, chair

L. Blitz   Photometric Evidence for a Bar   30 min

J. Binney   The Dynamics of the Bulge/Bar   30 min

K. Koyama   X-Ray Observations of the Galactic Bulge   30 min

1600   **Tea**

1630   **4.**   **Our Local Environment**   **N. Kassim**, chair

D. McCammon   The Hot Ionized Medium   30 min

R. Reynolds   The Warm Ionized Medium   30 min

D. York   Composition of the Local ISM   30 min

R. Warwick   The EUV source population and the local bubble   10 min

1830   **Informal evening poster session with light refreshments**

TUESDAY, OCTOBER 13, 1992

0830	5.	Large Scale IR Emission	E. Dwek, chair

E. Wright   Cold Dust in the ISM   30 min

M. Hauser   IR Emission from Stars and Dust   30 min

T. Nakagawa   Far-Infrared [CII] Line Survey of the Galaxy   10 min

S. Odenwald   The Cygnus-X Region: An IRAS View   5 min

1000	Coffee

1030	6.	Large Scale ISM Emission	F. Verter, chair

T. Dame   Distribution and Kinematics of the Interstellar Gas   30 min

C. Bennett   Cooling of the Interstellar Gas   30 min

D. Sanders   A Multitransition CO & ^{13}CO Survey of the Plane   5 min

1200	Lunch

1400	7.	Stellar Dynamics	I. King, chair

D. Spergel   The Shape of the Disk   30 min

M. Weinberg   Distribution of Stars in the Disk   30 min

R. Wyse   Dynamical Evolution of the Galaxy   30 min

E. Kokubo   Velocity Evolution of Disk Stars Due to Gravitational
                               Scattering by Clouds   6 min

E. Griv   Relaxation and Stability of the Disk of the Galaxy   6 min

1600	Tea

1630	8.	Energetic Particles and Fields	N. Gehrels, chair

R. Ramaty   Implications of Diffuse < 1 Mev Gamma Rays   25 min

H. Bloemen   Implications of Diffuse > 1 Mev Gamma Rays   25 min

W. Purcell   OSSE Observations of Diffuse Galactic γ-Ray Emission   10 min

R. Sunyaev   Observations of Diffuse Emissions by GRANAT   10 min

P. Sreekumar   What Can the Magellanic Clouds Tell Us About
Cosmic Rays in the Milky Way?   2 min

D. Hartmann   Galactic Gamma-Ray Emission from Pulsars   2 min

1830   **Conference Banquet**   **Speaker: M. Harwit**

WEDNESDAY, OCTOBER 14, 1992

0800   9.   The Outer Galaxy   **M. Roberts**, chair

M. Merrifield   Rotation, Scale Height and Mass   30 min

L. Sparke   The Warp of the Milky Way   30 min

P. Schechter   Evidence for Dark Matter   30 min

G. Byrd   Major Dynamical Events in the Local Group from
15×10^9 B.C. to 15×10^9 A.D.   10 min

1000   Coffee

1030   10.   The Corona and Halo   **R. McCray**, chair

C. McKee   Fountains, Bubbles and Hot Gas   30 min

R. Manchester   Globular Clusters and Pulsars   30 min

W. Waller   Worms or Froth? Fine Structure in Far-IR Milky Way   10 min

W. Reach   Galactic Worms: Infrared and Radio Observations   10 min

1200   Lunch

1330   11.   Summary Session   **S. Holt**, chair

S. Tremaine   Rapporteur: Structure and Dynamics   45 min

C. Heiles   Rapporteur: Structure and Energetics   45 min

Appendix B

List of Attendees

THE 3rd ANNUAL
ASTROPHYSICS CONFERENCE IN MARYLAND
"BACK TO THE GALAXY"
University of Maryland
Center of Adult Education
College Park, Maryland

12 - 14 October 1992

List of Participants

Acord, Jerry	University of Wisconsin-Madison
Adler, David	NRAO
Aizenman, Morris	National Science Foundation
Albert, Elise	U.S. Naval Academy
Allen, R.J.	STScI
Arendt, Rick	Applied Research Corporation
Arnaud, Keith	NASA/Goddard
Audley, Damian	UMD/NASA Goddard
Banday, Tony	USRA/NASA Goddard
Baranov, Alexander	Inst. for Theoretical Astronomy
Barnes, Bill	M.I.T.
Barrett, Paul	USRA/NASA-HEASARC
Baum, Stefi	STScI
Beckman, John	Instituto Astrofisica Canarias
Bennett, Chuck	NASA Goddard
Berriman, Bruce	General Sciences Corporation
Bicay, Michael	NASA Headquarters
Biello, Joseph	Columbia University
Binney, James	Dept. of Theoretical Physics
Bisnovatyi-Kogan, Gena	Space Research Institute
Blitz, Leo	University of Maryland

Bloemen, Hans	Space Research, Leiden
Bock, Jamie	UC/Berkeley
Boggess, Nancy	NASA Goddard
Boldt, Elihu	NASA/Goddard
Borkowski, K. J.	University of Maryland
Boulanger, Francois	IAS-Orsay
Bronfman, Leonardo	Universidad de Chile
Buss Jr., Richard	The Johns Hopkins University
Byrd, Gene	University of Alabama
Carey, Sean	Rensselaer Polytechnic Institute
Casey, Sean	NRC/NASA Goddard
Chen, Wesley	The Johns Hopkins University
Cheng, Ed	NASA Goddard
Cheung, Cynthia	NASA Goddard
Chi, L. K.	U. S. Naval Academy
Chiu, Hong-Yee	NASA Goddard
Christian, Eric	USRA/NASA Goddard
Clark, Frank	Phillips Laboratory
Clegg, Andrew	Naval Research Laboratory
Cline, Thomas	NASA Goddard
Corcoran, Michael	USRA/NASA Goddard
Corliss, Charles	Forest Hills Laboratory
Corliss, Edith	Forest Hills Laboratory
Dame, Thomas	Center for Astrophysics/JHU
Danly, Laura	STScI
Davidsen, Arthur	Center for Astrophysics/JHU
D'Amario, James	Harford Community College

De Amici, Giovanni	SSL-LBL
Deane, Jim	University of Hawaii
DeGeus, Eugene	University of Maryland
Dettmar, Ralf	ESA/STScI
Digel, Seth	Center for Astrophysics
Dinerstein, Harriet	University of Texas at Austin
Dixon, Van	The Johns Hopkins University
Doi, Yasuo	Inst. of Space & Astro. Science
Durouchoux, Plulitte	CE-SACLAY
Dwek, Eli	NASA Goddard
Elitzur, Moshe	University of Kentucky
Felten, James	NASA Goddard
Ferrara, Andrea	STScI
Fichtel, Carl	NASA Goddard
Fischer, Marc	UC/Berkeley
Frail, Dale	NRAO-VLA
Freudenreich, Henry	Hughes STX Corporation
Fridman, Alexei	Russian Academy of Sciences
Frost, Kenneth	NASA Goddard
Fuchs, Burkhard	Astronomisches Rechen-Institut
Gammie, Charles	University of Virginia
Gehrels, Neil	NASA Goddard
Gerhard, Ortwin	Landessternwarte, Heidelberg
Gezari, Dan	NASA Goddard
Gould, Andrew	Inst. for Advanced Study
Greyber, Howard	Greyber Associates
Griv, Evgenij	Ben-Gurion Univ. of the Negev

Gull, Ted	NASA Goddard
Gurksy, Herbert	NRL
Hanami, Hitoshi	Iwate University
Handa, Toshihiro	University of Tokyo
Hartman, Bob	NASA Goddard
Hartmann, Dap	Leiden Observatory
Hartmann, Dieter	Clemson University
Harwit, Martin	National Air & Space Museum
Hasan, Hashima	STScI
Hauser, Michael	NASA Goddard
Hawkins, Isabel	UC/Berkeley
Haywood, Misha	Observatoire de Besancon
Heaton, Harold	Johns Hopkins University/APL
Heiles, Carl	University of Hawaii
Herbstmeier, Uwe	Radioastronomisches Institute
Hinshaw, Gary	USRA/NASA Goddard
Hirao, Takanori	Nagoya University
Holt, Steve	NASA Goddard
Howe, John	University of Maryland
Hufnagel, Beth	Univ. California-Santa Cruz
Hunter, Chris	Florida State University
Hunter, Stan	NASA Goddard
Jackson, Peter	Hughes STX Corporation
Jahoda, Keith	NASA Goddard
Jalota, Lalit	USRA/NASA Goddard
Jonas, Justin	Rhodes University
Jones, Frank	NASA Goddard

Jones, Mark	Queen Mary & Westfield College
Kammeyer, Peter	U.S. Naval Observatory
Kassim, Namir	NRL
Kawada, Mitsunobu	Nagoya University
Kazanas, Demos	NASA Goddard
Kelsall, Thomas	NASA Goddard
Kerr, Frank	UMD/USRA
Khorujii, Oleg	Russian Academy of Sciences
Kimble, Randy	NASA Goddard
King, Ivan	UC Berkeley
Koch, Timothy	Univ. California-Santa Barbara
Kokubo, Eiichiro	University of Tokyo
Kondo, Masa-aki	Senshu University
Kondo, Yoji	NASA Goddard
Koyama, Katsuji	Kyoto University
Kruk, Jeffrey	The Johns Hopkins University
Kuchar, Thomas	Phillips Lab
Kuijken, Konrad	Center for Astrophysics
Kundic, Tomislav	Princeton University
Kuntz, K. D.	STScI
Kurfess, Jim	NRL
Kwok, Ping-Wai	NRC/NASA Goddard
Ladd, Ned	University of Hawaii
Lee, Geunho	The Johns Hopkins University
Leung, S. M.	University of Hong Kong
Lis, Darek	CALTECH
Loewenstein, Michael	USRA/NASA Goddard

Long, Kevin	SUNY Brockport
Lui, Ji-Cheng	The Johns Hopkins University
Lyakhovich, Valya	Russian Academy of Sciences
McCammon, Dan	University of Wisconsin
McCray, Richard	University of Colorado
McKee, Christopher	University of California
Machacek, Marie	Northeastern University
Madau, Piero	STScI
Madejski, Greg	NASA Goddard
Manchester, Richard	CSIRO/Australia Telescope National Facility
Mangus, John	NASA Goddard
Maran, Stephen	NASA Goddard
Marr, Jonathan	Haverford College
Marshall, Frank	NASA Goddard
Mather, John	NASA Goddard
Matsuhara, Hideo	Nagoya University
Mattox, John	Compton Observatory
Mead, Kathryn	Union College
Merrifield, Michael	C.I.T.A. McLennan Labs.
Metzger, Mark	MIT
Mitchell, John	USRA/NASA Goddard
Moffat, Anthony	Universite de Montreal
Moreno, Edmundo	Universidad Nacional de Mexico
Morris, Mark	UCLA
Mushotzky, Richard	NASA Goddard
Nakagawa, Takao	Inst. of Space & Astro. Science
Neff, Susan	NASA Goddard

Neufeld, David	The Johns Hopkins University
Norman, Michael	Univ. Illinos/Urbana-Champaign
Odenwald, Sten	BOMI
Oelfke, William	Univ. of Central Florida
Oka, Tomoharu	University of Tokyo
Ormes, Jonathan	LHEA/NASA Goddard
Ostriker, Jeremiah	Princeton University Obs.
Ozernoy, Leonid	NASA Goddard
Petre, Robert	NASA Goddard
Petrosian, Vahe	Stanford University
Petuchowski, Sam	NASA Goddard
Pfenniger, Daniel	Geneva Observatory
Philip, A.G. Davis	Van Vleck Obs.
Pier, Jeff	U.S. Naval Observatory
Polyachenko, Valerij	Russian Academy of Sciences
Praton, Elizabeth	University of Massachusetts
Purcell, William	Northwestern University
Ramaty, Reuven	NASA Goddard
Ratnatunga, Kavan	The Johns Hopkins University
Reach, William	NASA Goddard
Reynolds, Ronald	University of Wisconsin
Rho, Jeonghee	University of Maryland
Riegler, Guenter	NASA Headquarters
Rieke, Marcia	University of Arizona
Roberts, Morton	NRAO
Robin, Annie	Observatoire de Besancon
Rohlfs, K.	Ruhr University

Roman, Nancy	Hughes/STX
Rots, Arnold	USRA/NASA Goddard
Rubin, Vera	Carnegie Inst. of Washington
Sakamoto, Seiichi	University of Tokyo
Sanders, David	University of Hawaii
Schechter, Paul	MIT
Schlegel, Eric	NASA Goddard
Schweizer, Francois	Carnegie-DTM
Seitzer, Patrick	University of Michigan
Sellwood, Jerry	Rutgers University
Serlemitsos, Peter	NASA Goddard
Shafer, Rick	NASA Goddard
Shaw, Lisa	University of Hawaii, Manoa
Silberberg, Rein	USRA/NRL
Silverberg, Robert	NASA Goddard
Skard, John	Hughes STX Corporation
Smith, Bruce	NASA Ames Research Center
Smith, Howard	SI-NASM
Sodroski, Tom	Applied Research Corporation
Sokolowski, James	STScI
Sparke, Linda	Washburn Observatory
Sparks, William	STScI
Spergel, David	Princeton University Obs.
Spicer, Dan	NASA Goddard
Spiesman, Bill	NASA Goddard
Sreekumar, Parameswaran	USRA/NASA Goddard
Stecher, Ted	NASA Goddard

Stiller, Bertram	
Stone, Jim	University of Maryland
Straizys, Vytas	Union College
Sunyaev, R.	Academy of Science of the USSR
Swank, Jean	NASA Goddard
Sygnet, Jean Francois	CNRS-IAP
Szymkowiak, Andrew	NASA Goddard
Tewari, Krishna	Hughes STX Corporation
Thompson, Dave	NASA Goddard
Titarchuk, Lev	NASA Goddard
Townes, Charles	UC/Berkeley
Trapero, Joaquin	Inst. de Astrofisica de Canarias
Trasco, John	University of Maryland
Tremaine, Scott	CITA
Trimble, Virginia	UC-Irvine/UMD
Turbide, Luc	Uiversite de Montreal
Tyler, Pat	HEASARC/CSSI
Tylka, Allan	NRL
Tytler, David	UC/San Diego
Ueono, Munetaka	University of Tokyo
Van Woerden, Hugo	Kapteyn Institute
Vayner, Boris	
Velusamy, T.	Radio Astronomy Centre India
Verter, Fran	USRA/NASA Goddard
Wall, William	NASA Goddard
Waller, William	NASA Goddard
Warwick, Bob	University of Leicester

Watson, William	University of Illinois
Webber, William	New Mexico State University
Weinberg, Martin	University of Massachusetts
White, Nick	HEASARC
Whittle, Mark	University of Virginia
Williams, Harold	Montgomery College
Wilson, Thomas	NASA Johnson Space Center
Wright, Edward	UCLA
Wu, Chi-Chao	Computer Sciences Corp./STScI
Wyse, Rosie	The Johns Hopkins University
York, Donald	University of Chicago
Yuan, Chi	City College of New York
Zhang, Weiping	NASA Goddard
Zhao, Hong-Sheng	Columbia University

Appendix C

Physical
Constants

TABLE OF PHYSICAL CONSTANTS

CONSTANT	SYMBOL	MKS	CGS	OTHER
speed of light	c	$3.00 \cdot 10^8$ m/s	$3.00 \cdot 10^{10}$ cm/s	(2.997925)
electron charge	e	$1.60 \cdot 10^{-19}$ coul	$4.80 \cdot 10^{-10}$ esu	
Planck constant	h	$6.63 \cdot 10^{-34}$ J·s	$6.63 \cdot 10^{-27}$ erg·s	
	$\hbar$	$1.05 \cdot 10^{-34}$ J·s	$1.05 \cdot 10^{-27}$ erg·s	
	hc	$1.99 \cdot 10^{-25}$ J·m	$1.99 \cdot 10^{-16}$ erg·cm	
	$\hbar c$	$3.15 \cdot 10^{-26}$ J·m	$3.15 \cdot 10^{-17}$ erg·cm	200 MeV·fm
Boltzmann constant	k	$1.38 \cdot 10^{-23}$ J/K	$1.38 \cdot 10^{-16}$ erg/K	$8.6 \cdot 10^{-5}$ eV/K
	k/h	$2.08 \cdot 10^{10}$ s^{-1}/K	$2.08 \cdot 10^{10}$ s^{-1}/K	
	k/hc	67.5 m^{-1}/K	0.675 cm^{-1}/K	
Gravitational constant	G	$6.67 \cdot 10^{-11}$ N·m^2/kg^2	$6.67 \cdot 10^{-8}$ dy·cm^2/gm^2	
Gas constant	R	8.314 J/K·mole	$8.31 \cdot 10^7$ erg/K·mole	
Avogadro's number (= R/k)	N	$6.02 \cdot 10^{26}$ amu/kg	$6.02 \cdot 10^{23}$ amu/kg	$6 \cdot 10^{23}$ molecules/mole
electron mass	m_e	$9.11 \cdot 10^{-31}$ kg	$9.11 \cdot 10^{-28}$ gm	0.51 MeV
proton mass	M_p	$1.67 \cdot 10^{-27}$ kg	$1.67 \cdot 10^{-24}$ gm	938 MeV
neutron mass	M_n	$1.67 \cdot 10^{-27}$ kg	$1.67 \cdot 10^{-24}$ gm	939 MeV
pion mass (=270·m$_e$)	m_π	$2.46 \cdot 10^{-28}$ kg	$2.46 \cdot 10^{-25}$ gm	140 MeV
muon mass (=207·m$_e$)	m_μ	$1.89 \cdot 10^{-28}$ kg	$1.89 \cdot 10^{-25}$ gm	106 MeV
classical elect radius (=e^2/mc^2)	r_c	$2.82 \cdot 10^{-15}$ m	$2.82 \cdot 10^{-13}$ cm	
Compton wavelength (=h/mc)	λ_c	$2.43 \cdot 10^{-12}$ m	$2.43 \cdot 10^{-10}$ cm	0.02 Å

Quantity	Symbol	CGS	SI	Other
Thomson cross-section	σ_T	$6.65\cdot10^{-25}$ cm²	$6.65\cdot10^{-29}$ m²	
Planck length $(=\sqrt{\hbar G/c^3})$	l_{Pl}	$1.61\cdot10^{-33}$ cm	$1.61\cdot10^{-35}$ m	
Planck time $(=\sqrt{\hbar G/c^5})$	t_{Pl}	$5.39\cdot10^{-44}$ s	$5.39\cdot10^{-44}$ s	
Planck density $(=c^5/\hbar G^2)$	ρ_{Pl}	$5.16\cdot10^{93}$ gm/cm³	$5.16\cdot10^{96}$ kg/m³	
Bohr radius $(=\hbar^2/me^2)$	r_B	$0.53\cdot10^{-8}$ cm	$0.53\cdot10^{-10}$ m	0.5 Å
Fine structure constant $(=e^2/\hbar c)$	α	$7.30\cdot10^{-3}$	$7.30\cdot10^{-3}$	1/137
Bohr magneton $(=e\hbar/2m_ec)$	μ_B	$9.27\cdot10^{-21}$ erg/gauss	$9.27\cdot10^{-24}$ J/T	
Nuclear magneton $(=e\hbar/2M_pc)$	μ_N	$5.05\cdot10^{-24}$ erg/gauss	$5.05\cdot10^{-27}$ J/T	
Permittivity of vacuum	ε_o		$8.85\cdot10^{-12}$ fd/m	$1/4\pi\varepsilon_o=9.0\cdot10^{9}$
Permeability in vacuum	μ_o		$4\pi\cdot10^{-7}$ Hen/m	
Stefan-Boltzmann constant	σ	$5.67\cdot10^{-5}$ erg/s·cm²·K⁴	$5.67\cdot10^{-8}$ W/m²·K⁴	
Rydberg $(=m_ee^4/2\hbar^2)$	R_∞	$2.18\cdot10^{-11}$ erg	$2.18\cdot10^{-18}$ J	13.6 eV
1 amu		$1.66\cdot10^{-24}$ gm	$1.66\cdot10^{-27}$ kg	931.5 MeV
1 calorie		$4.19\cdot10^{7}$ erg	4.19 J	
1 year		$3.16\cdot10^{7}$ s	$3.16\cdot10^{7}$ s	
1 atmosphere		$1.01\cdot10^{6}$ dyne/cm²	$1.01\cdot10^{5}$ N/m², $1.01\cdot10^{5}$ Pascal	14.2 lbs/in², 760 Torr
1 eV		$1.6\cdot10^{-12}$ erg	$1.6\cdot10^{-19}$ J	
		$1.24\cdot10^{-4}$ cm	$1.24\cdot10^{-6}$ m	11,605 K
		10^4 gauss	1 Tesla	

ASTROPHYSICAL CONSTANTS

CONSTANT	SYMBOL	MKS	CGS	OTHER
astronomical unit	AU	$1.50 \cdot 10^{11}$ m	$1.50 \cdot 10^{13}$ cm	
	AU/year			4.74 km/s
parsec	pc	$3.09 \cdot 10^{16}$ m	$3.09 \cdot 10^{18}$ cm	3.26 LY
solar mass	$M_\odot$	$1.99 \cdot 10^{30}$ kg	$1.99 \cdot 10^{33}$ gm	
solar luminosity	$L_\odot$	$3.90 \cdot 10^{26}$ J/s	$3.90 \cdot 10^{33}$ erg/s	
solar effective temperature	$T_{eff\odot}$	5780 K	5780 K	
solar radius	$R_\odot$	$6.96 \cdot 10^{8}$ m	$6.96 \cdot 10^{10}$ cm	
Earth radius	$R_\oplus$	$6.38 \cdot 10^{6}$ m	$6.38 \cdot 10^{8}$ cm	
Earth mass	$M_\oplus$	$5.98 \cdot 10^{24}$ kg	$5.98 \cdot 10^{27}$ gm	
Earth density	$\rho_\oplus$	5520 kg/m^3	5.52 gm/cm^3	
Jansky	Jy	$1.0 \cdot 10^{-26}$ W/m^2•Hz	$1 \cdot 10^{-23}$ erg/s•cm^2•Hz	
Hubble constant	H_o	$3.24h \cdot 10^{-18}$ s^{-1}	$3.24h \cdot 10^{-18}$ s^{-1}	$100h$ km/s•Mpc
critical density (=$3H_o^2/8\pi G$)	ρ_o	$1.88h^2 \cdot 10^{-26}$ kg/cm^3	$1.88h^2 \cdot 10^{-29}$ gm/cm^3	
plasma frequency				$8.98\sqrt{n_e(\text{cm}^{-3})}$ kHz/gauss
radian				$57.29578° = 206{,}265''$
CMB photon density	n_γ	$4.15 \cdot 10^{5}$ m^{-3}	415 cm^{-3}	

Author

Index

Subject

Index